NOT
RETURNABLE
IF
OPENED

About the CD

The CD packaged on the facing page contains invaluable tools for the study and application of the principles of chemical analysis. Its semi-automated data tables and plotting routines make complex analyses easy to implement. Access to these powerful tools adds depth and practicality to your understanding of the chemical systems studied. For example, approximate solutions of complex problems in chemical equilibrium and titrimetric procedures are just a click away, and the effects of ionic strength on formal equilibrium constants can be instantly assessed. Below is a description of the contents and applications of the CD, by section.

Equation Solvers

The equation solver capabilities of scientific calculators are introduced. Instructions in the use of the equation solver function in Microsoft Excel are provided. A spreadsheet template enables the quick solution of quadratic equations.

Dynamic Data Tables

These tables are duplicates of the thermodynamic equilibrium and standard reduction potential tables in the appendix of the book, with one important addition. Entry of the ionic strength into the table results in an estimation of the formal constants for all table entries automatically.

Derivations

Complete derivations of equations for titration curves and other relationships are included in this section for study, variation, and emulation.

Titration Curves

Titration curves for monoprotic and polyprotic acids, complex forming reactions, precipitate forming reactions, and redox reactions can be predicted automatically using these spreadsheet templates. Appropriate equilibrium constants and concentrations are entered, and plots of the response versus either the volume of titrant or fraction titrated are generated automatically.

Log Concentration Plots

Log concentration plots enable the approximate solutions of very complex equilibrium problems to be quickly obtained. In such situations, they clearly indicate which of the several equilibria occurring are the most significant. Critical points on titration curves can be assessed quickly. This section contains spreadsheets in which the log concentration plots are automatically generated for monoprotic and polyprotic acids, complex forming reactions, precipitate forming reactions, electrode potentials, and redox reactions. These spreadsheets eliminate the tedium of drawing log concentration plots, so their full diagnostic power can be readily applied to all problems.

Dynamic Figures

Many figures in this book were created from spreadsheet plots of the function illustrated. This section contains the spreadsheets from which the figures marked with the CD icon were obtained. With these sheets, one can alter the parameters used for the book plots to demonstrate how the function will change under different circumstances. These sheets will be especially useful in the solution of various "what-if" problems and in deepening one's understanding of the functions illustrated.

Web Support (www.wiley.com/college/enke)

The Web support for this book can be found at

www.wiley.com/college/enke; it includes:

Solutions to the Practice Questions and Problems at the end of the chapters.

Access to these solutions requires a password, available to instructors of all classes using this book.

Additional Practice Questions and Problems and their solutions

More Practice Questions and Problems will be added frequently to this part of the website. They will be organized by chapter and section. This feature will assure instructors of a continuous supply of fresh problems for their classes. Worked-out solutions will also be available to instructors.

Related Laboratory Experiments

We will make available here as much information about the lab experiments implemented by the various adopters of the book as we can obtain.

Overhead Figures

In this section are printable copies of selected figures from this text. They may be used for the preparation of slides or overheads to aid in the illustration of the concepts presented in the book.

Round Table

This is a general place for the exchange of information among people studying and teaching with the book. The author will frequently review these discussions and will contribute where appropriate.

Additions and Errors

Based on feedback from readers and instructors, we will keep an ongoing list of errors detected. If there are discussions to be amplified or new material that has been developed, those materials will be included here, too.

The Art and Science
of Chemical Analysis

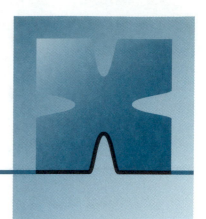

The Art and Science of Chemical Analysis

Christie G. Enke

Professor of Chemistry
University of New Mexico, Albuquerque

JOHN WILEY & SONS, INC.

New York ■ Chichester ■ Weinheim ■ Brisbane ■ Singapore ■ Toronto

Acquisitions Editor	David Harris
Senior Production Editor	Elizabeth Swain
Senior Marketing Manager	Charity Robey
Senior Designer	Karin Gerdes Kincheloe
Interior Designer	Lee Goldstein
Illustration Editor	Anna Melhorn
Cover Designer/Illustrator	Carol Grobe

This book was set in 10/12 Times Roman by UG / GGS Information Services, Inc. and printed and bound by Von Hoffmann. The cover was printed by Lehigh.

This book is printed on acid-free paper.

To order books or for customer service, call 1(800)-CALL-WILEY (225-5945).

Library of Congress Cataloging-in-Publication Data

Enke, Christie G., 1933-
    The art and science of chemical analysis/Christie G. Enke.
        p. cm.
    Includes index.
ISBN 0-471-37369-9 (cloth: alk. paper)
1. Chemistry, Analytic.   I. Title: Art and science of chemical analysis.   II. Title.
    QD75.2.E54 2001
    543—dc21

00-043399

To Bea

About the Author

Chris Enke's first book was "Electronics for Scientists" written with Howard Malmstadt, and in later versions with Stan Crouch as well. This book is the ninth he has authored or coauthored. Between these "bookends," he has been active in research and teaching. He is currently Professor of Chemistry at The University of New Mexico and Professor Emeritus at Michigan State University. His Ph.D. was earned at the University of Illinois. He has mentored over 60 Ph.D. students whose theses are in the areas of electroanalytical chemistry, computer-based instrumentation, optical spectroscopy, and mass spectrometry. Over 150 papers and book chapters have resulted from this research. He has received American Chemical Society awards in Chemical Instrumentation and Computers in Chemistry, the Distinguished Faculty Award at Michigan State University, and is an AAAS Fellow. Among his several inventions is the triple quadrupole mass spectrometer (with Rick Yost) for which they received the Distinguished Contribution Award from the American Society for Mass Spectrometry. Chris has served as President of ASMS, Chairman of the Computers in Chemistry Division of ACS, and has been a member of many professional society and journal advisory boards. This present book was written out of his love for the subject of analytical chemistry and in appreciation of congenial colleagues, stimulating students, and rewarding research.

Preface

This book began almost five years ago during a drive along the Turquoise Trail between Albuquerque and Santa Fe. Graham and Maria Cooks were with me. While Maria was admiring the scenery, which reminded her of her native Valencia province in Spain, Graham asked if I was content with the way we were teaching analytical chemistry. I was teaching the instrumental methods course that term and was feeling frustrated because there were far too many techniques to cover in one semester. In addition, there seemed to be little to connect the topics with each other. As we talked, it became clear that the challenge was to identify the unifying science that underlies the field of chemical analysis. If this were done, selected techniques would then illustrate the analytical principles, and an appreciation of the general approach to the solution of problems in chemical measurement could be gained.

To this end, I started with the concept of analytical chemistry as the science of chemical measurement. From this point, it is natural to ask how chemical measurements differ from physical measurements. One answer is that in chemical measurement, the substances we are trying to characterize are often intermingled with many other substances. To perform any measurement of the desired chemical (or chemical substructure), we must identify a quality that it has but which is not shared by other components in the sample. I refer to this quality as the *differentiating characteristic*. Examples are the ability to act as an acid, the ability to absorb yellow light, and the ability to affect an electrode potential.

Identifying a suitable differentiating characteristic is thus a necessary and fundamental part in the art and science of chemical analysis. Once the differentiating characteristic is selected, we must devise a "probe" for that characteristic, anticipate the nature of the sample's response to the probe, measure the response to the probe, and interpret the measurement data to obtain the desired information. This is a pattern common to all analytical methods. Further, the ways each of the above steps have been implemented for particular differentiating characteristics are fascinating stories involving the science of each of the steps and the creative ingenuity of analytical chemists: thus, the science and the art. For this reason, this book is organized by differentiating characteristic.

Four goals of chemical analysis have been identified: quantitation, identification, detection, and separation. Each of these goals requires the use of a differentiating characteristic, so all four goals are discussed in the context of most of the characteristics. This gives the book a broader view than a book focused solely on quantitation.

The question of which topics to include in an analytical text is always a difficult one. Texts too limited in scope have a correspondingly limited audience, and texts that try to be comprehensive become encyclopedic rather than interrelated or coherent. In traditional texts, the choices are among which techniques to include. In writing this book, the choices were which differentiating characteristics to cover. There is considerable overlap in the lists of potential techniques and useful differentiating characteristics, but the options are not equivalent and the consequences of the choices are not the same. As a priority, I chose those characteristics that include the material expected of the introductory analytical course in most curricula. These characteristics are acid-base reactivity, photon absorption, complexation reactivity, precipitation reactivity, electrode potentials, redox reactivity, and interphase partition. In terms of techniques, these chapters cover titrimetry, gravimetry, potentiometry, UV-visible and IR spectrometry, coulometry, extraction, distillation, and chromatography.

In organizing the subject by differentiating charcteristic, the distinctions between "wet methods" and instrumental methods are considerably blurred. For example, potentiometric and spectrometric methods for titrimetric end point determinations are introduced in the context of the reactivity used in the titration, and sensors based on antibody–antigen reactivity are covered in the chapter on analysis by biochemical reactivity. This blurring allows the course based on this text to be modern without compromising the important chemical fundamentals.

Beyond the "required" topics, some choices were made among additional topics. These choices were based on what I considered most important for students who might not take any further analytical courses. They include techniques based on photon emission (fluorescence, phosphorescence, and chemiluminescence) and techniques based on biochemcal reactivity. The inclusion of this latter chapter reflects the increasing application of such techniques as enzyme reactivity and immunoassay for analysis in both biological and non-biological systems.

Even though this book approaches the subject of chemical analysis from a nontraditional perspective, the knowledge obtained from a study of it meets the expectations of current standardized exams in chemistry and medicine for the course in quantitative analysis. The examples of methods developed around various differentiating characteristics include all the usual techniques and a substantial number of modern and instrumental techniques. Please see Appendix H for a cross-reference by technique. In the study of the basis for each characteristic, an understanding of the chemical equilibria involved is emphasized. The existence of a conceptual framework for the facts learned enhances relevance and retention of the material.

True to the concept of analytical chemistry as the science of chemical measurement, the book begins with a development of the measurement process itself. Making measurements, understanding the differentiating characteristic employed, and knowing how to interpret the data (technique, instrumentation, chemical principles, and chemometrics) are all seen as integral parts of the art and science of chemical analysis.

Each chapter supports concurrent laboratory experiments so that the laboratory experiences and the study of the text reinforce each other throughout the term. At the end of the course, the students are able to think creatively about choosing a differentiating characteristic for a particular analysis, implementing that characteristic, and solving difficulties and interferences that may be encountered. The structure provided for the solution of analytical problems enables and encourages analytical thought and should be very complementary to problem-based laboratories.

I am grateful to many people for their willingness to read and provide feedback on various parts of this book. They include Fritz Allen, Ma'an Amad, Nadja Cech, Terri Constantopoulos, Yanga Dijiba, Gary Eiceman, Ben Gardner, Peter Griffiths, Kelly Halle, Chuck Henry, Jack Holland, Carol Korzeniewski, Debra Dunaway-Mariano, Tom Niemczyk, Scott Nutter, Ignacio Villegas, Bea Reed, Curtis Shannon, Keith Vitense, George Wilson, and Vicki Wysocki. Nadja Cech prepared some of the example problems and solutions. The reviewers of the first edition included Todd L. Austell (University of North Carolina—Chapel Hill), Michael D. DeGrandpre (University of Montana), Thomas C. Farrar (University of Wisconsin—Madison), W. Ronald Fawcett (University of California—Davis), Asoka Marasinghe (Moorhead State University), John C. Schaumloffel (Washington State University), and Peter C. Uden (University of Massachusetts—Amherst).

Class testers for the Preliminary edition included Gary Eiceman (New Mexico State University), Peter Griffiths (University of Idaho), Chuck Henry (Mississippi

State University), Carol Korzeniewski (Texas Tech University), Curtis Shannon and
Eric Bakker (Auburn University), Keith Vitense (Cameron University), George Wilson
(University of Kansas), Douglas Beussman (Purdue University), and Thomas C. Farrar
(University of Wisconsin—Madison).

I want to also thank Graham Cooks for his support and for the many discussions
we have had regarding the overall approach taken in this book. Most importantly, my
wife Bea Reed, to whom this book is dedicated, provided constant encouragement and
proofreading for this project despite the havoc it wrought on our discretionary time for
almost three years.

It has been a pleasure to work with the Wiley staff during the preparation of this
book. David Harris, the Acquisitions Editor has given excellent advice and constant encouragement. He also skillfully guided the book into production. Cathy Donovan was always there to handle the many logistical tasks. The production staff has also been superb.
I have enjoyed knowing and working with the production editor, Elizabeth Swain, the illustration editor Anna Melhorn, the CD editor Martin Batey, the senior designer Karin
Kincheloe, and the copyeditor Connie Parks. Prior to this edition, Wiley published beta
and preliminary editions while this book was being class tested. Many thanks to Jay
Beck and Wiley Custom Publishing for doing a fine job during that phase.

CHRIS ENKE

September 2000
Albuquerque

Contents

Chapter 3
Acidity, Activity, and pH   61

Chapter 4
Analysis by Acid—Base Reactivity   90

Chapter 5
Analysis by Absorption of Light   138

Chapter 12
Analysis by Biochemical Reactivity   403

Background Materials   449

Appendices   461

Index   483

Chapter One

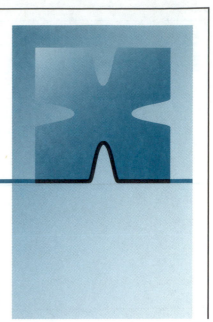

CHEMICAL ANALYSIS: WHAT, WHO, AND WHY

Broadly speaking, chemical analysis includes any aspect of the chemical characterization of a sample material. It is often the determination of the amount of one or more substances in a sample (the science we call quantitative analysis). Because of its significance in the fields of environmental monitoring, drug testing, and clinical diagnostics, quantitation is the most widely recognized area of chemical analysis. However, many analytical chemists are involved in the identification of unknown substances (unknown qualitative analysis) or in the determination of whether particular substances are present (targeted species detection). Furthermore, it is frequently the case that for accurate quantitative measurement or identification, the species of interest must be separated from the other materials in the sample. Chemists who study and apply the science of separation are thus essential to the field of analytical chemistry. Together, analytical techniques provide an invaluable function to the advance of many sciences. They can confirm the products of a chemical synthesis, identify the active substances in pharmaceutical research, and determine the exact structure of compounds of biological significance, to name just a few areas of application.

Chemical analysts serve industry in many areas including quality control, testing products for purity and correct composition, monitoring raw materials for adjustments in processing, and testing waste for levels and types of contamination. In addition, chemical analysis is an indispensable tool in virtually all phases of chemically related research, since it provides the methodologies that underlie all chemical measurements. In turn, chemical measurements provide data essential for informed decisions in many areas of commerce, politics, and choice of personal lifestyle. In this chapter, we will look at the general approach taken in the development and application of methods used in all four areas of chemical analysis, namely, quantitation, detection, identification, and separation.

Areas of chemical analysis and the questions they answer

- QUANTITATION: How much of substance X is in the sample?
- DETECTION: Does the sample contain any of the substance X?
- IDENTIFICATION: What is the identity of the substance(s) in this sample?
- SEPARATION: How can the species of interest be separated from the sample matrix for better quantitation and identification?

A. Chemical Quantities

The determination of the amount of a particular substance in a sample is called **quantitative analysis**. In essentially all cases, the substances of interest are in a matrix that includes many other substances. In many cases, the substance of interest is present in the sample in only minute or trace amounts. In chemical analysis, the substance to be assessed is often called the **analyte**. The challenge to the analytical methods developer is to devise a way to "see" just the analytes of interest and to determine the quantities of just those analytes without being influenced by the nature and amounts of the other materials present. Analytical chemists have devised many techniques to accomplish just that. Up to now, the study of analytical chemistry has been organized according to the techniques devised and employed by earlier analysts. However, a common approach underlies all techniques of quantitative analysis. The steps involved in this approach to the development of any quantitative technique are outlined in the following sections. Viewing quantitative analysis from the perspective of these common steps allows us to understand the general pattern shared by all techniques of quantitative analysis. We can then appreciate the creativity that has been involved in the development of analytical methods, and we will see how the approaches developed for one method have added to the kit of tools available to future analysts. As a preamble to the study of quantitative analysis, we will review the units chemists have devised to express the amounts of chemical substances.

UNITS INVOLVED IN THE EXPRESSION OF AMOUNT

The expression of the amount of material can take many forms. The fundamental unit of the amount of material is its mass. In chemical terms, an equally useful unit may be the number of moles of the elements or compounds in the sample. The number of moles is equal to the mass in grams divided by the **formula weight** (FW). The term **molecular weight** (MW) is also used for molecular species. The relationship between mass and FW or MW is only valid for pure materials of known chemical formula. A **mole** of a substance is Avogadro's number ($N_A = 6.022 \times 10^{23}$) of the smallest unit (atoms, molecules, ions, etc.) that retains the qualities of the substance. The concept of moles is extremely important in chemical analysis because the recognition of the analyte frequently takes place on the atomic or molecular level.

   When the analyte is dissolved in an appropriate liquid (frequently referred to as the **solvent**), it becomes uniformly dispersed in the solvent. The ratio of the amount of analyte (or **solute**) present in a given amount of solution is generally called the **concentration**. Several ways of expressing the concentration of dissolved species have been devised. Principal among expressions of concentration is the molar concentration C_M expressed in **molarity** (moles of analyte per liter of solution). The concentrations of very dilute analytes is sometimes given in parts per thousand, parts per million, or parts per billion (C_{ppt}, C_{ppm}, C_{ppb}). These refer to the weight fraction of analyte in the solution. Other expressions of concentration less used in analytical chemistry are the **molality** (moles of solute per 1000 g of solvent), weight percent, and volume percent.

Expressions of amount

- 1 mol of X = 6.022×10^{23} of the single X units

- The formula weight of X (FW_X) is the mass of 1 mol of X in grams

- The number of moles of X in m grams of X is m/FW_X

Expressions of concentration

Molarity: C_M = moles of X per L of solution

Parts per thousand: C_{ppt} = (grams of X per gram of solution) $\times$ 1000

Parts per million: C_{ppm} = (grams of X per gram of solution) $\times 10^6$

Weight fraction: $C_{w/w}$ = (grams of X per gram of solution)

Volume fraction: $C_{v/v}$ = (mL of X per mL of solution)

Molality: C_m = moles of X per kg of solvent

Study Questions, Section A

1. What are the four goals of chemical analysis?

2. How many units of a substance are contained in 1 mol of that substance?

3. How can you determine how much 1 mol of a substance weighs?

4. How can you determine the number of moles in a given mass of a substance?

5. A piece of pencil lead, which is pure carbon, weighs 1.0 mg. Calculate

   A. the moles of carbon in the pencil lead

   B. the number of carbon atoms in the pencil lead

6. A piece of iron contains 31.83 mol of iron atoms. How much does it weigh?

7. How many moles are there in 100.0 mL of mercury at 20 °C? (The density of mercury at 20 °C is 13.59 g/mL, and its atomic weight is 200.59 g/mol.)

8. What does it mean if a solution is said to be 0.36 molar in a particular substance?

9. Calculate the number of moles of NaCl in 50.0 mL of a 0.200 M NaCl solution.

10. Calculate the molarity of $CaCO_3$ in a solution prepared by placing 0.5471 g of $CaCO_3$ in a 100.0-mL volumetric flask and diluting to the mark with distilled H_2O. The formula weight of $CaCO_3$ is 100.09 g/mol. Assume the calcium dissolves completely.

11. What do the terms ppt, ppm, and ppb mean?

12. Calculate the concentration in ppm of $CaCO_3$ in the $CaCO_3$ solution in Question 10. Assume that 1 mL of solution weighs ~1 g.

13. Calculate the concentration in ppm of a solution that is 0.0300 M in HCl. The formula weight of HCl is 36.461 g/mol. Assume that 1 mL of solution weighs 1 g.

14. A water sample with a mass of 53.5 g contains 20.3 ppb of lead. Calculate the moles of lead in the solution.

Answers to Study Questions, Section A

1. The four goals of chemical analysis are quantitation, detection, identification, and separation.

2. A mole of a substance contains 6.022×10^{23} units of that substance. This number is called Avogadro's number and has the symbol N_A.

3. To determine the weight of a mole of a substance, you obtain the chemical formula of the substance and sum the elemental weights of all the atoms in the formula. This value is called the formula weight.

4. The relationship between formula weight, mass, and moles is FW = g/mole. Rearranging, moles = g/FW.

5. A. The formula weight of carbon, from the periodic table, is 12.01 g/mol. The mass in mg can first be converted to grams and then multiplied by the formula weight to find moles.

$$(1.0 \text{ mg})\left(\frac{1 \text{ g}}{1000 \text{ mg}}\right)\left(\frac{1 \text{ mol}}{12.01 \text{ g}}\right) = 8.3 \times 10^{-5} \text{ mol C}$$

   B. Multiplying by Avogadro's number converts moles to atoms.

$$(8.3 \times 10^{-5} \text{ mol})\left(\frac{6.022 \times 10^{23} \text{ atoms}}{1 \text{ mol}}\right) = 5.0 \times 10^{19} \text{ atoms}$$

6. The formula weight of iron from the periodic table is 55.85 g/mol; thus,

$$(31.83 \text{ mol})\left(\frac{55.85 \text{ g}}{1 \text{ mol}}\right) = 1.778 \times 10^3 \text{ g}$$

7. The volume can be converted to weight using density, and moles can be calculated from weight using the formula weight.

$$(100.0 \text{ mL})\left(\frac{13.59 \text{ g}}{1 \text{ mL}}\right)\left(\frac{1 \text{ mol}}{200.59 \text{ g}}\right) = 6.775 \text{ mol Hg}$$

8. The molarity of a solute in a solution is the number of moles of that solute there would be in exactly 1 liter of the solution. Thus, for a 0.36 M solution of some substance, there would be 0.36 mol of the substance in 1.000 liter of solution.

9. The expression 0.200 M (0.200 molar) means the solution contains 0.200 mol of NaCl per liter of solution. The number of moles in 50.0 mL of solution can be calculated by converting mL to L and multiplying by concentration.

$$(50.0 \text{ mL})\left(\frac{1 \text{ L}}{1000 \text{ mL}}\right)\left(\frac{0.200 \text{ mol}}{1 \text{ L}}\right) = 0.0100 \text{ mol NaCl}$$

10. Since molarity is moles per liter, it is first necessary to convert the grams of $CaCO_3$ to moles. The molarity is then calculated by dividing moles $CaCO_3$ by liters of solution.

$$(0.5471 \text{ g})\left(\frac{1 \text{ mol}}{100.09 \text{ g}}\right)\left(\frac{1}{100.0 \text{ mL}}\right)\left(\frac{1000 \text{ mL}}{1 \text{ L}}\right)$$

$$= 0.05466 \text{ M CaCO}_3$$

11. In these terms, the "pp" is an acronym for "parts per." The t, m, and b are thousand, million, and billion. These terms express the fraction of the sample weight that is contributed by the component the term is applied to.

12. From the equation in the margin on p. 2,

$$C_{ppm} = \frac{\text{g solute}}{\text{g solution}} \times 10^6$$

To use this equation, we need the mass of the solution; since we have assumed that 1 mL of solution weighs 1 g, the solution weighs 100.0 g, so

$$C_{ppm} = \frac{0.5471 \text{ g CaCO}_3}{100.0 \text{ g solution}} \times 10^6 = 5471 \text{ ppm CaCO}_3$$

Often, when dilute solutions are made in water, which weighs about 1 g/mL, C_{ppm} is calculated as mg/L. This calculation follows:

$$\frac{547.1 \text{ mg CaCO}_3}{0.1000 \text{ L solution}} = 5471 \text{ ppm CaCO}_3$$

It is important to realize that this way of calculating ppm involves an approximation because the exact mass of the solution is not 1 g/mL. More accurate calculations require using the exact density of the solution.

13. As in Question 12, if we assume that 1 mL of solution weighs 1 g, C_{ppm} is equal to mg/L solution. The concentration is already expressed in moles/L, so the C_{ppm} can be found by converting moles to mg. This is done using formula weight.

$$\left(\frac{0.0300 \text{ mol}}{1 \text{ L}}\right)\left(\frac{36.461 \text{ g}}{1 \text{ mol}}\right)\left(\frac{1000 \text{ mg}}{1 \text{ g}}\right)$$
$$= \frac{1.09 \times 10^3 \text{ mg}}{\text{L}} = 1.09 \times 10^3 \text{ ppm HC1}$$

14. Since $C_{ppb} = $ (g solute/g solution) $\times 10^9$, 20.3 ppb of lead means that the amount of lead in solution is 20.3×10^{-9} g lead/g solution. Multiplying the grams of solution by this value gives grams of lead. Moles are calculated from grams by dividing by formula weight (from the periodic table).

$$\left(\frac{20.3 \times 10^{-9} \text{ g Pb}}{\text{g solution}}\right)(53.5 \text{ g solution})\left(\frac{1 \text{ mol}}{207.2 \text{ g}}\right)$$
$$= 5.24 \times 10^{-9} \text{ mol Pb}$$

B. The Differentiating Characteristic

In performing a chemical analysis or separation, it is necessary for the analyte to have some property or quality by which it can be distinguished from the other material with which it is combined. The proverbial problem of finding needles in a haystack provides a good example. We can think of the various qualities of needles that are not shared by the hay stems. The needles have a metallic luster, they may respond to a magnetic field, they have very sharp points, they do not bend easily, and they do not burn. Each of these distinctive characteristics could be used as a basis for separation and/or quantitation. In the same way, when a chemical species is dissolved in a solution, vaporized in a gas, or mixed with other species in a solid, a distinguishing characteristic must be identified before an analysis can be performed. The analyst must consider the various physical and chemical properties of the analyte by which it can be differentiated from the matrix in which it exists.

It often happens that some other materials present in the sample share with the analyte the most distinctive characteristic the analyst can identify. In this case, a second differentiating characteristic must be used to separate the analyte from those materials that share the characteristic to be used for quantitation or detection. In this study, we will explore analytical methods based on the differentiating characteristics of acid–base reactivity, absorption of light, reactivity to form complexes and precipitates, partitioning between two phases, oxidation–reduction reactivity, ability to affect electrode potentials, characteristic emission of light when excited, and participation in biochemical reactions. These are just a few of the many distinguishing characteristics that have been exploited for chemical analysis, but they serve to introduce the fundamental concepts of analytical chemistry and provide the basis for learning valuable quantitative skills in the laboratory.

DEVISING A PROBE FOR THE DIFFERENTIATING CHARACTERISTIC

Once the differentiating characteristic that will serve as the basis for the determination of the amount of analyte has been chosen, it is necessary to devise a probe for that particular characteristic. For example, if we are going to use the fact that the analyte molecules absorb green light as the differentiating characteristic, we must devise a way to expose the sample to green light. If the fact that the needles respond to a magnetic field is used to find needles in haystacks, a means to apply a magnetic field to the stack must be devised. This process is diagrammed in Figure 1.1. In each of the characteristics we

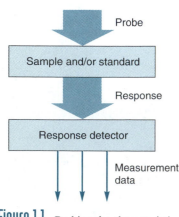

Figure 1.1. Probing the characteristic quality and measuring the response.

will study, we will see that analytical chemists have often created more than one useful way to probe each differentiating characteristic.

ANTICIPATING THE RESPONSE TO THE PROBE

Before the response to the probe can be measured, we must anticipate what form that response will take. It is in this area that much of the theory of chemical analysis is developed. A knowledge of how the analyte will respond to the stimulus of the probe aids in the perfection of the probe system, establishes exactly what must be measured to determine the analyte response, suggests the best methods for the interpretation of the measurement data, and reveals what the limitations and interferences of the technique are likely to be.

MEASURING THE RESPONSE TO THE PROBE

It is helpful to distinguish between the application of the probe and the measurement of the response to that probe because it helps us to realize that there may be several ways to measure the response of the analyte to any given probe. In the needles in a haystack example, we could count the needles that were attracted to a magnet, we could release the needles onto a scale to weigh them, or we could weigh the magnet before and after its probe of the haystack. The response of the analyte molecules capable of absorbing green light is to absorb some of the light impinging on them. To determine the extent of this absorption, the analyst must find a way to determine the amount of the light the sample has absorbed. This is where the measurement part of the analytical process takes place. In the case of the green light probe, the measurable quantity is the intensity of the green light that gets through the sample. Measuring this will still not tell us how much light was absorbed by the analyte. To know this, we must also measure the intensity of green light transmitted by a sample holder with no analyte present. Finally, we must find a way to relate the amount of green light absorbed with the amount of analyte in the sample. The sequence of operations performed in an analysis scheme is shown in the margin. These six steps will be a recurrent theme in our study of chemical analysis.

Steps in chemical analysis

1. Select a differentiating characteristic.

2. Devise a probe for the characteristic chosen.

3. Anticipate the response of the analyte(s) to the probe.

4. Measure the response to the probe.

5. Interpret the measurement data to obtain the desired information.

6. Assess the method for accuracy, precision, and interferences.

Study Questions, Section B

15. What are the six steps of chemical analysis?

16. You are asked to nondestructively determine the number of apples in a box of mixed fruit that includes also bananas, pears, and grapes. Identify each of the six steps of chemical analysis in the process you would follow.

Answers to Study Questions, Section B

15. The six steps of chemical analysis are selecting the differentiating characteristic, devising a probe for that characteristic, anticipating the response of the analyte to the probe, measuring the response to the probe, interpreting the measurement data, and assessing the limitations of the technique.

16. One must first choose a differentiating characteristic. I would choose shape (somewhat spherical or larger at the stem end) because some apples may have the same color as pears. The probe is a visual assessment of the shape. The anticipated response to the probe is that either the fruit fits the shape criterion or it doesn't. The measurement would be to count the pieces of fruit that match the shape criterion. The interpretation of the data is that the count number is equal to the number of apples in the box. The potential interferences are any pieces of fruit that are not apples but match the shape criterion used. If the box also contained oranges, mangoes, Asian pears, and pomegranates, for example, the differentiating characteristic chosen would not be sufficiently differentiating, and more or different criteria would have to be employed.

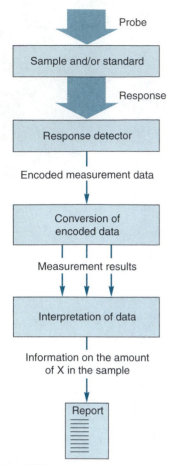

Figure 1.2. Process of quantitative analysis.

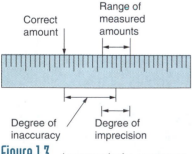

Figure 1.3. Accuracy is the agreement with the accepted value. Precision is the agreement among repetitive measurements.

C. Quantitation

The data obtained from measuring the response to the probe are related to the amount of analyte present, but they do not directly tell us how much is there. All measurements of probe response begin with an encoded signal or quantity that must be subsequently converted and interpreted to yield the desired analytical information (see Figure 1.2). Sometimes the interpretation is a simple multiplication by some predetermined proportionality constant. Quite often, a more complex relationship must be invoked. Frequently, data from more than one measurement are required in the interpretive step.

The measurement conditions and the relationships used in the interpretive step may have been empirically determined by previous analysts, they may follow laws discovered by earlier fundamental investigators, or they may have been developed and extended by analysts who specialize in data interpretation. Part of our study will be focused on the process of data interpretation. We will see how the response measurement data are related to the quantity of analyte both theoretically and in practical systems.

There are many areas in which we need reliable quantitative data. One of the most commonly encountered is that of clinical analysis. A conservative approach to health care leads to increased numbers and types of analyses. The cost factors limit the range of tests that will be ordered in a particular circumstance. Clearly, the development of rapid, reliable, and less expensive methods will be of great benefit. As such, a great deal of effort is being put in this direction. In the 1999 Application Reviews issue of *Analytical Chemistry,* 80 pages were devoted to various aspects of clinical chemistry. New immunoassay techniques (see Chapter 12) are supplementing the more traditional spectroscopic methods in many areas.[1]

ASSESSING ERRORS AND INTERFERENCE

One of the essential steps in the characterization of an analytical procedure is the determination of the quality of the results obtained by that method. There are a number of factors by which quality can be described and compared. The obvious ones applicable to any measurement are **accuracy** (correctness) and **precision** (related to reproducibility and resolution). In science, these terms have precise meanings, as illustrated in Figure 1.3. The assessment of these qualities involves an appreciation of the sources of inaccuracy and imprecision inherent in each step of the analytical procedure. Errors can arise at any step: the application of the probe, the measurement of probe response, and the interpretation of measurement data. Other figures of merit for a quantitative technique are the **detection limit** (the minimum detectable amount), the maximum measurable amount (where the probe or response systems reach some practical limit), and the **range** (the ratio of the maximum measurable amount to the detection limit).

Another very important consideration in chemical characterization is whether the entire probe response is due to the specific analyte sought. Other compounds that can act like the analyte and thus fool the probe and response analysis system are called **interferents**. Since the possibility for interference is usually present in real-life samples, the assessment of potentially interfering substances is important in the choice and application of an analytical technique. Where the presence or risk of interferences is unacceptable, the analyst has two choices. Either she can search for another differentiating characteristic less likely to be shared by other sample components, or she can use a combination of differentiating characteristics, one for separation and one for analysis.

---

[1]L. Schoeff, *Anal. Chem.* **1999**, *71*, 351R–355R.

17. What is the role of the differentiating characteristic in quantitation?

18. In scientific measurements, it is important to use the terms accuracy and precision accurately. Why is the last word in the previous sentence "accurately" rather than "precisely"?

19. A careful person uses a poorly calibrated scale to repeatedly determine the mass of an object. A much less careful person measures the mass of the same object with a recently calibrated scale. Discuss the differences in their likely results with respect to accuracy and precision.

17. Quantitation of a component of a mixture requires the ability to determine its amount independently of its surroundings. This requires a way to distinguish the analyte from the other species present. This characteristic is then probed, and the response to the probe is related to the amount of the component present.

18. A person using a term precisely will use it consistently, but not necessarily correctly. The person using a term accurately always conveys the universally accepted meaning. A person using a term inaccurately conveys misinformation and confusion.

19. The careful person would have more precise data than the less careful one, but the accuracy may not be good because of the poor calibration of the scale. The less careful person would probably produce a wider range of measurement values (poorer precision), but the average of his results could be more accurate because he was using a more accurate instrument.

D. Detection

In chemical detection, the information sought is the presence or absence of a particular substance (molecule or element) in the sample. If the substance is present, the question of exactly how much is present is usually pursued by a later quantitative analysis. Without a need for exact quantitation, species detection is generally simpler to carry out than normal quantitative analyses. This makes chemical detection useful when screening large numbers of samples for the substances of interest (for example, a particular mineral or drug metabolite). Despite the difference in the information sought, detection and quantitation have much in common. In detection, the analyst must still find a differentiating characteristic of the substance sought, devise a probe for that characteristic, and measure the response to that probe. The measurement results are then interpreted in terms of the presence or absence of the sought substance rather than its exact amount. A comparison of Figure 1.2 and 1.4 reveals the differences between quantitation and detection.

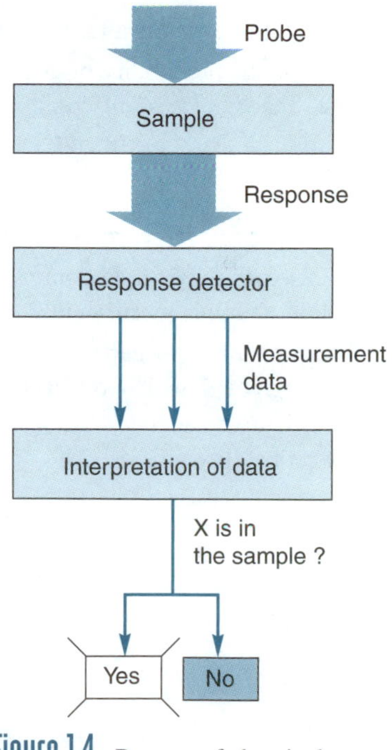

A current area of concern is the detection of infectious microorganisms in drinking water,[2] particularly *Cryptosporidium* and *Giardia*. The approaches under development include filtration and centrifugation for separation, use of monoclonal antibodies for specific reactivity, use of the polymerase chain reaction (PCR) for improvement in detection limit, and fluorescence and microscopy for identification and detection. The research in this area exemplifies the wide range of techniques available to the modern analyst and the great overlap with other disciplines such as biochemistry and microbiology.

Figure 1.4. Process of chemical detection.

ASSESSING CONFIDENCE IN THE RESULTS

The assessment of quality in the results of chemical detection is in the degree of certainty with which we can say that the substance sought is present or absent. Our confidence in the resulting answer depends on two factors: the possibility of interferents in

[2]For an extensive review, see S. Richardson, *Anal. Chem.* **1999**, *71*, 181R–215R.

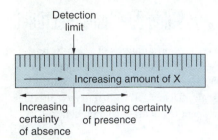

Figure 1.5. The detection limit is a crucial point relative to the certainty of detection analysis.

the sample and the detection limit of the technique. In the case of interferents, the considerations are the same as with quantitative analysis, but the consequences are different. Concluding that a substance is present when the measured response may have been due to a different substance (such a response is called a **false positive**) can lead to an incorrect decision regarding environmental safety, the fate of an entire production run, or the future of an Olympic athlete. A **false negative** is also possible. In this case, an interferent has suppressed the response of the analyte to the probe.

As shown in Figure 1.5, the **detection limit** of the technique necessarily forms the boundary between the conclusions of presence and absence. To be quite certain the substance is absent, the measured response must not contain the slightest sign of the sought substance's tested characteristic. On the other hand, to be quite sure that the substance is present, the measured response must be far enough above the detection limit to be certain there is no mistake.

The assessment of the suitability of a particular targeted species detection technique thus centers on the degree of discrimination between the analyte and potential interferents and the certainty that the response at the required detection level is well above the detection limit. An assessment of the confidence one could place in a particular result requires knowledge of statistics and the relative responses of the analyte and the interferents to the probed qualities.

Study Questions, Section D

20. In detection, what is a false positive?
21. Discuss the probability of generating a false positive relative to the choice of differentiating characteristic used.

22. Why are values of response for the analyte near the detection limit inconclusive for the analytical goal of detection?

Answers to Study Questions, Section D

20. A false positive is a determination that a substance is present in a sample when, in fact, it is not.
21. The probability of generating a false positive is related to the degree to which the differentiating characteristic is specific to the analyte. The more specific it is, the lower is the probability.

22. At the detection limit, the response is not sufficiently larger than the background or noise level to satisfactorily eliminate the possibility of the detection being a random error. Complete absence of a substance is not demonstrable. We can only say it is not more than the amount equal to the detection limit.

E. Identification

When the information desired is the identity of an unknown substance, the approach to the analysis differs from that of quantitation and detection. One thing we can do is to use a differentiating characteristic that provides different responses for different substances and then measure the value of that response. An example is the melting point temperature of the pure substance. If the melting points of all the compounds that could be in a sample are known, a measurement of the melting point of the compound can eliminate most of the compounds from the list. If more than one possibility remains, another differentiating characteristic (crystalline form, degree of solubility in ethanol, etc.) can be applied to narrow the possibilities still further. The goal, of course, is an unequivocal identification of the compound. This process of successive elimination is diagrammed in Figure 1.6.

Identification analysis is most easily performed on a pure substance. Separation of the substance from its original environment is thus a key part of the analysis. Even after purification, the determination of compound identity is a daunting prospect. There are over 10 million chemical compounds recognized, named, and numbered by the Chemical Abstracts Service of the American Chemical Society. Through synthesis and discovery, this number continues to grow by about 10% per year. Complete and certain identification usually follows the application of a series of powerful characterizing techniques. Those techniques with the greatest degree of discrimination narrow the list of possibilities by the greatest amount.

The technique of fluorescence analysis has been developed to the point of being able to detect a single fluorescent molecule.[3] The current challenge is to unambiguously identify the fluorescing molecule. The fluorescence probability, wavelength, polarization, and lifetime are differentiating characteristics that are being investigated. In one report, several of these factors are used in combination to provide a kind of unique fingerprint by which positive identification can be achieved.[4]

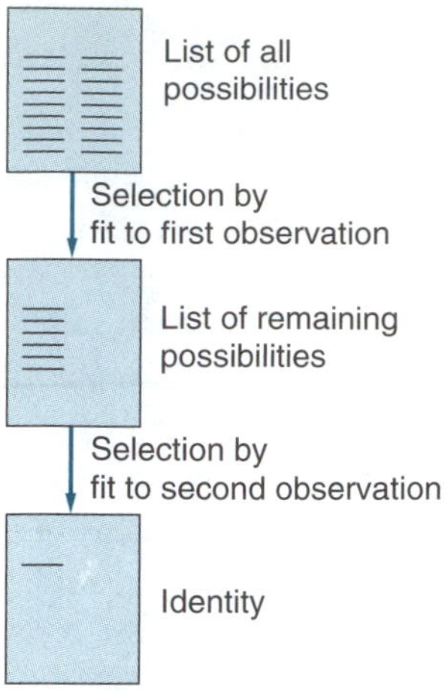

IDENTIFICATION BY DEDUCTIVE REASONING

Compound identification can sometimes be accomplished by deducing the structure of the compound from known information and rules that correlate the structure with that information. Many types of information can be used for identification, as indicated in Figure 1.7. Knowledge of the chemical formula of the compound is of great value in this process. The chemical formula for many organic compounds can be determined by CHN analysis (a technique that gives the ratios of the atoms of carbon, hydrogen, and nitrogen in the substance). Some types of very high-resolution mass spectrometry can also provide the chemical formula. Unfortunately, because of the possibility of isomers (molecules with the same amount of each element, but arranged differently), knowledge of only the chemical formula is not sufficient. Various spectroscopies such as infrared and nuclear magnetic resonance provide information about compound substructures and interatomic bonds, adding valuable information for the deductive process. The application of these techniques has become invaluable in the areas of chemical synthesis and the elucidation of biological mechanisms.

The quality of a compound characterization or identification analysis depends entirely on the confidence that can be ascribed to the result. Sometimes this can be a virtual certainty, particularly when several techniques provide confirming information. At other times, despite our best efforts and techniques, we are left with a range of possibilities rather than a single firm answer. The techniques of interpreting spectra to provide composition and structure information about unknown compounds continues to be an area of active research.

"ANALYZE THIS FOR ME!"

One of the most challenging situations in chemical analysis is the determination of the components in an unknown mixture. This process is sometimes referred to as **qualitative analysis**. Qualitative analysis may be performed to determine the formulation of a competitor's product, the impurities in drinking water or synthesized drugs, or the contaminants in a soil sample. Such a total qualitative analysis on a sample derived from nature is not practical since there may well be thousands of different compounds at levels above their detection limits. This point is caricatured, but without exaggeration, in Figure 1.8. In such cases, the question must be narrower, such as, "What metabolites of this drug are present?" or, "What compounds are causing the odor of this sample?" Even so, the analyst must often apply a variety of techniques to first separate and then identify the components of interest.

Figure 1.6. Identification by elimination of nonqualifying possibilities.

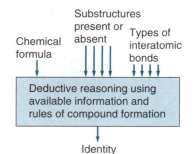

Figure 1.7. Compound identification by deduction.

Figure 1.8. Report on all compounds present in a petroleum sample.

[3]S. Nie, D. T. Chiu, and R. N. Zare, *Anal. Chem.* **1995**, *67*, 2849; Y. H. Lee, R. G. Maus, B. W. Smith, and J. D. Winefordner, *Anal. Chem.* **1994**, *66*, 4142.

[4]M. Prummer, C. G. Hubner, B. Sick, B. Hecht, A. Renn, and U. Wild, *Anal. Chem.* **2000**, *72*, 443–447.

23. What is the role of the differentiating characteristic in the identification of a compound by the elimination of non-qualifying possibilities?

24. What is the consequence of the list of all possibilities not being complete when employing the elimination technique for compound identification?

25. Why is knowledge of the chemical formula so helpful in the process of compound identification by induction?

26. An unknown salt is one of the salts listed in the following table. When reacted with H_2S in mild base and with $(NH_4)_2CO_3$, it forms a precipitate. Which of the salts can it be and, if more than one, what test will complete the identification?

Dissolved Salt	Precipitating Agent			
	HCl	H_2S (acid)	H_2S (base)	$(NH_4)_2CO_3$
Hg_2O_4	precipitate	soluble	soluble	precipitate
$FeSO_4$	soluble	soluble	precipitate	precipitate
$CaSO_4$	soluble	soluble	soluble	precipitate
K_2SO_4	soluble	soluble	soluble	soluble
$CdSO_4$	soluble	precipitate	precipitate	precipitate
$BaSO_4$	soluble	soluble	soluble	precipitate
$CuSO_4$	soluble	precipitate	precipitate	soluble

23. Each stage of elimination requires the use of a new differentiating characteristic. The more discriminating the characteristic, the more quickly the list converges.

24. There is a risk that the compound sought will be one of the missing ones. When this happens, no correct result is possible.

25. The rules of compound formation are based on the ways in which the elements are known to bond to each other. If the numbers of the atoms of each element are not known, applying these rules is more difficult, and the results are less conclusive.

26. Salts forming a precipitate with H_2S (base) and $(NH_4)_2CO_3$ are $FeSO_4$ and $CdSO_4$. The H_2S (acid) test resolves these two. The HCl test gives the same result for both.

F. Separation

We have already seen that separating one material from a mixture of materials is a very important part of chemical analysis. In particular, separations are used to remove analytes from other materials that might interfere with the intended analysis. Just as with quantitation and identification, separation processes exploit differentiating characteristics. In this book, we will see how the processes of separation and analysis are closely linked in principle and in practice. However, in separation, the differentiating characteristic must be one that enables the physical separation of the analyte from the other materials. Differentiating characteristics that have this capability are based on physical properties (such as size, mass, charge, or density), partition between phases (solubility, vaporization, or extraction), and chemical change (electrochemical deposition).

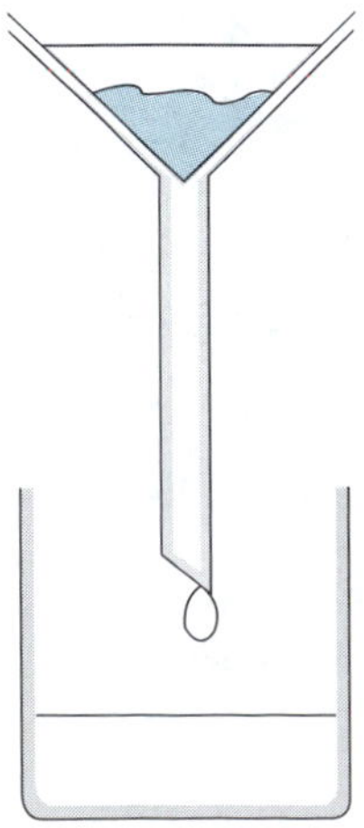

SINGLE-STEP SEPARATION

Where the differentiating characteristic is unique to the analyte, the separation can be completed in a single step, as in the case of a compound that can be precipitated and thus separated from all the compounds remaining in the solution. This single-step process is illustrated in Figure 1.9. Flotation depends on density as the differentiating characteristic. All components less dense than the liquid they are in will float on the liquid surface. All components more dense than the liquid will sink to the bottom of the container. Other examples include distillation of a compound that is volatile to separate it from ones that are not, extraction of a compound into an organic solvent to separate it from salts dissolved in an aqueous solution, and filtration of a solution to separate dissolved molecules from larger particles. Even though we are familiar with

Figure 1.9. Filtration separates the solvent and dissolved ions and molecules from the much larger particles.

the general nature of these processes, in the context of chemical analysis, we must be concerned with the completeness of the separation. In mining, for example, we may be content with recovering 95% of a metal from its ore, but such an incomplete extraction could cause an unacceptable error in a quantitative analysis.

SEPARATION BY DISPERSION

Analytical chemists are called on to quantitate and identify materials in increasingly complex mixtures. In fact, all natural samples become more complex as detection limits are improved. It is estimated that for every 10-fold reduction in the detection limit, there is a 10-fold increase in the number of detectable compounds in any natural sample (lake water, blood serum, soil, etc.). In such a situation, it is unlikely that there is a differentiating characteristic that is unique to the specific analyte of interest. Therefore, quantitative (complete) separation of a trace analyte from a natural sample can rarely be accomplished in a single step. In such cases, we use a differentiating characteristic for which the analytes differ by degree. A method is then devised that will sort out the compounds in the mixture according to the extent to which they exhibit the tested characteristic.

Figure 1.10 illustrates a popular chromatographic technique for dispersive separation. The separation medium is a strip of porous material (stiff paper or coated glass or plastic). A sample containing a mixture of analytes is placed in a dot near the bottom of the strip. The strip is then placed in a beaker with some solvent in the bottom. As the solvent wicks up the porous strip, it brings the analytes with it to a degree. The differences in velocity among the analytes depends on their relative preference for the solid or solution phases in the strip. Another example is fractional distillation, in which the most volatile component appears first followed by others in order of decreasing volatility. Note that this sorting must result in a physical dispersion in time or space in order to result in a physical separation. A number of ingenious separation techniques based on the dispersion concept have been developed. For each of the differentiating characteristics studied in this book, we will explore its potential to serve as a basis for separation. Because of its widespread application, we will pay particular attention to the dispersion technique called chromatography.

One of the most dramatic examples of a separation technique is the development of forensic DNA analysis. This powerful technique has essentially taken over the previous domain of blood type analysis.[5] The distinctive pattern of the digested DNA fragments as they are dispersed along a chromatographic gel has an extremely high degree of specificity.

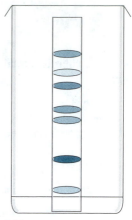

Figure 1.10. As the solvent in the bottom of the beaker travels up the strip, it carries solutes from the mixture placed at the bottom along with it. The difference in their velocities results in a physical separation along the strip.

Study Questions, Section F

27. What special quality must a differentiating characteristic have to be used for separation?

28. What are the major limitations of single-step separations?

29. How can analytes be separated when they differ only by the degree to which they express the differentiating characteristic?

Answers to Study Questions, Section F

27. The differentiating characteristic must enable the physical separation of the analyte from the rest of the materials in the sample. This often involves the analyte undergoing a transition between two phases to isolate it from the components that do not undergo this transition.

28. The major problem with single-step separations is that virtually all the analyte must undergo the separation process

while none of the components from which it is to be separated undergo this process at all. This is an unlikely scenario, especially with complex samples.

29. The separation technique must be developed as a dispersion technique in which the degree to which the compounds express the differentiating characteristic results in a dispersion in time, space, or both.

[5]T. A. Brettell, *Anal. Chem.* **1999**, *71*, 235R–255R.

G. Chemical Analysis in Science and Society

Areas where chemical analysis is central

- Environmental testing
- Human genome project
- Studies in drug delivery and metabolism
- Operating room monitoring
- Quality control in all chemical products
- Clinical laboratories and pathology
- Drug discovery through screening of active agents
- Studies of biological processes
- Assessment of geological resources

The areas of chemical characterization and chemical measurement are of great importance in science. The field of chemistry is heavily dependent on the ability of analysts to provide essential data for improved synthesis and quality control of chemical products and industrial waste. Chemical research in all areas depends on and is limited by the variety and quality of analytical techniques available. Chemical analysts provide the techniques that allow chemists to peer inside the workings of the chemical and biological systems under study. In addition, the rate of improvement and expansion of analytical techniques has a great impact on the progress of research in virtually all other areas of science, including pharmacy, pathology, toxicology, biology, geology, physics, and environmental science.

Chemical analysis also provides essential data in an increasingly technological society. It is widely used in modern forensics, it serves as a watchdog on the effects our technology is having on the environment, and it is a key player in detecting the abuses of technology in industry, politics, and sports, to name just a few examples. Advances in the detection limits and selectivity of chemical analyses have enabled the setting of testable standards for safety in water, food, the workplace, and home as well as assessment of the basis for the importance of these standards.

CHALLENGES AND REWARDS

Analytical chemists exercise many skills, much knowledge, and a great deal of ingenuity in the practice of their art. The selection of the distinguishing characteristic on which an analysis is to be based requires an intimate knowledge of the physical and chemical properties of the analyte as well as the other compounds in the sample matrix. The analyst must then be able to prepare the samples appropriately for the method selected, carry out the measurements under effective conditions, and perform the data interpretation to yield the desired information. In addition to such "on-line" problem solvers, other analytical chemists specialize in the development of new kinds of probes and measurements (instrumentalists), some are engaged in the development of new tools of data analysis (chemometricians), and some explore better methods of analyte separation (chromatographers or separations specialists). As indicated in Figure 1.11, people with all these areas of expertise make essential contributions to the field of analytical chemistry.

Many chemists are drawn into the field of analytical chemistry because they appreciate the satisfaction of providing accurate and reliable information that is then the basis for important decisions. They also enjoy the wide range and continuous variety of the challenges presented to them for solution. This stimulating field continues to demonstrate a rapid growth and development. Every advance in measurement technology, every new kind of chemical reaction that is discovered, and every advance in data processing capability offers new opportunities for better chemical analyses. Industry has long recognized the value of analytical chemists, and these professionals have been among the most highly sought graduates in all areas of science.

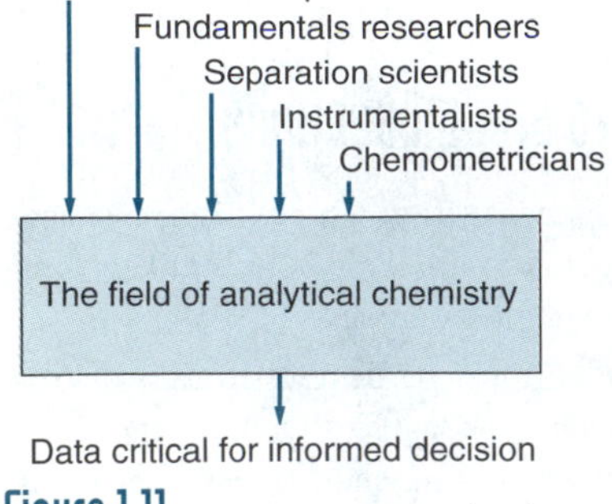

Even if the reader is not planning a career in science, an understanding of the approaches, capabilities, and limitations of analytical techniques is valuable in assessing the appropriate use of analytical data in corporate, social, and political decisions. If you are pursuing a study of any science, an understanding of the measurement process, particularly chemical measurements, will form an invaluable foundation for your future work.

Methods developers
Fundamentals researchers
Separation scientists
Instrumentalists
Chemometricians

The field of analytical chemistry

Data critical for informed decision

Figure 1.11. Many types of scientists contribute to the development and application of analytical chemistry.

THE EVOLUTION OF CHEMICAL ANALYSIS

The field of chemical analysis has gone though a number of stages or periods of development. In the early stages, analysis was based principally on chemical reactivity and selective solubility. The techniques of volumetric and gravimetric analysis were extensively developed during this period. These are the so-called **wet methods** of analysis. This approach predominated up to the **age of electronics**. The electronics era made oscilloscopes, strip-chart recorders, and sensitive signal amplifiers available to scientific experimenters. These capabilities enabled the development of the first sophisticated instruments for chemical analysis. An extensive love affair with **instrumentation** developed, as it seemed that all the measurement goals could be met without the need for the complex and time-consuming chemical methods. Analytical courses and textbooks are still largely divided between wet methods and instrumental analysis.

The introduction of inexpensive computing power brought on one of the most recent advances. **Computers** were first used for routine data analysis, then for data collection and instrument control, then for intelligent **automation**, and finally for sophisticated **chemometric data analysis**. It was hoped that what the instrumental methods lacked in specificity could be regained by mathematical analysis of the huge amounts of data the automated instruments could generate. This has happened to a remarkable extent, but every advance has been accompanied by the desire to push the limits of detection and discrimination still further. In the present era, where the detection capability for some techniques has reached the ultimate limit of single molecules, our greatest need is for differentiating characteristics that are more selective. An increased appreciation of the exquisite selectivity of many **biochemical reactions** has ushered in the most recent trend in chemical analysis. We are still exploring how to better utilize these new chemical techniques for both simple sensors and elegant complex analysis. In many ways, it would seem that the field has circled back to its chemical roots. This time, however, it has the powerful array of tools and techniques that have been developed in the intervening years.

Practice Questions and Problems

1. The four areas of chemical analysis are quantitation, detection, identification, and separation. Give an example of each kind of determination that you know about from your work or reading. Your examples do not have to be from chemistry class. They can be from biology, geology, physics, medical technology, clinical chemistry, pharmacy, etc. (or even from the news channel on TV).

2. In our laboratory, we are involved in some very sensitive measurements. We are currently making standard solutions as dilute as 10^{-10} M.

   A. If the molecular weight of the substance is 147 g/mol, how many grams of substance are required to make 100 mL of 10^{-10} M solution?

   B. What is the resulting concentration in ppm or ppb?

3. Consider a trip to the supermarket for food. Name and explain the differentiating characteristics (and the corresponding probe and response) you use when you choose

   A. bananas

   B. eggs

   C. chiles

4. A recent TV program on jewelry fraud told of the problem of distinguishing turquoise that is real (stone) from that which is fake (plastic). To tell which is which, suggest a differentiating characteristic, a probe for that characteris-

tic, and the expected responses to that probe. The method used by the gemologist was destructive to the fake turquoise only.

5. Which compounds listed in the table of Study Question 26 cannot be distinguished by the tests given and why?

6. You have heard the story about the person who, when asked to determine the number of people present, repeatedly neglected to count himself. Discuss the accuracy and precision of his results.

7. Discuss the reasons for any discrepancy you may have noticed between the temperature given in the weather report and the temperature you read at that same time on your outdoor thermometer. How much correlation between the two values have you come to expect?

8. Name the differentiating characteristic that is used in the following processes. For each, identify the method of probing that characteristic, the method of detecting the response to the probe, and the way the desired information is obtained.

   A. panning for gold

   B. detecting a gas leak

   C. ticketing a speeder

9. Find a recent article about a situation of personal or social significance in which a chemical analysis was involved.

Learn all you can about the technique employed, the differentiating characteristic employed, the detection limit, the method of interpreting the data, and the assessment of confidence in the result. Also indicate whether the chemical analysis was principally quantitation, detection, identification, or separation. Include your satisfaction with the depth of reporting and with the way in which the analytical data were used.

10. Define the following and give a scientific or everyday life example:

    A. quantitation

    B. detection

    C. identification

    D. separation

11. It is found that 0.020 L of a water sample contains 0.0097 mg of lead. What is the concentration of lead in moles per liter and in ppm?

12. Describe a technique that could be used to separate a mixture of salt, sand, and small Styrofoam pellets. List the characteristic used to differentiate each component from the others.

13. For an acid–base titration, describe

    A. the distinguishing characteristic of the analyte

    B. the method used to probe this characteristic

    C. the method used to measure the response to the probe

    D. the way in which the desired information (concentration of acid in an unknown sample) is obtained from the measurement data

14. A geode is a rock that is remarkably spherical and which has layers (from the outside in) of rock and quartz crystals. The center is air. Without breaking it open, how would you test whether a rock is a geode or not?

15. What is meant by the term, "detection limit?" What are the difficulties of achieving the four goals of quantitation, detection, identification, and separation when the analyte level is near the detection limit?

16. Suppose 20.00 mL of a solution that is 0.146 M in NaCl is mixed with 25.00 mL of a solution that is 0.573 M in KNO_3. When dissolved in water, both salts are completely dissociated into their constituent ions. What are the concentrations of the four ionic species in the final solution?

17. Consider 50.00 mL of a solution that is 0.0429 M in HCl that is mixed with 20.00 mL of a solution that is 0.106 M in HCl. What is the concentration of HCl in the final solution?

18. What is the molar concentration of glucose in a solution that is 34 ppm glucose. The molecular formula for glucose is $C_6H_{12}O_6$.

19. Explain why the differentiating characteristic is at the heart of achieving each of the four goals of quantitation, detection, identification, and separation when the sample is a mixture of many compounds.

20. Explain why the list of differentiating characteristics that would be effective for a given analysis depends on the nature of the other components of the sample in which the analyte is found.

21. A. Distinguish the techniques of quantitation and detection.

    B. Distinguish the techniques of identification and separation.

22. The detection limit for chlorobenzene C_6H_5Cl, with GC/MS (gas chromatography/mass spectrometry) is 100 fg.

    A. How many moles of the compound is this?

    B. If the volume of solution injected into the analyzer is 2 mL, what is the concentration of the chlorobenzene in this solution in moles/liter?

    C. For the solution in part B, what is the chlorobenzene concentration in ppm?

23. The reaction of glucose ($C_6H_{12}O_6$) with O_2 in water to form gluconic acid ($C_6H_{12}O_7$) and hydrogen peroxide (H_2O_2) is catalyzed by the enzyme glucose oxidase.

    A. Write a balanced equation for this reaction.

    B. If 3.00 mg of glucose is oxidized in 25 mL of solution, what concentration of H_2O_2 is created? (For a review of stoichiometry and equation balancing, see Background Materials, section A.)

24. Workers in the clean rooms at Intel can identify coworkers with whom they are familiar despite their identical, full-body suits and helmets. Faces are not visible through the helmet windows. What differentiating characteristics could be used for this identification? Could any of these be developed into an automated means of positive identification?

25. Do you think you will be able to pursue your career without performing or relying on chemical analyses? Name your intended profession and explain.

Suggested Related Experiments

1. Lab check-in.
2. Observations of processes and instances of chemical analysis in life and in the news.

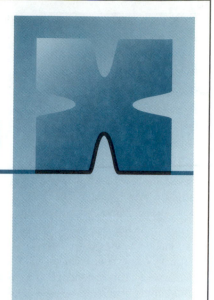

Chapter Two

THE ELEMENTS OF MEASUREMENT

B y now, measurement is an inherent part of our life, as are walking and talking. At first, it may seem odd to study a process we have been doing for so long. However, as this chapter demonstrates, there is more to measurement than meets the eye (literally), and there are many unexamined assumptions we have developed in the course of making so many measurements. Scientists should be rigorous in their study of nature. This must surely include rigor in the way we obtain the values used in our attempts to understand natural phenomena. In this chapter, a systematic view of the process of measurement is developed, and a precise vocabulary for the communication of our techniques and results is presented.

A. Measurement, Interpretation, and Observation

Before a measurement is made, two questions must be answered. They are "What is the characteristic or property we want to measure?" and "What are the units with which we want to express the amount of this property?" For example, we may want to measure the length of an object in centimeters or the temperature of the room air in degrees centigrade as in Figure 2.1. The measurement result will then be expressed as a number of the units of the measured property such as 3.8 cm or 27°C. From these common examples, **measurement** can be defined as the determination of the number of standard units of a property inherent in an object or system. *The result of a measurement is always a number combined with the units of the property measured.* Determinations made without quantitative measurement are more properly called **observations.** Examples of observations are "You look pale" or "It is getting warmer."

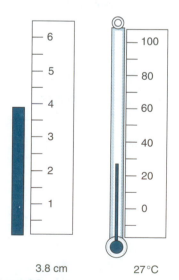

3.8 cm   27°C

Figure 2.1. Measurements result in numbers with units.

15

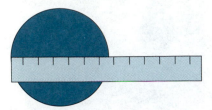

Figure 2.2. To determine the area of the circle, the accessible quantity is measured ($d = 6.8$ cm), and then the area is calculated from the relationship $A = \pi(d/2)^2$. In this case, $A = 36$ cm^2. In measurements, the desired information usually comes from a combination of measurement data and interpretive relationships.

MEASUREMENTS AND INTERPRETATIONS

It is important to distinguish between quantities measured directly and those calculated from the measurement results. For example, we may measure the diameter of a circle and then use this value to calculate its area. It is not strictly accurate to say that we have measured the area of the circle if we have determined it in this way. As shown in Figure 2.2, the determination of the area occurred in two steps: the **measurement** of the diameter and the **calculation** of the area. The measurement step produces a number of centimeters that is related to the true diameter in ways affected by the measurement process. The area calculation step produces an area in square centimeters by means of the relationship between the diameter and the area of a circle. This calculation is an **interpretation step** between what we have measured and what we wanted to know about the circle. Interpretation steps involve assumptions and limitations that are different from those of the measurement steps. Since the variety of the quantities and properties we want to know is very much greater than those we can measure directly, almost all scientific determinations include both measurement and interpretation steps. In fact, a principal part of the art of chemical measurements is recognizing what can be measured directly and then devising a method to obtain the desired information from those measurement data.

INFORMATION FROM SETS OF MEASUREMENT DATA

It is often necessary to make more than one measurement to obtain the desired information. For example, the area of a rectangle requires the measurement of both the length and width of the rectangle. Both data are used in the calculation of the area. Another example is the determination of the normal range of body temperatures in healthy people. This will require the measurement of the temperatures of many subjects and a statistical analysis of the measurement data. In this study, we will encounter many examples of interpretation steps that require rather large sets of measurement data to produce the desired information.

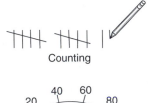

Counting

Scale reading

Digital display

Figure 2.3. Measurement numbers come from three sources.

SOURCES OF MEASUREMENT NUMBERS

We have seen that the basic measurement step produces a number that is the number of units of the measured quantity. This basic observation is useful in categorizing the measurement processes, because people have devised only three basic ways for obtaining measurement numbers. They are **counting**, reading from a numbered **scale**, and reading from a **digital display**, as shown in Figure 2.3. It follows, then, that *everything we measure must be converted to one of these representations*. Realizing this, one can only marvel at the ingenuity of the people who have invented methods to convert temperature, pressure, time, volume, amperes, voltage, mass, light intensity, and many other quantities into these few forms of numerical representation.

MEASUREMENT QUALITIES

absolute error = true value − measured value

relative error = |absolute error/measurement value|

The quality of a measurement may be assessed in several ways. The one we usually think of first is the **accuracy,** the smallness of the difference between the measured and true values. Accuracy of a measurement is usually established by applying the measurement procedure and device(s) to the measurement of a known standard for the measured quantity. The result of an accuracy test is described by the amount of the error involved. The **absolute error** is the absolute value of the difference between the true value and the measured value. A length measurement that is in error by 1.4 mm is high or low by 1.4 mm. The **relative error** is the fraction that the absolute error is of the measured amount. If an absolute error of 1.4 mm occurred in the measurement of a 1 meter standard length, the relative error would be 1.4 mm/1000 mm. The relative accuracy is often expressed as percent or as parts per thousand (ppt). For the example used, the relative accuracy would be 0.14% or 1.4 ppt.

Another measure of measurement quality is the **resolution,** that is, the smallest difference in the measured quantity that can be detected. For example, a scale with marks every millimeter may be able to be read to the nearest 0.2 mm. Thus, the resolution for the measurement would be 0.2 mm. Sometimes resolution is expressed in relation to the measured value. If a length of 1 meter were measured, the **relative resolution** would be 0.02% or 0.2 ppt.

A third means of determining the quality of a measurement is the **precision,** that is, the degree of agreement among repetitive measurements of the same quantity. When a measurement is made near the limit of resolution for that measurement, some imprecision may come from variations in reading the number from the scale. Imprecision can also come from variations in the measurement system that affect the relationship between the quantity measured and the number produced. Further, the quantity measured may vary somewhat with time. The variation in values from repetitive measurements of the same quantity is sometimes called **noise.**

Different types of measurements lead to a variety of distributions of differences in the values from repetitive measurements. The quantitative determination of precision for measurements that produce the so-called random or normal distribution is covered in sections H and I of this chapter.

Resolution is the smallest change in the measured quantity that can be detected.

The resolution for a scale is generally 1/10 or 2/10 of the value between the finest markings.

Precision is a measure of the reproducibility of repetitive measurements of the same quantity.

Study Questions, Section A

1. What is the difference between a measurement and an observation?

2. What two elements are required for any measurement result?

3. Why is interpretation so often a part of the process of determining a quantity?

4. Is it accurate to say that we are going to measure the speed of a runner with the use of a stopwatch and a set distance?

5. If a snail travels 5.00 cm in 300 seconds, what is its speed? What kinds of measurements are made in this determination?

6. What is the density in $g \, cm^{-3}$ of a spherical lead fishing sinker 4.00 mm in diameter that weighs 380 mg? What types of measurements are made in the determination of the sinker's density?

7. A person measures the length of line on a graph as 7.35 cm. The actual length of the line is 7.31 cm. Calculate the absolute and relative errors. Express the relative error in percent and in parts per thousand.

8. Discuss the possibility that the measurement referred to in Question 7 is limited by the resolution of the ruler used.

Answers to Study Questions, Section A

1. A measurement is quantitative. It is numerical. An observation is qualitative or comparative. It uses such words as increasing or more than, but without a numerical evaluation of the relative quantities.

2. A measurement includes the number of units of the quantity being measured and the name of those units.

3. Interpretation of measurement data is often part of the determination of a quantity because many of the quantities we want to know cannot be measured directly.

4. The measurement number comes from the stopwatch, so the measured quantity is necessarily time. To convert the time required to run a certain distance to velocity involves an in-

terpretation step. Thus, it is not accurate to say that we measure speed in this case.

5. Speed is the ratio of distance and time. The distance could be measured with a meter stick. In this application, the meter stick acts as a length-to-number of centimeters converter. The time can be measured with a stopwatch, a time interval-to-number of seconds converter.

$$\text{speed} = \frac{d}{t} = \frac{5.00 \text{ cm}}{300 \text{ sec}} = 0.0167 \text{ cm/sec}$$

6. The volume of the sphere is

$$V = \frac{4}{3}\pi r^3 = \frac{4}{3} \cdot 3.14 \cdot (2.00 \text{ mm})^3 = 33.49 \text{ mm}^3$$

Density is mass over volume.

$$\rho = m/V = 0.380 \text{ g}/3.35 \times 10^{-2} \text{ cm}^3 = 11.3 \text{ g cm}^{-3}$$

A measurement of length (radius, diameter, or circumference) is required. Measurement of the sphere's mass is also needed. A length-to-number converter is used for the length (ruler, caliper, or tape measure), and a scale or balance is used for the mass.

7. The absolute error is $|7.31 \text{ cm} - 7.35 \text{ cm}| = 0.04 \text{ cm}$. The relative error is $0.04 \text{ cm}/7.35 \text{ cm} = 0.005$. In percent, this is 0.5%. In parts per thousand, this is 5 ppt.

8. The error is only 0.04 cm or less than half a millimeter. The marks on many rulers are not fine enough to resolve small fractions of millimeters. Thus, it is very possible that the measurement error was due to limited resolution of the readout scale.

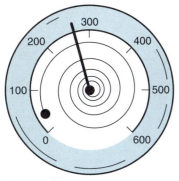

Figure 2.4. A therometer based on two conversion devices, the bimetallic coil that converts temperature to the angle of the pointer and the scale that converts the angle of the pointer to a number.

B. Elements of Measurement Systems

A device of some kind is used for all measurements other than counting. Measurement devices at all levels of complexity can be characterized and understood in terms of the **conversion devices** they employ. All the dials, scales, numerical displays, and strip-chart recorders from which we obtain our measurement data are readout conversion devices. They in turn obtain their information from other conversion devices that relate back to the quantity being measured. An example is the thermometer illustrated in Figure 2.4. It contains a bimetallic strip that causes the pointer to rotate as the temperature changes. This device is a temperature-to-angle converter. As the angle of the pointer changes, the number it points to also changes. This is the function of the printed scale. The scale is thus an angle-to-number converter. In combination, these two conversion devices convert the temperature to a related number.

All measurement systems can be similarly broken down into their component conversion devices. The concept of conversion devices thus becomes the core of the understanding of measurement systems. An advantage of this approach to the study of measurement systems is that it is not necessary to understand the mechanics or electronics involved in the conversion devices in order to adequately understand their function in the measurement process. In this section, the concept of the conversion device is explored, and several familiar measurement systems are viewed from this perspective.

CONVERSION DEVICES

The output quantity of a readout conversion device is a number.

The input quantity of an initial conversion device is the quantity being measured.

The input quantity of each conversion device must be the output quantity of the one that precedes it.

Conversion devices can be thought of in three categories: initial conversion devices, intermediate conversion devices, and readout conversion devices. **Readout conversion devices** produce the measurement number. They are the familiar scales we see on rulers, protractors, dials, and gauges such as thermometer scale in Figure 2.4. When we read the position of the pointer or measured object against these scales, we obtain the measurement number. We will refer to these as **scalar readout devices** or **analog readout devices.** Other forms of readout conversion devices produce a digital number, which can be displayed in a numerical readout or transferred to a computer. These are called **digital readout devices.**

Initial conversion devices convert the quantity that is being measured into another quantity. They are also sometimes called input conversion devices, **sensors** or **transducers.** Examples are thermistors that convert temperature to electrical resistance, pH electrodes that convert the solution pH to voltage, and phototubes that convert light intensity to electrical current. The bimetallic strip in the thermometer of Figure 2.4 is an example of an initial conversion device.

An **intermediate conversion device** converts one quantity to another as needed to match the initial conversion device to the readout conversion device. For example, to use a pH electrode with a scalar readout device, we would need an intermediate conversion device to convert the voltage from the pH electrode to the position of the pointer along the scale. A block diagram of a measurement system containing all three forms of conversion devices is shown in Figure 2.5. The output quantity of each conversion device is the input quantity of the next.

INPUT/OUTPUT RELATIONSHIPS

Every conversion device has a specific relationship between its input and output quantities. This is expressed mathematically as

$$Q_{out} = f(Q_{in}) \qquad\qquad 2.1$$

where f is some mathematical function. If the output is a linear function of the input, as is often the case,

$$Q_{out} = Q_{out}^{\circ} + KQ_{in} \qquad\qquad 2.2$$

where Q_{out}° is the output value when Q_{in} is zero and K is the proportionality constant between the input and output quantities. K has the units of output units $\times$ input units^{-1}. For the case of the pH electrode,

$$V = V_{pH=0} + K \times pH \qquad\qquad 2.3$$

When the input/output relationship can be expressed in a mathematical function such as Equation 2.3, the relationship is called the **transfer function.** The transfer function should always be expressed as the output quantity equals some function of the input quantity.

Sometimes we use conversion devices for which a theoretical relationship between the input and output quantities has not yet been established. Even in these cases, a relationship still exists; it just has to be established empirically (by experiment). To do so, one would plot values of the output quantity for known values of the input quantity. A plot of output values versus the input quantity is called a **working curve.** If a theoretical transfer function is known, this function can be fit to the working curve over the **operating range** of the conversion device. A working curve for a pH electrode is shown in Figure 2.6. The dotted line corresponds to the theoretical transfer function (Equation 2.3), and the heavy line corresponds to the experimentally determined response. The portion of the working curve over which the response is linear is called the **linear operating range.** Outside this range, the conversion device might still be useful, but the linear relationship can no longer be assumed.

The input and output quantities of a true conversion device are always expressed in different units. The ratio of the output units to the input units is the proportionality

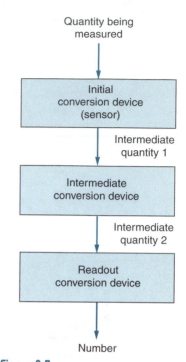

Figure 2.5. When a series of conversion devices is used, the output quantity of one must be the same as the input quantity of the next.

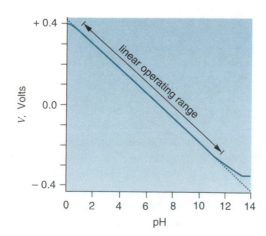

Figure 2.6. A working curve (solid line) compared with a plot of the transfer function (dotted line) for a pH sensor. The portion of the operating range over which the response is linear is indicated.

constant K in Equation 2.2. For the case of the pH electrode (Equation 2.3), the proportionality constant K is approximately -0.06 V/pH. The value of the proportionality constant is also called the **sensitivity** of the conversion device. The greater the change in the output quantity per unit change in the input quantity, the greater is the sensitivity. The sensitivity can also be obtained from the slope of the working curve in the linear operating range. Thus,

$$\text{Sensitivity} = \frac{\text{output for input } b - \text{output for input } a}{\text{input } b - \text{input } a} \frac{\text{output units}}{\text{input units}} \quad\quad 2.4$$

$$= \frac{\Delta(\text{output quantity})}{\Delta(\text{input quantity})} \frac{\text{output units}}{\text{input units}}$$

Study Questions, Section B

9. Name and define the three types of conversion devices.

10. What are the two kinds of readout conversion devices?

11. For the thermometer shown in Figure 2.4, what is the intermediate quantity between the initial and readout conversion devices?

12. A ruler converts a length to a number. What is the form of the transfer function for a ruler?

13. Define the sensitivity of a conversion device.

14. The output voltage of a thermocouple goes from 34 mV at 600 °C to 60 mV at 1000 °C. The response curve is essentially linear in this region. What is the sensitivity of the thermocouple in the region between 600 and 1000 °C?

15. Justify the statement that the sensitivity of a conversion device is equal to the slope of its working curve.

Answers to Study Questions, Section B

9. The three types of conversion devices are initial conversion devices, for which the input quantity is the quantity being measured; readout conversion devices, which provide the output number; and intermediate conversion devices, which provide any required conversions between the initial and readout conversion devices.

10. Readout conversion devices are either analog (numbers on a marked scale such as a ruler) or digital (a specific number in numerical form such as with a digital watch).

11. The initial conversion device converts temperature to angle. The readout conversion device converts angle to number. Therefore, the intermediate quantity that encodes the data is the angle of the pointer.

12. The output quantity is a number. The input quantity is a length. There is a linear relationship between the length and the number. Therefore, $\# = \#° + KL$, where $\#$ is the output number, $\#°$ is the output number when the length is 0, L is the length, and K is the proportionality constant between length and number. Since the ruler markings usually start at one end of the ruler, the quantity $\#°$ would usually be zero.

13. The sensitivity of a conversion device is the amount of change in the output quantity for a unit change in the input quantity.

14. The sensitivity, K, of the thermocouple is

$$\frac{\Delta \text{output}}{\Delta \text{input}} = \frac{60 \text{ mV} - 34 \text{ mV}}{1000 \text{ °C} - 600 \text{ °C}} = \frac{26 \text{ mV}}{400 \text{ °C}} = 6.5 \times 10^{-5} \text{ V/°C}$$

15. The working curve is a plot of the output quantity versus the input quantity. The slope of such a plot is thus Δoutput/Δinput, which is the definition of sensitivity.

C. Examples of Measurement Systems

THE ANALOG THERMOMETER

The familiar thermometer (shown in Figure 2.7) involving a bulb containing an expandable liquid and the stem from which the degree of expansion can be determined is a good example for an analysis of the relationship between the input quantity and the output number. The thermometer bulb, stem, and liquid form a conversion device by which the temperature of the liquid in the bulb is converted to the length of the liquid column in the stem. Since temperature is the quantity we are trying to measure, this is an initial conversion device. The transfer function has the form $l = f(T)$. This relationship is nearly linear over the operating range, so we can also write

$$l = l° + KT \qquad\qquad 2.5$$

where $l°$ is the length of the liquid column when $T = 0$. The term K is the sensitivity of the conversion device; in this case, K is the change in length for a unit change in temperature.

The quantity length can be converted directly to number with a linear scale. The linear scale is then used as a readout conversion device. It has a transfer function of the form $\# = f'(l)$. If the transfer function of the input conversion device is linear, the linear scale will also be linearly calibrated, so that

$$\# = \#° + K'l \qquad\qquad 2.6$$

where $\#°$ is the number marked on the scale where $l = 0$ and K' is the change in the number value per unit length of scale. If one were making the thermometer, it would be handy to know how to draw the scale. This can be determined by substituting Equation 2.5 into Equation 2.6 as follows.

$$\# = \#° + K'(l° + KT) = \#° + K'l° + K K'T \qquad\qquad 2.7$$

The condition for the scale number to read the temperature directly is that $\# = T$. This condition will be met for Equation 2.7 when

$$\#° = -K'l° \qquad \text{and} \qquad K' = 1/K \qquad\qquad 2.8$$

The first condition sets the zero reading correctly, and the second condition sets the calibration for nonzero temperatures. It is interesting that the K values for the two conversion devices are reciprocals. This will always be true when the scale is calibrated in the units of the input quantity. (see Example 2.1) It is also of interest that even if the transfer function of the initial conversion device were not linear, a direct reading could still be obtained if the transfer function of the readout device is its complement.

THE AMMETER

The moving coil ammeter (Figure 2.8) is a basic measurement device that measures electrical current as the position of a pointer against a curved scale. It, too, is a combination of two conversion devices. The first is the device that converts the magnitude of the current to the angle of the pointer. The second is the scale against which the pointer position is read. The current to be measured is in a coil of wire wound on a drum that

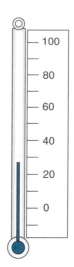

Figure 2.7. The thermometer as a combination of a temperature-to-length initial conversion device and a length-to-number linear scale readout conversion device.

For direct reading of the number of units of the measured quantity, the readout device transfer function must be the complement of the combined transfer functions of the previous conversion devices.

Example 2.1

For the thermometer of Figure 2.7, if $l° = 2$ cm, $K = 0.10$ cm deg^{-1}, and $\#° = -20$, for the number $\#$ to equal the temperature T, $K' = 1/K = 10$ cm^{-1} and $\#° = -10$ cm^{-1} $\times 2$ cm $= -20$. So the scale should begin at -20 and increase at 10 per centimeter.

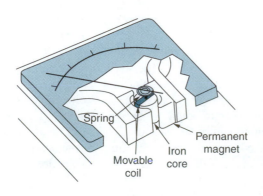

Figure 2.8. The moving coil ammeter is a combination of a current-to-angle conversion device (the meter movement) and an angle-to-number conversion device (the scale).

pivots on fine bearings. A magnetic field is produced that is proportional to the current through the coil. The magnetic field of the coil acts against the magnetic field of a permanent magnet to provide a rotating force between the coil and the magnet. Since the magnet is fixed and the coil is pivoted, the coil rotates. A spring provides a counteractive force so that the angle of rotation is proportional to the current through the coil. The transfer function for this device is $\theta = \theta° + Ki$, where θ is the angle of rotation, $\theta°$ is the angle when the current is zero, K is the sensitivity of the conversion device in degrees per ampere, and i is the current through the coil.

The curved scale is the output conversion device. It is ruled and labeled so that the value of the current can be read directly. The transfer function for the scale is $\# = \#° + K' \theta$. Just as in the case of the thermometer, the conditions for direct reading of the current ($\# = i$) are $\#° = -K' \theta°$ and $KK' = 1$.

It might seem as though high sensitivity in a conversion device would be the best thing. However, the maximum possible input quantity is equal to the readout device range times the conversion device sensitivity. For example, if the meter scale covered a range of 90 degrees and the meter movement had a sensitivity of 2 μA degree^{-1}, the maximum input current on the meter scale (called the **full-scale reading**) would be 180 μA. For a scalar readout, the resolution is a constant fraction of the range. If the pointer position can be read to 0.5 degree, the resolution of the current measurement would then always be 1/180 of the full-scale current value, or 1 μA for the example given. Thus, as we shall see, the sensitivity, range, scale length, reading resolution, and resolving power are all interrelated.

An increase in the sensitivity of a conversion device may improve resolution and precision, but it is likely to also decrease the range because of a limit in the maximum output value the conversion device can produce.

For decades, the basic ammeter provided the scalar or analog representation of the measurement values for many measured quantities. When the current meter is used for the measurement of quantities other than current, other conversion devices must be employed to convert the measured quantity to a related current.

THE ANALOG CLOCK

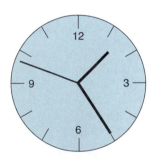

Figure 2.9. The analog clock is a combination of an initial conversion device that converts time to rotation of the hands and a readout conversion device (the dial) that converts rotation to number.

The addition of multiple hands moving at a constant ratio of rates can expand the range of the circular scale indefinitely.

The analog clock (Figure 2.9) is used to indicate the current relative position in the cycle called the day. In a 24-hour clock, the hour hand rotates once each day. The more familiar 12-hour clock has an hour hand that rotates twice each day, and we must rely on an auxiliary observation to determine whether the current half-cycle is ante meridiem (am) or post meridiem (pm). The heart of any analog clock is a kind of motor that can rotate the hour hand at a uniform rate and that can be calibrated to complete one or two rotations every day. The clock motor is an initial conversion device that converts time to rotation through its constant speed. The transfer function for this conversion device is thus $\theta = Kt$, where θ is the angle of rotation, t is time, and K is the sensitivity of the conversion device in degrees of rotation per unit of time. The circular scale of the clock must be divided into an integer number of input units if the readings after the first complete rotation are to have any meaning. Thus K must have the value of 360° per cycle. The cycle for the clocks we are familiar with is one-half day (12 hours).

For the hour hand of a 12-hour clock, K is not very large: 360 °/0.5 days = 720° per day. The resolution of time reading from the hour hand is sufficient to know when it is time for lunch, but not when it is time to be at your 9:00 am class. To measure the position in the daily cycle more precisely, the day was divided into smaller units, namely, hours, minutes, and seconds. The sensitivity for the hour hand in degrees per hour is 360°/12 hours = 30° per hour. For minutes, however, the sensitivity is only 360°/720 min = 0.5° per minute. Therefore, to resolve minutes, another pointer was added with a rotation speed of 360°/hour or 6° per minute. This hand is useful for reading the relative position within the cycle of each hour. The use of the same motor with gearing keeps the hands synchronized so that the minute hand goes through exactly 12 revolutions per revolution of the hour hand. The combination of these hands lengthens the effective scale greatly, so that there are $12 \times 360° = 4320°$ of rotation every twelve hours. Adding a hand that rotates once each minute (the second hand) provides still further resolution, now giving over a half million degrees of rotation per day. The possibility to provide scale expansion by multiple, coordinated pointers is easily accomplished with the circular scale.

The **stopwatch**, a device for measuring time interval, is a variation on the analog clock. It includes a mechanism for setting the hands to their zero position and a control for starting and stopping the motor. The time interval between the actuation of the start and stop controls is read by the position of the hands against the circular scale.

pH PAPER

It is interesting to consider pH paper as a measurement system. It is the kind of paper used to measure the pH of water in aquariums, pools, and spas. A test strip of the paper is dipped into the water. The resulting color of the strip is related to the pH of the water. To make the correlation between the color and the pH, one compares the color of the strip with a table printed on the dispenser containing the strips. In the table, there is a pH range associated with each color. The system is composed of two conversion devices. One is the paper, which is a pH-to-color converter. The second is the table that converts color to number. The table acts as a kind of scale for the conversion of color to number. If the color appears to be between those of adjacent table entries, it is possible to interpolate between the tabular numbers to obtain the pH reading. This would appear to be an example of a scale for which the quantities of length or angle are not involved.

Study Questions, Section C

16. Derive an equation relating the temperature to the output number for the bimetallic coil thermometer shown in Figure 2.4. Give the conditions required for the output number to be equal to the temperature in degrees.

17. What conversion devices are used in an analog clock?

18. For a current meter with a full-scale reading of 100 μA and an accuracy rating of 2% of full scale, calculate the absolute and relative accuracy of a current reading at one-third of full scale.

19. A clinical thermometer has a scale that is 44 mm in length. The range is from 96 to 106 °F. The finest markings are every 0.88 mm. What is the sensitivity of the temperature-to-length converter, and what is the resolution of the read-out scale markings?

20. An automobile speedometer has a rotational range of 270° and a speed range of 0 to 120 mph. What is the sensitivity of the speed-to-angle converter?

Answers to Study Questions, Section C

16. $\theta = \theta° + KT$ and $\# = \#° + K'\theta$

    $\# = \#° + K'(\theta° + KT) = \#° + K'\theta° + K'KT$

    For $\# = T$, $\#° + K'\theta° = 0$ and $K' = 1/K$.

17. The input conversion device is a motor with a constant speed of rotation, so it is a time-to-angle converter. The output conversion device is the scale on the clock face that converts the angle to a number.

18. The reading is $1/3 \times 100$ μA $= 33.3$ μA. The inaccuracy may be as large as 2% of full scale, or 0.02×100 μA $=$

2 μA. The absolute accuracy is 2 μA. The relative accuracy is then

$$2 \ \mu A/33.3 \ \mu A \times 100 = 6\% \text{ error}$$

19. The sensitivity is

$$44 \text{ mm}/(106 - 96) \ °F = 4.4 \text{ mm per } °F$$

The temperature change between each mark is thus 0.88 mm/4.4 mm °F^{-1} = 0.2 °F.

20. The sensitivity is 270 °/120 mph = 2.25 degrees per mph. The resolution is 0.2 mark $\times$ 2 mph/mark = 0.4 mph.

D. Characteristics of Scalar Readouts

We are all very familiar with the process of reading a number from a scale. The marks on the scale are usually uniformly spaced, and they are labeled with numbers often enough to easily determine the value represented by each mark. The measurement number is obtained from the value of the mark closest to the position of the pointer or the mark opposite the end of the object being measured. There are two general forms of the scale: the linear scale represented by the ruler and the circular scale represented by the clock (Figure 2.10). In either case, the number obtained is not automatically in numerical form as in counting or digital display; it must be obtained from the scale by an observer. In this section, we will look at the quantities represented by scales, the methods of reading numbers from scales, and the sources of imprecision and inaccuracy in scalar readouts.

LINEAR SCALES

Only the scalar expression of measurement number allows interpolation between the calibrated measurement increments.

The ruler is the fundamental device that has a linear scale. The marks on the ruler are spaced in uniform units of length such as millimeters. The physical dimension of an object is obtained by lining up one edge of the object against the zero mark on the scale and reading the value of the mark corresponding to the other edge of the object. If the edge of the object falls between two marks on the scale, the fraction of the interval included in the dimension can be estimated. This process is called **interpolation.**

The ruler is a readout conversion device. It is used to convert the property of **length** into a **number** of length units. *Length is the only input quantity this conversion device can use.* Therefore, the ruler is a complete measurement system for the measurement of length. Whenever a linear scale is used for the measurement of a quantity other than length, another conversion device is required along with the linear scale.

The accuracy of the conversion from length to number depends completely on the accuracy of the markings on the scale. The accuracy of a ruler is generally not expected to be better than the interval between adjacent marks. For accurate work, one would choose a ruler of specified accuracy made of material that is dimensionally stable against changes in temperature and humidity. Alternatively, one could periodically check the accuracy of a utility ruler against a carefully maintained object of standard length.

Devices with a linear scalar output generally have a measurement precision that is independent of the absolute measurement value. If the absolute precision is the same whether the measured value is large or small, it follows that the relative precision for scalar measurements will increase as the magnitude of the measured quantity decreases.

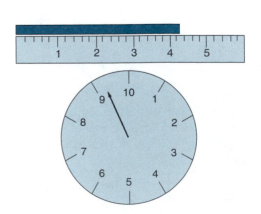

Figure 2.10. Scales are either linear or circular. The measurement number is obtained by observing the position of the indicator against the scale.

SCALAR READOUT CHARACTERISTICS

All conversion devices have a limited **range** of input or output values over which their operation is satisfactory, and it is important that each conversion device in a measurement system be operated within this range. The range of the readout device, however, puts a physical limitation on the largest possible measurement number. For the typical ruler, for example, the largest length that can be measured (in a single step) is 12 inches. Increasing the range of the measurement requires the use of a different device (such as a 20-foot tape).

The other characteristic is the **resolution,** the smallest detectable difference in the measurement number. The limit of resolution in reading a linear scale is generally one-fifth to one-tenth the distance between the closest markings on the scale. See Figure 2.11. The resolution of a scale reading depends greatly on the interpolation skill of the observer (see the next section), the resolution of the scale markings, and the fineness of the edge of the object being measured (or the thickness of the pointer). The values of repetitive measurements made near the limits of resolution for the ruler will have some variation.

INTERPOLATION ERRORS

An interesting phenomenon occurs with the reading of a scale when the process of interpolation between scale marks is employed. Apparently, the mind's eye tends to favor particular values. Laitinen[1] tested the least significant digit (the interpolated one) in 1500 student buret readings for bias toward particular numerals. In these readings, there should have been an equal representation of all the numerals in the interpolated (least significant) digit. In other words, the frequency for each numeral should have been 1500/10 = 150. A very strong bias appeared toward the lower digits (0–3), with all the numerals from 4 to 9 underrepresented, especially the number 7. See Table 2.1. The application of the χ^2 (chi-square) statistical test to the actual measurement values indicated that the test for a uniform distribution of the 10 numerals was improbable far past the 99.9 percentile level.

VERNIERS

The vernier is a device to improve on the precision and reduce the human error in an interpolated reading. The **vernier** is a movable secondary scale whose zero is lined up with the edge of the object being measured as shown in Figure 2.12. The intervals between markings on the vernier are 0.90 times the interval spacing in the main scale. The vernier marking that lines up with a mark on the main scale gives the correct interpolated value. The vernier thus greatly reduces the interpolation bias discussed above. Verniers are included on scales intended for the most precise measurements.

Figure 2.11. The resolution of a scale reading is 0.1 to 0.2 times the smallest marking interval. In this case, the resolution is 0.01 to 0.02 scale units.

Table 2.1
Frequency of Interpolated Numeral Readings for 1500 Readings

Terminal numeral	Frequency
0	212
1	212
2	229
3	166
4	124
5	107
6	110
7	81
8	134
9	125

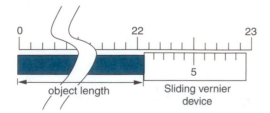

Figure 2.12. A vernier used in length measurement. The measured value is 22.05.

[1]H. A. Laitinen and W. E. Harris, *Chemical Analysis,* 2nd ed. McGraw-Hill, New York, 1975.

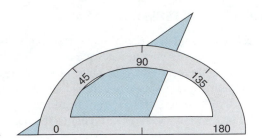

Figure 2.13. A protractor for measuring angle.

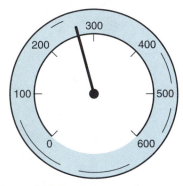

Figure 2.14. An oven thermometer with a circular scale.

SCALES IN ARCS OR CIRCLES

Scales arranged on an arc or a circle use a pointer that rotates so that it can follow the curve of the scale. A curved scale is thus very useful if the conversion device produces a rotation related to the input quantity. The fundamental device with a curved scale is the protractor used for the measurement of angle. In this application, the protractor (Figure 2.13) is an angle-to-number conversion device. The electrical current meter, described previously, is a good example of a measurement in which an arced scale is used as the readout conversion device. Examples of scales using only an arc of the circle are an automobile speedometer and the oven thermometer shown in Figure 2.14.

As we have seen in the clock example, scales arranged on a complete circle are especially useful for indicating relative position in a repetitive process or cycle.

Study Questions, Section D

21. What is the only quantity that can be measured by a linear scale?

22. Regardless of its markings, a circular or arc scale can only measure what quantity?

23. What is interpolation, and how does a vernier aid in this operation?

24. Readout conversion devices involving a scale are said to be analog readouts as opposed to digital readouts. How would you distinguish between these two forms of readouts?

Answers to Study Questions, Section D

21. A linear scale can only measure length.

22. The only quantity that can be measured by a scale on an arc or circle is angle.

23. Interpolation is the estimation of values on a scale that are between the marks. The vernier provides another scale that enables the interpolated value to be read directly.

24. The scalar or analog readout is provided by matching an object or pointer against the markings on the scale. Its nor-
mal operation involves an observer to obtain the output number. The value of the readout between the markings can be estimated by interpolation. With a digital readout, the output number is readily recorded without an observer, and no estimation of the values between the numbers presented is possible.

E.   Volume-to-Length Converters

It is very common to see a linear scale on devices used to measure volume. Examples are measuring cups, graduated cylinders, oil dipsticks, liquid-based thermometers, and volumetric flasks. As we now understand, such devices are a combination of a volume-to-length conversion device and a length-to-number conversion device. Because of their prevalence in chemical measurements, it is worth having a careful look at how these devices work and how they are used. The graduated cylinder is a good example for further study. The height of the column of liquid in the cylinder is related to its volume by the relationship $l = V/\pi r^2$ where l is the length of the liquid column, V is the volume of the liquid, and r is the interior radius of the cylinder.

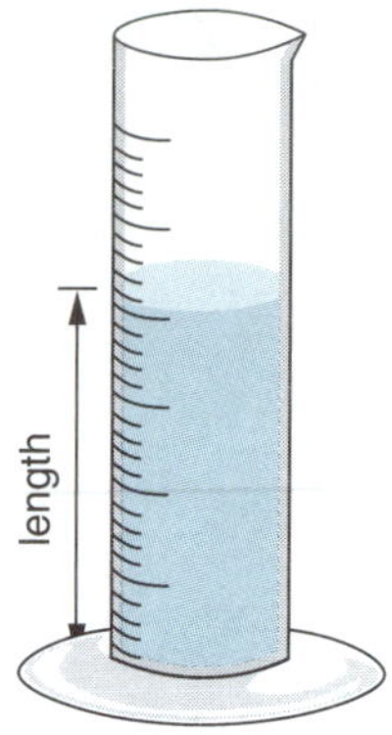

THE GRADUATED CYLINDER

Consider the graduated cylinder shown in Figure 2.15. The transfer function for this device will have the form $l = f(V)$. For a perfect cylinder, $V = \pi r^2 l$. Therefore, the transfer function is $l = V/\pi r^2$, and the sensitivity (output units × input units^{-1}) is $1/\pi r^2$ cm mL^{-1}. For this relationship to hold exactly, we must assume that the cylinder is a perfect geometric form, that r is constant throughout, and that the "0" length mark is exactly at the base of the cylinder.

For convenience, the scale on the cylinder is generally in volume units. The volume marks correspond to the volumes associated with each length according to the transfer function of the device. Before the techniques for precise glass fabrication had been developed, the marks on each device were obtained by the introduction of known standard volumes. This corresponds to the development of the scale from the working curve.

SCALE INDICATION BY LIQUID LEVEL

The art of using a graduated cylinder also involves the correct and consistent reading of the position of the top of the liquid column. This is more complex than measuring the length of a stick for two reasons. One is that the top of the liquid is curved because of its contact angle with the material of the cylinder, as shown in Figure 2.16. This curvature is called the **meniscus.** Practice has shown that the readings taken from the level in the central, flat portion of the liquid column are the most consistent. Thus, the cylinder markings are based on the central level. Another difficulty is that the markings are on the outside of the cylinder material and the liquid column is inside. The length scale and the object are thus separated by the thickness of the cylinder material. Viewing the liquid behind the scale at an angle other than orthogonal will result in an error called **parallax.** The basis for the parallax error is illustrated in Figure 2.17.

THE BURET

The function and operation of the buret (Figure 2.18) are similar to that of the graduated cylinder except that a valve is located at the bottom end of the cylinder and the "0" mark is located at the top of the scale. A buret measures the volume of liquid delivered through the valve, not the volume of liquid contained in the buret. In operation, the buret is filled with the liquid to be delivered in measured quantities. The buret volume that includes the valve and delivery tip must also be completely filled. For convenience, it is not necessary to fill the buret exactly to the zero mark. The volume delivered can be obtained by subtracting the initial scale reading from the final scale reading. Another aspect of the art of using a buret involves the delivery of the metered volume. In general, the dispensed volume will leave the tip of the buret below the valve in drops. Until the volume contained in the drop has become part of the solution in the receiving vessel, the scale reading will not accurately reflect the volume delivered. This error is significant: one drop is roughly 0.05 mL and the resolution of the buret reading is 0.01 mL or less (with interpolation). Skilled operators accomplish complete delivery by touching the side of the receiving flask to the buret tip to transfer any partial drop hanging there to the flask. This must be done into a waste flask before the first reading is taken and into the receiving flask before each delivered reading is taken.

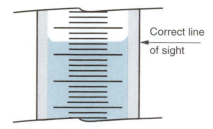

Figure 2.15. A graduated cylinder is a conversion device for converting volume to length combined with a scale for converting length to number.

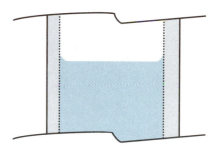

Figure 2.16. Meniscus formed at contact between the liquid and the container.

Figure 2.17. Reading the position of a liquid level without parallax. Reading from any other angle will put the bottom of the meniscus behind a portion of the scale that is above or below the correct position.

THE VOLUMETRIC FLASK

Flasks designed to contain a specific volume are called **volumetric flasks** (Figure 2.19). They have a mark near the top of the flask where the flat portion of the liquid surface should be when the liquid in the flask has the calibrated volume. The flask is shaped with a narrow neck so that the change in the solution level for a given change in solution volume (sensitivity) is relatively large. Obviously, the volume measurement cannot be exact if there are any droplets on the neck of the flask above the calibration mark. These can be avoided by beginning with a scrupulously cleaned flask. This admonition is basic to the accurate use of all volumetric glassware.

THE PIPET

The pipet (Figure 2.19) is a device designed to deliver a single specific volume. The liquid to be delivered is drawn into the pipet by the application of suction through the tube at the top until the liquid level is above the calibration mark. Then through partial release of the vacuum above the liquid, the liquid is delivered to a waste container until the liquid meniscus is exactly at the calibration mark. The pipet tip is then touched to the side of the waste container to remove any partial drop, and the measured volume is ready for delivery to the receiving vessel. After the solution has drained into the receiving flask, the tip is again touched to the side of the flask for partial drop removal. A bit of solution remains in the very tip of the pipet, but this should not be blown into the receiving vessel since this retention has been taken into account in the volume calibration process. Again, cleanliness is essential because a dirty pipet will lead to droplets of solution clinging to the pipet surface.

EFFECTS OF TEMPERATURE ON VOLUMETRIC MEASUREMENTS

There are two effects of temperature on the measurement of volume by volume-to-length conversion devices. One is due to the relationship between temperature and the calibrated volume(s) of the device. Thermal expansion of the device will increase the volume contained by the square of the change in device radius. The length of the scale will also increase, but only directly with the degree of expansion. Therefore, the volume contained for a given calibration mark is a function of temperature. Fortunately, volumetric ware is generally made of glass with a very low thermal coefficient of expansion, so this effect is below the resolution of the measurement unless extreme temperatures are involved. A greater effect can come from the expansion of the liquid being measured. This effect does not involve an error in the volume measured, just an inconsistency in the amount (mass) of material delivered or contained at different temperatures. To avoid being misled by this effect, remember that it is volume that is measured by the volumetric equipment within the limits described above. If the quantity of matter is the actual value of interest, this involves an additional interpretive step in which the temperature is involved. Where volume is being interpreted to represent a certain quantity of matter, careful attention must be paid to the temperature of the solutions at the time of the volume measurement.

The normal method for the calibration of volumetric glassware is to weigh the amount of water the device contains or delivers when filled to the mark. At 20 °C, the volume occupied by 1.000 g of water is 1.0028 mL. At other temperatures, the volume of one gram of water is different. When it is necessary to calibrate volumetric ware at a room temperature other than 20 °C, a correction factor must be applied. Table 2.2 contains the factors used. For example, a 50-mL volumetric flask was weighed dry and then filled to the mark with water. The temperature of the laboratory, flask, and water were 28 °C (warm day). The dry and filled masses of the flask were 28.185 g and

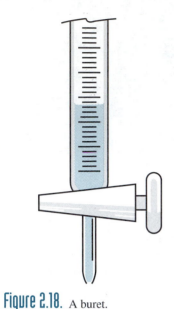

Figure 2.18. A buret.

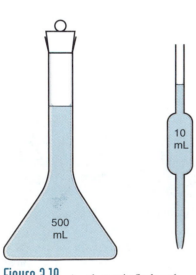

Figure 2.19. A volumetric flask and pipet.

Table 2.2
Volume of 1.000 g of Water at Various Temperatures

Temperature (T, °C)	Volume (mL)	Volume (mL) corrected to 20 °C
10	1.0013	1.0016
12	1.0015	1.0017
14	1.0018	1.0019
16	1.0021	1.0022
18	1.0024	1.0025
20	1.0028	1.0028
22	1.0033	1.0032
24	1.0037	1.0036
26	1.0043	1.0041
28	1.0048	1.0046
30	1.0054	1.0052

78.029 g. The actual volume of the flask at 28 °C is the product of the mass of the water times the volume per gram of water at 28 °C:

$$\text{Volume} = (78.029 - 28.185) \times 1.0048 = 50.083 \text{ mL}$$

However, taking the expansion of the flask also into account, the volume of the flask at 20 °C would be $(78.029 - 28.185) \times 1.0046 = 50.073$ mL

Study Questions, Section E

25. Volumetric glassware is a combination of what two conversion devices?

26. A graduated cylinder is being designed to have 1.00 cm between marks that represent 5.00 mL of volume. What should the inside diameter of the cylinder be?

27. Why is the surface of a liquid in a container not exactly on the same plane, and what part of the surface should be compared with the markings on volumetric glassware?

28. What errors occur in volumetric measurements when the observer's line of sight is not parallel to the surface of the liquid?

29. Which is a better measure of the amount of a liquid material, volume or mass, and why?

30. The initial reading of a buret containing water at 26 °C is 44.42 mL. The buret is used to deliver water to a weighing bottle with an empty weight of 2.6943 g. After delivery, the buret reading is 20.14 mL, and the weighing bottle weighs 26.9136 g. Assuming the weighing to be correct, what is the accuracy of the buret volume readings? Use the data from Table 2.2.

Answers to Study Questions, Section E

25. The shape of the volumetric glassware provides a volume-to-length conversion. The markings on the glassware are a scale that converts length to number.

26. $V = \pi r^2 l$, $\quad \dfrac{l}{V} = \dfrac{1}{\pi r^2} = \dfrac{1 \text{ cm}}{5 \text{ cm}^3}$, $\quad r^2 = \dfrac{5 \text{ cm}^2}{\pi}$,

$r = \sqrt{5/\pi} = 1.26$ cm, $\quad d = 2r = 2.52$ cm

27. The interaction of the liquid with the container material creates a meniscus where the surface of the liquid touches the container wall. The level at the container–liquid interface is often different from that in the middle of the liquid surface. The level at the middle of the liquid surface should be used for volumetric measurements.

28. A parallax error occurs owing to the fact that the markings on the glassware are closer to the observer than the liquid surface being measured. Because of the angle of viewing, the liquid surface will not appear to be behind the correct marking.

29. Mass is a better measure of the amount of material because it is always exactly related to the number of molecules of the substance present. The volume of a given amount of material is a function of its density, and the density of most materials changes with temperature.

30. The mass of the water delivered is

$$26.9136 \text{ g} - 2.6943 \text{ g} = 24.2193 \text{ g}$$

The volume of this mass water at 26 °C is

$$24.2193 \text{ g} \times 1.0043 \text{ mL g}^{-1} = 24.323 \text{ mL}$$

The nominal volume delivered is

$$44.42 \text{ mL} - 20.14 \text{ mL} = 24.28 \text{ mL}$$

The error is 0.04 mL in 24.28 mL or 0.165%.

F. Characteristics of Digital Readouts

Analog signals encode data as the voltage, current, charge, or power.

Digital signals encode integer numbers as combinations of HI and LO (1 and 0) signal levels.

Digital readout devices obtain the number they display in either of two ways. The number results either from a counting operation or from the conversion of an electrical voltage to a related number (which is also sometimes accomplished by counting). A conversion device that converts a signal voltage to a proportional digital number is called an **analog-to-digital converter** or ADC. The analog part of the term comes from the fact that the magnitude of the voltage signal is related to (or is analogous with) the value of the data represented. Analog also means that the signal voltage is continuously variable, that is, it can have fractional values and can change infinitesimally. Voltage is not the only electrical quantity that can form the basis for an analog signal; the magnitudes of current, charge, and power can also be used to represent measurement data. (For a review of electrical quantities and their relationships, please see Background C.) However, nearly all ADCs assume that the analog input signal is in the voltage form. Many of the commonly used sensors produce an analog signal related to the quantity they are sensing. Because the trend is toward use of the digital computer for processing measurement data, the conversion of analog data to digital form is essential and common to virtually all computer-based instruments.

Table 2.3
Values of 2^n

n	2^n
2	4
4	16
6	64
8	256
10	1024
12	4096
14	16,384
16	65,536
20	1,048,576
24	16,777,216
28	268,435,456
32	4,294,967,296

THE NUMBER REGISTER

The output number in all digital readout devices is held in a number memory device called a **register.** The number register may be based on either the binary or decimal number system. In the **binary** system, the only numerals used are 0 and 1. A **binary number** then is a string of 1's and 0's, such as 10010111. Each digit in a binary number is called a **bit** (short for binary digit). In the example given, there are 8 bits. The rightmost bit is called the **least significant bit** (LSB) and has a value of zero or one. The next bit to the left has a value of zero or two (decimal), the next zero or four, and so on. An 8-bit number has $2^8 = 256$ different values or states, so an 8-bit register has 255 steps between its minimum value (0) and its maximum value (255). In general, an n-bit register can contain any value from 0 to $2^n - 1$. Referring to Table 2.3, a 10-bit register can contain any whole number from 0 to 1023; a 20-bit register can contain any value to over a million.

    Binary registers are most often used in computer-based ADCs. The computer mathematically converts the binary number to decimal form before displaying the result. Decimal registers with a display capacity of three to eight decimal digits are generally used in small instruments. A decimal register uses sets of four binary registers, one set of 4 bits for each decimal digit. A 4-bit register can contain any number from 0 to 15, but in this application, all the values above 9 are skipped. Thus, each 4-bit section can represent all 10 values one decimal digit can have. A register arranged in this way is called a **binary-coded decimal register.** Thus, for a register with five decimal digits (values from 00000 to 99999), five sections of 4 bits each, or 20 bits altogether, are required. These same 20 bits in a binary register could contain all the numbers from 0 to 1,048,575.

THE STEPWISE FORM OF THE TRANSFER FUNCTION

The form of the digital output signal of an ADC is an integer number. Digitally encoded data share the integer-based characteristic of count data; no fractional values are represented. A change of 1 in the least significant bit or digit is the smallest change one can make to a digitally encoded number. You have certainly seen digital displays that appear to show decimal fractions such as 17.35 V. This displayed value is actually a combination of the digital number 1735 and additional information about the correct position of the decimal point. For the ADC, as the input voltage slowly increases, the output number will increase in unit steps. The working curve of the ADC shows this stepwise quality (Figure 2.20). Voltage changes in the ranges between the steps in the output number (such as the one shown at the output number 9) will not cause any change in the output number.

In terms of conversion devices, the ADC is a readout conversion device because its output quantity is a number. Ideally, the output number is linearly related to the input voltage. The sensitivity of the ADC is the change in output number per 1-V change in the input. For the 4-bit converter for which the working curve is shown in Figure 2.20, it is 15 steps/V.

RANGE AND RESOLUTION

A critical characteristic of an ADC is the number of steps that it takes before reaching the full-scale output number. The number of bits in its register determines the maximum output number available from the ADC and, therefore, its maximum resolution. Thus, an ADC with a 10-bit register (called a 10-bit converter) has 1023 steps and can represent all the values from 0 to 1023. The resolution of such a converter will be 1 ppt (1 part in 1023) with respect to the full-scale value. One can also express the resolution in terms of the magnitude of the input voltage change required to cause a unit change in the output number. This can be obtained by dividing the voltage required for full-scale output by the number of values the ADC can represent (see Example 2.2). If greater resolution is required for the measurement, one must either reduce the full-scale voltage of the ADC or use an ADC with more bits.

Example 2.2

A 12-bit converter can represent 2^{12} or 4096 different values. If the input voltage range is 10.0 V, the input voltage change corresponding to each step in the output number is 10.0 V/4095 steps = 2.44 mV/step. If a smaller resolution is required, one needs to have a converter with more bits in the register or one needs to decrease the range.

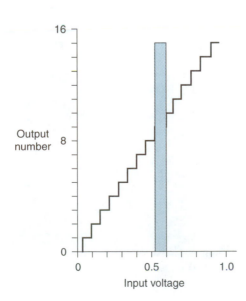

Figure 2.20. Working curve for a 4-bit ADC.

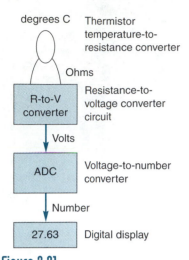

Figure 2.21. A digital thermometer based on a thermistor transducer and an ADC.

MEASUREMENTS INVOLVING ADCS

The quantity converted to number by the ADC is voltage. Therefore, other intermediate conversion devices are required for the measurement of any other quantities. In Chapters 3 and 8 we will see that the electrodes used to measure solution pH produce a voltage that is related to the pH. Thus the electrodes are a pH-to-voltage converter. When we combine the electrodes with an ADC and digital display, we have a digital pH meter. The relationship between the input pH and the output number is dependent on the transfer functions of the electrodes and the ADC. Complete digital measurement systems include an initial conversion device (also called a sensor or transducer) that converts a physical or chemical quantity such as pH or light intensity into an electrical signal. Sensors have been developed for sensing temperature, light, mass, humidity, pressure, strain, pH, and many other qualities and properties. Temperature sensors are often temperature-to-resistance converters. When an ADC is used with such a converter for temperature measurement (as in Figure 2.21), a resistance-to-voltage converter must be interposed between the temperature transducer and the ADC. The transfer functions of all three conversion devices affect the resulting relationship between the input temperature and the output number. The development of effective transducers and other conversion devices has been an important aspect of measurement science.

THE DISCRETE NATURE OF THE DATA

The result of each analog-to-digital conversion is a number. To record the change in a measured quantity over a period of time, it is now customary to connect the output of an ADC to a computer and store the successive numbers in the computer memory. From there, with the appropriate program, a plot of the variation in the quantity versus time can be produced. Such a plot is shown in Figure 2.22. The plot is not a continuous line of recorded values as from a strip-chart recorder (an analog device). It is a plot of separate points taken at different times. To produce the plot, the plotting program must be able to read both the Y-axis value (the conversion value) and the X-axis value (the sampling time) for each point. Thus each point is, in fact, a data pair. The acquisition of individual points along a continuous function is called **sampling.** One must arrange to sample the input voltage often enough to present the recorded changes in sufficient detail. However, selecting too high a sampling rate can create an unnecessarily large data file.

There are immense advantages to the digitization of data. The data can be coupled directly to a computer for automatic processing. It can be stored and manipulated without error. It can be transmitted reliably to remote locations. However, it is impor-

Figure 2.22. A plot of successive digital measurements. The integer resolution and the discrete nature of each data point are evident.

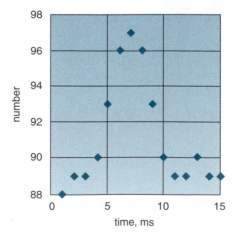

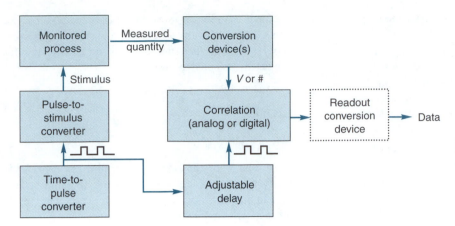

Figure 2.23. In stimulus–response measurements, the data collection is timed relative to the stimulus. A single point may be taken after each stimulation, or the stimulus may trigger a series of data conversions.

tant to keep in mind the limitations of its discrete nature. A digital measurement is quantized in magnitude (reflecting the stepwise quality of the numerical output), and it represents the measured value at only a single point in time.

STIMULUS–RESPONSE MEASUREMENTS

The basis of many analytical instruments and data processing subsystems is the correlation of a stimulus applied to the experiment and the response to that stimulus. This involves a timing operation so that the response can be monitored at the desired time following the application of the stimulus. The responses to neurological stimuli are often studied in this way. In many instruments, the stimulus may be the injection of the sample or the initiation of a chemical reaction. In photometric instruments, the light source may be pulsed and the time and nature of the sample response determined. A general block diagram of a measurement system of this type is shown in Figure 2.23. Many examples of this kind of measurement system will be encountered in this study.

Study Questions, Section F

31. What is the principal difference between analog electrical signals and digital electrical signals?

32. Why is the working curve for an analog-to-digital converter a step function as shown in Figure 2.20?

33. A 14-bit ADC has a range of ± 10.00 V. What is the resolution, in volts, of this conversion measurement?

34. It is desired to have the output number of the ADC in Question 33 be exactly equal to the magnitude of the input voltage in millivolts. This is to be accomplished by adjusting the input voltage range of the ADC. To what range should it be adjusted?

35. Many sensors have an output quantity of current or charge. How would these sensors be used with an ADC?

36. Derive the relationship between the input temperature and the output number for the temperature measurement system shown in Figure 2.21. The relationship between temperature and resistance for the thermistor is $R = 10^4/T$ in ohms. The voltage output of the resistance-to-voltage converter is directly proportional to the input resistance $(V = KR)$.

37. In the data set plotted in Figure 2.22, what limits the resolution of the data on each axis?

Answers to Study Questions, Section F

31. An analog signal encodes the data as a continuously variable voltage, current, charge, or power. A digital signal encodes the data as combinations of HI and LO signal levels that can represent numerical data.

32. The working curve for an analog-to-digital converter is a step function because the output number can only change in unit values, that is, it must step from one value to the next as the input quantity is increased. All values of the

input quantity between the steps produce the same output value.

33. The 14-bit converter will have $2^{14} = 16,384$ steps. These are divided over the voltage range of -10.00 to $+10.00$ V, covering 20.00 V. Thus, 20.00 V/16,384 steps = 0.00122 V/step. Since the input voltage must change 1.221 mV for the output number to change, this is the resolution of the measurement.

34. The ADC has 16,384 steps. Each step is to be equal to 1.000 mV. Thus, the full range will be 16,384 steps $\times$ 1.000 mV/step = 16,384 mV. If this is to be equally divided between positive and negative voltages, the full range would be $\pm 16,384/2$ mV = $\pm 8,192$ mV = ± 8.192 V.

35. The input quantity for most ADCs is voltage. For sensors with an output quantity other than voltage, an intermediate conversion device is required to convert the sensor output quantity to a related voltage.

36. $$\# = K'V, \qquad V = KR, R = 10^4/T.$$

Therefore,

$$\# = K'K \times 10^4/T \qquad \text{or} \qquad T = 10^4 KK'/\#.$$

37. The resolution on the Y-axis is determined by the sensitivity of the measurement system (change in output number per unit change in input value). On the X-axis the resolution is determined by the sampling rate.

G. Null Measurements and Conversion Devices

One of the ways to make a conversion device is through the comparison of the input quantity with an adjustable quantity of the same type. The comparison enables the adjustable quantity to be made equal to the input quantity. The result of the conversion is the magnitude of the adjustable quantity. A familiar example of this technique is the balance shown in Figure 2.24 used for measurement of mass. This approach is called **null measurement** when the output number is that of the quantity being measured. The concept of null measurement is shown in block form in Figure 2.25. The quantity to be measured is balanced directly against an **adjustable standard quantity,** and the standard quantity is varied until the difference between the standard and unknown quantities is zero. The **difference detector** (also called a **null detector**) consists of conversion devices so that the magnitude of the difference can be observed by the operator and a null or zero difference between the quantities can be achieved. In this kind of measurement, only the null value of the difference detector readout must be true. The actual measurement number is obtained from the amount of the variable standard required to bring the measurement to null.

DOUBLE-PAN BALANCE

To understand null measurements more deeply, one takes a careful look at the quantity that is actually being compared. As an example, we will look further at the double-pan balance. The sample and the standard masses are each placed in weighing pans hung from opposite ends of an arm. This arm is supported by a pivot in its middle. Each pan exerts a downward torque on the arm about the pivot point. When the torques on each side of the arm are equal, the arm angle indicator points to the "0" mark. The torque T applied to each side of the pivot is a function of the weight distribution of the arm, the

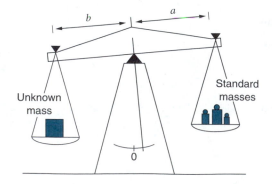

Figure 2.24. Double-pan balance. When the standard masses are equal to the unknown mass, the indicator points to zero. The result of the measurement is the value of the standard masses.

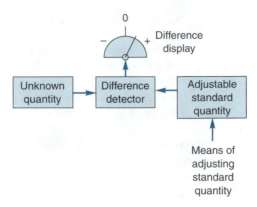

Figure 2.25. Null measurement system. The standard quantity is adjusted until the difference detector shows no difference between the standard and unknown quantities. The output number comes from the standard quantity, not the difference detector.

weight of the pan, support, and contents, and the distance between the pan suspension point and the pivot point of the arm. Specifically, for the right and left sides of the arm,

$$T_{rt} = T_{rt\,arm} + ag(m_{pan} + m_{support}) + ag(m_{standard}) \qquad 2.9$$

$$T_{lft} = T_{lft\,arm} + bg(m_{pan} + m_{support}) + bg(m_{sample}) \qquad 2.10$$

where a and b are the distances between the pivot point and pan suspension, g is the gravitational constant, and m is mass. The balance contains an adjustment that can equalize the torques when both pans are empty. When this adjustment has been made,

$$T_{rt\,arm} + ag(m_{pan} + m_{support}) = T_{lft\,arm} + bg(m_{pan} + m_{support}) \qquad 2.11$$

In application, the sample is placed in the left pan and standard masses of various values are placed in the right pan. The position of the arm angle indicator tells the sign of the difference in the right and left torques. Using this information, the operator adjusts the magnitude of the standard masses until the arm angle indicator points as close to zero as possible. How close to a zero (null) reading the operator can get depends on the smallest available change in the standard mass value, $\Delta m_{min\,std}$. At this point (within $\Delta m_{min\,std}$ and assuming Equation 2.11 is still true),

$$T_{rt} = T_{lft} \qquad \text{and} \qquad ag(m_{standard}) = bg(m_{sample}) \qquad 2.12$$

In the original double-pan balances, an adjustment allowed a and b to be exactly equal (determined by demonstrating a null reading when two identical masses were placed in each pan). With $a = b$ and g canceling out in Equation 2.12, at the null point,

$$m_{sample} = m_{standard} \qquad 2.13$$

The measurement value is obtained by observing the value of the standard mass required to obtain the null point.

NULL MEASUREMENT CHARACTERISTICS

Several general characteristics of null measurements can be illustrated by the example of the double-pan balance. First, a difference detector is required in order to achieve the null point. This difference detector is actually a conversion device that converts torque difference into pointer angle. It would be even better if it converted mass difference into pointer angle since mass is what we want to measure, but the zero and arm length adjustments bring us close to that ideal. Second, the transfer function of the difference conversion device does not need to be known, calibrated, or constant. The off-null response, however, must be sensitive enough to indicate a mass change of $\Delta m_{min\,std}$. The only point on the difference detector scale that is used in the final measurement is

In null measurements:

- A conversion device is used for the difference detector.

- The sensitivity of the difference detector should be sufficient to resolve the smallest desired change in the adjustable standard.

- The zero indication from the difference detector should be accurate and constant.

- Neither the linearity nor the calibration of the difference detector transfer function is required.

- Factors that affect the difference detector response function often cancel out at null.

- The accuracy and resolution depend on the adjustable standard.

- Automated nulling can often be used to increase speed and reduce tedium of measurement.

the zero point. This point must be accurate and constant. It can be checked and adjusted frequently by placing objects of identical mass in each pan. Third, with a functioning and properly adjusted balance, the accuracy of the mass measurement depends entirely on the accuracy of the standard masses employed. The resolution of the measurement in normal operation would be $\Delta m_{min\ std}$, but some methods of interpolation have been developed to estimate the sample mass to fractions of the smallest increment of the standard. The measurement precision depends greatly on operator skill and care, and on the condition of the balance, but ideally, it would be less than $\Delta m_{min\ std}$.

The null measurement approach has many advantages. The calibration and accuracy do not depend on the transfer function of the conversion device; they depend on standards of the same quantity that is being measured. The direct comparison of the unknown quantity and standard quantities often eliminates some assumptions inherent in the devices used for direct measurement. For example, in measuring mass with the spring scale, the relationship between force (the quantity actually affecting the pointer position) and mass involves the gravitational constant, g. Under conditions where the value of g might be different from that when the scale was calibrated, this must be taken into account. With the double-pan balance, the effect of g cancels out, as seen in Equation 2.12.

The principal disadvantage of null measurements is the tedium involved in achieving the null condition when the standard value must be manually adjusted. However, modern methods of automation have eliminated the slow manual process of standard adjustment for many null measurement techniques. In an automated null measurement, the difference output from the difference conversion device would be connected to an automatic device for adjusting the standard quantity. (Refer to Figure 2.25.) This kind of feedback operation for adjusting the standard to keep the difference output at zero is sometimes called a **servo mechanism.** This adjustment mechanism would also keep updated the displayed value of the amount of the standard applied. This readout provides the numerical output of the measurement.

NULL CONVERSION DEVICES

Another way to look at the null comparison mechanism is as a method to achieve a conversion device. The double-pan balance is a mass-to-number converter where the resulting number inherently has the units of the measured quantity, mass. Since it produces a number, the balance is an example of a readout conversion device. In fact, the null balance concept can be used for any of the three conversion devices, initial, intermediate, or readout. A conceptual diagram for a general conversion device based on the null principle is shown in Figure 2.26. The difference between the values of the

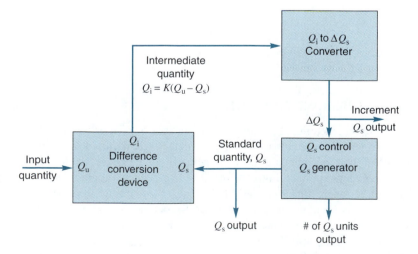

Figure 2.26. The null balance principle is used for the construction of many automated conversion devices. A variety of outputs has added to the versatility of this approach.

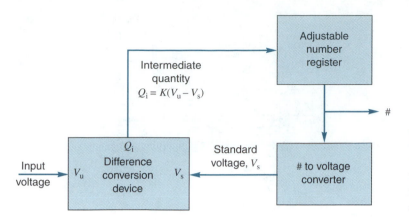

Figure 2.27. An ADC based on the null measurement approach. The number in the register used to control the number-to-voltage converter is the readout number when the system has come to balance.

input and standard quantities is used (through the feedback loop) to control the amount of the standard quantity. This process automatically keeps the intermediate quantity Q_i to nearly zero. The output can be taken from several places depending on the application. The matching quantity Q_s can be obtained at the Q_s output. This mode of operation does not perform a conversion since the units of input quantity Q_u and Q_s are the same. However, there are often advantages in connecting the Q_s output rather than the input quantity to further conversion devices. An example is the voltage follower circuit often used to isolate input sensors from readout conversion devices. Another output is the signal used to control the value of Q_s. The rate at which Q_s is incremented can be a useful signal. The current-to-frequency converter is based on this principle. For a numerical output, the Q_s generator can provide a number equal to the number of Q_s units it is applying at the Q_s output. A version of the ADC uses this concept.

One of the methods most often used to design an ADC is shown in the block diagram of Figure 2.27. The output of the difference detector is used to adjust a number in the number register. If the difference is low, the number is decreased; if high, it is increased. The number output is also used to set the standard voltage, V_s. The change in V_s reduces the difference between V_s and input voltage V_u. This cycle continues until the difference is reduced to zero. At this time, the number in the register is directly proportional to the input voltage.

THE ELECTRONIC ANALYTICAL BALANCE

An electronic balance (Figure 2.28) is an excellent illustration of many of the principles developed in this chapter. The shaft supporting the sample pan is attached to an electromagnetic coil. The coil is placed in the field of a permanent magnet so that the

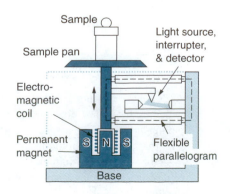

Figure 2.28. In the modern electronic balance, the force caused by the measured weight is balanced by the upward force exerted by the electromagnetic coil. The magnitude of the current in the coil required to achieve this balance is related to the weight of the object.

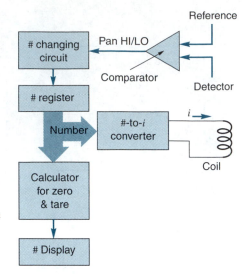

Figure 2.29. Block diagram illustrating the operation of the electronic balance. The pan position detector raises or lowers the number, which in turn changes the current through the coil. When the pan is returned to the null position, the number is related to the weight on the pan. This diagram is a specific instance of the general digital null measurement described in Figure 2.26.

greater the current in the coil, the greater is the upward force asserted by the coil. The shaft is attached to the base via a flexible parallelogram that allows vertical motion only. The balance of the forces caused by the sample and the coil is detected by an optical system consisting of a light source, a light detector, and an interrupter. The light interrupter is attached to the supporting shaft.

The circuit that provides an automatic balance between the two forces is shown in Figure 2.29. The current through the coil is proportional to the number in the number register. The position of the shaft is detected by the optical interrupter system. The light detector output is compared with a reference value. If the detector output is too high, the coil is lifting the pan too much. The pan HI/LO signal then commands the number-changing circuit to lower the number in the register. This decreases the current. This process is repeated until the reference and detector values are equal. This electronic feedback system for automatically keeping the system at null is an example of a servo system. One advantage of using the current in the coil to achieve the force balance is that the current can be quickly adjusted electronically.

The electronic balance can now be seen to be a weight-to-number converter. Because mass is not the quantity used to create both forces that are compared in the electronic balance, the effect of g does not cancel out. Therefore, the electronic balance must be recalibrated whenever it is moved to a new location.

The working curve between mass and current is determined by calibration with standard masses. This is done by loading a standard mass on the pan and pushing the calibrate button. When weighing an object, the number in the register at null is acquired by the calculator, which processes data from the zero, calibration, and tare measurements to produce the number displayed.

Study Questions, Section G

38. What is the role of conversion devices in null measurements?

39. In a null measurement, what part of the measurement system provides the readout number?

40. What aspects of the standard quantity affect the accuracy and precision of the readout number?

41. What are the most important qualities of the difference detector?

42. What quantity is being compared in the double-pan balance?

43. A double-pan balance has a set of standard masses in which the smallest mass is 1 mg. The balance beam pointer moves along a scale 4 cm in length (0 ± 2 cm) with markings every millimeter. The sensitivity of the null detector is roughly 5 mm/mg. What is the limit of resolution of the mass measurement of this system, and what is the limiting factor?

44. The standard mass set used with a double-pan balance has standard masses of 1, 2, 2, and 5 for each decade of mass from 1 mg to 10 g. What is the accuracy required for each standard weight used in the weighing of a 14.394 g object to an accuracy of 1 mg?

45. Trace the feedback loop by which the automated null system of Figure 2.26 maintains the difference detector output to be zero.

46. Riding a bicycle involves the automatic maintenance of the quantities of balance and direction. Compare the process of staying vertical and pointed in the right direction with the operation of a servomechanism.

47. Calculate the precision required in the current-adjusting circuitry and the digital readout for an electronic balance to read mass up to 100 g to a precision of 0.1 mg.

Answers to Study Questions, Section G

38. Conversion devices are used in the difference detector because the output of the difference detector has different units than that of the quantities being compared.

39. The readout number comes from the number of standard units used to balance the input quantity.

40. The accuracy of the readout depends on the accuracy of the standard units used in matching the input quantity. The precision of the readout number can be limited by the smallest increment by which the standard quantity can be changed.

41. The difference detector must have a zero output when the difference between the input quantities is zero. If the difference detector is not sensitive enough to give a sensible output for the smallest increment of standard quantity, the difference detector will limit the attainable resolution.

42. The double-pan balance compares the torque produced on either side of the balance beam.

43. The null detector will move 5 mm for a mass difference equal to the finest change available in the standard mass. Since a movement of 5 mm is easily detected, the factor limiting the resolution is the smallest increment in the standard mass. If the null detector sensitivity is calibrated by observing the deflection for a 1 mg change in mass, it can be used to estimate the sample mass to a fraction of a milligram. If the null position could be read to 0.2 mm, the mass resolution would then be 0.2 mg ÷ 5 mm/mg = 0.04 mg.

44. The standard weights required are 10 g, 2 g, 0.1 g, 0.2 g, 0.02 g, 0.05 g, and 2 mg. The accuracy specification (in %) for each is (1 mg/std mass) × 100. The requirements are, respectively, 0.01%, 0.05%, 1%, 0.5%, 5%, 2%, and 50%.

45. The output of the difference detector is connected to the system that converts a nonzero signal to a command to change the standard quantity. The changed standard quantity is applied to the difference detector to reduce the difference. This continues until the difference is as close to zero as it can get.

46. The rider detects an off-balance condition and applies a steering and posture correction to restore balance. This process is continuously applied to keep the difference between attitude and verticality as small as possible. Similarly, the rider senses an incorrect direction and applies a steering correction to reduce the difference between the path of the bike and the intended direction as close to zero as possible.

47.
$$0.1 \times 10^{-3} \text{ g} \div 100 \text{ g} = 10^{-6}$$

The precision required in the current-adjusting circuitry is thus 0.0001% or 1 part per million. The maximum reading is 100.0000 g, so six full decimal digits and one partial digit are required in the readout.

H. Measurement Accuracy and Precision

MEASURES OF PRECISION

If a measurement is repeated, but identical results are not obtained, a question arises about which of the measured values to use. If all values were obtained with equal care, there is no basis for choosing among them. In this case, we will use all of them by calculating their average value. The **average** is the sum of the values measured divided by the number of values averaged. This operation is expressed by

$$\bar{x} = \frac{\sum_{i=1}^{N} x_i}{N} \qquad\qquad 2.14$$

where x_i represents each of the individual values, $\bar{x}$ is the average value, and N is the number of values averaged. The average value is also sometimes called the **mean,** or the **arithmetic mean.**

Example 2.3

To calculate the mean and the average deviation

The sum is calculated:

$$x_1 = 32.14 \text{ mL}$$
$$x_2 = 32.11 \text{ mL}$$
$$x_3 = 32.08 \text{ mL}$$
$$x_4 = 32.16 \text{ mL}$$
$$\text{Sum} = 128.49 \text{ mL}$$

The average or mean is determined:

Average $= \bar{x} = \text{Sum}/N = 128.49 \text{ mL}/4$

$$= 32.12 \text{ mL}$$

The deviations are calculated and averaged:

$$d_1 = |32.14 - 32.12| = 0.02 \text{ mL}$$
$$d_2 = |32.11 - 32.12| = 0.01 \text{ mL}$$
$$d_3 = |32.08 - 32.12| = 0.04 \text{ mL}$$
$$d_4 = |32.16 - 32.12| = 0.04 \text{ mL}$$
$$\text{Sum} = 0.11 \text{ mL}$$

Average deviation $= 0.11 \text{ mL}/4 = 0.03 \text{ mL}$

Example 2.4

To calculate the standard deviation

Using data from Example 2.3, the squares of the deviations are summed:

$$d_1^2 = 4 \times 10^{-4}$$
$$d_2^2 = 1 \times 10^{-4}$$
$$d_3^2 = 16 \times 10^{-4}$$
$$d_4^2 = 16 \times 10^{-4}$$
$$\text{Sum} = 37 \times 10^{-4}$$

The sum is divided by $N - 1$ to get s^2:

$$\text{Sum}/(N - 1) = (37 \times 10^{-4})/3 = 12 \times 10^{-4} = s^2$$

Take the square root to get s:

$$s = (12 \times 10^{-4})^{1/2} = 0.035 \text{ mL}$$

Calculation of the RSD allows evaluation of measurement precision. For Example 2.4,

$$\text{RSD} = (0.035/32.12) \times 1000 = 1.1 \text{ ppt}$$

The difference d_i between an individual measurement and the mean value is called the **deviation from the mean,** and it is expressed as the absolute value (i.e., it is always a positive number). Thus, $d_i = |x_i - \bar{x}|$. The larger the deviations from the mean, the less precise is the measurement (see Example 2.3). It is helpful to have a means of assessing and comparing degrees of precision. The average of the deviations from the mean (**average deviation**) has been used, but a related value with greater statistical significance is the **standard deviation.** The standard deviation s for a small set of data is calculated from Equation 2.15:

$$s = \sqrt{\frac{\sum\limits_{i=1}^{N} (x_i - \bar{x})^2}{N - 1}} = \sqrt{\frac{\sum\limits_{i=1}^{N} d_i^2}{N - 1}} \qquad 2.15$$

This process is illustrated in Example 2.4.

The significance of the calculation of s is that it provides a means to determine the expected range of deviations for the measurement process that led to the specific values for which s was calculated. For example, if the distribution of deviations is known to follow the normal random measurement error, we can say that 95% of the measured values are expected to fall within a range of $\pm 2s$ of the mean value $\bar{x}$.

The portion of Equation 2.15 within the square root sign contains the sum of the squares of the individual deviations from the mean. If this sum were divided by N, the average of the squares of the deviations would be obtained. When $N - 1$ is used, the calculated value of s is larger. This is the penalty for having to estimate the standard deviation from a small set of data. As the number of measurements increases, the effect of subtracting 1 from N decreases. When N is 25 or larger, the 1 in the denominator can be ignored. The value of the mean is also affected by an increased number of measurements. It becomes closer to the value it would have if the measurement were repeated hundreds or thousands of times. The symbol for the mean of a very large data set is μ. The formula for the standard deviation σ of larger data sets is

$$\sigma = \sqrt{\frac{\sum\limits_{i=1}^{N} (x_i - \mu)^2}{N}} \qquad 2.16$$

The values for s and σ calculated from the above equations are the **absolute standard deviations**. If the standard deviation has been calculated in order to determine the degree of precision obtained in a particular measurement process, it is more usefully expressed as the fraction of the measured value. The standard deviation divided by the average measured value is called the **relative standard deviation** (RSD). To express the RSD in parts per thousand,

$$\text{RSD} = (s/\bar{x}) \times 1000 \quad \text{ppt} \qquad \text{or} \qquad \text{RSD} = (\sigma/\mu) \times 1000 \quad \text{ppt} \qquad 2.17$$

If the RSD is expressed in percent rather than parts per thousand (obtained by multiplying by 100 instead of 1000 in Equation 2.17), it is called the **coefficient of variation** (CV). In quantitative chemical analysis, a common goal is the achievement of an RSD of 1 ppt. The range of quantitative RSD extends to 10 ppt (CV = 1%). The CV range from 1 to 10% is sometimes referred to as "semiquantitative."

CONFIDENCE INTERVALS

When one is working with a limited data set, the average value of a measurement is only an estimate of the actual mean value one would obtain with a very large set of values. The possible differences between the true mean and the limited set average increases with increasing standard deviation. It also increases with a decreasing number of data points to average. The uncertainty about the true value of the measurement result is usually expressed as the average value and an interval that could contain the true mean. This expression has the form $\mu = \bar{x} \pm U$ where U is the maximum probable difference between the average and true mean values. The two values of μ obtained from this expression are called the **confidence limits.** The difference between the two values of μ is the **confidence interval.** This is equal to $2U$.

The value of U can be calculated from the standard deviation and the number of samples that went into the average value. Two formulas are used, depending on whether the true standard deviation, σ, is known or just s:

$$\mu = \bar{x} \pm t_\mathrm{p} \frac{\sigma}{\sqrt{N}} \qquad\qquad 2.18$$

$$\mu = \bar{x} \pm t \frac{s}{\sqrt{N}} \qquad\qquad 2.19$$

where t_p and t are values that one looks up in tables for that purpose. Equations 2.18 and 2.19 are identical except for the value of t to be used. Where σ is known, the confidence interval depends only on the confidence level required. Table 2.4 gives the value of t_p for several different levels of confidence. If one wishes to set the confidence limits such that they will include 95% of the likely values, a value of 1.960 is used for t_p in Equation 2.18. In statistical analysis, the quantity $s/(N)^{1/2}$ is sometimes called the **standard error of the mean** (SEM).

It is very rare, for a single sample, to have enough replicate data to determine σ, and if one did, μ would be known. The situation where σ is known but μ is not can occur when a given technique is run repeatedly in the same laboratory with only a few replicates of each sample. Much more likely is the situation where the confidence limits must be obtained for a small data set. As you might expect, the smaller the data set, the less certain we can be of the confidence interval. Thus, the confidence limits increase with decreasing size of the data set. When using Equation 2.19, then, the value of t depends on both the confidence level and N, the number of data points in the set. For applicability in a variety of situations, the t table does not indicate N directly, but rather a quantity called the **degrees of freedom.** For a set of replicate measurement data, the degrees of freedom, or D.F., is equal to $N - 1$.

The critical t values for a few combinations of confidence level and degrees of freedom are in Table 2.5. If triplicate results were available for a sample, and the confi-

Table 2.4
Values of t When σ Is Known

Confidence level, %	t_p
50	0.674
90	1.645
95	1.960
99	2.576
99.5	2.807

Table 2.5
Critical Values for t

D.F.	Confidence Level, %					
	50	90	95	99	99.5	99.9
1	1.00	6.31	12.71	63.66	127.32	636.62
2	0.82	2.92	4.30	9.93	14.09	31.60
3	0.76	2.35	3.18	5.84	7.45	12.92
4	0.74	2.13	2.78	4.60	5.60	8.61
5	0.73	2.01	2.57	4.03	4.77	6.87
6	0.72	1.94	2.45	3.71	4.32	5.96
7	0.71	1.90	2.37	3.50	4.03	5.41
8	0.71	1.86	2.31	3.36	3.83	5.04
9	0.70	1.83	2.26	3.25	3.69	4.78
10	0.70	1.81	2.23	3.17	3.58	4.59
15	0.69	1.75	2.13	2.95	3.25	4.07
20	0.69	1.73	2.09	2.85	3.15	3.85
∞	0.67	1.65	1.96	2.58	2.81	3.29

Example 2.5

To calculate the confidence limits the standard deviation is calculated.

Using the numerical example begun with Example 2.3, the average was 32.12 mL. The value of s was determined in Example 2.4 to be 0.035 mL. There are four values in the data set. Therefore, at the 95% confidence level with 4 samples, $t = 3.18$ and,

$$\mu = 32.12 \text{ mL} \pm 3.18 \frac{0.035}{2} \text{ mL}$$

$$\mu = 32.12 \pm 0.06 \text{ mL}$$

The 95% confidence limits are 32.06 mL and 32.18 mL.

Example 2.6

Measurement numbers are expressed to avoid presenting any digits past the first uncertain digit. If a measurement and its uncertainty were 11.824 ± 0.037 V, the 2 is the first uncertain digit. Therefore, the number is expressed as 11.82 ± 0.04 V.

Example 2.7

Calculator generation of nonsignificant digits

One wishes to calculate the number of moles of $CaCO_3$ in a 42.18 g sample of pure $CaCO_3$. The MW of the $CaCO_3$ is $40.078 + 12.011 + 3(15.999) = 100.086$ g mol^{-1}. When the weight of the sample is divided by the MW the calculator could display 0.421437564. Since there were only four digits in the weight of the sample, which is also the number with the least number of significant digits, the answer should be rounded to four digits (i.e., 0.4214 mol).

dence limits are to be set at the 95% level, the value of t to be used in Equation 2.19 is 4.30. This is to be compared with the value of 1.960 that would be used in the case where σ is known. The penalty for the small data set and the lack of information about σ is a factor of over 2 in the size of the confidence interval. You can also see in Table 2.5 that as the value of N increases, the value for t at the 95% confidence level approaches 1.96. The values in the row for an infinite number of degrees of freedom are the same as those for t_p in the table of t values when σ is known (see Table 2.4 and Example 2.5).

The tedium associated with confidence interval calculations has largely been eliminated by modern calculators and by the spreadsheet, statistical, and plotting programs available for computers.

SIGNIFICANT FIGURES

One of the many uses of the confidence interval is that it informs us how many of the figures obtained in our measurement (or calculation of the average measured value) are significant. For example, if the average value of a set of measurements is calculated to be 154.4762 g and the confidence interval of the measured values is (0.0264 g, the confidence interval indicates that there is an uncertainty of almost 0.03 g in the measurement. With such an uncertainty in the hundredths place, the numerals in the places further to the right are not relevant. The general rule is to report all the digits that one is certain of plus the first uncertain digit, that is 154.47 ± 0.03 g. Even if the confidence interval is not reported, all digits past the first uncertain digit should be removed (see Example 2.6).

Simple removal of the nonsignificant digits is easy when they are to the right of the decimal point as in the examples given. It is another matter when they are to the left. For example, if a weight is reported as 1500 g, one does not know if the weight is known to the unit gram level or to the hundred gram level because the zeros have the task of holding the place value.[2] Only if the weight were given as 1500.0 would we know the precision implied because the "point zero" is unnecessary except to indicate precision at the one-tenth gram level. To avoid the problem of uncertain precision in whole numbers with zeros in the less significant digits, use scientific notation. For this example, the weight could be reported as 1.5×10^3 g, 1.50×10^3 g, 1.500×10^3 g, 1.5000×10^3 g, etc.

One of the consequences of the digital revolution has been the proliferation of the reporting of nonsignificant figures. A conscious effort is required to avoid writing down the multidigit answers produced by even simple calculators. A modest HP 27S calculator will display a number that is 16 digits long. It is easy to start out with data that may be valid to one part in 1000 and end up with a calculated value with six or eight digits. When writing the final answer, we must always remember to trim away the nonsignificant numbers (see Example 2.7).

Instruments with digital displays can also be the source of numbers that contain nonsignificant figures. It is desirable for the instrument resolution to be smaller than the precision of the measurement. However, whenever this is the case, the readout number is likely to have insignificant digits. The only way we can know how many digits are significant is through repetitive measurements and the calculation of the confidence interval.

MEASUREMENT PRECISION IN DATA PROCESSING

Each measurement has a certain precision associated with it. In the course of obtaining the desired information from the measurements that we make, we often have to use several measurement values. The question then arises, "What is the precision of the

[2]For a fascinating discussion of the importance of zeo, see R. Kaplan, *The Nothing That Is, A Natural History of Zero*. Oxford University Press, Oxford, 2000.

final result?" The precision of the result of a calculation involving several data values can itself be calculated if the precisions of the several data values are known.

The way in which the precisions combine to produce the precision of the result depends on the mathematical relationship used in the data calculation. For example, if the data are *added* or *subtracted* to get the final result, the standard deviation of the sum is given by

$$\text{For } y = a + b + c, \qquad s_y = \sqrt{s_a^2 + s_b^2 + s_c^2} \qquad 2.20$$

The implementation of this formula may be made with either the standard deviation or the confidence interval (see Example 2.8). If the standard deviations are used, then the standard deviation of the result is obtained. Notice that the confidence interval for the result is worse than that for any of the values used in the calculation.

The operations of addition and subtraction are the only ones for which the standard deviations of confidence interval can be used directly. For all other operations, such as multiplication, division, roots, logs, and powers, the relative standard deviation or confidence interval must be used:

If the data are *multiplied* or *divided* in calculating the final result, the same relationship applies, but the relative standard deviations must be used.

$$\text{For } y = \frac{ab}{c}, \qquad \frac{s_y}{y} = \sqrt{\left(\frac{s_a}{a}\right)^2 + \left(\frac{s_b}{b}\right)^2 + \left(\frac{s_c}{c}\right)^2} \qquad 2.21$$

For the operations of *power, log,* and *antilog* on a single data point, the relationships between the standard deviation of the data and the result are

$$\text{For } y = a_x, \qquad \frac{s_y}{y} = x\frac{s_a}{a} \qquad 2.22$$

$$\text{For } y = \log_{10} a, \qquad s_y = \frac{s_a}{2.303a} \qquad 2.23$$

$$\text{For } y = \text{antilog}_{10} a, \qquad \frac{s_y}{y} = 2.303 s_a \qquad 2.24$$

For various combinations of the operations, it is necessary to take the calculation in steps (see Example 2.9). For example, if the data calculation to get the result is $y = \log_{10}[(a + b)/c]$, one would first calculate the standard deviation for $a + b$, then for that result over c, and finally for the log of the quotient.

Example 2.9

If $y = x \log c$, $x = 4.3 \pm 0.15$, and $c = 6.2 \times 10^2 \pm 84$, then

$$\log c = 2.79 \pm \frac{84}{2.303 \times 620} = 2.79 \pm 0.06$$

$$\frac{s_{x \log c}}{x \log c} = \sqrt{\left(\frac{s_x}{x}\right)^2 + \left(\frac{s_{\log c}}{\log c}\right)^2}$$

$$\frac{s_{x \log c}}{12.0} = \sqrt{\left(\frac{0.15}{4.3}\right)^2 + \left(\frac{0.06}{2.79}\right)^2}$$

$$= \sqrt{1.23 \times 10^{-3} + 4.8 \times 10^{-4}} = 0.041$$

$$s_{x \log c} = 12.0 \times 0.041 = 0.49$$

$$x \log c = 12.0 \pm 0.5$$

Example 2.8

Calculating a new confidence interval when adding or subtracting data

If $R = A + B - C$ and

$$A = 23.42 \pm 0.03$$
$$B = 43.65 \pm 0.05$$
$$C = 37.83 \pm 0.06$$

then $R = 29.24$.

The confidence interval for R is then

$$\text{CI} = \sqrt{(0.03)^2 + (0.05)^2 + (0.06)^2}$$
$$= 0.084$$

Therefore, $R = 29.24 \pm 0.084$ which, when rounded to remove nonsignificant digits, is $R = 29.24 \pm 0.08$.

48. How do you obtain the average or mean value of a set of data?

49. What is meant by the deviation of an individual data value that is in a set of values?

50. How is the standard deviation of the data from repetitive measurements obtained?

51. The following set of data was obtained for the number of seeds in a sunflower. Calculate the mean, standard deviation, and average deviation of the data. Number of seeds: 525, 527, 527, 530.

52. The following data were obtained for the amount of calcium ion in a water sample: 769.5, 770.1, 770.2, 770.8, 771.3, 771.5, 772.1, 772.1, 772.2, 772.2, 772.2, 772.3, 772.3, 772.3, 772.4, 772.5, 772.6, 772.7, 772.7, 773.0, 773.0, 773.1, 773.2, 773.5, and 774.1 ppm. For any 5, 10, and 25 of these points, calculate and compare the mean, s, and σ.

53. How does the relative standard deviation differ from the standard deviation?

54. What is the coefficient of variation?

55. Calculate the RSD and CV for the five-point calculation performed in Question 52.

56. Relationships were given in Equations 2.20 through 2.24 for calculating the standard deviation of the result when mathematical operations are performed on the mean values. All the operations except two require the use of the relative standard deviation rather than the standard deviation. What are the only operations that can be performed on the standard deviation directly?

57. The following data were collected, and, from replicate measurements, the confidence intervals (at the same confidence level) were determined

$$A = 47.72 \pm 0.002$$
$$B = 0.836 \pm 0.003$$
$$C = 9.759 \pm 0.008$$

Calculate R and the confidence interval for R if $R = A + B + C$.

58. Using the data in Question 57, calculate R and the confidence interval for R if $R = (B + C)/A$.

59. Calculate R and the confidence interval for R if $R = \log D$, where $D = 4.58 \times 10^{-4} \pm 0.04 \times 10^{-4}$.

Answers to Study Questions, Section H

48. One sums the data values and divides by the number of data values. The equation for this operation is

$$\bar{x} = \frac{\sum\limits_{i=1}^{N} x_i}{N}$$

49. The deviation of a data value is the absolute difference between it and the mean of the data set. The equation expressing this is $d_i = |x_i - \bar{x}|$.

50. The standard deviation is the sum of the squares of the deviations after it has been divided by one less than the number of data values. The equation for this operation is

$$s = \sqrt{\frac{\sum\limits_{i=1}^{N}(x_i - \bar{x})^2}{N-1}} = \sqrt{\frac{\sum\limits_{i=1}^{N} d_i^2}{N-1}}$$

51. The mean is calculated from Equation 2.14.

$$\bar{x} = \frac{\sum\limits_{i=1}^{N} x_i}{N} = \frac{525 + 527 + 527 + 530}{4} = 527.25$$

The standard deviation is calculated from Equation 2.15.

$$s = \sqrt{\frac{\sum\limits_{i=1}^{N}(x_i - \bar{x})^2}{N-1}}$$

$$= \sqrt{\frac{(2.25)^2 + (0.25)^2 + (0.25)^2 + (2.75)^2}{4-1}} = 2.1.$$

Note that the standard deviation equation for a small data set is used here. If a data set of more than 25 values was used, Equation 2.16 could have been applied in place of Equation 2.15, the only difference being that the denominator is N instead of $N - 1$. The average deviation is the average of the deviations from the mean:

$$\bar{d} = \frac{\sum\limits_{i=1}^{N} |x_i - \bar{x}|}{N}$$

$$= \frac{|525-527.25| + |527-527.25| + |527-527.25| + |530-527.25|}{4}$$

$$= 1.37$$

52.

# of data points	$\bar{x}$	s	σ
5	772.1	1.3	1.1
10	772.2	1.3	1.2
25	772.2	1.1	1.1

To make the above table, the given number of data points were chosen randomly from the data set. Note that s and σ approach each other as the data set gets larger. This is why we can use σ instead of s for data sets with more than 25 values.

53. The standard deviation is an absolute amount, that is, the actual magnitude of the deviation in the units of the quantity being measured. The relative standard deviation is the fraction of the measured value that is uncertain. It is obtained by dividing s or σ by the mean of the measured value. The equations for this operation are $RSD = s/\bar{x}$ or σ/μ expressed as percent ($\times$ 100) or ppt ($\times$ 1000).

54. The coefficient of variation is the relative standard deviation expressed in percent.

55. $RSD = s/\bar{x} = 1.3/772.1 = 0.0017$ for the five-point case. Multiply by 100 to get $\%RSD = CV = 0.17\%$ or by 1000 to get $RSD = 1.7$ ppt.

56. The only operations that can be performed on the standard deviation directly are addition or subtraction. Multiplication, division, roots, and powers all require use of the relative standard deviation.

57. $R = 47.72 + 0.836 + 9.759 = 58.32$

$CI = \sqrt{(0.002)^2 + (0.003)^2 + (0.008)^2} = 0.009$

Thus $R = 58.32 \pm 0.01$, from which we see that the largest confidence interval dominates the resulting confidence interval.

58. $R = (0.836 + 9.759)/47.72 = 0.2220$

The relative deviation for $(B + C)$ is

$$\sqrt{(0.003)^2 + (0.008)^2} = 0.009$$

The relative deviation for $(B + C)/A$ is

$$\sqrt{\left(\frac{0.009}{10.595}\right)^2 + \left(\frac{0.002}{47.72}\right)^2} = 0.001$$

The confidence interval is $CI = 0.001 \times 0.222 = 0.0002$. Thus $R = 0.2220 \pm 0.0002$.

59. $R = \log(4.58 \times 10^{-4}) = -3.339$

The absolute deviation for R is

$$\frac{0.04 \times 10^{-4}}{2.303 \times 4.58 \times 10^{-4}} = 0.0038$$

The relative deviation is $\dfrac{0.0038}{3.339} = 1.1 \times 10^{-3}$.

Thus $R = -3.339 \pm 0.001$.

I. Comparing Means and Deviations

A common occurrence in the analytical laboratory is the need to compare the results obtained with what is considered the true value. Is the mean value obtained truly different from the accepted value, or is the difference within the margin of error for the test? Another question that we often need to answer is whether there is any significant difference in the results obtained with different laboratories, methods, or operators, either in the means or in the standard deviations. Such questions are tests of whether the differences are real or are within the expected variance. The test begins with the **null hypothesis,** which proposes that the data belong to the same set, i.e., that there is no statistical difference between them. Then the probability of this hypothesis being true is determined using accepted statistical approaches. The statistical tests for several different comparisons are described in this section. From them, the limit of detection for a technique can be quantitatively determined, and tests for the rejection of outlying data values can be developed.

COMPARING THE MEAN WITH AN ACCEPTED VALUE

To determine if the mean of the test set is statistically different from an accepted value, we just have to find out if the accepted value falls within the confidence limits of the test set. One can go about this in either of two ways. The first is to calculate the confi-

Example 2.10

To compare the mean with an accepted value by the confidence limit method

The confidence interval for the volume measurements we have been working with was calculated in Example 2.5 to be $\mu = 32.12 \pm 0.06$ mL. These confidence limits were calculated at the 95% confidence level. If the true value were considered to be 32.20 mL, the true value would not be included in the confidence limits, so we conclude that the data and the accepted value are not the same. (The null hypothesis has been rejected.) We are at least 95% sure of this conclusion.

dence limits for the data set and see if the accepted value falls within the limits or not. If it does not, it means that the measured data do not agree with the accepted value and that the difference between them is unlikely to occur at the confidence level chosen when the confidence limits were calculated (see Example 2.10).

If the accepted value falls outside the confidence limits for the measurement, it means that the precision is better than the accuracy for the measurement. In other words, the results are consistent, but they are wrong. When this occurs, we say that there is **systematic error** in the method. The same error is occurring reproducibly. This can come from a consistent error in technique, a poorly calibrated instrument, or an error in the method of data interpretation.

The second method of comparison reveals more easily the level of probability that the null hypothesis is true. For this approach, a rearrangement of Equation 2.19 is used. For the difference between μ (the accepted value) and $\bar{x}$ (the mean of the data set) to be within the confidence limits,

$$|\mu - \bar{x}| \le t \frac{s}{\sqrt{N}} \qquad 2.25$$

A further rearrangement, where $t_{\exp}$ is now the value of t for these data, gives

$$|\mu - \bar{x}| \frac{\sqrt{N}}{s} = t_{\exp} \qquad 2.26$$

The value of $t_{\exp}$ is calculated and compared with the critical values for t in Table 2.5. This process is called student's t-test after the mathematician Student who developed the table of t values. The confidence levels at which the null hypothesis can be asserted (the data set includes the true value) are those for which the critical value of t is greater than $t_{\exp}$ (see Example 2.11).

In the case where the true standard deviation of the method is known, the same approach is used, but beginning with Equation 2.18. In this case, the equation used to calculate t for the data set is

Example 2.11

Comparing the mean with an accepted value by the t calculation method

Continuing with the data used in Example 2.10, the value of t for the data set is calculated using Equation 2.26:

$$|32.20 - 32.12| \frac{2}{0.035} = 4.6$$

Now the values of t from Table 2.5 are consulted for the line where $N - 1 = 3$. The calculated value is greater than the value of 3.18 at the 95% level and smaller than the value of 5.84 at the 99% level. Therefore, the probability that the data mean and the accepted value are different is greater than 95% but less than 99%.

$$|\mu - \bar{x}| \frac{\sqrt{N}}{\sigma} = t_{\exp} \qquad 2.27$$

and the table used for comparison is that for t_p (Table 2.4).

COMPARING THE MEANS OF TWO DATA SETS

The process of comparing the means of two sets of experimental data is very similar to that for comparing the mean with an accepted value. The principal difference is that the possible variance of both means must be considered. In this comparison, to use a single value for the standard deviation, the standard deviations of the two data sets must be pooled. If the standard deviations for both sets have already been calculated, then obtain the value of s_{pooled} from

$$s_{\text{pooled}} = \sqrt{\frac{s_1^2(N_1 - 1) + s_2^2(N_2 - 1)}{N_1 + N_2 - 2}} \qquad 2.28$$

where s_1 and N_1 are the standard deviations and number of data points in set 1 and s_2 and N_2 are the same quantities from the second data set. If the standard deviations have not already been determined, calculate the sum of the squares of the deviations for each set and use this value instead of the $s^2(N - 1)$ term in Equation 2.28.

Now a value for t_{exp} can be calculated for the combined data set and compared to that in the critical t table. This is accomplished with Equation 2.29.

$$\frac{|\bar{x}_1 - \bar{x}_2|}{s_{pooled}} \sqrt{\frac{N_1 N_2}{N_1 + N_2}} = t_{exp} \qquad 2.29$$

For the null hypothesis to be true (the means of the data sets are not statistically different), the value for t_{exp} must be smaller than the critical value from the table. In using the table, the degrees of freedom are $N_1 + N_2 - 2$. If one of the data sets has a value of N that is very much larger than the other is, Equation 2.29 reduces to Equation 2.26 where N and s are taken from the set with the smaller number of repetitions (see Example 2.12).

Example 2.12

Comparing the means of two data sets

Another experimenter repeating the same measurement described in Example 2.3 obtained the following data: 32.18, 32.13, 32.09, and 32.23 mL. The mean and standard deviation for these data are 32.16 mL and 0.06 mL. Comparing the data, one can clearly see that the values from the second experimenter are higher than those of the first. The question is whether it is improbably higher given the deviations in both sets. The value for s_{pooled} is calculated to be

$$s_{pooled} = \sqrt{\frac{(0.035)^2 (4 - 1) + (0.061)^2 (4 - 1)}{4 + 4 - 2}}$$

$$s_{pooled} = 0.050$$

The calculated value for t, from Equation 2.29 is

$$\frac{|32.12 - 32.16|}{0.050} \sqrt{\frac{4 \times 4}{4 + 4}} = 1.1$$

This is the value of t at the 70% confidence level. At a more usual level of 90%, t is 1.94. Thus the means of the two data sets cannot be distinguished with a sufficiently high level of certainty.

LIMIT OF DETECTION

A very important quantity in chemical analysis is the limit of detection. The **limit of detection** is the smallest concentration or amount that can be detected with reasonable certainty for a given analytical procedure.[3,4] Such a definition clearly requires statistical evaluation of the data to be performed. The question is whether the determined amount could statistically belong to the set of data with no analyte present. An often-used criterion is that the measured amount should be three times the standard deviation of the amount measured. Application of this criterion makes it almost 99.9% certain that the measured value is not a part of the set of measurements made on a sample devoid of analyte. This criterion is based on the assumption that the deviations follow the standard error function at this low level of determination. Thus, the criterion actually

[3]"Nomenclature, Symbols, Units and Their Usage in Spectrochemical Analysis—II," *Spectrochim. Acta B* **1978**, *33B*, 242.

[4]G. L. Long and J. D. Winefordner, *Anal. Chem.* **1983**, 55, 713A–724A.

Example 2.13

Calculating a detection limit
An instrument gives a reading that is known to change by 1000 per milligram of analyte in the sample. Eight repetitive readings were taken from samples presumed to contain no analyte. The mean and standard deviation of those readings were 57.6 and 4.8. Duplicate sample measurements are to be taken. From Equation 2.30, the minimum difference between the background and the detection limit is thus 3.5 × 4.8 × $(10/16)^{1/2}$ = 1.3. Since there are 1000 output units per milligram, this corresponds to 1.3 × (1 mg/1000) = 1.3 μg.

represents a best case, with real situations providing somewhat lower levels of confidence.

Another, less stringent definition is that the mean of the measured value should be different from the mean of the background measurements at the 99% confidence level. To apply this criterion, Equation 2.29 is rearranged as

$$\bar{x}_{dl} - \bar{x}_b = ts_b \sqrt{\frac{N_s + N_b}{N_s N_b}} \qquad 2.30$$

The subscripts s and b refer to the sample and the blank. This equation calculates the average value for the sample measurements at the detection limit $\bar{x}_{dl}$. As with Equation 2.29, the degrees of freedom to use when looking up t should be $N_s + N_b - 2$. Equation 2.30 assumes that the standard deviation for the sample measurements is going to be the same as those for the blank. At sample values so close to the blank values, this may be a reasonable assumption. Notice that if a very large number of data points has been used for the calculation of the blank, the number of sample replications does not significantly affect the result. Thus a careful determination of the blank statistics can enable detection with relatively few repetitions of the measurement (see Example 2.13).

COMPARING THE DEVIATIONS OF TWO DATA SETS

You may have noticed that the values for the standard deviations in the two sets of data used in the means comparison example are quite different (i.e., 0.035 and 0.061 mL). To answer the question of whether the second experimenter is less precise than the first, one can employ the **F-test,** which compares the variances of the two sets. The **variance** of a data set is equal to the square of the standard deviation. The critical value of F is the number that the ratio of the variances of the two measurement sets is unlikely to exceed. The first step is to take the ratio of the variances to calculate the experimental F value for these two sets:

$$F_{exp} = \frac{s_1^2}{s_2^2} \qquad 2.31$$

Example 2.14

Comparing the variances of two data sets
Continuing the use of the two data sets applied in the examples thus far, the two sets had standard deviations of 0.035 and 0.061 mL, and each set consisted of four values. To apply the F-test, the ratio of the variances are calculated and the value compared with that in the F table.

$$F_{exp} = (0.061)^2/(0.035)^2 = 2.9$$

The degrees of freedom for both sets is $4 - 1 = 3$. The critical value of F at the intersection of 3 and 3 is 9.3. Since F_{exp} would exceed the critical value 5% of the time, no difference is indicated.

The larger value is used in the numerator so that the calculated value is greater than 1. The value for F_{exp} is then compared with a table of values such as Table 2.6. To use the table, one must find the intersection of the degrees of freedom in each set. Table 2.6 is set at the confidence level of 95%. For a different confidence level, a table for that level must be used (see Example 2.14).

REJECTION OF DATA

Occasionally, a data set will include a value that seems to be far removed from the other values in the set. Such a data point is called an **outlier.** If the data point truly does not belong in the set, through some error in transcribing or other procedural error in measurement, it is a good idea to remove it, as it can severely affect the calculated mean and standard deviation of the remaining set. A note in the lab book such as "overshot the end point" or "recalibration required after this reading," combined with the error and suspected direction, is ample reason for its rejection. When such a clear indication is not available, one looks to statistical justification for verification that the outlier does not belong in the data set. One cannot make this decision on the basis of confidence limits. The confidence limits will include or not include the outlier depending on whether the outlier was used in the calculation. Two approaches are presented here. You will notice that they will often not reject values that we sense intuitively

Table 2.6
Critical Values of *F* at the 95% Confidence Level

D.F., denominator	D.F., Numerator								
	1	2	3	4	5	6	7	8	12
1	161	200	216	225	230	234	237	239	244
2	18.5	19.0	19.2	19.3	19.3	19.3	19.4	19.4	19.4
3	10.1	10.0	9.3	9.1	9.0	8.9	8.9	8.8	8.7
4	7.7	7.0	6.6	6.4	6.3	6.2	6.1	6.0	5.9
5	6.6	5.8	5.4	5.2	5.1	5.0	4.9	4.8	4.7
6	6.0	5.1	4.8	4.5	4.4	4.3	4.2	4.2	4.0
7	5.6	4.7	4.4	4.1	4.0	3.9	3.8	3.7	3.6
8	5.3	4.5	4.1	3.8	3.7	3.6	3.5	3.4	3.3
9	5.1	4.3	3.9	3.6	3.5	3.4	3.3	3.2	3.1
10	5.0	4.1	3.7	3.5	3.3	3.2	3.1	3.1	2.9
15	4.5	3.7	3.3	3.1	2.9	2.8	2.7	2.6	2.5
20	4.4	3.5	3.1	2.9	2.7	2.6	2.5	2.5	2.3
∞	3.8	3.0	2.6	2.4	2.2	2.1	2.0	1.94	1.8

must be wrong. This is partly due to the many differences between intuition and statistical predictions, but it is mostly due to the meager basis that exists for throwing out a data point without other evidence.

The *Q*-test[5] test is based on the difference between the outlier and the nearest data point. The ratio of this distance to the total range of data points is calculated:

$$Q_{exp} = \frac{|x_{out} = x_{nearest}|}{x_{max} - x_{min}} \qquad 2.32$$

The experimental value for *Q* is then compared with the critical values found in a table. Critical values for *Q* at the 95% confidence level are shown in Table 2.7. For rejection of the outlier, Q_{exp} must be larger than Q_{crit}. We can see what a stringent test this is. For a triplicate measurement, the difference between the two close points would have to be less than 3% of the total spread of data points. Even with a larger number of points, the distance to the outlier must be almost half the total spread.

A troublesome aspect of the *Q*-test is that it depends so much on the value of the next closest point. A test that is based on the whole data set is the *z*-score test.[6] The *z*-score for an outlier is based on the difference between the outlier and the mean, relative to the standard deviation:

$$z = \frac{|x_{out} - \bar{x}|}{s} \qquad 2.33$$

Table 2.7
Table of Critical *Q* Values at the 95% Confidence Level

N	Q_{crit}
3	0.97
4	0.83
5	0.71
6	0.63
7	0.57
8	0.53
9	0.49
10	0.47

---

[5] R. B. Dean and W. J. Dixon, *Anal. Chem.* **1951**, *23*, 626.

[6] W. P. Gardner, *Statistical Analysis Methods for Chemists.* The Royal Society of Chemistry, Cambridge, 1997.

If the z-score is greater than 2, it is possible that the outlier does not belong in the set. If the z-score is greater than 3, it is likely that the outlier should be rejected. No confidence levels are provided for this decision, but the $z > 2$ or $z > 3$ criterion is not easily met as the mean and standard deviation are calculated with the outlier in the set (see Example 2.15).

Example 2.15

Tests for rejection of an outlying data point

We will add another data point to the set we added in Example 2.12. The new set now has five points. They are 32.14, 32.11, 32.08, 32.16, and 32.30 mL. The last point is the added "outlier." The mean and standard deviation of the first four points were determined in Example 2.5 to be 32.12 mL and 0.035 mL.

The mean and standard deviation of the set with the outlier added are 32.16 and 0.085 mL.

$$Q = (32.30 - 32.16)/(32.30 - 32.08)$$

$$Q = 0.636$$

This value is smaller than the critical value of Q for a data set of five which is 0.71. Therefore, the rejection is denied on the basis of the Q-test. The z-score is

$$z = 0.14/0.085 = 1.65$$

This value is smaller than 2, so the rejection is denied on the z-score as well.

Much better, where possible, is the practice of repeating the experiment. As the number of data points increases, the confidence in rejecting the outlier improves, assuming the new data points do not merely confirm that the deviations are large.

Study Questions, Section I

60. Define null hypothesis, systematic error, detection limit, and outlier.

61. Identify the statistical table used for each of the following tests: determination of the confidence interval, comparing two means, comparing two standard deviations, determining the detection limit, rejecting an outlier.

62. Identical soil samples were sent out to two testing labs to determine the level of polychlorinated biphenyl (PCB) contamination. They were concocted samples, each containing 267 ppb PCB. The labs were asked to make eight repetitive determinations and report the mean and standard deviation. Lab A reported 249 ppb with a standard deviation of 13 ppb. Lab B reported 283 ppb with a standard deviation of 35 ppb.

    A. Does either of these labs appear to have a systematic error? (That is, is the difference between the measured and true values significant at the 95% confidence level?)

    B. Is there a significant difference (95% level) between the means of the two labs?

    C. Is there a significant difference (95% level) between the standard deviations of the two labs?

    D. Estimate the limit of detection of each lab.

63. After running 10 blank determinations for Pb in drinking water, for which the readout average was 487 with a standard deviation of 53, a water sample was run. Duplicate measurements averaged 534.

    A. Has lead been detected in this sample at the 95% confidence level?

    B. If the instrument response is 150 readout units per ppm, estimate the detection limit in parts per million of Pb.

64. A. A student determining the percentage of a weak acid in a sample by titration obtained the values 45.36, 45.39, and 46.04. Having made no note that would justify the rejection of the last value, she tried the Q-test for rejection of the outlier. Was rejection justified?

    B. Not wishing to turn in the average of the three data points, she repeated the experiment two more times. This time the results were 45.34 and 45.37. What value should she submit for the result? Use both the Q-test and the z-score test to confirm your conclusion.

60. The null hypothesis in statistics is the assumption that two data sets are identical (in whatever quality is being tested). A systematic error is one that is reproducible and thus affects the accuracy but not the precision of the technique. The detection limit is the smallest amount of analyte that can be determined with a specified level of confidence that it is there. An outlier is a data point relatively far removed from the closest value and from the sample mean.

61. The t table is used for determining the confidence interval, for comparing two means, and for determining the detection limit. The F table is used for comparing two standard deviations, and the Q table is sometimes used for rejection of an outlier.

62. A. Equation 2.26 is used to calculate t_{exp} for each of the data sets. For lab A,

$t_{exp} = (267 - 249)(8)^{1/2}/13 = 3.92$. For lab B, $t_{exp} = 1.29$. From Table 2.5, for D.F. $= 8 - 1 = 7$, $t = 2.37$. Thus, lab A appears to have a systematic error (even at the 99% confidence level), but lab B does not.

B. We must first calculate s_{pooled} from Equation 2.28:

$s_{pooled} = \{[(13)^2(8 - 1) + (35)^2(8 - 1)]/(8 + 8 - 2)\}^{1/2} = 26.4$

Now t_{exp} is calculated from Equation 2.29:

$t_{exp} = (283 - 249)(64/16)^{1/2}/26.4 = 2.58$

In looking up the critical value for t, D.F. $= 14$. This value for D.F. is not in Table 2.5, but either the value for D.F. $= 15$ can be used or the value can be estimated as 2.15 by interpolation. The value of t_{exp} is greater than the critical value; thus the null hypothesis is rejected, and there is a difference in the means between the two labs.

C. Equation 2.31 is used to calculate the experimental value for F:

$F_{exp} = (35)^2/(13)^2 = 7.25$

From Table 2.6, $F = 3.8$. Thus, the standard deviations are significantly different.

D. Since the standard deviation of the background is not known, we must use the criterion of three times the standard deviation to estimate the detection limit. For lab A, this is 39 ppm; for lab B, it is 105 ppm.

63. a. Equation 2.30 is used, solving for the detection limit:

$$\bar{x}_s = \bar{x}_b + ts_b \sqrt{\frac{N_s + N_b}{N_s N_b}}$$

$$\bar{x}_s = 487 + 2.23 \times 53 \sqrt{\frac{2 + 10}{2 \times 10}} = 578$$

Since the readout number at the detection limit is larger than the readout number for the sample, it cannot be said that Pb was detected in the sample.

B. The readout number that is in excess of the blank is $578 - 487 = 91$ readout units. With a sensitivity of 150 readout units (ru) per ppm, the detection limit is 91 ru/(150 ru ppm^{-1}) = 0.61 ppm.

64. A. Applying Equation 2.32, the value for Q_{exp} is $(46.04 - 45.39)/(46.04 - 45.36) = 0.96$. This value is just less than the critical value of Q, which for three data points is 0.97. Therefore, the outlier should not be rejected.

B. With the new data, the value for Q_{exp} is 0.93. For five data points, the critical value is 0.71, so rejection is justified. The average is then recalculated with the outlier removed. The result is 45.37%. To confirm this with the z-score, the average and mean are calculated for the set of five values. They are 45.50 and 0.30. From Equation 2.33, $z = 0.54/0.30 = 1.8$. This value is less than 2, so the z-score could not be said to confirm the conclusion. This indicates that the z-score test is more conservative than the Q-test.

J. Least Squares Method for Linear Plots

In the development of working curves or calibration plots for techniques or conversion devices, we are frequently called on to connect the individual measured points with a line that represents the expected output/input relationship. Where the response is a linear function of the input quantity, the form of the working curve is expected to be a straight line. If the measurement of the response had perfect precision, one need only draw the line through the points. A perfectly straight line would go through them all. More realistically, we must choose where to draw the line, making a compromise among the deviations of the measured points from the straight-line value. The **least-squares method** finds the equation for the line for which the cumulative deviations (squared) from the linear value is a minimum.

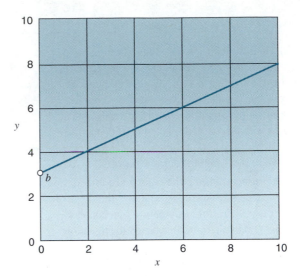

Figure 2.30. The plot of a straight line for which the intercept (b) is 3 and the slope ($\Delta y/\Delta x$) is $\frac{1}{2}$.

THE STRAIGHT LINE

The general form of the equation for a straight line is

$$y = mx + b \qquad\qquad 2.34$$

where y is the value along the Y-axis, x is the value along the X-axis, m is the slope of the line ($\Delta y/\Delta x$), and b is the Y-intercept (the value of y when $x = 0$). A linear plot for which $m = 0.5$ and $b = 3$ is shown in Figure 2.30. The slope of 1/2 is verified by observing that when y increases by 1, x increases by 2.

From Equation 2.34, we see that only two parameters are required to determine the position of a line. They are the slope m and the intercept b. From this it follows that the goal of the least squares method is to obtain the values of m and b for the line that has the least variance from the measured values. The nature of the deviation of a point from the line is shown in Figure 2.31, where the measured value is y_i and the corresponding line value is y_l. In this development, we will assume that the precision with which the x values are known is much greater than that for the y values. This is normally the case with calibration curves where the x values are derived from the standards and the y values are the measured response.

LINEAR REGRESSION

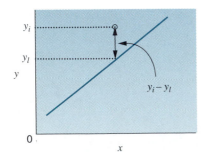

Figure 2.31. The deviation of the measured point from the line is determined along a line of constant x.

The process of determining the straight line for which the deviations are minimized is called **linear regression.** The value of the deviation d_i of each point is $y_i - y_l$. The value of this difference is called the **residual.** The value for y_l (from the equation for a straight line) is $mx_i + b$. Therefore,

$$d_i^2 = (y_i - y_l)^2 = (y_i - mx_i - b)^2 \qquad\qquad 2.35$$

The quantity to be minimized in the calculation of m and b is the sum of the squares of the residuals for all the x_i, y_i data pairs. The derivation of the equations that accomplish minimization involves finding the minimum in the derivative of Equation 2.35 for the sum of all the data points. Many standard statistics texts[7] include this derivation. Only the results are presented here:

[7]C. A. Bennett and N. L. Franklin, *Statistical Analysis in Chemistry and the Chemical Industry,* John Wiley & Sons, New York, 1954.

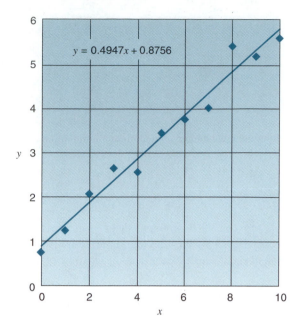

⊙ **Figure 2.32.** A least squares line calculated from the indicated data points and plotted by the spreadsheet program Excel.

$$m = \frac{\sum x_i y_i - \dfrac{\sum x_i \sum y_i}{N}}{\sum x_i^2 - \dfrac{(\sum x_i)^2}{N}}$$

2.36

$$b = \frac{\sum y_i}{N} - m \frac{\sum x_i}{N}$$

2.37

The manual implementation of these equations is daunting. Fortunately, many calculators and most spreadsheet programs have included the functions for automatic calculation of m and b from the data set. Figure 2.32 is a plot of the least squares line fit through the data points shown. The spreadsheet program automatically displayed the parameters of the straight-line equation. If this were a calibration plot, the value of the unknown could be calculated from the response (y value) from a rearrangement of the straight-line equation:

$$x_s = \frac{y_s - b}{m}$$

2.38

where y_s is the response value and x_s is the analytical result.

PRECISION OF THE REGRESSION RESULTS

When the plot of Figure 2.32 was constructed, a random number generator with a normal error distribution was used to determine the deviations from a straight line. The selected slope for the line was 0.50, and the intercept was 1.0. The random variations about these values had a mean of 0 and a standard deviation of 0.3. It is interesting to compare these "true" parameters with the values produced by the regression analysis. First, consider why there should be any difference. The reason is that the deviations were sufficiently large and the number of points sufficiently small that there is an uncertainty in the prediction of the slope and the intercept. The magnitude of the deviations can best be seen from a plot of the residuals, such as that shown in Figure 2.33.

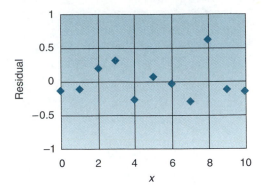

⊙ **Figure 2.33.** a plot o f the residuals of the data in Figure 2.32.

If the residuals are caused by random error, their deviations will follow the Gaussian error curve. This fact enables the measured values to be pooled to calculate the standard deviation of the y values. The equation for this operation is

$$s_y = \sqrt{\sum y_i^2 - \frac{(\sum y_i)^2}{N} - \frac{\left(\sum x_i y_i - \frac{\sum x_i \sum y_i}{N}\right)^2}{\sum x_i^2 - \frac{(\sum x_i)^2}{N}}} \qquad 2.39$$

and, once again, we can be grateful for the automatic calculation of this value. This quantity is also called the **standard deviation of regression** and given the symbol s_r. If regression were perfectly accurate, the standard deviation should match that used in the random generation of the deviations from the ideal line. The value of s used in the random generation was 0.30, and the value of s_y calculated by the regression analysis package was 0.289. Again, the difference is due to the uncertainty introduced by the random error and the relatively small data set.

The standard deviations for the values of m and b can be calculated from the data values. Table 2.8 was taken directly from the spreadsheet produced when the regression function was applied to the data plotted in Figure 2.32. The standard deviations for the values of m and b are called the **standard error of the estimate** in Table 2.8. From this value, the confidence interval and limits for the slope and intercept can be calculated using

$$m_{\text{true}} = m_{\text{reg}} \pm t s_m \qquad 2.40$$
$$b_{\text{true}} = b_{\text{reg}} \pm t s_b \qquad 2.41$$

in which the subscripts true and reg signify the true values and those obtained from the regression analysis. The value for t to be used in this equation is that for which the degrees of freedom are $N - 2$. When this calculation is performed, one will have the confidence interval as well as the confidence limits. The latter are already calculated in Table 2.8 for the 95% confidence level. (Any other level could have been specified.)

Table 2.8
Coefficient Statistics for the Data of Figure 2.32

	Coefficients	Standard error	Lower 95%	Upper 95%
Intercept b	0.875628	0.16249777	0.508033	1.243224
Slope m	0.494731	0.02746714	0.432596	0.556866

Note also that these equations are different from the equations used to calculate the confidence interval of the mean (Equations 2.18 and 2.19) in that the root N is taken as 1. This is because the s for the slope and the intercept is the standard error for a single parameter.

The equations by which s_m and s_b are calculated are

$$s_m = \sqrt{\frac{s_y^2}{\sum x_i^2 - \frac{(\sum x_i)^2}{N}}} \qquad 2.42$$

$$s_b = s_y \sqrt{\frac{\sum x_i^2}{N \sum x_i^2 - (\sum x_i)^2}} \qquad 2.43$$

Table 2.9

Regression Statistics for the Data of Figure 2.32

Multiple R	0.986411
R^2	0.973007
Adjusted R^2	0.970008
Standard error	0.288078
Observations	11

Additional statistics are available from the regression analysis. They are shown in Table 2.9, also copied from the spreadsheet regression of the data in Figure 2.32. The first value, multiple R, is also called the **correlation coefficient.** This is a measure of the goodness of the fit of the data to the function specified. In this case, the function was the straight-line function. The correlation coefficient, for which the symbol is often r, can have any value from -1 to $+1$. If the value is -1 or $+1$, the fit of the data to the equation is perfect. The positive value of the correlation coefficient indicates that the dependent variable increases as the independent variable increases. A negative value indicates a correlation of the opposite slope. A value of 0.98 for the correlation coefficient indicates a good fit. If the value were greater than 0.99 (or less than -0.99), the fit would be considered excellent. A correlation coefficient between 0.9 and 0.95 is thought to be only fair for scientific data.

The square of the correlation coefficient will always have a positive value between 0 and 1. The closer the value is to 1, the better the data fit the specified function. This quantity is called the **coefficient of determination** and often has the symbol R^2 or r^2. The equation for the calculation of R is

$$R = \frac{N \sum x_i y_i - \sum x_i \sum y_i}{\sqrt{[N \sum x_i^2 - (\sum x_i)^2][N \sum y_i^2 - (\sum y_i)^2]}} \qquad 2.44$$

The coefficient of determination will always be smaller than the correlation coefficient. The coefficient of determination is sometimes called the unadjusted R^2. This is because this value can be adjusted to take into consideration the degrees of freedom involved in the data used for its calculation. The value of R^2 adjusted will always be smaller than R^2, but the difference becomes very small as the value of N increases. The coefficient of determination is the parameter most often used as an indication of goodness of fit. Figure 2.34 shows two plots for which the true equation of the line is the same as in Figure 2.32 ($y = 0.50x + 1.00$). However, the values of s_y are large in plot A and small in plot B. Note the great effect of the magnitude of the residuals on the R^2 values and the accuracy of the coefficients for the line.

The standard error given in Table 2.9 is the **standard error of regression.** Again, because it is for a single parameter, the standard deviation and the standard error are the same. This value was calculated using Equation 2.39, and as such, it represents the standard deviation in the values of y.

The example carried through in this discussion has been that for a straight-line calibration curve. Regression analyses can also be performed for responses that are exponential, power, logarithmic, and polynomial functions of the standard quantity. For each model tried, the resulting R^2 values will give an indication of its validity.

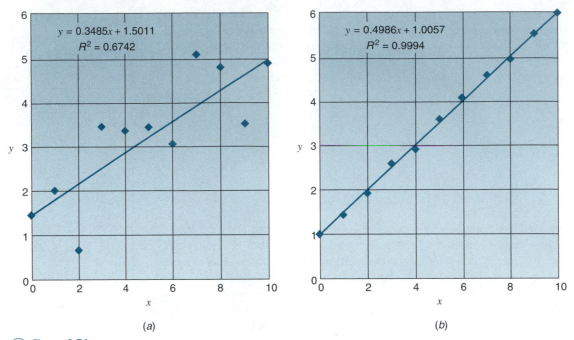

Figure 2.34. These two plots are based on the same line equation as Figure 2.32. In A, $s_y = 0.85$. In B, it is 0.04.

PRECISION OF VALUES CALCULATED FROM THE WORKING CURVE

Having determined the values of the slope and intercept of the working curve by linear regression, we now want to use Equation 2.38 with these values to calculate the value for the quantity of analyte. This is called predicting the value of x based on a measurement of y. Since there is uncertainty in the values of the slope and intercept of the working curve, there will be uncertainties in the values calculated for the unknowns. Assuming that the precision for the unknown determination is the same as that for the standards, the standard deviation in the results calculated from the working curve can be anticipated. Two forms for this equation are

$$s_x = \frac{s_y}{m} \sqrt{\frac{1}{M} + \frac{1}{N} + \frac{(\bar{y}_{meas} - \bar{y}_{calib})^2}{m^2 \left[\sum x_i^2 - (\sum x_i)^2/N\right]}} \qquad 2.45$$

$$s_x = \frac{s_y}{m} \sqrt{\frac{1}{M} + \frac{1}{N} + \frac{(\bar{y}_{meas} - \bar{y}_{calib})^2}{m^2 \sum (x_i - \bar{x}_i)^2}} \qquad 2.46$$

where M is the number of data points in the determination of the measured y, N is the number of data points used in the determination of the calibration curve, the y bars are the averages of the values measured for y for the unknown and for the calibration curve, and the x_i values are those used for the calibration curve. From Equation 2.45, one can go on to predict the confidence interval for the predicted values of x. The equation for this is

$$x_{true} = \bar{x} \pm t s_x \qquad 2.47$$

where the value of t is obtained from Table 2.5 for the appropriate level of confidence (see Example 2.16).

Example 2.16

Calculation of the confidence interval for predicted values of x
Equation 2.46 was evaluated for the three plots of Figure 2.34A, 2.32, and 2.34B with values of s_y of 0.85, 0.29, and 0.04. It was assumed that triplicate measurements were made with an average y value of 3.00. The resulting values of s_x were 1.04, 0.25, and 0.036. The value of t for 95% confidence and D.F. $= N - 2 = 9$ is 2.26. From Equation 2.47, the confidence intervals for the three plots are

$$x_{\text{true}} = 4.3 \pm 2.4$$
$$x_{\text{true}} = 4.3 \pm 0.57$$
$$x_{\text{true}} = 4.00 \pm 0.08$$

The true value was 4.00.

Two characteristics of the standard deviation can be seen from Equation 2.45. One is that if the number of data points used to make the calibration curve greatly exceeds the replicate measurements for each sample, the term $1/M$ in the equation can be ignored. The second is that the numerator in the last term in the radical will be minimized when the measured value of y is near the average of the values used to make the calibration curve. The uncertainty of the predicted values of x then increases the further one gets from the average of the values used for calibration. This is because the uncertainty in the slope of the line has a greater effect on the line position as one moves away from the center region of the line.

Study Questions, Section J

65. Why is it desirable to apply a least squares fit and linear regression to the data points collected for a calibration curve?

66. What is meant by the deviation of a point on the calibration plot?

67. How can statistics be used to evaluate the points on a calibration plot even when each point is the result of only one measurement?

68. What is the significance of the standard deviation of regression?

69. How can the confidence intervals for the slope and intercept be obtained from the standard error of the estimate for them?

70. What are the correlation coefficient and the coefficient of determination, and what are the ideal values for each of them?

71. In the calculation of the confidence interval for predicted values of x, the standard deviation of the predicted x values is greater than that for s_y/m. Is this coincidental, or would you expect this always to be true? Explain.

72. A calibration plot was made from the data at the end of this question. The units used for the concentration were parts per million. From these data, calculate the following:

A. the standard deviation of regression

B. the slope, its standard deviation, and its 95% confidence interval

C. the intercept, its standard deviation, and its 95% confidence interval

D. the coefficient of determination

E. the concentration of an unknown for which the triplicate responses measured were 60.355, 61.199, and 60.747

F. the confidence interval for the unknown

G. an estimate of the detection limit at the 95% confidence level

Concentration	Response
0.00	3.66
1.00	17.66
2.00	30.12
3.00	41.99
4.00	60.17
5.00	71.50
6.00	87.81
7.00	100.31
8.00	115.16
9.00	128.71
10.00	144.59

65. The least squares fit defines the line from which predicted values of analyte can be calculated given the measured response. The linear regression evaluates the goodness of the fit to a straight line and provides data that can be used for a calculation of the confidence intervals for the predicted values of x.

66. The deviation of a point on the calibration plot is the difference between its y value and the y value of the line at the same value of x.

67. The residuals, which are the differences of each point from the best straight line, have a random distribution. These can be minimized to obtain the best line, and the remaining differences can be used to determine the standard deviations of the slope, the intercept, and the values of y.

68. The standard deviation of regression is the standard deviation of the y values from the best straight line. The larger the value of s_n, the less certain we are of the parameters of the line.

69. The confidence interval is $\pm t$ times the standard error of estimate. The degrees of freedom to use for the value of t is $N - 2$.

70. The correlation coefficient is the value R calculated from Equation 2.44. When its value is 1.0 or -1.0, the data perfectly fit the predicted equation for the input–output relationship. The coefficient of determination is the square of R, so it is always positive. The ideal value (perfect fit) of R^2 is 1.0.

71. The standard deviations for the predicted values of x are always greater than s_y/m because of the additional uncertainty in the measured values of y for the sample.

72. The following answers were taken directly from an Excel spreadsheet analysis except where indicated.

    A. The standard deviation of regression is 1.387 (given as standard error in the regression statistics table).

    B. The slope is 14.07, its standard deviation is 0.132, and the 95% confidence limits are 13.78 and 14.37. Divide

Answers to Study Questions, Section J

the difference between the confidence limits by 2 to get the confidence interval. From this, $m_{true} = 14.07 \pm 0.30$.

C. The intercept is 2.508, its standard deviation is 0.782, and the 95% confidence limits are 0.739 and 4.278. Divide the difference between the confidence limits by 2 to get the confidence interval. From this, $b_{true} = 2.5 \pm 1.8$.

D. The coefficient of determination (R^2) is 0.9992.

E. The mean of the responses is 60.767. Equation 2.38 was used to calculate the predicted value of concentration:

$$\text{Conc} = (60.767 - 2.5)/14.07 = 4.1 \text{ ppm}$$

F. Equation 2.46 was used to calculate the standard error for x. First, the square of the deviations was calculated for the x calibration values. The function named DEVSQ in the Excel spreadsheet was used for this. The result was 110. The value for M is 3, and for N, it is 11. The mean value of y used in the calibration is 72.881. With these values, s_x was calculated to be 0.064. The value of t for 95% confidence and D.F. = $11 - 2 = 9$ is 2.26. From Equation 2.47, $x_{true} = 4.14 \pm 0.14$ ppm.

G. There are not enough data for zero concentration (a blank) to use Equation 2.30 to calculate the detection limit. If one used the criterion of three times the standard deviation of the measurement, $3 \times s_x$ is 0.19 ppm. Another approach could be to use the standard deviation in the value for zero concentration from the calibration curve. The standard error in the intercept is 0.78 response units. Three times this value is 2.3 response units. To get this in terms of parts per million, divide the response units by the slope (response units per ppm) to get 0.17 ppm. Rounding up slightly from these criteria, one would safely put the detection limit at 0.2 ppm.

Practice Questions and Problems

1. It is said that the only things we can measure without the use of conversion devices are things we can count. Debate this statement by citing exceptions or prove it with an explanation.

2. For one full day, make a note of each instance in which you gain numerical information by your own observation. For each instance, identify the source of the measurement number (counting, scale reading, or digital display), the units associated with the measurement number, and the resolution of the measurement.

3. The measurement of the diameter of a cylinder yielded 2.38 ± 0.04 cm, and the length measurement was 14.83 ± 0.05 cm. Calculate the volume of the cylinder including the uncertainty.

4. In calibrating a 10 mL pipet, a student obtained weights of delivered water of 10.014, 10.025, 10.018, 10.009, 10.020, 10.010, 10.015, 10.008, 10.012, 10.019, and 10.014 g. The temperature was 25 °C. Calculate the accuracy of the pipet and the precision of the student's technique.

5. In terms of the measurements and interpretations involved, describe and explain the process of determining the percent fat in a person by their weight and the volume of water he or she displaces.

6. The scale on your buret reads from 0 to 50 mL, and the inner diameter of the buret is 1.00 cm.

   A. What is the length of the scale?

   B. What is the transfer function of the buret in length/volume?

   C. What is the transfer function of the scale markings in number/length?

7. The inner diameter of your buret is 1.0 cm, and the inner diameter of the 10 mL pipet is 3.0 mm in the region of the volume mark. The sensitivity of a volume-to-length converter is the change in length for a given change in volume. The pipet is how many times more sensitive than the buret?

8. An oven thermometer with a 270 degree circular scale has a range of 0 to 500 °F. The scale has a mark every 20 degrees. The thermometer can be calibrated to read correctly at 212 °F by dipping it in boiling water. The accuracy rating, when calibrated, is ±5% of the difference between the reading and 212 °F.

   A. What is the sensitivity of the temperature-to-angle conversion device?

   B. What is the approximate resolution of a reading, in degrees Fahrenheit?

   C. For a reading of 400 °F, which is limiting, the resolution or the accuracy? Why?

   D. Name and describe the conversion devices involved in this measurement.

   E. Give the transfer function for each conversion device identified in part D.

9. Perhaps you have seen graduated liquid containers that have a volume shaped like an inverted cone, wider at the top than at the base.

   A. Describe the spacing between the markings on this device assuming each interval represents an equal difference in volume.

   B. Sketch the shape of the working curve for this volume-to-length conversion device.

10. For a clock, what is the angle-to-number conversion of the dial in terms of number/angle?

11. List definitions for the following (include equations when appropriate):

    A. measurement

    B. accuracy

    C. precision

    D. resolution

    E. average/mean

    F. deviation

    G. standard deviation

    H. coefficient of variance

12. A goal of many experiments in the quantitative analysis laboratory is the determination of the amount of the analyte to within 1 ppt.

    A. If the amount of analyte is proportional to the volume reading of a buret (as in a titration) and the buret volume is 34.73 mL, how accurate must the volume reading be in order to achieve the desired goal? If the major markings in the sketch of a buret in Figure 2.18 are 0.5 mL apart, is any interpolation required to achieve the needed resolution in the volume reading?

    B. In the case of a titration, the accuracy of the result frequently depends on two volume readings, one for the calibration of the titrant and one for the titration of the unknown. In this case, how accurate must the volume readings be to achieve the desired accuracy goal? Assume that the average volumes and the inaccuracies of both readings are the same.

    C. If a sample weight is determined to be 0.4592 g, what is the allowable inaccuracy of weighing if the goal of 1 ppt accuracy is to be met?

13. Explain the fundamental difference between scalar and digital readout devices, emphasizing the characterization of one as an analog device and the other as digital.

14. Make a block diagram similar to Figure 2.21 for the use of a Pt resistance thermometer (resistance is a function of temperature) with the readout provided by a moving coil current meter.

15. Draw a block diagram of a system for the measurement of current that has an analog-to-digital converter (ADC) for the readout device.

16. The pH electrode you use in the lab is a conversion device that has a change in output voltage of 0.060 V per unit change in pH. Its output is to be connected to an analog-to-digital converter (ADC) for readout of the pH. It is desirable to read pH directly for pH = 0 to pH = 14 with a resolution of 0.01 pH units.

    A. What is the required input voltage range for the ADC?

    B. If the ADC has a binary register, how many bits are required? (As a point of information, a digital calculator that converts the binary number to the related pH is also required.)

17. A. Explain the role of the conversion device(s) in a null measurement.

    B. Explain why the electronic analytical balance is not a true null measurement of mass.

18. What is the resolution of a 12-bit ADC that has a voltage range of −1.0 to +1.0V?.

19. Define a conversion device (see Chapter 2). Tell whether the following measurements require conversion devices. If

they do, list the conversion devices involved and the input and output units of the data involved.

    A. measuring a person's height

    B. measuring an angle

    C. measuring the temperature of water

    D. measuring the time it takes someone to run 500 yards

20. Draw a block diagram of a pH meter using the following components: ADC, pH electrode, and digital display. Use Figure 2.21 as an example. List the ways in which the data are encoded between each conversion device.

21. A. Define null measurement.

    B. Discuss whether or not titration is a null measurement.

22. List and discuss the errors involved in measuring the length of an object with a linear scale (ruler).

23. A. Give the name and function of the three types of conversion devices used in measurement systems.

    B. For a measurement that you make this week, describe the measurement process in terms of the conversion devices involved and the way in which the measurement data were interpreted to obtain the information you sought.

24. Students in an analytical chemistry laboratory analyzed samples for chloride ion two ways. One method involved weighing the precipitate formed by the reaction of Cl^- with Ag^+. The other method was titration with standard Ag^+ solution. The results of repetitive measurements are listed below. The units are mg of Cl.

Gravimetric: 34.0, 34.4, 34.6, 34.8, 34.9, 34.9, 35.0, 35.0, 35.0, 35.1, 35.1, 35.2.

Volumetric: 34.5, 34.6, 34.7, 34.7, 34.8, 34.8, 34.8, 34.9, 34.9, 35.0, 35.1, 35.2, 35.4.

    A. Calculate the mean and standard deviation for both data sets.

    B. Determine if there is any basis for rejecting any of the values in either set. If so, recalculate the means and standard deviations excluding the rejected point(s).

    C. Determine if the difference in the means of the two sets is statistically significant (at the 95% confidence level).

25. Explain very carefully the difference between the detection limit and the sensitivity of a measurement.

26. A. How many bits are required in an ADC that is used to measure to the nearest 0.1% of full scale?

    B. If one wanted all readings above one-tenth of full scale to have a resolution of at least 0.1%, how many bits would be required?

    C. If the full-scale input voltage range of the converter in part B is 0.00 to 10.0 V, what is the resolution of the converter in volts?

27. A. What are the quantities that are being balanced in a modern electronic balance?

    B. How is the balanced quantity related to the mass of the object being weighed (in general)?

28. A calibration curve was obtained for the determination of an analyte. The response of the technique was taken as a function of the number of milligrams of standard quantity for the analyte. The resulting data table was

mg	Response
0	0.308
5	2.008
10	3.787
15	5.932
20	8.042
25	9.288
30	10.381
35	11.999
40	13.738
45	15.443
50	17.748

From these data, use the spreadsheet regression function to calculate

    A. the standard deviation of regression

    B. the slope, its standard deviation, and its 95% confidance interval

    C. the intercept, its standard deviation, and its 95% confidence interval

    D. the coefficient of determination

    E. the amount of an unknown for which the triplicate responses were 10.472, 10.480, and 10.419.

    F. the confidence interval for the amount of analyte in the sample

    G. an estimate of the detection limit at the 95% confidence level

Suggested Related Experiments

1. Report on the resolution, range, and minimum detectable amount for several pieces of laboratory apparatus (balance, buret, meter stick, etc.)

2. Arrange a number of burets filled to different levels and have all members of the class read all of them. Compare results, calculate the precision, and test for bias in the interpolated values.

3. Become familiar with available statistical analysis tools (calculator and spreadsheet).

Chapter Three

ACIDITY, ACTIVITY, AND pH

The first of the differentiating characteristics to be applied in this study of chemical analysis is the quality of acidity. In this chapter, the concept of acidity will be explored in some depth. We will review just what it is that makes a substance an acid or a base and how acids and bases react with each other and with the solvent in which they are dissolved. With a firm grasp of the nature of acidity and the concept of pH, we will see how the pH electrode can be used to probe this quality, how its response can be measured, and what can be learned from the measurement of pH.

A. Acids, Bases, and Their Reactions

A substance that is acting as an **acid** is donating a proton (hydrogen ion) to another substance. One that is acting as a **base** is accepting a proton. The reactions we will study take place in a solution, but gas-phase proton exchange reactions are also well known.

CONJUGATE ACID–BASE PAIRS

When a chemical species acts as an acid and gives up a proton, it is changed into a species that can accept a proton, that is, a base. This reaction is shown in Equation 3.1.

$$\text{Acid}_1 \rightleftharpoons \text{H}^+ + \text{Base}_1 \qquad 3.1$$

The acid and base forms of the species, differing by just one proton in composition, are called a **conjugate acid–base pair.** The base form of a conjugate acid–base pair can accept a proton to produce the acid form (the reverse of the reaction in Equation 3.1).

Many species can lose or accept a hydrogen ion (proton). A species that loses a proton is acting as an acid. Having done so, it has become the conjugate base of the acid. The charge on the acid form is always one unit more positive than that on its conjugate base.

The reaction shown in Equation 3.1 does not normally occur by itself. The base form of another species usually accepts the proton released by the acid. The acceptance of the proton by the second species is shown in reaction 3.2.

$$Base_2 + H^+ \rightleftharpoons Acid_2 \qquad 3.2$$

Again, the reaction in Equation 3.2 does not occur by itself but in conjunction with another species acting to provide the proton needed. Thus, the reactions of Equations 3.1 and 3.2 are only partial reactions. We will call them **proton half reactions** since they only describe half of the overall process that occurs. When Equations 3.1 and 3.2 are combined to show the acidic form of the first species reacting with the basic form of the second, the overall or total acid–base reaction results.

$$Acid_1 \rightleftharpoons H^+ + Base_1$$
plus $$Base_2 + H^+ \rightleftharpoons Acid_2$$
gives $$Acid_1 + Base_2 \rightleftharpoons Acid_2 + Base_1 \qquad 3.3$$

In the total acid–base reaction of Equation 3.3, we see that two conjugate acid–base pairs are involved, the acid form of one donating its proton to the basic form of the other.

The total acid–base reaction has no protons as reactants or products. They cancel when the two proton half reactions are correctly summed to give a balanced reaction.

RELATIVE STRENGTHS OF ACIDS AND BASES

Conjugate acid–base pairs can be arranged in order of their relative strengths in acid–base reactions as in Table 3.1. The species that releases the most energy by giving up a proton is the acid form in the upper left corner of the table ($HClO_4$). Similarly, the species that releases the most energy by accepting a proton is in the lower right corner of the table (O^{2-}). A total acid–base reaction is formed by combining two of the half reactions in the table so that the acid form in one of the half reactions reacts with the base form in another. An example would be

$$CH_3COOH + NH_3 \rightleftharpoons NH_4^+ + CH_3COO^- \qquad 3.4$$

Notice that in this instance, the product acid form (NH_4^+) is a weaker acid than the reactant acid form (CH_3COOH) and that, correspondingly, the product base formed (CH_3COO^-) is a weaker base than the reactant base (NH_3). Because the stronger acid is reacting with the stronger base to form the weaker products, the reaction will proceed to the right until the concentrations of the product species are greater than the concentrations of the reactant species. This will be the case for any total reaction in which an acid form from one half reaction reacts with a base form from a half reaction lower than it in Table 3.1. Correspondingly, a total reaction in which an acid form reacts with a base form in a half reaction higher than the one it is in, the reaction will not proceed sufficiently to the right to produce a predominance of reaction products.

Table 3.1
Conjugate Acid–Base Pairs

Acid form $\rightleftharpoons$ H$^+$ + Base form

increasing acid strength →

$HClO_4 \rightleftharpoons H^+ + ClO_4^-$
$HI \rightleftharpoons H^+ + I^-$
$HBr \rightleftharpoons H^+ + Br^-$
$HCl \rightleftharpoons H^+ + Cl^-$
$H_2SO_4 \rightleftharpoons H^+ + HSO_4^-$
$C_2H_5OH_2^+ \rightleftharpoons H^+ + C_2H_5OH$
$H_3O^+ \rightleftharpoons H^+ + H_2O$
$H_2C_2O_4 \rightleftharpoons H^+ + HC_2O_4^-$
$HSO_4^- \rightleftharpoons H^+ + SO_4^{2-}$
$H_3PO_4 \rightleftharpoons H^+ + H_2PO_4^-$
$HNO_2 \rightleftharpoons H^+ + NO_2^-$
$HF \rightleftharpoons H^+ + F^-$
$HC_2O_4^- \rightleftharpoons H^+ + C_2O_4^{2-}$
$CH_3COOH \rightleftharpoons H^+ + CH_3COO^-$
$H_2CO_3 \rightleftharpoons H^+ + HCO_3^-$
$H_2PO_4^- \rightleftharpoons H^+ + HPO_4^{2-}$
$NH_4^+ \rightleftharpoons H^+ + NH_3$
$HCO_3^- \rightleftharpoons H^+ + CO_3^{2-}$
$HPO_4^{2-} \rightleftharpoons H^+ + PO_4^{3-}$
$H_2O \rightleftharpoons H^+ + OH^-$
$C_2H_5OH \rightleftharpoons H^+ + C_2H_5O^-$
$OH^- \rightleftharpoons H^+ + O^{2-}$

increasing base strength ↓

EQUILIBRIUM CONSTANTS

Equations 3.3 and 3.4 are written as equilibrium reactions with a double arrow connecting the reactants and products. This double arrow does not just symbolize that the reaction is free to proceed in either direction; it means that both the forward and reverse reactions are actually occurring simultaneously. The true state of **equilibrium** is achieved when the rates of the forward and reverse reactions are equal. To achieve equilibrium, the reaction has proceeded to the point where the concentrations of the products are sufficient to produce a reverse reaction rate equal to the forward reaction rate. When equilibrium is achieved, the reactant and product concentrations are *not* equal, but there is a known relationship between the concentrations of the reactants and products. The relationship is called the **formal equilibrium constant expression.** For the reaction shown in Equation 3.3, the formal equilibrium constant expression is

$$K'_{eq} = \frac{[Acid_2][Base_1]}{[Acid_1][Base_2]} \qquad 3.5$$

where K'_{eq} is the **formal equilibrium constant** and the brackets indicate the molar concentrations of the species. The general form of the equilibrium expression for the generic reaction $aA + bB \rightleftharpoons cC + dD$ is

$$K'_{eq} = \frac{[C]^c[D]^d}{[A]^a[B]^b} \qquad 3.6$$

When K'_{eq} is large (>100) we say that the reaction is **complete** because virtually all of the reactants have been converted to the product forms. For reactions formed from the half reactions in Table 3.1, the greater the difference in the acid strengths of the reactant and product acids, the greater will be the magnitude of K'_{eq}.

The equilibrium constant of the total reaction will be greater than 1 if the reactant acid is stronger than the product acid. In such a case, the reactant base is necessarily stronger than the product base.

Study Questions, Section A

1. What is the principal quality of an acid?

2. What is the principal quality of a base?

3. What is a conjugate acid–base pair?

4. Is it possible for a species to be both an acid and a base?

5. Why is the reaction $Acid_1 \rightleftharpoons H^+ + Base_1$ called a proton half reaction?

6. What is the form of a complete acid–base reaction?

7. Write the expression for the formal equilibrium constant of a complete acid–base reaction.

8. Acetic acid (CH_3COOH) is mixed with ammonia (NH_3) in water.

   A. Write and balance the reaction that occurs.

   B. Using Table 3.1, decide whether the reaction favors reactants or products.

   C. Will the equilibrium constant, K'_{eq}, be greater or less than 1?

Answers to Study Questions, Section A

1. An acid can transfer a proton to another species.

2. A base can accept a proton from another species.

3. A conjugate acid–base pair is two species that differ by only one exchangeable proton.

4. A species can be both an acid and a base if it can give up a proton to another species but also accept a proton from another species. In other words, it is a species that is not in its most acid form nor in its most basic form.

5. The reaction of Equation 3.1 is called a proton half reaction because it involves the loss or gain of a proton between the acid and base forms of the conjugate pair. It is a half reaction because it does not include the species that is accepting the proton (if the reaction is going to the right) or donating the proton (if the reaction is going to the left).

6. A complete acid–base reaction has both the proton donor and the proton acceptor acting as reactants with the conjugate forms of each of these as products.

$$Acid_1 + Base_2 \rightleftharpoons Acid_2 + Base_1$$

7. The equilibrium constant for an acid–base reaction has the form

$$K'_{eq} = \frac{[Acid_2][Base_1]}{[Acid_1][Base_2]}$$

8. A. The proton half reactions for acetic acid and ammonia from Table 3.1 are as follows:

$$CH_3COOH \rightleftharpoons H^+ + CH_3COO^-$$
$$NH_4^+ \rightleftharpoons H^+ + NH_3$$

Since the starting materials are CH_3COOH and NH_3, the second reaction can be reversed and the overall reaction written as follows:

$$CH_3COOH \rightleftharpoons H^+ + CH_3COO^-$$
$$\underline{NH_3 + H^+ \rightleftharpoons NH_4^+}$$
$$CH_3COOH + NH_3 \rightleftharpoons NH_4^+ + CH_3COO^-$$
$$\text{stronger} \quad \text{stronger} \quad \text{weaker} \quad \text{weaker}$$

The overall reaction will proceed to the right.

B. Acid–base reactions will form mostly products if the reacting acid is stronger than the product acid. Since CH_3COOH is a stronger acid than NH_4^+ (it is higher on Table 3.1) the reaction will favor products.

C. From Equation 3.5, the equilibrium constant can be written as follows:

$$K'_{eq} = \frac{[NH_4^+][CH_3COO^-]}{[CH_3COOH][NH_3]}$$

Since we know from part B that the reaction favors products, it is apparent that the numerator of the K'_{eq} equation is greater than the denominator. Consequently, K'_{eq} will be greater than 1. This is true for any reaction that favors products.

B. Acids and Bases in Water

Acid–base reactions often take place in a solution. In such cases, the solvent molecules are frequently involved. Water is the principal solvent discussed here since it is so commonly used.

REACTIONS OF ACIDS WITH WATER

In Table 3.1 of conjugate acid–base pairs, there is an entry ($H_3O^+ \rightleftharpoons H^+ + H_2O$) in which water is the base form of the acid H_3O^+. When an acid species is dissolved in water, it can transfer a proton to water molecules to produce the acid form of the water (H_3O^+, sometimes called the **hydronium ion**) and the base form of the acid. For example,

$$HCl + H_2O \rightleftharpoons H_3O^+ + Cl^- \qquad 3.7$$

Since HCl is above H_3O^+ in the table of acid strengths, this reaction proceeds predominately to the right, and its equilibrium constant is greater than 1. The net effect of this reaction is to produce the base form of HCl and the acid form of water. It also produces a cation and an anion from uncharged reactants. If the reaction of Equation 3.7 has an equilibrium constant of 100 or more, the reaction is complete. An acid that reacts completely with water is said to be a **strong acid** in water.

Consider the reaction of acetic acid with water.

$$CH_3COOH + H_2O \rightleftharpoons H_3O^+ + CH_3COO^- \qquad 3.8$$

Acids above H_3O^+ in Table 3.1 may be strong acids in water. Those below will be weak acids in water.

In this case, CH_3COOH is below H_3O^+ in Table 3.1; thus the reaction proceeds only partially to the right, and its equilibrium constant is less than 1. Acids with an equilibrium constant of less than 1 for the reaction with the solvent are called **weak acids**. All acids below H_3O^+ in the table of conjugate acid–base pairs (Table 3.1) are weak acids in water.

The equilibrium constant expression of Equation 3.5 for the reaction of an acid with water is written as follows.

$$K'_a = \frac{[H_3O^+][Base_1]}{[Acid_1][H_2O]} = \frac{[H_3O^+][Base_1]}{[Acid_1]} \qquad 3.9$$

Example 3.1

The formal equilibrium constant expression for the reaction of acetic acid with water is

$K'_a = [H_3O^+][CH_3COO^-]/[CH_3COOH]$

The subscript a denotes that it is the equilibrium constant for an acid reacting with water. Water is not included with the reactants in the denominator of the equation because it is the solvent and as a (nearly) pure substance has a value in the equilibrium expression of 1 (see Example 3.1). Sometimes the protonated water reaction product is written as H^+ rather than H_3O^+. This is often done for the sake of convenience, but it has sometimes led to confusion between acid–water reactions that involve proton exchange and simple dissociation reactions in which a salt forms solvated ions on dissolution.

When a strong acid is dissolved in water, the molar concentration of the H_3O^+ formed is equal to the number of moles of acid added per liter of solution. It is interesting to note that the acidic species that now exists in that solution is H_3O^+ regardless of which strong acid was dissolved in the water. This is called the **leveling effect.** The strongest acid that can exist in water is the hydronium ion H_3O^+. If a stronger acid is needed for a reaction, a solvent whose protonated form is a stronger acid than H_3O^+ should be selected. For example, from the position of protonated ethanol ($C_2H_5OH_2^+$) in Table 3.1, we see that a strong acid dissolved in ethanol will produce a solution of greater acidity than when that same acid is dissolved in water.

When a weak acid is dissolved in water, the molar concentration of H_3O^+ produced can be very much less than the number of moles of acid added per liter of solution. In the case of acetic acid ($K_a' \approx 10^{-5}$), a 1 molar solution (about the same concentration as in vinegar) will have an H_3O^+ concentration of only 4×10^{-3} M. The acetic acid is about 99.7% unreacted with the water. In this solution, two acids exist: the unreacted acetic acid and the small concentration of H_3O^+. Since acetic acid is a much weaker acid than H_3O^+, the acid strength of a 1 M solution of acetic acid is much less than that of a 1 M solution of a strong acid such as HCl.

Different weak acids have different equilibrium constant values for their reaction with water. The approximate K_a' values of several weak acids in water are given in Table 3.2. Those with a higher value of K_a' will react more completely with the water when dissolved in it. Methods for calculating the extent of the reactions and the concentrations of all the species in the solution are discussed in the next section. A more complete table of equilibrium constants for weak acids is found in Appendix B.

The reaction

$$HCl + H_2O \rightleftharpoons H_3O^+ + Cl^-$$

is a reaction in which the dissolving HCl is protonating the water solvent. This is not the same as a salt dissociation reaction such as $KCl_{solid} \rightleftharpoons K^+_{aq} + Cl^-_{aq}$ in which the salt dissolves by dissociation into its constituent ions.

RELATIONSHIP BETWEEN [H₃O⁺] and [OH⁻] in Water

We can see that water exists in the table of conjugate acid–base pairs in two places: once for the H_3O^+/H_2O pair and once for the H_2O/OH^- pair. From these two entries, a total reaction can be written in which the acid form of the first pair reacts with the base form of the second pair as follows.

$$H_3O^+ + OH^- \rightleftharpoons 2H_2O \qquad\qquad 3.10$$

$$H_3O^+ \rightleftharpoons H^+ + H_2O$$
$$H_2O \rightleftharpoons H^+ + OH^-$$

The equilibrium constant for this reaction is very large, about 10^{14}. The value in considering this reaction and its equilibrium constant lies in its prediction of a relation-

Table 3.2

Approximate K_a' Values for Some Weak Acids in Water

Reaction	$\approx K_a'$	[H₃O⁺]*
$H_2C_2O_4 + H_2O \rightleftharpoons H_3O^+ + HC_2O_4^-$	6×10^{-2}	0.2 M
$HSO_4^- + H_2O \rightleftharpoons H_3O^+ + SO_4^{2-}$	1×10^{-2}	0.1 M
$H_3PO_4 + H_2O \rightleftharpoons H_3O^+ + H_2PO_4^-$	7×10^{-3}	8×10^{-2} M
$HF + H_2O \rightleftharpoons H_3O^+ + F^-$	7×10^{-4}	3×10^{-2} M
$HC_2O_4^- + H_2O \rightleftharpoons H_3O^+ + C_2O_4^{2-}$	5×10^{-5}	7×10^{-3} M
$CH_3COOH + H_2O \rightleftharpoons H_3O^+ + CH_3COO^-$	2×10^{-5}	4×10^{-3} M
$H_2CO_3 + H_2O \rightleftharpoons H_3O^+ HCO_3^-$	5×10^{-7}	7×10^{-4} M
$H_2PO_4^- + H_2O \rightleftharpoons H_3O^+ + HPO_4^{2-}$	2×10^{-7}	4×10^{-4} M
$NH_4^+ + H_2O \rightleftharpoons H_3O^+ + NH_3$	6×10^{-10}	2×10^{-5} M
$HCO_3^- + H_2O \rightleftharpoons H_3O^+ + CO_3^{2-}$	5×10^{-11}	7×10^{-6} M
$HPO_4^{2-} + H_2O \rightleftharpoons H_3O^+ + PO_4^{3-}$	5×10^{-13}	7×10^{-7} M

*Approximate [H₃O⁺] for 1 M HA.

ship between the concentrations of H_3O^+ and OH^- species in aqueous solution. For this purpose, the reaction is usually written the other way about,

$$2H_2O \rightleftharpoons H_3O^+ + OH^- \qquad 3.11$$

which demonstrates that even in pure water, there will be a small but finite concentration of H_3O^+ and OH^- ions. The formal equilibrium constant for reaction 3.11 is

$$K'_w = [H_3O^+][OH^-] \qquad 3.12$$

Equation 3.12 demonstrates a reciprocal relationship between $[H_3O^+]$ and $[OH^-]$. Thus, when $[H_3O^+]$ is high, $[OH^-]$ will be correspondingly lower, and vice versa.

where K'_w is called the **ion product constant** for water. It is also sometimes called the **autoprotolysis constant** or the solvent **disproportionation constant.** The value of K'_w is 1.0×10^{-14} in pure water at 25 °C.

Equilibrium constants often vary greatly with temperature, and K'_w is no exception. Values of K'_w for a few temperatures are shown in Table 3.3. An even greater variability of the concentration of protonated solvent at neutrality occurs with changing solvent. For example, the ion product constant of ethanol is 3×10^{-20}.

Table 3.3

K'_w Values versus Temperature

Temperature, °C	K'_w
0	0.114×10^{-14}
25	1.01×10^{-14}
50	5.47×10^{-14}
100	49×10^{-14}

REACTIONS OF BASES WITH WATER

When a base (proton acceptor) is dissolved in water, a water molecule can donate a proton to it. As seen in Table 3.1, when water is acting as an acid, its base form is OH^-, the **hydroxide ion.** The total reaction of a base with water is

$$H_2O + Base_1 \rightleftharpoons Acid_1 + OH^- \qquad 3.13$$

The formal equilibrium constant expression is

$$K'_b = \frac{[Acid_1][OH^-]}{[H_2O][Base_1]} = \frac{[Acid_1][OH^-]}{[Base_1]} \qquad 3.14$$

where is K'_b the formal base equilibrium constant. Again, the term for water, which would appear in the denominator, is assumed to be 1 since it is the solvent.

If a base that is a stronger base than OH^- is dissolved in water, the reaction in Equation 3.13 proceeds predominately to the right, and the equilibrium constant is greater than 1. When the equilibrium constant is 100 or greater, the reaction is essentially complete and we say that the base is a **strong base** in water. However, after the reaction of the strong base with water, the basic species that exists in the solution is hydroxide ion, regardless of the strength of the original base. Thus, the leveling effect exists at the basic extreme of the relative acidity scale just as it does on the acidic end. From this, we see that H_3O^+ and OH^- define the range of acid and base strengths available in water. Other solvents have smaller and larger ranges, which can be either more acidic or more basic than that of water.

H_3O^+ and OH^- are the strongest acid and base species that can exist in aqueous solution. Any stronger acids or bases dissolved in water react with the water to form H_3O^+ or OH^-.

If a base weaker than OH^- is dissolved in water, the reaction of that base with water will be incomplete. An example is an ammonia solution.

$$H_2O + NH_3 \rightleftharpoons NH_4^+ + OH^- \qquad 3.15$$

Example 3.2

The formal equilibrium constant expression for the reaction of NH_3 with water is

$$K'_b = [NH_4^+][OH^-]/[NH_3]$$

The equilibrium constant for this reaction (see Example 3.2) is much less than 1 ($K'_b \approx 10^{-5}$) so the ammonia remains mostly unreacted. A base with a base strength that is less than that of OH^- is said to be a **weak base** in water. The resulting solution contains two basic species: the unreacted base and OH^-.

Values of equilibrium constants for the reactions of bases with water are sometimes found in the literature and in tables, but it is far more common to express the relationship between the acid and base forms as the acid form reacting to give up a proton. Since the reaction of a base form of a conjugate acid–base pair with water produces the acid form, the reaction of a base with water is closely related to the reaction of its acid form with water. Comparing the equilibrium constant expressions 3.9 and 3.14 demonstrates this point. Rearranging Equation 3.9 to obtain the ratio $[Acid_1]/[Base_1]$

$$[Acid_1]/[Base_1] = [H_3O^+]/K_a' \qquad 3.15$$

and then substituting this in the expression for K_b' (Equation 3.14) gives

$$K_b' = \frac{[Acid_1][OH^-]}{[Base_1]} = \frac{[H_3O^+][OH^-]}{K_a'} = \frac{K_w'}{K_a'} \qquad 3.16$$

which rearranged is

$$K_w' = K_a' K_b' \qquad 3.17$$

This expression allows an easy conversion between values of K_b' and K_a' depending on which constant is available and which is needed. Equation 3.17 also illustrates how intimately involved the solvent ion product constant is in the relationship between K_b' and K_a'. It is clear that a base with a high value for K_b' will have a correspondingly low value for K_a'.

Example 3.3

The K_b' for NH_3 in water at 25 °C is 1.75×10^{-5}. Thus the K_a' for NH_4^+ in water at 25 °C is $1.0 \times 10^{-14}/1.75 \times 10^{-5} = 5.7 \times 10^{-10}$.

LOGARITHMIC CONCENTRATION EXPRESSIONS

The molar concentration of H_3O^+ in a 1 M solution of strong acid is 1 M. From Equation 3.17, $[OH^-]$ will be 10^{-14} M in that solution. In a solution of 1 M strong base, $[OH^-]$ will be 1 M and $[H_3O^+]$ will be 10^{-14} M. This huge range of reasonable concentrations is most conveniently expressed as the logarithm of the concentration. (For a review of logarithms, please see Background B.) Since the values of $[H_3O^+]$ and $[OH^-]$ are usually less than 1 and therefore their logs less than 0, a convention was developed to use the *negative* logs of their concentrations. The negative log function is abbreviated "p." Thus,

$$p[H_3O^+] \equiv -\log[H_3O^+] \qquad 3.18$$

This same "p" function is also useful for equilibrium constants, so $pK_{eq}' = -\log K_{eq}'$.

Equation 3.2 can be expressed using the "p" function:

$$pK_w' = p[H_3O^+] + p[OH^-] \qquad 3.19$$

where pK_w' has the value of 14 for pure water at 25 °C. Equation 3.19 can be plotted to show the values of $\log[H_3O^+]$ and $\log[OH^-]$ as a function of $p[H_3O^+]$, as shown in Figure 3.1. The OH^- line is obtained from the equation $\log[OH^-] = -(14 - p[H_3O^+])$. Such a plot is called a **logarithmic concentration plot.** With it, one can visually estimate the value of $\log[OH^-]$ for each value of $p[H_3O^+]$. If $p[H_3O^+]$ is 5.5 ($[H_3O^+] = 3 \times 10^{-6}$ M), a **solution line** is drawn at that $p[H_3O^+]$. Solution lines are always vertical lines at the solution $p[H_3O^+]$. All the **solution points** fall on the solution line. Horizontal lines drawn left from the solution points give the respective log concentrations for the species. Thus $\log[OH^-]$ is seen to be -8.5 ($[OH^-] = 3 \times 10^{-9}$ M). In this plot, the acidity of the solution increases toward the left (i.e., with decreasing values of $p[H_3O^+]$).

At the point in the logarithmic concentration diagram where $[H_3O^+]$ and $[OH^-]$ are equal (where the species lines intersect), the solution is said to be **neutral,** that is, neither acidic nor basic. On the plot, we see that this occurs where $\log[H_3O^+]$ and

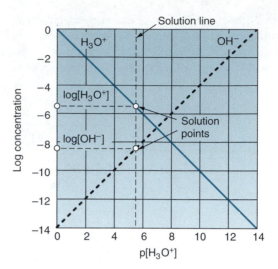

Figure 3.1. Log concentration plot giving the log concentrations for H_3O^+ and OH as a function of $p[H_3O^+]$. The solution line is drawn at a constant $p[H_3O^+]$, and the concentrations of each species correspond to the intersection of the solution line with the species lines. The log concentration value is read from the Y-axis.

$log[OH^-]$ are both -7 and $p[H_3O^+] = 7$. However, this is true only for an aqueous solution at 25 °C. In a solution at 50 °C, pK'_w is 13.26. This would cause the line for $[OH^-]$ to shift to the left in the logarithmic concentration diagram. The $p[H_3O^+]$ of the neutral solution is $13.26/2 = 6.63$.

OTHER AMPHIPROTIC SOLVENTS

Solvents other than water can exist in more acidic and more basic forms than the solvent molecule. Such solvents are called **amphiprotic.** All amphiprotic solvents undergo autoprotolysis. Several such solvents and their autoprotolysis constants, K'_{solv}, are given in Table 3.4. For each of these solvents, the more acid form is the solvent molecule with one more hydrogen atom and a single plus charge. Similarly, the more basic form is the solvent molecule with one less hydrogen atom and a single negative charge. For liquid ammonia (NH_3) the acidic form is NH_4^+ and the basic form is NH_2^-. For methanol (CH_3OH) the acidic form is $CH_3OH_2^+$ and the basic form is CH_3O^-. For methanol, the autoprotolysis expression would be

$$K'_{solv} = [CH_3OH_2^+][CH_3O^-]$$

Just as with water, a strong acid in a given solvent will react to donate all its protons to the solvent. To do so, it must be a stronger acid than the protonated solvent. If the solvent is more acidic than water, strong acids will be leveled at a stronger acid strength in the solvent than in water. An acid that is a strong acid in water may only be a weak acid in the more acidic solvent. Similarly, solvents more basic than water will be leveled by strong bases to a more basic solution than an aqueous solution of OH^-.

Table 3.4
Solvent Autoprotolysis Constants

Solvent	K'_{solv}
Water	1×10^{-14}
Methanol	2×10^{-17}
Ethanol	3×10^{-20}
Acetic acid	3.6×10^{-15}
Formic acid	6×10^{-7}
Sulfuric acid	1.4×10^{-4}
Ammonia	1×10^{-33} (at -50 °C)
Ethylenediamine	5×10^{-16}

9. What is the reaction that accompanies the dissolution of an acid in water?

10. What determines whether an acid is a strong or weak acid in water?

11. Why is dissociation not an accurate term to describe the reaction that occurs when an acid is dissolved in water?

12. Perchloric acid ($HClO_4$) is a stronger proton donor than HCl. Yet, a 0.1 M water solution of these acids has the same acid strength. Why is that true, and what determines the acid strength of these solutions?

13. What is autoprotolysis?

14. What is the equilibrium constant expression for the autoprotolysis constant of water?

15. What is the concentration of OH^- in water at 25 °C if the concentration of H_3O^+ is 3.6×10^{-4} M?

16. Write the autoprotolysis reactions for liquid ammonia and for acetic acid.

17. What is the reaction that accompanies the dissolution of a base in water?

18. If 3.00 g NaOH is dissolved in 500.0 mL of water, what will the molar concentration of OH^- be? The formula weight of NaOH is 40.00 g/mol.

19. What is the equilibrium constant expression for the reaction of Question 17, and how is the equilibrium constant related to K_a'?

20. The K_a' for acetic acid in water is 1.8×10^{-5} at 25 °C. Calculate the K_b' for acetate ion in water at that same temperature.

21. Is $NaNH_2$ a stronger base when dissolved in water or ethanol? Why?

22. A. If the concentration of OH^- in a solution of water at 25 °C is 3.0×10^{-4} M, what is the concentration of H_3O^+ in that solution?

    B. If the water temperature is increased to 50 °C, what is the concentration of H_3O^+?

    C. What is $[H_3O^+]$ at 25 °C if the solution is 80% ethanol and 20% water instead of pure water? The K_{solv}' for 80% ethanol and 20% water at 25 °C is 1.23×10^{-16}.

23. When each of the following substances is dissolved in water, does it act as an acid or a base? Write the reaction between the compound and water and its corresponding equilibrium expression.

    A. $NaHC_2O_4$

    B. $NaHCO_3$

24. If $[H_3O^+]$ is 7.4×10^{-8} M, what is $p[H_3O^+]$?

25. Draw a logarithmic concentration diagram for $C_2H_5OH_2^+$ and $C_2H_5O^-$ in ethanol. You can either create this plot manually or use the plotting program in the CD.

9. Part or all of the acid reacts with the water to form its conjugate base and protonated water.

10. If the acid is a stronger proton donor than protonated water, the reaction will be predominately to form products, and the acid will be considered a strong acid in water. If protonated water is a stronger proton donor than the acid, the reaction will proceed only slightly, and the acid will be considered a weak acid in water.

11. Dissociation implies the mere separation of the proton from the acid to form the conjugate base. This is only a proton half reaction. The dissolved acid reacts to donate protons to the water, with the water acting as a proton accepter.

12. Both $HClO_4$ and HCl are both strong acids in water. Having a higher proton donor capability than protonated water, they react completely to form a 0.1 M solution of H_3O^+. Thus, the acid strength of both solutions is that of a 0.1 M solution of H_3O^+. Their acidity has been leveled to the same point by their reaction with the water.

13. When a solvent can act as an acid and a base, some small fraction of it will undergo proton exchange with itself and thus be in the more acidic and basic forms.

14. $K_w' = [H_3O^+][OH^-]$

15. From Equation 3.12,

$$[OH^-] = 1.0 \times 10^{-14}/3.6 \times 10^{-4} = 2.8 \times 10^{-11} \text{ M}$$

16. $$2NH_3 \rightleftharpoons NH_4^+ + NH_2^-$$
$$2HOAc \rightleftharpoons H_2OAc^+ + OAc^-$$

    where Ac is a frequently used abbreviation for CH_3CO

17. The water transfers protons to some or all of the dissolved base, which increases the concentration of OH^- in solution and creates some of the conjugate acid of the dissolved base.

18. Since NaOH is a strong base, it reacts completely in water. Therefore, all of the NaOH placed in solution will form OH^-:

$$NaOH + H_2O \rightleftharpoons Na^+ + OH^- + H_2O$$

The concentration of OH^- is equal to the concentration of NaOH, so

$$(3.00\ g)\left(\frac{1\ mol}{367.41\ g}\right)\left(\frac{1}{500.0\ mL}\right)\left(\frac{1000\ mL}{1\ L}\right) = \frac{0.150\ mol}{L}$$

$$\frac{0.150\ mol}{L} = C_{NaOH} = [OH^-]$$

19. The equilibrium constant expression is

$$K_b' = \frac{[\text{Acid}_1][\text{OH}^-]}{[\text{Base}_1][\text{H}_2\text{O}]} = \frac{[\text{Acid}_1][\text{OH}^-]}{[\text{Base}_1]}$$

The relationship between K_b' and K_a' is

$$K_w' = K_a' K_b'$$

20. From Equation 3.17, $K_a' K_b' = K_w' = 1.0 \times 10^{-14}$ at 25 °C.

$$K_a' = \frac{K_w'}{K_b'} \qquad K_b' = \frac{1.0 \times 10^{-14}}{1.8 \times 10^{-5}} = 5.6 \times 10^{-10}$$

21. NaNH_2 is a stronger base in ethanol because, from Table 3.1, $\text{C}_2\text{H}_5\text{O}^-$ (the base formed when NaNH_2 is dissolved in ethanol) is a stronger base than OH^- (the base formed when NaNH_2 is dissolved in water).

22. A. $[\text{OH}^-]$ is related to $[\text{H}_3\text{O}^+]$ by Equation 3.12,

$$K_w' = [\text{H}_3\text{O}^+][\text{OH}^-]$$

At 25 °C, the value of K_w' is 1.0×10^{-14}, so

$$1.0 \times 10^{-14} = [\text{H}_3\text{O}^+](3.0 \times 10^{-4})$$

$$[\text{H}_3\text{O}^+] = 3.3 \times 10^{-11} \text{ M}$$

B. At 50 °C, the value of K_w' is 5.47×10^{-14} (from Table 3.3), so

$$[\text{H}_3\text{O}^+] = \frac{5.47 \times 10^{-14}}{3.0 \times 10^{-4}} = 1.8 \times 10^{-10} \text{ M}$$

C. The relationship between the concentrations of the protonated and unprotonated solvent molecules holds true in any solvent. In a mixed solvent, the protonated and deprotonated forms will be those that are most easily protonated and deprotonated. In a water–ethanol mixture, it will be H_3O^+ and OH^- because these are weaker acids and bases than the protonated and deprotonated ethanol. They are simply related by K_{solv}', the autoprotolysis constant for that solvent. Therefore,

$$K_{solv}' = [\text{H}_3\text{O}^+][\text{OH}^-]$$

$$[\text{H}_3\text{O}^+] = \frac{1.23 \times 10^{-16}}{3.0 \times 10^{-4}} = 4.1 \times 10^{-13} \text{ M}$$

23. A. Since HC_2O_4^- is a stronger acid than water (see Table 3.1) it will act as an acid when mixed with water, and water will act as a base. The reaction is

$$\text{HC}_2\text{O}_4^- + \text{H}_2\text{O} \rightleftharpoons \text{H}_3\text{O}^+ + \text{C}_2\text{O}_4^{2-}$$

The acid equilibrium expression is written using Equation 3.9:

$$K_a' = \frac{[\text{H}_3\text{O}^+][\text{C}_2\text{O}_4^{2-}]}{[\text{HC}_2\text{O}_4^-]}$$

Note that water is not included in the expression since it is assumed to be a pure solvent.

B. Since HCO_3^- is a weaker acid than H_3O^+, it will act as a base in water, and water will act as an acid:

$$\text{HCO}_3^- + \text{H}_2\text{O} \rightleftharpoons \text{H}_2\text{CO}_3 + \text{OH}^-$$

The base equilibrium expression is written using Equation 3.14:

$$K_b' = \frac{[\text{H}_2\text{CO}_3][\text{OH}^-]}{[\text{HCO}_3^-]}$$

24. The term "p" stands for "the negative log of." Therefore, $p[\text{H}_3\text{O}^+] = -\log(7.4 \times 10^{-8}) = 7.13$.

25. Plot $p[\text{EtOH}_2^+]$ versus $\log[\text{EtO}^-]$. The line for EtOH_2^+ is obtained from $\log[\text{EtOH}_2^+] = -p[\text{EtOH}_2^+]$. The line for ETO^- is obtained from $\log[\text{EtO}^-] = -(pK_{solv}' - (p[\text{EtOH}_2^+]))$. pK_{solv}' for ethanol is obtained from Table 3.4. The $p[\text{EtOH}_2^+]$ of a neutral solution of ethanol is where the lines intersect.

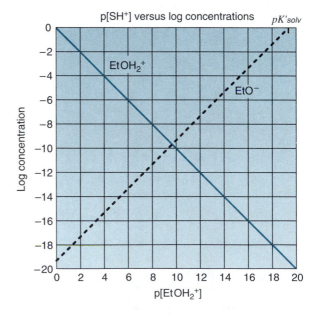

C. Concentrations, Activities, and pH

In general, the chemical reactivity (usually called the chemical activity) of a species in solution increases with its concentration. However, this increase is often not exactly proportional to the change in concentration. In addition, the chemical activity of an ionic species can be affected by the presence of other ions in solution. An understanding of the relationship between the chemical activity of a species and the amount of that species is essential if we are to use chemical reactions as a basis for chemical analysis. This section begins with the development of the concept and definition of chemical activity and proceeds to explore its relationship to chemical concentration. We will also see how the fundamental definition of pH is related to the chemical activity of H^+ in solution.

CHEMICAL POTENTIAL

In an earlier section of this chapter, we defined the equilibrium state as one in which a reversible reaction is proceeding in the forward and reverse directions at the same rate. Consider the reaction

$$HF + H_2O \rightleftharpoons H_3O^+ + F^- \qquad 3.20$$

When equilibrium has been reached, the rate at which hydrofluoric acid molecules protonate water is equal to the rate at which hydronium ions protonate fluoride ions. Thus, equilibrium is a dynamic state; the reaction not only *may* proceed in either direction, it is *actively going* in both directions simultaneously and equally.

For a chemical reaction system at equilibrium, there is no change in composition so no reaction would appear to be occurring. In fact, the reaction may be very rapid but going in both directions at equal rates.

A reaction system comes to an equilibrium condition when the energy of the system cannot decrease further by proceeding in either direction. If the reaction proceeds in the forward direction, reactant species are lost and product species are gained. As more product is formed, the energy to make still more product increases. In addition, the energy gained by the loss of the already depleted reactant decreases. At some point, the increasing energy to form more products will equal the declining energy gained by the loss of reactants. For the reaction to proceed beyond that point, the system energy would have to increase. Thus, it remains in this equilibrium state with no net change in composition.

If one artificially disturbs the equilibrium state by, for instance, adding more product (as in adding F^- to the system of Equation 3.20), the system would lose energy by the reverse reaction to lose product and create reactant molecules, but only to the point where this is no longer favorable. Thus, there is a minimum in the system energy for each combination of reactant and product concentrations. The system automatically seeks this minimum by forward or reverse reaction until the equilibrium state is achieved. This energy minimum is shown schematically in Figure 3.2.

The increase in system energy accompanying the addition of an infinitesimal amount of species X (in J mol^{-1}) is called the **chemical potential** of species X. It has the symbol μ_X. If the sum of the chemical potentials of the reactants is greater than the sum of the chemical potentials of the products, the reaction will proceed in the forward

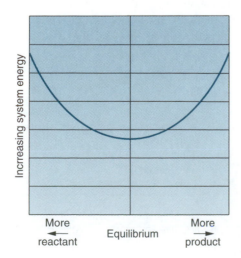

Figure 3.2. The system energy is a minimum at equilibrium. The reaction will proceed toward equilibrium from either side.

direction. If it is less, the reverse reaction will occur. At equilibrium, the combined chemical potentials of the reactants and the products will be minimized and equal. Thus, for the reaction of Equation 3.20, at equilibrium,

$$\mu_{HF} + \mu_{H_2O} = \mu_{H_2O^+} + \mu_{F^-} \qquad 3.21$$

In the next section, Equation 3.21 will be used to obtain the equilibrium constant expression. Before doing so, we must develop the relationship between the chemical potential of a species and its concentration.

CHEMICAL ACTIVITY

In general, the chemical potential of a species will increase with increasing concentration. More specifically, the increase is roughly in proportion to the log of the concentration. This relationship is not exact, so a quantity called the **chemical activity** has been defined for which this is true.[1] The relationship that defines chemical activity is

$$\mu_X = \mu_X^\circ + RT \ln \frac{a_X}{a_X^\circ} = \mu_X^\circ + RT \ln a_X \qquad 3.22$$

where R is the molar gas constant in J mol^{-1} K^{-1}, T is the temperature in K, μ_X° is the chemical potential of species X in its **standard state,** a_X is the chemical activity of X, and a_X° is the chemical activity of X in the standard state. The quantity a_X° for any species is exactly equal to 1, so the equation is usually written as shown after the second equals sign. The units in which the activity is expressed cancel with those of the implied standard state. From Equation 3.22, for a species in its standard state, $a_X = 1$ and $\mu_X = \mu_X^\circ$ as expected.

For solid substances and solvents, the **standard state** is pure material, and the activity is expressed in the mole fraction of the species in that solid or solvent. For gases, the standard state is the pure gas at 1 atmosphere pressure and 273 K (standard temperature and pressure, or STP), and the activity is expressed in the partial pressure of the species in the gas phase. For solutes whose concentration is expressed in moles per liter, the standard state is an ideal 1 molar solution, and the activity has the units of molarity. These standard states and activity units are summarized in Table 3.5.

For solids and solvents, the mole fraction is 1 when the material is pure and 0 when there is none present. The chemical activity is also 1 when the material is pure and 0 when there isn't any. If the concentration (in mole fraction) is equal to the activity at mole fractions of both 0 and 1, why aren't they also equal at all fractional mole fractions? The answer is that when the mole fraction is less than 1 or more than 0, the system is a mixture, and the species in the mixture can interact with each other. The degree to which they are mutually attractive or repulsive changes their availability for chemical reaction and thus affects their chemical activity.

Table 3.5
Standard States of Materials

Material	Standard State	Units for Activity
Solids	Pure solid	Mole fraction
Liquids	Pure liquid	Mole fraction
Gases	Pure gas at STP	Partial pressure
Solutions	"Ideal" 1 M solution	Molarity

[1]G. N. Lewis and M. Randall, *Thermodynamics,* 2nd ed., revised by K. S. Pitzer and L. Brewer. McGraw-Hill, New York, 1961.

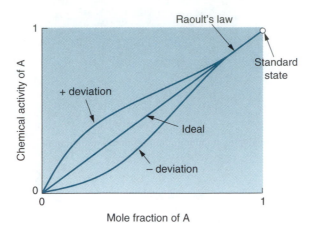

Figure 3.3. Raoult's law predicts a linear relationship between chemical activity and mole fraction. Positive and negative deviations from this rule are caused by the mutual attraction or repulsion between the solute and the matrix.

Figure 3.3 is a plot of the activity of a species in three matrices versus the mole fraction of the species. In one matrix (the lower line) the solute A interacts positively with the solvent or matrix it is in. This interaction with the matrix molecules makes the A molecules less active for other chemical interactions than they otherwise would be. At high mole fractions of A, the mixture begins to approach pure A, and the availability of the matrix species for interaction decreases. Thus, the activity line approaches the linear relationship as the mole fraction approaches 1. In the middle line, the degree of interaction of A molecules with the matrix molecules is exactly the same as that the molecules have among themselves. In this situation, the presence of the matrix or solvent has no effect on the activity of A, and the activity is exactly proportional to the mole fraction. Such an ideal solution conforms to **Raoult's law,** which is stated

$$a_X = f_X \qquad\qquad 3.23$$

where f_X is the mole fraction of X.

The top line in Figure 3.3 has a positive deviation from Raoult's law. This occurs when there is less attraction between the solute and the matrix than the matrix molecules have for each other. This relative repulsion of the A molecules increases their chemical activity for other interactions. These same considerations apply to gaseous mixtures between zero partial pressure and pure gas. The extent of the deviation in gases, however, is generally much smaller.

For relatively dilute solutions, the relationship between the concentration and activity is shown at the lower left corner of Figure 3.3. For all nonideal solutions, the activity is not equal to the mole fraction in this region. Nevertheless, the activity may be approximately proportional to the concentration at low concentrations. That is,

$$a_X = k_H C_X \qquad\qquad 3.24$$

where k_H is a constant and C_X is the molar concentration of X.

A solution for which Equation 3.24 is true is said to follow **Henry's law.** If the Henry's law constant k_H is equal to 1, the solution also follows Raoult's law. Thus, a solute at low concentrations may behave ideally with respect to Henry's law, but still have a positive or negative deviation relative to Raoult's law.

Henry's law ideal behavior occurs at very low concentrations where the solute molecules or ions are so widely dispersed that their entire interaction is with solvent molecules. As the concentration of solutes increases in this very dilute region, their chemical activity increases in proportion to their concentration, as shown in Figure 3.4. At higher concentrations, the possibility of interaction among the solutes increases, and this can affect the way the chemical activity changes with increasing concentra-

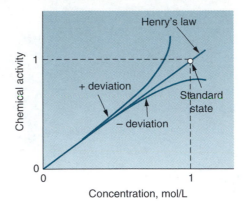

Figure 3.4. Henry's law states that the chemical activity is proportional to the concentration. At low concentrations, this is generally true. However, most species exhibit positive or negative deviations from this general rule at larger concentrations.

tion. Solute/solvent combinations can be classified by whether their deviation from the Henry's law linear relationship between activity and concentration is positive, negative, or zero.

ACTIVITY COEFFICIENTS

A term called the **activity coefficient** has been devised to describe the extent of the deviation from Henry's law. Thus,

$$a_X = \gamma_X[X] \qquad \text{or} \qquad \gamma_X = a_X/[X] \qquad\qquad 3.25$$

where a_X and γ_X are the activity and activity coefficient of species X. For solutions with a positive deviation from Henry's law, γ_X is greater than 1; for a negative deviation, γ_X is less than 1; for an ideal solution, $\gamma_X = 1$.

For ionic solutes, the deviation from the Henry's law line is always very strongly negative ($\gamma_X < 1$). (See Figure 3.5.) In very dilute solutions, the solute ions interact only with the surrounding water molecules. Changing the concentration in this range does not affect the environment of the ions very much. At higher concentrations, the attractive and repulsive forces of the other ions in the solution affect each of the ions much more. An "ionic atmosphere" of oppositely charged ions begins to form around each of the ions, making the environment more attractive to the ions. This has the effect of decreasing their activity for chemical interaction with other species. Note that all ionic solutes will enhance this effect for all the ionic species in the solution.

In 1923, P. Debye and E. Hückel[2] developed a model for the effect of increased ionic concentrations on the chemical activity of ions. This model is based purely on the

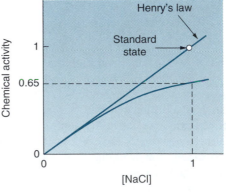

Figure 3.5. The deviation from Henry's law always begins in the negative direction for ions dissolved in water. The activity of the salt NaCl is only 0.65 when the concentration is 1 M. For ions, the differences between concentration and activity are significant even at relatively low concentrations.

[2]P. Debye and E. Hückel, *Phys. Z.* **1923,** *24,* 185.

interactions among the ions owing to their electric charge. From the model, they derived an equation for the activity coefficient of an ion as a function of the concentrations of all ions in the solution. The simplest form of this relationship is called the Debye–Hückel limiting law (DHLL):

$$-\log \gamma_X = A z_X^2 \sqrt{S}$$
$$\text{In water at } 25°C, \quad -\log \gamma_X = 0.5091 z_X^2 \sqrt{S} \qquad 3.26$$

The term A is called the **Debye–Hückel coefficient.** It has the value of $A = 1.842 \times 10^6/(DT)^{3/2}$ L mol^{-1}, where D is the dielectric constant of the solvent and T is the temperature in kelvins. z_X is the number of electron charges on the ion, and S is the ionic strength of the solution.

The **ionic strength** depends on the total ionic concentration but is also strongly affected by the charges on the dissolved ions in the following way.

$$S = \frac{1}{2}\sum_{i=1}^{n}[i]z_i^2 = \frac{1}{2}([A]z_A^2 + [B]z_B^2 + [C]z_C^2 + \cdots) \qquad 3.27$$

For a solute that dissolves to form two singly charged ions (such as NaCl), S is equal to the molar concentration of the salt. For a salt with a doubly charged cation and a singly charged anion,

$$S = \frac{1}{2}[M(2)^2 + 2M] = 3M$$

and for a salt with doubly charged cation and anion, the ionic strength is 4 times the concentration. Table 3.6 provides the ratios of ionic strength to concentration for several salt types. The factors assume that the salt is completely dissociated. This may not always be the case.

According to the DHLL, a singly charged ion in a solution with an ionic strength of 0.01 would have an activity coefficient of 0.89. The activity coefficients of doubly and triply charged ions in the same solution would be 0.63 and 0.35, respectively. Only at an ionic strength of less than 0.0001 would the activity of a singly charged ion be within 1% of its concentration. From these examples and from the curves of the plot in Figure 3.6, we can see that the activity of an ion is much lower than its concentration even at this relatively low ionic strength. (The ionic strength of blood is typically 0.05, and for seawater it is 0.46.) Unfortunately, the approximations involved in the derivation of the DHLL make the law increasingly inaccurate for solutions with an ionic strength above 0.01.

Table 3.6
Ionic Strength Factor for Various Salt Types

Type	Example	S/C
1:1	NaCl	1
1:2	Na$_2$CO$_3$	3
2:1	Ba(NO$_3$)$_2$	3
2:2	BaSO$_4$	4
3:1	AlCl$_3$	6
3:2	Fe$_2$(SO$_4$)$_3$	15
3:3	LaPO$_4$	9

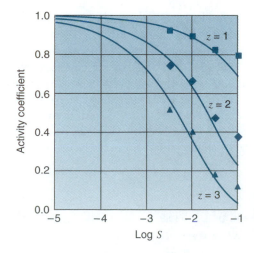

Figure 3.6. Actual and calculated activity coefficients for three charge types. The points are measured values. The lines were calculated using the DHLL.

The range of ionic strengths for which the activity coefficient can be calculated can be extended somewhat using the more complete Debye–Hückel equation (DHE). This equation is

$$-\log \gamma_X = \frac{A z_X^2 \sqrt{S}}{1 + B\mathbf{a}_X \sqrt{S}}$$

In water at 25 °C, $$-\log \gamma_X = \frac{0.5091 z_X^2 \sqrt{S}}{1 + 3.29 \mathbf{a}_X \sqrt{S}}$$ 3.28

where $\mathbf{a}_X$ is the approximate diameter of the hydrated ion (in nanometers) and $B = 502.9/(DT)^{1/2}$ nm^{-1} where D is the dielectric constant of the solvent. Values of $\mathbf{a}_X$ vary from 0.25 nm for Ag^+ and NH_4^+ to 1.1 nm for Ce^{4+} and Sn^{4+}. (See Appendix A for a table of $\mathbf{a}_X$ values.)

Calculations of the activity coefficient of a sodium ion in a solution of ionic strength 0.01 by the DHLL and the DHE yield 0.89 and 0.90. At an ionic strength of 0.1, the calculated values deviate much more (0.69 and 0.77). Where they are different, the value obtained from the DHLL may be higher or lower than that from the DHE, depending on the value of $\mathbf{a}_X$. For larger diameter ions, the use of the DHE at the higher ionic strengths has an even greater effect. The range of usefulness of the DHE is, however, limited to ionic strengths of 0.1 or less because above this value, the proximity of the ions allows species-specific interactions to become important.

For many ions in water, the product of $\mathbf{a}_X$ and 3.3 is approximately 1. This prompted Guntleberg[3] to suggest the following equation:

$$-\log \gamma_X = \frac{A z_X^2 \sqrt{S}}{1 + \sqrt{S}}$$ 3.29

which has a range of application somewhat between that of the DHLL and the DHE. More recently, the Guntleberg variation has been empirically extended by Davies[4] as shown in Equation 3.30. The Davies equation is useful in aqueous solutions to ionic strengths up to 0.5.

$$-\log \gamma_X = \frac{A z_X^2 \sqrt{S}}{1 + \sqrt{S}} - 0.2S$$ 3.30

In practical terms for analytical chemistry, we will emphasize the use of concentration equations wherever possible. However, some sensors are responsive to chemical activity rather than concentration, so it is very important to distinguish between them. Another important factor we should not ignore is the temperature dependence of both the A and B terms in the Debye–Hückel equation. At temperatures much different from 25 °C, this can have a significant effect on the values calculated for γ. Even greater will be the effect of changing solvents; the dielectric constant D for ethanol is only 25 compared to a value of 78 for water.

For the protonated water molecule, the ionic diameter is 0.9 nm. For this ion, the activity coefficients, calculated with the DHE for ionic strengths of 0.001, 0.01, and 0.1, are 0.97, 0.91, and 0.83, respectively. The values of $\gamma_{H_3O^+} = a_{H_3O^+}/[H_3O^+]$ calculated by the DHLL, Davies, and DHE equations are plotted as a function of ionic strength in the graph in Figure 3.7. As we will learn in the next section, the pH electrode responds to the activity of the hydrogen ion, not its concentration. At even modest ionic strengths, there is a large difference between the activity and concentration of H_3O^+.

---

[3]E. Guntleberg, Z. *Phys. Chem.* **1926**, *123*, 199.

[4]C. W. Davies, *Ion Association.* Butterworths, London, 1962.

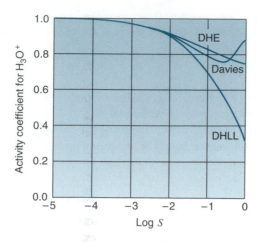

Figure 3.7. The values of the activity coefficient predicted by the DHE, DHLL, and Davies equation differ significantly for ionic strengths above 0.01. For the DHLL, the ion diameter is that for the H_3O^+ ion.

MEAN IONIC ACTIVITIES

Interestingly, the chemical activity of a single ionic species cannot be measured. This is because all measurements of activity are those of the dissolved salt (both cation and anion parts of the salt). The quantity measured is the mean ionic activity $a_\pm$. The relationship between the mean ionic activity and the mean ionic concentration is $a_\pm = \gamma_\pm C_\pm$, where $\gamma_\pm$ is the mean ionic activity coefficient. For a simple 1:1 salt such as NaCl,

$$a_\pm = (a_{Na^+} a_{Cl^-})^{1/2}, \qquad \gamma_\pm = (\gamma_{Na^+} \gamma_{Cl^-})^{1/2}, \qquad C_\pm = C_{NaCl}$$

For the more complex salt $CaCl_2$, the mean ionic activity coefficient is

$$a_\pm = (a_{Ca^{2+}} a_{Cl^-}^2)^{1/3}, \qquad \gamma_\pm = (\gamma_{Ca^{2+}} \gamma_{Cl^-}^2)^{1/3}, \qquad C_\pm = (4C_{CaCl_2}^3)^{1/3}$$

For the general salt formula $A_m B_n$,

$$a_\pm = (a_{A^{z_A}}^m a_{B^{z_B}}^n)^{1/m+n}, \qquad \gamma_\pm = (\gamma_{A^{z_A}}^m \gamma_{B^{z_B}}^n)^{1/m+n}, \qquad C_\pm = [(mC_{A_mB_n})^m (nC_{A_mB_n})^n]^{1/m+n} \quad 3.31$$

From a measurement of the mean ionic activity for known concentrations of salt, the mean ionic activity coefficient for the salt in that solution can be calculated. The results of this calculation can then be compared with the mean ionic activity calculated from combining the separate activity coefficients obtained from the DHLL and DHE. These comparisons have provided validation of the DHLL for low ionic strength solutions. They have also been the basis for an empirical determination of the appropriate size parameter when the DHE is used. Because of the consistency between the experimental results and the Debye–Hückel theory, we have come to trust the calculated activity coefficients in low ionic strength solutions.

An important consequence of the ability to calculate the chemical activity of ions in low ionic strength solutions of acids is that the activity of the hydronium ion can be calculated. The "p" function of the hydronium ion activity, $p(a_{H_3O^+})$, is often called the pH of the solution. However, as we shall see in the next section, the official definition of pH is based on a quantity we can actually measure. In this text, we will be careful to use $p[H_3O^+]$ when we mean the negative log of the hydronium ion concentration, $p(a_{H_3O^+})$ when we mean the negative log of the hydronium ion activity, and pH when referring to the quantity measured by with the pH electrode.

Mean ionic activities can be measured for ionic solutes. Single species ionic activities cannot. Therefore, only the mean ionic activity coefficient can be determined experimentally.

ACTIVITY COEFFICIENTS OF NEUTRAL SPECIES

The effect of ionic strength on the activity of neutral species is much less than that for dissolved ions. An effect of increased ionic strength is to increase the polar nature of the solution. This decreases the degree of interaction of nonpolar solutes with the solvent and increases their chemical activity. The activity coefficient of neutral species follows the simple relationship

$$\log \gamma_0 = k_0 S \qquad\qquad 3.32$$

where k_0 is a constant. The value for k_0 can vary from 0 for solvated ion pairs to 0.2 or more for large organic molecules. For species such as CO_2 and NH_3, it is about 0.11. Since k_0 and S are both positive values, the value for the activity coefficient for neutral species will always be greater than 1. Further, the activity coefficient will increase rather than decrease as the ionic strength increases.

Study Questions, Section C

26. What is the chemical activity of a species in its standard state?

27. What is the chemical potential, and how is it used in the exact description of a chemical reaction at equilibrium?

28. Name the standard states for solids, gases, liquids, and solutions whose concentrations are expressed in moles per liter.

29. What is the quality of a solution that obeys Henry's law? How is this different from one that obeys Raoult's law?

30. What is the activity coefficient?

31. What is the value of the activity coefficient if the activity of a 0.300 M solution is 0.24?

32. Why is the activity coefficient for dissolved ions in water usually less than 1? Why is the activity coefficient for dissolved neutral molecules 1 or greater?

33. Calculate the ionic strength of a solution that is 0.100 M in Na_2SO_4 and 0.0100 M in KCl.

34. To what level of ionic strengths do the DHLL, the DHE, and the Davies equation provide a reasonable estimate of the activity coefficient?

35. Calculate the activity coefficient of Na^+ in an aqueous solution at 25 °C that is 0.010 M in NaCl and one that is 1.0 M in NaCl using

   A. The DHLL

   B. The DHE

   C. The Davies equation

   D. What error in the actual activity of the solutions is involved if the activity coefficient γ is assumed to be 1?

36. Why is it not possible to measure the activity of a single ionic species in solution?

37. The activity coefficients of K^+ and SO_4^{2-} in a 0.5 M K_2SO_4 solution are 0.238 and 3.21×10^{-3} respectively. Calculate the mean activity coefficient for this solution.

38. The measured mean activity coefficient for a 0.80 M solution of $CaCl_2$ is 0.0265, and the calculated activity coefficient of Ca^{2+} is 7.00×10^{-4}. What is the activit, of coefficient of Cl^-?

39. Distinguish among the terms pH, $p[H_3O^+]$, and $p(a_{H_3O^+})$.

Answers to Study Questions, Section C

26. Since chemical activity is the activity of a species relative to its standard state, the activity when in the standard state is 1.

27. The chemical potential of $X\mu_X$ is the increase in system energy in J mol^{-1} that results from a tiny increase in the amount of X. At equilibrium, the sum of the chemical potentials of the reactants is equal to the sum of the chemical reactants of the products.

28. The standard states are pure solid, pure gas at STP, pure liquid, and *ideal* 1 M solution.

29. If a solution obeys Henry's law, the activity of the solute is directly proportional to its concentration. If it obeys Raoult's law, the activity of the solute is directly proportional to the mole fraction over the entire range from 0 to 1.

30. The activity coefficient γ_X is the factor by which the concentration of a solute X should be multiplied to equal its activity a_X.

31. From Equation 3.25,

$$\gamma_X = a_X/[X] = 0.024/0.030 = 0.80$$

32. As the concentration of the ions in water increases, the attractive forces between oppositely charged ions increase. This positive interaction between the ions decreases their activity for other reactions. Conversely, neutral molecules are less solvated as the ionic concentration increases. This increases their activity for other reactions.

33. It is helpful to first write the reactions for the dissolution of the salts in water.

$$Na_2SO_{4(s)} \rightarrow 2Na^+_{(aq)} + SO_4^{2-}_{(aq)}$$
$$KCl_s \rightarrow K^+_{(aq)} + Cl^-_{(aq)}$$

Note that two sodium ions are released into solution for every one Na_2SO_4 molecule; thus, the concentration of Na^+ will be twice that of Na_2SO_4.

$[Na^+] = 2(0.100) = 0.200$ M, $[SO_4^{2-}] = 0.100$ M,
$[K^+] = [Cl^-] = 0.0100$ M

The ionic strength S is calculated from Equation 3.27:

$$S = \tfrac{1}{2}([Na^+]z^2_{Na^+} + [SO_4^{2-}]z^2_{SO_4^{2-}} + [K^+]z^2_{K^+} + [Cl^-]z^2_{Cl^-})$$
$$= \tfrac{1}{2}[0.200(1)^2 + 0.100(2)^2 + 0.0100(1)^2 + 0.0100(1)^2]$$
$$= 0.310.$$

34. The DHLL can provide a good approximation of the activity coefficient to ionic strengths of 0.01. The DHE extends this to 0.1. The Davies equation works well up to 0.5.

35. A. The ionic strength of the 0.010 M solution is calculated using Equation 3.27:

$$S = \tfrac{1}{2}\{[Na^+](1)^2 + [Cl^-](-1)^2\}$$
$$= \tfrac{1}{2}(0.010 + 0.010) = 0.010 \text{ M}$$

Applying the DHLL (Equation 3.26) gives an activity coefficient of

$$-\log \gamma_{Na^+} = 0.5091(1)^2 \sqrt{0.010} = 0.051$$
$$\gamma_{Na^+} = 10^{-0.5091} = 0.89$$

For the 1.0 M solution,

$$S = \tfrac{1}{2}(1.0 + 1.0) = 1.0$$
$$-\log \gamma_{Na^+} = 0.5091(1)^2 \sqrt{1.0}$$
$$\gamma_{Na^+} = 0.31$$

B. Using the DHE (Equation 3.28) for the 0.010 M solution gives an activity coefficient of

$$-\log \gamma_{Na^+} = \frac{0.5091(1)^2 \sqrt{0.010}}{1 + 3.29\,\alpha_{Na^+} \sqrt{0.010}}$$

Using 0.42 nm for α_{Na^+} from Appendix A, $\gamma_{Na^+} = 0.90$. For the 1.0 M solution,

$$-\log \gamma_{Na^+} = \frac{0.5091(1)^2 \sqrt{1.0}}{1 + 3.29\alpha_{Na^+} \sqrt{1.0}} = 0.214$$
$$\gamma_{Na^+} = 0.61$$

C. Using the Davies equation (Equation 3.30) for the 0.010 M solution gives an activity coefficient of

$$-\log \gamma_{Na^+} = \frac{0.5091(1)^2 \sqrt{0.010}}{1 + \sqrt{0.010}} - 0.2 \times 0.010$$
$$\gamma_{Na^+} = 0.90$$

For the 1.0 M solution,

$$-\log \gamma_{Na^+} = \frac{0.509(1)^2 \sqrt{1.0}}{1 + \sqrt{1.0}} - 0.2 \times 1.0$$
$$\gamma_{Na^+} = 0.0546$$

From this example, we see that using the DHLL instead of the DHE gives an answer only slightly lower than the DHE at a NaCl concentration of 0.010 M but gives a much lower answer at a NaCl concentration of 1.0 M. The Davies equation gives a very high value at the higher ionic strength. Probably none of the calculations for the 1 M case are trustworthy, since 1 M is beyond the upper limit for accurate calculations by any of the methods

D. Using the more accurate DHE calculations, the 0.010 M solution has an activity of $0.010 \times 0.90 = 0.0090$. This value is off 10% from the 0.010 value for the activity if activity coefficients are ignored. The situation is much worse at the higher concentration. For the 1.0 M solution, the activity was calculated to be 0.61 versus 1.0. The error, in this case, is almost a factor of 2.

36. Ionic species must be in solution as a salt with both cations and anions. The measured activity is then a combination of the activities of the cationic and anionic species.

37. From Equation 3.31 for the salt A_mB_n,

$$\gamma_\pm = (\gamma^m_{A^{z_A+}}\gamma^n_{B^{z_B-}})^{1/m+n}.$$

$$\gamma_{K_2SO_4} = (\gamma^2_{K^+}\gamma^1_{SO_4^{2-}})^{\frac{1}{2+1}} = [(0.238)^2(3.21 \times 10^{-3})]^{1/3}$$

$$\gamma_{K_2SO_4} = 0.0567$$

38. $\gamma_{CaCl_2} = (\gamma_{Ca^{2+}}\,\gamma_{Cl^-}^{-2})^{1/3}$, $\gamma^3_{CaCl_2} = \gamma_{Ca^{2+}}\,\gamma^2_{Cl^-}$

$$\sqrt{\frac{\gamma^3_{CaCl_2}}{\gamma_{Ca^{2+}}}} = \gamma_{Cl^-}, \qquad \gamma_{Cl^-} = \sqrt{\frac{0.0265^3}{7.00 \times 10^{-4}}}$$

$$\gamma_{Cl^-} = 0.163$$

39. pH is the quantity measured by a pH meter, $p[H_3O^+]$ is the negative log of the H_3O^+ concentration, and $p(a_{H_3O^+})$ is the negative log of the H_3O^+ activity.

D. Equilibrium Constants

Equilibrium constants are essential to the study of analytical methods involving chemical reactions, that is, all techniques in which one or another kind of chemical reactivity is probed. They are used to determine the extent of a reaction when two reactants are mixed, and they provide the fundamental basis for relating the activities (or concentrations) of all the reacting species in a solution.

THE EQUILIBRIUM STATE

When a reaction is at equilibrium, the reason that the forward and reverse reaction rates are equal is that there is no net energy driving the reaction in either direction. This was illustrated in Figure 3.2. The mathematical statement of the equilibrium condition is that the sum of the chemical potentials of the reaction products is exactly equal to the sum of the chemical potentials of the reactants. This was written for a specific case in Equation 3.21. We write it again here for a general reaction, $A + 2B \rightleftharpoons P + 2Q$,

$$\mu_A + 2\mu_B = \mu_p + 2\mu_Q \qquad 3.33$$

Equation 3.33 is the exact and universally applicable statement of the equilibrium condition for the reaction given.

In this example, we see that the stoichiometric coefficients of the reactants and products are included to provide the correct ratios of chemical potentials.

When the equation for chemical potential (Equation 3.22) is substituted into Equation 3.33,

$$\mu_A^\circ + RT\ln a_A + 2\mu_B^\circ + 2RT\ln a_B = \mu_p^\circ + RT\ln a_p + 2\mu_Q^\circ + 2RT\ln a_Q \quad 3.34$$

Like terms may be collected to give:

$$\frac{-(\mu_p^\circ + 2\mu_Q^\circ - \mu_A^\circ - 2\mu_B^\circ)}{RT} = \ln a_p + 2\ln a_Q - \ln a_A - 2\ln a_B \qquad 3.35$$

The lefthand side of Equation 3.35 is all constants. The term for the differences in the standard chemical potentials (in the parentheses) is ΔG°, the standard free energy change for the reaction. Using this term and combining the log terms,

$$\frac{-\Delta G^\circ}{RT} = \ln \frac{a_p a_Q^2}{a_A a_B^2} = K_{eq}^\circ \qquad 3.36$$

For the general reaction $aA + bB + \cdots \rightleftharpoons pP + qQ + \cdots$,

$$K_{eq}^\circ = \frac{a_p^p a_Q^q \cdots}{a_A^a a_B^b \cdots}$$

from which we see that

$$K_{eq}^\circ = \frac{a_p a_Q^2}{a_A a_B^2} \qquad 3.37$$

where K_{eq}° is called the **thermodynamic equilibrium constant.** The form of this equilibrium constant is the same as that for the formal equilibrium constant except that species activities are used instead of their concentrations.

THERMODYNAMIC AND FORMAL EQUILIBRIUM CONSTANTS

When values or tables of equilibrium constants are given, they are generally the thermodynamic equilibrium constants. The reason for this convention is that the thermodynamic equilibrium constant is a more fundamental quantity for the specified reaction. The values of the often more useful formal equilibrium constants, written as the concentration products of the products and reactants, vary with the concentrations of other solutes. As we have seen, this is especially true for reactions involving ionic species. Let's take the reaction $HOAc + H_2O \rightleftharpoons H_3O^+ + OAc^-$ as an example. (HOAc is an abbreviation for acetic acid, CH_3COOH.) Applying Equations 3.33 and 3.34 to this reaction,

$$\mu_{H_3O^+} + \mu_{OAc^-} = \mu_{HOAc} + \mu_{H_2O}$$

$$\frac{-(\mu_{H_3O^+}^\circ \mu_{OAc^-}^\circ - \mu_{HOAc}^\circ - \mu_{H_2O}^\circ)}{RT} = \ln \frac{a_{H_3O^+} a_{OAc^-}}{a_{HOAc} a_{H_2O}} \qquad 3.38$$

The water is a nearly pure solvent, so its activity relative to its standard state (pure water) is essentially 1. Thus the thermodynamic equilibrium constant expression is

$$K_{eq}^{\circ} = \frac{a_{H_3O^+} a_{OAc^-}}{a_{HOAc}} \qquad 3.39$$

which has exactly the same form as the formal equilibrium constant expression (K_a' in Equation 3.9), here written explicitly for the reaction of acetic acid with water:

$$K_a' = \frac{[H_3O^+][OAc^-]}{[HOAc]} \qquad 3.40$$

THE EFFECT OF IONIC STRENGTH ON K_a'

The two kinds of equilibrium constants we have studied so far are clearly related in form. We also understand that the thermodynamic equilibrium constant is fundamental to the reaction but that it is not so useful when the concentrations rather than the activities of the species in the reaction mixture are known or sought. Furthermore, the formal equilibrium constant cannot be looked up for all circumstances because it depends on the nature of the other solutes in the solution. Fortunately, we can use the DHLL to calculate an approximate value for K_a' if K_a° and the ionic strength are known. Figure 3.8 is a plot of the actual values of the K_a' for acetic acid (the diamonds) and the value calculated by a simple relationship based on the DHLL. We start by substituting $\gamma_x[X]$ for each activity in the thermodynamic equilibrium expression (from Equation 3.25). Equation 3.39 then becomes

$$
\begin{aligned}
K_a^{\circ} &= \frac{\gamma_{H_3O^+}[H_3O^+]\gamma_{OAc^-}[OAc^-]}{\gamma_{HOAc}[HOAc]} \\
&= \frac{\gamma_{H_3O^+}\gamma_{OAc^-}}{\gamma_{HOAc}}\frac{[H_3O^+][OAc^-]}{[HOAc]} = \frac{\gamma_{H_3O^+}\gamma_{OAc^-}}{\gamma_{HOAc}}K_a' \qquad 3.41
\end{aligned}
$$

So,

$$K_a' = \frac{\gamma_{HOAc}}{\gamma_{H_3O^+}\gamma_{OAc^-}}K_a^{\circ} \qquad 3.42$$

In order to substitute the equation for the activity coefficients for each of the ionic species into Equation 3.42, it is convenient to first take the log of each term.

$$\log K_a' = \log K_a^{\circ} + \log \gamma_{HOAc} - \log \gamma_{H_3O^+} - \log \gamma_{OAc^-} \qquad 3.43$$

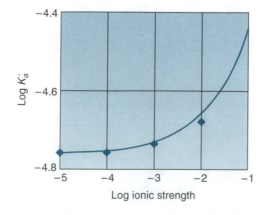

Two forms of the equilibrium expression

The two forms of the equilibrium expression exemplified by Equations 3.39 and 3.40 are sometimes called the activity equilibrium constant and the concentration equilibrium constant.

Figure 3.8. The solid line follows the values for log K_a' for acetic acid calculated with Equation 3.44. The data points are the actual values. Good agreement is achieved to an ionic strength of 0.01.

When the equation for $-\log \gamma_X$ from the DHLL (Equation 3.26) for each species is substituted in Equation 3.43,

$$\log K_a' = \log K_a^\circ + 0 + 2(0.51\sqrt{S}) = \log K_a^\circ + 1.02\sqrt{S} \qquad 3.44$$

where S is the ionic strength of the solution and the value of A for aqueous solutions at 25 °C is used. A simple relationship for K_a' in terms of K_a° and S results. (Note that for the neutral species at moderate concentrations, the activity coefficient is 1.) From this relationship we can see that for dilute solutions, when S is very low, $K_a' = K_a^\circ$. As the ionic strength increases, $\log K_a'$ increases somewhat. The amount of that increase is shown in the plot in Figure 3.8. A similar plot for the autoprotolysis constant of water is shown in Figure 3.9.

Not all formal equilibrium constants will increase with increasing ionic strength and not all to the same degree. It depends on where the ionic species appear in the equilibrium expression and what their charge is. For example, the reaction of NH_4^+ with H_2O gives $H_3O^+ + NH_3$. In this case, the relationship between K_a' and K_a° is

$$K_a^\circ = \frac{\gamma_{H_3O^+}\gamma_{NH_3}}{\gamma_{NH_4^+}} K_a' \qquad 3.45$$

Then taking the log of all terms and substituting from the DHLL for all the $\log \gamma_X$ terms (taking 0 for $\log \gamma_{NH_3}$ because NH_3 has no charge),

$$\log K_a' = \log K_a^\circ - (0.51)\sqrt{S} + (0.51)\sqrt{S}$$
$$\log K_a' = \log K_a^\circ \qquad 3.46$$

So for the ammonia case, K_a' and K_a° are equal over the range of ionic strengths for which the DHLL is valid.

USING FORMAL EQUILIBRIUM CONSTANTS

There are many advantages to using formal equilibrium constants for the calculation of equilibrium concentrations of reactants and products. In the next chapter, we will see that it is actually more accurate to do so. We only need to remember that while the thermodynamic constants are a function of temperature and solvent, the formal constants are also a function of ionic strength. When the formal constants are not available, they can be estimated as above using the single, simple equation developed below.

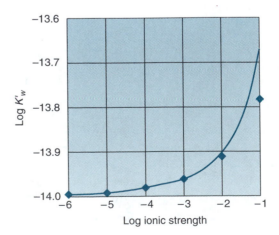

Figure 3.9. Measured and calculated values for the autoprotolysis constant of water. At high ionic strengths, what happens to the p[H_3O^+] of a neutral solution?

Consider the general reaction

$$aA^{z_A} + bB^{z_B} + \cdots \rightleftharpoons pP^{z_P} + qQ^{z_Q} + \cdots \qquad 3.47$$

for which the lowercase letters are the stoichiometric coefficients and the z's are the charges for each species. The equilibrium constant is then

$$K_{eq}^{\circ} = \frac{[P^{z_P}]^p[Q^{z_Q}]^q \cdots}{[A^{z_A}]^a[B^{z_B}]^b \cdots} \times \frac{\gamma_{P^{z_P}}^p \, \gamma_{Q^{z_Q}}^q \cdots}{\gamma_{A^{z_A}}^a \, \gamma_{B^{z_B}}^b \cdots} = K_{eq}' \frac{\gamma_{P^{z_P}}^p \, \gamma_{Q^{z_Q}}^q \cdots}{\gamma_{A^{z_A}}^a \, \gamma_{B^{z_B}}^b \cdots} \qquad 3.48$$

Taking the log of both sides

$$\log K_{eq}^{\circ} = \log K_{eq}' + \log \frac{\gamma_{P^{z_P}}^p \, \gamma_{Q^{z_Q}}^q \cdots}{\gamma_{A^{z_A}}^a \, \gamma_{B^{z_B}}^b \cdots}$$

$$\log K_{eq}' = \log K_{eq}^{\circ} - p \log \gamma_{P^{z_P}} - q \log \gamma_{Q^{z_Q}} - \cdots + a \log \gamma_{A^{z_A}} + b \log \gamma_{B^{z_B}} + \cdots \quad 3.49$$

$$\log K_{eq}' = \log K_{eq}^{\circ} + A\sqrt{S}\,(pz_P^2 + qz_Q^2 + \cdots - az_A^2 - bz_B^2 - \cdots)$$

$$\log K_{eq}' = \log K_{eq}^{\circ} + NA\sqrt{S} \quad \text{where } N = (pz_P^2 + qz_Q^2 + \cdots - az_A^2 - bz_B^2 - \cdots)3.50$$

Using Equation 3.50 with the value of N in the appendix tables

Equation 3.50 allows the simple estimation of a formal equilibrium constant from the thermodynamic constant and the ionic strength. The parameters in the parentheses sum to a simple, small integer. This integer is a constant for a given reaction. The parameter N has been calculated for many of the equilibrium constants given in the appendices.

Study Questions, Section D

40. What is the difference in form and significance of the thermodynamic equilibrium constant expression and the formal equilibrium constant expression?

41. Write the formal and the thermodynamic equilibrium constant expressions for the reaction of HSO_4^- with CH_3COO^-. See Table 3.1 to get the products for this reaction.

42. How does Equation 3.42 demonstrate that the formal equilibrium constant could be predicted from the DHLL for solutions with modest ionic strength?

43. Obtain an expression for the value of K_w' in terms of K_w° and the ionic strength S at 25 °C.

44. Obtain an expression for the value of K_a' in terms of K_a° and the ionic strength S for the reaction $HPO_4^{2-} + H_2O \rightleftharpoons H_3O^+ + PO_4^{3-}$ at 25 °C.

45. Look at the table of K_a° values in Appendix B. In this table is a coefficient N. Using the solution to Question 44, show how this coefficient was obtained for the K_a° of HPO_4^{2-}.

46. Verify that the coefficient N in Appendix B for the loss of the most acidic proton of aspartic acid is correct.

Answers to Study Questions, Section D

40. The thermodynamic equilibrium constant expression involves the activities of the reactants and products. As such, it is a true constant that will apply at all levels of concentration. The problem is that we do not know the activities or the activity coefficients for many practical situations. The formal equilibrium constant expression involves the concentrations of the products and reactants. Because the activity is a function of concentration, the formal equilibrium constant will vary with solution conditions, particularly the ionic strength. The value for the formal equilibrium constant can be experimentally determined for a given operating condition.

41. The balanced equation is

$$HSO_4^- + CH_3COO^- \rightleftharpoons CH_3COOH + SO_4^{2-}$$

The formal and thermodynamic equilibrium expressions are

$$K_{eq}' = \frac{[CH_3COOH][SO_4^{2-}]}{[HSO_4^-][CH_3COO^-]}$$

$$K_{eq}^{\circ} = \frac{a_{CH_3COOH}\, a_{SO_4^{2-}}}{a_{HSO_4^-}\, a_{CH_3COO^-}}$$

42. Equation 3.42 shows that the formal and thermodynamic equilibrium constants are related through a term containing only activity coefficients. The activity coefficients can be estimated with the DHLL for solutions where the ionic strength is not above 0.01.

43. The reaction is $2H_2O \rightleftharpoons H_3O^+ + OH^-$. Applying Equation 3.49,

$$\log K_w' = \log K_w^{\circ} + A\sqrt{S}(1 \cdot 1^2 + 1 \cdot 1^2)$$

$$= \log K_w^{\circ} + 1.02\sqrt{S}$$

44. $\log K_a' = \log K_a^\circ + A\sqrt{S}(1 \cdot 1^2 + 1 \cdot 3^2 - 1 \cdot 2^2)$
    $= \log K_a^\circ + A\sqrt{S}\,(6)$
    $= \log K_a^\circ + 3.06\sqrt{S}$

45. The coefficient for the term $A\sqrt{S}$ in the solution to Question 44 is 6. This is the same as the value of N for the K_a° of HPO_4^{2-} in Appendix B.

46. The reaction of $HAsp^+$ with water is $HAsp^+ + H_2O \rightleftharpoons Asp + H_3O^+$. Using Equation 3.50,

$$N = (1 \cdot 0^2 + 1 \cdot 1^2 - 1 \cdot 1^2 - 1 \cdot 0^2) = 0$$

This is the same value entered in the table in Appendix B.

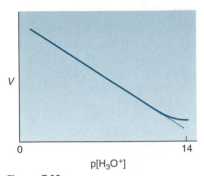

Figure 3.10. The response of the glass electrode to changes in $p[H_3O^+]$ is essentially linear until the H_3O^+ concentration is extremely low. The slope of the response curve is a function of the solution temperature and the presence of other salts.

E. pH Electrode and the Definition of pH

Early in the twentieth century, it was discovered that an electrical potential was developed between solutions on two sides of a thin glass membrane if the solutions differed in their acidity. Soon after, a conversion device based on this phenomenon was developed. The difference in electrical potential (voltage) between the glass electrode and a reference electrode is related to the acidity of the solution in which they are dipped. The glass electrode became a standard laboratory tool some years before the mechanism for this phenomenon was known. Of course, when the operation of the device was better understood, the development of improved devices quickly followed. In this chapter, the glass electrode will be treated as a sensor without detailed explanation. The basis for its operation is discussed in detail in Chapter 9.

THE pH ELECTRODE

The transfer function (working curve) for the glass electrode/reference electrode pair can be obtained by plotting the potential produced versus the concentration of a simple strong acid (such as HCl) dissolved in the solution. It is observed that the voltage decreases approximately 60 mV per 10-fold increase in $[H_3O^+]$. Thus, if we plot the sensor voltage versus the log $[H_3O^+]$, we get a very nearly straight line. Because of the wide use of the "p" function, the response is usually plotted with respect to $p[H_3O^+]$ or $p(a_{H_3O^+})$ as shown in Figure 3.10. When the electrode response is plotted versus $p[H_3O^+]$, the plot deviates from a straight line at lower acid concentrations (higher $p[H_3O^+]$) and has a lower slope when other salts are added to the solution. This ionic strength effect suggests that the electrode response is more directly related to the chemical activity of H_3O^+ than to its concentration. As we shall see in Chapter 9, there is a fundamental reason for this conclusion.

The linear portion of the glass electrode working curve follows the equation

$$E_{pH} = E_{const} + \frac{RT}{F} \ln a_{H_3O^+} = E_{const} - 2.303 \frac{RT}{F} p(a_{H_3O^+}) \qquad 3.51$$

where E is the symbol for **electromotive force**, **electrical potential difference**, or **voltage** and F is the Faraday constant (see Appendix G). At 25 °C, $2.303RT/F = 0.0592$ V so that the output potential changes 59.2 mV for each order of magnitude change in $a_{H_3O^+}$. The quantity $2.303RT/F$ is so widely used, we will give it a symbol V_N. The V reminds us that it has the units of volts. At other temperatures, the value of V_N and thus the slope of the working curve are different. (See Table 3.7.)

Although the effect of the hydronium ion activity on the *change* in the output voltage of the pH electrode is well established, the value of E_{const} can vary from one electrode to another. This prevents the direct conversion of the pH electrode voltage to $p(a_{H_3O^+})$ without calibration with a solution of known $a_{H_3O^+}$. If the pH electrode that is to be used in the measurement is first immersed in a solution of known (standard) $a_{H_3O^+}$, the measured potential will be

$$E_{pH,\,std} = E_{const} - V_N p(a_{H_3O^+,\,std}) \qquad 3.52$$

Table 3.7

Temperature Dependence of pH Electrode Sensitivity

Temperature, °C	V_N
0	0.0542 V
25	0.0592 V
50	0.0641 V
100	0.0741 V

whereas the potential when immersed in the unknown solution will be

$$E_{pH, \, unk} = E_{const} - V_N p(a_{H_3O^+, \, unk})$$ 3.53

When Equations 3.52 and 3.53 are combined to eliminate E_{const},

$$p(a_{H_3O^+, \, unk}) = p(a_{H_3O^+, \, std}) + \frac{E_{pH, \, std} - E_{pH, \, unk}}{V_N}$$

$$p(a_{H_3O^+, \, unk}) = p(a_{H_3O^+, \, std}) + \frac{E_{pH, \, std} - E_{pH, \, unk}}{0.0592} \qquad (\text{at } 25\,°C)$$ 3.54

pH STANDARDS

Equation 3.54 forms the basis for the measurement of $a_{H_3O^+}$ in unknown solutions, but it would not be useful if we could not readily obtain solutions of known $a_{H_3O^+}$ to use as standards. For this we are indebted to the work of Roger Bates. While he was at the National Bureau of Standards (now NIST, National Institute of Standards and Technology), he painstakingly measured responses of the glass electrode in solutions whose $a_{H_3O^+}$ could be reliably calculated with the DHLL. Then he compared these with the response in solutions of stable $a_{H_3O^+}$ that could be used as practical standards. From these measurements, the set of currently available standard pH solutions were developed[5] (see Table 3.8), and the $a_{H_3O^+}$ of the standard solutions was determined as accurately as possible over a wide range of temperatures.

THE DEFINITION OF pH

Because $-\log a_{H_3O^+}$ cannot be measured, the accepted definition of pH has come to be that which is measured with a glass electrode and pH meter that has been calibrated with one of the pH standard solutions. A few of these standards are given in Table 3.8. On this basis, the current definition of pH, as recommended by the International Union of Pure and Applied Chemistry, is the following version of Equation 3.54.

$$pH_{unk} = pH_{std} + \frac{E_{pH, \, std} - E_{pH, \, unk}}{V_N}$$ 3.55

When we use the term "pH" in this book, we will be referring to the value measured using a glass electrode relative to a standard pH solution.

Table 3.8

pH Assigned to Some Standard pH Solutions*

Temperature °C	"pH 4"	"pH 7"	"pH 10"
0	4.003	6.894	10.317
15	3.999	6.900	10.118
25	4.008	6.865	10.012
40	4.035	6.838	9.889
60	4.091	6.836	NA
90	4.205	6.877	NA

*"pH 4" = 0.05 molal $KHC_8H_4O_4$ (potassium hydrogen phthalate, sometimes called KHP; "pH 7" = 0.025 molal KH_2PO_4 and 0.025 molal Na_2HPO_4; "pH 10" = 0.025 molal $NaHCO_3$ and 0.025 molal Na_2CO_3. NA = not available.

[5]R. G. Bates, *J. Res. Natl. Bur. Stand.* **1962**, *66A*, 79.

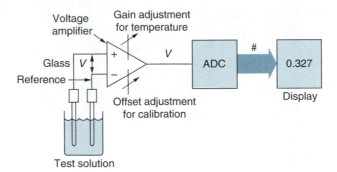

Figure 3.11. Block diagram of a digital pH meter. Gain and offset adjustments allow the digital display to read directly in pH units.

THE MEASUREMENT OF pH

The measurement of the output voltage of the glass electrode conversion device is accomplished with an instrument called a **pH meter.** A normal voltmeter cannot measure the pH electrode output voltage accurately because of the very high resistance of the glass membrane. The pH meter includes a special kind of amplifier that can respond to the electrode voltage without expecting that voltage source to supply any significant current. pH meters also include a readout device, usually a digital display that includes an analog-to-digital converter (ADC).

Figure 3.11 is a block diagram of the pH measurement system. In this sketch, the glass and reference electrodes are shown as separate units to emphasize that a voltage difference is being measured. In order for the digital display to read directly in pH units, the voltage difference between the pH and reference electrodes must be processed electronically. At 25 °C, the electrode voltage will change 59.2 mV for every 10-fold change in $a_{H_3O^+}$. The resolution of most digital pH meters is 0.01 pH units. Therefore, the numerical readout should change by 100 for a 59.2 mV change in the input voltage. However, for measurements of pH in solutions at other temperatures, the electrode voltage change per pH unit change is different. For this reason, all pH meters have an adjustment for the solution temperature. This adjustment controls the gain (output volts per input volt) of the amplifier. Usually this adjustment is marked in degrees Celsius. Calibration of the pH scale is accomplished by immersing the pH electrode in a standard pH solution and adjusting the amplifier offset (output volts for zero input volts) to read the exact pH of the standard solution.

Earlier pH meters used a moving coil current meter for the readout device. A block diagram of such an instrument is shown in Figure 3.12. A combination electrode is shown in this sketch. Such a device includes the glass and reference electrodes in a single unit. As in the digital display pH meter, a voltage amplifier processes the output voltage of the combination electrode. A voltage-to-current converter must follow the amplifier to provide the type of signal encoding required by the current meter readout.

When making pH measurements, it is desirable to use a standard solution as close as possible to the pH range of your measurements. The temperature of the pH

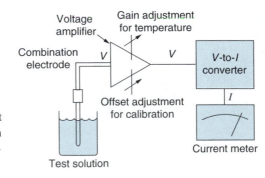

Figure 3.12. The conversion devices for a current meter output are different from those of the digital meter. The amplifier with the gain and offset adjustments works the same way in both instrument types.

standard should be the same as that of the solution you will be testing. The pH meter should be set to the solution temperature, and the calibration should be to the pH of the standard at that temperature. The pH values of the standard solutions assume that water is the solvent being used. pH electrodes indicate pH changes or differences quite well in other solvents, but without a standard solution for the specific solvent (or solvent combination) you are using, the absolute pH will not be known.

The concept of pH is widely used in chemical, biological, environmental, and geological science. The accurate definition of pH has been a major factor in its wide acceptance, as has the ability to measure it so readily. The measurement of pH, following the scheme introduced in Chapter 1, is outlined in the margin.

In terms of chemical analysis, pH measurements do not measure the concentration of acid present (because of the difference between the activity and concentration of the acid), nor do they indicate the species of acid present. In the next chapter, however, we shall see how pH measurements can be used to follow acid–base reactions in such a way as to accomplish the goals of quantitation and identification.

Analytical scheme for the measurement of pH

DIFFERENTIATING CHARACTERISTIC: The activity of H_3O^+ produced in the solvent by the acid (or base)

PROBE: pH electrode

RESPONSE: Change in the voltage between the pH and reference electrodes

MEASUREMENT OF RESPONSE: Measure the electrode voltage for unknown and standard solutions

INTERPRETATION OF DATA: Use Equation 3.55

Study Questions, Section E

47. What are the input and output quantities for the conversion device called the pH electrode?

48. According to the working curve of Figure 3.10, how does the output voltage of the pH electrode change as the pH increases? As the acidity increases?

49. In the transfer function for the glass pH electrode, the slope of the working curve is reliable, but the absolute value of the voltage as a function of pH is not. How, then, can we use this sensor to make accurate measures of pH?

50. What is the currently accepted definition of pH?

51. What is the effect of temperature on the measurement of pH?

52. Why are the voltage amplifiers in Figures 3.11 and 3.12 not conversion devices?

53. What two controls exist on a pH meter, and what is the function of each?

54. The voltage of a pH electrode (with respect to a reference electrode) was 0.226 V when dipped into a pH standard of 6.865. In a solution of unknown pH, its voltage was 0.025 V. What is the pH of the unknown solution? Both solutions were at 25 °C at the time of the measurement.

55. A student who had not studied this chapter calibrated his pH meter with a pH = 4.008 standard at 25 °C and then measured the pH of a 40 °C solution as 6.052.

    A. Assuming he set the temperature dial on the pH meter to 40 °C, what error is involved?

    B. What error is involved if he left the temperature setting at 25 °C?

56. What are the pitfalls of using a pH measurement to calculate the concentration of acid in an unknown solution?

Answers to Study Questions, Section E

47. The input quantity is pH, and the output quantity is voltage. The glass pH electrode is thus a pH-to-voltage converter.

48. The output voltage decreases as the $p[H_3O^+]$ increases. Since the $p[H_3O^+]$ and the pH are roughly related, the output voltage decreases as the pH increases. An increase in pH is actually a decrease in acidity, so the output voltage increases as the acidity increases.

49. Since the slope is reliable, the change in voltage is a true reflection of the change in pH. By using a pH standard, the working curve can be calibrated, and the difference in voltage that accompanies the change from the standard to the unknown solutions is a reliable measure of the *difference* in their pH values.

50. The currently accepted definition of pH is that which is obtained by measurement with a glass pH electrode calibrated against an accepted standard.

51. As the temperature changes, the change in electrode voltage for a given change in pH changes. This change in slope is predictable, so a measurement of temperature allows for the correction of this effect.

52. The input and output quantities of a voltage amplifier are both voltage. Since there is no conversion of the input quantity to a different output quantity, the voltage amplifier is not a conversion device.

53. The controls on a pH meter are the temperature (slope) and the calibration (offset). Their function is to set the slope of the working curve to that appropriate for the temperature of

the solution and to set the pH reading to the correct value when the electrode is dipped into the standard solution.

54. Using Equation 3.55,

$$pH_{unk} = pH_{std} + \frac{E_{pH,\,std} - E_{pH,\,unk}}{0.0592}$$

$$= 6.865 + \frac{0.226 - 0.025}{0.0592} = 10.260$$

55. A. According to Table 3.8, the pH 4 standard has a pH of 4.035 at 40 °C. Had the student standardized the meter with the standard at 40 °C, the unknown reading would have been higher (keeping the same difference) by $4.035 - 4.008$. The correct pH would then be 6.079.

B. At 40 °C, the term $2.303RT/F$ equals 0.0622 V, instead of 0.0592 V at 25 °C. Thus, the difference in pH be-

tween the unknown and standard calculated by the meter using the factor 0.0592 is wrong. The correct pH difference will be

$$(6.052 - 4.008)\frac{0.0592}{0.0622} = 1.945$$

When this difference is added to the correct standard value of 4.035, the result is 5.980.

56. The first pitfall is that the pH is more closely related to the activity of H_3O^+ than to its concentration. In solutions of high ionic strength, the difference between activity and concentration can be quite large. The second pitfall is that the dissolved acid might not be a strong acid. In this case, the concentration of the acid could be very much greater than the concentration of H_3O^+ in the solution.

Practice Questions and Problems

1. A. Describe the relationship between the two forms in a conjugate acid–base system.

   B. Give an example.

2. What does it mean that one acid is stronger than another?

3. A. What is the strongest acid that can exist in water, and why is this so?

   B. What is the leveling effect?

4. A. Write the reaction that will occur when $H_2C_2O_4$ is mixed with NaF in water.

   B. Write the equilibrium constant for the reaction in part A.

   C. What will happen if the $H_2C_2O_4$ is mixed with HCO_3^- instead? Write the reaction.

5. Autoprotolysis occurs in ethanol as well as in water.

   A. What two proton half reactions in Table 3.1 are involved in the ethanol autoprotolysis?

   B. Combine the two reactions in part A to obtain the K'_{solv} expression for ethanol.

6. The K'_b for 1-naphtholate ion is 2.17×10^{-5}. What is the K'_a for 1-naphthol?

7. An aqueous solution is 0.01 M in $FeCl_3$, which is completely dissociated.

   A. Calculate the ionic strength of this solution.

   B. Using the DHLL (Equation 3.26), calculate the activity coefficient for each of the ionic species.

8. Solve for the relationship between $K°_a$ and K'_a (similar to Equations 3.49 and 3.50) for the acid HPO_4^{2-}.

9. Consider the measurement of pH with a pH electrode.

   A. How much does the electrode voltage change for a unit change in pH?

   B. For a pH measurement using a digital meter, briefly describe the conversion devices employed and their transfer functions.

10. A solution is 0.0010 M in H_2SO_4. At that concentration, both H_2SO_4 and HSO_4^- are strong acids. Calculate

    A. The ionic strength of the solution.

    B. The activity of the SO_4^{2-} ion.

    (The A term in the DHLL is 0.509.)

11. The autoprotolysis constant of glacial (pure) acetic acid is 3.6×10^{-15} at 25 °C.

    A. Use the spreadsheet program to draw the log concentration plot. Label the axes and lines appropriately for the species involved.

    B. What are the strongest bases and acids that can exist in glacial acetic acid? Why?

    C. Does glacial acetic acid have a wider or narrower range of acid strength available than water? Why?

    D. Is the strongest acid in acetic acid stronger or weaker than the strongest acid in water? Why?

    E. What is the pH of a neutral solution in glacial acetic acid? Is this more acidic or less acidic than a neutral water solution? Why?

    F. Comment on the limitations of pH as a measure of acid strength when different solvents are involved.

12. The table in Appendix B has a value of N for each K'_a value given.

    A. Explain the use of the value N in terms of predicting the effect of ionic strength on K'_a.

    B. Justify the values of N given for the K'_a's of arginine and citric acid.

    C. What are the limitations of the use of N in this table?

13. A. Why are the deviations from Henry's law generally negative for ions in water?

    B. The constant A in the Debye–Hückel equation is generally given as 0.509. Name the solvent conditions that cannot be changed without changing this value.

    C. Why does a doubly charged ion contribute more to the ionic strength per mole than a singly charged one? Why do you think the effect is equal to the second power of the charge rather than the first?

    D. Comment on the usefulness of the DHLL in predicting the formal equilibrium constant under typical analytical conditions.

    E. Comment on the usefulness of thermodynamic equilibrium constants for the analysis of solution composition under typical analytical conditions.

14. A. What is the current accepted definition of pH? Why is it more useful than $-\log a_{H_3O^+}$?

    B. Why is the temperature of the measurement solution and the standard buffer so important in the measurement of pH?

15. For the reaction of H_2CO_3 and Na_2CO_3:

    A. Write a balanced chemical reaction. Consult Table 4.1 for the product(s).

    B. Write the equilibrium constant expression for the reaction.

    C. Indicate whether you expect the reaction to proceed significantly or not and why.

16. For the reaction of H_2CO_3 and water:

    A. Write a balanced chemical reaction. Consult the Table 4.1 for the product(s).

    B. Write the equilibrium constant expression for the reaction.

    C. Indicate whether you expect the reaction to proceed significantly or not and why.

    D. Indicate whether H_2CO_3 is a strong acid in water or not and why you think so.

    E. Name an amphiprotic solvent in which H_2CO_3 will be a stronger acid than it is in water and one in which it will be weaker and explain how you made your choice.

17. A. What is an acid–base couple? How do the members of the couple differ from one another? Give an example.

    B. Is the statement that every acid–base couple has a K_a and a K_b true or false? Explain.

18. A. Prove that the product of the concentrations of the deprotonated solvent molecules and the protonated solvent molecules is a constant.

    B. What factors will this constant depend on? (What solution factors will cause it to vary?)

19. The thermodynamic ion product constant (autoprotolysis constant) for water is 1.0×10^{-14} in pure water at 25 °C.

    A. Explain why the formal ion product constant for pure water at 25 °C has the same value.

    B. Calculate the autoprotolysis constant for an aqueous solution of 0.01M NaCl. Use the DHLL to make this calculation.

20. In the Table of K_a's in Appendix B, why are some of the values of N positive, some negative, and some zero?

21. What is the quantity V_N in this text and how is it related to the sensitivity of the pH electrode as a conversion device?

Suggested Related Experiments

1. Become familiar with the operation of a pH meter and the care of the electrodes and standards.

2. Compare the readings of a pH meter for several solutions with the results when using pH paper.

3. Measure the pH of a solution of a strong acid in water versus concentration. Relate the difference between concentration and antilog pH to the expected activity coefficient.

4. Measure the pH of a 10^{-4} M strong acid solution versus ionic strength. Relate the damage in pH versus ionic strength to the change in activity coefficient.

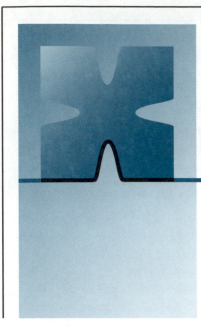

Chapter Four

ANALYSIS BY ACID—BASE REACTIVITY

An excellent way to probe the characteristic acidity of an analyte is to react it with a base. From such a reaction, it is possible to learn a great deal about the acidic analyte. For example, if we could learn how much base is required to react with the acid, we would know how much acid was present. This is generally accomplished by gradually adding a known concentration of base to the unknown solution and then determining exactly how much base was added when the last of the acid has reacted. When the reaction is carried out in this way, it is called a **titration**. The pH electrode introduced in the last chapter is a very convenient way to follow the course of a titration. As we shall see, the pH values during the titration provide an indication of the strength of the analyte acid, which can also aid in the identification of the acid. Similarly, basic analytes can also be studied by their reaction with an acid.

A. Equilibrium Concentrations

When a weak acid is dissolved in water, it reacts partially with the water according to the reaction

$$\text{HA} + \text{H}_2\text{O} \rightleftharpoons \text{H}_3\text{O}^+ + \text{A}^- \qquad 4.1$$

Solutions of acids or bases in water contain both forms of the conjugate acid–base pairs as well as protonated and deprotonated water (H_3O^+ and OH^-).

where HA and A^- are symbols for any conjugate acid–base pair. The concentration of the acid form that remains unreacted and the concentrations of the H_3O^+ and A^- produced depend on the K_a' of the acid and its initial concentration. If we are to use acid–base reactivity as a differentiating characteristic, we must be able to assess the extent of the reaction and calculate the concentrations of all the species in an acid–base solution.

The first step is to write the chemical expression for the reaction of the acid with the solvent, as in Equation 4.1. The equilibrium constant expression for this reaction is

$$K'_a = \frac{[H_3O^+][A^-]}{[HA]} \qquad 4.2$$

If Equation 4.2 is rearranged to separate the $[H_3O^+]$ term,

$$[H_3O^+] = K'_a \frac{[HA]}{[A^-]} \qquad 4.3$$

we see that for every value of $[H_3O^+]$, there is a specific ratio of the concentrations of the acid and base forms of the acid and that this ratio depends on the value of K'_a. When the two forms have equal concentrations, $[H_3O^+] = K'_a$. If the acid form predominates, $[H_3O^+] > K'_a$, and if the majority of the couple is in the base form, $[H_3O^+] < K'_a$.

Looking at Equation 4.3, consider the following.

1. *What is the $[HA]/[A^-]$ ratio when $[H_3O^+] = K'_a$?*

2. *What is the predominant form of the conjugate acid–base pair when $[H_3O^+]$ is greater than K'_a?*

THE FRACTION OF THE CONJUGATE PAIR IN EACH FORM

From Equation 4.3 we can derive the fraction of the acid that is in each form as a function of $[H_3O^+]$. This fraction is generally given the term α (alpha). Thus,

$$\alpha_{HA} = \frac{[HA]}{C_T} \quad \text{and} \quad \alpha_{A^-} = \frac{[A^-]}{C_T} \qquad 4.4$$

where C_T is the total concentration of both forms of the acid:

$$C_T = [HA] + [A^-] \qquad 4.5$$

Equation 4.5 is called a **mass balance** equation. It recognizes that the total amount of species A must remain constant even though the portion of it in either form may change.

To obtain an expression for α_{HA} in terms of the values of K'_a and $[H_3O^+]$, first solve Equation 4.2 for [HA] and substitute it twice for [HA] as follows.

$$\alpha_{HA} = \frac{[HA]}{C_T} = \frac{[H_3O^+][A^-]}{K'_a([HA] + [A^-])} = \frac{[H_3O^+][A^-]}{[H_3O^+][A^-] + K'_a[A^-]}$$

$$= \frac{[H_3O^+]}{[H_3O^+] + K'_a} = \frac{1}{1 + K'_a/[H_3O^+]} \qquad 4.6$$

A similar expression can be obtained for α_{A^-} (after solving Equation 4.2 for $[A^-]$).

$$\alpha_{A^-} = \frac{[A^-]}{C_T} = \frac{K'_a[HA]}{[H_3O^+]([HA] + [A^-])} = \frac{K'_a[HA]}{[H_3O^+][HA] + K'_a[HA]}$$

$$= \frac{K'_a}{[H_3O^+] + K'_a} = \frac{1}{1 + [H_3O^+]/K'_a} \qquad 4.7$$

A plot of the fraction of each species as a function of $p[H_3O^+]$ is shown in Figure 4.1 for an acid with a pK'_a of 1.0×10^{-5}. As this plot demonstrates, at higher values of $p[H_3O^+]$, the acid–base pair is virtually entirely in the basic form, and at lower values of $p[H_3O^+]$, it is essentially all in the acidic form. The value at which the two forms have equal concentrations is the value of $p[H_3O^+]$ that is equal to pK'_a. Changing the value of pK'_a simply shifts the crossover point to correspondingly higher or lower val-

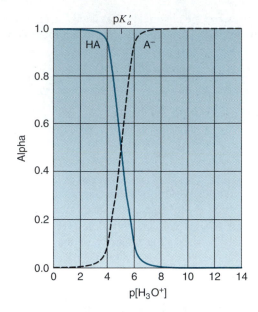

Figure 4.1. The form that predominates switches from the acid form to the base form where $p[H_3O^+] = pK_a'$.

ues of $p[H_3O^+]$. It is also interesting to note that at only one $p[H_3O^+]$ unit higher or lower than pK_a', the fraction of the total concentration that is in the dominant form has reached a value of approximately 90% and the minor component, 10%. At two $p[H_3O^+]$ units away from the crossover $p[H_3O^+]$, the minor form is only 1% of the total; 99% of the couple is in the dominant form.

SOLVING FOR [H₃O⁺] IN SOLUTIONS OF ACIDS

If we assume that the reaction of Equation 4.1 is the only one occurring when an acid is dissolved in water, we can readily solve for the H_3O^+ concentration that will result. We begin with the mass balance equation written as

$$C_{HA} = [HA] + [A^-] \qquad\qquad 4.8$$

where C_{HA} is called the **analytical concentration** of HA. The analytical concentration is the concentration of a species added regardless of the form or forms it may take after it is dissolved. The term C_X is used for the analytical concentration of species X. If the solution were made by dissolving some HA and some NaA in the solvent, the total concentration would be both the sum of the analytical concentrations C_{HA} and C_{A^-} and the sum of the equilibrium concentrations [HA] and [A⁻].

$$C_T = C_{HA} + C_{A^-} = [HA] + [A^-] \qquad\qquad 4.9$$

In the case of dissolving only HA in the solution, C_{A^-} is equal to 0, and C_{HA} is equal to C_T so $C_T = C_{HA} = [HA] + [A^-]$.

Now we are able to make appropriate substitutions for [HA] and [A⁻] in Equation 4.3. From the stoichiometry of the reaction in Equation 4.1, equal amounts of H_3O^+ and A⁻ are formed by this reaction. This means that $[A^-] = [H_3O^+]$. Substituting $[H_3O^+]$ for [A⁻] in Equation 4.8, we get $[HA] = C_{HA} - [H_3O^+]$. Now,

$$[H_3O^+] = K_a'\frac{[HA]}{[A^-]} = K_a'\frac{C_{HA} - [H_3O^+]}{[H_3O^+]} \qquad\qquad 4.10$$

Equation 4.10 is quadratic in $[H_3O^+]$, so it must be solved by approximation or by the use of the **quadratic formula**. The rearrangement best suited for the approximation approach is

$$[H_3O^+] = \sqrt{K_a'(C_{HA} - [H_3O^+])} \qquad 4.11$$

and that for using the quadratic formula is

$$[H_3O^+]^2 + K_a'[H_3O^+] - K_a'C_{HA} = 0 \qquad 4.12$$

The Quadratic Formula

The value of x *can be solved for a quadratic equation of the form* ax² + bx + c = 0, *where* a, b, *and* c *are numbers. The equation for the solution is called the quadratic formula. It is*

$$x = \frac{-b \pm \sqrt{b^2 - 4ac}}{2a}$$

For Equation 4.12, a = 1, b = K_a', *and* c = $- K_a'C_{HA}$.

An example will serve to illustrate both approaches. Assume that a solution is 0.010 M in an acid for which the K_a' is 1×10^{-4}. Equation 4.11 is solved neglecting the $[H_3O^+]$ (relative to C_{HA} in the square root). From this $[H_3O^+] \approx (10^{-4} \times 10^{-2})^{1/2} = 1.0 \times 10^{-3}$ M. Now we check to see if $[H_3O^+]$ really is negligible with respect to C_{HA}. If $[H_3O^+] = 1.0 \times 10^{-3}$ M, it is one-tenth the value of C_{HA}. This is small, but not negligible. So the value of 1.0×10^{-3} M for $[H_3O^+]$ is used in the square root side, and Equation 4.11 is solved again. Now $[H_3O^+] \approx (1.0 \times 10^{-4} \times 9 \times 10^{-3})^{1/2} = 9.5 \times 10^{-4}$ M. Then this value is used again in Equation 4.11 and the next approximation calculated. This time $[H_3O^+]$ is again 9.5×10^{-4} M, which means that the **successive approximations** have converged and the solution has been found. In general, the substitutions are repeated until the result no longer changes at the level of precision desired.

The use of the quadratic formula yields the correct answer in a single but more complex step. For the problem of an acid dissolved in water, the quadratic formula is

$$[H_3O^+] = \frac{-K_a' + \sqrt{K_a'^2 + 4K_a'C_{HA}}}{2} \qquad 4.13$$

When the data of the example are put in Equation 4.13, the result is $[H_3O^+] = 9.5 \times 10^{-4}$ M.

Spreadsheet Version of Equation 4.13

Equation 4.13 can be easily solved using the spreadsheet implementation. Data for K_a' *and* C_{HA} *are entered. The* $[H_3O^+]$ *and* $p[H_3O^+]$ *are calculated automatically. On the CD, look in Equation Solvers, Quadratic Solver Template to find this sheet.*

SOLVING FOR [H₃O⁺] IN SOLUTIONS OF BASES

When the base form of a conjugate acid–base pair is dissolved in water, the reaction is

$$A^- + H_2O \rightleftharpoons HA + OH^- \qquad 4.14$$

The mass balance equation is now $C_{A^-} = [A^-] + [HA]$, and, from the stoichiometry of the reaction in Equation 4.14, $[HA] = [OH^-]$. Since it is generally the H_3O^+ concentration we wish to calculate, we must also use the autoprotolysis equation (Equation 3.12), $K_w' = [H_3O^+][OH^-]$. When these substitutions are made into Equation 4.3,

$$[H_3O^+] = K_a'\frac{[HA]}{[A^-]} = K_a'\frac{K_w'/[H_3O^+]}{C_{A^-} - K_w'/[H_3O^+]} \qquad 4.15$$

From Equation 4.15, the rearrangement most useful for the approximation method is

$$[H_3O^+] = \sqrt{\frac{K_w'}{C_{A^-}}(K_a' + [H_3O^+])} \qquad 4.16$$

and the form most useful for solution by the quadratic formula is

$$C_{A^-}[H_3O^+]^2 - K_w'[H_3O^+] - K_a'K_w' = 0 \qquad 4.17$$

As an example, let us use the salt of the weak acid used for the acid calculations. This would be a 0.010 M solution of NaA where Na is any cation used to form the salt of the base form A^-. Substituting $C_{A^-} = 0.010$, $K'_w = 1.0 \times 10^{-14}$, and $K'_a = 1.0 \times 10^{-4}$ into Equation 4.16 and ignoring $[H_3O^+]$ relative to K'_a, we get $[H_3O^+] = 1.0 \times 10^{-8}$ M. Then we check to see that $[H_3O^+] << K'_a$ and it is by four orders of magnitude. In this case, the answer was obtained on the first try. This is clearly a much simpler solution than the quadratic formula, but for confirmation, we will do that too. For solutions of the base form of the conjugate pair, the quadratic formula is

Spreadsheet Version of Equation 4.18

Equation 4.18 also has a spreadsheet implementation. Data for K'_a, K'_{solv} and C_T are entered. The $[H_3O^+]$ and $p[H_3O^+]$ are calculated.

$$[H_3O^+] = \frac{K'_w + \sqrt{K'^2_w + 4C_{A^-}K'_wK'_a}}{2C_{A^-}} \qquad 4.18$$

When the substitutions from the example are made into Equation 4.18 the result is $[H_3O^+] = 1.0 \times 10^{-8}$ M as expected.

So far, we have ignored any other acid–base reactions that could be going on in the solution at the same time. One that is certainly occurring is the autoprotolysis of the solvent. In the next section, we will learn how to take this reaction into account and under what circumstances it is important to do so.

Study Questions, Section A

1. Obtain a formula for the ratio of acid to base forms of a conjugate acid–base pair in terms of $[H_3O^+]$ and K'_a from the formal equilibrium constant expression.

2. Lactic acid ($CH_3CHOHCOOH$) solution has a K'_a of 1.38 $\times 10^{-4}$. Calculate the ratio of the acid to base forms of lactic acid in a solution with a $p[H_3O^+]$ of 4.00.

3. Write the mass balance equation for a 3.0×10^{-4} M solution of acetic acid (HOAc).

4. A solution of iodic acid (HIO_3) has a $p[H_3O^+]$ of 2.00. The K'_a of the acid is 1.7×10^{-1}. Use Equations 4.6 and 4.7 to

calculate the fraction of acid in its acidic and basic forms in the solution.

5. Use the spreadsheet on the CD to make an alpha plot for iodic acid (see Question 4). Confirm the results obtained in the solution to Question 4.

6. Calculate the $[H_3O^+]$ and the $p[H_3O^+]$ of a 0.000300 M aqueous solution of benzoic acid. Assume that $K'_a \approx K°_a$.

7. Calculate the $[H_3O^+]$ and the $p[H_3O^+]$ of a 0.002 M aqueous solution of dimethylamine, $(CH_3)_2NH$. Assume that $K'_a \approx K°_a$.

Answers to Study Questions, Section A

1. $K'_a = \dfrac{[H_3O^+][A^-]}{[HA]}, \qquad \dfrac{K'_a}{[H_3O^+]} = \dfrac{[A^-]}{[HA]}$

from which we see that if $[H_3O^+] > K'_a$, the acid form will predominate, and if $[H_3O^+] < K'_a$, the base form will predominate.

2. We write the acid–base reaction of lactic acid with water as follows:

$CH_3CHOHCOOH + H_2O \rightleftharpoons H_3O^+ + CH_3CHOHCOO^-$

From Equation 4.2, the equilibrium expression is

$$K'_a = \frac{[H_3O^+][CH_3CHOHCOO^-]}{[CH_3CHOHCOOH]}$$

Rearranging yields

$$\frac{[CH_3CHOHCOOH]}{[CH_3CHOHCOO^-]} = \frac{[H_3O^+]}{K'_a}$$

$$[H_3O^+] = 10^{-pH_3O^+} = 10^{-4.00} = 1.0 \times 10^{-4}$$

$$\frac{[CH_3CHOHCOOH]}{[CH_3CHOHCOO^-]} = \frac{1.0 \times 10^{-4}}{1.38 \times 10^{-4}} = 0.72$$

3. From Equation 4.9,

$$C_{HAc} = 3 \times 10^{-4} = [HOAc] + [OAc^-]$$

4. $\alpha_{HIO_3} = \dfrac{1}{1 + K'_a/[H_3O^+]}$

$$= \frac{1}{1 + (1.7 \times 10^{-1})/(1.0 \times 10^{-2})} = 0.056$$

$$\alpha_{IO_3^-} = \frac{1}{1 + [H_3O^+]/K'_a}$$

$$= \frac{1}{1 + (1.0 \times 10^{-2})/(1.7 \times 10^{-1})} = 0.94$$

Note that the fractions of the acid in its acidic and basic forms do add up to 1.00.

5. From the alpha plot, we see that at $p[H_3O^+]$ 2.00 the fraction in the acid form is about 0.06 and the fraction in the basic form is about 0.94.

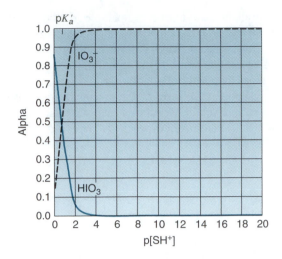

6. Using the spreadsheet, Equation 4.13, or the quadratic formula and Equation 4.12, $[H_3O^+] = 4.04 \times 10^{-4}$ M. This concentration is not negligible with respect to C_{HA}, so the first approximation of Equation 4.11 will not be accurate. The $p[H_3O^+] = -\log[H_3O^+] = 3.39$.

7. This is a fairly concentrated solution of a rather strong base, so it is possible that Equation 4.16 will work on the first approximation, that is, that $[H_3O^+] = K_w'K_a'/C_T)^{1/2} = 4 \times 10^{-12}$ M. We see that this is not negligible with respect to K_a' (1.68×10^{-11}), so the longer solution is required. When the spreadsheet or the quadratic equation (Equation 4.18) is used, the answer is $[H_3O^+] = 2.66 \times 10^{-12}$ M, and $p[H_3O^+] = -\log [H_3O^+] = 11.58$.

B. Exact Equilibrium Expressions

In the previous section, we solved for the equilibrium concentrations of the acid and base forms when an acid or base was dissolved in solvent. In doing so, we considered only the equilibrium for the reaction of the acid or base with the solvent. There is another related equilibrium occurring in such solutions. It is the autoprotolysis of the solvent. There are circumstances where the contribution of the autoprotolysis equilibrium will affect or even dominate the equilibrium concentrations that exist. To obtain an exact solution, we must consider all related system equilibria, as shown by the approach illustrated in the margin. We begin by writing all the related reactions and their equilibrium constants. This is illustrated for the case of an acid being dissolved in water. First, there is the acid–solvent reaction (repeated from Equation 4.1), for which the equilibrium expression (repeated from Equation 4.2) is as shown. In addition, there is the autoprotolysis of water (Equation 3.11), for which the equilibrium expression is Equation 3.12.

The mass balance equation(s) is always specific for the given problem conditions. The first one shown is the simple mass balance for the forms of the acid added. The second one comes from a combination of the two reaction stoichiometries as follows. When the reaction of Equation 4.1 proceeds to the right, it produces equal amounts of H_3O^+ and A^-. Therefore, we could substitute $[H_3O^+]$ for $[A^-]$ in the mass balance equation to obtain $C_{HA} = [HA] + [H_3O^+]$. However, we would be forgetting the amount of H_3O^+ that comes from the autoprotolysis of water. From Equation 3.11, we can see that for every H_3O^+ produced by the water autoprotolysis, an OH^- ion is also produced. If we then subtract the OH^- concentration from the H_3O^+ concentration, we will have the concentration of H_3O^+ due to the reaction of the acid with the water. Now the mass balance equations can be written

$$[A^-] = [H_3O^+] - [OH^-], \qquad C_{HA} = [HA] + ([H_3O^+] - [OH^-]) \qquad 4.19$$

The equilibrium concentration of HA from Equation 4.19 can be substituted into a rearranged form of the equilibrium constant expression (Equation 4.3) as follows

$$[H_3O^+] = K_a'\frac{[HA]}{[A^-]} = K_a'\frac{C_{HA} - ([H_3O^+] - [OH^-])}{([H_3O^+] - [OH^-])} \qquad 4.20$$

The complication in Equation 4.20 is the terms in parentheses, that is, the difference between the concentrations of H_3O^+ and OH^-. We know from the autoprotolysis equi-

---

An Approach to Solving Problems with Multiple Equilibria: The example of HA in water

1. Define the problem:

    An acid is dissolved in water.

2. Write all the reactions and their equilibrium expressions:

    $$HA + H_2O \rightleftharpoons H_3O^+ + A^- \qquad 4.1$$

    $$K_a' = \frac{[H_3O^+][A^-]}{[HA]} \qquad 4.2$$

    $$2H_2O \rightleftharpoons H_3O^+ + OH^- \qquad 3.11$$

    $$K_w' = [H_3O^+][OH^-] \qquad 3.12$$

3. Write the mass balance equations:

    $$C_{HA} = [HA] + [A^-] \qquad 4.8$$

    $$[A^-] = [H_3O^+] - [OH^-]$$

4. Combine to obtain an equation in known constants, the analytical concentrations, and the sought equilibrium concentration:

    See Equation 4.20 for the result.

librium equation that these concentrations can only be comparable in a nearly neutral solution. Below $p[H_3O^+] = 6$, the OH^- concentration will be negligible with respect to the H_3O^+ concentration. It is inconceivable that the OH^- term could predominate in a solution that was made by adding an acid to water. If the solution is acidic enough to neglect $[OH^-]$, Equation 4.20 becomes identical to Equation 4.10.

Now consider the situation when the base form of the conjugate acid–base pair HA/A^- has been dissolved in water. This would necessarily involve a salt of the base form such as NaA. When the salt of A^- is dissolved in water, A^- reacts as a base with the water.

$$A^- + H_2O \rightleftharpoons HA + OH^- \qquad\qquad 4.21$$

The mass balance for this situation is

$$C_{A^-} = [A^-] + [HA] \qquad\qquad 4.22$$

Following the same reasoning used to develop Equation 4.20, we first note that the OH^- concentration would be equal to the HA concentration (from Equation 4.21) except for the contribution to $[OH^-]$ that comes from the autoprotolysis of water. This, in turn, will be equal to $[H_3O^+]$, so $[HA] = [OH^-] - [H_3O^+]$. From Equation 4.22, $[A^-] = C_{A^-} - [HA]$. Now Equation 4.23 can be written for $[A^-]$, and the equations for $[A^-]$ and $[HA]$ can be substituted into the equilibrium expression as shown in Equation 4.24.

$$[A^-] = C_{A^-} + ([H_3O^+] - [OH^-]) \qquad\qquad 4.23$$

$$[H_3O^+] = K'_a \frac{[HA]}{[A^-]} = K'_a \frac{-([H_3O^+] - [OH^-])}{C_{A^-} + ([H_3O^+] - [OH^-])} \qquad\qquad 4.24$$

When a base is dissolved in water, we expect the solution to be basic, that is, $[H_3O^+] < [OH^-]$. If the solution is sufficiently basic that $[H_3O^+]$ can be neglected ($p[H_3O^+]$ over 8), Equation 4.24 simplifies to Equation 4.15 as we expect.

Finally, we will consider the case when both the acid and base forms have been dissolved. Using the mass balance Equations 4.19 and 4.23 in the equilibrium expression results in the equation

$$[H_3O^+] = K'_a \frac{C_{HA} - ([H_3O^+] - [OH^-])}{C_{A^-} + ([H_3O^+] - [OH^-])} \qquad\qquad 4.25$$

Equation 4.25 is a general and exact equation that can be used to solve for the concentrations of the species present when known amounts of acids and/or their conjugate bases have been dissolved in water. This equation is extremely useful in finding the exact $p[H_3O^+]$ of buffer solutions. It is usually solved by making approximations (neglecting $[OH^-]$ with respect to $[H_3O^+]$ or vice versa) on the basis of one's knowledge of the solution conditions. For example, if only the acid form was added, assume that $[OH^-]$ is negligible. Similarly, if only the base form was added, assume that $[H_3O^+]$ can be neglected. If both acid and base forms were added in roughly comparable amounts, assume that $[H_3O^+]$ is approximately equal to K'_a.

An equation that contains only $[H_3O^+]$, C_{HA}, C_{A^-}, and K'_a can be obtained from Equation 4.25 by substituting $K'_w/[H_3O^+]$ for $[OH^-]$. It is

$$[H_3O^+]^3 + (C_{A^-} + K'_a)[H_3O^+]^2 - (K'_a C_{HA} + K'_w)[H_3O^+] - K'_a K'_w = 0 \qquad 4.26$$

Equation 4.26 can be evaluated by solver programs, or Equation 4.25 can be solved by successive approximations. (See Example 4.1.) You can see how mathematically complex a system can become when we do not know which of the simpler equations might apply. In Section E of this chapter, we will show how log concentration diagrams can be used to directly obtain approximate solutions and to clearly indicate which species can be neglected.

Example 4.1

Find the p[H₃O⁺] of a solution made up of 0.0001 M HOCl and 0.0030 M NaOCl

The K_a' for HOCl is 3.0×10^{-8}.

When the analytical concentrations of the starting materials are applied to Equation 4.25,

$$[H_3O^+] = 3.0 \times 10^{-8} \frac{1.0 \times 10^{-3} - ([H_3O^+] - [OH^-])}{3.0 \times 10^{-3} + ([H_3O^+] - [OH^-])}$$

Since the concentrations of the acid and base forms are nearly equal, we expect that $[H_3O^+]$ is nearly equal to K_a'. If this is true, the $[H_3O^+]$ and $[OH^-]$ are both negligible compared to the analytical concentrations. Thus,

$$[H_3O^+] \approx 3.0 \times 10^{-8} \frac{1.0 \times 10^{-3}}{3.0 \times 10^{-3}} = 1.0 \times 10^{-8} \text{ M}$$

and $p[H_3O^+] = 8.0$. The approximation is verified.

If the analytical concentrations had been 1.0×10^{-5} M for the HOCl and 3.0×10^{-5} M for the NaOCl, the approximation is no longer valid. Taking the result from the higher concentration solution as an approximation, neglect the $[H_3O^+]$ in the parentheses and use 1×10^{-6} M for $[OH^-]$. This gives

$$[H_3O^+] = 3.0 \times 10^{-8} \frac{1.0 \times 10^{-5} + 1.0 \times 10^{-6}}{3.0 \times 10^{-5} - 1.0 \times 10^{-6}}$$

from which $[H_3O^+] = 1.14 \times 10^{-8}$ M and $p[H_3O^+] = 7.94$. The approximations are verified, and no further iterations are necessary.

At still lower analytical concentrations, the result will approach that of a neutral solution. The solvent autoprotolysis becomes increasingly significant.

Study Questions, Section B

8. The challenge in solving acid–base equilibria is to find the simplest equation for which the solution will be valid. Calculate the equilibrium p[H₃O⁺] for each of the following aqueous solutions using the simplest equation possible:

   A. $C_{HA} = 10^{-2}$ M, $K_a' = 10^{-6}$

   B. $C_{HA} = 10^{-4}$ M, $K_a' = 10^{-2}$

   C. $C_{HA} = 10^{-6}$ M, $K_a' = 10^{-11}$

   D. $C_{A^-} = 10^{-6}$ M, $K_a' = 10^{-2}$

   E. $C_{A^-} = 10^{-4}$ M, $K_a' = 10^{-13}$

   F. $C_{A^-} = 10^{-2}$ M, $K_a' = 10^{-8}$

9. Find the exact p[H₃O⁺] of a 1.0×10^{-3} M solution of phenol, for which $K_a' = 1.0 \times 10^{-10}$.

Answers to Study Questions, Section B

8. A. This is a solution of an acid at a concentration well above 10^{-7} M where autoprotolysis must be considered. Apply Equation 4.11, neglecting the $[H_3O^+]$ in the parentheses.

   $$[H_3O^+] \approx (10^{-6} \times 10^{-2})^{1/2} \approx 10^{-4} \text{ M and } p[H_3O^+] = 4.$$

   B. This is a low concentration of a relatively strong acid. Neglecting the $[H_3O^+]$ in the parentheses of Equation 4.11 gives a nonsense answer. When Equation 4.13 is applied, it leads to the answer $p[H_3O^+] = 4$. The weak acid has reacted almost completely with the water at this low concentration.

   C. This is a very low concentration of a very weak acid. It is likely that the autoprotolysis of water will be the dominant effect. The simplified form of Equation 4.11 yields

   $$[H_3O^+] \approx (10^{-11} \times 10^{-6})^{1/2} \approx 10^{-8.5} \text{ M}.$$

This suggests the solution became quite basic from the addition of an acid. Since this does not make sense, the correct answer is $p[H_3O^+] = 7$ from the autoprotolysis of water. In fact, if Equation 4.20 were solved exactly, the $p[H_3O^+]$ would be very slightly less than 7.

D. This is a very low concentration of a very weak base. Again, the autoprotolysis of water will be likely to dominate. If the simplified form of Equation 4.16 is applied, the result is

$$[H_3O^+] \approx [(10^{-14}/10^{-6}) \times (10^{-2})]^{1/2} \approx 10^{-5} \text{ M}.$$

The result of a very acidic solution from adding a weak base is a sure sign that the base is too weak to affect the neutral $p[H_3O^+]$ established by the autoprotolysis. Thus the correct answer is $p[H_3O^+] = 7$.

E. Here we have a low concentration of a relatively strong base. We might expect it to have reacted quite completely with the water to produce a solution with $[OH^-] \approx C_{A^-}$. One can confirm this by substituting $[H_3O^+] = 10^{-10}$ M in the parentheses in Equation 4.16 and solving for $[H_3O^+]$ to see if that is true. (It is.) Another approach would be to solve Equation 4.18. The end result is $p[H_3O^+] = 10$.

F. The concentration is quite high and the strength of the base only moderate. Autoprotolysis should not make a significant contribution, so try the simplified form of Equation 4.16:

$$[H_3O^+] \approx (10^{-12} \times 10^{-8})^{1/2} \approx 10^{-10} \text{ M}$$

From this, $p[H_3O^+] = 10$.

9. We can begin by trying to neglect the autoprotolysis of water. The solution to Equation 4.13 is $[H_3O^+] \approx 3.16 \times 10^{-7}$ M. The OH^- concentration at this value of $[H_3O^+]$ is not negligible. Therefore, Equation 4.25 must be used. The application of a solver program to Equation 4.26 gave $p[H_3O^+] = 6.48$.

Another method is to use successive approximations with Equation 4.25. Begin with the value of the estimate obtained from the application of Equation 4.13. For $[H_3O^+] = 3.16 \times 10^{-7}$ M, the $[OH^-]$ is 3.16×10^{-8} M. When these values are substituted into the right-hand side of Equation 4.25, the result is $[H_3O^+] = 3.51 \times 10^{-7}$ M. This value is then used in the right-hand side of Equation 4.25 to find a new result for $[H_3O^+]$, which is then used again in the right-hand side of Equation 4.25 until the result equals the initial $[H_3O^+]$. This process takes several iterations. The final result is $[H_3O^+] = 3.31 \times 10^{-7}$ M, for which $p[H_3O^+] = 6.48$.

C. Quantitation by Acid—Base Titration

In this section, we explore ways to use the action of an analyte as an acid or base to determine how much of the analyte is present. An obvious probe for the differentiating characteristic of acidity is its reaction with a base. In the technique called **titration**, the base is added incrementally to determine exactly how much base is required to react with all the acid present.

In terms of measurement principles, the technique of titration is a classic example of a null measurement. A standard quantity (moles of base or acid) is added incrementally until the amount added is exactly sufficient to react with the moles of acid or base present. The reagent added to react with the analyte is called the **titrant**. A **detector** is required so that the null point can be determined. Remember that in a null measurement, the measured quantity is obtained from the amount of standard added at the null point, not from the detector response. The purpose of the detector is to provide an indication that the null point has been reached. In this case, the detector must be able to determine the point at which the difference between the amount of acid in the sample and the amount of base added is zero. A block diagram of this process is shown in Figure 4.2, and the analytical scheme is outlined in the margin.

MAKING THE STANDARD TITRANT

One of the most common methods for the addition of standard base is to make a base solution of known concentration and measure the volume of it that is added to the analyte solution. In this way, the moles of titrant added is simply the concentration (mol/L) times the volume (L). The accuracy of the standard thus depends on both the accuracy with which the standard solution concentration is known and the accuracy of the measurement of the volume added. The usual goal for accuracy in an acid–base titration is 0.1% or 1 ppt.

Analytical Scheme for Acid–Base Titration

Differentiating Characteristic
Ability to react with a base.

Devise a Probe:
Add measured amounts of a basic reactant to a solution of the acid.

Anticipate the Response:
The acid reacts with the base, consuming the acid and raising the solution pH.

Measure the Response:
Follow the increase in pH as the base is added. Determine the exact volume of base added when the amount of added base equals the amount of acid originally present.

Interpret the Data:
Use the equivalence point volume of the base and its concentration to calculate the amount of acid originally present.

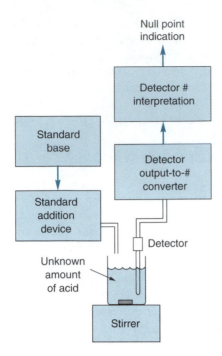

Null point
indication

Detector #
interpretation

Standard
base

Detector
output-to-#
converter

Standard
addition
device

Detector

Unknown
amount
of acid

Stirrer

Figure 4.2. A block diagram for the titration of an acid with a base. The detector output must be interpreted to determine the amount of standard base that was added up to the equivalence point.

Acid–base titrants are generally solutions of hydrochloric acid (HCl) or sodium hydroxide (NaOH). Strong acid and base titrants are used to achieve the highest possible equilibrium constant for the titration reaction. When the titrant concentration is determined by diluting an accurately known amount of acid or base to a known volume, it is called a **determinate preparation.** When a titration itself is used to establish the titrant concentration, the titrant is said to have been **standardized**. The difficulty with determinate preparation is identifying a form in which the HCl or NaOH can be accurately weighed. The composition of a constant-boiling HCl solution (at about 20 weight %) is very reproducible and has been well characterized. An amount of the constant-boiling HCl can be weighed out and diluted to its final volume. The dilution of standard solutions is carried out with a pipet and volumetric flask. Pure NaCl or KCl can also be weighed out and dissolved. The resulting solution is run through an ion exchange column to replace the K^+ or Na^+ with H_3O^+. With care, this can be accomplished exactly and completely (or, as analytical chemists often say, "quantitatively"). The resulting solution is then diluted to an exact volume. From the mass of the standard material weighed, the formula weight of the standard material, and the final solution volume, the concentration of standard acid can be calculated.

There are no methods for the determinate preparation of standard solutions of NaOH. Solid NaOH generally has too much included water and Na_2CO_3 for its weight to be accurately converted to moles of NaOH. Thus, NaOH is weighed to provide an approximate solution concentration that is later standardized to obtain the exact concentration. The main difficulty in the preparation of NaOH solutions to be used as titrants is the elimination of as much of the carbonate as possible. The usual technique is to begin with a very concentrated NaOH solution, in which the Na_2CO_3 is only slightly soluble. The insoluble salt is then filtered out just prior to dilution to the final volume.

Standardization is the usual practice for the preparation of standard acids and bases. In this technique, an acidic or basic material that is available in very pure form is weighed out to produce a solution with an accurately known amount of analyte present. Materials that are suitable for this purpose are called **primary standards**. The primary standard material used in the standardization of acids is Na_2CO_3. From the volume of acidic titrant required to exactly react with the Na_2CO_3 (to form H_2CO_3),

Example 4.2

Standardizing a titrant

A solution was made to be approximately 0.1 M in NaOH. This solution was used to titrate weighed amounts of KHP (FW = 204.22 g mol^{-1}).

g, KHP	$V_{titrant}$, mL	$M_{titrant}$
0.7225	34.85	0.10152
0.6832	32.93	0.10159
0.7346	35.38	0.10167

The molarity of the titrant was calculated from the relationship

$$M_{titrant} = \text{g KHP}/(V_{titrant} \times 204.22)$$

The average value of the molarity is 0.1016 mol L^{-1}.

the weight of Na_2CO_3 titrated, and the number of moles of HCl that react with each mole of Na_2CO_3 (exactly 2), the concentration of standard acid can be calculated. The primary standard used for the standardization of basic titrants is potassium hydrogen phthalate ($KHC_8H_4O_4$, sometimes abbreviated KHP). It reacts in a 1:1 ratio with the NaOH in the standard solution (see Examples 4.2 and 4.3). Primary standard materials are usually heated before weighing to eliminate any adsorbed water. Care must be taken to avoid the decomposition that can result from overheating.

ADDING KNOWN VOLUMES OF STANDARD SOLUTION

The resolution of a null measurement is determined by the smallest amount by which the standard quantity can be accurately changed. The most common device for adding standard solution is a buret. The volume levels in a standard 50-mL buret can be read to 0.01 mL. This amount then, times the molar concentration, gives the number of moles of titrant at the limit of resolution of the volume measurement. This number can be made smaller by use of a buret with finer graduations or by use of a standard titrant with lower concentration. Standard titrant solution is poured into the buret, filling it almost to the top reading. The solution level is then recorded after making sure the valve and tip of the buret are also filled with the standard solution. Increments of standard solution are then added (coarsely at first) while watching the detector output (indicator color or electrode response). As the point of equivalence between the moles of titrant added and the moles of analyte present approaches, the increments are reduced with the aim of measuring the buret level at the exact null point. The null point level alone may be recorded, or the buret levels and corresponding indicator responses may be recorded at many points during the titration. The advantages of the latter practice will be demonstrated later in this chapter.

The optimum concentration for the titrant is that which will use 60 to 80% of the buret volume to reach the null point. If too small a volume is delivered, the reading resolution limit of 0.01 mL can reduce the relative precision. If the volume required exceeds the buret capacity, the buret must be refilled, which adds two more volume readings to the process.

The classic buret is not the only method for titrant delivery to the reaction mixture. In more automated systems, a syringe is used to dispense the titrant liquid. Modern syringes come in a wide range of sizes from 50 mL to 5μL. The inside diameter of the syringe tube is very uniform, and the plunger fits tightly against the walls. The result is a very reproducible volume of solution delivered per centimeter of distance moved by the plunger. The volume markings on the barrel are never used for quantitative work. Rather, the plunger is moved by a precision screw, and the fractional turns of the screw, which are related to the volume delivered, are recorded. Two common types of titrant delivery systems based on the syringe are the digital buret and the syringe infusion pump.

Digital burets come in two varieties. In the manual one, a knob is turned to advance the plunger screw. A digital display counts the fractional rotation of the knob. The syringe plunger and barrel are an integral part of an assembly that replaces the screw cap on a reagent bottle containing the titrant. This very simple device can be quite effective when titrating with an indicator or to a set end point. The trend is toward the availability of an electrical connection by which the digital display information can be transferred to a computer for automated data logging. Precision to 0.01 mL is claimed for most digital burets. Single fill delivery capacities of 25 mL and 50 mL are the most common.

More sophisticated titrant delivery systems are incorporated into digital titration systems such as the one shown in Figure 4.3. The syringe is filled from the stock solution container by withdrawing the syringe plunger. The digital display is then zeroed and the fill valve returned to the dispense position. The complete titration system includes a clamp to hold a sensing electrode and a computer for control and data inter-

Example 4.3

Calculating the amount of the species titrated

Aliquots ($V = 50.00$ mL) of an acid solution were titrated with the base standardized in Example 4.2. Repetitive titration volumes were 26.25, 26.22, and 26.24 mL. The average of these is 26.24 mL. The amount of acid in 50.00 mL of solution is

$$0.02624 \text{ L} \times 0.1016 \text{ mol L}^{-1}$$
$$= 2.666 \times 10^{-3} \text{ mol}$$

The concentration of the acid in the solution is

$$2.666 \times 10^{-3} \text{ mol}/0.05000 \text{ L}$$
$$= 0.05332 \text{ M}$$

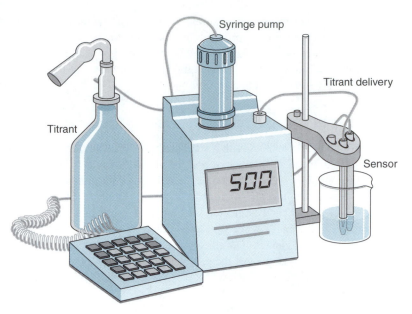

Figure 4.3. A digital titration system. The panel displays the volume delivered. The volume increments, which are often 0.01 mL, can be dispensed at a constant rate or on demand. The titration systems generally combined with computer data acquisition and automatic determination of end point volume.

pretation. Several modes of titrant delivery and end point detection can be programmed into the system for maximum versatility.

A syringe-based infusion pump (such as that shown in Figure 4.4) is just a device for clamping a standard syringe in place so that a screw-driven block can push on the syringe plunger. A stepper motor drives the screw over a wide range of user-settable rates. In addition, most infusion pumps are set up for computer control of the titrant delivery. Operation of the syringe pump for titrations is similar to that of the digital buret, except that the titrant delivery occurs continuously at the set rate or the rate may be varied by the same computer that is logging the response from the titration mixture sensor. With such an automated system, the titrant delivery rate could be automatically slowed in the region of the equivalence point.

Measurement of the amount of titrant delivered is not limited to volume. Several ingenious techniques have been devised. In one, tiny droplets of uniform size are generated on demand. The volume delivered to the reaction mixture is proportional to the number of droplets generated.[1] Another clever method doesn't measure volume at all.

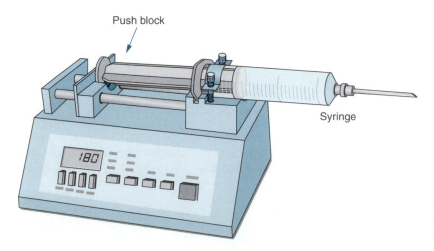

Figure 4.4. A syringe-based infusion pump. Any commercial syringe is placed in the holder and the drive rate selected. A very wide range of solution delivery rates are possible with this simple device.

[1]Steele, A. W. and Hiefje, G. M. *Anal. Chem.*, *56*, 2884 (1984).

The reaction vessel is placed on a top-loading balance. The increase in the weight of the reaction vessel as the titration proceeds is due only to the weight of the titrant added. In an earlier method, a **weight buret** is used, weighing the buret before the titration and at the end point.[2] This approach has the advantage that the effect of temperature on the calibration of the buret is eliminated.

DETECTING THE RESPONSE TO THE TITRANT

In the titration of an acidic analyte with a standard base, the solution will become less acidic as more titrant is added. In other words, the pH will increase. At the equivalence point (where the number of moles of titrant is exactly the right number to complete the reaction with the analyte), the pH often changes quite rapidly with respect to titrant volume. This change in pH can be detected with a pH electrode and pH meter or with a chemical indicator. Other methods could also be devised to follow the course of this reaction, but the pH electrode and the indicator methods are explored here first. To do so, we will first consider the extent of the pH change and how it is related to the equivalence point. In this consideration, we will look first at the situation where the analyte is a strong acid or base.

What is needed is an equation that predicts the change in $p[H_3O^+]$ as a function of the fraction of acid titrated. This can be developed from the equilibrium equations involved and some other relationships that are true for the titration situation. Since the acid to be titrated and the titrant are both strong, there is no K_a' or K_b' equation. However, the autoprotolysis equation for water is needed.

$$K_w' = [H_3O^+][OH^-] \qquad 4.27$$

The initial concentration of acid is C_i. Since the strong acid reacts completely with the water, the concentration of the base form of the acid is also C_i. Thus, at the beginning, $[A^-] = C_i$. As the titration proceeds, A^- will be diluted by the added titrant. We will assume that the concentration of the titrant is similar to that of the reactant. This is the usual situation because it keeps the amount of titrant needed in the desirable range for buret delivery.

The point at which the amount of titrant added is exactly equal to the amount of analyte in the sample is called the **equivalence point**. At this point, the fraction titrated, θ, is exactly 1.00. If enough titrant has been added to react with exactly 50% of the acid, $\theta = 0.50$, and so on. As the titration proceeds (θ increases), A^- is diluted:

$$[A^-] = C_i\left(\frac{1}{1 + \theta}\right) \qquad 4.28$$

Next, we will obtain an equation for the sodium ion that is introduced to the solution from the NaOH titrant:

$$[Na^+] = C_i\left(\frac{\theta}{1 + \theta}\right) \qquad 4.29$$

Finally, we will introduce a new equation that is often very useful. It is the **charge balance equation**. It simply states that the sum of the concentrations of the cations (times their charge) must equal the sum of the concentrations of the anions (times their charge). For the present situation, the charge balance equation is

$$[H_3O^+] + [Na^+] = [A^-] + [OH^-] \qquad 4.30$$

[2]Kolthoff, I. M. and Furman, W. H. *Volometric Analysis*, Vol II Practical Principles, John Wiley & Sons, New York, 1929.

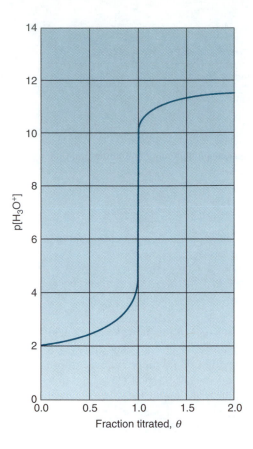

Figure 4.5. A titration plot in which $p[H_3O^+]$ is monitored while standard base is added to a 0.01 M solution of strong acid.

When Equations 4.27 through 4.30 are combined to eliminate all terms except constants, $[H_3O^+]$, and θ, the following equation results.[3]

$$[H_3O^+]^2 - C_i\left(\frac{1 - \theta}{1 + \theta}\right)[H_3O^+] - K_w' = 0 \qquad 4.31$$

From this equation, the value of $p[H_3O^+]$ a function of θ was plotted in Figure 4.5 for C_i $= 10^{-2}$ M. Such a plot of $p[H_3O^+]$ versus volume of titrant or fraction titrated is called an **acid–base titration curve**. As we see from this plot, there is a dramatic increase in $p[H_3O^+]$ where $\theta = 1$, that is, where exactly enough titrant has been added to react with the analyte. This dramatic increase is the basis for most methods of equivalence point detection. An experimental plot of pH versus titrant volume could be obtained by recording the reading of a pH meter for each increment of titrant added. Drawing a vertical line through the rapidly rising portion of the pH meter response will intersect the volume axis (X-axis) at the titrant volume at which equivalence was reached. Note that the desired information (amount of acid present in the sample) is obtained from the titrant volume at the equivalence point, not from the reading of the pH meter.

USING AN INDICATOR

Another approach is to use a chemical **indicator**. An acid–base indicator is a weak acid for which the acid and base forms have different colors in solution. A very small amount of this indicator is added to the solution to be titrated. When the $p[H_3O^+]$ of

---

[3]The derivation of Equation 4.31 for the titration of a strong acid with a strong base was included to illustrate how such equations are developed. In general, text space will not be taken for such derivations. However, complete derivations can be found on the accompanying CD.

Table 4.1

Some Acid–Base Indicators

Common Name	pK_a'	Color*
Thymol blue	1.6	R/Y
Methyl yellow	3.3	R/Y
Methyl orange	4.2	R/O
Methyl red	5.0	R/Y
Bromothymol blue	7.1	Y/B
Phenol red	7.4	Y/R
Cresol purple	8.3	Y/P
Thymol blue	8.9	Y/B
Phenolphthalein	9.7	C/R
Thymolphthalein	9.9	C/B

*Colors are listed as acid form/base form.
B = blue, C = colorless, O = orange, P = purple, R = red,
Y = yellow.

The end point of a titration is the point at which the amount of titrant is experimentally assumed to be equivalent of the amount of sample being titrated.

The equivalence point of a titration is the point at which the amount of titrant delivered is exactly equivalent to the amount of sample being titrated.

These two points may differ as a result of systematic and random errors.

the solution is equal to the pK_a' of the indicator weak acid, the indicator is split equally between the acid and base forms. At a higher $p[H_3O^+]$ it will be predominately in its base form, and at lower $p[H_3O^+]$ it will be in its acid form. Thus when the $p[H_3O^+]$ of the solution being titrated reaches the $p[H_3O^+]$ equal to the indicator pK_a', the solution will be in the middle of its transition from the acid color to the base color. A number of indicators are available, so one can generally find one with a pK_a' for which the color change occurs near the $p[H_3O^+]$ of the equivalence point solution. Table 4.1 is a list of some indicators.

When an indicator is used, the titrant is added in small increments in the region where the indicator color is changing. When the color change has just occurred, the titration is stopped and the delivered volume is recorded. This volume is called the **end point volume**. If the wrong indicator has been chosen or if the titration is not carefully done, the end point volume can be different from the equivalence point volume, and an erroneous concentration will be calculated. In this text, the term **equivalence point volume** will be reserved for the titrant volume at which the titrant and titrated species are present in exactly equivalent amounts. The term end point volume will be used for the experimentally determined approximation of the equivalence point volume.

The range of $p[H_3O^+]$ over which the color change for an indicator occurs can be seen from the alpha plot of Figure 4.1. At one $p[H_3O^+]$ unit less than $pK_{a,ind}'$, the indicator will be 90% in the acid form. Similarly, at one $p[H_3O^+]$ unit greater than $pK_{a,ind}'$, the indicator will be 90% in the base form. Further changes in $p[H_3O^+]$ have little effect on the composition of the indicator as far as the color is concerned. Thus, the color change for most indicators occurs over a range of two $p[H_3O^+]$ units, centered on the $p[H_3O^+]$ that is equal to $pK_{a,ind}'$. From the titration curve shown in Figure 4.5 for the titration of 0.01 M strong acid, any indicator with a $pK_{a,ind}'$ between 5 and 9 would appear to serve the purpose. However, the titration curve shown does not have sufficient resolution along the fraction titrated axis to assess an accuracy to 0.1%. Let us look at an expansion of the curve in the region of the equivalence point. From Figure 4.6, we can see that for 0.1% precision (the area between $\theta = 0.999$ and $\theta = 1.001$) the $pK_{a,ind}'$ of the indicator should be between 5.5 and 8.5.

The equivalence point in the titration of a strong acid with a strong base is at a $p[H_3O^+]$ equal to $\sqrt{K_w'}$. (It is found by setting $\theta = 1$ in Equation 4.31.) For a solution at room temperature, this will be 6.998. Therefore, the optimum indicator for this titration is one that has a color change centered at $p[H_3O^+] = 7.0$. Among the indicators listed in Table 4.1, the best choice would be bromothymol blue, with phenol red as an acceptable alternative.

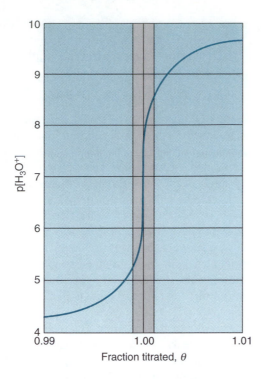

⊙ Figure 4.6. This expanded version of Figure 4.5 shows the narrow range of $p[H_3O^+]$ that exists for the equivalence point to be determined to 0.1%.

Achieving the goal of finding the end point volume with a precision of 0.01 mL requires some skill. As the end point is near, very small increments of titrant must be added and then the solution stirred while the indicator color is observed. Fortunately, the solution itself gives us a clue regarding the impending end point. In the region where the titrant enters the solution, the local excess of titrant causes the indicator color change to occur. Before the end point, this color will dissipate on stirring. As the end point approaches, the time required for complete dissipation of the color change increases. In this region, the operator uses increasingly small increments of titrant, taking care to rinse any titrant adhering to the tip of the buret into the solution. If the end point volume is exceeded without recording the exact end point volume, we say the end point has been overshot and the titration is spoiled.

USING A pH METER

When a pH meter is used to determine the equivalence point in a titration, either of two methods can be used. In one, the correct equivalence pH is calculated, and the titrant is added until this value of pH is reached. The pH meter is thus used as an end point indicator. The technique of achieving the end point volume is the same as with the chemical indicator. The other approach is to record the pH value and titrant volume at many points in the titration, especially in the region of the equivalence point. These points define an experimental titration curve. These data points can then be fit to the theoretical titration curve equation to mathematically determine the end point volume. The advantages of this technique are that it is not necessary to have a data point at the exact equivalence point, and random errors in reading the titrant volume are averaged over many volume readings. The disadvantage is that automated data logging is desired to provide many data points without the tedium of manual entry.

Another advantage to the curve fitting approach is that, if the correct algorithm is used for the titration curve, the agreement between the determined end point volume and the true equivalence point volume can be very good. This is in contrast to the end point techniques for which the end point indication might not occur at the exact equivalence point. Examples of situations in which the end point volume may not be the same as the equivalence point volume are the incorrect choice or use of indicator, incorrect calculation of the expected equivalence point pH, and the expectation that the equivalence point will occur at the inflection point of the titration curve.

CONCENTRATION EFFECT IN STRONG ACID–BASE TITRATION

The degree of change in $p[H_3O^+]$ before and after the equivalence point in the titration of strong acids and bases depends on the initial concentration of the analyte. Again assuming the concentration of the titrant to be similar to that of the analyte, the titration curves for various concentrations of strong acid analyte are shown in Figure 4.7. From this we see that the magnitude of the break decreases dramatically as the analyte concentration decreases. This decreases the range of $p[H_3O^+]$ over which the titrant volume is within 0.1% of the equivalence point volume. Given that the range of $p[H_3O^+]$ over which an indicator color change occurs is about two $p[H_3O^+]$ units, a 10^{-3} M solution of analyte is about the least concentrated one could accurately titrate with an indicator. Using the technique of fitting the titration curve to the data, this concentration limit could probably be extended to about 10^{-5} M. Below this concentration, the $p[H_3O^+]$ change over the entire titration is too small to make titration a useful quantitative method in aqueous solutions.

The log concentration plot of the autoprotolysis reaction of water introduced in Figure 3.1 can be used to provide an estimate of the magnitude of the $p[H_3O^+]$ change one could expect in the titration of a strong acid or base. This is shown in Figure 4.8. If a solution of about 10^{-2} M strong acid is to be titrated, the initial $p[H_3O^+]$ will be about 2. When titrated with a similarly concentrated base, the $[OH^-]$ when the fraction titrated is 2.0 will be 3×10^{-3} M, or $p[H_3O^+] = 11.5$. Thus the total change in $p[H_3O^+]$ will be from 2.0 to 11.5, as shown in Figure 4.8 by the length of the line for a log concentration of -2. Now if the initial concentration of the strong acid is 10^{-4}, the

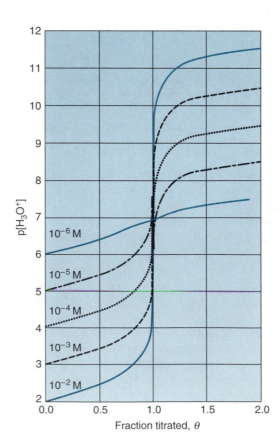

◎ Figure 4.7. The initial concentration of acid to be titrated has a great effect on the magnitude of the change in $p[H_3O^+]$ in the equivalence point region.

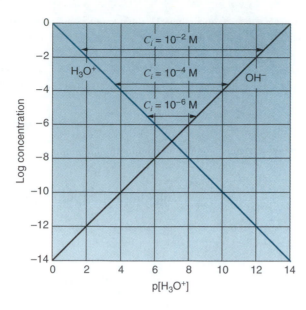

⊚ **Figure 4.8.** The autoprotolysis log plot for water can be used to predict the magnitude of the $p[H_3O^+]$ change for the titration of strong acids or bases given their initial concentration in aqueous solution.

$p[H_3O^+]$ change is from 4.0 to 9.5, as shown. This simple plot makes it clear why the titration break decreases with decreasing initial concentration. It is also clear why acid–base titrations become increasingly impractical as the initial concentration approaches 10^{-7} M.

The use of the autoprotolysis log plot to predict the magnitude of the titration break for different concentrations of acid or base analyte underscores the significance of the solvent in this determination. A solvent with a higher value of autoprotolysis constant would provide a smaller break in pH for the titration of the same analyte concentration. For example, if the solvent were formic acid ($pK'_{solv} = 6.2$), a 0.1 M analyte would produce only a modest break. On the other hand, there are several useful solvents for which the autoprotolysis constant is much lower than water and which, therefore, would allow acid–base titrations with much lower concentrations of analyte.

Study Questions, Section C

10. In the titration of an acid, what two quantities are used in the calculation of the amount of acid present? If one wanted to know the concentration of the acid, what additional fact is required?

11. In an acid–base titration, what is the probe of the differentiating characteristic?

12. Titration is a null measurement technique. What is the adjustable standard, what is being compared, what is the difference detector, and how is the amount of adjustable standard at the null point determined?

13. How is the exact concentration of the standard titrant known?

14. What quantities need to be measured in performing the standardization of the titrant?

15. You have a strong base solution in your buret that is nominally 0.05 M. You wish to standardize this solution with weighed out amounts of potassium hydrogen phthalate (KHP). The buret has a capacity of 20.00 mL, and you want to perform the standardization titration without refilling the buret. Approximately what weight of KHP should you dissolve in the titration container?

16. How many milliliters of 0.01206 M strong base titrant will be required to titrate a 20.00 mL aliquot of a solution that contains 2.717×10^{-2} mol L^{-1} of acetic acid?

17. The change in what quality of the solution is the basis for all methods of acid–base equivalence point detection?

18. Use Equation 4.31 to calculate the equivalence point pH for the titration of 0.100 M HCl and for 0.0001 M HCl.

19. In choosing an indicator for an acid–base titration, why does one select an indicator that has a pK'_a near the pH of the equivalence point solution?

20. What is the distinction between the equivalence point volume and the end point volume for a titration?

21. Using Equation 4.31, calculate the relative error in the fraction titrated when a 0.001 M solution of strong acid is titrated with a strong base using thymol blue as an indicator (yellow to blue transition).

22. What are some advantages in using a pH electrode and pH meter to record the titration curve for the determination of the titration end point?

23. Why does the degree of change at the equivalence point for the titration of a strong acid with a strong base decrease with decreasing concentration of acid titrated?

24. If an acid-base titration were carried out in a solvent for which the autoprotolysis constant is $10^{-19.5}$ (as it is for pure ethanol), what would be the minimum acid concentration for practical titration with a strong base, assuming a need for a change of 3 in $p[SH^+]$ over the course of the titration? (SH^+ is the protonated solvent molecule.) You can use a spreadsheet to make a plot similar to that of Figure 4.8.

Answers to Study Questions, Section C

10. The amount of the acid present is obtained from the volume of the standard titrant at the equivalence point and the concentration of the standard titrant. The titration determines the total amount of the analyte. To determine its original concentration, one would have to know the original sample volume.

11. The differentiating characteristic is the ability to undergo an acid–base reaction. If the analyte is an acid, the probe is to add a basic reactant. If the analyte is a base, an acid is added.

12. The adjustable standard is the volume of the standard titrant added to the sample. The amount (number of moles) of the titrant is being compared with the amount of the analyte. The difference detector is the equivalence point detection mechanism (pH measurement or indicator). There must be a way to measure the volume delivered. The equivalence point volume is determined by measuring the volume at an end point or by calculating the equivalence point volume from a recorded titration curve.

13. The concentration of the standard titrant is known by making the solution from a known amount of reactant to an exact volume (determinate preparation) or by standardizing the titrant by using it to titrate a known amount of primary standard.

14. To standardize the titrant, one must know the amount of primary standard being titrated (from weighing) and the volume of the titrant delivered at the equivalence point.

15. We expect $0.02000 \text{ L} \times 0.05 \text{ mol L}^{-1} = 1 \times 10^{-3}$ mol of base titrant, maximum. The equivalent amount of KHP is 1×10^{-3} mol $\times 204.22$ g mol$^{-1} = 0.204$ g. Since this is the maximum amount, a slightly smaller quantity, say between 0.16 and 0.18 g would be ideal.

16. We expect 2.717×10^{-2} mol L$^{-1} \times 0.02000$ L $= 5.434 \times 10^{-4}$ mol of acetic acid to be titrated. The amount of strong base is 5.434×10^{-4} mol$/0.01206$ mol L$^{-1} = 45.06$ mL.

17. As the sample is titrated, the pH of the solution changes. If the analyte is an acid, the pH increases as the titration proceeds.

18. If $\theta = 1$ in Equation 4.31, the second term becomes 0 and $[H_3O^+]^2 = K'_w$ for both starting concentrations. Thus,

$$[H_3O^+] = \sqrt{K'_w} = \sqrt{10^{-14}} = 10^{-7}$$

and $p[H_3O^+] = 7$.

19. The indicator will change color when the fractions in each form change significantly. Since the indicator is a weak acid, the fraction in the acid and base forms will depend on the pH (see Figure 4.1). The pH at which the ratio of the fractions in each form changes most dramatically is equal to the pK'_a of the weak acid.

20. The equivalence point volume is the volume at which the amount of titrant was exactly the right amount to react with all the analyte. The end point volume is the experimental approximation of the equivalence point.

21. From Table 4.1, the pK'_a for thymol blue (yellow to blue transition) is 8.9. At its transition, $[H_3O^+] = 10^{-8.9} = 1.26 \times 10^{-9}$ M. For the titration of a strong acid with a strong base, the $[H_3O^+]$ value at the equivalence point is 1.00×10^{-7} M. Equation 4.31 can be rearranged and the value for $[H_3O^+]$ substituted.

$$\left(\frac{1-\theta}{1+\theta}\right) = \frac{[H_3O^+]^2 - K'_w}{C_i[H_3O^+]}$$

$$= \frac{-10^{-14}}{1.26 \times 10^{-12}} = -7.94 \times 10^{-3}$$

Various values of θ can then be tried until this equation is satisfied. In this way θ is found to be 1.016. This means that the true equivalence point is exceeded by 1.6%.

22. Some advantages of using a pH electrode and meter are as follows. There is no need to select an indicator in advance. The change in pH that occurs at the end point will be recorded for any end point pH value. Variations in assessment of the indicator color at the end point are eliminated, as is the amount of titrant required to change the color of the indicator. The use of many recorded pH versus volume points improves the precision of the result.

23. As the concentration of the analyte acid at the beginning of the titration decreases, the initial pH will increase. When titrating a smaller amount of acid, a lower titrant concentration is used, resulting in a lower value of the pH in the region past the equivalence point.

24. Draw the log concentration diagram for a solvent with an autoprotolysis constant of $10^{-19.5}$. From this plot, find the concentration for which the difference between the lines for $[(S - H)^-]$ and $[SH^+]$ is three "p" units. The graph is shown below. The log of the concentration is seen to be -8.3, from which the concentration is 5×10^{-9} M.

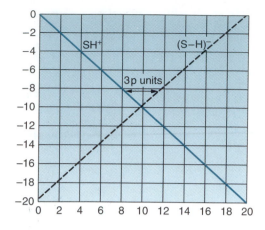

D. Weak Acids, Weak Bases, and Buffers

The analysis of weak acids and bases is similar to that for the strong acids and bases except for the additional complication that a significant fraction of the dissolved acid or base can remain unreacted with the solvent. This fact affects the value of the pH of the analyte solution and the shape of the pH volume curves that occur during titration.

TITRATION OF WEAK ACIDS

We will first develop the form of the titration curve for the titration of a weak acid. The equations that apply are the equilibrium constants for the weak acid and the autoprotolysis of water.

$$K_a' = \frac{[H_3O^+][A^-]}{[HA]} \qquad \text{and} \qquad K_w' = [H_3O^+][OH^-] \qquad \text{4.2 and 3.12}$$

We can no longer assume that all the dissolved acid is in the basic form, so we invoke the mass balance equation

$$C_i = [HA] + [A^-] \qquad \text{4.32}$$

which simply states that all the acid dissolved must be in one form or the other. The charge balance condition of Equation 4.30 is also used. These equations are combined as in the derivation of Equation 4.31 for the case of a titration of a strong acid. The resulting equation[4] is cubic in $[H_3O^+]$. An example of a plot of the weak acid titration equation is shown in Figure 4.9 for the titration of a 0.100 M solution of a weak acid that has a K_a' of 1.0×10^{-5}. This plot was produced by the titration curve spreadsheets on the CD.

Take a moment to compare the weak acid titration curve of Figure 4.9 with that of the strong acid (Figure 4.5). Despite the higher concentration of the weak acid, the break in the titration curve is much smaller than that for the strong acid. This has the effect of reducing the potential precision of the titration and makes the choice of indicator more critical. In addition, there is an inflection point at the half-titrated point. At the halfway point in the titration, the solution behaves as though it had been made up of equal concentrations of the acid and base forms of the conjugate pair. Referring to

---

[4]For the details of this derivation, consult the derivations section of the accompanying CD.

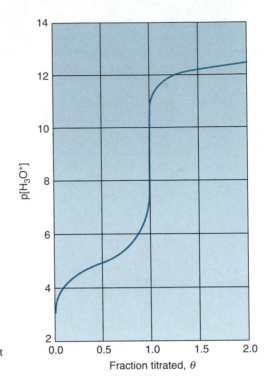

⊙ **Figure 4.9.** The titration curve for a weak acid flattens out to $p[H_3O^+] = pK'_a$ at the mid-titration point ($\theta = 0.5$).

Equation 4.25, if C_{A^-} and C_{HA} are equal and $[H_3O^+]$ is large compared to $[OH^-]$, $[H_3O^+]$ is equal to K'_a. For most weak acid titrations, this is true, and thus *the $p[H_3O^+]$ value at the half-titrated point can be used to determine the pK'_a of the weak acid.*

The titration curves of weak acids are also affected by the initial concentration of the acid. To illustrate this, the titration of three concentrations of an acid with a K'_a of 10^{-5} are shown in Figure 4.10. Not only does the magnitude of the break decrease with

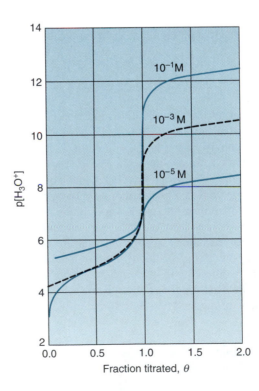

⊙ **Figure 4.10.** The effect of initial concentration on the magnitude of the $p[H_3O^+]$ break at the equivalence point for a weak acid with $K'_a = 10^{-5}$.

decreasing concentration, but the $p[H_3O^+]$ value at the equivalence point also changes. At concentrations of 10^{-1} M and 10^{-3} M, the inflection at the half-titrated point remains at the value of pK_a'. For the titration at 10^{-5} M concentration, this is no longer the case. In fact, the titration curve looks much like that of a strong acid, with the $p[H_3O^+]$ value at the equivalence point nearly equal to 7. This is due to the large fraction of the weak acid that has reacted with the water. As the concentration decreases, the fraction of a dissolved weak acid that reacts with the solvent increases. In this case, the reaction is 62% complete, making the weak acid behave very much like a dilute solution of a strong acid.

The calculation of specific points on a titration curve allows an estimation of the complete curve by simply joining the points with a curve of the familiar **S** shape. The $p[H_3O^+]$ before the titration has begun can be calculated from Equations 4.11, 4.12, or 4.13 if the acid concentration is well above $(K_w')^{1/2}$. We will take the example of the titration of a 10^{-3} M solution of an acid that has a K_a' of 1.0×10^{-5} with a 1.0×10^{-3} M solution of strong base. Using the result of Equation 4.13, $p[H_3O^+] = 4.0$. At the 50% titration point, half the HA has been titrated to A^-. From Equation 4.3, the $p[H_3O^+]$ would be equal to pK_a'. From Figure 4.10, we see this is true for an initial acid concentration of 10^{-3} M, but it is not true for an initial concentration of 10^{-5} M. At this low concentration, the autoprotolysis reaction has a significant effect, and Equation 4.25 or 4.26 must be used.

At the equivalence point, exactly enough strong base has been added to react with the amount of HA initially present. The solution composition is the same as that of a solution of NaA. Therefore, we can use Equations 4.16, 4.17, or 4.18 to solve for the $p[H_3O^+]$ at the equivalence point. The value of C_{A^-} used is half the initial acid concentration to take into account the dilution caused by the addition of an equivalent amount of titrant. Using Equation 4.18, the calculated $p[H_3O^+]$ is 7.85. Again, this matches the value in the plot of Figure 4.10. The final, easily calculated point is that at 200% titrated ($\theta = 2$). At this point, the OH^- concentration is one-third that of the titrant concentration (because the solution volume is now three times that of the original acid). For a titrant concentration of 1.0×10^{-3} M, the $[OH^-]$ at the 200% point is 3.3×10^{-4} M and the $p[H_3O^+]$ is 10.5, as shown in the plot.

If the initial acid concentration is much lower than 10^{-3} M, the more complex Equation 4.25 must be used to calculate these points. In the next section, we will use simply constructed log concentration plots to quickly obtain estimates of the four critical titration curve points. This graphical approach is valid under all conditions.

BUFFER SOLUTIONS

The titration curves of a weak acid have two inflection points. One is near the equivalence point, where the slope of the titration curve is a maximum. The other inflection point is at the mid-titration point, where the slope of the titration curve is a minimum. At this point, the degree of change in $p[H_3O^+]$ as a function of the amount of acid (or base) added is a minimum. This kind of solution is called a **buffer solution** because the presence of the nearly equal concentrations of the conjugate acid and base forms establishes a $p[H_3O^+]$ value nearly equal to pK_a' and resists changes in $p[H_3O^+]$ from that value. The equilibrium constant (as in Equation 4.3) can be used to calculate the ratio of the acid and base forms required to establish a given value of $p[H_3O^+]$. Similarly, the $p[H_3O^+]$ established by a given ratio of acid and base form concentrations can be calculated.

When it is desired to carry out a reaction at a specific $p[H_3O^+]$, a buffer solution is used. It is particularly important to use a buffer when the reaction itself may involve protons and thus may work to change the $p[H_3O^+]$ of the solution. A buffer solution can be rated by the molar concentration of strong acid or base it would take to change the p[H_3O^+] by a particular small amount. This rating is called the **buffer capacity** or the **buffer index**. In terms of a formula, the buffer capacity, β, is

$$\beta = \frac{\partial C_{base}}{\partial p[H_3O^+]} = -\frac{\partial C_{acid}}{\partial p[H_3O^+]} \quad \frac{mol}{L \cdot pH \; unit} \qquad 4.33$$

where the symbol ∂ means "very small change in." For an acid–base couple, the buffer capacity has been derived to be[5]

$$\beta = 2.303\left\{\frac{C_T K_a'[H_3O^+]}{([H_3O^+] + K_a')^2} + \frac{K_w'}{[H_3O^+]} + [H_3O^+]\right\} \qquad 4.34$$

A plot of the buffer capacity of a 0.01 M solution of an acid–base couple with a K_a' of 10^{-6} is shown in Figure 4.11. This curve was plotted using the buffer capacity spreadsheet for monoprotic acids on the CD. The peak in the middle of this plot is due to the buffer capacity of the weak acid couple. To achieve a p[H_3O^+] of less than 4.0 or greater than 10 (for this case), it is necessary to add strong acid or strong base to the solution containing the weak acid couple. The buffer capacities at p[H_3O^+] values below 4 and above 10 are due to the strong acid and base. The buffer capacity from the weak acid couple is the sharp maximum at a p[H_3O^+] equal to the pK_a' of the couple. This corresponds to the minimum slope in the titration curve, which also occurs at that same p[H_3O^+].

The calculation of the buffer capacity can be greatly simplified in most situations. First, the second and third terms within the braces of Equation 4.34 are [OH^-] and [H_3O^+]. Unless the buffer pH is nearly neutral, one of these terms will be insignificant with respect to the other. If equal quantities of the acid and base forms have been used in the buffer so that the buffer capacity is at its maximum for the total buffer concentration, [H_3O^+] = K_a'. When this substitution is made in Equation 4.34,

$$\beta_{max} = 2.303\left(\frac{C_T}{4} + \frac{K_w'}{[H_3O^+]} + [H_3O^+]\right) \qquad 4.35$$

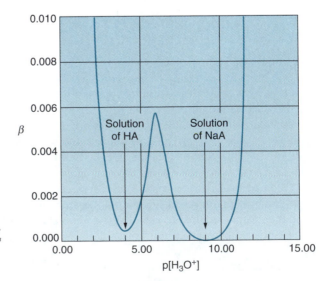

Figure 4.11. The result of Equation 4.34 is plotted for a 0.0100 M solution of an acid–base couple with a K_a' of 1.0×10^{-6}. The buffer capacity falls off steeply at values of p[H_3O^+] either side of pK_a'.

[5]J. N. Butler, *Ionic Equilibrium*, p. 134. John Wiley & Sons, Inc., New York, 1998.

For virtually any reasonable concentration of buffer with $p[H_3O^+]$ in the region of 3 to 11, $C_T/4$ will be larger than the other two terms in the parentheses of Equation 4.35. In this case, the maximum buffer capacity is simply

$$\beta_{max} = 0.576C_T \qquad 4.36$$

(See Example 4.4.)

The effective buffer region for a given weak acid couple extends only approximately one $p[H_3O^+]$ unit either side of $p[H_3O^+] = pK_a'$. To calculate the $p[H_3O^+]$ that will be established by any given mix of acid and its salt, use Equation 4.25. For most realistic concentrations, the terms in the parentheses in Equation 4.25 can be neglected and $[H_3O^+]$ calculated very simply. Once $[H_3O^+]$ has been calculated, the buffer capacity can be determined with Equation 4.34.

Note that the formal equilibrium constant has been used throughout this discussion of buffers. If the thermodynamic equilibrium constant is used, the actual buffer pH and the pH of maximum buffer capacity will be shifted because of the effect of the ionic strength. For many practical buffer systems, the formal values are available. Where they are not, the effect of the ionic strength can be estimated from the N values given in the K_a° table of Appendix B. (See Example 4.5.) Maintaining a constant pH is very important for most biochemical reactions. Buffer systems favored by biochemists for such applications are listed in Table 12.2.

TITRATION OF WEAK BASES

The approach used to predict the curves for the titrations of weak acids can be used for the titration of weak bases. It is more convenient to use the K_a' expression for the equilibrium between the conjugate acid–base forms than the K_b' expression. The exact equation for the titration curve can be derived from the equilibrium, mass balance, and charge balance equations as before. The complete derivation of the titration curve is given in the derivations section of the accompanying CD.

Plots of the titration curves for several concentrations of a weak base with a K_a' of 1×10^{-8} are shown in Figure 4.12. As expected, the solutions start out basic and be-

Example 4.4

The maximum buffer capacity for the buffer system plotted in Figure 4.11 is calculated using Equation 4.36.

$$\beta_{max} = 0.576 \times 0.0100 = 5.76 \times 10^{-3}$$

This value is seen to correspond with the peak value in Figure 4.11.

Example 4.5

From the table in Appendix B, the pK_a° of benzoic acid is 4.202. The value of N is 2.

$$pK_a' = pK_a^\circ - NAS^{1/2}$$

Since N has a positive value, the value of pK_a' will decrease as the ionic strength increases. For an ionic strength of 0.1, the decrease would be approximately 0.3 so that the $p[H_3O^+]$ at the maximum buffer capacity would be approximately 3.9.

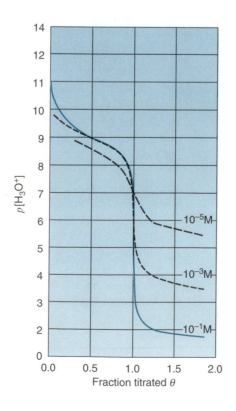

Figure 4.12. The similarity between the titration curves for a weak acid and a weak base can be seen by comparing this figure with Figure 4.10. Again, the midpoint inflection occurs at the pK_a' of the base being titrated.

come more acidic as the acid titrant is added. The equivalence point is marked by a break (more rapid change) in the $p[H_3O^+]$ versus the fraction titrated curve. Again, as the analyte concentration decreases, the break decreases. Note also that the optimum pK_a' for an indicator becomes greater as the equivalence point $p[H_3O^+]$ increases.

The weak base titration curve can be predicted from the application of Equation 4.25 or its appropriate simplifications. At the 0% titration point, C_{A^-} is equal to the initial base concentration. At the 50% point, C_{A^-} and C_{HA} are equal. At the 100% point, C_{HA} is equal to the initial base concentration times the dilution factor, and at 200%, the H_3O^+ concentration is equal to about one-third the titrant concentration. Just as with the weak acid, as the concentration of the base being titrated decreases, the relative contribution of the water autoprotolysis increases and the $p[H_3O^+]$ at the equivalence point approaches 7.0. The weaker the base, the smaller will be the break at the equivalence point for a given concentration. Remember, the weakest bases are those for which the pK_a' of the conjugate acid–base pair is lowest.

Study Questions, Section D

25. Why is there a smaller break in the titration of a weak acid than a strong one of the same concentration?

26. Using approximate equations, predict the titration curve for 0.100 M formic acid, HF, by calculating the $p[H_3O^+]$ at the 0, 50, 100, and 200% titration points. The pK_a' of formic acid is 3.74.

27. What information can be obtained from the pH at the 50% titration point of a weak acid?

28. Explain how the titration curve of a weak acid exhibits the optimum buffer composition and the $p[H_3O^+]$ at which this buffer will work best.

29. A. What is the composition and $p[H_3O^+]$ of the optimum buffer composed of NH_4^+ and NH_3 (total concentration of 0.10 M)? The K_a' for NH_4^+ is 5.70×10^{-10}.

    B. Calculate the buffer capacity of this buffer.

    C. Calculate the number of moles of acid a 100 mL solution of this buffer can withstand before the $p[H_3O^+]$ changes by 0.2 units.

30. Predict the titration curve for a titration of 1.00×10^{-4} M arginine with a strong acid of the same concentration. Arginine reacts with water to form $ArgH^+$. The pK_a' for $ArgH^+$ is 9.0.

Answers to Study Questions, Section D

25. The break is smaller because the pH of the weak acid sample solution is higher than that of a strong acid of the same concentration. The smaller the K_a' of the acid, the higher is the initial concentration and the smaller is the break.

26. The acid being titrated is relatively strong and the solution is relatively concentrated. Therefore, the simplest equations should apply without significant error. At 0%, the solution is just that of HF in water. From Equation 4.11,

    $$[H_3O^+] = (K_a'C_{HF})^{1/2} = (1.82 \times 10^{-4} \times 0.100)^{1/2}$$
    $$= 4.27 \times 10^{-3} \text{ M}, \quad p[H_3O^+] = 2.37$$

    At 50%, $p[H_3O^+] = pK_a' = 3.74$. For 100%, use Equation 4.16, neglecting $[H_3O^+]$. Thus,

    $$[H_3O^+] = (K_w'K_a'/C_{A^-})^{1/2} = (10^{-14} \times 1.82 \times 10^{-4}/0.0500)^{1/2}$$
    $$= 6.03 \times 10^{-9} \text{ M}, \quad p[H_3O^+] = 8.22$$

    At 200%, $[OH^-] = 0.100/3 = 3.33 \times 10^{-2}$ M and $p[H_3O^+] = 12.50$.

27. At the 50% titration point of a weak acid, the concentrations of the acid and base forms of the species being titrated are equal. The pH when this is true is equal to the pK_a' of the weak acid being titrated. This only works, however, when the contribution of the autoprotolysis of the solvent to the H_3O^+ concentration is negligible.

28. The titration curve shows the least change in $p[H_3O^+]$ versus volume of titrant at the 50% titration point. The acid and base forms of the species being titrated have equal concentrations at the 50% titration point. This is the optimum buffer composition. The optimum $p[H_3O^+]$ for this buffer is then where $p[H_3O^+] = pK_a'$.

29. A. The point of maximum resistance to change in $p[H_3O^+]$ occurs when the NH_4^+ and NH_3 concentrations are equal. Therefore, the composition of the buffer will be $[NH_4^+] = [NH_3] = 0.10/2 = 0.050$ M. From Equation 4.3, if the acidic and basic forms have equal concentrations, $p[H_3O^+] = pK_a' = 9.24$.

    B. Since the buffer composition is optimum, we can use Equation 4.36 for β_{max}.

    $$\beta_{max} = 0.576 \times 0.10 = 5.8 \times 10^{-2}$$

C. If acid is added, the $p[H_3O^+]$ will decrease, so the solution to part A is used in the calculation of the new conjugate pair concentrations for a $p[H_3O^+]$ of $9.24 - 0.20 = 9.04$.

$$\frac{[NH_4^+]}{[NH_3]} = \frac{[H_3O^+]}{K_a'} = \frac{9.12 \times 10^{-10}}{5.70 \times 10^{-10}} = 1.60$$

We also know that $[NH_4^+] + [NH_3] = 0.10$ M, so $[NH_3] = 0.10 - [NH_4^+]$. Substituting for $[NH_3]$, $[NH_4^+] = 1.60 \ (0.10 - [NH_4^+]) = 0.160 - 1.60[NH_4^+]$.

From this, $[NH_4^+] = 0.0615$ M. The amount of the *increase* in $[NH_4^+]$ is due to the amount of acid added. Thus,

$$\frac{(0.0615 - 0.050) \text{ mol/L} \times 100 \text{ mL}}{1000 \text{ mL/L}} = 1.15 \times 10^{-3} \text{ mol}$$

Another way to estimate this is from the buffer capacity calculated in part B, since the buffer capacity is the molar concentration of acid added per unit change in pH. For a β_{max} of 5.76×10^{-2} mol L^{-1} pH^{-1} the number of moles is 5.76×10^{-2} mol L^{-1} $pH^{-1} \times 0.2$ pH $\times 0.1$ L $= 1.15 \times 10^{-3}$ mol. Because of the change in buffer capacity as a function of pH, this solution will only work for pH values close to the buffer pH.

30. The solution at 0% titrated is that of 1.00×10^{-4} M base. From Equation 4.16,

$$[H_3O^+] = (10^{-14} \times 1 \times 10^{-9}/1.00 \times 10^{-4})^{1/2}$$
$$= 3.2 \times 10^{-10} \text{ M}, \qquad p[H_3O^+] = 9.5$$

The solution at 50% titrated is equal concentrations of $ArgH^+$ and Arg, so

$$p[H_3O^+] \approx pK_a' = 9.0$$

From Equation 4.25, we can see that this answer is not completely correct, since $[OH^-]$ at this $p[H_3O^+]$ is only 0.1 of C_T. The exact solution will be slightly more acidic than the approximate one.

The solution at 100% titrated is that of a 5.00×10^{-4} M solution of $ArgH^+$. From Equation 4.11, this will be

$$[H_3O^+] = (K_a' C_{ArgH^+})^{1/2} = (1 \times 10^{-9} \times 5.00 \times 10^{-5})^{1/2}$$
$$= 2.2 \times 10^{-7} \text{ M}, \qquad p[H_3O^+] = 6.7$$

The solution at 200% titrated is $C_T/3$ in excess strong acid titrant.

$$[H_3O^+] = 3.33 \times 10^{-5} \text{ M}, \qquad p[H_3O^+] = 4.5$$

E. Logarithmic Concentration Plots

Recall the alpha plot for an acid–base conjugate pair shown in Figure 4.1. For values of $p[H_3O^+]$ more than 2 units away from pK_a', the minor component's fraction of the total concentration can no longer be obtained from the plot, as it is less than 1%. This situation can be overcome by the simple ploy of plotting the log of alpha versus $p[H_3O^+]$ instead. The log alpha plot in Figure 4.13 clearly shows the concentration of

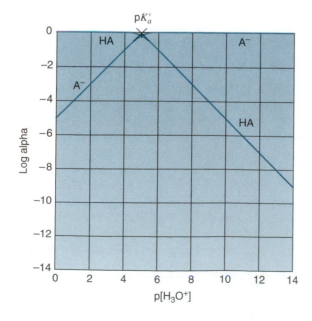

Figure 4.13. A plot of log α versus $p[H_3O^+]$ shows the fractions of each form at $p[H_3O^+]$ values well away from pK_a'.

the minor component at all values of $p[H_3O^+]$. It is also convenient because it is easily constructed with graph paper and a straightedge. One simply places an $\times$ at the intersection of $\log \alpha = 0$ and $p[H_3O^+] = pK'_a$. From this point, a horizontal line is drawn in each direction for the major component. The minor component $\log \alpha$ values appear on lines that begin at the $\times$ and proceed downward at unit slopes in both directions. For completeness, curved lines can be drawn to join the major and minor component sections of each line. These two curves intersect 0.3 $\log \alpha$ units below the $\times$, corresponding to α values of 0.5. Just as in the case of the linear α plot, a change in the pK'_a of the acid being plotted simply moves the $\times$ (and the intersection point under it) to higher or lower values of $p[H_3O^+]$. Spreadsheets for automatic plotting of log alpha plots are provided on the accompanying CD.

CONSTRUCTING LOGARITHMIC CONCENTRATION DIAGRAMS

The $\log \alpha$ plot can be modified to provide the actual concentration of the species given knowledge of C_T, the total concentration of both forms. Though these plots are as simple to construct as the $\log \alpha$ plot, for the sake of rigor, we will derive the equations for the lines that make up these plots. Solving Equation 4.6 for [HA] and taking the logarithm of both sides of the equation yields

$$[HA] = C_T\alpha_{HA} = C_T\left(\frac{1}{1 + K'_a/[H_3O^+]}\right)$$

$$\log[HA] = \log C_T - \log\left(1 + \frac{K'_a}{[H_3O^+]}\right)$$

4.37

Similarly, an equation for the log of [A$^-$] can be obtained.

$$[A^-] = C_T\alpha_{A^-} = C_T\left(\frac{1}{1 + [H_3O^+]/K'_a}\right)$$

$$\log[A^-] = \log C_T - \log\left(1 + \frac{[H_3O^+]}{K'_a}\right)$$

4.38

Now the equations for the lines of the remaining species (H_3O^+ and OH^-) are needed. For H_3O^+, it is the identity $\log[H_3O^+] = -p[H_3O^+]$. For OH^-, it is from the autoprotolysis equation. (A log plot for these species was previously developed in Figure 3.1).

$$\log[OH^-] = p[H_3O^+] - pK'_w$$

4.39

Equations 4.37 through 4.39 were used in the spreadsheet program to obtain the logarithmic concentration diagram shown in Figure 4.14. In this example, the pK'_a for the acid–base pair is 5.0 and C_T is 0.01 M. This plot is very powerful because it represents the equilibrium concentrations of all species in the acid–base–solvent system as a function of $p[H_3O^+]$. For example, the concentrations of all species at $p[H_3O^+] = 10.0$ are found along the vertical line corresponding to $p[H_3O^+] = 10$. They are seen to be $[HA] = 10^{-7}$ M, $[OH^-] = 10^{-4}$ M, and $[A^-] = 10^{-2}$ M.

Logarithmic concentration diagrams are easily constructed on graph paper. An $\times$ is placed at the intersection of $\log C_T$ and the $p[H_3O^+]$ value that is equal to pK'_a. This intersection is called the **system point**. Horizontal lines are drawn to the left and right of the system point along the value of $\log C_T$. Lines are drawn downward at unit slope from the system point. A curved connection is made between the lines through a point 0.3 log units beneath the system point, and the lines are labeled. The lines for H_3O^+ and OH^- are added as developed in Figure 3.1.

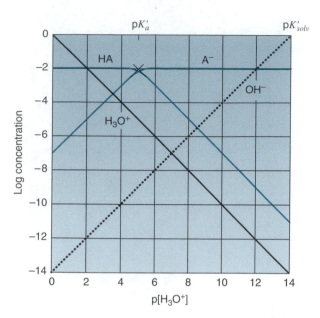

⊘ **Figure 4.14.** The equilibrium concentrations of the acid and base forms of both the weak acid and the solvent for any value of $p[H_3O^+]$ can be seen in the log concentration plot.

The log concentration plots are most easily constructed by entering the data values in the appropriate spreadsheet. The spreadsheets are easily located by category, the two data values are entered, and the plot is produced automatically. A press of a button provides a printout.

SOLVING PROBLEMS WITH LOG PLOTS

Logarithmic concentration plots can be a very effective aid in the solution of a variety of problems regarding acid–base systems. They are a simple alternative to the use of the equations we have developed in earlier sections, they can provide approximate solutions that are useful in choosing valid assumptions for exact solutions, and they can provide information on expected titration curve points virtually by inspection. We will begin our appreciation of log plots with the solution of a few problems of the type earlier solved algebraically.

The log concentration plot of Figure 4.14 reveals the concentrations of all the acid species in a solution of an acid-base couple with $K_a' = 10^{-5}$ and $C_T = 10^{-2}$ as a function of $p[H_3O^+]$. To use this plot to solve a problem related to this system, we need an additional equation that describes the starting conditions of the problem. When this equation is combined with the graphed component values, the solution point can be found. The equation that is the key to the application of the log plot is a relationship called the **proton balance equation**. This equation simply states that, relative to the starting materials, the concentrations of the more protonated species must be equal to the concentrations of the less protonated species. One can think of this as the law of conservation of protons. For example, if the starting materials are HA and H_2O, the only more protonated species is H_3O^+, and the less protonated species are A^- and OH^-. Therefore, the proton balance equation for this situation is

$$[H_3O^+] = [A^-] + [OH^-] \qquad\qquad 4.40$$

The $p[H_3O^+]$ in a 10^{-2} M solution of weak acid ($pK_a' = 5$) can then be found on the logarithmic concentration plot where Equation 4.40 is true. This process is illustrated in Figure 4.15. The solution is at the intersection of the line for H_3O^+ with that of the A^- line or the OH^- line. The intersection chosen is the one that gives the larger value for $[H_3O^+]$. In the case of the logarithmic concentration plot shown, the intersec-

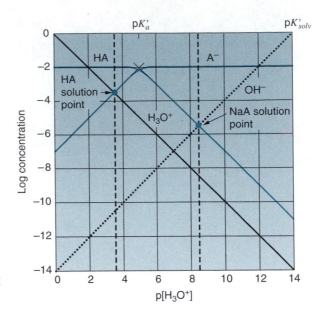

Figure 4.15. Here the log concentration plot is used in conjunction with a proton balance equation to solve for the concentrations of all species given the initial solution conditions. Solution points for 0.010 M acid and 0.010 M salt are shown.

Proton balance equations

To obtain the proton balance equation for a given starting solution, list all the acid–base species in the solution. For a monoprotic acid in water, this list will be

$$HA, H_3O^+, H_2O, A^-, OH^-$$

Cross off the starting materials (HA and H_2O for a solution of the acid and A^- and H_2O for a solution of a base).

Arrange the remaining species on either side of the equals sign depending on whether they are more protonated or less protonated than the starting materials.

Do not try to memorize proton balance equations. Just obtain them from the starting materials as needed.

---

tion that satisfies Equation 4.40 is where the line for H_3O^+ crosses that for A^-. The $p[H_3O^+]$ value at this intersection is 3.5. A dashed solution line can then be drawn vertically through the solution point. At $p[H_3O^+] = 3.5$, the concentrations of the species are $[H_3O^+] = 3 \times 0^{-4}$ M, $[HA] = 1 \times 10^{-2}$ M, $[A^-] = 3 \times 10^{-4}$ M, and $[OH^-] = 3 \times 10^{-11}$ M. The concentration of OH^- is thus seven orders of magnitude less than the concentration of A^- at this $p[H_3O^+]$. This huge difference completely justifies neglecting $[OH^-]$ relative to $[A^-]$ in choosing the intersection that satisfies Equation 4.40.

The same logarithmic concentration plot can be used to find the $p[H_3O^+]$ of a solution made by dissolving NaA in water at 10^{-2} M. The proton balance equation for this situation is

$$[HA] + [H_3O^+] = [OH^-] \qquad\qquad 4.41$$

(If you're not sure why Equation 4.41 is correct, please follow the procedure in the margin note.)

To find the $p[H_3O^+]$ value for the solution of the salt of the base form, we look again for the position on the log plot (Figure 4.15) for which the proton balance equation (Equation 4.41) is true. Here it is convenient to follow the line for OH^- from its high concentration value (in the upper right-hand corner) to the line for either one of the species mentioned on the other side of the proton balance equation. In the example we are using, this is the line for HA. At this intersection, the $p[H_3O^+]$ is 8.5. Again, a dashed solution line is drawn vertically through the solution point. The concentrations of the other species at that $p[H_3O^+]$ are $[A^-] = 10^{-2}$ M, $[HA] = 3 \times 10^{-6}$ M, and $[OH^-] = 3 \times 10^{-6}$ M. Before accepting this solution, we need to make sure the proton balance equation (Equation 4.41) is satisfied. At $p[H_3O^+] = 8.5$, $[HA]$ is 1000 times greater than $[H_3O^+]$, so this solution is correct to 1 part in 1000 or 0.1%.

Logarithmic concentration plots are also an excellent tool to use when the simplifying assumptions are not valid or are not known. In fact, the plots give an immediate indication when that is the case. For example, consider an acid with a pK'_a of 3 and a C_T of 10^{-5} M. The log concentration plot for this system is shown in Figure 4.16. When the proton balance equation for HA dissolved in H_2O is invoked ($[H_3O^+] = [A^-] + [OH^-]$), the solution point is seen to be at the intersection of the H_3O^+ and A^- lines. At this intersection, the concentration of A^- is equal to C_T. In other words, the reaction between the weak acid and water is essentially complete just as if the acid were a strong acid in water. From this example we can see that if the system point lies significantly to the left or below the H_3O^+ line on the log concentration diagram, the acid will behave as

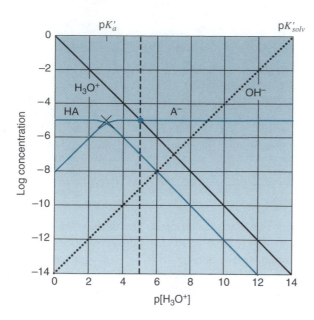

Figure 4.16. Here the log concentration plot demonstrates that when the system point is below and to the left of the H_3O^+ line, the reaction of the dissolved acid with the solvent is essentially complete.

a strong acid, producing an H_3O^+ concentration equal to C_T. Use the log concentration diagram to also prove that if the system point lies significantly below or to the right of the OH^- line, the dissolved acid has little effect on the solution $p[H_3O^+]$.

As we have seen, the key to using the log concentration plot to solve acid–base system problems is the correct development and application of the proton balance equation. A complication can arise in the application of the logarithmic concentration diagram when the smaller of the terms on the side of the proton balance equation with two terms cannot be neglected. An example of this situation is seen in Figure 4.17 for the case when pK'_a equals 10 and C_T is 10^{-3} M. The proton balance equation ($[H_3O^+]$ $= [A^-] + [OH^-]$) provides an approximate solution point at the intersection of the H_3O^+ and A^- lines where $p[H_3O^+] = 6.5$. However, at that point, $[OH^-]$ is one-tenth of $[A^-]$ and thus not a negligible amount. A better approximation can be gained by considering that, from the proton balance equation, the value $[H_3O^+]$ must be the sum of $[A^-]$ and $[OH^-]$. This would mean the $p[H_3O^+]$ at the actual solution point would be a bit lower than that at the intersection, say 6.4. The approximate concentrations

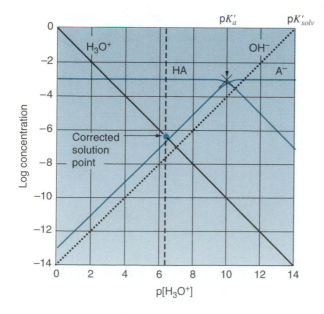

Figure 4.17. When the solution point falls too closely to a minor term in the proton balance equation to neglect it, an approximate solution can still be obtained. The more accurate solution point is shifted slightly to the left of the first approximate solution point at $[H_3O^+] = [A^-]$.

Table 4.2
Rules of Thumb for Various System Point Locations

1. To the left of the H_3O^+ line: Acid form reacts almost completely with the solvent. Base form reaction with the solvent is negligible.
2. To the right of the OH^- line: Base form reacts almost completely with the solvent. Acid form reaction with the solvent is negligible.
3. Well above both the H_3O^+ line and OH^- line: The simplified forms of Equations 4.11 and 4.16 can be used.
4. Below the intersection of the H_3O^+ and OH^- lines: Dissolved acid or base has little effect on $p[H_3O^+]$.

provided by the logarithmic concentration diagrams are often sufficient, but for exact work, we will need to use the rigorous relationships developed in the previous sections. Even then, the approximate solutions provided give a good value from which successive approximations can begin, and they indicate which terms in the equations cannot be safely neglected. Some rules of thumb are given in Table 4.2.

OBTAINING TITRATION CURVE POINTS FROM LOG PLOTS

Log concentration plots can be used to provide an estimate of the shape of the titration curve. They indicate the approximate $p[H_3O^+]$ at the beginning of the titration, the halfway point, the equivalence point, and the point of 100% excess titrant. This process is illustrated in Figure 4.18 for the titration of a 10^{-2} M solution of an acid with a pK_a' of 10^{-5}. First, the log concentration plot is constructed. Then, the conditions at each of the prediction points are assessed. It is convenient to line up a titration plot with the log concentration diagram so that the points can be easily moved from one plot to the other. This is best accomplished by rotating the log concentration plot 90° counterclockwise, as shown in Figure 4.18.

At the beginning of the titration illustrated in Figure 4.18, the solution point fits the proton balance equation $[H_3O^+] = [A^-] + [OH^-]$, and a line is drawn along that

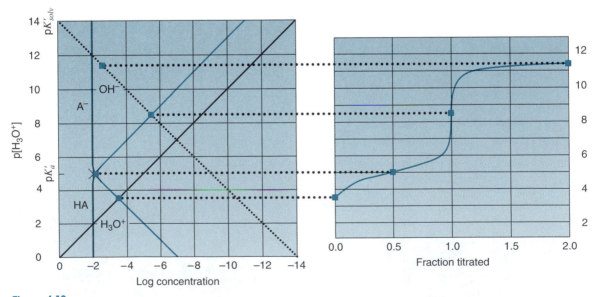

Figure 4.18. Rotating the log concentration plot for an acid–base system allows the $p[H_3O^+]$ values for 0, 0.5, 1, and 2 fraction titrated points to be transferred directly to a titration curve. The correlation between the points on the log plot and the titration curve are better appreciated.

$p[H_3O^+]$ to the $\theta = 0$ point. At the halfway point, $[HA] = [A^-]$. The equivalence point composition is that of a solution of the base form of the titrated acid. The proton balance equation is $[HA] + [H_3O^+] = [OH^-]$. When $\theta = 2$, the solution has been diluted threefold, so the final $[OH^-]$ is one-third the concentration of the titrant. Assuming a titrant concentration equal to that of the analyte, the final $p[OH^-]$ is half a p unit less than the $p[H_3O^+]$ of the titrant. The use of these diagrams allows the feasibility of a titration to be quickly assessed and the approximate $p[H_3O^+]$ value at the equivalence point to be readily determined.

From correlation of the points on the log concentration plot with the points on the titration curve, several generalizations can be made. One is that the break at the equivalence point will decrease with increasing values of K_a' and decreasing initial concentration. Another is that if the system point lies to the right of the OH^- line, there will be no break at all. If the system point lies on or below the H_3O^+ line, the acid acts as a strong acid, and no information regarding the acid K_a' is available. Finally, as the acid concentration approaches 10^{-7} M, the break begins to disappear completely.

The weak base titration curve can be predicted from the logarithmic concentration plots just as in the case of the weak acid. Figure 4.19 is the combined log concentration and titration curve plot for the titration of 10^{-2} M base with a K_a' of 10^{-8}. At the beginning of the titration, the solution is that of the salt of the base form of the acid. The proton balance equation for this condition is $[HA] + [H_3O^+] = [OH^-]$. The approximate $p[H_3O^+]$ for this condition is 10. At the midpoint in the titration, the concentrations of A^- and HA are nearly equal. The $p[H_3O^+]$ at the intersection of those two lines is 8. At the equivalence point, the solution composition is that of the acid form, for which the proton balance equation is $[H_3O^+] = [A^-] + [OH^-]$, and the $p[H_3O^+]$ is 5. At $\theta = 2$, the excess 0.01 M titrant has been diluted threefold, so the $p[H_3O^+]$ is about 2.5.

Several generalizations can be made from correlation of the points on the log concentration plot with the points on the titration curve. The break at the equivalence point will decrease with decreasing values of K_a' and decreasing initial concentration. If the system point lies below the H_3O^+ line, there will be no break. If the system point lies on or to the right of the OH^- line, the base acts as a strong base, and no information regarding the base K_a' is available. The break will disappear completely as the base concentration approaches 10^{-7} M.

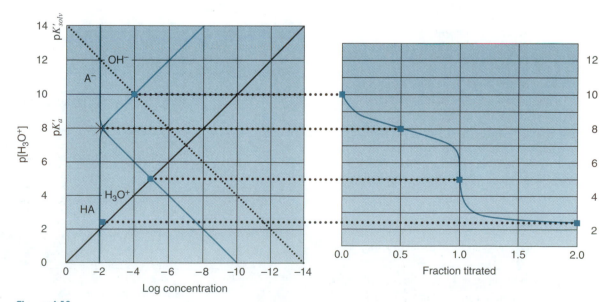

Figure 4.19. Base titration curves can also be predicted with the use of the log concentration plot.

Study Questions, Section E

31. Compare Figures 4.13 and 4.14. What are the principal differences between the log alpha plot and the log concentration plot?

32. Writing the correct proton balance equation is the key to the use of log concentration plots to solve problems in acid–base equilibria. $NaCH_3COO$, the salt of the conjugate base of acetic acid, is dissolved in water to make a 0.01 M solution. Write the proton balance equation for the solution. (Hint: The salt dissolves completely to form Na^+ and CH_3COO^-. The CH_3COO^- then reacts with the water.)

33. Write proton balance equations for these solutions:

    A. 0.01 M HNO_2

    B. 0.01 M $NaNO_2$

34. Use the spreadsheet plotting program to obtain the log concentration plot for a 3×10^{-3} M solution of hypochlorous acid.

35. Use the spreadsheet plotting program to solve the following problems:

    A. What is the $p[H_3O^+]$ of a 0.01 M solution of CH_3COONa?

    B. What is the $p[H_3O^+]$ of a 0.01 M solution of CH_3COOH?

36. Use the spreadsheet plotting program to find the $p[H_3O^+]$ of a 0.0010 M solution of phenol. Use 10.0 as the pK_a'.

37. If it is desired to have a break of 4 pH units between the 50% and 200% titration points for a titration, what are the minimum analyte concentrations for which this would be true for the systems in Figures 4.18 and 4.19?

38. Using a log concentration plot, predict the titration curve for 0.100 M formic acid, HF. The pK_a' of formic acid is 3.74.

39. Use a log concentration plot to predict the titration curve for a titration of 1.00×10^{-4} M arginine with a strong acid of the same concentration. Arginine reacts with water to form $ArgH^+$. The pK_a' for $ArgH^+$ is 9.0. Compare your answer with that obtained in Question 30.

Answers to Study Questions, Section E

31. The differences between the log alpha plot and the log concentration plot are as follows: in the latter, the Y-axis is concentration rather than α, the actual concentrations of the species are plotted, and the lines for H_3O^+ and OH^- are included.

32. The $NaCH_3COO$ dissociates in water to form Na^+ and CH_3COO^-. The CH_3COO^- reacts with water as follows:

$$CH_3COO^- + H_2O \rightleftharpoons CH_3COOH + OH^-$$

Also occurring in solution is the autoprotolysis of water:

$$H_2O \rightleftharpoons H_3O^+ + OH^-$$

The proton balance equation is the sum of everything in solution less protonated than the starting material equated to everything in the solution more protonated than the starting material.

The starting materials are H_2O and CH_3COO^-. Species more protonated are H_3O^+ and CH_3COOH.

Species less protonated are OH^-. The proton balance equation is

$$[H_3O^+] + [CH_3COOH] = [OH^-]$$

Note that CH_3COO^-, the starting material, does not appear in this equation.

33. The starting materials for part A are H_2O and HNO_2. All the species in the solution are H_2O, H_3O^+, NO_2^-, and

OH^-. Write an equation with all the species more protonated than the starting materials on the left and those less protonated than the starting materials on the right.

A. $[H_3O^+] = [NO_2^-] + [OH^-]$

B. $[HNO_2] + [H_3O^+] = [OH^-]$

34. See the plot.

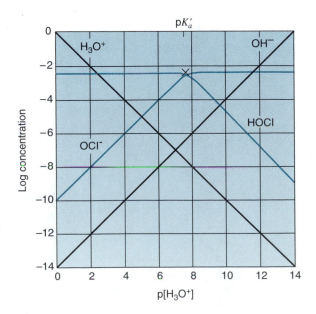

35. The log concentration diagram can be used to solve these problems.

A. The proton balance equation is

$$[H_3O^+] + [CH_3COOH] = [OH^-]$$

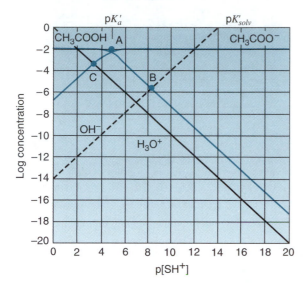

The solution point is found by starting on the line representing OH^- (since it is alone on its side of the equation) and following it until it intersects with that of CH_3COOH (point B). Here the $[H_3O^+]$ is negligible, so the $p[H_3O^+]$ at this point is the $p[H_3O^+]$ of the solution. The pH is ~8.4.

B. The proton balance equation is

$$[H_3O^+] = [CH_3COO^-] + [OH^-]$$

The solution is found by starting on the line representing H_3O^+ (since it is alone on its side of the equation) and following it until it intersects with that of CH_3COO^- (point C). Here the $[OH^-]$ is negligible, so the $p[H_3O^+]$ at this point is the $p[H_3O^+]$ of the solution. The $p[H_3O^+]$ is ~3.3.

36. A log concentration plot is draw using the point (10.0, −3.0) as the system point. [Remember, log(0.0010) = −3.0.] The proton balance equation for phenol (abbreviated PhOH) in water is

$$[H_3O^+] = [PhO^-] + [OH^-]$$

Following the H_3O^+ line, we intersect the PhO^- line at point A. However, a check of the validity of the proton balance equation for this point reveals that while $[PhO^-]$ > $[OH^-]$, the $[OH^-]$ is one-tenth the value of $[PhO^-]$. An approximate solution can still be found that reduces the 10% error. This is done by considering where, on the H_3O^+ line, we would need to be for $[H_3O^+]$ to equal

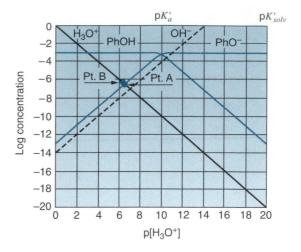

the sum of $[PhO^-]$ and $[OH^-]$, as the proton balance equation states. This will be true at the point where $[H_3O^+]$ is about 10% higher than it is at point A. Thus, the adjusted solution point is point B. The $p[H_3O^+]$ at this point is ~6.4.

37. The 50% $p[H_3O^+]$ of the system in Figure 4.18 will remain at approximately $p[H_3O^+]$ = 5 down to an analyte concentration of just over 10^{-5} M. The $p[H_3O^+]$ at the 200% point will decrease along the OH^- line as lower concentrations of titrant are used. The minimum concentration would then be that at which the OH^- line crosses the line for $p[H_3O^+]$ = 5 + 4 = 9. This concentration is approximately 10^{-5} M. When dilution from the titrant is considered, the initial analyte concentration would have to be at least 2×10^{-5} M.

In the case of the system in Figure 4.19, the 50% titration pH will remain at $p[H_3O^+]$ = 8 to an analyte concentration just above 10^{-6} M. The minimum concentration of analyte would be that which would put the 200% titration point at $p[H_3O^+]$ = 8 − 4 = 4. This would be true for an analyte concentration of 10^{-4} M if dilution were ignored. Thus, a more exact answer would be 2×10^{-4} M.

38. The log concentration diagram is drawn first with a blank titration curve adjacent. The solution points for several conditions along the titration curve are found. First, at the 0% titrated point, the solution is HF. From the proton balance equation for this condition, the solution point is at the intersection of the H_3O^+ line and the F^- line. The $p[H_3O^+]$ for this point is transferred to the 0% titrated point on the titration curve plot. At $\theta = 0.5$, the HF and F^- concentrations are equal. At the equivalence point, the condition is that of a solution of F^-, and the solution point is the intersection of the OH^- and HF lines. When $\theta = 2.0$, the OH^- concentration is half the initial concentration of HF. These points are then joined by the familiar sigmoid curve as shown.

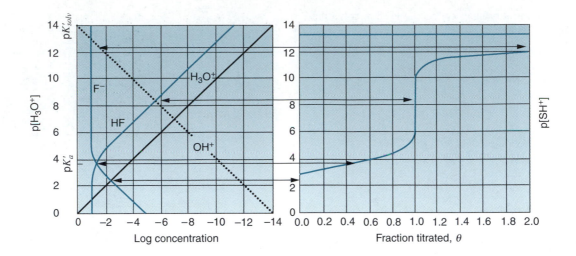

39. The process is the same as that in Question 38, and the resulting curve is shown. The proton balance equations used for the 0, 50, and 100% titrated points are $[ArgH^+] + [H_3O^+] = [OH^-]$, $[ArgH^+] = [Arg]$, $[H_3O^+] = [OH^-] + [Arg]$. In this last case, some approximation is necessary because the $[OH^-]$ is not negligible. The plot confirms the $p[H_3O^+]$ values of 9.5, 9.0, 6.7, and 4.5 obtained in Question 30 for the 0, 50, 100, and 200% titration points for this system.

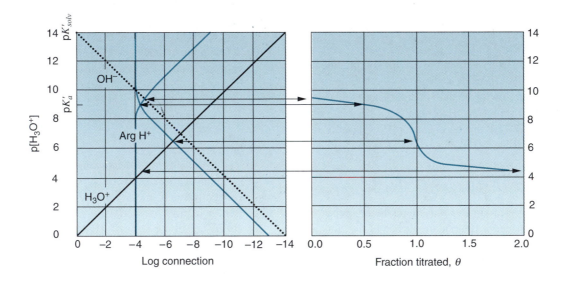

F. Quantitation in Polyprotic Systems

Many species have several degrees of protonation. These species are said to be **polyprotic**. We have already observed that water is such a system in which the neutral water can either gain or lose a proton, making a total system with three degrees of protonation (H_3O^+, H_2O, and OH^-). Other examples are phthalic acid, phosphoric acid, and alanine. Phthalic acid has two carboxylic acid groups and can donate two protons to basic reactants. Such an acid is called **diprotic** or **dibasic**. The proton loss undergone by a polyprotic acid is generally treated as successive losses of single protons. Thus (abbreviating the phthalate dianion as Pth^{2-}),

$$H_2Pth + H_2O \rightleftharpoons H_3O^+ + HPth^-$$
$$HPth^- + H_2O \rightleftharpoons H_3O^+ + Pth^{2-}$$

The equilibrium constants for these two reactions are

$$K'_{a1} = \frac{[H_3O^+][HPth^-]}{[H_2Pth]} \qquad \text{and} \qquad K'_{a2} = \frac{[H_3O^+][Pth^{2-}]}{[HPth^-]} \qquad 4.42$$

where the subscript 1 generally refers to the first proton loss, 2 refers to the second, and so on. In such systems, a species that is the base form of one conjugate acid–base pair may be the acid form for another. Such systems appear at several places in Table 3.1 and Appendix B. There is a separate entry (and separate K'_a) for each conjugate pair.

The general reactions and equilibrium constants for a triprotic acid such as phosphoric acid are

$$H_3A \rightleftharpoons H_3O^+ + H_2A^-$$
$$H_2A^- \rightleftharpoons H_3O^+ + HA^{2-} \qquad 4.43$$
$$HA^{2-} \rightleftharpoons H_3O^+ + A^{3-}$$

and

$$K'_{a1} = \frac{[H_3O^+][H_2A^-]}{[H_3A]}$$

$$K'_{a2} = \frac{[H_3O^+][HA^{2-}]}{[H_2A^-]} \qquad 4.44$$

$$K'_{a3} = \frac{[H_3O^+][A^{3-}]}{[HA^{2-}]}$$

Although the change in charge of the acid–base forms shown in Equations 4.43 and 4.44 are general, the absolute value of the charge is not followed in every case. For instance, for alanine, the most protonated form is the protonated neutral molecule, which we will abbreviate $AlaH^+$. For this species, the reactions are

$$AlaH^+ + H_2O \rightleftharpoons H_3O^+ + Ala$$
$$Ala + H_2O \rightleftharpoons H_3O^+ + (Ala - H)^-$$

ALPHA PLOTS FOR POLYPROTIC SYSTEMS

The fraction of the total system concentration in each form can be derived from the equilibrium constant equations and the mass balance equation. The equilibrium constant equations follow the form given in Equations 4.44. For the diprotic acid H_2A, the mass balance equation is $[H_2A] + [HA^-] + [A^{2-}] = C_{HA}$ where C_{HA} is the total concentration of all species. The two expressions for K'_a and the mass balance equation are solved to give the three alpha equations shown.

$$\alpha_{H_2A} = \frac{[H_3O^+]^2}{[H_3O^+]^2 + K'_{a1}[H_3O^+] + K'_{a1}K'_{a2}} \qquad 4.45$$

$$\alpha_{HA^-} = \frac{K'_{a1}[H_3O^+]}{[H_3O^+]^2 + K'_{a1}[H_3O^+] + K'_{a1}K'_{a2}}$$

$$\alpha_{A^{2-}} = \frac{K'_{a1}K'_{a2}}{[H_3O^+]^2 + K'_{a1}[H_3O^+] + K'_{a1}K'_{a2}}$$

The linear alpha plot shown in Figure 4.20 is for phthalic acid. This acid has pK'_a values of 2.95 and 5.41. As we might expect, at very acidic $p[H_3O^+]$'s, the system is entirely in the most acidic form, H_2A. Similarly, at very basic $p[H_3O^+]$ values, the system is in the most basic form, A^{2-}. At a $p[H_3O^+]$ equal to pK'_{a1}, the fraction of the system in the HA^- and H_2A forms are almost equal, as expected from the equilibrium

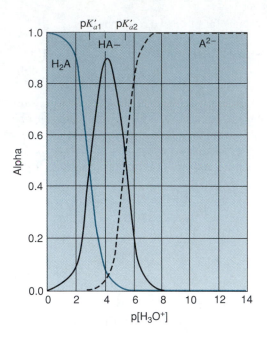

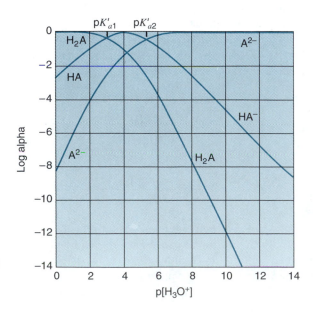

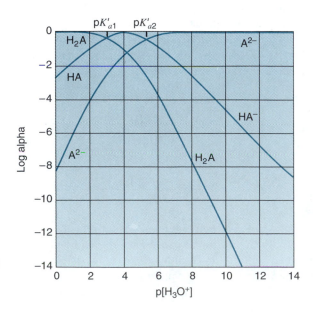**Figure 4.20.** The alpha plot for phthalic acid gives the fraction of each form as a function of $p[H_3O^+]$. The $p[H_3O^+]$ values at which two forms are at equal concentrations are approximately equal to the pK_a' values.

expression for K_{a1}'. Another crossover point occurs at the $p[H_3O^+]$ equal to pK_{a2}'. Here the fractions in the HA^- and A^{2-} forms are approximately equal. Between the two crossover points, the fraction of the system in the HA^- form has a maximum. This maximum will occur at the $p[H_3O^+]$ equal to the average of the $p[H_3O^+]$ values at the two crossover points (in other words, exactly between them). Because the two pK_a' values are relatively close together, the fraction of the system in the HA^- form never reaches 1.0. This is seen quite clearly in Figure 4.20.

The plot of $\log \alpha$ is better for obtaining the fractional values of the system components when these fractions are less than 1%. The log alpha plot for phthalic acid is shown in Figure 4.21. At first glance, comparison of this plot with that of the monoprotic acid shown in Figure 4.13 suggests that it is simply the two plots for the individual K_a' values superimposed. This is very nearly true. The system points are where $p[H_3O^+]$ is equal to each pK_a', and the slope of the alpha lines leading away from those points is 1 (a $\log \alpha$ change of 1 for each unit change in $p[H_3O^+]$). However, where the

Figure 4.21. The log alpha plot is better for assessing the fractions of the several forms when the fractions are very small.

alpha line crosses under the next system point, the slope changes to two (a log α change of 2 for each unit change in $p[H_3O^+]$).

LOGARITHMIC CONCENTRATION DIAGRAMS

The construction of a log concentration plot for a polyprotic system follows simple rules similar to those for the monoprotic system. We begin with a log–log plot that contains the lines for H_3O^+ and OH^-. Then the system points are placed where the line for log C_{HA} intersects the values of $p[H_3O^+]$ that are equal to each pK'_a. From the system points, the log concentration lines fall away at a slope of 1. However, when a log concentration line passes under another system point, the slope of that line increases by 1. The log concentration diagram for a 0.10 M solution of the phosphate system is shown in Figure 4.22. (For phosphate, $pK'_{a1} \approx 2.15$, $pK'_{a2} \approx 7.20$, and $pK'_{a3} \approx 12.35$.) The transitions of the dominant forms are seen to progress from the most acidic (H_3PO_4) at low values of $p[H_3O^+]$ to the most basic (PO_4^{3-}) at high $p[H_3O^+]$'s.

It is in the polyprotic systems that the log concentration plots show one of their greatest virtues. Such systems are more difficult than monoprotic systems to solve algebraically, but the determination of all species concentrations from the log plot is not more complex. One simply writes and applies the proton balance equation for the solution condition. For example, if a solution is made by dissolving NaH_2PO_4 to a concentration of 0.10 M, the proton balance equation is

$$[H_3O^+] + [H_3PO_4] = [HPO_4^{2-}] + 2[PO_4^{3-}] + [OH^-] \qquad 4.46$$

Note that the PO_4^{3-} concentration is multiplied by 2 because two protons were lost from the starting material to make it. The $p[H_3O^+]$ of this solution is found by following the H_3O^+ line from its highest value. The first intersection is with the line for H_3PO_4, which is on the same side of Equation 4.46 as H_3O^+. This means that from that point on $[H_3PO_4]$ is greater than $[H_3O^+]$. From this point the line for H_3PO_4 will be followed. The next intersection is with the line for HPO_4^{2-}. This intersection becomes the tentative solution point. The concentrations of the other species on the right-hand side of Equation 4.46 are truly negligible at this point, but $[H_3O^+]$ is seen to be about one-tenth of $[H_3PO_4]$ and so is not completely negligible. The actual $p[H_3O^+]$ is then a bit higher than that of the intersection, say about 4.7. The intersection of the line for $p[H_3O^+] = 4.7$ with all the other lines gives the concentrations of all the species in solution.

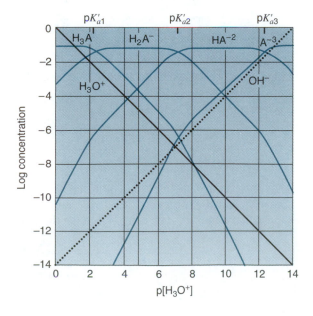

⊙ **Figure 4.22.** The log concentration plot for a polyprotic acid is similar to the log alpha plot, except that the units are now concentration and the lines for H_3O^+ and OH^- are included.

Polyprotic proton balance equations

Follow the procedure for the monoprotic system. There will be more species to list. For example, for a triprotic system, the list of species is

H_3A, H_2A^-, HA^{2-}, A^{3-}, H_3O^+, H_2O, OH^-

Cross off the starting materials (H_3A and H_2O for a solution of the most protonated form).

Arrange the remaining species on either side of the equals sign depending on whether they are more protonated or less protonated than the starting materials.

If they are more than one proton different from the starting material, use the number of protons gained or lost as a coefficient.

For H_3A in water, the equation is

$$[H_3O^-] = [H_2A^-] + 2[HA^{2-}] + 3[A^{3-}] + [OH^-]$$

The exact $p[H_3O^+]$ for the solution of NaH_2PO_4 can now be found more simply because one can see from the log concentration diagram which equilibria and which species are significant. The relevant reactions and equilibrium constants are those that contain the concentrations of the significant species (H_3O^+, H_3PO_4, $H_2PO_4^-$, and HPO_4^{2-}). These will be considered in terms of the reactions of the dissolved NaH_2PO_4.

$$H_2A^- \rightleftharpoons HA^{2-} + H_3O^+ \qquad K'_{a2} = \frac{[H_3O^+][HPO_4^{2-}]}{[H_2PO_4^-]}$$

$$H_2A^- + H_3O^+ \rightleftharpoons H_3A + H_2O \qquad \frac{1}{K'_{a1}} = \frac{[H_3PO_4]}{[H_3O^+][H_2PO_4^-]}$$

$$H_2O \rightleftharpoons H_3O^+ + OH^- \qquad K'_w = [H_3O^+][OH^-]$$

The proton balance equation (neglecting the PO_4^{3-}) is

$$[H_3O^+] + [H_3PO_4] = [HPO_4^{2-}] + [OH^-]$$

The equilibrium expressions are used to produce an equation that contains only $[H_2PO_4^-]$, $[H_3O^+]$, and constants:

$$[H_3O^+] = K'_{a2}\frac{[H_2PO_4^-]}{[H_3O^+]} + \frac{K'_w}{[H_3O^+]} - \frac{[H_3O^+][H_2PO_4^-]}{K'_{a1}}$$

Solving this expression for $[H_3O^+]$,

$$[H_3O^+] = \sqrt{\frac{K'_{a1}(K'_{a2}[H_2PO_4^-] + K'_w)}{K'_{a1} + [H_2PO_4^-]}}$$

From the log concentration plot (Figure 4.22), we can see that at the solution point, $[H_2PO_4^-] = C_{HA}$. This observation allows the substitution of 0.10 M for $[H_2PO_4^-]$ in the above equation. The calculated value of $[H_3O^+]$ is then 2.05×10^{-5} M ($p[H_3O^+] = 4.69$).

TITRATION CURVES IN POLYPROTIC SYSTEMS

Titration curves for polyprotic systems can be derived in the same way as those for monoprotic systems.[6] We begin with the charge balance equation for the titration of a dibasic acid with NaOH.

$$[H_3O^+] + [Na^+] - 2[A^{2-}] - [HA^-] - [OH^-] = 0$$

The Na^+ concentration is $C_i \theta/(1 + \theta)$ as in Equation 4.29. The acid system concentration C_{HA} is $C_i/(1 + \theta)$. This time, we will use the alpha equations (Equations 4.45) times C_{HA} for the concentrations of A^{2-} and HA^-. Thus,

$$[H_3O^+] + C_i\frac{\theta}{1 + \theta} - \frac{C_i}{1 + \theta}\left(\frac{2K'_{a1}K'_{a2} + K'_{a1}[H_3O^+]}{[H_3O^+]^2 + K'_{a1}[H_3O^+] + K'_{a1}K'_{a2}}\right) - \frac{K'_w}{[H_3O^+]} = 0 \quad 4.47$$

When Equation 4.47 is solved for θ, the titration curve can be plotted. Curves for the titration of 10^{-1} M, 10^{-3} M, and 10^{-5} M phthalic acid are shown in Figure 4.23. Recall that the pK'_a values for phthalic acid are 2.95 and 5.41. In the titration of the most con-

---

[6]Complete derivations of the titration curves for polyprotic acids and bases are found on the CD. Spreadsheets for automatic plotting of polyprotic titration curves are also on the CD.

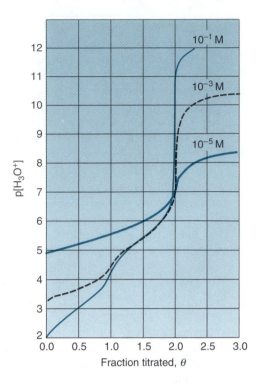

Figure 4.23. The titration curve for a diprotic acid such as phthalic acid will generally have two breaks, one for each of the protons titrated. At low concentrations, these breaks may become indistinct.

centrated solution, two breaks in the curve are clearly seen. One is at the first equivalence point (when enough NaOH has been added to convert all the H_2Pth to $HPth^-$ and $\theta = 1$); the second is at $\theta = 2$ when enough NaOH has been added to convert all the $HPth^-$ to Pth^{2-}. The second break is much steeper than the first. The size of the breaks and the $p[H_3O^+]$ values at the inflection points can be readily estimated from a log concentration plot.

A combination of the titration curve and log concentration diagram for the titration of 0.10 M phthalic acid is shown in Figure 4.24. The initial $p[H_3O^+]$ is obtained from the proton balance equation for a solution of H_2Pth ($[H_3O^+] = [HPth^-] + 2[Pth^{2-}] + [OH^-]$). The solution to this equation is seen at about $p[H_3O^+] = 2$ where $[H_3O^+] = [HPth^-]$. When the H_2Pth is half-titrated to $HPth^-$, $[H_2Pth] = [HPth^-]$ and $p[H_3O^+] = 2.95$. When the H_2Pth is fully titrated to $HPth^-$, the proton

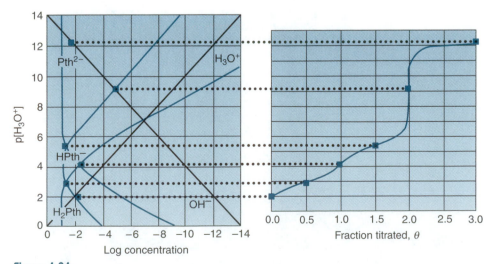

Figure 4.24. The log concentration plot is again very helpful in assessing the magnitude of breaks in the titration curve.

balance equation is the same as if the salt NaHPth had been dissolved. Thus $[H_3O^+]$ + $[H_2Pth]$ = $[Pth^{2-}]$ + $[OH^-]$. The solution point is at the intersection of the H_2Pth and Pth^{2-} lines. Here the $p[H_3O^+]$ is 4.18 (the midpoint between the two K_a' values). The reason for the shallow break at the first equivalence point is clear; the $p[H_3O^+]$'s at 0, 0.5, and 1.0 fraction titrated are only separated by about 1 $p[H_3O^+]$ unit.

The second part of the titration curve is more pronounced. The $p[H_3O^+]$ at the 1.5 fraction titrated point is where $[HPth^-]$ = $[Pth^{2-}]$. The $p[H_3O^+]$ at the second equivalence point is that of a solution of Na_2Pth, for which the proton balance equation is $[H_3O^+]$ + $2[H_2Pth]$ + $[HPth^-]$ = $[OH^-]$. Taking dilution into account, the OH^- concentration at fraction titrated = 3 is the original $C_{H_2A}/4$.

From this overall titration curve, we can see that titration to the first equivalence point could not be accomplished precisely with an indicator, though the second equivalence point could be found quite readily. The shapes of the titrations curves for phthalic acid in more dilute solutions can also be anticipated from the log concentration diagram. Imagine the phthalate system lines shifted to the right to a log C_T of −3. At this concentration, the system point for pK_{a1}' will be very close to the H_3O^+ line, substantially raising the $p[H_3O^+]$ at both the 0.5 and 1.0 fraction titrated points. This reduces the break still further, as seen in Figure 4.23. At 10^{-5} M phthalic acid, H_2Pth behaves as a strong acid, reacting completely with the solvent. Even the second system point at pK_{a2}' is near the H_3O^+ line, so that the reaction of $HPth^-$ with water is also significant. The resulting titration curve shows no break at the first equivalence point. Only the atypical shape of the titration curve gives a hint that both protons of a weak dibasic acid are being titrated. Without some care, it would be easy to mistake this titration for a monoprotic acid with a pK_a' of about 5.5 and at twice the concentration of the H_2Pth actually being titrated.

TITRATIONS OF ACID MIXTURES

Mixtures of acids or bases can sometimes be titrated. The success of this depends on the presence of distinct break points for each of the acid or base species present. A common situation might be the mixture of a dibasic acid and its salt, such as carbonic acid and $NaHCO_3$. For the carbonate system, the pK_a' values are 6.35 and 10.33. The log concentration plot for C_{HA} = 10^{-2} M is shown in Figure 4.25. A mixture of compa-

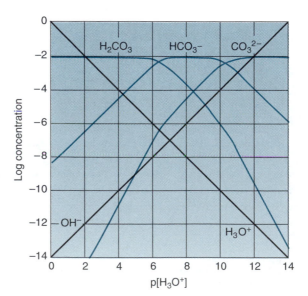

Figure 4.25. This log concentration plot for the carbonic acid system is useful in assessing the feasibility of titrations to determine the amount of each form of the acid.

rable concentrations of H_2CO_3 and $NaHCO_3$ will have an initial $p[H_3O^+]$ between 4 and 8. One way to analyze this mixture would be to divide the unknown sample into several parts called **aliquots**. The mixture in one part could be titrated with a strong acid to convert all the HCO_3^- to H_2CO_3. The mixture in another part could be titrated with a strong base to CO_3^{2-}. In the latter titration the initial H_2CO_3 would lose two protons and the initial HCO_3^- one proton. The amount of base required to titrate the H_2CO_3 could be subtracted from the total end point volume to provide the amount of titrant used on the HCO_3^-. Many mixtures of acids or bases and their salts can be analyzed by this approach. (See Example 4.6.)

Another common situation is a mixture of a strong acid or base with a weak acid or base. An example could be a mixture of NaOH and sodium acetate ($pK_a' = 4.76$). This mixture could be titrated with a strong acid, which would react first with the OH^- and then with the acetate anion. Assuming sufficient initial concentrations of each component, there will be a clear break at the equivalence point for the OH^- titration and another for the titration of acetate anion to acetic acid. The feasibility of this example is aided greatly by the fact that there is a substantial difference in the base strengths of OH^- and acetate anion. Again, the suitability of such titrations can be readily assessed using log concentration diagrams. The initial $p[H_3O^+]$ would be principally established by the concentration of OH^-. At the equivalence point of the OH^- titration, the composition of the solution would be that of the acetate anion. If the acetate system concentration were 10^{-2} M, this would be at a $p[H_3O^+]$ of about 8.3. (How do we know this?)[7] Assuming the titration started at about a $p[H_3O^+]$ of 12 or so, the equivalence point detection for the OH^- titration is not difficult. Continuing on to titrate the acetate ion, the midpoint $p[H_3O^+]$ for this titration will occur at $p[H_3O^+] = 4.76$, the equivalence point at about $p[H_3O^+] = 3.4$, and the 200% point at about $p[H_3O^+] = 2$. These changes will enable a very practical titration of the mixture. Of course, the titrant volume related to the amount of acetate present is the difference between the first and second equivalence point volumes.

Example 4.6

A 20.00 mL mixture of H_2CO_3 and $NaHCO_3$ was titrated with a 0.0500 M solution of strong acid. This required 24.73 mL to reach the end point at $p[H_3O^-] \approx 3.5$.

In this titration the HCO_3^- was titrated to H_2CO_3. The amount of $NaHCO_3$ in the sample is thus 0.0500 M $\times$ 24.73 mL = 1.237 mmol. The concentration is $1.237/20.00 = 0.0618$ M.

Another 20.00 mL aliquot was titrated with a 0.1000 M solution of a strong base. This required 34.69 mL to reach an end point at $p[H_3O^+] \approx 11.7$. In this titration, both the H_2CO_3 and the $NaHCO_3$ are titrated to CO_3^{2-}. The total amount of titrant used was 34.69 $\times$ 0.1000 = 3.469 mmol. Of this, 1.237 mmol was required to titrate the $NaHCO_3$. Thus, $3.469 - 1.237 = 2.232$ mmol was used to titrate the H_2CO_3. The amount of $H_2CO_3 = 2.232/2 = 1.116$ mmol. The concentration is $1.116/20.00 = 0.0558$ M.

Study Questions, Section F

40. Among the entries in the table of K_a° values in Appendix B, which of the acids with names beginning with "A" are polyprotic?

41. How many species and conjugate acid–base pairs are there in a system that has three equilibrium constants?

42. The alpha equations for a diprotic acid system are given in Equations 4.45. Write the alpha equations for a triprotic acid system. You do not need to derive them. They continue the same trend developed between Equations 4.6 and 4.7 and Equations 4.45. Hint: The first equation is

$$a_{H_3A} = \frac{[H_3O^+]^3}{[H_3O^+]^3 + K_{a1}'[H_3O^+]^2 + K_{a1}'K_{a2}'[H_3O^+] + K_{a1}'K_{a2}'K_{a3}'}$$

43. Draw a log alpha plot for glutamine, HG. The pK_a' values for HGH^+ and HG are approximately 2.17 and 9.01. You can use the spreadsheet plot for your answer.

44. A. Draw the log plot for the glutamic acid (H_2Gl) system at 1.0×10^{-3} M. The pK_a' values for glutamic acid are approximately 2.23, 4.42, and 9.95. The lowest pK_a' is for H_2GlH^+.

   B. Estimate the $p[H_3O^+]$ for a solution of 1.0×10^{-3} M H_2Gl.

   C. Calculate the exact solution to part B.

45. From the log plot made for Question 44, predict the success of the titration of each of the species in the glutamic acid system at a concentration of 1.0×10^{-3} M with a strong acid or strong base.

46. What are the requirements for the separate quantitation of a mixture of two or more acids by titration?

47. It is quite common for relatively concentrated solutions of NaOH to be contaminated with Na_2CO_3. If this standard is

---

[7]See Figure 4.15 for the log plot of a 0.010 M solution of an acid with the nearly same pK_a' as acetic acid. The solution of NaA has a $p[H_3O^+]$ of about 8.3.

used to titrate a strong acid, there is more than the expected difference between the phenolphthalein and methyl orange end points. At the phenolphthalein end point, the carbonate is in the HCO_3^- form, and at the methyl orange end point, the carbonate is in the carbonic acid form

(H$_2$CO$_3$). The end point volumes for a titration of 30.00 mL of 0.1000 M HCl with a nominal 0.1000 M NaOH standard were 30.32 and 30.94 mL for methyl orange and phenolphthalein, respectively. Calculate the concentrations of NaOH and Na$_2$CO$_3$ in the standard titrant.

Answers to Study Questions, Section F

40. The polyprotic acids are those with more than one entry in the table. The polyprotic acids with names beginning with "A" are alanine, arginine, ascorbic acid, asparagine, and aspartic acid.

41. The system will have one conjugate pair for each equilibrium constant. There will be one more species than there are conjugate pairs, so that a system with three equilibrium constants will have four species and three conjugate pairs.

42. For the triprotic system,

$$\alpha_{H_3A} = \frac{[H_3O^+]^3}{[H_3O^+]^3 + K'_{a1}[H_3O^+]^2 + K'_{a1}K'_{a2}[H_3O^+] + K'_{a1}K'_{a2}K'_{a3}}$$

$$\alpha_{H_2A^-} = \frac{K'_{a1}[H_3O^+]^2}{[H_3O^+]^3 + K'_{a1}[H_3O^+]^2 + K'_{a1}K'_{a2}[H_3O^+] + K'_{a1}K'_{a2}K'_{a3}}$$

$$\alpha_{HA^{2-}} = \frac{K'_{a1}K'_{a2}[H_3O^+]^{2-}}{[H_3O^+]^3 + K'_{a1}[H_3O^+]^2 + K'_{a1}K'_{a2}[H_3O^+] + K'_{a1}K'_{a2}K'_{a3}}$$

$$\alpha_{A^{3-}} = \frac{K'_{a1}K'_{a2}K'_{a3}}{[H_3O^+]^3 + K'_{a1}[H_3O^+]^2 + K'_{a1}K'_{a2}[H_3O^+] + K'_{a1}K'_{a2}K'_{a3}}$$

43. Place an $\times$ at the intersections of the pK'_a values and log α = 0. From each $\times$ draw lines with unit slope to the left, changing to a slope of 2 when passing under the other $\times$. Draw lines along the right axis skipping the region around each $\times$. Join the sloped lines with the vertical lines by a curve that goes 0.3 log units underneath each $\times$. Label the

lines according to the species they represent. The most basic species has the highest α at the highest p[H$_3$O$^+$], and so on.

44. A. Follow the instructions for drawing the log concentration diagram, placing the system points at the intersections of the pK'_a values and -3.0.

B. For the starting materials of H$_2$Gl and water, the proton balance equation is

$$[H_2GlH^+] + [H_3O^+] = [HGl^-] + 2[Gl^-] + [OH^-].$$

The tentative solution point is at the intersection of the [H$_3$O$^+$] and [HGl$^-$] lines. This is at a p[H$_3$O$^+$] of 3.7. However, at this point, the value of [H$_2$GlH$^+$] is not negligible, so the estimate of p[H$_3$O$^+$] will have to be reduced somewhat, say to 3.8.

C. For the exact solution, we can see that [Gl^{2-}] is negligible and that, therefore, the most basic equilibrium need not be considered. We can also see that the value of [H$_2$GlH$^+$] is one-tenth that of [H$_3$O$^+$]. Now the second pK'_a can be used to find the p[H$_3$O$^+$].

$$K'_{a2} = \frac{[H_3O^+][HGl^-]}{[H_2Gl]}$$

$$[H_2Gl] = \frac{[H_3O^+][HGl^-]}{K'_{a2}}$$

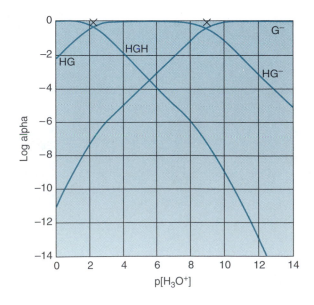

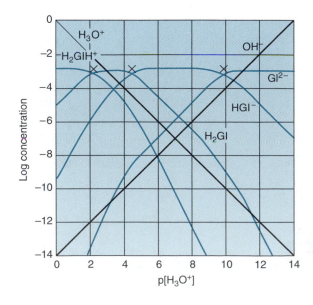

From the proton balance equation, neglecting insignificant concentrations, we have

$$[H_2GlH^+] + [H_3O^+] = [HGl^-] = 1.1[H_3O^+]$$

We will now substitute in the mass balance equation:

$$[H_2GlH^+] + [H_2Gl] + [HGl^-] = 1.0 \times 10^{-3} \text{ M}$$

$$0.1[H_3O^+] + \frac{1.1[H_3O^+]^2}{K'_{a2}} + 1.1[H_3O^+] = 1.0 \times 10^{-3} \text{ M}$$

This equation can be solved by successive approximations starting with the value of $[H_3O^+] = 1.59 \times 10^{-4}$ M from the graphical estimate of $p[H_3O^+] = 3.8$. This gives a sum that is slightly less than 1.0×10^{-3}, so the $[H_3O^+]$ is increased to 1.66×10^{-4} M, which results in a sum within 4 ppt of the right value. Thus $p[H_3O^+] = 3.78$.

45. Titration of the most basic species, Gl^{2-}, with a strong acid will have an initial $p[H_3O^+]$ of about 10.5, a midpoint $p[H_3O^+]$ of about 9.9, and an equivalence point $p[H_3O^+]$ of about 7.2. Since the midpoint in the titration of the HGl^- produced is at a $p[H_3O^+]$ of about 4.4, this titration should work well.

    In the titration of the HGl^- species, the initial $p[H_3O^+]$ is about 7.0. If HGl^- is titrated with an acid, the midpoint $p[H_3O^+]$ would be about 4.4, but there will be no significant break to the 200% point at $p[H_3O^+] = 3.7$. No break is observed for the formation of H_2GlH^+ because, at 10^{-3} M, it is completely reacted with the water.

Essentially the same problem exists if the HGl^- is titrated with a base. The titration of H_2GlH^+ with a strong base will begin at $p[H_3O^+] = 3.0$, change almost imperceptibly at the midpoint, and have an equivalence point $p[H_3O^+]$ of about 3.8. This break is too small. The titration of H_2Gl, on the other hand, will begin at about 3.8, have a midpoint at about 7, and rise to about 9.5 at the 200% point. This titration will work well. Note that the H_2GlH^+ could be titrated with a strong base if there were no H_2Gl in the solution and the titration were taken to the H_2Gl equivalence point (requiring 2 mol titrant for each mole of H_2GlH^+).

46. The acids must have significantly different K'_a values and still be strong enough to produce a reasonable break at the weakest acid equivalence point.

47. At the equivalence points, the millimoles of acid and base are equal. If x is the concentration of NaOH and y is the concentration of Na_2CO_3, at the methyl orange end point,

$$30.00 \times 0.1000 = 30.32(x + 2y)$$

and at the phenolphthalein end point,

$$30.00 \times 0.1000 = 30.94(x + y)$$

With these two equations, the values of x and y can be solved. From this, $[Na_2CO_3] = 1.99 \times 10^{-3}$ M and $[NaOH] = 0.950$ M.

G. Other Analyses by Acid—Base Reaction

The areas of chemical analysis are quantitation, detection, identification, and separation. In the above sections, we have seen how acid–base reactivity is useful in quantitative analysis. In this section we will consider how other types of analytical information can be obtained from acid–base reactivity.

IDENTIFICATION BY ACID–BASE REACTIVITY

We have studied how the amount of base required to react with an unknown acid and vice versa can be used to determine the amount of that acid or base. This same technique can be used to learn other facts about the acid or base being titrated. For example, if the weight of the unknown acid or base dissolved in the titrated solution is known, the equivalence point volume will allow us to calculate the molecular weight of the unknown substance. Knowing the molecular weight may be very useful in identifying the substance titrated. If the entire titration curve is obtained, and if a weak acid or base was titrated, the pK'_a of the acid–base conjugate pair can also be determined. This will be known from the $p[H_3O^+]$ at the half-titration point in simple cases or from a fitted parameter in more complex situations. The combination of the pK'_a and the molecular weight can often greatly reduce the number of possibilities in the search for the identity of the unknown substance. When these techniques are used for unknown identification, one must consider that the unknown may be polyprotic and that the unknown acid or base may be an intermediate form (conjugate base of a stronger acid as well as conjugate acid of a stronger base). The shape of the titration curve can often indicate whether the unknown is polyprotic, and the initial $p[H_3O^+]$ is often a clue about whether the unknown is an intermediate form in a polyprotic system.

DETECTION BY ACID–BASE REACTIVITY

The concept of detection is often related to the analysis of trace components in complex mixtures. In this area of analytical application, acid–base reactivity is of limited value. We have seen that the ability to follow the acid–base reaction (at least with a pH meter) becomes severely hampered as the analyte concentration approaches 10^{-6} M, even for strong acids and bases. On the other hand, we have seen that most acids and bases behave as if they were strong (completely reacted with the water) at those low concentrations. Detection by the glass electrode and pH meter are similarly limited to those regions where the effect of the acid or base on the pH exceeds that of the auto-protolysis of the solvent. Interference is another factor that must be considered in trace component detection. An acid interferent at the 10^{-5} M level will not affect the detection of an acid at the 10^{-3} M level, but as the detection limit approaches the concentrations of the interfering compounds, the interference problem increases. Unfortunately, it is exactly at these lower levels that more interfering substances are likely to be found. Since virtually all acids and bases have completely reacted with the water at very low concentrations, they can no longer be distinguished from each other by their K_a''s. In general, detection and identification of an acid or base at a concentration below 10^{-5} M will be based on some other differentiating characteristic of the analyte.

SEPARATION BY ACID–BASE REACTIVITY

The acid–base reactions we have studied do not result in a separation of the analyte from the solution in which it is dissolved. Thus acid–base reactivity, by itself, is not the basis for physical separation of analytes from other components in a mixture. Acidity can, however, have a great effect on solvent extraction, as we shall see in Chapter 11. On the other hand, it is important to note that for the purposes of analysis, separation is not necessary if the analyte has been correctly distinguished from all potentially interfering substances. For example, it would not be necessary to separate a weak acid analyte from a strong acid in the mixture if they can be resolved by analysis of the titration curve. As analysts have been confronted with the need to resolve increasingly complex mixtures, the use of mathematical analysis of all the response data has gained greatly in sophistication and effectiveness.

Practice Questions and Problems

1. Hypochlorous acid, HOCl, has a K_a' in water of 3.0×10^{-8}.

   A. At what value of $p[H_3O^+]$ will the concentrations of the acid and base forms be equal?

   B. What will be the predominate form at a $p[H_3O^+]$ that is +2 units greater than your answer to part A?

   C. Draw a log concentration diagram for a solution that is 0.01 M in HOCl.

   D. Write the proton balance equation for the solution in part C and estimate the $p[H_3O^+]$ of this solution using the log plot.

   E. Write the proton balance equation for a solution of 0.01 M NaOCl and estimate the $p[H_3O^+]$ of this solution using the log plot.

   F. Calculate the exact $p[H_3O^+]$ for the solutions in parts D and E and compare your results with the estimates obtained from the log plots.

   G. Over what concentration range will the reaction between HOCl and H_2O to form H_3O^+ and OCl^- be essentially complete?

2. You now want to find out the exact concentration of an HOCl solution that you expect is about 0.01 M.

   A. Briefly describe the essential steps in preparing to titrate this sample.

   B. What will be the $p[H_3O^+]$ values at the 50, 100, and 200% titrated points? Use the log concentration diagram to obtain your answers, but explain how you got them.

C. Which of the indicators listed in Table 4.1 would you think most suitable for this titration and why?

D. If you desire a change in $p[H_3O^+]$ of at least 3 units between the 50 and 200% titrated points, what is the lowest concentration of HOCl that you can titrate? How did you obtain this answer?

3. Malonic acid, H_2Ma, has K_a' values of 1.4×10^{-3} and 2.0×10^{-6}.

A. Draw a log concentration plot for a 0.10 M solution of malonic acid.

B. Estimate the $p[H_3O^+]$ of the solution of part A.

C. Draw an approximation of a titration curve for the titration of this solution with 0.1 M NaOH. Label the curve with the $p[H_3O^+]$ values at all points for which estimated values are obtainable from the log plot. Comment on the suitability of the two possible equivalence points as end points in the titration.

4. This course is based on the concept that chemical analysis requires the identification and exploitation of a differentiating characteristic. Thus far, the only differentiating characteristic we have studied is acid–base reactivity. For the titration method of obtaining quantitative information about an acidic sample,

A. What is the probe that tests this characteristic?

B. What is the analyte's response to this probe?

C. How is the response measured?

D. How are the measurement data interpreted to reveal the quantity of acid in the sample?

5. A. Draw a log concentration plot for a 0.10 M solution of ascorbic acid (vitamin C). The pK_a' values for ascorbic acid are 4.3 and 11.8. Label all the concentration lines. For labeling, use H_2Asc, $HAsc^-$, and Asc^{2-}.

B. For the solution in part A, write the proton balance equation, mark the solution point on the graph with a B, and estimate the $p[H_3O^+]$.

C. Assume a portion of the solution in part A is to be titrated with 0.10 M NaOH.

(i) Estimate the $p[H_3O^+]$ when the first proton of the ascorbic acid is half-titrated. Mark this point on the plot with the legend 50%.

(ii) Estimate the $p[H_3O^+]$ at the equivalence point for the titration of the first proton. Write the equation you used to find this point. Mark this point on the plot with the legend 100%.

(iii) Comment on the suitability of phenolphthalein as an indicator for this titration. $pK_{ind} = 9.7$.

D. Estimate the degree of change that would occur during the titration of the second proton of the ascorbic acid and comment on the practicality of titrating the second proton of ascorbic acid.

6. The monoprotic acid benzoic acid (C_6H_5COOH) has a K_a' of 6.28×10^{-5}.

A. Draw a log concentration diagram for 10^{-1} M benzoic acid. Clearly label all lines with the species they represent. Label the system point. From the log concentration diagram, determine the pH of a solution with a 1.0×10^{-1} M analytical concentration of benzoic acid. Label the solution point.

B. Using the same log concentration diagram, what is the pH of a solution with 1.0×10^{-1} M analytical concentration of NaC_6H_5COO? ($C_6H_5COO^-$ is the conjugate base of benzoic acid.)

C. Label the solution point.

D. Apply Equation 4.20 with the appropriate approximations to determine the exact pH of the solution in part A.

E. From the log concentration diagram, what must the concentrations of H_3O^+, OH^-, $C_6H_5COO^-$, and C_6H_5COOH be for a solution to have a $p[H_3O^+]$ of 5?

7. Use the log concentration diagram to find the critical points on the titration curve ($p[H_3O^+]$ as a function of θ) for the titration of benzoic acid with NaOH. List the approximate $p[H_3O^+]$'s of the solution at $\theta = 0, 0.5, 1$, and 2. From Table 4.1, what acid–base indicator would work for this titration?

8. Draw a log concentration diagram for 1.0×10^{-2} M arsenic acid (H_3AsO_4), with $K_{a1}' = 5.8 \times 10^{-3}$, $K_{a2}' = 1.1 \times 10^{-7}$, and $K_{a3}' = 3.2 \times 10^{-12}$. Using the log concentration diagram, what will the approximate $p[H_3O^+]$ of a 1.0×10^{-2} M solution of Na_2HAsO_4 be?

9. What are the advantages of using an entire titration curve (collected with a pH meter) to determine the equivalence point of a titration rather than an indicator? What are the advantages of using an indicator?

10. In an acid–base titration, what factors affect the accuracy of the result? What factors affect the precision?

11. What two factors in the titration of a weak acid with a strong base have an impact on the magnitude of the break at the equivalence point? How do changes in each of these factors affect the magnitude of this break?

12. In Table 4.2, rules of thumb for system point location, explain why $pH = (K_w')^{1/2}$ if the system point falls below the intersection of the H_3O^+ and OH^- lines.

13. These questions have to do with the distribution of forms of acids and bases when dissolved in a solvent.

A. When a species that has a conjugate acid or base form is dissolved in water, some of it will take the conjugate form. True or false and why?

B. How is the $p[H_3O^+]$ at which the conjugate forms are roughly equal in concentration related to the K_a' of the conjugate pair?

C. As the $p[H_3O^+]$ becomes higher than the value in question B, which form will predominate and why?

D. For which species are the concentrations plotted on an acid–base log-concentration plot assuming a monoprotic acid? From which basic equations are the lines for each of the species obtained?

E. Figure 4.14 is a plot of an acid with a K_a' of 1×10^{-5} dissolved in water. What aspects of this plot would change if the same acid were dissolved in a different solvent?

14. Write the proton balance equations for the following solutions.

A. NH_4^+ in water

B. HF in water

C. CH_3COO^- in water

D. HSO_4^- in water (Consider all forms.)

E. PO_4^{3-} in water (Consider all forms.)

F. H_2CO_3 in water (Consider all forms.)

15. Draw the log concentration plot for a 0.10 M solution of HF in water.

A. Graphically determine the $p[H_3O^+]$ of this solution.

B. Graphically determine the $p[H_3O^+]$ of an HF solution that is 1×10^{-5} M. Graphically determine the fraction of the HF–F^- conjugate pair that is in the F^- form in this solution.

16. In general chemistry, we may have been taught that for the reaction of disolving an acid HA in water (HA + H_2O ⇌ H_3O^+ + A^-), we could say that equal amounts of H_3O^+ and A^- were formed so that the equilibrium expression could be written

$$K_a' = \frac{[A^-][H_3O^+]}{C_{HA} - [H_3O^+]} \approx \frac{[H_3O^+]^2}{C_{HA}}$$

from which

$$[H_3O^+] \approx \sqrt{K_a' C_{HA}}$$

A. Based on Equation 4.11, what must be true for the above relationship to be valid?

B. Use equations to find the exact solution to the HF solutions of Question 15, parts A and B.

17. These questions have to do with the titration of a weak acid with a standard base solution.

A. What is the method and purpose of standardizing the base solution with which you are titrating?

B. If you overshoot the end point in the standardization titration, will your result for the amount of weak acid be high or low and why?

C. If you do not completely dry your weak acid before the analysis, will your result for the amount of weak acid be high or low and why?

D. From Table 4.1, choose an indicator that would be very good for the titration of L-ascorbic acid. Give the reason for your choice.

18. A. Create a titration curve for the titration of 25.0 mL of a solution that is approximately 1×10^{-3} M L-ascorbic acid. Use a titrant concentration that is roughly equal to the expected analyte concentration.

B. Create a log concentration plot for the tiration of a solution that is approximately 1×10^{-3} M L-ascorbic acid.

C. Mark on the log plot the solution points for the 0%, 50%, 100%, and 200% titration points. Compare the $p[H_3O^+]$ at each of these points with those predicted on the titration curve.

19. The pH of human blood is between 7.35 and 7.45. The blood also contains phosphate and carbonate. For these questions, assume that the K_a' values in the table of Appendix B are equal to the K_a' values.

A. For a pH of 7.4, what are the fractions of each of the major forms for the carbonate system?

B. For a pH of 7.4, what are the fractions of each of the major forms for the phosphate system?

C. Why are there only two major forms in each case?

D. For which of these two systems is the assumption about the equivalence of the thermodynamic and formal equilibrium constants worse and why?

20. The amino acids from which peptides are composed themselves include an amine group (—NH_2) and a carboxylic acid (—COOH). The pK_a' of the carboxylic acid group is in the range of 1.8 to 2.5. In its basic form, the carboxylic acid group is —COO^-. The pK_a' of the amine group is in the range of 9.0 to 10.7. In its acidic form, it is —NH_3^+.

A. In what $p[H_3O^+]$ range would the acidic form of the carboxylic acid predominate?

B. In what $p[H_3O^+]$ range would the basic form of the amine group predominate?

C. In a neutral solution ($p[H_3O^+] \approx 7$), which forms of the carboxylic acid and amine groups will predominate?

D. For those amino acids that have no other acidic or basic groups (amino acids with nonpolar side chains), what is the charge on the amino acid in a neutral solution?

E. Over what range of $p[H_3O^+]$ values will 99% of the amino acids with nonpolar side chains be in the 0 charge state?

F. The pK_a' values for the amino acid valine are 2.3 and 9.7. What will be the $p[H_3O^+]$ of a 0.1 M solution of valine?

Suggested Related Experiments

1. Titration of KHP to standardize NaOH titrant, using standard buret, pH meter, and indicator.

2. Titration of unknown amount of weak or strong acid to get original concentration.

3. Titration of unknown weak acid to determine K'_{a1}, molecular weight, and identity of acid.

4. Titration of weak base, salt of dibasic acid, or mixture of salt and acid or base.

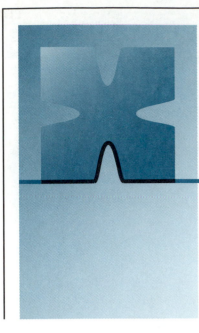

Chapter Five

ANALYSIS BY ABSORPTION OF LIGHT

The selective absorption of light is what gives color to our world. The light from the sun and most lamps is white light, that is, it contains light of many colors. Objects illuminated by white light may not reflect or transmit all the colors of the illuminating light equally. Thus, the light coming from the object has a different color balance than the illuminating light. Our eyes perceive this difference as color. Atoms or molecules in the illuminated object are responsible for the selective absorption of the illuminating light, so different materials absorb the light differently. In other words, the way in which objects reflect or transmit light is a function of their composition. Any relationship between composition and an observable quantity can become the basis for a method of analysis. In this chapter, we will see how the selective absorbance of light by solutions can be used in quantitation, detection, identification, and yes, even separation (analytically speaking).

A. Colored Solutions and White Light

We are all familiar with colored solutions such as grape juice, orange soda, and Easter egg dye. A colored solution looks colored because the distribution of the colors of light coming through the solution is different from that in the light that is illuminating the solution. For example, grape juice appears purple because it transmits the purple colored light from the illuminating source more effectively than it does some of the other colors of light.

To understand how transmission of light can depend on its color, it is helpful to recall that light behaves both as a continuous **electromagnetic wave** and as a particle of essentially pure energy. The fundamental quality that distinguishes light of one color from all the others is the frequency of its **electromagnetic radiation**, v. (See Figure 5.1.) This **frequency** has the units of cycles per second, or hertz (Hz). Light that our eyes perceive as purple has a frequency of about 7×10^{14} Hz. At the other end of the **visible spectrum** (the range of electromagnetic radiation that we can see) is red light with a frequency of about 4×10^{14} Hz.

THE WAVE NATURE OF LIGHT

When light acts as a wave passing a single plane, one can imagine the field strength of the wave at that plane, going through one cycle every $1/v$ seconds. This time is the **period** of the wave oscillation. One could also imagine stepping back a bit to see an entire cycle of the oscillation, as shown in Figure 5.2. The space taken up by one cycle (called the **wavelength**) is the distance the wave moves in one period. Of course, this distance also depends on the velocity of the wave. The **velocity of light** in a vacuum, c, is 2.998×10^8 m/s. The wavelength, λ, is thus

$$\lambda = \frac{\text{velocity}}{\text{frequency}} = \frac{c}{v} \qquad 5.1$$

in meters. In the visible part of the spectrum, wavelengths range from purple light (430 nm) to red light (700 nm).

The visible part of the spectrum is an extremely small part of the total range of possible (and detectable) frequencies of electromagnetic radiation. (See Figure 5.3.) At longer wavelengths (lower frequencies) than visible light is the infrared (IR) region (0.70–300 μm), and at frequencies higher than blue light is the ultraviolet (UV) region (180–430 nm). Light in these regions, even though invisible to the human eye, can be detected by some other animals and by some of the same kinds of light-detecting instruments designed for the visible region. Electromagnetic radiation at frequencies higher than the ultraviolet range and lower than the infrared range are not called light. They are **X rays** and **gamma rays** on the high frequency side and **microwaves** and **radio waves** on the low frequency side.

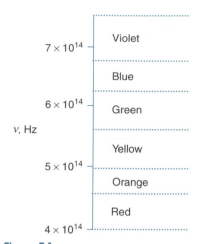

Figure 5.1. The different frequencies of light are perceived as different colors. The red end of the spectrum has the lowest frequency.

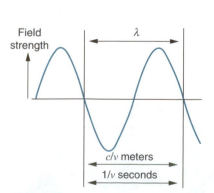

Figure 5.2. A traveling wave with frequency v, velocity c, and wavelength λ

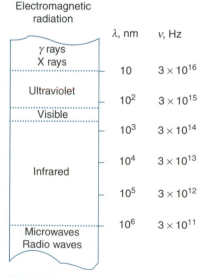

Figure 5.3. In all tests devised thus far to determine if light is a particle or a wave of electromagnetic energy, the answer is both, even simultaneously.

Example 5.1

Frequency, Energy and Wavelength of Light

What is the frequency and color of light that has a wavelength of 630 nm in a vacuum? What is the energy of 1 photon of this light?

From Equation 5.1,

$$v = \frac{c}{\lambda} = \frac{3.00 \times 10^8 \, m \, s^{-1}}{6.30 \times 10^{-7} \, m}$$
$$= 4.76 \times 10^{14} \, Hz$$

From Figure 5.1, the color of this frequency of light is orange.

From Equation 5.2,

$$\mathscr{E}_{photon} = hv$$
$$= 6.63 \times 10^{-34} \, Js$$
$$\times 4.76 \times 10^{14} \, s^{-1}$$
$$= 3.16 \times 10^{-19} J.$$

White light contains light of all the visible frequencies.

An object appears colored when it absorbs some of the illuminating frequencies of light more effectively than others. The color balance of the reflected light gives the object its color.

THE PARTICLE NATURE OF LIGHT

Light also behaves as a particle of energy with essentially no mass. Particles of light are called photons. The energy (in joules) of a photon is also related to the frequency of the wave by Planck's constant, h (which is 6.626×10^{-34} joule-seconds). Thus

$$\mathscr{E}_{photon} = hv \qquad\qquad 5.2$$

(See Example 5.1.) The individual photon energy is the basis for the phenomenon of light absorption. When the photon energy matches an allowed energy transition within the material through which the photon is passing, the energy (the photon) can be absorbed. Photons cannot lose just part of their energy through normal absorption; it is an all-or-nothing phenomenon. This means that the energy of a photon is normally unchanged between its inception and its total absorption. It also means that the absorbing molecule must be able to absorb the entire energy of the photon. Putting it another way, the photon energy must be equal to an allowed energy transition in the absorbing species. This factor makes different species absorb light of different energies.

SOURCES OF LIGHT

We know that light emanates from hot objects. We see red light coming from fires and stove tops. We see white light coming from the filaments in lightbulbs and from the sun. Such light is generated by the conversion of heat into photons. The heat excites the atoms or molecules of the hot object, which then become de-excited by the emission of a photon. The distribution of the energies of the photons thus produced is a function of the distribution of the energies of the excited atoms or molecules in the material. A plot of the intensities of several light sources as a function of photon energy is shown in Figure 5.4. Such a plot is called a **spectrum**. **Spectra** (the word for more than one spectrum) may be plotted as intensity versus frequency, wavelength, or energy. In this case, the intensity distribution is a continuous, smooth function of the photon energy. The light emanating from such an object has photons of many energies or waves of many frequencies. If the light has a substantial representation of light of all the colors, we say the light is **white light**. Perfectly white light would have exactly equal intensities of all the frequencies, that is, the spectrum would be a horizontal line. Perfectly white light sources are very rare. Most light sources have a little color. For example, the light of an incandescent lamp has a higher proportion of red frequencies than blue and so may appear somewhat reddish in contrast to sunlight.

We are used to observing objects that are illuminated by approximately white light. Light of all colors strikes the object. If the object is opaque, some of the light will be reflected and some will be absorbed. The fraction of the light reflected and absorbed generally depends on the frequency of the light and on the material that reflects it. This causes the distribution of frequencies in the reflected light to be different from that of the illumination. It is the same case with transparent objects such as colored glass or solutions. The illuminating light has all the colors of light, but they are not transmitted

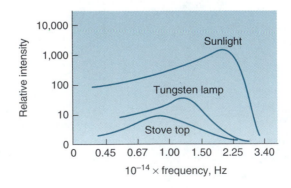

Figure 5.4. Different sources of visible radiation have different spectral distributions. The more red the appearance of the light, the less high frequency radiation it contains.

Table 5.1

Colors, Wavelengths, and Complements*

Light Color	λ, nm	Complement
Violet	400–450	Yellow-green
Blue	450–480	Yellow
Green-blue	480–490	Orange
Blue-green	490–500	Red
Green	500–560	Purple
Yellow-green	560–580	Violet
Yellow	580–600	Blue
Orange	600–650	Green-blue
Red	650–750	Blue-green

through the object with equal efficiency, that is, some of the colors of illuminating light are partially absorbed. The color imbalance of the emerging light (relative to the illuminating light) makes the solution appear colored.

LIGHT ABSORBING SOLUTIONS

In the visible region, water is transparent; it transmits all of the visible wavelengths of light passing through it. What makes an aqueous solution appear colored, then, is the selective absorption of light by a component dissolved in the solution. A solution of potassium permanganate ($KMnO_4$) appears purple. This is because the purple light from the source is absorbed less than light of purple's complementary color (yellow-green). In other words, the solution appears purple because it absorbs light in the yellow-green part of the spectrum. (See Table 5.1.) Different dissolved substances absorb light in different regions of the spectrum and thus appear differently colored. We noticed this in Chapter 4 when working with the colored indicators in acid–base titrations. As the indicator changes from its acid form to the conjugate base form, the color of light it absorbs changes. The fact that we can tell lemonade from root beer by the color of the solution suggests that we could develop methods to identify dissolved substances by the colors of light they absorb. In this chapter, we will see how this has been done.

Color change is not the only effect induced by light absorbers. We have also observed that if we dissolve more of a colored substance in a solution, the color of the solution becomes more intense. The increase in color intensity is due to an increase in the selective absorption of light by the solution. This relationship between the amount of the light being absorbed and the concentration of the dissolved substance suggests that a method of quantitative analysis could also be based on this phenomenon. To accomplish either of these analytical goals, we must have a way to measure the degree to which light of a particular wavelength is being absorbed by a solution.

When light travels through material other than a vacuum, its velocity may be decreased (Figure 5.5). Its frequency remains constant, so the wavelength of the light

Color helps us identify objects; it can be used for molecules, too.

The intensity of the color is related to the concentration of the absorbing species.

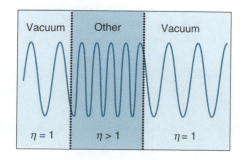

Figure 5.5. The velocity of light is decreased in media other than a vacuum. The frequency stays the same, but the wavelength is reduced.

Table 5.2

Index of Refraction for Some Materials

Material	η
Air	1.003
Water	1.333
Fused silica	1.458
Quartz	1.544
Sapphire	1.769
Benzene	1.500

through such a medium will be decreased while it is in it. This change in velocity and wavelength between two transparent media is the basis for the optical phenomenon of **refraction**. The ratio of the velocity of light in a vacuum (c) to its velocity in another medium (v) is called the **index of refraction**, η. Thus,

$$\eta = c/v \qquad\qquad 5.3$$

Table 5.2 lists some materials and their index of refraction for light at 589 nm. Generally, the index of refraction of a material is different for different wavelengths of light. From Table 5.2 and Equation 5.3 we see that the velocity of light through water is only three-fourths of its velocity in air. This reduction in velocity while the frequency remains constant causes a decrease in wavelength, as shown in Figure 5.5. (See Example 5.2.)

Example 5.2

Effect of Refractive Index on Wavelength

What is the wavelength of light that appears yellow in air (at 590 nm) when it is traveling through water?

From Equation 5.1,

$$v = c/\lambda$$
$$= 3.00 \times 10^8 \text{ m s}^{-1}/5.90 \times 10^{-7} \text{ m}$$
$$= 5.08 \times 10^{14} \text{ s}^{-1}.$$

From Equation 5.3, v in water =

$$c/\eta = 3.00 \times 10^8 \text{ m s}^{-1}/1.33.$$
$$= 2.26 \times 10^8 \text{ m s}^{-1}.$$

From Equation 5.1,

$$\lambda \text{ in water} = \frac{2.26 \times 10^8 \, ms^{-1}}{5.08 \times 10^{14} \, s^{-1}}$$

$$= 445 \, nm$$

Study Questions, Section A

1. Which color of light has the higher frequency, red or blue?

2. Which has the higher energy, a photon of red light or a photon of blue light?

3. What is the wavelength of light that has a frequency of 5.00×10^{14} Hz? Into what range of the light spectrum does it fall?

4. Calculate the energy of
   A. A photon of red light (l 5 700 nm)
   B. A photon of purple light ($\lambda = 430$ nm)
   C. A mole of purple photons

5. What is a spectrum, and what is the form of the spectrum of white light?

6. What is the phenomenon that makes a solid yellow object appear yellow?

7. Is it the wavelength or the frequency of light that changes as it enters into a different substance?

8. What is the velocity of light passing through quartz if the quartz has a refractive index of 1.54 for that wavelength of light?

1. From Figure 5.1, blue light has a higher frequency than red light.

2. Since energy is proportional to frequency (Equation 5.2) and blue light has the higher frequency, the blue photon will have the higher energy.

3. Using Equation 5.1, $\lambda = c/v$ where $c = 3.00 \times 10^8$ m/s,

$$\lambda = \frac{3.00 \times 10^8 \text{ m/s}}{5.00 \times 10^{14} \text{ s}^{-1}} = 6.00 \times 10^{-7} \text{ m}$$

We know 6.00×10^{-7} m $= 600$ nm, so the light is in the visible region of the spectrum.

4. A. From Equation 5.2, $\mathscr{E} = hv$ and from Equation 5.1, $v = c/\lambda$. Combining the equations gives, $\mathscr{E} = hc/\lambda$, where h (Plank's constant) is 6.626×10^{-34} J $\cdot$ s, so

$$\mathscr{E} = \frac{(6.626 \times 10^{-34} \text{ J} \cdot \text{s})(3.00 \times 10^8 \text{ m/s})}{700 \times 10^{-9} \text{ m}}$$

$$= 2.83 \times 10^{-19} \text{ J/photon of red light}$$

Such small amounts of energy can better be expressed with electron volts, where 1 eV $= 1.602 \times 10^{-19}$ J.

$$\mathscr{E} = 2.83 \times 10^{-19} \text{ J} \cdot \frac{1 \text{ eV}}{1.602 \times 10^{-19} \text{ J}} = 1.77 \text{ eV}$$

B. For a photon of purple light,

$$\mathscr{E} = \frac{(6.626 \times 10^{-34} \text{ J} \cdot \text{s})(3.00 \times 10^8 \text{ m/s})}{430 \times 10^{-9} \text{ m}}$$

$$= 4.62 \times 10^{-19} \text{ J}$$

$$\mathscr{E} = 4.62 \times 10^{-19} \text{ J} \left(\frac{1 \text{ eV}}{1.602 \times 10^{-19} \text{ J}} \right) = 2.88 \text{ eV}$$

Note that red light has a longer wavelength and a lower energy than purple light.

C. To find how much energy a mole of photons has, we must multiply the energy of one photon by the number of photons per mole.

$$4.62 \times 10^{-19} \text{ J} \frac{6.022 \times 10^{23} \text{ photons}}{\text{mol}} = 2.78 \times 10^5 \text{ J/mol}$$

5. A spectrum is a plot of light intensity versus the wavelength or frequency of the light. White light contains all the visible wavelengths of light, so its spectrum would have significant (ideally, equal) intensities from 430 to 700 nm.

6. The object will appear yellow if the wavelengths of light that have the color of the complement of yellow are selectively absorbed by the object. The colors that are not absorbed are reflected. Using Table 5.1, we see that if the object absorbs blue light (more than the light of the other colors) it will appear yellow.

7. It is the wavelength of light that depends on the refractive index of the material through which it is passing. The energy of the light remains constant, and since the energy is proportional to the frequency, the frequency cannot change.

8. From Equation 5.3, we have $\eta = c/v$, so

$$v = \frac{c}{\eta} = \frac{3.00 \times 10^8 \text{ m/s}}{1.54} = 1.95 \times 10^8 \text{ m/s}$$

B. Measuring the Absorption of Light

The general approach for chemical analysis is to choose a differentiating characteristic for the species to be analyzed, develop a probe for that characteristic, anticipate the response of the analyte to that probe, measure the response to that probe, and interpret the results of the measurement in terms of the information desired. A flowchart of this process is shown in Figure 5.6. In this chapter, the differentiating characteristic discussed is the selective absorption of light. In this section, we will develop a probe for this characteristic, see how the response to this probe is measured, and determine how the probe response can be interpreted to reveal the fraction of the light shining on a sample that is absorbed by it.

DEVELOPING THE PROBE

A test for the differentiating characteristic of the selective absorption of light will necessarily involve the use of light. Specifically, we will want a source of light to test for absorption, and we will have to arrange the sample so that light from the

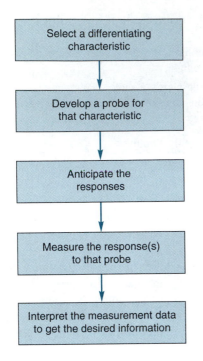

Figure 5.6. The scheme for the application of light absorption for chemical analysis.

source can shine through it. Three mechanisms for doing this are shown in Figure 5.7. In Figure 5.7a, the liquid sample is in a transparent container through which the light shines. In Figure 5.7b, the light is introduced through the entrance optical fiber (light pipe) that is part of a dip probe assembly. This assembly includes the entrance fiber, a mirror, and the exit fiber. The light from the entrance fiber is transmitted through a short length of sample to a mirror, which reflects some of the light back through the sample solution and into the exit fiber. Figure 5.7c is a flow cell through which the sample solution is pumped. Light from the source is focused on the entrance window, is transmitted through a length of sample, and exits the far window for detection. Each of the three methods of exposing the sample to the probe illumination represents different modes of operation. In Figure 5.7a, the sample is brought to the instrument in a special cell designed for absorbance measurement. In Figure 5.7b, the optical sensing probe is brought to the sample and dipped into the sample solution. In Figure 5.7c, the optical measurement is introduced into a flowing stream of sample for continuous monitoring of the solution composition.

Another important aspect of the probe is the nature of the light chosen to transmit through the sample. The differentiating characteristic being probed is the *selective* absorption of light, that is, the preferential absorption of light of certain colors. This will require that the light source produce light in the wavelength region of the selective absorption. In the case of a sample that appears yellow, the light source must emit light in the blue part of the spectrum (from 435 to 480 nm) to probe for light absorption in that region.

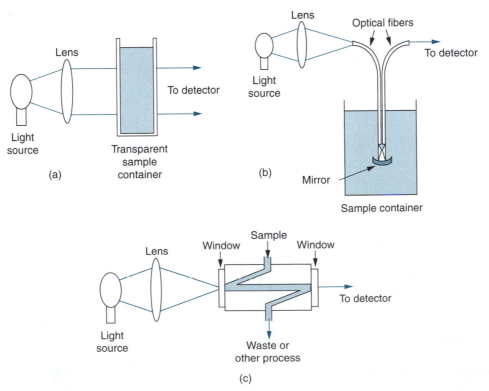

Figure 5.7. Three common methods of transmitting the probe illumination through a sample solution.

ANTICIPATING THE RESPONSE TO THE PROBE

We anticipate that the response of the analyte to the probe of illumination is to absorb some of the light. More specifically, we expect the light of particular wavelengths to be absorbed more than that of others. In this stage of the analysis, a thorough understanding of the phenomenon being used for the differentiating characteristic is very helpful. This understanding aids in the design and application of the probe and in the interpretation of the data. In the case of light absorption, we will see that we need to be able to distinguish between the light absorbed by the sample and that lost in the light path and the sample container. Further, we need to develop a relationship between the intensity of the light transmitted and the concentration of the absorbing analyte in the sample. The ways in which this has been accomplished are described in the following sections.

It is a truism that the more we understand about the phenomenon underlying the differentiating characteristic, the more precisely and intelligently we are able to apply it in chemical analysis. Thus, exploration of these phenomena has become a very important part of the ongoing research in analytical chemistry.

DETECTING THE RESPONSE TO THE PROBE

The anticipated response to the probe is a diminution in the intensity of light of the absorbed wavelengths passing through the sample. To detect this response, we must arrange to measure the intensity of the light that has passed through the sample. To do this, we allow the transmitted light to fall on a photodetector and measure the detector output. A **photodetector** is a conversion device that converts the intensity of the light to a related electrical signal. Many kinds of photodetectors have been developed. Photodetectors are based on the phenomenon of photon absorption. In detectors for visible light, the photon energy is used to separate an electron from its atom. This charge separation can result in a measurable electrical current. A specific example is the photodiode, which converts light intensity to a related electrical current. (See Figure 5.8.) Light intensity is measured in units of radiant power, P (ergs per square centimeter per second). Additional conversion devices must then be employed to obtain a number related to the light intensity. In modern instruments with a digital readout, the current is converted to a related voltage, which is then connected to an ADC. For the old standard readout, the moving coil ammeter, a current amplifier, was used to raise the detector output current to a level compatible with the ammeter sensitivity. Block diagrams of these systems are shown in Figure 5.9.

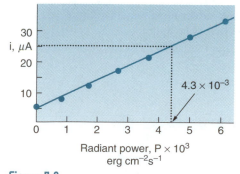

Figure 5.8. Most photodetectors produce a current that increases linearly with the detected light intensity.

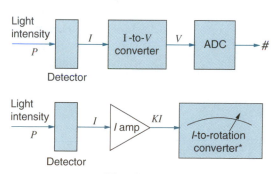

*Experimenter and scale perform rotation-to-number conversion.

Figure 5.9. Block diagrams of light detection systems. The conversions that relate the number output to the light intensity for the digital readout system are as follows:

$$I = K_1P + C_1, \qquad V = K_2I + C_2, \qquad \# = K_3V + C_3,$$
$$\# = K_1K_2K_3P + K_3K_2C_1 + K_3C_2 + C_3,$$
$$\# = K_1K_2K_3P + C_4$$

What is the relationship between the light intensity and the number we observe from the meter pointer or the digital display? From arguments developed in Chapter 2, we know that it depends on the transfer function of each conversion device employed. The block diagrams of Figure 5.9 are drawn to reveal each conversion device and the units of the quantity encoding the data at each point. This method of diagramming a measurement system is strongly recommended because all the encoding forms and conversion devices are identified. If the transfer functions of all the conversion devices are linear (output = $K \times$ input + C), the overall transfer function from light intensity P to number has the same form. The overall light measurement sensitivity (the change in output number for a unit change in the input light intensity) will be the product of the sensitivities (K values) of each of the conversion devices involved.

OBTAINING THE FRACTION OF LIGHT TRANSMITTED

Having a way to measure the amount of light transmitted by the sample does not directly answer the question of what fraction of the illuminating light was absorbed by the sample. To accomplish this, we could compare the intensities of light hitting the detector when the sample is present and when it is absent. When making such a comparison measurement, it is important to minimize the differences that are due to effects other than the one being measured. If we remove the sample from the light path altogether, we remove not only the sample, but the sample container and the medium in which the sample may have been dissolved. (See Figure 5.10.) To the extent that these components could affect the intensity of the transmitted light, their effect will be added to that of the analyte molecules. Therefore, to determine how much light the sample molecules alone absorb, we leave the sample container and the solvent in the light path and remove only the analyte molecules. The solution used in this measurement is called a **blank**. A **blank solution** and its container should be as much like the analyte solution and its container as possible.

The spectroscopist's term for the fraction of the light transmitted by a sample (relative to the blank solution) is the **percentage transmittance** or simply %T. To a first approximation, the %T is the ratio of the readout number for the sample, N_s, to the readout number for the blank, N_b (times 100). However, this interpretation would assume not only that the transfer functions of all the conversion devices are linear, but that their intercepts were zero (i.e., that the C_4 value in the overall transfer function of Figure 5.9 is 0). For sensitive detectors, this is not true. There is always a small output current even when the detector is exposed to no light at all. The output of an unilluminated detector is called the **dark current**. This effect must be taken into account for accurate light intensity measurements, especially at low light levels. To do so, we must make a separate *dark* measurement. This is accomplished by placing a shutter in the

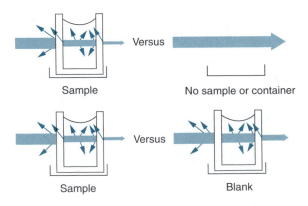

Figure 5.10. Using a blank solution to separate the analyte absorbance from other light loss mechanisms.

light path and recording the readout number when the shutter is closed (Figure 5.11). We will call this number N_d. The light intensity measurement is corrected by subtracting N_d from N_s or N_b. The accurate relationship for $\%T$ is then

$$\%T = \frac{N_s - N_d}{N_b - N_d} \times 100 \qquad\qquad 5.4$$

We can see that we don't measure $\%T$ directly; we measure the light intensities related to the data N_s and N_b and then use Equation 5.4 along with N_d to calculate the $\%T$. We also see that the usefulness of this relationship depends on a linear transfer function for all the conversion devices employed in the light intensity measurement process and that the source light intensity is the same for measurements of N_b and N_s.

As the concentration of the absorbing substance increases, we expect the $\%T$ to decrease. This will be true as long as the illuminating light source includes light in the wavelength region of the analyte absorbance, but what if it includes many other colors of light as well? Presumably, light outside the absorbance region of the analyte will pass through to the detector unattenuated. The presence of this light will increase N_s and the $\%T$. In this way, we see that $\%T$ will be a function not only of the absorber concentration, but also the distribution of wavelengths from the light source. To minimize this effect, efforts are made to restrict the probe light to the wavelength region absorbed by the particular analyte being tested. A wavelength-selecting device is placed in the light path to avoid the measurement of light intensities that are outside the desired wavelength region. Note from Figure 5.12 that it does not matter whether the wavelength-selecting device is placed between the source and sample or between the sample and detector. Several methods of accomplishing wavelength selection are discussed in later sections.

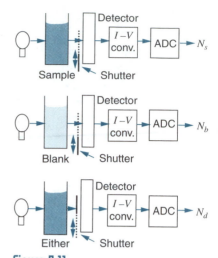

Figure 5.11. Arrangements for the three measurements required to calculate $\%T$.

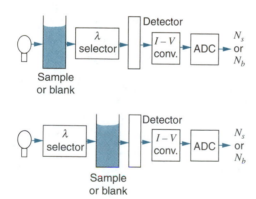

Figure 5.12. A wavelength selecting device is used to limit the probe illumination to those wavelengths that are absorbed by the analyte.

Study Questions, Section B

9. What is the nature of the probe that tests for the characteristic of light absorption?

10. What is the response of the analyte to this probe?

11. How is the response to the probe measured?

12. What are the input and output quantities of the usual photodetector?

13. Calculate the sensitivity of a light intensity measurement system that utilizes a detector with the response function illustrated in Figure 5.8, a current-to-voltage converter with an output of 0.0100 V/μA, and a 12-bit ADC with a full scale response of 0 to 10.00 V.

14. Why is the use of a blank solution rather than no sample cell or an empty sample cell so important?

15. What are the quantities N_s, N_b, and N_d?

16. A spectrometer reads 27 when the shutter is closed, 3056 with a blank solution, and 2375 with the sample in place. What is the $\%T$ for this sample?

9. The probe for light absorption is light of the wavelength that is absorbed.

10. The response of the analyte to the probe is the absorption of some of the light at certain wavelengths.

11. The intensity of the light transmitted through the sample compartment is measured with and without the analyte present.

12. Most photodetectors convert radiant power in erg cm^{-2} s^{-1} to current in amperes.

13. From the plot of Figure 5.8, the response of the detector is seen to be

$$\frac{(25 - 5) \times 10^{-6}\ \text{A}}{4.3 \times 10^{-3}\ \text{erg cm}^{-2}\ \text{sec}^{-1}} = \frac{4.65 \times 10^{-3}\ \text{A}}{\text{erg cm}^{-2}\ \text{sec}^{-1}}$$

The conversion factor for the ADC is

$$\frac{2^{12}\ \text{units}}{10.00\ \text{V}} = \frac{409.6\ \text{units}}{\text{V}}$$

Overall, the conversion factor (sensitivity) is

$$\frac{4.65 \times 10^{-3}\ \text{A}}{\text{erg cm}^{-2}\ \text{sec}^{-1}} \times \frac{1.00 \times 10^{4}\ \text{V}}{\text{A}} \times \frac{409.6\ \text{units}}{\text{V}}$$

$$= \frac{1.9 \times 10^{4}\ \text{units}}{\text{erg cm}^{-2}\ \text{sec}^{-1}}$$

14. The fraction of the incident light reaching the detector is a function of the cell material and surfaces, the refractive index of the liquid inside the cell, and any absorption of the solvent. The use of a blank solution to measure the unabsorbed light intensity reduces these effects.

15. The quantities N_s, N_b, and N_d are the numerical outputs of the light intensity measurement when the sample is in place, when the blank is in place, and when the shutter is closed.

16. Using Equation 5.4,

$$\%T = \frac{2375 - 27}{3056 - 27} \times 100 = 77.52\%$$

C. Relating Light Absorption to Concentration

We have shown that we expect %T to be a function of the concentration of the analyte and that we have a way to determine the %T (by a combination of light intensity measurements and calculation). This expectation of a relationship between %T and concentration for an absorbing analyte can be tested by determining the %T of a number of samples of known concentration. A plot of the data thus obtained is called a **working curve**. A working curve for a solution of β-carotene is shown in Figure 5.13. The first thing we notice is that the %T does not decrease linearly with increasing concentration. The achievement of a linear relationship between an instrument output and the quantity being determined used to be a major goal for instrument developers. This is because only linear readout devices (ammeters and strip-chart recorders) were available. Our fixation with this quality, however, has outlived its usefulness. With the computing power available in even the simplest instruments, accommodation of a nonlinear work-

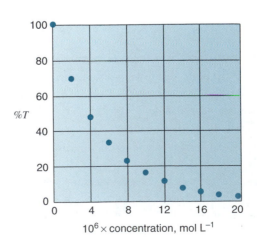

Figure 5.13. Working curve for %T versus concentration of β-carotene.

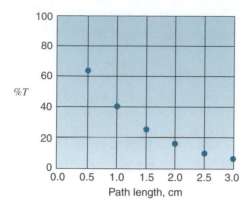

Figure 5.14. The %*T* does not decrease linearly with increasing concentration or path length.

ing curve is trivial. Nevertheless, determination of the mathematical function of the working curve is both interesting and useful.

RATIONALIZING THE WORKING CURVE

The result that the %*T* is not cut in half when the concentration is doubled is counter to our intuition. After all, if the number of absorbing species has doubled, why has not the number of absorbed photons doubled with it? What would happen if, instead of doubling the concentration, one doubled the length of the sample container along the light path? In this experiment, too, the number of absorbing species through which the light must pass is doubled. The results of this experiment are shown in Figure 5.14. Again, the relationship is not linear. What is happening? An insightful question would be, "What is the difference between the second centimeter of solution and the first?" There is less light entering the second centimeter of solution because of the light that was absorbed in the first part. In fact, this must be happening all through the solution. A less intense light illuminates the solution farther away from the front of the sample container. The farther back solution absorbs the same *fraction* of the light it gets as the front part of the sample, but since it gets less light, it absorbs less light. This process is illustrated in Figure 5.15.

A model for the absorption of light through the sample can now be constructed. It involves dividing the sample into an infinite number of infinitely thin slices through which the light passes. The amount of light passed on to each slice from the previous one is reduced by the same fraction. The magnitude of this fraction depends not only on the concentration of the absorber but also on the capture cross section. The **capture cross section** is the area surrounding the absorber that the photon must hit in order to be absorbed. The magnitude of the capture cross section varies greatly from one absorbing species to another. The cumulative effect of the fractional reduction of all the slices is then derived. Mathematically, this derivation involves the use of integral calculus. If you would like to see the complete derivation, it is in the derivations section of the accompanying CD. The result of this derivation is

$$\log \frac{P_0}{P} = \varepsilon b C \equiv A \qquad 5.5$$

where P_0 is the radiant power incident on the sample, P is the radiant power emerging from the sample, C is the molar concentration of the absorbing species, b is the length of the light path through the sample, and ε is the **molar absorptivity** or **molar absorption coefficient** (related to the capture cross section). This relationship is called **Beer's law**. In it, a new term, A, has been defined. The term A is called the **absorbance** of the sample. The absorbance is greater for solutions with a lower percentage transmittance.

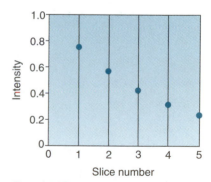

Figure 5.15. The sample solution is divided into five slices. Each slice absorbs 25% of the light reaching it. The dot at the rear of each slice represents the relative intensity of the light that got through the slice.

The molar absorptivity, ε is specific to the λ of the illumination and to the absorbing species.

A rearrangement of Equation 5.5 reveals the form of the relationship between the emerging light intensity and the concentration of the sample.

$$P = P_0 \times 10^{-\varepsilon bC} \qquad\qquad 5.6$$

From this relationship we can see that the intensity of the emerging light falls off exponentially with increasing concentration and sample thickness. This corresponds to the general shape of the operational working curve. However, to more exactly compare the predictions of this theory with the working curve, we must relate the quantities used in the model with the quantities we can measure. A restatement of Beer's law in terms of the measured quantities is

$$\log\frac{N_b - N_d}{N_s - N_d} = \varepsilon bC = A \qquad\qquad 5.7$$

Comparing this relationship with that of Equation 5.4, we see that

$$\log\frac{N_b - N_d}{N_s - N_d} = \log\frac{100}{\%T} = \varepsilon bC = A \qquad\qquad 5.8$$

The Beer's law model deals only with the phenomenon of absorbance by the sample constituents. In practice, this must be reconciled with the other reasons for light loss in the illumination of the sample and the collection of light from it. The light intensity measured with the blank solution is taken to be the measurable intensity with which the sample was illuminated.

Absorbance is a very convenient quantity: it is directly proportional to the concentration (as well as to the path length and molar absorptivity,) and it is readily calculated from measurable quantities.

THE ABSORBANCE-CONCENTRATION RELATIONSHIP

On the basis of the model from which Beer's law was developed, we expect a linear relationship between absorbance and concentration of analyte. For this reason, practical working curves are usually plotted in this way. (See Figure 5.16.) The slope of the working curve (the sensitivity) is equal to εb. The greater the molar absorptivity of the analyte and the greater the light path through the sample, the more the absorbance changes with concentration.

There are upper and lower limits to the practical range of measurable absorbance. The plot of absorbance versus $\%T$ in Figure 5.17 reveals the basis for these limits. At an absorbance of 1, the transmittance is 10%, at 2, it is 1%, and at 3 it is only 0.1%. The amount of light transmitted through the sample decreases rapidly

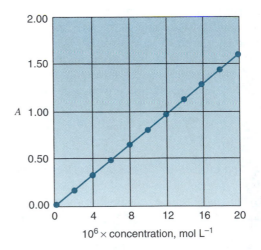

Figure 5.16. An ideal working curve.

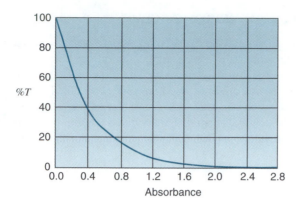

Figure 5.17. At values of A above 2, the %T becomes extremely small. Any other light reaching the detector can cause a significant error.

with increasing absorbance in this region. This poses some difficulties for the detection of the reduced light intensities. However, a greater problem is that there is always a small amount of **stray light** hitting the detector. This light may be reflected or incident light that did not go through the sample, or it may be light of wavelengths less absorbed by the sample that has gotten through the wavelength selection device. At very low levels of transmittance, the detector response to the stray light can become significant, so that the detected light no longer decreases by a factor of 10 with each unit increase in absorbance. As a result, the working curve becomes nonlinear, as shown in Figure 5.18. The amount of stray light depends greatly on the design of the spectrometer, with cost increasing rapidly as the stray light specification goes down. Most analysts avoid working at high absorbance levels by accurately diluting their sample.

At very low absorbances, the transmittance is approaching 100%. At an absorbance of 0.01, only 2.3% of the light is absorbed by the sample. The difference between the light transmitted by the blank and that by the sample is becoming very small. The precision of the measurement based on this small difference declines rapidly in this absorbance range. The dependence on an accurate blank measurement becomes greater. Even with matched cells for the sample and blank solutions, the differences between them and the difficulty in achieving exactly the same cell position limit the precision in this range. When absorbance is low, spectroscopists use longer path length cells and/or a fixed cell through which the sample and blank are alternately pumped.

For the reasons given earlier, the best range for absorbance measurements is in the region between 0.1 and 1. With care and expensive instrumentation, this range can be extended from 0.01 to 3.

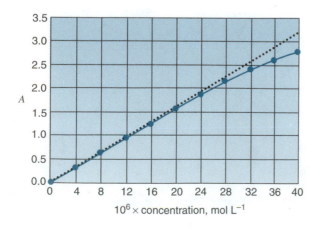

Figure 5.18. A working curve for a spectrometer with 0.1% stray light.

CALIBRATION OF THE WORKING CURVE

Earlier in this section, we saw how the working curve was rationalized and how, in the linear portion of the curve, there is a simple relationship between the absorbance and concentration. Since the absorbance can be determined by a combination of measurement and calculation, one could, if one knew b and ε, calculate the concentration. In practice, this is rarely done. Instead, one simply determines the absorbance of a known concentration of the analyte. From this, the value of $b\varepsilon$ can be determined for the specific measurement conditions used. In the linear region of the working curve, one can use the equation

$$\frac{A_{std}}{C_{std}} = b\varepsilon = \frac{A_{unk}}{C_{unk}}, \qquad \text{so} \qquad C_{unk} = C_{std}\frac{A_{unk}}{A_{std}} \qquad\qquad 5.9$$

where C_{std} and C_{unk} are the concentrations of the absorbing analyte in the standard and unknown solutions and A_{std} and A_{unk} are the absorbances of the standard and unknown solutions. The graphic in Figure 5.19 illustrates this method. A safer method is to use several standard solutions that comfortably bracket the range of absorbances of the unknown solutions. The results of the standardization can be used to verify the linearity of response and to improve the precision of the determination of $b\varepsilon$.

Another method of calibration, where the response for a standard solution cannot be determined separately, is to use the **standard addition** technique. In this approach, a known amount of the analyte is added to the unknown solution. The absorbance of this solution is determined before and after the standard addition. From the absorbance increase due to the increased analyte concentration, the original concentration can be determined. With standard addition, the standard solution concentration is $C_{unk+std}$, which is $(C_{unk}V_{unk} + C_{std}V_{std})/(V_{std} + V_{unk})$ where V_{unk} and V_{std} are the volumes of the sample and the added standard solution of concentration C_{std}. When combined with a variation of Equation 5.9 ($C_{unk} = C_{unk+std}A_{unk}/A_{unk+std}$), an equation for C_{unk} can be obtained.

$$C_{unk} = \frac{C_{std}V_{std}}{\dfrac{A_{unk+std}}{A_{unk}}(V_{std} + V_{unk}) - V_{unk}} \qquad\qquad 5.10$$

The standard addition method requires that Equation 5.9 apply, so both linearity and zero intercept are essential. An advantage of the standard addition method of calibration is that it ensures that the matrix of the standard is the same as that of the sample.

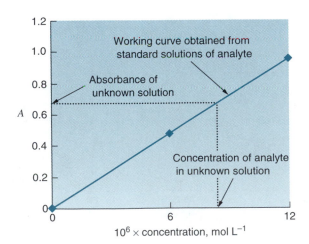

Figure 5.19. A linear working curve for an absorbance determination. The calibration approach does not require that the curve be linear or have a zero intercept, but the standard addition method does.

Working curve obtained from standard solutions of analyte

Absorbance of unknown solution

Concentration of analyte in unknown solution

A

$10^6 \times$ concentration, mol L^{-1}

A way to use photometry in a null measurement is to use a photometer to monitor the progress of a titration. Any characteristic reactivity of the analyte can be probed by a reactant. If any of the reactants or the product has an absorbance, the equivalence point can be determined photometrically.

Study Questions, Section C

17. Why is the fraction of the light absorbed not a linear function of the concentration of absorbing analyte?

18. What is the absorbance, and how is it related to $\%T$?

19. What is the absorbance of the solution in Question 16?

20. A solution has an absorbance of 0.237. What is its percentage transmittance?

21. A 0.188 M solution of cobalt nitrate is placed in a cuvette which is 1.00 cm long and illuminated with light of a 510.7 nm wavelength. The measured absorbance of the solution is 0.9158. What is the molar absorptivity of the solution?

22. A dip probe is placed in a solution of chromium nitrate, and the measured absorbance is 0.435. The absorbance of the same solution when a cuvette with a 1.00 cm path length is used is 0.810. What is the path length of the dip probe?

23. Keeping in mind the distinction between measurement and interpretation introduced in Chapter 2, is it appropriate to say one measures absorbance? What are the quantities that are actually measured?

24. The working curve in absorption spectroscopy is a plot of what two functions?

25. What limits the reasonable range of sample absorbance over which good accuracy and precision can be obtained?

26. Four standard solutions are made up, each having a known concentration of aspirin. The absorbances of these solutions are shown below. What is the concentration of an unknown aspirin solution analyzed under the same conditions with an absorbance of 0.1832?

[Standard].	Absorbance
1.000×10^{-4}	0.0820
2.000×10^{-4}	0.1636
3.000×10^{-4}	0.2427
5.000×10^{-4}	0.3862

27. Iron can be analyzed by UV spectrometry by reacting it with an excess of o-ophenanthroline to form $Fe(Phen)_3^{2+}$, a complex that absorbs UV light. In an analysis, 10 mL of standard. 0.050 mg/mL $Fe(Phen)_3^{2+}$ is added to 40 mL of a solution containing an unknown amount of $Fe(Phen)_3^{2+}$. The absorbance of the unknown before addition of the standard is 0.212, and after adding the standard it is 0.738. What is the concentration of $Fe(Phen)_3^{2+}$ in the unknown?

Answers to Study Questions, Section C

17. The fraction of the light absorbed by the first part of the solution decreases the light available to the solution deeper into the sample container. This produces a nonlinear absorbance between the absorbance and the concentration of absorber.

18. The absorbance of a solution is a measure of its light absorbing capability. It is a quantity that is proportional to the concentration of absorber, the length of the light path through the solution, and the molar absorbance of the absorbing species. Numerically, it is equal to log $(100/\%T)$.

19. $A = \log \dfrac{100}{\%T} = \log \dfrac{100}{77.52} = 0.1106$ or

$$A = \log\left(\dfrac{N_b - N_d}{N_s - N_d}\right) = \log\left(\dfrac{3056 - 27}{2375 - 27}\right)$$

$$= 0.1106$$

20. $A = \log \dfrac{100}{\%T}, \qquad 10^A = \dfrac{100}{\%T},$

$\%T = \dfrac{100}{10^A}, \qquad \%T = \dfrac{100}{10^{0.237}} = 57.9\%$

21. Using Equation 5.8, $A = \varepsilon bC$

$$\varepsilon = \dfrac{A}{bc} = \dfrac{0.9158}{(1.00 \text{ cm})(0.188 \text{ M})} = 4.87 \ cm^{-1} \, \text{M}^{-1}$$

22. This problem can be solved using the ratio

$$\dfrac{1.00 \text{ cm}}{0.810} = \dfrac{x \text{ cm}}{0.435}$$

Thus,

$$x = 0.537 \text{ cm}$$

23. Absorbance is not measured. It is calculated from the measurement of two light intensities, N_s and N_b, and the detector dark current.

24. The working curve is a plot of sample absorbance versus analyte concentration.

25. At high absorbances, the stray light reaching the detector limits the precision and accuracy. At low concentrations, the absorbance is so small that there is little difference between the transmittance of the sample and the blank. This, too, limits both precision and accuracy.

26. To find the concentration of the unknown, it is necessary to graph the calibration curve (absorbance versus concentration) and determine its best-fit line through a least squares analysis. The data plot is shown here.

    Least squares analysis gives an equation for the best-fit line of the data as follows:

$$y = (7.582 \times 10^2)x + 1.0126 \times 10^{-2}$$

Since absorbance is the dependent variable (it varies as a function of concentration), this equation can be written as

$$A = (7.582 \times 10^2) C + 1.0126 \times 10^{-2}$$

Solving for concentration gives

$$C = \frac{A - 1.0126 \times 10^{-2}}{7.582 \times 10^2}$$

To find the concentration of the unknown, we need only plug its absorbance into the above equation and solve.

$$C = \frac{0.1832 - 1.0126 \times 10^{-2}}{7.582 \times 10^2} = 2.283 \times 10^{-4} \text{ M}$$

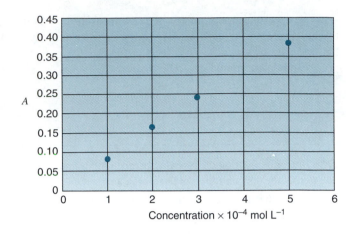

27. We need Equation 5.10,

$$C_{unk} = \frac{C_{std} V_{std}}{\dfrac{A_{unk+std}}{A_{unk}} (V_{std} + V_{unk}) - V_{unk}}$$

After inserting the data from the problem into Equation 5.10, the result is

$$C_{unk} = \frac{0.050 \text{ mg mL}^{-1} \times 10 \text{ mL}}{\dfrac{0.738}{0.212} (40 \text{ mL} + 10 \text{ mL}) - 40 \text{ mL}}$$

$$= 0.0037 \text{ mg/mL}^{-1}$$

The concentration of $Fe(Phen)_3^{2+}$ in the unknown is therefore 0.0037 mg mL^{-1}.

D. Instruments for Absorption Measurements

Complete instruments based on the measurement principles described earlier are very common. They range from simple photometers to sophisticated spectrophotometers. Their components and operation are described in this section.

PHOTOMETERS

A block diagram of an instrument called a **photometer** is shown in Figure 5.20. It includes a light source, a wavelength selecting device, a means of passing the light through the sample, a light intensity measuring system including a light detector, and a method for calculating %T. Most simple instruments perform the %T calculation by means of adjustments of the instrument parameters. One adjustable parameter is the readout number when the shutter is closed. This offset adjustment is set so that the readout number is zero when the shutter is closed. Internally, this offset correction applies an adjustable compensating current to nullify the dark current from the detector. Then the sensitivity of the current-to-voltage converter is adjusted so that the readout is 100% or 1.000 when the blank solution is in place. When these adjustments have been made, Equation 5.4 becomes

$$\%T = \frac{N_s}{100} \qquad \text{or} \qquad \frac{N_s}{1.000} \times 100 \qquad\qquad 5.11$$

which allows the readout number to be equal to %T.

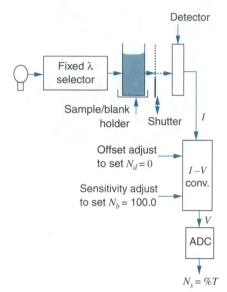

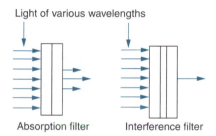

Figure 5.20. Photometers have a fixed probe wavelength and adjustments that enable a numerical readout directly in %T.

For photometers, the wavelength-restricting device most often used is an **optical filter**. The simplest optical filter, called an **absorption filter**, is a piece of colored glass. An absorbing material is dissolved in the glass or in a gel sandwiched between protecting layers of glass, as shown in Figure 5.21. The nature of the absorber is to transmit only a particular range of wavelengths. Absorption filters are available in the range of 400 to 2000 nm. In application, the wavelength of transmission of the filter is chosen to match the wavelength of absorption of the analyte.

Another type of optical filter is the **interference filter**. In this device, a very thin layer of transparent dielectric material is placed between glass sheets that have a partially reflecting metallic coating. The thickness of the dielectric is carefully controlled to be half the wavelength of the light to be transmitted. Note that the wavelength of the light when in the dielectric will generally be shorter than it is in air. When light impinges on one side of the filter, the first metallic layer reflects some of it. The part that gets through this layer may be transmitted or reflected by the second metallic layer. The reflected part can be repeatedly reflected between the metallic layers until it is transmitted through one of them. If the internally reflected light emerges in the same phase as the originally transmitted light, it will reinforce this light. If it is of random phase, it will tend to cancel out.

The reinforcement will only happen if the internally reflected light was in the dielectric for an integer number of wavelengths. The wavelength (in air) of the reinforced transmission will be $\lambda = 2t\eta/n$ where t and η are the thickness and refractive index of the dielectric and n is the integer number of wavelengths corresponding to $2t$ in the dielectric. The band of transmitted wavelengths for the interference filter is much narrower than for the absorption filter (see Figure 5.22), but the interference filter can transmit several such bands (for different values of n). When an interference filter is combined with an absorption filter, the undesired transmittance bands can be removed. Interference filters can be obtained for operation throughout the UV, visible, and IR regions of the spectrum.

Another option for the photometer is the use of a source that inherently emits light over a narrow band of wavelengths. Such monochromatic sources are lasers and light-emitting diodes.

Filter and monochromatic source photometers are used in situations where simplicity is desired and where the absorbance wavelength to be tested does not need to be changed very often. For general-purpose laboratory use, changing the filter or light source for each different analyte absorbance can be tedious. In this application, the

Light of various wavelengths

Absorption filter   Interference filter

Figure 5.21. Optical filters are very simple devices for limiting the probe wavelength.

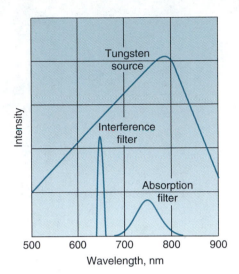

Figure 5.22. Optical filters can be used with a white light source to provide a limited band of wavelengths. Filters are available for many different wavelengths. Changing the probe wavelength requires changing the filter.

ability to continuously vary the wavelength of the probe illumination is very desirable. Work on tunable monochromatic sources is progressing at a great rate, but they are currently limited by either the available intensity, wavelength range of operation, or both. Thus, the classic spectrometer, described next, is still the basis of most general-purpose laboratory absorption instruments.

SPECTROMETERS

A **spectrometer** is an absorbance measurement instrument that uses a wavelength dispersion device for the test wavelength selection. The two wavelength dispersion devices in common use are the prism and the grating. Both devices are illuminated by a beam of parallel polychromatic light, and in each case, the angle of the light emerging is a function of its wavelength. This spatial dispersion of the wavelengths of light allows the selection of any region of the spectrum as the probe for the light absorption measurement. When the selected region is a relatively narrow band of wavelengths, the wavelength selection device is called a **monochromator**. Light path diagrams of prism and grating based monochromators are shown in Figures 5.23 and 5.24. The prism device relies on the refraction of light that occurs at each air–prism interface. For dispersion to occur, the refractive index of the prism must change over the range of wavelengths of light to be dispersed. Since the light is transmitted through the prism, the wavelength range available with the prism monochromator is necessarily limited to the wavelengths for which the transmittance of the prism material is high. Fused silica or quartz prisms work well through the UV, visible, and near IR regions (160–3200 nm). For IR work, a NaCl prism is used.

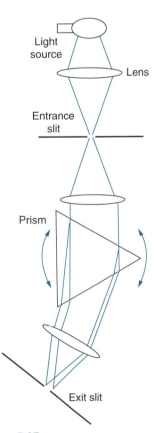

Figure 5.23. Optical diagram for a prism monochromator.

In the case of the grating, the dispersion is achieved by the constructive interference of light reflected from narrow parallel grooves in the grating surface. (For a more detailed discussion of this phenomenon, see the derivation of Bragg's Law on the CD.) The angle of constructive reflection is a function of the density of grooves in the grating surface and the wavelength of the light being reflected. Unlike the prism, the relationship between the angle and wavelength is linear. Gratings are available for use over the entire range of absorbance measurements. Their only drawback is that, as for the interference filter, light of 1/2, 1/3, 1/4, etc., the wavelength of the principal constructive interference will also be reflected at the same angle. These are called higher order reflectances. In the regions where the source may produce such wavelengths, they are eliminated from the light path with an optical filter. Most modern spectrometers are based on the grating approach.

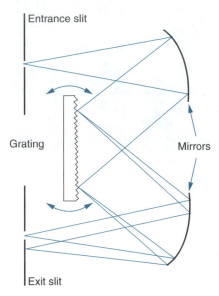

Entrance slit

Grating

Mirrors

Exit slit

Figure 5.24. Optical diagram for a grating monochromator.

The quality of the monochromator part of a spectrometer is judged by three parameters. One is the narrowness of the wavelength range of light they transmit. The range of wavelengths used as the test probe for absorbance is called the **bandwidth**. Narrow bandwidths in prism and grating monochromators are achieved by the use of narrow slits at the entrance and exit of the monochromator. It follows that the narrower the bandwidth, the less intense the probe illumination will be. The second parameter is the **stray light** specification. This is the fraction of the transmitted illumination that is composed of wavelengths other than those in the desired bandwidth. As disussed earlier, stray light is not affected by the sample in the same way as is light of the desired probe wavelengths. The effect of stray light on the measurement process is shown in Figure 5.18. The third parameter is the efficiency of the monochromator, expressed in terms of the fraction of the source light within the desired bandwidth that appears at the exit. Good efficiency is especially important when narrow bandwidths are required.

Important figures of merit are bandwidth of exit light, fraction of the exit light that is outside this bandwidth (stray light), and intensity of the exit light relative to the power of the source (efficiency).

FLOW INJECTION ANALYSIS

When a flow cell such as that in Figure 5.7c is used, the sample must be introduced into the liquid stream that is flowing through the cell. This is accomplished by a sampling valve, as shown in Figure 5.25. Two streams are involved: the sample stream and the solvent stream. The object is to introduce a plug of sample into the solvent stream.

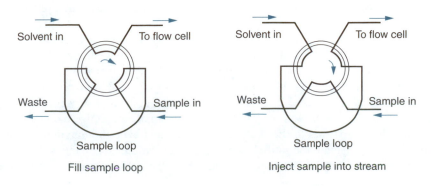

Solvent in   To flow cell

Waste   Sample in

Sample loop

Fill sample loop

Solvent in   To flow cell

Waste   Sample in

Sample loop

Inject sample into stream

Figure 5.25. A sampling valve for use with flow injection analysis. In one position of this valve, the sample loop is filled; in the other, the solution in the sample loop is sent to the flow cell.

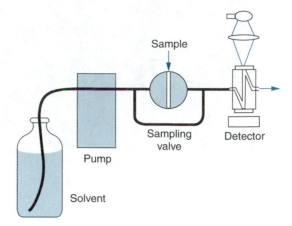

Figure 5.26. A schematic of a simple flow injection measurement system. A plug of sample solution is introduced into the stream of fluid pumped through the flow cell.

In the fill position of the valve, (Figure 5.25, left) the sample stream fills the tubing comprising the sample loop while the solvent flows directly to the flow cell. Rotating the valve 60° clockwise rearranges the flow pattern. Now the solvent flow goes through the sample loop and on to the flow cell, and the sample stream goes directly to waste. When the valve is in this position, the sample stream is usually turned off after the loop is filled. Meanwhile, the sample solution that was in the sample loop is on its way to the flow cell.

A more complete flow injection analysis (FIA) system is shown in Figure 5.26. It includes a solvent reservoir, a pump, the sampling valve, and the flow cell. A commonly used symbol for the sampling valve is used in this diagram. The pump is often a **peristaltic pump** that works by massaging flexible tubing to move the solution inside it along. The solvent stream is used to pass the sample plug on to the flow cell for the determination of absorbance. Initially, the plug of sample has constant composition throughout its length. If the solution absorbance were determined after a very short flow, as suggested in the diagram, the absorbance versus time plot would have the shape of a square peak with a peak width equal to the length of time required to pump the sample plug through the flow cell.

In most flow injection analyses, a variety of operations are performed on the sample after injection. These can include mixing with one or more reagents, heating, and adding a time delay for the reactions to be complete. In these applications, the length of tubing between the sampling valve and the absorbance detection cell can be a meter or more. Over that length of tubing, several effects will cause the edges of the initial sample plug to spread out. This effect is called **dispersion**, and the phenomena that cause it include the drag of the walls on the pumped solution and the diffusion of species from higher to lower concentration regions of the solution. In the regions where the sample has spread out to adjacent solvent, the sample is diluted by the solvent, and the absorbance is thus decreased over that of the original sample.

The concentration profile of the sample plug for a series of injection volumes is shown in Figure 5.27. The injection volume is changed by changing the length of the tubing in the sample loop of the sampling valve. The longer the loop, the larger is the sample plug injected. The dilution of the edges of the sample plug is apparent from the profiles in Figure 5.27. For the largest injection volume used, the sample in the middle of the plug was not diluted. For the smaller injection volumes, however, the dilution extends to the center of the plug, and even the maximum concentration is smaller than the initial sample concentration. A quantity called the **dispersion coefficient**, D, has been defined. It is the concentration of the sample

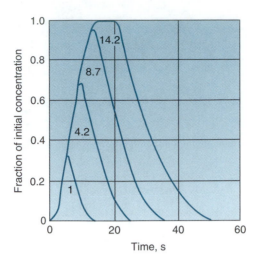

Figure 5.27. Concentration profiles of the sample plug for several different volumes of sample injected. The numbers by each peak are the ratios of the injection volumes.

plug at the time of sample determination relative to the undiluted sample concentration. Thus,

$$D = \frac{C_s^\circ}{C_s} \qquad\qquad 5.12$$

where C_s is the concentration of the analyte at the time of the absorbance determination and C_s° is the undiluted analyte concentration. If the absorbance at the peak value of the sample plug is used,

$$D_{max} = \frac{C_s^\circ}{C_{s,max}} \qquad\qquad 5.13$$

If the dispersion coefficient is in the range of 1 to 2, the system is said to have *limited dispersion*. Systems with values of D from 2 to 10 are said to have *medium dispersion. Large dispersion* values are those over 10. The dispersion coefficient decreases with increasing sample volume up to a maximum of 1, after which it plateaus as shown in Figure 5.27. For a given injection volume for which D_{max} is more than 1,

$$D_{max} = Kl^{1/2}r^2 \qquad\qquad 5.14$$

where l is the length of the tubing between the injector and detector and r is the radius of the tubing used. From Equation 5.14 we see that the greatest effect on D_{max} is from the radius of the tubing. Doubling the tubing length will only increase D_{max} by a factor of 1.4, but halving the tubing diameter will decrease D_{max} by a factor of 4. The sample volume for which the value of D_{max} is 2.0 is defined as $S_{1/2}$. The value of $S_{1/2}$ decreases with decreasing tubing diameter.

It is customary to calibrate FIA systems by the injection of a series of standard solutions. Multiple injections of the same sample are often done, as shown by the response plot of Figure 5.28. The repetitive responses provide data on the precision of the determination and on whether the samples are free of carryover from one sample to the next. If the analyte concentration at the detector has not dropped to an insignificant amount by the time the next sample peak has arrived, the measured value will be affected by the concentration of the previous sample. When carryover is occurring, the responses of the repetitive samples will follow a trend. The first value of each series with increasing concentration will be lower than the others; it will be higher than the others when decreasing standard concentrations are used. From the time scale of this

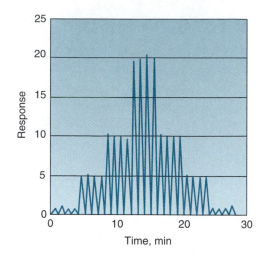

Figure 5.28. The results of a calibration run with an FIA system. Four repetitive samples of each standard were run in both increasing and decreasing concentration directions.

plot, one can see that FIA systems are capable of running many samples in a very short time. They are widely used in situations where a large number of the same types of samples are run routinely. Prime examples are environmental laboratories determining chlorine levels in drinking water and clinical laboratories performing lipid profiles for blood cholesterol.

In this discussion, we assumed that the analyte had a detectable absorptivity. In most FIA applications, the absorbing species must be derived from the analyte by one or more reactions. These reactions are also carried out in the flowing system. Examples of FIA analyses will be discussed throughout this book in the context of the specific chemistry and detection systems involved.

Study Questions, Section D

28. What is the principal difference between a photometer and a spectrometer?

29. The adjustments on a photometer are dark and %T. In terms of Equation 5.4, what do these adjustments do?

30. An interference filter is optimized to pass a narrow band of wavelengths around 682 nm. What other wavelength regions will have a significant transmittance?

31. Which is currently used in spectrometers more often, gratings or prisms?

32. Define "bandwidth" and "stray light."

33. What is the purpose of the sampling valve in flow injection analysis?

34. What is dispersion in flow injection analysis?

35. When the sample was placed in the flow cell directly, an absorbance of 0.26 was determined. The peak absorbance from an injection was determined to be 0.04. What is the value of D_{max}, and does this system have limited, medium, or large dispersion?

36. If the tubing in the system used for question 35 is reduced from 3 mm inner diameter (i.d.) to 2 mm i.d., what absorbance value for the sample would be expected?

Answers to Study Questions, Section D

28. The photometer uses an optical filter to limit the wavelengths of the probe illumination. The spectrometer uses a wavelength dispersion device with a light slit to accomplish this goal.

29. The dark adjustment is made with the shutter closed. It effectively sets the value of N_d to zero. The %T adjustment is made with the blank solution in place. It sets the value of N_b to 1.00. This done, $N_s \times 100$ is now %T as in Equation 5.11.

30. Given $n\lambda = 2t\eta$, for $n = 1$, $2t\eta = 682$ nm. For $n = 2$ and $n = 3$,

$$\lambda_2 = 682/2 = 341 \text{ nm}, \qquad \lambda_3 = 682/3 = 227 \text{ nm}$$

31. Gratings are more often used. They can be replicated inexpensively and provide a linear relationship between wavelength and angle of reflectance.

32. Bandwidth is the range of wavelengths transmitted through the wavelength selection device when set for the

desired test wavelength. Narrow bandwidths are generally desired. Stray light is the fraction of the transmitted illumination that is outside the specified bandwidth. Low stray light is desired.

33. The sampling valve inserts a plug of the sample solution in the flowing stream of solvent.

34. Dispersion is the dilution of the sample plug at the leading and trailing edges of the plug. It is caused by diffusion and unequal flow velocity at the center and near the wall of the tubing.

35. Since the absorbance is proportional to the concentration, we can use the absorbance ratio in Equation 5.13 to calculate D_{max}. Thus,

$$D_{max} = 0.26/0.040 = 6.5$$

Since D_{max} is between 2 and 10, the system is a medium dispersion system.

36. The value of D_{max} will be decreased by the ratio of the square of the radii, or 2.25. This will increase the concentration and therefore the absorbance by the same factor, so the expected absorbance would be $0.04 \times 2.25 = 0.09$.

E. The Absorbance Spectrum

One way to determine the colors of light a particular analyte absorbs is to measure the absorbance as a function of the wavelength or frequency of the illuminating radiation. The plot resulting from such an experiment is called an **absorbance spectrum**. The absorbance spectrum for β-carotene in the visible region of the spectrum is shown in Figure 5.29. The absorbance spectrum indicates which wavelengths of light the sample absorbs and which it does not. From the spectrum, it is possible to deduce the color we would perceive the object to be. We also see from the spectrum that as the absorbance of a sample is a function of the wavelength of light, so will be the molar absorptivity, ε. Therefore, when the molar absorptivity for a compound is listed, the wavelength of light for which that value is true must also be specified. It is usually given for the wavelength for which the absorbance is a maximum in the region of the absorbance peak.

RATIONALIZING THE ABSORBANCE SPECTRUM

The absorbance spectrum of a compound results from the fundamental qualities of its atoms and the chemical bonds between them. Therefore, the spectrum can provide useful information regarding the chemical composition and structure of the analyte. A molecule, atom, or ion can absorb a photon if the energy of the photon matches an allowed change in the energy state of the species. The energies of ultraviolet and visible photons are matched to changes in the energy states of the bonding electrons in the absorbing species. The ultraviolet spectrum of benzene vapor is shown in Figure 5.30. The energies of the absorption peaks correspond to known electron energy transitions within the benzene molecule. The photon energy axis of UV and visible spectra is generally plotted as the wavelength in nanometers. Shorter wavelengths correspond to higher energies.

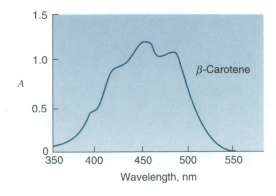

Figure 5.29. The change in ε as a function of probe wavelength gives rise to a spectrum of characteristic shape.

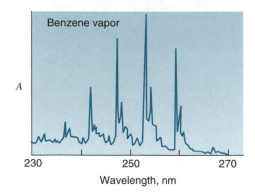

Figure 5.30. Absorbance peaks correspond to the allowed transitions between energy levels of the absorbing molecule.

An interesting phenomenon occurs when the spectrum of a benzene solution is taken. Instead of the sharp absorption peaks at specific photon energies, the absorption bands are much broader, as shown in Figure 5.31. The interactions between molecules in the liquid state increase the allowed energy transitions to an apparent continuum of energy states rather than the few distinct states allowed by the individual molecules. This phenomenon of absorbance-band broadening is common in all dissolved or liquid-state absorbers.

Photons in the infrared part of the spectrum do not have enough energy to produce changes in the electronic structure of an analyte. The energy changes associated with infrared photon absorption are associated with the bonds between atoms. The distance between each pair of bonded atoms has a preferred value. Energy is required to compress or stretch this distance. If it is compressed, the atoms will tend to push apart. The energy of this repulsion makes them farther apart than the optimum distance, so then an attractive force takes over. This sequence is repeated indefinitely; in other words, the bond distance has an oscillatory nature and the bond is said to be vibrating. Vibrations in the bond distance between atoms are often called the stretching motion. The energy that started the vibration is stored in the vibrating bond. The more energy that is stored, the more intense the vibration. Interestingly, the allowed energy states for vibrations are also limited to specific values that depend on the particular atoms involved.

Other oscillations in the relative positions of bonded atoms can occur. These are rotations of an atom around the bond axis and changes in the angle between two bond axes associated with a single atom. The IR spectrum of a compound reveals all of these allowed energy absorptions. An example is the IR spectrum of 2-methylpentane shown in Figure 5.32. The photon energy axis in IR spectra is generally plotted as the **wavenumber** $\bar{v}$ (in cm^{-1}), which is the reciprocal of the wavelength. Thus, $\bar{v} = 1/\lambda$. The larger the value of $\bar{v}$, the higher the energy of the photon.

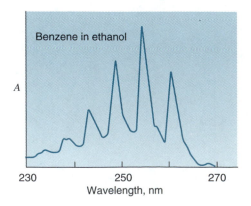

Figure 5.31. Intermolecular interactions in the liquid state provide a continuum of allowed energy states. The spectrum loses much of its distinctiveness.

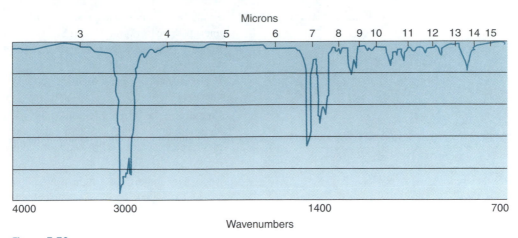

Figure 5.32. Infrared spectrum of 2-methylpentane.

   In the infrared, spectra can contain considerable detail regarding the structure of the analyte molecule. The vibrational, bending, and stretching energy states that absorb the IR photons are little influenced by intermolecular interactions. Therefore, solid and dissolved analytes retain the narrow absorption peaks characteristic of a limited band of allowable energy transitions. Not all stretching, bending, and rotational modes can be activated by IR absorption. Only those that result in a change in the dipole moment of the molecule will result in an IR absorption peak. Fortunately, the electron distribution in most compounds is sufficiently unsymmetrical to enable the absorption of infrared radiation.

Study Questions, Section E

37. What is an absorbance spectrum?

38. What is the relationship between the absorbance spectrum and the fundamental characteristics of the absorbing species?

39. Calculate the energy in electron volts and in wavenumbers of the following wavelengths of light absorbed by benzene.

   A. 229 nm (the UV region)

   B. 6250 nm (the IR region).

40. What types of energy changes are associated with the infrared region of the spectrum?

41. What is the wavenumber of a light wave, and how is it related to the energy of the light wave?

Answers to Study Questions, Section E

37. An absorbance spectrum is a plot of the absorbance of a sample versus the wavelength (or wavenumber) of light.

38. The energy of the absorbed light corresponds to specific allowed energy transitions within the absorbing species. Therefore, the energies of the wavelengths at which a sorption peaks occur are related to molecular composition and structure.

39. A. In Question 4, we used the relationship between $\mathscr{E} = hc/\lambda$ energy in joules and wavelength. Apply-

ing this relationship to the wavelength of absorption gives

$$\mathscr{E} = \frac{(6.626 \times 10^{-34}\ \text{J} \cdot \text{s})(3.00 \times 10^{8}\ \text{m/s})}{229 \times 10^{-9}\text{m}}$$

$$= 8.68 \times 10^{-19}\ \text{J}$$

$$\mathscr{E} = (8.68 \times 10^{-19}\ \text{J})\left(\frac{1\ \text{eV}}{1.602 \times 10^{-19}\ \text{J}}\right) = 5.42\ \text{eV}$$

The energy in wavenumbers is simply the reciprocal of the wavelength, or

$$\bar{v} = \frac{1}{\lambda} = \frac{1}{229 \text{ nm}} = 0.00437 \text{ nm}^{-1}\left(\frac{1 \text{ nm}}{10^{-7} \text{ cm}}\right)$$
$$= 4370 \text{ cm}^{-1}$$

B. In the same way,

$$\mathscr{E} = \frac{hc}{\lambda} = \frac{(6.626 \times 10^{-34} \text{ J} \cdot \text{s})(3.00 \times 10^8 \text{ m/s})}{6250 \times 10^{-9} \text{ m}}$$
$$= 3.18 \times 10^{-20} \text{ J}$$

$$\mathscr{E} = (3.18 \times 10^{-20} \text{ J})\left(\frac{1 \text{ eV}}{1.602 \times 10^{-19} \text{ J}}\right) = 0.199 \text{ eV}$$

$$\bar{v} = \frac{1}{\lambda} = \frac{1}{6250 \text{ nm}} = 1.60 \times 10^{-4} \text{ nm}^{-1}\left(\frac{1 \text{ nm}}{10^{-7} \text{ cm}}\right)$$
$$= 1600 \text{ cm}^{-1}$$

From these calculations we see that UV light is higher in energy than IR light. UV light has a lower wavelength than IR light, since wavelength is inversely proportional to energy. UV light has a higher wavenumber than IR light, since wavenumber is directly proportional to energy.

40. Energies in the infrared region of the spectrum are associated with the energies in interatomic bonds. Specific energies are associated with bond stretching, bending, and rotating.

41. The wavenumber $\bar{v}$ of a light wave is the reciprocal of its wavelength: $\bar{v} = 1/\lambda$. Longer wavelengths correspond to lower energies, so the larger the value of $\bar{v}$, the higher is the energy.

F. Obtaining an Absorbance Spectrum

An absorbance spectrum is always obtained by a separate determination of the absorbance for each wavelength sampled. That is, N_s and N_b must be measured separately at each wavelength. This is because the intensity of the light source will be a function of wavelength, as are the sensitivity of the detector and the transmittance of the blank solution. A necessary function of a spectrometer that can be used to obtain the absorbance spectrum of a sample is the ability to separately test each wavelength of the light whose transmittances through the sample and blank are to be determined. Two ways to accomplish this are described below.

SCANNING SPECTROMETERS

The earliest and still some of the most precise spectrometers are called scanning spectrometers because they produce a spectrum by scanning through the desired wavelength range, measuring N_s and N_b and calculating A as the wavelength is varied. A plot of A versus. λ is produced. The block diagram of a scanning spectrometer is shown in Figure 5.33. Light from a light source that contains all the wavelengths of interest is sent through a **monochromator**. The monochromator is used to select the wavelength of light to be tested and reject all the others. The light emerging from the monochromator is said to be **monochromatic**, that is, of one color. It isn't strictly of just one wavelength, though. Rather, it is composed of a band of wavelengths, the narrowness of which depends on the quality of the monochromator. The width of wavelengths transmitted by the monochromator is called the **bandwidth**. As we shall see, the bandwidth of the monochromator affects the accuracy of absorbance measurements in wavelength regions of changing absorption. The selected wavelength band is scanned systematically through the wavelengths of interest by a motor rotating the wavelength selection device in the monochromator. The narrower the bandwidth, the better will be the wavelength resolution of the spectrum that can be produced

From the monochromator, the light passes alternately (or simultaneously) through the sample and the blank solutions on its way to the detector. The need to measure N_b and N_s at every wavelength necessitates a mechanism for the rapid alternation of the light intensity measurement between the paths through the sample and the blank solutions. A variety of ways to accomplish this alternation has been developed, includ-

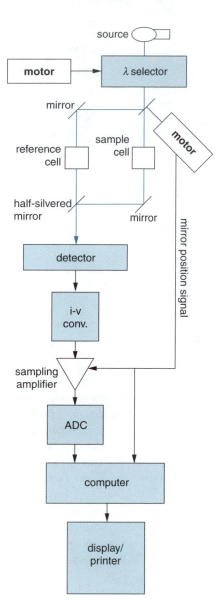

Figure 5.33. A scanning spectrometer must have a way of continuously changing the wavelength of the probe illumination and a way to measure N_b and N_s at each wavelength value.

ing moving the light beam, moving the cells that hold the test solutions, and alternating the path impinging on the detector.

In the example illustrated in Figure 5.33, the beam is sent alternately through the sample and reference cells by a rotating disk with transparent and reflecting segments. The output from the detector is converted to a number related to the light intensity for each light path position at each wavelength of light. Because the detector dark current is relatively constant, it is not necessary to measure it at each wavelength. Rather, the dark current is compensated (offset to zero) by a manual (or automatic) adjustment when a shutter in the light path is closed. This process is needed only infrequently.

The numbers related to the intensity measurements are digitized in modern instruments, and the absorbance for each wavelength is calculated by computer. The computed value is stored in computer memory along with the associated wavelength for displaying and plotting the spectrum. Scanning spectrometers come in a wide variety and range of capabilities. They differ mainly in the mechanism for sample/blank alternation, the quality of the monochromator employed, the wavelength range over which they operate, and the nature of the controls and readouts employed.

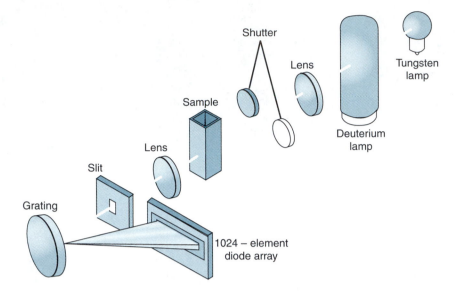

Figure 5.34. Optical diagram of the HP 8453 diode array spectrophotometer. Courtesy of Hewlett Packard Company, Palo Alto, CA.

ARRAY DETECTOR SPECTROMETERS

An increasingly popular variation on the classic spectrometer is the **array detector spectrometer**. In these instruments, the monochromator exit slit and photodetector are replaced by an array of photodiode **detector elements,** each of which detects light over a particular bandwidth. The set of detector elements is called a **photodiode array**. A block diagram of such an instrument is shown in Figure 5.34. The photodiode array is exposed to the light transmitted through the sample or blank solutions, much as one would expose a photographic film. The exposure creates a buildup of electrical charge in each element of the array corresponding to the amount of light falling on that element during the exposure. At the end of the exposure, the charge on successive elements is converted to a related voltage and read out as a series of values related to the position of the element in the array. This process can be repeated many times each second. Continuous and dramatic improvement in the sensitivity and dynamic range of array detectors has made the characteristics of this type of spectrometer competitive with the scanning spectrometer for many applications.

The great advantage array spectrometers have over the scanning type is that they simultaneously measure the light intensity at each wavelength region impinging on a detector element. This obviates the need for scanning the monochromator and makes much more efficient use of the light transmitted through the sample. To obtain a spectrum, the calculation of Equation 5.4 must be made for each detector element separately. With the shutter closed, the dark reading for each element is obtained. These dark readings are different for each wavelength because each detector element has slightly different characteristics. With the blank solution in place, the complete set of N_b values is obtained for the spectrum. (See Figure 5.35.) Then the sample is placed in the light path, and the set of N_s values is obtained. The computer then calculates the absorbance value for each detector element and plots the absorbance spectrum using the wavelength associated with each detector element for the X-axis values. It is not necessary to collect the dark and blank spectra for each sample spectrum as along as the drift in these values is within the required precision. Therefore, once the dark and blank values have been collected, the sample cell can be left in the light path, and the spectrometer can collect sample spectra as fast as the detector expose and read cycle will allow.

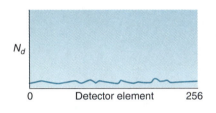

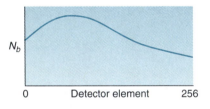

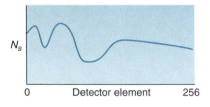

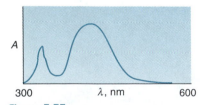

Figure 5.35. These curves illustrate the three sets of data taken in the array detector spectrometer from which the absorbance can be calculated at the wavelength associated with each detector element. Outputs from a 256-element detector are shown.

The ability to collect complete sample spectra at rates of 10 spectra per second or more makes the array spectrometer uniquely suited for situations in which the sample composition is changing with time. Examples are in the areas of titrations, kinetics measurements, and chromatographic detection. Obtaining complete spectra as a function of time or titrant volume creates a voluminous, three-dimensional data set, such as

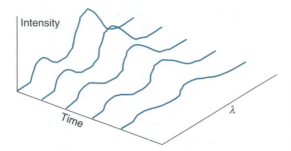

Figure 5.36. A cartoon of spectra taken at regular time intervals. The spectral generation rate determines the resolution available along the time axis. There are several practical limits to the time resolution attainable.

that of Figure 5.36, comprised of hundreds of spectra. Such an extensive data set was not practical until inexpensive computer power and data storage were available. Now that such data sets can be gathered routinely, the wealth of analytical information they contain is beginning to be explored and utilized.

The principal disadvantage of array detector spectrometers is the limited number of detector elements along the range of wavelengths sampled. For example, if the detector has 1024 elements and the wavelengths dispersed along those elements range from 190 to 850 nm, the bandwidth associated with each element will be (850 − 190)/1024 = 0.64 nm. For an array detector spectrometer, this figure becomes the effective bandwidth of the system. Although this is small compared to the absorption peak widths in liquid spectra, it may be limiting in some situations. In contrast, the bandwidth of the illumination reaching the detector in many scanning spectrometers can be as little as 0.01 nm. Better spectral resolution is possible with the array detector spectrometer if the detector elements are used to cover a narrower spectral range. To achieve 0.01 nm resolution with 1024 elements, the total available spectral range would only be 1024 × 0.01 = 10.24 nm. Another disadvantage is the dynamic range available within one spectrum. With a scanning spectrometer, the system can accommodate great differences in the measured light intensity in different parts of the spectrum by dynamically adjusting the sensitivity of the detector and amplifier. This is more difficult to accomplish when the signal from all wavelength elements is measured the same way. A recent improvement in array detectors can allow selective adjustment of the exposure time for each element to extend the effective dynamic range.

FOURIER TRANSFORM SPECTROMETERS

Many spectrometers that operate in the infrared region of the spectrum use a Michelson interferometer in conjunction with a broadband IR emitting light source. (See Figure 5.37.) The interferometer contains a beam splitter that divides the incoming light into

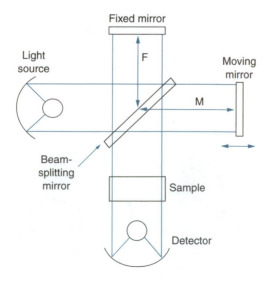

Figure 5.37. Because of their simplicity, high spectral generation rate, and low-cost computing power, Fourier transform IR instruments have replaced scanning IR instruments for many applications.

two paths (F and M in Figure 5.37). A moving mirror changes the length of the M path before the two beams are recombined at the exit. The differences in the path lengths provide constructive and destructive interference of the various wavelengths of light at the output. The light intensity is measured as a function of the position of the mirror (path length difference). When the two paths are of exactly equal length, all wavelengths constructively interfere at the exit, and the light intensity is high. At other mirror positions, constructive interference will occur for those wavelengths for which the path difference is some multiple of the wavelength of the light. In the same way, destructive interference will occur for those wavelengths equal to $n_{odd}/2$ times the path difference, where n_{odd} is any odd integer. As shown in Figure 5.38, taking the Fourier transform of the intensity versus mirror position data yields an intensity versus wavelength plot, in other words, a spectrum. However, since the Fourier transform spectrometer is essentially an IR instrument and IR spectra are normally plotted versus wavenumber, that is the way they are presented here. (For a mathematical explanation of the operation of the Fourier transform spectrometer, see the derivation section of the accompanying CD.)

To acquire a spectrum from a sample, the sample is placed in the light path between the interferometer and the detector. The result of the Fourier transform is a spectrum of the absolute intensity of the transmitted light as a function of wavenumber. To calculate the absorbance versus wavelength, the spectrum of the light intensity transmitted through a blank is required. Then Equation 5.4 is used to calculate the %T for each wavenumber.

The Fourier transform spectrometer shares many of the advantages of the array detector spectrometer. Data to generate a complete spectrum are obtained for each excursion of the mirror. Since the mirror can be cycled through its excursions several times per second, the spectral generation rate can be high. Even though only a single-channel detector is used, all the wavelengths in the source are being sampled at every position of the mirror. In the Fourier transformation, every point in the acquired data set contributes to the intensity determination at every wavelength. Thus, there is highly efficient use of the transmitted light. The wavelength resolution (effective bandwidth of the probe illumination) depends on the mechanical and optical precision of the interferometer. Mirror and mirror movement tolerances must be substantially less than one wavelength of the shortest wavelength in the spectral range probed. At 1000 nm (the far red end of the visible spectrum) the required mirror flatness and movement tolerances are less than 1/1000 of a millimeter. Compared to array detector spectrometers,

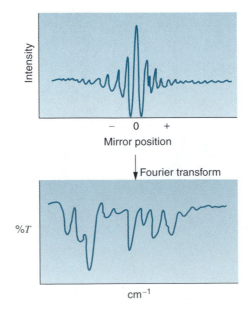

Figure 5.38. The data set of intensity versus mirror position can be transformed into a spectrum of %T versus wavenumber.

the Fourier transform spectrometer offers no advantage in the UV and visible spectral regions. In the IR and far IR regions, however, it has become the most common form of spectrometer. As with the array detector spectrometer, the availability of inexpensive computing power (for quickly calculating the Fourier transform) has allowed an originally esoteric laboratory technique to enter the arena of mainstream applications.

ATTENUATED REFLECTANCE SPECTROSCOPY

When a light wave encounters a boundary between two substances of significantly different refractive index, it can be reflected back into the initial material. The likelihood of reflectance increases with the difference in refractive index and with a decrease in the angle with which the beam approaches the interface (a more glancing angle giving high reflectance). An optical element designed to implement such reflections is called a **waveguide**. A typical waveguide shape with associated light path is shown in Figure 5.39.

Since the property of reflectance is so much affected by the difference in the refractive indices between the two materials, it follows that the light, in some way, has probed the material on the other side of the boundary from which it is being reflected. We now understand that the reflected wave imparts an electromagnetic field into the medium on the other side of the boundary. This field is called an **evanescent wave**. The depth of the penetration of this field is very small, of the order of one wavelength of the reflected illumination. Further, the intensity of this field decreases rapidly with distance from the boundary. Of great analytical significance is the fact that absorption of the evanescent wave can occur. In other words, absorbing molecules, close to the surface of the waveguide, can cause less of the wavelength they absorb to be reflected back into the waveguide. This effect is small, but after many reflections it can become significant. A comparison of the transmitted light intensity with and without the analyte(s) contacting the waveguide at each wavelength enables the generation of an absorption spectrum for the components of the sample. This spectroscopic method is called **attenuated total reflectance** (ATR).

The sensitivity of the ATR technique can be increased by placing a layer of specific absorbent or adsorbent on the waveguide reflecting surface. If the analyte has a relatively high affinity for the absorbent or adsorbent, the concentration of analyte at the waveguide surface will be increased and thus the number of sample molecules exposed to the evanescent wave will be larger. A flow cell implementation of this concept is illustrated in Figure 5.40. An additional differentiating characteristic (greater selectivity) is provided by the absorbing layer. Molecules not attracted to this layer are dispersed throughout the solution, and even if they absorb light, they will have little effect on the spectrum.

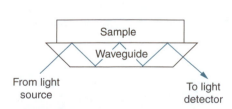

Figure 5.39. An optical waveguide with multiple internal reflections. The sample is placed on one of the waveguide surfaces, where it can absorb selected wavelengths of the source illumination.

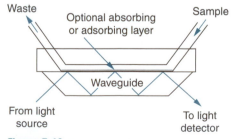

Figure 5.40. An optical waveguide with a flow-through sampling arrangement.

As indicated, the depth of penetration of the evanescent wave is roughly equal to one wavelength of light. Figure 5.3 gives an indication of the sampling depth for various regions of the spectrum. In the visible and ultraviolet regions of the spectrum, the depth is 750 nm or less. If IR radiation is used, the penetration depth is greatly increased. The ATR method is often combined with Fourier transform IR analysis for this reason. Detection limits in the parts per billion range have been achieved for a variety of gases and small organic molecules.[1]

Study Questions, Section F

42. In a scanning spectrometer, what measurements must be taken at each wavelength?

43. Considering Figure 5.10, what problems could arise when using separate containers for the blank and sample in the double-beam scanning spectrometer?

44. In Figure 5.33, why is it necessary to have the motor on the beam-switching mirror send information about its position to the sampling amplifier and the computer?

45. How does the array detector spectrometer avoid scanning the wavelength?

46. How is the information about the blank transmittance and the dark current obtained with the array detector spectrometer?

47. What is the principal advantage of the array detector spectrometer for obtaining the spectra of a sample stream of changing composition?

48. An array detector spectrometer is being used to follow a chemical reaction. The spectral generation rate is set to 10 spectra per second. The spectrometer has a 1024-element array, and a 16-bit absorbance value is recorded. The reaction is followed for 20 minutes. If the intensity data occupy 60% of the total digital storage space required for these data, calculate the size of the resulting data set in megabytes. (A byte is 8 bits.)

49. Some array detector spectrometers use separate arrays for the UV and visible portions of the spectrum. Assuming that the UV detector monitors the region from 190 to 440 nm with 512 elements and the visible detector covers the region from 400 to 850 nm with 2048 elements, calculate the detector resolution limit for each region.

50. Justify the statement, "Only when the path lengths M and F (see Figure 5.37) in the interferometer are equal will all wavelengths of light constructively interfere."

51. How is information about the blank obtained in a Fourier transform IR spectrometer?

52. What is the purpose of the waveguide in attenuated total reflectance spectroscopy?

53. Compare the depth of penetration of the evanescent wave between green light and 1000 wavenumber illumination.

Answers to Study Questions, Section F

42. The values for N_s and N_b must be measured at each wavelength. The value for N_d need not be taken so often since it is not a function of wavelength.

43. The optical behavior of the two sample compartments might not be identical. They could differ in reflectivity, scattering, or path length.

44. The sampling amplifier needs to sample the detector output when the beam is totally in the reference path or the sample path and not in between. The number from the ADC can be either N_s or N_b depending on the mirror position. The data interpretation step, performed in the computer, requires this information.

45. Scanning is avoided by using an array of detectors spread across the dispersion axis of the dispersing device. Each detector element detects the intensity of one narrow section of the spectrum.

46. The dark readings are obtained for all detector elements when the shutter is closed. A blank solution must be put in the sample compartment to obtain the blank information for each element.

47. The stream can be run through the sample compartment and the spectra obtained continuously. The spectral generation rate is very much faster than attainable with a scanning spectrometer, so much more rapid changes in composition can be followed.

48. Since the array detector records absorbance at every element, the entire amount of storage space used by each spectrum will be the product of the bits per absorbance value and the number of elements in the array.

(1024 elements)(16 bits/element) = 16,384 bits/spectrum

[1]Niemczyk, T., Haaland, D. and Han, L. *Applied Spectroscopy* **53**, 381 (1999) and **53**, 390 (1999).

The total number of spectra collected is

$$\frac{10 \text{ spectra}}{\text{sec}} \times 20 \text{ min}\left(\frac{60 \text{ sec}}{\text{min}}\right) = 12{,}000 \text{ spectra}$$

The amount of digital storage space taken up *by just the intensity data* is then

$$\left(\frac{16{,}384 \text{ bit}}{\text{spectrum}}\right)(12{,}000 \text{ spectra})\left(\frac{1 \text{ byte}}{8 \text{ bits}}\right) = 2.46 \times 10^7 \text{ bytes}$$

The intensity data occupy only 60% of the total storage space. The rest of the storage space in the file is occupied by commas, spaces, etc., which make the data readable. We calculate the total size of the data set based on that of the intensity data as follows (where S is the size of the data set):

$$S \times 0.60 = 2.46 \times 10^7 \text{ bytes}$$

$$S = 4.10 \times 10^7 \text{ bytes}.$$

So the entire data set takes up 41 megabytes of storage space.

49. Remember from Chapter 2 that resolution is defined as the smallest difference in the measured quantity that can be detected. We can calculate the resolution in the two regions of the spectrum by dividing the wavelength range by the number of spectral elements. For the UV region,

$$\text{detector resolution} = \frac{440 - 190 \text{ nm}}{512} = 0.488 \text{ nm}$$

For the visible region

$$\text{detector resolution} = \frac{850 - 400 \text{ nm}}{2048} = 0.220 \text{ nm}$$

For this particular spectrometer, the resolution in the visible region is significantly better than that in the UV range.

50. When M and F are equal lengths, all wavelengths will constructively interfere because they all have the same path length. When M and F are different, and the difference is equal to a multiple of the light wavelengths, constructive interference will occur, but this can be true for only a few wavelengths at any value of the path length difference.

51. A blank solution is placed in the sample compartment and the 100% transmittance spectrum is obtained.

52. The waveguide contains the probe illumination and provides the reflecting surface against which the sample is placed.

53. From Figure 5.1, green light has a frequency of 6×10^{14} Hz. Ignoring the decrease in velocity it has through the waveguide, the wavelength will be $3 \times 10^8 \text{ m s}^{-1}/6 \times 10^{14}$ $\text{s}^{-1} = 5 \times 10^{-7}$ m. Light with a wavenumber of 1000 cm^{-1} has a wavelength of 1×10^{-3} cm or 1×10^{-5} m. Therefore the penetration of the IR light is $1 \times 10^{-5} \text{ m}/5 \times 10^{-7}$ m = 20 times greater than the green light.

G. Spectral Precision and Accuracy

The absorbance spectrum of a compound or solution usually has regions of high and low absorptions. Information regarding the amount of absorbing material in the sample lies in the accuracy of the absorbance measurement in the regions of peak absorption. Information regarding the identity of the absorber lies in the variations of absorption as a function of wavelength (the "shape" of the spectrum). In either case, accuracy in the absorbance values obtained and in the wavelengths at which they are measured is essential to the interpretability of the data. The influences of stray light and the absorbance value on the accuracy and precision of the absorbance determination have already been discussed. Another factor to be considered is the effective spectral bandwidth relative to the features in the spectrum.

An absorbance measurement will be inaccurate if the sample absorbance changes significantly over the effective bandwidth of the probe illumination. Consider an absorbance measurement at a wavelength corresponding to a straight side of an absorption peak, as diagrammed in Figure 5.41. At one side of the bandwidth the absorbance is 0.620, and at the other side it is 0.660. These correspond to %T values of 23.99 and 21.88, respectively. With a symmetrical distribution of intensities within the bandwidth, we will obtain a value for the average %T over the bandwidth. This is 22.94 for the example. The absorbance corresponding to this %T is 0.639. The exact value for the absorbance in the center of the bandwidth sampled is 0.640. For a narrower bandwidth of probe illumination, the error would be decreased. For this reason, quantitative absorbance measurements are best performed at the peak absorption wavelengths, where the change of absorbance over the effective bandwidth is minimized. However, if the absorption peak

Information regarding the concentration of an absorbing species depends on the accuracy of the absorbance values obtained. The interpretation of absorbance data to determine the identity of the absorbing species depends on obtaining undistorted spectra. Excessive probe illumination bandwidth leads to errors in both peak absorbance values and distortions in the spectral shape.

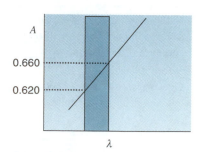

Figure 5.41. When the absorbance changes across the bandwidth of the probe illumination, the average absorbance is not equal to the absorbance in the center of the bandwidth.

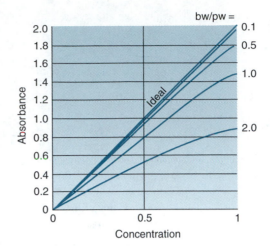

Figure 5.42. The effect of the ratio of bandwidth (bw) to peak width (pw) on the slope and linearity of the working curve.

is not substantially wider than the effective bandwidth of the probe, the error will still exist. The effect of having a bandwidth comparable to the absorption peak is a significant curvature in the working curve and a significant decrease in the slope. This effect is illustrated in Figure 5.42. The wider the bandwidth relative to the peak width, the greater is the curvature and the lower is the absorbance for a given concentration.

The shape of the absorbance curve can also be distorted by excessive bandwidth in the probe illumination. Unless the bandwidth is small compared to the features (peaks, valleys, and edges) of the spectrum, these features will be broadened. These distortions will interfere with the techniques of data analysis used to interpret such spectra according to the composition of the sample. As we have seen, dissolved and liquid samples do not produce sharp features in the UV and visible regions of the spectrum. Therefore, moderately narrow bandwidths will usually provide good data. On the other hand, the IR region of the spectrum contains a wealth of detail, more and more of which is revealed as the wavelength resolution of the measurement system increases.

Study Questions, Section G

54. What are the consequences of a spectral bandwidth in the probe illumination over which the absorbance of the sample is not constant?

55. How would you experimentally determine whether the bandwidth of your spectrometer is causing the type of error discussed in this section?

Answers to Study Questions, Section G

54. The absorbance determination will be in error, and the absorbance of the sample may not be a linear function of the concentration.

55. One way is to take a spectrum at several concentrations and see if the calculated absorbance remains proportional

to the concentration throughout the spectrum. Another way is to compare spectra taken at two bandwidths. If they are the same, then both bandwidths are sufficiently small. If they are different, then at least the larger one involves errors.

H. Identification by the Absorption of Light

Absorbance spectra of compounds taken in the visible and UV regions of the spectrum tend to have broad absorption peaks of 100 nm or more. For single compounds, the peak shape is very reproducible. Some examples are shown in Figure 5.43. Although the spectra are not dramatically distinctive, it is clear that one can readily see the dif-

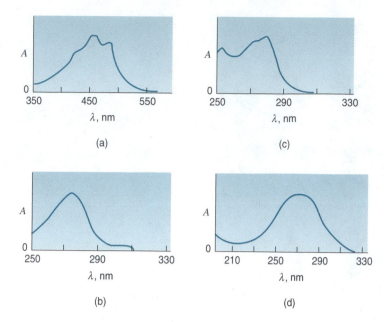

Figure 5.43. Absorbance spectra for four different compounds illustrating their distinctiveness as a basis for identification.

ferences in the spectra of the several compounds. The compounds do not need to be fundamentally different in order to have significant spectral differences. Consider the case of an acid–base indicator. Its color in solution changes dramatically with only the addition or removal of a proton.

Spectra of the indicator Congo red are shown in Figure 5.44 for different values of pH. Solutions of Congo red change from red to blue as the pH increases through the value of pK_a ($\approx$ pH 4). These spectra reveal several very important points. One is that minor variations in the form of a molecule can have a significant effect on its spectrum. Other examples of changes in form are dimerization, complexation, and solvation with different solvents. The second point is that, after standardization, one could use the spectrum of the indicator to reveal the pH of the solution it is in (in the region where pH $\approx pK_a$). The third point is that when an analyte can undergo a reaction in the sample solution, the absorbance–concentration relationship is likely to not be linear. This is because the fraction of the analyte in each of its forms is not likely to be a linear function of the total concentration of analyte in the sample. For example, if dimerization (the combining of two similar molecules) occurs, the fraction of the analyte in the dimer form will increase with increasing concentration. When dissolving a weak acid in water, the fraction of the acid in its base form will decrease with increasing concentration. These nonlinearities are sometimes called **chemical deviations** from Beer's law.

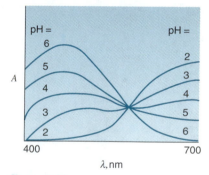

Figure 5.44. Absorbance at several pH values for Congo red. The different forms a species can have in solution frequently have quite different spectra.

COMPOUND IDENTIFICATION BY SPECTRAL MATCHING

If one had a library of spectra that included the spectrum of the unknown, there is a good chance that the compound could be identified by the quality of the match between the unknown and library spectra. One difficulty with this approach is the effect of concentration on the appearance of the spectrum. The spectrum of $KMnO_4$ at several concentrations is shown in Figure 5.45.[2] If the library and unknown spectra were taken at different concentrations, the match is much harder to make. An improvement on the use of library spectra is to compare plots of the log of the absorbance. For these spectra, the features maintain their relative magnitude and shape over a wider range of concentrations.

Library matching of UV and visible spectra is of limited value because of the general lack of distinctive features in the typical spectrum. Matching could be success-

[2]M. G. Mellon, *Analytical Absorption Spectroscopy*, pp. 104–106. John Wiley & Sons, New York 1950.

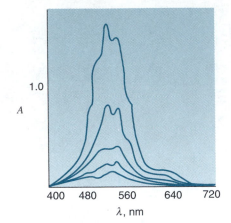
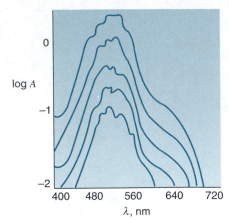

Figure 5.45. The appearance of absorbance spectra can change as a function of concentration, here as shown, for $KMnO_4$. A plot of log A versus wavelength enables a better comparison of spectral shape.

ful over a limited range of compounds with absorptions in different areas of the spectrum, but as soon as the library of possible compounds reaches a few dozen, the differences between some of the spectra become too small to enable a positive identification of the compound. This is not the case in the IR region. There, the wealth of detail contained in the spectra provides a great deal of discriminating power for the process of compound identification. (Remember, the energy states involved in IR absorption are not "smeared out" from the proximity of other molecules in a condensed phase.) Two methods of compound identification are used in the IR. They are spectral matching and structural group identification.

DEDUCTION OF MOLECULAR STRUCTURE AND COMPOSITION

In the IR region, particular absorption bands are associated with particular bond motions between particular atoms or substructures. Some of these correlations are shown in Table 5.3. The appearance of absorption bands in the spectrum of an unknown compound is indicative of the presence of the species in that bonding system. Each additional piece of information about the composition or structure of a compound brings the investigator that much closer to the complete description of the unknown. Note that a compound identity may be given in many ways; one way is as the molecular formula, such as $C_5H_{10}O$. This description, however, is not complete since there are dozens of compounds, many with quite different characteristics, that share that same formula.

Table 5.3
Stretching Frequencies for C—H and C═O

Molecule	Wavenumber, cm^{-1}
C—H Stretching	
CH_4	3020
$C_2H_2Cl_2$	3089
C_6H_6	3099
CH_3OH	2977
C_2H_2	3287
C═O Stretching	
CH_3COCH_3	1715
CH_3CHO	1729
$C_2H_5COC_2H_5$	1720
$HCOOH$	1729
CH_3COOH	1718

These are called **compositional isomers**. More rigorous is the structural representation, such as $C_2H_5COC_2H_5$. This descriptive form is more definitive in that it indicates the other atoms each atom is bonded to. Also definitive is the name of the compound, when the naming follows the conventions.

For all but the simplest compounds, deductive interpretation of the IR spectrum is not sufficient for a complete identification. In such a case, there are two alternatives. One is to obtain more information about the compound through the use of other techniques, and the other is to use a digitally encoded library of IR spectra and a spectral matching program. The spectral matching program may use any of several different algorithms to find the library spectrum that most closely matches the unknown compound. When it has finished going through the library looking for likely "hits," the program will print a list of all the most likely candidates in order of the **match factor** or **similarity index** assigned to each candidate. It is interesting to ask why a spectral matching program can be more successful in compound identification than the deductive approach when they both use the same data. One opinion is that the deductive approach can use only those spectral features for which we have rules that relate them to compound structure. The matching program uses all the features in the spectrum without concern for whether they have been rationalized in terms of structure. If this is true, it suggests that there is more information in the IR spectrum than we currently have rules to use for structural determination.

Study Questions, Section H

56. Develop the analytical scheme around the differentiating characteristic for the process of compound identification by absorption spectroscopy.

57. The molar absorptivity of a monoprotic weak acid in its protonated form is 6.32×10^2 L/mol · cm, and for its basic form ε is 1.325×10^4 L/mol · cm. Calculate and plot the working curve (absorbance versus total concentration) for the acid when it is dissolved in pure water with the analytical concentration of acid being 0.00, 5.0×10^{-6}, 1.0×10^{-5}, 2.0×10^{-5}, and 5.0×10^{-5} M. The K_a' for the acid is 1.0×10^{-5}. Use 1 cm as the path length of the spectrometer.

Answers to Study Questions, Section H

56. The differentiating characteristic is a distinctive spectrum in one or more of the wavelength regions used for spectroscopy. The probe is light that includes all the wavelengths over which the spectrum is distinctive. The response to the probe is the characteristic selective absorption of certain wavelengths of light. The measurement of the response is to measure the sample and blank transmittances over the wavelength region of interest. The interpretation of the data is to calculate the resulting absorbance spectrum and to compare it with the absorbance spectra of known compounds to obtain a match. Alternatively, one can interpret the absorption bands in terms of the substructures known to absorb in those regions and then infer the structure of the analyte molecule.

57. First calculate the concentrations of A^- and HA at each concentration. Start with Equation 4.13 for $[H_3O^+]$. This is a valid equation because the solution is quite acidic. This equation is then solved for the several concentrations. From the proton balance equation, $[H_3O^+] = [A^-] + [OH^-]$, again, the $[OH^-]$ can be ignored so that $[A^-] =$ $[H_3O^+]$. [HA] will be $C_A - [A^-]$, and the absorbance can be calculated from the equation

$$A = \varepsilon_{HA}b[HA] + \varepsilon_{A^-}b[A^-]$$

The resulting plot and values are shown. Note that the working curve is nonlinear due to the differing fractions of the acid in each form as the concentration changes.

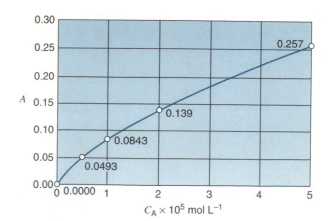

I. Separation by the Absorption of Light

We have seen in this chapter how the selective absorption of light can be used in the identification and quantitation of many types of compounds. At first glance, it would seem that separation (one of the four goals of chemical analysis) could not be achieved through this differentiating characteristic. Of course, if one thinks only of physical separation, this is true. However, from an analytical perspective, if a mixture of compounds can be resolved into its components and these can be identified or quantitated, an analytical differentiation has been achieved. We will see, in this section, how the spectrum of a mixture of compounds can often be resolved to reveal the makeup of the mixture.

MIXTURE SPECTRA

The discussion thus far has assumed that the analyte is the only compound in the sample that has a significant absorption within the probe bandwidth. When compounds other than the analyte can cause a response to the probe chosen, those compounds are **interferents** in the determination of the analyte. Their presence can cause an incorrect result in the quantitation of the analyte and can change the shape of the absorbance spectrum, decreasing the reliability of analyte identification. In general, either the interfering substances must be separated from the sample by a prior step, or the contribution of their response must be separated from that of the analyte. When a spectrum includes the responses of more than one analyte, it is called a **mixture spectrum**, as in Figure 5.46. The methods of resolving the mixture into the compounds that contribute to the response are called **multicomponent analysis**. Several of these are described in the following sections.

RESOLUTION BY SIMULTANEOUS EQUATIONS

The problem of interference exists when the measured response is a function of more than one compound in the sample. If two compounds in a sample absorb light of the same wavelength, a probe at that wavelength alone will not be able to reveal the part of the response due to just one of the compounds. In the simplest case, the absorbances due to each compound are additive, so that

$$A_{M,\lambda_1} = \varepsilon_{M,\lambda_1} b C_M$$
$$A_{N,\lambda_1} = \varepsilon_{N,\lambda_1} b C_N$$
$$A_{\text{Mixture},\lambda_1} = \varepsilon_{M,\lambda_1} b C_M + \varepsilon_{N,\lambda_1} b C_N$$

5.15

where $A_{\text{Mixture},\lambda_1}$ is the absorbance of the sample at λ_1, b is the cell path length, $\varepsilon_{M,\lambda_1}$ is the molar absorptivity of compound M at λ_1, C_M is the concentration of compound M in the sample, and the other terms have corresponding meanings for the compound N. The concentrations of M and N cannot be obtained from this equation. This is because there is

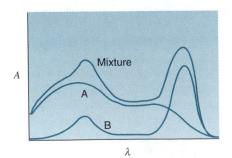

Figure 5.46. The spectrum of a mixture is the sum of the spectra of the individual components. Quantitation and identification are both made more difficult when the absorbance bands overlap.

only one equation and, even if the molar absorptivities of M and N are known, there are still two unknowns. However, the chances are that at a different wavelength, the molar absorbances of the two compounds will be different, and so also will be the total absorbance. From the measurement at a second wavelength, we can write another equation.

$$A_{\text{Mixture},\lambda_2} = \varepsilon_{\text{M},\lambda_2} bC_{\text{M}} + \varepsilon_{\text{N},\lambda_2} bC_{\text{N}} \qquad 5.16$$

Now, assuming the molar absorptivities of both compounds are known at both wavelengths, we have two equations and just two unknowns. These equations can then be solved by the method of simultaneous equations to obtain the concentrations of each of the components.

In practice, the method of simultaneous equations is implemented by first determining the absorbance of known concentrations of each pure compound (standard solutions) at the two wavelengths. From Equation 5.15, solving for C_{N} yields

$$C_{\text{N}} = \frac{A_{\text{Mixture},\lambda_1} - \varepsilon_{\text{M},\lambda_1} bC_{\text{M}}}{\varepsilon_{\text{N},\lambda_1} b} \qquad 5.17$$

This equation is used to substitute for C_{N} in Equation 5.16.

$$A_{\text{Mixture},\lambda_2} = \varepsilon_{\text{M},\lambda_2} bC_{\text{M}} + \varepsilon_{\text{N},\lambda_2} b\left(\frac{A_{\text{Mixture},\lambda_1} - \varepsilon_{\text{M},\lambda_1} bC_{\text{M}}}{\varepsilon_{\text{N},\lambda_1} b}\right) \qquad 5.18$$

Equation 5.18 is then solved for C_{M} to get

$$C_{\text{M}} = \frac{A_{\text{Mixture},\lambda_2} \varepsilon_{\text{N},\lambda_1} - A_{\text{Mixture},\lambda_1} \varepsilon_{\text{N},\lambda_2}}{b(\varepsilon_{\text{M},\lambda_2}\varepsilon_{\text{N},\lambda_1} - \varepsilon_{\text{N},\lambda_2}\varepsilon_{\text{M},\lambda_1})} \qquad 5.19$$

The value for C_{M} calculated from Equation 5.19 is then used in Equation 5.17 to obtain C_{N}.

For the best results, the two wavelengths chosen should provide the maximum discrimination between M and N. This will be at those wavelengths for which the difference in the ratios $\varepsilon_{\text{M},\lambda_1}b/\varepsilon_{\text{N},\lambda_1}b$ and $\varepsilon_{\text{M},\lambda_2}b/\varepsilon_{\text{N},\lambda_2}b$ is the greatest.

If determination of the absorbance at two wavelengths is sufficient to calculate the concentrations of two absorbers, then three components could be resolved by the measurements at three wavelengths, and so on. The application of the simultaneous equation technique involves two major assumptions. One is that all the absorbers present in the sample are known and that standard spectra for each of them have been obtained. Another is that there is no interaction between the analytes that affects their composition or structure. Such interactions that allow the amount of one analyte to affect the response of the system to another analyte are called **matrix effects**. Matrix effects are among the most challenging problems confronting the analyst.

In practice, when more than two components are involved, spectrometers, rather than photometers, are used for the measurements so that entire absorbance spectra can be readily obtained. To provide the required data, there needs to be as many standard spectra as there are absorbing components in the mixture and one spectrum of the unknown sample. These spectra contain information concerning the standard and sample responses at many wavelengths. In other words, dozens or even hundreds of equations could be written, one for each wavelength sampled in the standard and sample spectra. When the number of wavelengths sampled exceeds the number of absorbing components in the sample, the analysis is said to be **overdetermined**. A repeated theme of this text is that the collection of extra data and its use in the achievement of the analytical goals can lead to a greater precision of result and a decreased dependency on a theoretical response curve.

The methods for dealing with the very large data sets involved when full spectra are used for multicomponent analysis have been worked out by analytical

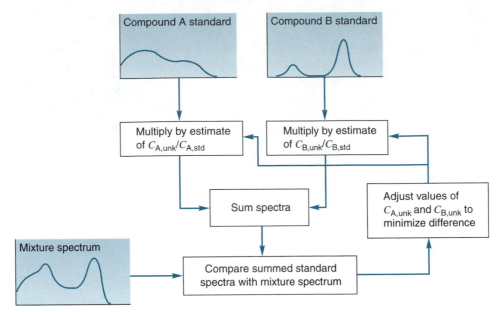

Figure 5.47. A method of using whole spectra for multicomponent analysis.

chemists working in the area of **chemometrics**. The data manipulations involve matrix algebra and are only practical when a computer is used for the calculations. Their implementation requires the efficient acquisition of the spectral data by a computer and the availability of the appropriate data analysis program. The results achieved by the application of chemometric methods to spectroscopic data are truly remarkable. The presence of an unexpected component can be discovered and its spectrum generated. The requirement of linear addition of absorbances (the assumption of no matrix effects) can be avoided. All regions of discrimination among the mixture components are used, so detection limits for the components are greatly improved.

The method of using full spectra in multicomponent analysis diagramed in Figure 5.47 is called the **classical least squares** (CLS) method or sometimes the **K-matrix calibration**. It is inherently a multiple linear regression. In this case, the inclusion of the absorbances of both components for many wavelengths will lead to a better precision in the determination of the concentrations of both components. Other approaches can take advantage of the extra data to avoid the assumption that the absorbances are exactly additive. The development of such techniques, made practical by inexpensive computers and friendly application programs is an exciting area of analytical research.

Study Questions, Section I

58. If a mixture contains four species of unknown concentration, what absorbance determinations are required to obtain the concentrations of all of them?

59. What is the advantage of taking more than the minimum amount of data to resolve a mixture?

60. A sample containing cobalt and chromium was analyzed by UV/VIS spectrometry. Two standards, one containing 0.0500 M Cr and one containing 0.0188 M Co, were also analyzed. The analysis was done at two wavelengths, $\lambda_1 =$ 408.4 nm and $\lambda_2 = 510.7$ nm. Using the data below, calculate the concentrations of cobalt and chromium in the unknown mixture. The cell pathlength is 1 cm.

	A_{λ_1}	A_{λ_2}
Co	0.0847	0.9158
Cr	0.8026	0.2705
Mix	0.3229	0.6935

58. Absorbance determinations of all four standard solutions are required at four different wavelengths. In addition, the sample absorbance must be determined at the same four wavelengths. This makes 20 absorbance determinations in all.

59. One advantage will be improved precision in the component determinations. Another is that unanticipated effects, such as the presence of an unexpected absorber or matrix effects among the analytes, can be observed and sometimes resolved.

60. To solve this problem, which has two unknowns, we must simultaneously solve Equation 5.19 to obtain the concentration of one of the components and then Equation 5.17 to obtain the concentration of the other. Before applying these equations, it is necessary to find the values of $\varepsilon_{N\lambda_1}$, $\varepsilon_{N\lambda_2}$, $\varepsilon_{M\lambda_1}$ and $\varepsilon_{M\lambda_2}$. This is done using the absorbances of the standards and Equation 5.8:

$$A = \varepsilon bC, \varepsilon = \frac{A}{Cb}, \text{ or, since } b = 1.00$$

in this situation, $\varepsilon = \dfrac{A}{C}$.

The molar absorptivities can therefore be calculated as follows. For cobalt at λ_1 (where M represents Co), the molar absorptivity is equal to the absorbance of the standard at λ_1 divided by its concentration:

$$\varepsilon_{M\lambda_1} = \frac{0.0847}{0.0188} = 4.51 \frac{L}{mol \cdot cm}$$

$$\varepsilon_{M,\lambda_2} = \frac{0.9158}{0.0188} = 48.71 \frac{L}{mol \cdot cm}$$

For chromium:

$$\varepsilon_{N\lambda_1} = \frac{0.8026}{0.050} = 16.052 \frac{L}{mol \cdot cm}$$

$$\varepsilon_{N\lambda_2} = \frac{0.2705}{0.050} = 5.41 \frac{L}{mol \cdot cm}$$

Now we can solve for the concentrations in the mixture: For the cobalt,

$$C_M = \frac{0.6935 \times 16.052 - 0.3229 \times 5.41}{1.00(48.71 \times 16.052 - 5.41 \times 4.51)}$$

$$= 0.0124 \text{ M}$$

For the chromium concentration,

$$C_N = \frac{0.3229 - 4.51 \times 1.00 \times 0.0124}{16.052 \times 1.00}$$

$$= 0.0166 \text{ M}$$

Practice Questions and Problems

1. A. Using Table 5.1, calculate the mid frequency of orange light and compare that with the mid frequency of violet light.

   B. What wavelength of light would you use to probe the concentration of a solution that appears blue?

   C. The velocity of light through a solution is smaller than it is in air or in vacuum. Does this result in a change of the frequency or the wavelength of light? Explain.

2. A. In the digital measurement of light intensity (Figure 5.9), what are the units of K_1, K_2, K_3, and the overall conversion, $K_1 K_2 K_3$?

   B. If we desired to measure a light intensity that is 5% of the full range to a precision of 0.1%, how many bits would the ADC have to have?

3. A. In making a light transmittance measurement, why is it necessary to use a blank solution?

   B. What is the %T of a solution for which the measured values of N_d, N_b, and N_s were 24, 8492, and 6741?

   C. How does a photometer arrange to have a readout that indicates %T directly? What needs to be reset if the wavelength is changed to a new value?

   D. Why is absorbance a more useful function for chemical analysis than %T?

   E. What is the absorbance of the solution in part B?

   F. If the absorbance of a 0.00235 M solution of that same analyte in a similar sample holder was determined to be 0.374, what is the concentration of the analyte in the solution of part B?

4. A. List the methods for limiting the bandwidth of the light that is used to probe for analyte absorbance.

   B. Explain why it is important to make the probe bandpass as narrow as it is practical to do.

   C. How could you tell if the bandpass is narrow enough for a particular determination?

5. What makes a solution appear colored? What makes a red shirt red?

6. Light source x emits light of higher energy than light source y.

   A. Which has the longer wavelength?

   B. Which has the higher frequency?

7. If the wavelength of a photon is 500 nm, what is its energy? What is the energy of a mole of such photons? What color is the light?

8. What are the absorbance and percentage transmittance of a sample if the numbers read by a spectrometer are $N_d = 32$, $N_b = 9870$, and $N_s = 5467$?

9. Why is the fraction of the light transmitted not directly proportional to the number of light absorbers in the sample?

10. In an experiment, 10.0 mL of 0.040 M $KMnO_4$ solution is added to 50.0 mL of a solution containing some amount of $KMnO_4$. The absorbance of the solution before addition of the standard is 0.56, and after the addition of the standard it is 0.78. What is the concentration of $KMnO_4$ in the unknown solution?

11. What steps would be required to obtain an absorbance spectrum (A versus 50 different wavelength values) using a photometer such as the Spectronic™ 20 (block diagram in Figure 5.20)?

12. What is occurring when a molecule absorbs a photon with a wavelength of 450 nm? Why would some molecules not absorb photons of this wavelength?

13. Explain the principal differences between a scanning spectrometer and an array detector spectrometer with regard to how each point in the absorbance spectrum is obtained.

14. Explain why the working curve for the solution in question 57 is not linear.

15. Relate detector wavelength resolution in an array detector spectrometer with monochromatic bandpass in a scanning spectrometer.

16. When measured with a 1-cm cell, a 3.5×10^{-5} M solution of species A exhibited absorbances of 0.148 and 0.624 at 443 and 679 nm, respectively. A solution of species B gave absorbances of 0.567 and 0.083 for a 4.58×10^{-5} M solution of species B under the same circumstances. The absorbances for a mixture of A and B were 0.484 at 443 nm and 0.829 at 679 nm. Calculate the concetrations of A and B in the mixture.

17. An array detector spectrometer uses a 4096-element array to cover the wavelength range from 200 to 900 nm.

    A. Calculate the maximum wavelength resolution attainable with this spectrometer.

    B. Because the resolution calculated in part A is more than necessary given the broad peaks of liquid absorbers in the UV and visible light ranges, the values obtained from each set of four detectors are averaged. Discuss the advantages of this approach over just using a 1024-element array for the detector.

18. Use Figure 5.3 to determine tha range of optical frequencies in the infrared region of the spectrum. If the mirror in

a Michelson interferometer has a linear motion of 3 cm/sec, what is the range of detector signal frequencies that will encompass the infrared region of the spectrum? Consult the FTIR derivation on the CD.

19. A scanning spectrometer is being used to obtain spectra from a flowing system in which the concentrations are changing with time. In order to adequately sample these changes, 5 spectra per minute are needed. The spectra should cover the wave length range from 300–500 nm to a resolution of 1 nm.

    A. Calculate the amount of time available to obtain a value at each wavelength. Assume equal time spent measuring N_b and N_s. Also assume that the wavelength drive mechanism can return to the starting wave length in 3 seconds.

    B. Compare the result of calculation (A) with the amount of time each detector can spend to acquire a value for N_s with an array detector spectrometer.

20. A. How many waves of the most energetic ultraviolet light would fit in one wavelength of the least energetic infrared radiation?

    B. What is the energy of a single photon of green light? For the units, consult the table of useful constants and conversions in Appendix G.

    C. Green light is passing through a medium of refractive index of 1.3 What are the wavelength, frequency, and velocity of the light in this medium?

    D. What color would a green sweater that was illuminated with only purple light appear?

21. A. What are the three measurements used to determine the $\%T$ of light through a solution?

    B. Why are all three of these measurements required?

    C. Is it appropriate to say that one measures $\%T$? Why or why not?

22. A. In a determination of $\%T$, why is it better to use the light source of the wavelengths absorbed by the sample rather than white light?

    B. Why are the entrance and exit slits necessary in the bandwidth limiting devices that use a prism and a grating?

23. A. The measurements of blank, sample, and dark currents for a particular absorption measurement were 3.754 nA, 2.236 nA, and 34 pA. Calculate the $\%T$ and the absorbance for this sample.

    B. A 10 mg/mL standard solution of this same absorber yielded detector currents of 3.726 nA, 1.857 nA, and 34 pA. What is the analyte concentration in the solution measured in part A?

    C. If the working curve begins to deviate from a straight line at values of A above 1.5 (see Figure 5.18), what is the value of the current below which the linear relationship used in your answer to part B would not be valid?

24. What is the wavenumber of green light (see Figure 5.1)?

   B. What is the wavenumber of light that has a wavelength of 8×10^{-6} M?

   C. In what region of the spectrum is the light of Part B above?

25. A. What is the principal difference in the wavelength selection devices that are used for scanning and array detector spectrometers?

   B. Could you use a filter as a wavelength selector in either the scanning or array detector spectrometers? Why or why not?

   C. Is a Fourier-transform spectrometer more like an array detector spectrometer or a scanning spectrometer? Why?

26. A. Is IR or UV-Vis spectroscopy better suited for the identification of molecular structure? Why?

   B. Does your answer apply to both matching and deductive determination or just one of them? Why?

27. A. If a 4.73×10^{-5} M solution of substance A gave an aborbance of 0.385 at λ_1 with a 1-cm cell, what is the value of its molar absorptivity?

   B. The molar absorptivity of substance A at λ_2 was 629. For substance B, the molar absorptivities at λ_1 and λ_2 were 3,620 and 14,715, respectively. An unknown solution gave an absorbance of 0.365 at λ_1 and 0.832 at λ_2 with a 1-cm cell. What are the concentrations of A and B in the unknown solution?

Suggested Related Experiments

1. Any photometric experiments relating absorbance and concentration.

2. Measuring spectra of indicators in each of their color states, perhaps titrating the indicator to show mixtures of states.

3. Observing deviations from Beer's law from changes in α (fraction of analyte in the absorbing form) as a function of analyte concentration.

4. Determining concentrations of separate components in a mixture using absorbance measurements at several wavelengths.

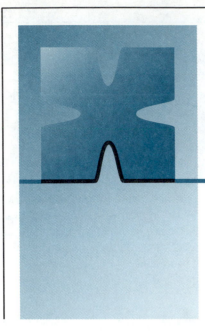

Chapter Six

ANALYSIS BY PHOTON EMISSION

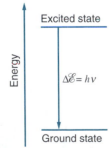

Figure 6.1. Atoms or molecules with excess internal energy (excited state) can lose some or all of this energy through photon emission. The emitted photon wavelength is determined by the energy difference between the initial and final states.

Atoms and molecules can be induced to emit photons. There are a variety of analytical methods based on this phenomenon because the quantity of photons emitted and their wavelengths can provide information regarding the quantity and identity of the species present. Photon emission is one of the ways an atom or molecule can lose energy. Therefore, for an atom or molecule to emit a photon, it must not be in its minimum energy state. The minimum energy state is called the **ground state.** Species that are not in the ground state are said to be in an **excited state.** Some excited state atoms or molecules can return to the ground state or a less excited state by the emission of a photon, as shown in Figure 6.1. The energy of the photon emitted, $h\nu$ from Equation 5.2, is equal to the difference in the energies of the initial and final states. Since the energies of all the possible states are characteristic of an atom or molecule, so are the wavelengths of the emitted photons.

To probe this phenomenon, one must arrange to put some of the atoms or molecules to be analyzed in an excited state. This generally requires that the analyte species absorb energy from an energy source. This chapter is organized by the type of species excited (atoms or molecules) and the manner of the excitation process. Three methods of excitation are in common use. They are thermal excitation, photon excitation, and excitation from chemical reactivity.

Following the now-familiar analysis scheme outlined in Figure 6.2, the response to the excitation probe is photon emission. The measurement of the response to the probe is thus the determination of the intensity of the emitted radiation. This is done by wavelength-selective photon detection. The information regarding the sample composition is then obtained from the photon intensity at the various wavelengths sampled. The way each of the excitation and emission phenomena have been developed into analytical methods and the way the data from them are interpreted to provide analytical information are explored in the sections to follow. One of the most

In the figure, the energy level diagram shows:
- Excited state
- $\Delta\mathscr{E} = h\nu$
- Ground state
- Energy (vertical axis)

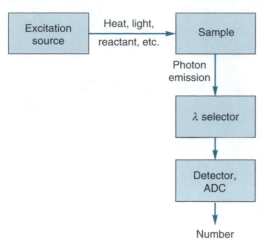

Figure 6.2. The general scheme for analysis by photon emission is to probe the differentiating characteristic (the ability to emit photons of characteristic wavelength from an excited state) by inducing excitation. The emitted photons are then sorted by wavelength and detected. The wavelengths and intensities of the emission are related to the quantity and identity of the emitting species in the sample.

exciting aspects of this study is that the story does not end with this chapter; new techniques and methods continue to be discovered and developed. They are providing important new tools for the chemical exploration of life (in biological studies) and the universe (in astronomy).

A. Photon Excitation of Molecular Species

Since the absorption and emission of a photon by a molecule are both related to the availability of energy states in that molecule, it is helpful to review what those energy states are. In Chapter 5, we outlined several regimes for the energy states. Those energy levels with the greatest differences between them are the electronic energy levels. Transitions among these states generally involve photons in the ultraviolet and visible regions of the spectrum. Other energy states are those of the interatomic bonds in a molecule. These include oscillations in the bond angle and length (called **vibrations**) as well as **rotations** about the bond axis. In a molecule, these two systems of energy levels are superimposed so that a molecule can be in its ground electronic state with excited vibrations or in an excited electronic state with either excited or ground state vibrations. A rough schematic of these energy levels is shown in Figure 6.3. The electronic states are designated S_0, S_1, etc., with S_0 being the ground electronic state.

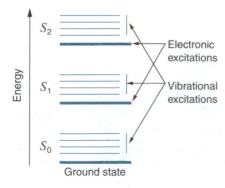

Figure 6.3. Vibrational levels associated with each electronic level.

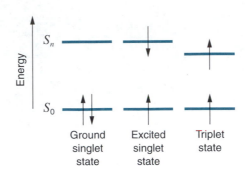

Figure 6.4. Relation of spin states in the singlet and triplet excited states.

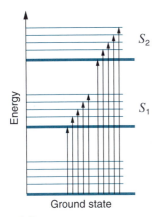

Figure 6.5. Energy transitions from the ground state to combinations of excited electronic and vibrational states.

PHOTON ABSORPTION

In a molecule, the same electronic energy state can be shared by only two electrons. To do so, the two electrons must be in opposite spin states called **up** and **down.** The **spin state** is indicated by the direction of arrows drawn crossing the energy level in energy level diagrams. Two electrons in the same energy level are called **paired** electrons. In the ground electronic state, all the electrons will be paired in the lowest energy states. The next higher state may contain an unpaired electron if the molecule has an odd number of electrons. On absorbing a photon corresponding to an electronic energy transition, an electron in a filled energy state will be promoted to a higher, unfilled level. This will create an unpaired electron in the new energy state and probably another one in the level from which it came. (See Figure 6.4.) If the promoted electron retains its spin state on excitation, the spins of the two unpaired electrons will remain opposed. This excited state is called the **excited singlet state.** If the spin of the promoted electron is reversed, it will be parallel to that of the other unpaired electron. This state is called the **triplet state.** The available energy levels for triplet states are different from those of the excited singlet state.

Photon excitation may occur between most of the energy levels available, both electronic and vibrational. Therefore, a molecule in the ground state may be excited electronically and vibrationally by the same photon absorption. The several possible energy transitions from the electronic and vibrational ground state are shown schematically in Figure 6.5. Where the absorbance bands corresponding to each absorbance wavelength overlap, the whole set is observed as one broad absorbance band. The result of photon absorption is the creation of a molecule that has more internal energy than most of the other molecules in the system. Next, we will look at the various ways by which the molecule can lose this "excess" energy. Each loss mechanism enables or prevents several forms of emission spectrometry.

SCATTERING OF THE EXCITATION RADIATION

The photon absorption process occurs in an incredibly short time. In fact, this time is essentially that of one period of the illuminating wave. In the visible region, this corresponds to approximately 10^{-15} second. Generally, the absorbed emission is reemitted very quickly. If it does so instantly, the initial direction of the radiation is maintained by constructive interference, and no effect on light transmission through the material is apparent. If the reemission is just a bit delayed, there is no constructive interference, and the emitted photon can go in any direction. This process is called **scattering.** Scattering results in a loss in transmission through the material. Most often, the scattered photon has the same energy as the excitation photon. Occasionally, the vibrational energy of the molecule can change while the excitation energy is present. In this case, the reemitted photon can be altered in energy by a vibrational transition. This process is called **Raman scattering.** As we shall see, Raman scattering can be used to obtain information about the vibrational energy levels available in the sample. Only a very

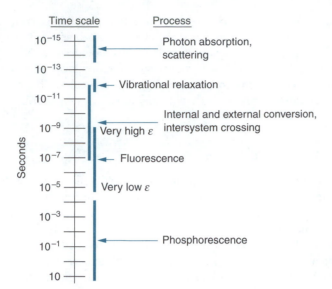

Figure 6.6. The time scale of various mechanisms for relaxation of excited states.

small fraction of the incident radiation undergoes Raman scattering. The rest is either not absorbed or remains in the molecule for a longer time.

VIBRATIONAL RELAXATION

If the photon remains absorbed, its energy is initially in the electronic and/or vibrational state resulting from the absorption process. From this state, a number of processes can occur. Each of them has its own time scale, as shown in the timeline in Figure 6.6. If the absorbing molecule is in a condensed phase (solid or liquid), the vibrations of its bonds are felt by the nearby molecules. Since the neighboring molecules are in lower energy states, they can quickly absorb this energy. Thus, within about 10^{-12} second, the vibrational excitation is brought to the ground *vibrational* state. In IR spectrometry, this process would be noted as an absorption of the vibrational excitation. However, if the excitation were electronic as well, the molecule would then be in the ground vibrational state of an excited electronic state. Fluorescence and phosphorescence are two of the ways that the remaining energy can be lost.

If the ground vibrational level of an excited electronic state overlaps the vibrationally excited state of a lower electronic state, the molecule can undergo an **internal conversion** to the latter state. (See Figure 6.7.) If the vibrational excitation exceeds the

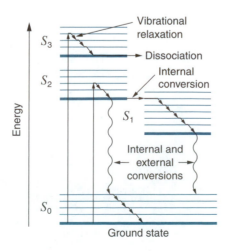

Figure 6.7. Several paths for the relaxation of internal energy.

energy of one of the molecular bonds, the molecule can undergo **dissociation,** breaking apart at one of its weaker bonds. A more likely event is the rapid loss of the vibrational excitation through vibrational relaxation. This would leave the molecule in its new, lower electronic state and in its ground state vibration level. Electronic excitation can also be lost through collisions with neighboring molecules. This process is called **external conversion.** Energy lost to vibrational modes of neighboring molecules appears in the system as heat.

MOLECULAR FLUORESCENCE

If an electronically excited molecule maintains an electronic excitation for as long as 10^{-9} second, there is an increasing chance that it may lose some of its energy by photon emission. **Fluorescence** is the emission of a photon from a molecule in the excited singlet state. In the condensed phase, emission will always occur from the ground vibrational state of the excited electronic state. The resulting state of the molecule will necessarily be a lower electronic state, but higher vibrational states are possible. This range of possibilities gives rise to a band of photon emission energies, as shown in Figure 6.8. If a molecule in the singlet excited state is going to fluoresce, it will generally do so in much less than a microsecond. Thus, special instrumentation is required to observe the time between the absorption of the exciting photon and the emission of the fluorescence radiation. In general, the lifetime of the excited state is inversely proportional to the molar absorptivity at the exciting wavelength. The quicker the fluorescence process, the more efficient it is because the longer the molecule remains in the excited state, the more likely it is that it will lose its energy by a nonradiative process.

Molecules vary greatly in the likelihood that they will lose their absorbed photon energy by fluorescence. The fraction of excited molecules that fluoresce is called the **quantum efficiency** for fluorescence. Molecules that have a usably high quantum efficiency for fluorescence often have multiple, conjugated double bonds. Examples are aromatic compounds and compounds that contain aromatic rings and chains. This includes all the polycyclic aromatic compounds, riboflavin, and proteins. The chemical structure of the compound affects both the quantum efficiency and the wavelength of the fluorescence. For example, among the polycyclic aromatic compounds, the wavelength of fluorescent emission increases with the number of rings in the compound as follows: benzene (278 nm), naphthalene (321 nm), anthracene (400 nm), and naphthacene (480 nm). Heterocyclic compounds in which the conjugated ring is interrupted by a nitrogen, sulfur, or oxygen (e.g., pyridine, pyrrole, thiophene, furan) do not fluoresce unless they are fused to a completely conjugated ring. Functional groups substituted on the conjugated systems have an effect on the quantum efficiency. Electron

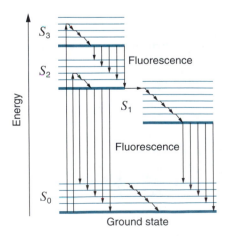

Figure 6.8. The energy transitions leading to fluorescence.

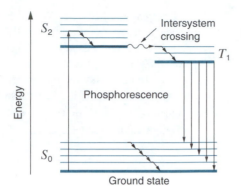

Figure 6.9. The energy states leading to phosphorescence.

donating groups such as $-NH_2$, $-OH$, and $-OCH_3$ increase the quantum efficiency, whereas electron withdrawing groups such as halides, carboxylic acids, and $-NO_2$ decrease it. If the substituent group can have an acidic and basic form, the measured fluorescence can be greatly affected by the pH of the solution, as the wavelength and quantum efficiency of the two forms are likely to be different. For example, the fluorescence of aniline and phenol are strongly affected by pH.

PHOSPHORESCENCE

It is possible for some molecules to make a transition from an excited singlet state to an excited triplet state, as shown in Figure 6.9. Although this is a low-probability process, it can occur if there is an overlap between the energy level of the excited singlet state and a vibrational mode of the triplet state. This process is called **intersystem crossing.** The vibrational excitation will be quickly lost through vibrational relaxation, but an excited triplet electronic state will remain. Transitions from this state have a low probability, so the molecule can remain in this state for a comparatively long time. A likely mechanism for energy loss from this state is photon emission. Photon emission from the triplet state is called **phosphorescence.** The delay between excitation and emission is much greater for phosphorescence than for fluorescence, as can be seen in the timeline of Figure 6.6.

Compounds that exhibit phosphorescence are even rarer than those that fluoresce because of the need for a triplet state accessible by rapid intersystem crossing and the competition of other energy loss mechanisms during the long time between excitation and emission. Many of the same factors that favor fluorescence also favor phosphorescence. In addition, the existence of heavier atoms in the molecule (e.g., halogens, sulfur, phosphorus) can enhance the probability of phosphorescence over fluorescence. Compound types for which phosphorescence spectrometry has been used include nucleic acids, enzymes, pesticides, and amino acids. The quantum efficiency for phosphorescence can be increased by cooling the sample.

When photon emission is stimulated by photon excitation, the process is called **photoluminescence.** In the next sections, we will see how the phenomenon of photoluminescence has been developed into the analytical techniques of fluorescence, phosphorescence, and Raman spectrometry.

Study Questions, Section A

1. What distinguishing characteristic is the basis for analysis by photon emission?

2. What is the probe discussed in this section for the characteristic named in Question 1?

3. What are the two states that can result from photon excitation of a molecule?

4. What is the time frame (in seconds) for the absorption of a photon that would appear blue to our eyes?

5. An absorbed photon can be very quickly emitted. What are the differences between scattering and transmission of an absorbed photon?

6. What is the time scale of vibrational relaxation in a condensed phase? What state will the absorber molecule be in after vibrational relaxation (assuming normal temperatures)?

7. What are internal and external conversions, and what is their time scale?

8. What are the principal differences between fluorescence and phosphorescence?

9. A light detector is set up to detect light transmitted through a sample cell. In the absence of the analyte, the detector current was 147.3 nA. With the analyte present,

the current was 122.9 nA. With the analyte present and the detector set up to intercept 6.2% of the fluorescent radiation, the current was 0.47 nA. The detector dark current was 0.02 nA. What is the quantum efficiency of the analyte molecules? Assume that the detector output current is proportional to the light intensity striking its surface.

10. Which of these two molecular structures is likely to have the higher fluorescence quantum efficiency and why?

Answers to Study Questions, Section A

1. Analysis by photon emission is based on the ability of excited atoms or molecules to emit photons.

2. The probe for analysis by photon emission is excitation of the analyte by photon absorption.

3. The excited singlet state (opposing unpaired spins) and the triplet state (parallel unpaired spins) are the two states that can result from photon excitation of a molecule.

4. Absorption of a photon occurs in the time of one period of the exciting radiation. For blue light, the wavelength is about 465 nm (from Table 5.1). From Equation 5.1,

$$\frac{1}{\nu} = \frac{\lambda}{c} = \frac{465 \times 10^{-9}\ \text{m}}{3.00 \times 10^{8}\ \text{m/s}} = 1.55 \times 10^{-15}\ \text{s}$$

5. In transmission, the absorbed photon is reemitted immediately. The direction of the photon path is unaltered. In scattering, the photon is absorbed for a time corresponding to at least one period of vibrational motion. The direction of the emitted photon is random. The wavelength of the emitted photon may be changed as well.

6. The time scale of vibrational relaxation is of the order of 10^{-12} second. After vibrational relaxation, the molecule is in the ground vibrational state of whatever electronic state it arrived at after excitation.

7. An internal conversion is the transition between the ground vibrational level of an excited electronic state and a comparable excited vibrational state of a lower

electronic state. An external conversion is the transfer of electronic state energy to another proximate molecule. The result is a lowering of the electronic excitation level. Both conversions result in a decrease of the electronic excitation and occur on a time scale of 10^{-12} to 10^{-7} second.

8. Fluorescence is the loss of electronic excitation by photon emission from the excited singlet state. Phosphorescence is the loss of electronic excitation by photon emission from the excited triplet state. Since the lifetime of the triplet state is normally much greater than that of the singlet state, the time scale for phosphorescence is much greater than that for fluorescence. Fluorescence generally occurs within a microsecond of excitation.

9. The absorbed light intensity is proportional to the difference in the detector currents with and without analyte: 147.3 − 122.9 nA = 24.4 nA. (The dark current corrections are negligible.) The measured fluorescence intensity is proportional to 0.47 − 0.02 nA = 0.45 nA. The total fluorescence intensity is then 0.45 nA × (1/0.062) = 7.3 nA. The quantum efficiency is 7.3 nA/24.4 nA = 0.30.

10. Phenol (left-hand structure) will have a greater quantum efficiency than chlorobenzene because the electron donating property of the −OH group enhances the benzene quantum efficiency while the electron withdrawing quality of the chlorine reduces it.

B. Fluorescence and Phosphorescence Spectrometry

The phenomenon to be probed in fluorescence spectrometry is the photon emission from molecules in an excited electronic state. The probe is the source of light used to excite the molecules. To become excited, a molecule must absorb an excitation photon. Therefore, the wavelength of the exciting illumination must correspond to an absorbance band of the molecules to be excited. The sample holder is arranged so that the

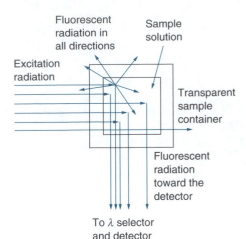

<figure>Figure 6.10. Stimulation and observation of fluorescence with the sample in a cuvette.</figure>

sample can intercept the excitation radiation. A common arrangement is shown in Figure 6.10. The incident radiation illuminates one face of a transparent sample container. Molecules in the sample then absorb some of this radiation. Excited molecules that fluoresce do so in a random direction. Thus, the fluorescent radiation is emitted in all directions. Only part of this radiation is effectively collected and detected. For simplicity, the distribution of fluorescence directions is shown for only one of the absorptions in the sketch and for only two dimensions.

As shown in Figure 5.15, the intensity of the excitation radiation decreases with increasing distance into the sample container because of the absorption of the light by the sample molecules. Thus, the sample molecules farthest from the illuminated face of the container have a lower probability of becoming excited. This effect, called **primary absorbance,** increases with decreasing percentage transmittance of the sample.

The fluorescent radiation must travel through the sample for some distance before leaving the sample container, and the wavelengths of the fluorescent radiation necessarily correspond to energy level differences in the fluorescing molecules. Therefore, the sample molecules will have a significant molar absorptivity for the fluorescent radiation, and some of the fluorescence will be absorbed before it leaves the sample container. This phenomenon is called **secondary absorbance.** As we will see, these phenomena affect the relationship between the intensity of the detected fluorescence and the concentration of fluorescing molecules in the sample.

Measurement of the response to the probe is the measurement of the sample fluorescence. Fluorescent radiation is collected from one of the faces of the sample container. This is most often done from a face that is orthogonal to the face used for excitation, but measurements of same-surface fluorescence (usually called **front-surface fluorescence**) are also used. (See Figure 6.11.)

It is a given that the wavelength of the fluorescent radiation cannot be shorter than that of the excitation radiation because there can be no transitions resulting from the excitation that are of greater energy than the energy of the absorbed photon. In fact, we see that vibrational relaxation and internal conversion processes only reduce the energy level from which the fluorescence takes place relative to the initial excited state. For this reason, the excitation and emission wavelength selectors operate in somewhat different parts of the spectrum.

A typical fluorescence instrument is shown in block diagram form in Figure 6.12. If filters are used along with a simple lamp source, the instrument is called a **fluorometer.** If monochromators are used for excitation and fluorescence wavelength selection, the instrument is called a **spectrofluorometer.** Some spectrofluorometers are also made with a wavelength dispersive analyzer, such as a grating and an array detector. These have the advantages that were discussed in Chapter 5 for array detector

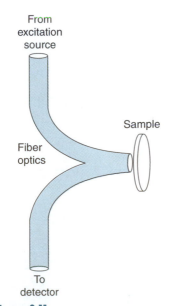

<figure>Figure 6.11. An arrangement using light pipes for front-surface fluorescence measurements.</figure>

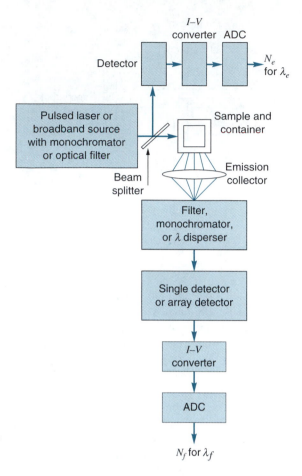

Figure 6.12. Schematic diagram of a complete fluorescence instrument.

spectrometers. The intensity of the resulting light is measured with a system consisting of a photometric detector (usually a light intensity-to-current converter), a current-to-voltage converter, and an ADC voltage-to-number converter. The number N_f that results from the measurement is directly related to the light intensity reaching the detector. This number is specific for the wavelength setting of the fluorescence monochromator. If a range or wavelengths is scanned or if an array detector is used, there will be a whole set of N_f values, one for each wavelength sampled. You will have noticed in Figure 6.12 that there is another separate system measuring the intensity of the excitation light. This is generally required because of drifts in the intensity of the excitation source. Since the source intensity is not easily held constant, it is measured as N_e and the effect of its value taken into account in the data interpretation step.

QUANTITATION

The data from a fluorescence instrument can be interpreted on several levels. On the most basic level, single wavelength bands are used for the excitation and fluorescence. Single values for N_f and N_e are obtained for each sample to provide quantitative information on the fluorescing species concentration. If secondary absorbance is ignored, the fluorescence intensity will be proportional to the quantity of excitation light absorbed by the sample:

$$N_f = KN_e\left(1 - \frac{\%T}{100}\right) = KN_e(1 - 10^{-A}) \qquad 6.1$$

where A is the absorbance of the fluorescing species at the excitation wavelength and K is a proportionality constant that includes the fluorescence quantum efficiency, the

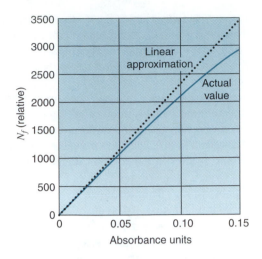

⊙ **Figure 6.13.** Deviations from a linear working curve plotted from Equation 6.1.

fraction of the fluorescent radiation impinging on the detector, the ratio of excitation intensities in the sample excitation and excitation detector paths, and scattering losses of excitation and emission radiation in the sample container. For very small values of A, Equation 6.1 reduces to[1]

$$N_f = 2.303 K N_e A = 2.303 K N_e \varepsilon b C \qquad 6.2$$

where ε is the molar absorptivity of the fluorescing species at the excitation wavelength, b is the length of the sample compartment, and C is the analyte concentration.

From Equation 6.2, one would conclude that the fluorescence intensity is proportional to the concentration of the fluorescing species. At the same time, we know that this is true only if the absorbance is much smaller than 1. The values of N_f calculated from Equations 6.1 and 6.2 are plotted in Figure 6.13. The error in assuming a linear response curve is approximately -1% for each 0.01 absorbance unit. The secondary absorbance, though usually a smaller effect, increases the deviation. There are two remedies to this situation. The one most often used is to keep the absorbance of the solution of the order of 0.01 absorbance units or less. The high sensitivity of fluorescence makes it an excellent technique for very low concentrations of analyte, where the absorbance is naturally low. The second remedy is to place a third detector behind the cell to measure the transmitted excitation light. From this value, the absorbance can be calculated (using Equation 5.8) and then Equation 6.2 can be used to calculate the concentration of the fluorescing species. In either case, a standard solution of analyte is used to determine the value of K (see Example 6.1).

Fluorescence is often used with fixed excitation and emission wavelength selectors to determine the concentration of a specific analyte or set of analytes with similar fluorescence characteristics. In such instruments, quantitation is accomplished by the application of Equation 6.2. The constant $K\varepsilon b$ is determined by the use of a standard solution. Remember that the linear relationship between fluorescence intensity and concentration only works for very high values of transmittance, of the order of 98% or higher. One would not expect very high sensitivity for fluorescence spectrometry with such a small fraction of the excitation wavelength absorbed; however, photons at the emission wavelength arise uniquely from the fluorescence process, so their detection indicates the presence of the fluorescing analyte. Since detecting the presence of a few

[1]From the relationship between natural logs and base 10 logs, $10^{-A} = e^{-2.303A}$. From the Taylor expansion, $e^{-x} = 1 - x$ for values of x much smaller than 1. Therefore, $1 - 10^{-A} = 1 - e^{-2.303A} = 1 - (1 - 2.303A) = 2.303A$.

■ **Example 6.1**

A fluorometer was used to determine the concentration of NADH (nicotinamide adenine dinucleotide). A blank sample gave a N_f reading of 48. The unknown sample gave a reading of 976, and a 5.0×10^{-7} M standard solution gave a reading of 2963.

Subtract the blank from both sample and standard readings. From Equation 6.2, concentration is proportional to fluorescence, so $C_{unk} = (5.0 \times 10^{-7})(976 - 48)/(2963 - 48) = 1.6 \times 10^{-7}$ M.

If the volume of the fluorescence cell was 0.010 mL, the total amount of NADH in the test solution was 1.6×10^{-7} mol L^{-1} $\times$ 10^{-5} L $= 1.6 \times 10^{-12}$ mol.

photons at the emission wavelength is much easier than detecting the loss of a few photons at the absorbance wavelength, fluorescence spectrometry can provide extremely low detection limits. In some dramatic experiments, detection of a single molecule has been accomplished.[2] For highly fluorescing species, detection limits in the 10^{-7} M range are common.

One of the problems in quantitative fluorescence spectrometry is that of quenching. Some species in the solution, notably molecular oxygen, can provide an energy loss mechanism through collisional absorption of the excited state energy; this process is called **quenching.** The quantum efficiency for fluorescence will depend on the concentration of the quenching species in the solution. If the sample solution contains a different concentration of quenching species from the standard solutions, the calculated concentration will be wrong. This disturbance of the analysis by other constituents is called **matrix interference.** To avoid it, one must either eliminate the quenching species prior to analysis or use the standard addition method of quantitation.

The use of a pulsed excitation source such as a pulsed laser allows the dimension of time to be added to the detection process. For example, the emission due to fluorescence can be distinguished from that due to scattering and phosphorescence because they occur on different time scales relative to the excitation incident. Sampling the detector output a fraction of a microsecond after the end of the laser pulse will eliminate the detection of scattered light. Ending the detection after 10 microseconds will prevent the detection of most of the phosphorescence. The fluorescence lifetime can be measured by recording the intensity versus time after a pulsed excitation. These data can provide much useful information about the excitation and fluorescence processes.

Fluorescence is widely used in flowing systems. One need only use a flow-through cell with the standard apparatus to accomplish this. Fluorescence is one of the most sensitive means of detecting compounds separated by liquid chromatography. Many compounds of biological importance fluoresce naturally. Those that do not can often be chemically modified by attaching a fluorescing subgroup to the analyte molecules before the chromatogram is run.

The degree of fluorescence quenching depends on the concentration of the quenching species. One should always be on the lookout for such relationships between effect and composition. Sensitive oxygen detectors based on measuring the extent of fluorescence quenching have been developed.

EXCITATION AND EMISSION SPECTRA

In addition to quantitation, identification information can be obtained from the excitation and emission spectra. The excitation spectrum is obtained by setting the emission monochromator at the wavelength of a fluorescence emission and scanning the excitation monochromator. The result is a plot of the emission intensity as a function of the excitation wavelength. Such a plot is shown by the solid line of Figure 6.14. The emission spectrum is obtained by setting the emission monochromator to one of the wave-

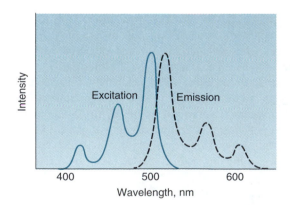

Figure 6.14. Comparison of the excitation and emission spectra of a compound.

---

[2]S. Nie, D. T. Chiu, and R. N. Zare, *Anal. Chem.* **1995,** *67,* 2849; Y.-H. Lee, R. G. Maus, B. W. Smith, and J. D. Winefordner, *Anal. Chem.* **1994,** *66,* 4142.

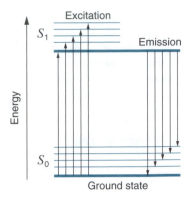

Figure 6.15. An energy level explanation of why emission energies are lower than excitation energies for the same molecule.

lengths that is effective in exciting the molecule to fluorescence emission and plotting the emission intensity as a function of the emission wavelength. The dashed line in Figure 6.14 is an emission spectrum

As expected, the emission spectrum is centered around a longer wavelength (lower energy) than the excitation spectrum. Excitation and emission spectra for the same compound often look like mirror images of each other. This is because the absorbance wavelengths for the emission radiation correspond to the transitions from the ground electronic and vibrational states to the vibrationally excited electronic state. (See Figure 6.15, which is a composite of Figures 6.5 and 6.8.) Thus, the *longest* excitation wavelength corresponds to the pure electronic transition. The more energetic transitions correspond to higher vibrational excitations. The vibrational excitation is quickly lost by vibrational relaxation, so the emission spectrum is from the excited electronic state to various vibrational excitations of the ground electronic state. As we see, the *shortest* wavelength of the emission spectrum corresponds to the pure electronic transition. To the extent that the vibrational states of the two electronic states are similar, the emission and excitation spectra proceed similarly in each direction from the wavelength corresponding to the electronic state transition.

With two wavelength variables in the fluorescence measurement, a complete plot of the fluorescence intensity and the two wavelengths involves three dimensions. Such a plot is shown in Figure 6.16.[3] A reasonably efficient way to obtain such a complete

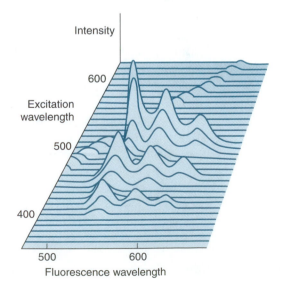

Figure 6.16. The three-dimensional plot of emission and excitation wavelengths.

---

[3]D. W. Johnson, J. P. Callis, and G. C. Christian, *Anal. Chem.* **1977**, *49*, 749A.

set of response data is to use an array detector spectrofluorometer for the emission and scan the excitation monochromator. From the standpoint of compound identification, the increased amount of data in the three-dimensional data set (over the emission spectrum alone) does not increase the level of discrimination from similar compounds significantly. This is because of the similarity in energy transition information provided by the two wavelength axes.

PHOSPHORESCENCE MEASUREMENTS

The apparatus for phosphorescence spectrometry is much the same as that for fluorescence spectrometry. However, because so many of the compounds that phosphoresce also fluoresce, observation of the emission that is strictly phosphorescence requires a pulsed excitation source with delayed detection of the emission. Since the delay times are in the millisecond region, even mechanical shutters can be used for this purpose. Emission wavelengths are still longer than those of fluorescence (relative to the excitation wavelength) because of the lower energy of the triplet state. Phosphorescence spectrometry is, in general, not as sensitive as fluorescence. Its main advantage is its greater selectivity, since so few compounds exhibit phosphorescence. Enhancement of phosphorescence can often be achieved by drying the sample to eliminate the solvent. Samples in the solid state have greatly reduced collisional quenching. Solid samples can best be used in a system that collects the emission radiation from the same surface impinged on by the excitation.

In application, preparation of the sample is important for the removal of quenching substances and for the choice of solvent. Detection limits for room temperature phosphorescence tend to run in the low nanograms for such favorable substances as adenine, cocaine hydrochloride, salicylic acid, sulfanilamide, and tryptophan. Low temperature phosphorescence from solution has achieved detection limits of the order of 10^{-8} M for such molecules as tryptophan and the pesticide DDT, and even lower for pyridine, sulfapyridine, sulfamethazine, and sulfamerazine.[4]

Study Questions, Section B

11. What is fluorescence?

12. What is used to induce fluorescence?

13. Why are the fluorescing molecules not distributed evenly throughout the illuminated region of the sample volume?

14. Give two reasons why only part of the fluorescence radiation is detected.

15. What is the difference between a fluorometer and a spectrofluorometer?

16. Why is it necessary to measure the excitation source intensity as well as that of the fluorescence?

17. For a standard solution of analyte of 0.00100 M, the measured excitation and fluorescence intensities were 10,538 and 2,029. For an unknown solution of the same analyte, the measured values were 11,319 and 1,438. What is the concentration of the analyte in the unknown solution?

18. Confirm the statement that the error in assuming a linear response curve involves approximately a 1% error for each 0.01 absorbance unit of the sample.

19. Name two processes by which matrix interference can occur with fluorescence spectroscopy.

20. Explain why the wavelength of the fluorescence emission must always be greater than the excitation wavelength.

21. Describe the timing of excitation and emission measurement required for phosphorescence spectrometry.

[4]J. D. Ingle and S. R. Crouch, *Spectrochemical Analysis*. Prentice-Hall, Englewood Cliffs, New Jersey, 1988.

11. Fluorescence is the emission of a photon from a molecule in the excited singlet state.

12. Electronic excitation of the analyte is produced by photon absorption.

13. The excitation radiation is absorbed by the sample, so the intensity of the excitation radiation decreases with increasing distance from the illuminated side of the sample container. This is called primary absorbance. This decrease in excitation intensity results in a decrease in fluorescence intensity.

14. Some of the fluorescence radiation is absorbed by the sample molecules before it can exit the sample container. This is called secondary absorbance. In addition, molecules fluoresce in all directions. To detect all the fluorescence emitted from the sample, the container would have to be completely surrounded by the detector.

15. The fluorometer uses filters to select the wavelengths of excitation and fluorescence detection. The spectrofluorometer uses monochromators for wavelength selection.

16. Excitation sources are not generally stable enough to provide a constant output. Taking the ratio of the fluorescence and excitation intensities removes the effects of this variation.

17. Rearrange Equation 6.2 to isolate the constants.

$$\frac{N_{f,std}}{N_{e,std}C_{std}} = 2.303K\varepsilon b = \frac{N_{f,unk}}{N_{e,unk}C_{unk}}$$

Then, for the unknown solution,

$$C_{unk} = C_{std}\frac{N_{f,unk}N_{e,std}}{N_{e,unk}N_{f,std}}$$

$$= 0.00100\ \frac{1,438 \times 10,538}{11,319 \times 2,029}$$

$$= 6.60 \times 10^{-4}\ \mathrm{M}$$

18. The error involves the difference between Equations 6.1 and 6.2. In the linear equation, $2.303A$ has replaced $(1 - 10^{-A})$ in the more exact equation. When there is a 1% error, the ratio of these two values will be 0.99. When $A = 0.0100$, $2.303A$ is 0.02303 and $(1 - 10^{-A})$ is 0.02276. The ratio of these two values is 0.988.

19. One process by which matrix interference can occur is when an interfering analyte has a significant fluorescence emission at the measured wavelength. The other is when an interferent acts to quench the fluorescence of the analyte by providing an alternate way for it to lose its excitation energy.

20. The excited state energy of the analyte will be equal to the energy of the exciting photon. Some of this energy is lost in the conversion to the ground vibrational level of the excited singlet state. The photon emitted in going to a lower energy state must have less energy than the original excitation. Photons with lower energies have longer wavelengths.

21. The excitation source must be turned off so that the fluorescence radiation ceases. In 10 microseconds or less (see the time table of Figure 6.6.), the remaining emission will be phosphorescence, and the emission intensity can be measured.

C. Raman Spectrometry

Raman scattering[5] occurs as a result of the interaction of the electric field of the electromagnetic light wave with the electrons in the sample molecules. If the vibration of a bond in that molecule results in a varying polarizability of the molecule, some of the incident radiation may be briefly absorbed. It can then be reemitted from the same or from a different vibrational state of the absorbing bond. If the vibrational level of absorption and emission are the same, the excitation and emission wavelengths are the same, and the only effect has been a scattering of the incident light called **Rayleigh scattering.** However, if the excited and emitting vibrational levels are different, there will be a corresponding shift in the emission wavelength relative to the excitation wavelength. The magnitude of the shift corresponds to the energy difference between the two vibrational levels.

The energy shift effect is shown schematically in Figure 6.17. From the diagram, we can see that excitations from the ground vibrational state will result in

---

[5]Raman scattering was discovered by the Indian physicist C. V. Raman in 1928. In 1931, he was awarded the Nobel Prize for his discovery and fundamental exploration of this process.

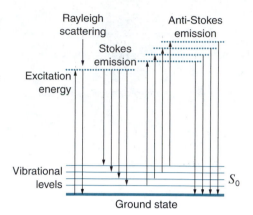

Figure 6.17. Schematic of the energy levels involved in Raman radiation.

Raman scattered radiation of a longer wavelength (lower energy) than the excitation radiation. The peaks of light intensity at these wavelengths are called **Stokes lines.** Excitation of an excited vibrational state with return to the ground state is also possible. Such radiation has wavelengths shorter than the excitation radiation, and the resulting light intensity peaks are called **anti-Stokes lines.** The resulting spectrum is shown in Figure 6.18. Because of differences in the populations of the initial states, the anti-Stokes lines are generally less intense than the Stokes lines. The shifts from the excitation radiation wavelength are identical for the Stokes and anti-Stokes lines. They both correspond to the vibrational energy levels for those vibrations that are Raman active, that is, those for which the molecular polarizability changes during the vibration. These wavelength shifts correspond exactly to the IR absorption wavelengths for the same vibrations. Thus, much the same information is gained through Raman and IR spectrometry. However, there are some differences because not all IR active vibrations are Raman active and vice versa. In the most extreme case, molecules with a center of symmetry have no vibrations that are both Raman and IR active. Raman and IR spectra of mesitylene are compared in Figure 6.19. The complementary nature of the spectra are clear from the number of peaks in each technique not shared by the other.

One form for the cell apparatus of a Raman spectrometer is shown in Figure 6.20. As in normal fluorescence measurements, the emitted radiation is sampled at right angles to the excitation light path. Since the efficiency of the Raman effect is very

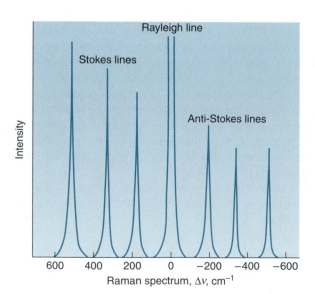

Figure 6.18. A typical Raman spectrum in which the information is in the magnitude of the wave number shift from the Rayleigh line.

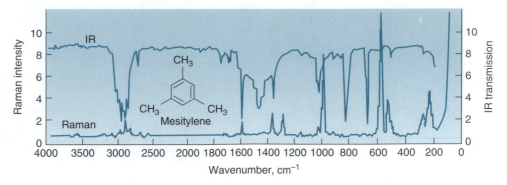

Figure 6.19. Infrared and Raman spectra of mesitylene. Courtesy of Perkin-Elmer Co., Norwalk, Connecticut. Note the similarities and differences in the data from these two techniques.

low (Rayleigh scattering is much less likely than transmission, and Raman scattering is less likely than Rayleigh scattering), collecting mirrors and lenses are used to increase the quantity of light sent to the detector. Sample cells are very small tubes that may hold as little as 10^{-8} L.

Choice of the excitation wavelength is critical. The efficiency of Raman scattering increases dramatically with increasing photon energy. However, using photons in the blue or UV regions of the spectrum greatly increases the likelihood of fluorescence. Since the fluorescence intensity is likely to be much greater than the Raman radiation, increasing the photon energy is counterproductive. Reducing the excitation energy (increasing the wavelength) decreases the opportunities for fluorescence. A compromise between Raman efficiency and background radiation reduction is generally optimum in the region of 800 nm. Therefore, light in this wavelength region is most often used for Raman spectrometry. Another problem for photon detection is that the Raman radiation is shifted only slightly from the excitation wavelength. Eliminating interference from the excitation source requires a very narrow band source and a very good quality emission monochromator. The former requirement is met with a laser source. Lasers are now the universal source for Raman spectrometry because they provide both high intensity and narrow-band emission.

Raman spectrometry is often used in conjunction with IR spectrometry to provide identification and structural information about organic compounds. The wave-

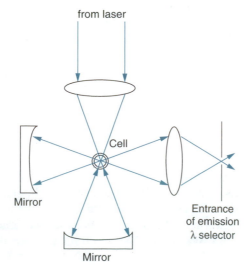

Figure 6.20. Form for a Raman sample cell. The cell volume can be a microliter or less.

lengths associated with such Raman active groups as —C—S—, —S—S—, —C—C—, —C=C—, etc., have been assigned to aid this process.

22. What is the requirement for Raman scattering?

23. Why is there an energy shift in Raman scattering?

24. How is it that the emitted photon could have a higher energy (shorter wavelength) than the exciting photon?

25. What are the names of the emission lines that are of longer length than the excitation radiation?

26. Which has the higher probability, the emission of wavelengths longer than the excitation wavelength or shorter?

27. Light of the excitation wavelength can also be scattered by the process of Rayleigh scattering. Which is the more probable scattering mechanism, Rayleigh or Raman?

28. Since the difference in the wavelength of the incident and Raman scattered radiation is due to changes in the vibrational states of the sample molecules, what kind of absorption spectroscopy provides similar information?

29. Why does the wavelength of the incident radiation in Raman spectrometry matter when the information is in the wavelength shift between the incident and emitted radiation?

30. How is the problem of the close proximity of the emission wavelengths to that of the excitation wavelength met in modern Raman spectrometers?

22. Raman scattering requires that the molecule contain a bond for which the vibration of that bond will result in a change in the polarizability of the molecule.

23. The energy shift is due to a difference in the vibrational state of a bond between the absorption and emission of the excitation radiation.

24. A molecule in an excited vibrational state can absorb the incident photon and emit it from the ground vibrational state. The overall energy transition of emission is thus greater than that of the absorption.

25. Emission wavelengths that are longer than those of the excitation are called Stokes lines. Anti-Stokes lines have shorter wavelengths than the exciting radiation.

26. The Stokes lines generally have a higher intensity than the anti-Stokes lines because there will generally be a larger fraction of the absorbing molecules going to a higher vibrational state during the absorption than to a lower one.

27. Scattering that does not involve a coincident change in the vibrational state of the molecule is far more common than

one that does. Therefore, the intensity of the Rayleigh scattered light is far greater than that of the Raman scattered emission.

28. The information contained in the Raman spectrum is similar to that obtained by IR absorption spectroscopy. Since the two techniques respond differently to various bond types, the information obtained is complementary rather than duplicative.

29. There is a trade-off between the higher Raman efficiency of shorter wavelength excitation and the higher level of fluorescence radiation that can occur when higher excitation energies are used. The compromise excitation wavelength generally used is 800 nm.

30. To perform Raman spectrometry one needs a source whose radiation bandwidth is much less than the smallest shift one wants to measure. Lasers provide such a narrowband source. Then, to measure a weak Raman light intensity in proximity to the strong excitation radiation, an excellent monochromator with low transmission of stray radiation is required.

D. Excitation from Chemical Reactivity

Some chemical reactions result in products that are in the excited state. If these excited state products can lose their internal energy through fluorescence, the reaction mixture glows from the fluorescence emission. This phenomenon is the source of the light from fireflies and the glowing wands and necklaces sold at fairs. The fluorescence emission that results from a chemical reaction is called **chemiluminescence** (CL). To emit a photon in the visible or near IR region of the spectrum, a molecule must be electronically excited. For this energy to have come from a chemical reaction, the ΔG for the reaction must be greater than 40 kcal mol^{-1}. Then a significant fraction of the excited molecular products of this reaction must avoid losing their internal energy by means other than fluorescence. As might be imagined, these conditions are rarely met. This is good from the standpoint of specificity, but it also means that CL has a limited range of application.

In a CL reaction, the fluorescence can come from the final reaction product or from an intermediate in the reaction process. To use CL as the discriminating characteristic for chemical analysis, the analyte either is the limiting reactant in a CL reaction, reacts to form a reagent for a CL reaction, or serves as a catalyst for the CL reaction. The remaining reagents required by the reaction are then the probe by which the CL is induced. The response to the probe is the CL reaction and the emission of fluorescence radiation. The intensity of the radiation is measured, perhaps as a function of time or wavelength. From these data, one may be able to confirm the presence or determine the amount of an analyte.

CHEMILUMINESCENCE MEASUREMENTS

Systems for the measurement of chemiluminescence can be very simple because the only light involved is the fluorescence radiation to be measured. A typical apparatus is shown in Figure 6.21. The reaction occurs in a lighttight box. The sample is introduced to the reagent solution, which is quickly mixed. Some of the resulting fluorescence radiation is collected and focused on a light detector. The output from this detector is converted to a related number. The fluorescence will continue to be emitted as long as the reaction continues. However, the reaction rate and thus the light intensity will decrease with time as the reactant gets used up. Since the light intensity is time dependent, the data acquisition is generally timed with the sample injection.

Chemiluminescence measurements can also be made in flowing systems. The reagents are added to the sample stream before the reaction mixture flows past the light detector. The length of the flow tube between the reagent mixer and the detector and the flow rate determine the reaction time after which the fluorescence is detected. Some gaseous reactions are chemiluminescent. Sample and reagent gases are leaked into a reaction chamber, which is kept at a constant pressure by a pump controller. Fluorescence radiation that escapes through a window in the chamber can be detected.

The intensities of light from chemiluminescence reactions can be very low. The fraction of the excited reaction products that fluoresces is 20% at best and generally much lower. In addition, the fraction of the luminescent photons intercepted by the measurement system is small. To effectively measure very low light fluxes, it is often necessary to reduce the dark signal (N_d) by cooling the detector and/or using photon counting techniques. In either case, a reasonable measurement precision may require substantial measurement times.

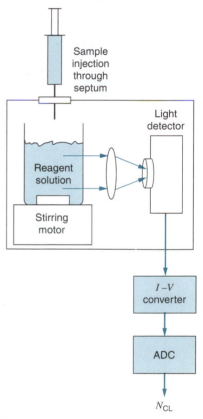

Figure 6.21. The apparatus for chemiluminescence measurements is contained in a lighttight box.

CHEMILUMINESCENCE APPLICATIONS

Virtually all of the CL reactions are oxidation reactions. A few examples are shown in the margin. The oxidant is often O_2, H_2O_2, or O_3. The reaction of ozone (O_3) with NO produces an excited NO_2 molecule; with ethylene (C_2H_4), the excited product is HCHO. These reactions are carried out in the gas phase and are used for the determination of NO and O_3, respectively. Since all nitrous oxides can be converted to NO by passing them over hot carbon, this is a relatively easy way to carry out this common air-quality determination.

$$NO + O_3 \rightarrow O_2 + NO_2{}^* \rightarrow O_2 + NO_2 + \text{light}$$

$$2O_3 + C_2H_4 \rightarrow 2O_2 + 2HCHO^* \rightarrow 2O_2 + 2HCHO + \text{light}$$

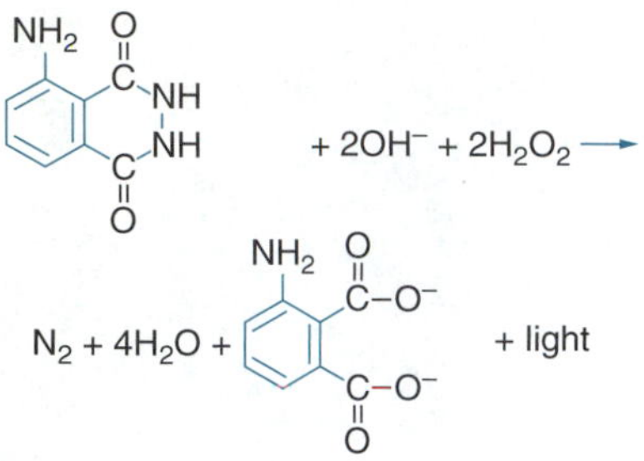

Figure 6.22. Reaction scheme for luminol chemiluminescence.

Fireflies glow because of bioluminescence:

$$LH_2 + E + ATP \rightleftharpoons E:LH_2:AMP + PP$$

$$E:LH_2:AMP + O_2 \rightleftharpoons E + L=O + CO_2$$

$$+ AMP + light$$

where E is firefly luciferase, LH_2 is firefly luciferin, ATP is adenosine triphosphate, AMP is adenosine monophosphate, PP is pyrophosphate, and L=O is oxyluciferin.

Compounds containing sulfur can be determined by combustion to form SO. The SO is then reacted with O_3 to form excited SO_2. The chemiluminescence emitted as excited SO_2 fluoresces is then detected. A sulfur detector based on this approach has a sensitivity of less than 0.5 pg of sulfur for one second of detection.[6]

In solution, molecules that can be oxidized into excited states that fluoresce have the familiar conjugated bond systems that enable photon excitation fluorescence. An example is luminol. As shown in Figure 6.22, in basic solution this reagent reacts with peroxide to form 3-aminophthalate, which is the fluorescing species. This reaction is used for the determination of the peroxide produced by the reaction of an oxidase enzyme on its substrate. Transition metal ions and compounds containing them (such as hemes and vitamin B12) enhance the rate of the reaction of luminol with peroxide. Since the increased rate results in an increased luminescence intensity, this effect can be used for the determination of these substances.

Some enzymatic reactions are luminescent. This phenomenon, a subclass of chemiluminescence, is sometimes called **bioluminescence.** An interesting reaction is the one causing the glow produced by fireflies. It is even more interesting in this context because of its analytical application. The reaction scheme shown in the margin indicates the many species that are involved in this reaction. Mg^{2+} is required as a catalyst. In the process, ATP is converted to AMP, and firefly luciferin is converted to oxyluciferin. This reaction is very specific and very efficient. In the laboratory, it is used as a basis for the determination of ATP. Since there is a strong correlation between the amount of ATP and the quantity of active biological cells, this is an important biological assay. The detection limit can be as low as 10^{-14} g for a 10 μL sample.[7]

Study Questions, Section D

31. Up to this point in this chapter, the probe for the distinguishing characteristic of photon emission has been excitation by photon absorption. What is the probe for the method called chemiluminescence?

32. What combination of distinguishing characteristics are required in an analyte for which chemiluminescence is a practical technique?

33. In the apparatus shown in Figure 6.21, what would limit the frequency of sample injection?

34. What is the oxidant used in the luciferin reaction?

35. In carrying out a measurement, would you choose to measure the chemiluminescence intensity at a specific time after sample injection or to integrate the intensity over a specific period of time following injection?

36. For the measurement of Question 35, would you expect the resulting measurement number to be proportional to the concentration or total amount of the analyte?

Answers to Study Questions, Section D

31. The probe for chemiluminescence is a chemical reaction that results in product that is in an excited electronic state.

32. The analyte must be able to participate as an essential reactant in a reaction that produces an excited state molecule that loses energy by fluorescence. The analyte molecule need not be the species that becomes excited; it could be a catalyst or other reactant in the excitation reaction.

33. The next sample should not be injected until the luminescence intensity from the previous sample has become essentially zero. The time required for this to occur would depend on the rate of the luminescence reaction. For very fast reactions, this could depend on the rate of mixing the reactants.

34. The oxidant in the luciferin reaction is O_2. With the aid of the enzyme and AMP, the luciferin is oxidized to excited

[6]P. L. Burrow and J. W. Birks, *Anal Chem.* **1997,** *69,* 1299–1306.

[7]J. D. Ingle and S. R. Crouch, *Spectrochemical Analysis,* p. 484. Prentice-Hall, Englewood Cliffs, New Jersey, 1988.

oxyluciferin, which can lose its excitation energy through fluorescence.

35. It would be preferable to average or integrate the intensity over an interval of time following injection of the sample. The longer measurement interval will allow more photons to be detected, which will improve the precision of the resulting number. Variations in mixing and sample introduction would also be averaged out.

36. The sample is injected onto the reaction mixture. Assuming quick dispersion throughout the reaction volume, the light intensity would be proportional to the concentration of the analyte in the reaction mixture. This is equal to the moles of analyte divided by the reaction mixture volume. Thus, the response would be proportional to total amount of analyte, not to the concentration of the analyte in its original solution.

E. Thermal Excitation and Atomic Emission

The use of heat as a source of excitation for photon emission is the basis of some of the earliest methods of trace metal analysis. The yellow color in a flame is due to the presence of sodium in the material that is burning. Excited atoms of other metals create the colorful lights observed in fireworks. Virtually all materials will emit photons when heated. The red glow of hot coals and the radiance of lightbulb filaments are familiar examples. The radiation emitted by solid materials often covers a very broad spectral range. If a very large number of closely spaced energy bands is available, the spectral distribution will match the distribution of excitation energies in the material. This is called **blackbody radiation**. The energy distribution of blackbody radiation, formulated by Planck, is a function of the temperature of the material. This function is shown in the plot in Figure 6.23 for three temperatures corresponding to the temperatures of a gas/air flame (1800 K), an acetylene/air flame (2500 K), and an acetylene/oxygen flame (3300 K). The higher the temperature, the higher will be the energy at the maximum in the energy distribution and, as we see, the higher will be the intensity of the radiation.

Because the emission spectrum from solids is broadband and much more a function of the temperature than the material, the emission spectrum from excited solids is not generally useful for chemical analysis. It is used, however, for the remote sensing of temperature in everything from furnaces to distant stars. For the spectrum to have analytical value, it is desirable for the energy bands represented in the emission to be characteristic of the material. This is accomplished in the vapor phase where the molecules or atoms are not in intimate contact with each other. There are far fewer energy levels available to molecules and atoms in the vapor phase, so the wavelengths emitted will be limited to these values. Similarly, atoms do not have the bond vibration and rotation levels that molecules have, so their emission spectra tend to have much narrower bands (peaks) of emission.

By the above arguments, we have established that excited vapor phase atoms and molecules can emit radiation at wavelengths characteristic of the species. To take ana-

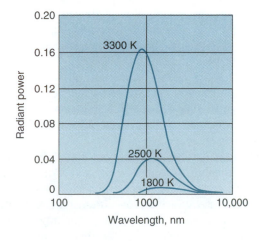

Figure 6.23. As the temperature of a blackbody radiator increases, the intensity and the average photon energy increase.

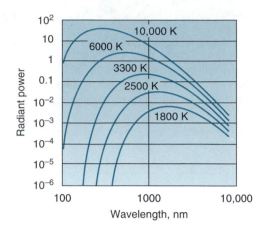

Figure 6.24. When heat is used to excite photon emission, the temperature greatly affects the range of energy levels that can be excited.

lytical advantage of this distinguishing characteristic, it is necessary to get the sample molecules and atoms into the vapor phase and to excite them. In this section, we will explore the methods by which heat is applied to the sample to accomplish these goals. Three forms of heating are in current use. They are the flame, the furnace, and the inductively coupled plasma. The temperature of the sample must get high enough to enable electronic excitation. An idea of how hot that would be can be obtained from the wavelength distribution of blackbody radiators. A log intensity plot similar to that given in Figure 6.23 is shown in Figure 6.24. The difference in the shape of the distribution curves results from plotting the log of the radiant power. To excite a sample to an energy level that would emit a 600 nm photon by raising its overall temperature, the temperature would have to be high enough for significant blackbody radiation at that wavelength. From the plot, we can see that this would be 2000 K or more. To excite characteristic radiation at lower wavelengths, an even higher temperature is required.

One of the consequences of using such high temperatures for excitation is that most of the molecules in the sample will decompose. If the temperature is sufficiently high, all the intermolecular bonds will be broken, so that the vapor then contains the atoms making up the original molecules. These, then, will be the excited species that provide the characteristic radiation. For this reason, this type of analysis is called **atomic emission analysis** or **elemental emission analysis.** In fact, it is desirable to avoid molecules in the vapor phase, as their broader emission spectra can interfere with the very narrow bandwidth of the atomic emissions. Molecular species can occur in the excitation region of the heating device both from the incomplete decomposition of the sample and from the combination of atoms in the vapor. The molecules formed by atomic recombination are called **refractory molecules** because they necessarily have a decomposition temperature higher than that of the excitation region. Refractory molecules are generally metallic oxides.

Atomic emission analysis involves the measurement of the emission intensity of at least one of the emission wavelengths of each element of interest. The sample is introduced into the heat source in a manner appropriate to the source. A portion of the radiation emitted from the source is focused on the entrance to a monochromatic or array spectrometer to obtain the desired intensity–wavelength information (see the block diagram in Figure 6.25). The very narrow bandwidths of the emission peaks for most elements makes the use of high resolution spectrometers advantageous. High resolution monochromators with narrow slits are used with single detectors. Where simultaneous, multielement intensity measurements are desired, several single detectors arranged behind separate exit slits may be used.

The linear diode array detector used for absorbance and fluorescence spectrometry does not offer enough elements to resolve the sometimes closely spaced emission peaks. Therefore, for array detection, a special monochromator that disperses the spectrum in two dimensions is used. Such a monochromator is called an **Eschelle monochromator.** Eschelle monochromators use a principal grating that is designed to operate efficiently at very high orders of reflectance. Normally this would cause an unaccept-

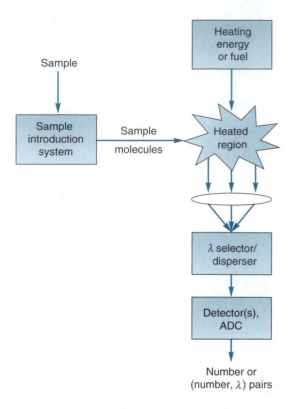

Figure 6.25. Block diagram of an atomic emission spectrometer.

able overlap of a wide range of the spectrum, but in the Eschelle monochromator, the reflection orders are dispersed in a second direction by a prism or another grating. This disperses a wide range of wavelengths over a two-dimensional area. In conjunction with a two-dimensional array detector (such as that used in a TV or digital camera), a million or more detector elements are available for the wavelength range covered.

The data obtained from atomic emission spectroscopy is emission intensity at the wavelength(s) corresponding to the element(s). Interpretation of these data into the amount of each element contained in the sample is generally accomplished by first constructing a calibration curve. This must be performed for each element because different elements have very different efficiencies of excitation and emission. There are other complications. One is self-absorption in which atoms in the ground state of the emitting element can absorb some of the emitted radiation before it has left the heated region. Vapor in the cooler outer parts of the heating device can absorb some of the emission from the excited atoms in the hotter central portion. In addition, some elements in the sample can contribute to the formation of refractory compounds that can remove some of the atoms from the emission process. An example is the suppressive effect of sulfates and phosphates on the intensity of calcium emission. These oxyacids promote the formation of refractory CaO. The presence of lanthanum can decrease this effect by reducing the degree of oxidation. Some complexing agents can have the same effect. Such agents are called **releasing agents,** but their action points out that the emission intensity of a particular element can be significantly affected by the concentrations of other elements in the sample. Such effects are called **matrix effects,** and they make quantitation much more difficult.

The remainder of this section is devoted to discussion of the three principal methods of providing the excitation heat. Each has advantages for particular types of samples.

FLAME EXCITATION

Flame is the oldest of the thermal excitation methods, and it is still one of the most widely used. As with all the thermal techniques, the flame is used to decompose the sample into atoms and excite the atoms to emission. With flame excitation, both gaseous and liquid samples can be used. The major design elements of a flame excita-

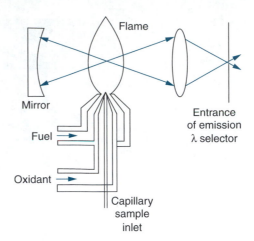

Figure 6.26. In the flame method of thermal excitation, the sample is introduced at the base of the flame.

tion source are the means of sample introduction into the flame, the type of fuel/oxidizer used for the flame, and the method of collecting the light emitted from the heated region. In the very simple design shown in Figure 6.26, the sample is introduced through a metal capillary that restricts the flow of the sample. The flow of fuel and oxidant past the upper tip of the capillary draws the sample up the capillary from the sample container in which it is placed. Acetylene is a common fuel; when it is used with air as the oxidant, the flame temperature is 2500 K in its hottest region. With oxygen as the oxidant, the maximum temperature is 3300 K owing to the lack of the energy-absorbing N_2. If operated with more than the stoichiometric amount of fuel, the flame is a reducing atmosphere. This has the effect of decreasing the formation of the refractory oxides.

The temperature of the flame is great enough to provide reasonably efficient excitation for emissions that occur in the low energy end of the visible spectrum (red, orange, and yellow). Thus, this method of analysis is used for alkali and alkaline earth metals, which have the requisite low energy electronic states. The simple flame apparatus lends itself to routine applications such as the determination of Na, K, and Li in the clinical laboratory. Although the use of exotic fuels and oxidants (oxyacetylene and NO_2) has increased the flame temperature high enough for the determination of over 60 elements, other techniques of atomic spectroscopy described later in this chapter are generally preferred.

PLASMA EXCITATION

The advantages of higher flame temperatures are offset by the difficulties and dangers of using increasingly exotic fuels and oxidizers. This led investigators to consider other sources of energy for the excitation process. Electric arcs and sparks were long used for emission spectroscopy because of the very high levels of energy they could provide. In these applications, the source of the energy is the conversion of electric power to heat in a region of conducting gas called a **plasma.** To create a plasma, some molecules of the air must become ionized (lose an electron). In the presence of a strong electric field, the electrons released will gain enough energy to impart ionizing energy to a majority of the molecules with which they collide. The electrons released from these molecules will ionize other molecules and so on in a rapidly cascading process. Soon a conducting region of the gas is created, with the electrons and ions serving as the mobile charge carriers. This process is the basis of arcs, sparks, and lightning.

The energy levels in a plasma can be very high, as seen in Table 6.1. A significant fraction of the molecules can be in the ionized state. Electron neutralization of ions can release a great deal of energy in the form of heat and light. Both these cause molecular and ionic excitation through energetic collisions and light absorption. The effective temperature depends on the gaseous species involved, the duration of the

Table 6.1

Energy Levels in Plasmas

Source	Temperature	Batch or Continuous
Spark	Up to 40,000	Batch
Arc	3000 to 8000	Batch
DC plasma	≈6000	Continuous
Inductively-coupled plasma	≈6000	Continuous

plasma, and the rate of energy loss through the escape of light, heat, and excited molecules and ions. In spark excitation, the duration of the plasma is very brief. The sample is generally the surface of one of the electrodes between which the spark occurs.

The plasma can be maintained as long as electric power continues to be supplied. If the electric current is supplied by a charged capacitor, the conducting plasma quickly discharges the capacitor, and the power supplied drops below the level required for maintenance. This is the process in photoflash lamps and in sparks. If a supply of power equal to the rate of energy loss can be maintained, the process can be continuous. The continuous plasma sources that have been developed thus far are the **DC plasma** (direct current plasma) and the **inductively coupled plasma** (ICP). Because the latter has much more widespread application, this discussion will focus on the inductively coupled plasma.

In an ICP, a plasma of argon ions and electrons is maintained by the continuous supply of energy in the form of a radio-frequency (RF) magnetic field. The rapid oscillation of the field accelerates the electrons in alternate directions. Electrons with high kinetic energy collide with argon atoms to form argon ions and more electrons. The plasma is only partially contained by the ICP structure, so a continuous supply of argon gas is required. A sketch of a typical source is shown in Figure 6.27. The tem-

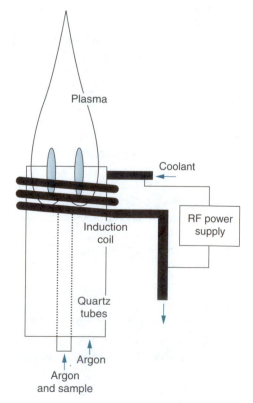

Figure 6.27. The inductively coupled plasma source obtains very high temperatures through the use of microwave excitation of argon ions and electrons.

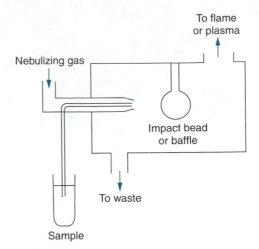

Figure 6.28. For introduction to the ICP, the sample solution must be nebulized into an aerosol. Shown here is one of many designs that accomplish this function.

perature of the argon ion plasma is as high as 10,000 K at the core (the shaded parts of the plasma). The portion of the plasma that provides the emission spectrum (above the tube) has a temperature closer to 5000 K. The sample is introduced as an aerosol mist mixed with the argon in the center tube. Laser ablation can be used for the introduction of solid particles into the gas phase.

CREATING THE AEROSOL

An aerosol is created from a liquid sample by a process called **nebulization.** Essentially, nebulization is the process of dividing the sample solution into an atmosphere of fine droplets called an **aerosol.** This can be accomplished by spraying from a fine nozzle, sometimes with the aid of an auxiliary gas around the nozzle. Larger droplets are separated from the aerosol by baffles in the gas flow. The smallest droplets will follow the gas flow around the baffle while the larger ones will hit the baffle. In some designs, the collision of the droplets has sufficient force to produce some smaller droplets. Some of the sample solution flows down the face of the baffle and returns to waste. The sample aerosol flows into the ICP or other heater, where the remaining solvent evaporates, the molecules dissociate into atoms, and the atoms get excited.

A typical **nebulizer** system is shown in Figure 6.28. This system uses a concentric gas flow to aid the breakup of the sprayed solution into finer droplets. Other designs use ultrasonic or piezoelectric devices to aid droplet breakup. The size of the droplets that remain in the mist delivered to the ICP is about 10 μm. Unfortunately, only a small fraction of the sample is transformed into aerosol. The remainder is drained away. In the design shown in Figure 6.28, the flow of the nebulizing gas past the end of the sample capillary creates a pressure reduction that draws the sample up the capillary tube. This is called **aspiration.**

Atomic emission analysis is well suited to analysis in flowing systems. The analyte solution can often be introduced directly into the flame or nebulizer. The technique is necessarily destructive, as the sample is consumed in the flame. Investigators have demonstrated some interesting advantages in using atomic spectroscopy for chromatographic detection. In effect, it can be used as a detector that is specific for compounds containing the analyte element.

Study Questions, Section E

37. In this section, a new probe for the distinguishing characteristic of photon emission from an excited species is introduced. What is the nature of this probe?

38. Approximately what temperature would be required for a blackbody to emit 10% of its peak radiation at the visible wavelength of 500 nm? Estimate this value from Figure 6.23.

39. Why are there few molecular species that can emit in the visible region after having been thermally excited?

40. What types of chemical species can exist at the thermal excitation temperatures?

41. Why is the technique of thermal excitation considered a technique for elemental analysis?

42. What information is available from atomic emission spectroscopy?

43. The two most common devices for thermal excitation in atomic emission spectroscopy are the flame and the inductively coupled plasma. Compare these two devices with respect to the temperatures achievable and the range of elements for which they can be used.

44. Name and describe the process of sample introduction into the flame or ICP.

Answers to Study Questions, Section E

37. The probe introduced in this section is heat. The intention is to apply enough heat to bring some or most of the sample atoms to such a high temperature that they would be in an electronically excited state.

38. First draw an imaginary vertical line on Figure 6.23 at a wavelength of 500 nm. This would be at a point 70% of the distance from the 100 nm line to the 1000 nm line. At 1800 K, the emission at 500 nm is 10^{-3} of the peak emission. At 2500 K, the emission at 500 nm is 0.1 of the peak emission. Therefore, the temperature of the blackbody would have to be at least 2500 K.

39. When thermally excited, the molecule is raised to the excitation temperature. The temperatures required to achieve significant electronic excitation are higher than the decomposition temperature of most molecules.

40. The species that can exist at the thermal excitation temperatures are atoms and refractory molecules. The latter are generally simple metal oxides.

41. Thermal excitation is an elemental analysis method because the characteristic emission is that of the excited species, which are almost completely atoms. Almost all molecular information is lost in the thermal decomposition of the sample.

42. The information obtained from atomic emission spectroscopy is the identification of the various elements that were present in the original sample (from the wavelengths of the emission) and the relative abundances of the elements present (from the intensities of the emission lines for each element).

43. The maximum flame temperature is between 2500 and 3300 K depending on whether air or oxygen is used as the oxidant. The ICP temperature is typically 6000 K. The higher temperature of the ICP results in much better efficiency of excitation for species that emit in the bluer regions of the spectrum.

44. The sample must be in a fluid form. If a gas, it can be introduced directly. If a liquid, it must be divided into fine droplets, that is, an aerosol, prior to introduction. This process is called nebulization. If a solid, the sample must first be dissolved and then introduced through a nebulizer or directly vaporized by laser ablation.

F. Photon Excitation of Atomic Species

In the atomic emission systems described above, the sample molecules are decomposed to their elemental state by the heat of the flame or plasma. At the temperatures involved, some of the atoms become electronically excited and emit characteristic radiation. In all of the systems, the fraction of the analyte atoms that are in the excited state is extremely small. In other words, almost all of the atoms are in the ground electronic state. Since they are atoms, there are no vibrational states available. With the development of efficient systems for atomization, and with the realization that the flame or plasma environment contains a predominance of ground state analyte atoms, the techniques of atomic absorption and atomic fluorescence spectroscopy have been developed. These techniques are simply the application of fluorescence and absorbance techniques, previously described, to a sample that is in the hot atomic environment.

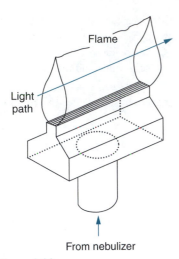

Figure 6.29. Another use for the flame and the nebulizer is to atomize the sample into the light path of a spectrometer. The analyte is then characterized by absorption rather than emission.

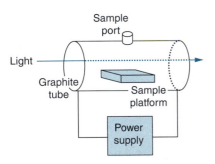

Figure 6.30. The furnace has also been used to convert the aerosol to its atomic constituents for spectroscopic determination.

FLAMES FOR USE WITH PHOTON EXCITATION

Burners used for atomic absorption spectroscopy are designed to provide as long a light path through the flame as possible. Thus the shape of the flame is arranged to be long and narrow. Furthermore, the burners are made to accommodate sample introduction in the form of an aerosol. A typical atomic absorption burner is shown in Figure 6.29.

FURNACES FOR SAMPLE ATOMIZATION

With the techniques of atomic absorption and atomic fluorescence, it is not necessary to heat the analyte molecules to the point of electronic excitation; just molecular decomposition is required. This fact has led to the development of furnace heaters that can produce the desired atomic vapor from a variety of sample types. One version of an atomic furnace is shown in Figure 6.30. The sample is injected through the injection port and placed on the sample platform. If the sample is a drop of liquid, the furnace is heated moderately to evaporate the solvent. Heating is accomplished by resistive heating of the graphite itself. This provides a uniform application of power along the length of the graphite tube. With the dry sample now on the sample platform, additional power is applied to heat the furnace to a temperature that will decompose any organic components in the sample (400–1100 °C). The rate of heating is determined by the amount of power applied to the graphite tube. The desolvation and organic decomposition phases of the process take only some tens of seconds to accomplish. Then full power is applied for a few seconds to heat the sample to 2000 to 3000 °C as quickly as possible. This is the atomization step in which the inorganic components of the sample are converted to atomic vapor. Heating rates in excess of 1000 °C/s are in common use.

The furnace atmosphere in this step is a bath of argon to prevent oxidation of the graphite. Since the furnace is not completely closed, the atomic vapor will eventually be swept out of the furnace with the bath gas. Thus, the atomic sample is only present for a brief time during which the spectroscopy must be accomplished. Furthermore, the analyte concentration in the furnace atmosphere is not constant over this period. These disadvantages are offset by the fact that the absence of combustion gases avoids the formation of some of the refractory molecules. In addition, the efficiency of atomization can be higher than in either a flame or a plasma.

ATOMIC FLUORESCENCE SPECTROSCOPY

Atoms in the vapor state can be excited to fluorescence. Researchers have taken advantage of this to demonstrate and explore this method of analysis. The probe for fluorescence is a light source of a wavelength that will cause electronic excitation of the analyte atoms. The expected response is a fluorescence emission at a longer wavelength. The response is measured by the use of a monochromator and detector. Atomic fluorescence was expected to have advantages over atomic absorption (described next) for low concentrations where absorbance methods are less reliable. This is because the presence of a small quantity of fluorescence light can be more easily measured than the absence of the same amount of the transmitted light. Unfortunately, lower detection limits were demonstrated for only a few elements. This may be due to the greater background radiation that exists in the flame or furnace, or it may be due to a lower fluorescence efficiency when atomic species are involved. In any case, the great success of atomic absorption spectroscopy (AAS) and the widespread availability of AAS equipment have discouraged the further development of atomic fluorescence spectroscopy.

ATOMIC ABSORPTION SPECTROSCOPY

As the name of the technique suggests, the differentiating characteristic exploited in atomic absorption spectroscopy is the ability of vapor phase atoms to absorb light of specific wavelengths. The probe would then be light of the wavelengths that can be absorbed. The absorbing wavelengths correspond to the electronic energy levels of the analyte atoms and ions. The measured decrease in the illuminating intensity is related to the concentration of the analyte atoms in the vapor according to the methods developed in Chapter 5. The atoms in the sample are converted to atomic vapor by a flame or furnace as described above.

One of the major factors contributing to the success of atomic absorption spectroscopy was the development of the hollow cathode lamp light source. The absorption bands of atomic vapor can be extremely narrow owing to the absence of the vibrational, rotational, and stretching energy levels associated with interatomic bonds. It is thus a challenge to obtain a light source with a bandwidth as narrow as the absorption bands. As discussed in Chapter 5, illumination with a bandwidth broader than the absorption band leads to serious deviations from the desired linear calibration plot. The bandwidth of light from even a good monochromator may be 100 times greater than the absorption bandwidth. The breakthrough concept that solved this problem was to use the analyte's atomic emission as the excitation radiation. Such emission automatically has the correct wavelength and can be made to have a bandwidth even narrower than the atomic absorption.

The device used to create the appropriate atomic emission is the hollow cathode lamp shown in Figure 6.31. The glass envelope is filled with neon or argon at a pressure of a few thousandths of an atmosphere. A dc power supply of a few hundred volts is applied through a series resistor to the contacts (positive to the anode and negative to the cathode). The applied voltage is sufficient to cause ionization of the neon or argon filler gas. Electrons released by the ionization are attracted to the anode. Filler gas cations are attracted to the cathode and can strike it with sufficient energy to release electrons from the cathode metal. Some of these electrons will combine with filler gas cations in the cathode region. This reaction creates an excited neon or argon atom that can then lose this energy by characteristic photon emission. This is the process involved in the orange glow of neon bulbs. The glowing region is near the surface of the cathode. In the case of the hollow cathode lamp, the cathode is made of, or coated with, the intended analyte metal. Metal atoms are also sometimes released from the cathode surface by the energetic collision of filler gas cations. These atoms, if not already excited, can become so by collision with energetic electrons or excited filler gas atoms. The excited metal atoms can also lose their energy by characteristic emission. The metal atom light emission region also remains close to the cathode surface.

The cathode has the shape of a hollow tube in order to concentrate the light emitting region and to maximize the chance that the atomic metal vapor will redeposit on the cathode surface rather than on other interior surfaces of the lamp. Hollow cathode lamps are available for virtually all of the metallic elements. Lamps are also made with cathode surfaces having a mixture of metals so that emission characteristic of a set of metals is produced. The light from all hollow cathode lamps includes the characteristic emission of the metals in the cathode, the atoms of the filler gas, and some unavoidable impurities. The fact that the emission lines are so narrow allows relatively easy selectivity of each appropriate wavelength for the intensity measurement.

The overall arrangement for atomic absorption spectroscopy is shown in Figure 6.32. Light from the hollow cathode lamp passes a shutter and then goes through the flame longitudinally. A monochromator is used to select the region of the particular hollow cathode lamp emission wavelength used. The monochromator bandwidth should be approximately 1 nm or less for the required selectivity and background rejection. The light transmitted through the monochromator is detected with a photomultiplier tube light intensity-to-current converter. The detection circuit is completed with

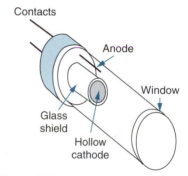

Figure 6.31. The hollow cathode lamp is a very narrow bandwidth emitter for just exactly the right wavelength for the analyte for which it was designed.

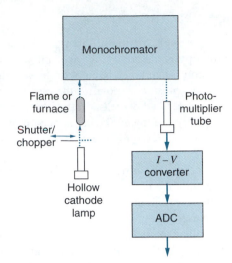

Figure 6.32. Block diagram of an atomic absorption spectrometer for elemental analysis.

a current-to-voltage converter and an ADC. Three detector output readings are required for the calculation of the atomic absorbance. One is the 0%T reading N_0 obtained with the source lamp blocked. One is the total sample reading N_s taken with the source on and the sample solution aspirating into the flame. The third is the total reference reading N_r taken with the source on and a blank solution aspirating into the flame. The quantity N_0 is subtracted from N_s and N_r to obtain the inverse of %T. The absorbance A is $A = \log(1/T)$. The overall relationship between the measured detector output values and the sample absorbance is

$$A = \log \frac{100}{\%T} = \log \frac{N_r - N_0}{N_s - N_0} \qquad 6.3$$

In practice, the values for the absorbance of source illumination are obtained by operating the source shutter as a continuous light-beam chopper. The change in detector output as the source light is turned on and off gives the value of $N_s - N_0$ or $N_r - N_0$ repetitively, depending on whether the sample or reference solution is aspirating into the flame. Just as in molecular absorption spectroscopy, there is a linear relationship between the absorbance A and the analyte concentration at low to moderate values of A. It is always a good idea to create a working curve to demonstrate the region over which this is true for each type of sample run.

Atomic absorption spectroscopy is a very widely used technique. It is capable of providing analysis for almost all of the elements in the periodic table.[8] Detection limits with the flame are often in the few nanograms per milliliter range. Where the furnace is applicable, the detection limits are often up to two orders of magnitude lower. The advantages are a quick and sensitive elemental analysis. The principal disadvantage is that there is no information available regarding the way the element was incorporated in the original sample, that is, in which compounds or even in which oxidation states.

Regarding analysis in flowing systems, the furnace-based techniques are limited to batch analyses and thus do not provide continuous data on the sample composition. Atomic absorption analysis can be used for continuous analysis, however, and it has been applied in chromatographic detection where having element-specific detection can be an advantage.

---

[8]R. J. Gill, *Am. Lab.* **1993**, *November*, 24F.

Study Questions, Section F

45. What is the purpose of the heat in the techniques of atomic spectroscopy?

46. Why is the temperature required for atomic absorption spectroscopy lower than that for atomic emission spectroscopy?

47. The lower heat requirement for atomic absorption spectrometry over atomic emission spectroscopy enables the use of a furnace for the source of heat for sample decomposition. How does the use of the oven alter the types of samples that can be used and their mode of introduction?

48. What is the favored source of the probe illumination for atomic absorption spectroscopy?

49. What is the purpose of the shutter and of the monochromator in an atomic absorption spectrometer?

50. Compare Equation 6.3 with Equation 5.8 and draw analogies among the terms involved.

51. Calculate the concentration of a solution of Cd^{2+} if the values for $N_r - N_0$ and $N_s - N_0$ are 27,473 and 16,429 for the unknown solution and 27,285 and 14,527 for a solution known to be 500 ppm in Cd^{2+}.

Answers to Study Questions, Section F

45. The purpose of the heat is to decompose the sample into its atomic constituents and introduce them into the gas phase.

46. A lower heat is required because the sample only needs to be decomposed; the atoms do not have to be excited to their electronic excited state.

47. Solids and liquids can both be placed on the sample platform of the atomic furnace, whereas only liquid samples could be used with the flame and ICP. The sample is introduced into the furnace in batch mode, whereas the sample was introduced into the flame and ICP continuously.

48. The favored source for the probe illumination in atomic absorption spectroscopy is the hollow cathode lamp. It is an automatic and inexpensive way to obtain a narrow-band source of the exact wavelengths absorbed by the targeted atomic vapor.

49. The shutter allows a light intensity reading to be taken with the light from the hollow cathode lamp off. This gives a value for the background illumination coming from the flame. The monochromator is used to obtain

wavelength selectivity for the light detected. Without it, too much background radiation from the flame would be included in the light intensity measurement.

50. The two equations have exactly the same form which allows a direct comparison of the terms. From this it is seen that N_0 in the case of atomic absorption is analogous to N_d in the case of molecular absorbance. Both are the reading of the detection circuit when the sample probe illumination is turned off. N_d comes from the dark current of the detector; N_0 comes from the light made by the flame, ICP, or furnace. The quantity N_s is the same in both equations. The quantities N_b and N_r both refer to the detector signal when the analyte is absent from the cell.

51. $A_{unk} = \log \dfrac{27,473}{16,429} = 0.2233$

$A_{std} = \log \dfrac{27,285}{14,527} = 0.2738$

$C_{Cd^{2+}} = C_{std} \dfrac{A_{unk}}{A_{std}} = 500 \dfrac{0.2233}{0.2738} = 408 \text{ ppm}$

Practice Questions and Problems

1. A. Among the lowest of the electronic excited states for an organic molecule (with no metallic elements) is that for molecules with a carbon–sulfur double bond. Its energy level is 4.32×10^{-19} J. What is the color and wavelength of a photon that can cause this electronic excitation?

   B. What wavelengths of light will be useful for the electronic excitation of organic molecules?

   C. Why is electronic excitation required for analysis by photon emission?

2. What are the several processes by which an excited molecule can lose its energy?

3. A travel clock has numerals and hands that emit light at night if they have been exposed to light the previous day. A common additive to laundry soap is called a brightener. It emits light in the blue end of the spectrum after being electronically excited. The shirts washed with such a soap do not glow in the dark unless illuminated by an ultraviolet light. For each example, explain whether the light emission is phosphorescence or fluorescence and why you think so.

4. Fluorescence is often used to locate peptides that have been separated by thin-layer chromatography. When illu-

minated with ultraviolet light, the peptides emit photons in the visible region. Peptides are chains of amino acids. Which of the amino acids, alanine or phenylalanine, is likely to have the stronger fluorescence response and why? Both amino acids have amine and carboxylic acid substructures. Where alanine has a methyl group, phenylalanine has a benzene ring attached to the methyl group.

5. Describe the analytical scheme for the technique of fluorescence.

6. A working curve was obtained for the fluorescence determination of a particular peptide. The ratios of N_f/N_e were as follows: 1.0 mM, 0.000343; 5 mM, 0.00172; 25 mM, 0.00858; 125 mM, 0.0403.

   A. At what approximate concentration will the deviation from linearity be as large as 1%?

   B. What is the concentration of the peptide solution for which the ratio N_f/N_e measured was 0.00574?

7. Do you think self-absorbance might be a problem for phosphorescence measurements? If so, how would the use of front surface excitation and detection (as in Figure 6.11) reduce this effect?

8. Raman spectra are frequently recorded from 0 to 4000 wavenumbers (cm^{-1}). Over what wavelength range will the Stokes and anti-Stokes lines appear if the excitation is a diode laser at 782 nm?

9. In a chemiluminescence determination of ATP, the photons counted over a 2 minute period following sample injection were 7,926. The count for 2 minutes following the injection of a solution containing 3.6 mmol of ATP were 10,385. When no sample was injected, the count over a 2 minute period was 384. How much ATP was present in the sample?

10. Using Figure 6.23, explain why the temperature of a hot object can be determined by the apparent color of the light it emits. Also consider why Edison had such a difficult time discovering a suitable filament material for a lightbulb that was to give illumination containing all the visible spectrum.

11. If the detector current is measured with a constant infusion of sample into the nebulizer for atomic emission spectrometry, is the total amount of analyte or the analyte concentration related to the measured detector current and why?

12. The concentration of Pb in tap water was determined with emission spectrometry using an ICP. An internal standard of Yt was used to factor out variations in the ICP and nebulizer. The Pb emission was measured at 405.8 nm and that of Yt at 242.2 nm. The detector current at those wavelengths when pure water was introduced were 2.8 nA and 2.6 nA, respectively. A sample containing 0.300 μg/mL each of Pb and Yt produced detector currents of 39.4 nA and 42.7 nA at 405.8 nm and 242.2 nm, respectively. The

tap water solution with 0.300 μg/mL of Yt added gave detector currents of 23.7 nA and 44.9 nA at 405.8 nm and 242.2 nm, respectively. What was the concentration of Pb in the tap water?

13. If one used an oven for sample decomposition with atomic absorption spectrometry and measured the transmission intensity versus time over the entire sample decomposition, then calculated the area under the absorbance versus time peak, would the resulting number be related to the total amount of analyte or to the concentration of analyte in the sample placed in the oven?

14. The concentration of Fe in engine oil is sometimes used as an indication of the extent of wear of the moving parts of the engine. Atomic absorption spectrometry is a common method for performing this test. The oil is diluted with a solvent and nebulized into an atomizing flame. Calculate the concentration of Fe in the engine oil if 1 mL of oil is diluted into 100 mL of solvent to form the sample. The values of $N_s - N_0$ and $N_r - N_0$ for the sample are 24,395 and 43,698, while those for a standard solution containing 100 ppm of iron are 29,718 and 43,486.

15. Why is phosphorescence a less likely result of molecular excitation than fluorescence?

16. A sample has a transmittance of 99.9%. The absorbing analyte in this sample has a quantum efficiency for fluorescence of 1.3%. The fluorescence apparatus detects 3% of the emitted radiation. A pulsed laser is used as a light source. How many photons per laser pulse are required if it is desired to detect 1000 fluorescence photons from each laser pulse?

17. Are the conditions of Question 16 such that primary or secondary absorption needs to be considered? Why or why not?

18. In the discussion of fluorescence quenching, it is mentioned in a side note that this phenomenon can be used as a method for the quantitation of the quenching agent. It is desired to use the fluorescence of a rubidium complex, readily quenched by O_2, as a method for the quantitation of O_2 in aqueous solution. Assume that this is a new idea and that you do not know the relationship between the degree of quenching and the O_2 concentration. Describe a procedure for developing a working curve and determining the O_2 concentration in an unknown sample.

19. Use Figure 6.15 to explain why the three-dimensional fluorescence plot of Figure 6.16 does not contain much more discriminating information than the emission spectrum alone.

20. Why is it desirable to use a narrow bandwidth source such as a laser when doing Raman spectrometry?

21. Why is a higher atomized sample temperature required for atomic emission spectrometry than for atomic absorption or atomic fluorescence spectrometry?

Suggested Related Experiments

1. Determination by fluorescence.
2. Determination by atomic absorption.
3. Determination by Raman spectroscopy.
4. Experiment with ICP emission.
5. Experiment with chemiluminescence.

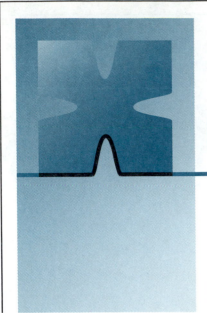

Chapter Seven

ANALYSIS BY COMPLEXATION REACTIVITY

The ability to participate in a complex formation reaction is a very useful differentiating characteristic. Since the number of compounds and ions that can form complexes is very large, many species can be analyzed taking advantage of this quality. Complexation can also be a quite specific characteristic. That is, the number of species that can form complexes with any specific complexing agent can be few. This property is very useful in reducing the number of possible interferences for quantitation and identification. The methods by which complex forming reactions can be used in chemical analysis are similar to those employed in acid–base reactivity. We must carry out the reaction in order to probe for this characteristic. Quantitative analysis can be accomplished by measuring the amount of reactant required to completely react with the analyte (in other words, titration) or by determining the amount of complex formed when the complexing agent is in excess. The basis for qualitative analysis and separation depends on the selectivity of the complexation reaction employed. We begin our study of the application of complex formation reactions in chemical analysis with a brief review of the nature of this type of reaction and the equilibria involved.

A. Complexation Reactions

The Reaction

$$Cu^{2+} + 4NH_3 \rightleftharpoons Cu(NH_3)_4^{2+}$$

is a typical complex formation reaction.

A **complexation reaction** is the reaction between two or more separately intact substructures. The word **complex**, used to describe the product, is used in the sense of an entity which itself made up of complicated parts. An example is the familiar copper–ammonia complex, $Cu(NH_3)_4^{2+}$. This complex is formed by the bonding of a cupric ion with four ammonia molecules. When the formula for the product is written, the identity of the nonelemental subgroups that make up the complex is kept within the parentheses. This is because, within the greater complex, these subgroups maintain their structural integrity. If the complex were to come apart, ammonia (not nitrogen and hydrogen in some other combination) would be a product. Complexation reactions can involve very simple reactants, as in the complexation of Cd^{2+} with I^- (to form CdI_6^{4-}), or they may involve very complex reactants, as in the reaction of the antibiotic gramicidin with K^+.

BONDS FORMED IN COMPLEXATION

The complexes formed from complexation reactions have a well-defined stoichiometry, that is, the reactants react in exact, small whole-number combinations. This fact suggests a reaction based on specific bond formation rather than a general kind of attraction. In fact, the bonds formed in complexation reactions are generally covalent bonds, though ionic bonds are known. Very often, a metal ion is involved. When this is the case, the role of the metal ion is that of the **coordination center**. The species that bond to the coordination center in a complexation reaction are called **ligands**. It is common for several bonds to be formed between the coordination center and one or more ligands. The number of ligand bonds formed around the coordination center is called the **coordination number**. In the cases of $Cu(NH_3)_4^{2+}$ and CdI_6^{4-}, the coordination numbers are 4 and 6, respectively. Specific coordination centers tend to have a given coordination number with which they are associated.

Complexes dissociate into the original ligands and coordination center.

$$Fe(CN)_6^{3-} \rightleftharpoons Fe^{3+} + 6CN^-$$

In solution, a coordination center will generally satisfy its coordination number by forming a complex with the solvent. Thus, in aqueous solution, the Cu^{2+} ion is present as $Cu(H_2O)_4^{2+}$. The complexation reaction is then actually a displacement reaction in which the ligands in the final complex replace the water ligands associated with the metal ion. This fact implies that the nature of the solvent can play a significant role in complex formation reactions.

Writing a complexation reaction this way

$$Cu(H_2O)_4^{2+} + 4NH_3 \rightleftharpoons Cu(NH_3)_4^{2+} + 4H_2O$$

illustrates the displacement of the solvent coordination with that of the ligand.

COMPLEXING SPECIES

Species that act as coordination centers include virtually all the transition metals and rare earth elements. In addition, some ligands are capable of complexing the alkaline earth metals (Ca^{2+} and Mg^{2+}). One of the chemical coups of the last two decades is the discovery of ligands (crown ethers) that effectively bind even the alkali metals Na^+ and K^+. Ligands come in a great variety. Many anions act as ligands. Among these are all the halides (Cl^-, Br^-, I^-), cyanide (CN^-), hydroxide (OH^-), oxalate ($C_2O_4^{2-}$), sulfate (SO_4^{2-}), thiocyanate (SCN^-), and thiosulfate ($S_2O_4^{2-}$). In addition, there are a variety of amines including ammonia. Some of the ligands are able to form multiple bonds with a single coordination center. Such ligands are referred to as **bidentate** (two bonds), **tridentate** (three bonds), **tetradentate** (four bonds), and **pentadentate** (five bonds). The term dentate refers to the concept that the ligand with multiple bonds is able to "bite" the central ion. The complex formed with a multidentate ligand is called a **chelate**. The antibiotic nonactin completely envelops a potassium ion with eight bonds.

Some ligands can satisfy more than one of the coordination center's coordination number Multidentate ligands with up to eight coordination sites are known.

REACTION RATES

Bond formation for complexes can be either fast or slow. A measurement of the average time for an H_2O ligand to be exchanged for another has been measured for many metal ions in aqueous solution. The times ranged from 10^{-9} sec for all the alkali and alkaline earth ions to 10^6 sec for Cr^{3+}. The actual bond formation times are greatly dependent on both the coordination center and the ligand. Quantitation by titration would require reaction times of less than one second to be practical. On the other hand, a reaction rate method may be able to gain extra discrimination if the reaction rates among potential interfering species differ significantly.

1. What is a complexation reaction?

2. The reactants in complexation reactions have specific names. What are they?

3. What is the meaning of the term "coordination number"?

4. What is a chelate?

1. A complexation reaction is the stoichiometric association of separate entities to form a complex ion or molecule that can dissociate into the entities that formed it.

2. The reactants in a complexation reaction are the coordination center and the ligands.

3. The coordination number of the coordination center is equal to the number of ligand bonds formed around it.

4. Some ligands can form multiple bonds with the coordination center. Such ligands are called chelating agents. The resulting complex is called a chelate.

B. Equilibrium Concentrations

As we have seen with the acid–base reactions, the equilibrium concentrations of the species involved in reactions are related through the formal equilibrium constants. The nature of complexation equilibrium expressions and the methods of using them for the determination of equilibrium concentrations are developed in this section.

FORMATION OF A 1:1 COMPLEX

To begin, we will consider a complexation reaction for which the stoichiometry is that of a single coordination center reacting with a single ligand. An example is the reaction of Cd^{2+} with $C_2O_4^{2-}$ to form CdC_2O_4. Complexation reactions are generally written in the direction of the formation of the complex. Thus, for the formation of CdC_2O_4 we write

$$Cd^{2+} + C_2O_4^{2-} \rightleftharpoons CdC_2O_4$$

For which the equilibrium expressions are

$$K_f^\circ = \frac{a_{CdC_2O_4}}{a_{Cd^{2+}}a_{C_2O_4^{2-}}} \quad and \quad K_f' = \frac{[CdC_2O_4]}{[Cd^{2+}][C_2O_4^{2-}]}$$

The subscript f is used to identify the equilibrium constant as that for a complex formation reaction. As before, the thermodynamic constant K_f°, in terms of activities, is a true constant, while the formal equilibrium constant K_f', in terms of concentrations, depends on the ionic strength of the solution.

Now let us consider the general case for a 1:1 complex formation reaction. We will use the normal convention of M for the metallic coordination center ion and L for the ligand. The charge on the ions in these general expressions is usually omitted, but it should be included in all specific cases. In this development, we will recognize that the coordination center M, in aqueous solution, is already coordinated with some number of water molecules. Therefore, the reaction is more exactly written as

The general form for a complex formation reaction is

$$M + nL \rightleftharpoons ML_n$$

where M is the coordination center and L is the ligand.

$$M(H_2O)_n + L \rightleftharpoons M(H_2O)_{n-1}L + H_2O \qquad 7.1$$

The thermodynamic equilibrium constant expression for reaction 7.1 is

$$K_f^\circ = \frac{a_{M(H_2O)_{n-1}L}a_{H_2O}}{a_{M(H_2O)_n}a_L} \qquad 7.2$$

In this expression, the activity of the water can be taken as 1 in dilute solutions, and the solvent ligands can be taken for granted in the formulas for the coordination center and the complex. When these steps are taken, the more familiar, simplified reaction and equilibrium expressions result.

$$M + L \rightleftharpoons ML \qquad 7.3$$

$$K_f^\circ = \frac{a_{ML}}{a_M a_L} \quad \text{and} \quad K_f' = \frac{[ML]}{[M][L]} \qquad 7.4$$

To obtain expressions for the equilibrium concentrations, we will use the formal equilibrium constant expressions. You will recall, from the treatment of acid–base reactions, that it is desirable to have both the equilibrium and mass balance expressions in terms of concentration. The relationship between the two equilibrium expressions, under conditions where the DHLL is valid, was given in Equation 3.49.

From reaction 7.3, we can see that the coordination center, M, is either free (solvated) or bonded with the ligand L. There are thus two forms for the species M: M and ML. We can then write a mass balance equation for the species M for which the total concentration (in both forms) is C_M.

$$C_M = [M] + [ML] \qquad 7.5$$

Now formulas for the fraction of C_M that is in each form can be derived. From the equilibrium expression,

$$[ML] = K_f'[M][L] \qquad 7.6$$

so

$$\alpha_M = \frac{[M]}{C_M} = \frac{[M]}{[M] + [ML]} = \frac{1}{1 + K_f'[L]} \qquad 7.7$$

and

$$\alpha_{ML} = \frac{[ML]}{C_M} = \frac{[ML]}{[M] + [ML]} = \frac{K_f'[L]}{1 + K_f'[L]} \qquad 7.8$$

These equations reveal that the fraction of M that is in each form depends on K_f' and [L]. A plot of this dependence is shown in Figure 7.1 for a simple complex with a K_f' of 1.0×10^8. From this plot and from the equilibrium constant expression, we can see that the concentrations of the two forms are equal when $p[L] = -pK_f'$.

The similarity between this plot and the alpha plot of a monoprotic acid (Figure 4.1) is unmistakable. This is due to the similarity of the equilibrium expressions. The

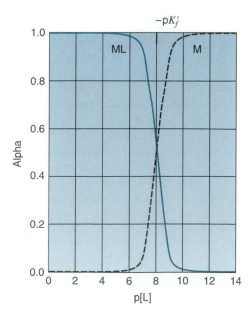

Figure 7.1. The fraction of M in the free (M) and complexed (ML) forms as a function of p[L]. For $p[L] < -pK_f'$, the coordination center is nearly completely complexed.

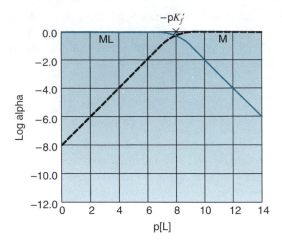

Figure 7.2. The log of the fraction of M and ML as a function of p[L]. The values of the fractions complexed and uncomplexed far from p[L] = $-pK_f'$, are easily obtained from the plot.

only difference is the convention of writing complexation reactions as the complex formation and acid–base reactions as the acid deprotonation. A plot of the logarithm of alpha versus p[L] can also be made. As Figure 7.2 shows, the log alpha plot better reveals the concentrations of the minor species at values of p[L] far from $-pK_f'$. This plot is readily constructed for any complexation system by placing an $\times$ at the point where $\alpha = 1$ and p[L] = $-pK_f'$. Then two horizontal lines are drawn either side of the $\times$ for the major component α values, and two angled lines are drawn at unit slope from the $\times$ for the minor component α values. A curved line intersecting 0.3 log α units under the $\times$ joins the major and minor lines for each species. From this plot, the value of α for either species can be obtained at any value of p[L]. For example, at a p[L] of 4.0, the fraction of the total M in the uncomplexed form is 1×10^{-4}, and the fraction in the complexed form is essentially 1.0. To make sure that at least 99% of the coordination center is in the complexed form, the ligand concentration must be at least 1×10^{-6} M (in excess).

SOLVING FOR CONCENTRATIONS IN COMPLEXATION SYSTEMS

The equilibrium concentrations of species involved in a complexation reaction are solved using a combination of the equilibrium constant expression (Equation 7.4) and mass balance equations. The mass balance equations are

$$C_M = [M] + [ML] \qquad\qquad 7.9$$

and

$$C_L = [L] + [ML] \qquad\qquad 7.10$$

where C_M and C_L are the analytical concentrations of M and L.

One of two situations generally exists in the solutions for which the concentrations are to be solved. One is when one of the complexing reagents is added in excess (more than enough to react with the other completely). The other is when the concentrations of uncomplexed M and L are equal. The latter is the condition that exists when pure ML is dissolved or when the equivalence point of a complexometric titration of M or L has been reached. We will first look at the situation where there is an excess of one reactant.

Assume that an excess of ligand has been added to a solution of M. In such a case, one would normally like to know the concentration of M that remains uncomplexed. Equation 7.4 is solved for [M] and substitutions are made from the mass balance equations, Equations 7.9 and 7.10.

$$[M] = \frac{[ML]}{K_f'[L]} = \frac{C_M - [M]}{K_f'(C_L - C_M + [M])} \qquad\qquad 7.11$$

Equation 7.11 is a quadratic equation that has the form

$$K_f'[M]^2 + [1 + K_f'(C_L - C_M)][M] - C_M = 0 \qquad 7.12$$

which can readily be solved by means of the quadratic formula. However, a simplification is often applicable without significant error. The simplification is possible when one can assume that the concentration of uncomplexed M is small relative to the analytical concentration of M ($[M] << C_M$) and also that $[M]$ is small compared to the excess concentration of L ($[M] << C_L - C_M$). This will be the case when the complex formation constant and/or the excess concentration of L is large. When this condition is met,

$$[M] \approx \frac{C_M}{K_f'(C_L - C_M)} \qquad 7.13$$

When Equation 7.13 is compared with the rearranged form of the equilibrium expression in the first part of Equation 7.11, one can see that another way to state the assumptions is that essentially all the M is complexed ($[ML] = C_M$) and that the unreacted L is equal to its excess concentration. (See Examples 7.1 and 7.2.)

Example 7.2

How many millimoles of L must be added to 10 mL of a 0.1 M solution of M to ensure that the M is 99% complexed? The formation constant for ML is 1×10^3.
    Equation 7.13 can be rearranged:

$$K_f'(C_L - C_M) \approx \frac{C_M}{[M]} = \frac{0.99}{0.01}$$

$$(C_L - 0.1) \approx \frac{99}{1 \times 10^3} \approx 0.1 \text{ M}$$

$$C_L \approx 0.2 \text{ M}$$

Thus, mmol L = 0.2 M $\times$ 10 mL = 2 mmol.

Example 7.1

Calculate the concentration of uncomplexed M when 50 mL of a 0.03 M solution of L is mixed with 50 mL of an 0.01 M solution of M. The formation constant for ML is 1×10^5.
    The values of C_L and C_M are calculated from the dilution as 15 mM and 5 mM. When these values are used in Equation 7.13, the result is

$$[M] \approx \frac{5 \times 10^{-3} \text{ M}}{1 \times 10^5 (15 - 5) \times 10^{-3} \text{ M}}$$
$$= 5 \times 10^{-6} \text{ M}$$

When these same values are used in the spreadsheet solution of Equation 7.12, the result is 4.99 $\times$ 10^{-6} M, indicating that the simplification was justified in this case.

The situation when there is an excess of M is exactly analogous. The exact, quadratic, and simplified equations for the case of solutions that contain more M than L are

$$[L] = \frac{[ML]}{K_f'[M]} = \frac{C_L - [L]}{K_f'(C_M - C_L + [L])} \qquad 7.14$$

$$K_f'[L]^2 + [1 + K_f'(C_M - C_L)][L] - C_L = 0 \qquad 7.15$$

$$[L] \approx \frac{C_L}{K_f'(C_M - C_L)} \qquad 7.16$$

Note that, in these cases, L is the minor component and we are solving for the concentration of it that is uncomplexed.
    The condition where there is no excess of either component is simply solved. Since $[M] = [L]$, the equilibrium expression can be written

$$[L] = \frac{[ML]}{K_f'[L]} = \frac{C_{ML} - [L]}{K_f'[L]} \quad \text{and} \quad [M] = \frac{[ML]}{K_f'[M]} = \frac{C_{ML} - [M]}{K_f'[M]} \qquad 7.17$$

Rearranging Equation 7.17,

$$[L] = \sqrt{\frac{C_{ML} - [L]}{K_f'}} \quad \text{and} \quad [M] = \sqrt{\frac{C_{ML} - [M]}{K_f'}} \qquad 7.18$$

■■ ••••••••••••••••
Example 7.3

What is the concentration of un-complexed M in a solution that is 1.5 mM in ML? The formation constant of ML is 3.0×10^4.

When Equation 7.19 is used, the result is $[M] = (1.5 \times 10^{-3}/3.0 \times 10^4)^{1/2} = 2.24 \times 10^{-4}$ M. This value is not negligible compared to 1.5×10^{-3} M, so successive approximations or the quadratic formula will have to be used. In either case, the result is $[M] = 2.19 \times 10^{-4}$ M.

Considering significant figures, all results are 2.2×10^{-4} M.

••••••••••••••••••••••••

These equations can be solved using the quadratic formula or can be simplified (where the concentration of uncomplexed M and L is much smaller than the analytical concentration of ML). The simplified equation is

$$[L] = [M] \approx \sqrt{\frac{C_{ML}}{K_f'}} \qquad 7.19$$

which can be easily solved and then confirmed by making sure the calculated value of $[M]$ or $[L]$ is much smaller than C_{ML}. (See Example 7.3.)

FORMATION OF HIGHER ORDER COMPLEXES

In many complexation reactions, the coordination number is high enough for more than one ligand to bond with the coordination center. A familiar example of this is the complexation of Cu^{2+} by ammonia in which four ammonia molecules can be complexed by the Cu^{2+} coordination center. However, the formation constant for the addition of the second ammonia is not the same as that for the first. The third is different from the second, and so on. This makes sense since the second ammonia is not bonding to the same species as the first one did. Normally, but not always, the formation constants for successive additions of ligand are progressively smaller. This leads to the phenomenon called **stepwise formation** of higher order complexes. A separate reaction and equilibrium expression is required for each step. Thus, for a system in which four ligands (L) can add to a single coordination center (M) we have

$$M + L \rightleftharpoons ML \qquad K_{f1}' = \frac{[ML]}{[M][L]}$$

$$ML + L \rightleftharpoons ML_2 \qquad K_{f2}' = \frac{[ML_2]}{[ML][L]}$$

$$ML_2 + L \rightleftharpoons ML_3 \qquad K_{f3}' = \frac{[ML_3]}{[ML_2][L]} \qquad 7.20$$

$$ML_3 + L \rightleftharpoons ML_4 \qquad K_{f4}' = \frac{[ML_4]}{[ML_3][L]}$$

ALPHA PLOTS

The fraction that is in each form can be derived from the equilibrium constant equations and the mass balance equation.

These are the general forms for the fraction of M in each complex form. If less than four formation constants are involved, use zero for the values of the extra constants. If more than four formation constants are needed, add the additional terms and equations following this pattern.

$$\alpha_M = \frac{1}{1 + K_1'[L] + K_1'K_2'[L]^2 + K_1'K_2'K_3'[L]^3 + K_1'K_2'K_3'K_4'[L]^4}$$

$$\alpha_{ML} = \frac{K_1'[L]}{1 + K_1'[L] + K_1'K_2'[L]^2 + K_1'K_2'K_3'[L]^3 + K_1'K_2'K_3'K_4'[L]^4}$$

$$\alpha_{ML_2} = \frac{K_1'K_2'[L]^2}{1 + K_1'[L] + K_1'K_2'[L]^2 + K_1'K_2'K_3'[L]^3 + K_1'K_2'K_3'K_4'[L]^4} \qquad 7.21$$

$$\alpha_{ML_3} = \frac{K_1'K_2'K_3'[L]^3}{1 + K_1'[L] + K_1'K_2'[L]^2 + K_1'K_2'K_3'[L]^3 + K_1'K_2'K_3'K_4'[L]^4}$$

$$\alpha_{ML_4} = \frac{K_1'K_2'K_3'K_4'[L]^4}{1 + K_1'[L] + K_1'K_2'[L]^2 + K_1'K_2'K_3'[L]^3 + K_1'K_2'K_3'K_4'[L]^4}$$

The linear alpha plot shown in Figure 7.3 is for the complexation of Tl^{3+} with Br^-. For this system, the $-pK_f'$ values are 8.15, 6.15, 4.36, and 2.85. At these same values of $p[L]$, the fractions of the species that are involved in the equilibrium for that pK_f' are approximately 0.5. From this plot, the $p[L]$ at which the fraction of each species will be maximum can be seen. For example, to create a solution in which the

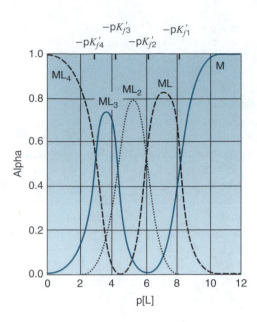

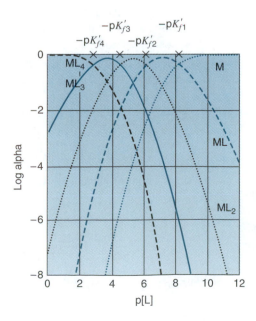

⊘ **Figure 7.3.** The alpha plot for a stepwise complex formation reaction shows the relative abundance of all species versus
p[L]. This plot is for the stepwise complexation of Tl^{3+} with
Br^-.

ML_3 form is at its maximum concentration, the p[L] should be adjusted to about 3.5.
The reason that alphas for ML, ML_2, and ML_3 never reach 1.0 is the closeness in the
values of the successive formation constants.

The closeness in the successive formation constants makes this kind of complex
formation reaction undesirable for quantitation by titration. The only two forms that
are quantitatively produced at any value of p[L] are the most complexed form and the
uncomplexed form. A titration of M with L to form ML_4 produces a gradual increase in
p[L] or a gradual decrease in p[M] versus volume of titrant as the complexation goes
through the formation of all the intermediate complex forms. The lack of a sharp
change in concentration makes the determination of the equivalence point quite difficult. Stepwise complex formation systems are useful for quantitation only if a large excess of the ligand is added to a solution of a coordination center analyte. The excess
should be such that all the analyte is completely complexed to a single complex form.
The concentration of complex is then determined spectrometrically. The purpose of the
complexing agent, in this case, is to produce a form of the analyte that has a high molar
absorptivity and/or a distinctive absorbance spectrum.

⊘ **Figure 7.4.** With the log alpha plot, lower values of
alpha can be determined. This plot is for the same system as
Figure 7.3.

As before, the log alpha diagram is more revealing of the concentrations of the various species at low values of alpha. This plot, for the Tl^{3+} system with Br^-, is shown in Figure 7.4. The graph you see was created in a spreadsheet programmed with Equations 7.21. It can also be constructed by placing an $\times$ at the intersection of the p[L] value that is equal to $-pK_f'$ and $\log \alpha = 0$ for each value of K_f'. From these points, the concentration lines move away to the left and right with slopes of 1 and -1. However, when these points pass under another system point, their slopes increase by 1. Because the system points are so close in this case, the slope changes sometimes occur before a section of constant slope has been established. For example, the M line, after passing under the system point of p[L] = 8.15 would have a slope of 1, but it passes under the p[L] = 6.15 system point while still in its curved section. From thence, its slope is 2 only briefly for when it passes under each remaining system point, its slope increases.

The alpha plots and log alpha plots for stepwise formation complexes are most easily obtained from the spreadsheet plotting programs contained on the accompanying CD. From them, it is easy to see the maximum fraction possible for each species in the series and the p[L] at which this occurs. One can also readily observe the relative abundance of all species at any given p[L]. The calculation of equilibrium concentrations of all species for given solution conditions is also greatly facilitated with log concentration plots. See Section F for the methods used.

Study Questions, Section B

5. Write the reaction, the thermodynamic equilibrium constant expression, and the formal equilibrium constant expression for the formation of a 1:1 complex between M and L.

6. What are the equations that relate the fraction of the coordination center that is in the complexed and uncomplexed forms as a function of the ligand concentration and the formal formation constant?

7. What is the relationship between the ligand concentration and the formal formation constant when the concentration of the complex is equal to the concentration of the uncomplexed coordination center?

8. Nitrilotriacetate, $N(CH_2CO_2^-)_3$, forms a 1:1 complex with many metal ions. Its formation constant with Ag^+ is 1.4×10^5.

   A. What is the concentration of Ag^+ in a solution in which the total silver concentration is 4.7×10^{-4} M and the total nitrilotriacetate concentration is 1.0×10^{-3} M?

   B. If a 0.0010 M solution of nitrilotriacetate is to be 99.9% complexed, what would the analytical concentration of silver have to be?

   C. What is the concentration of uncomplexed silver and nitrilotriacetate in a solution for which the analytical concentration of the complex is 5×10^{-2} M?

9. A complex ML absorbs light at a given wavelength but M and L do not. Plot absorbance versus analytical concentration of ML (C_{ML}) at ML concentrations of 1.0×10^{-3} M, 3.5×10^{-3} M, 5.0×10^{-3} M, 8.0×10^{-3} M, and 1.0×10^{-2} M. Use 60.72 L mol^{-1} cm^{-1} for the molar absorptivity of ML, 2.35 as $\log K_f'$, and 1.0 cm as the cell path length.

10. What is stepwise formation of higher order complexes?

11. What would the equations corresponding to Equations 7.21 be for a stepwise complex formation where the highest complex formed is ML_2?

12. Looking at the log alpha plot of Figure 7.4, what is the concentration of L for which the ML form is a maximum?

Answers to Study Questions, Section B

5. The general form of the reaction is

$$M + L \rightleftharpoons ML$$

where all three species are dissolved in the solvent. The equilibrium expressions for this reaction are

$$K_f^\circ = \frac{\alpha_{ML}}{\alpha_M \alpha_L} \quad \text{and} \quad K_f' = \frac{[ML]}{[M][L]}$$

6. The fractions of the total M (C_M) that are in each form are

$$\alpha_M = \frac{[M]}{C_M} = \frac{1}{1 + K_f'[L]}$$

$$\alpha_{ML} = \frac{[ML]}{C_M} = \frac{K_f'[L]}{1 + K_f'[L]}$$

7. When [M] = [ML], p[L] = $-pK_f'$.

8. A. The ligand is in excess. Using the quadratic formula to solve Equation 7.12, one gets 3.7×10^{-4} M. Under these circumstances, only a small fraction of the silver is complexed.

B. Equation 7.16 can be used since we know that the ligand is to be essentially completely complexed. Solving Equation 7.16 for $C_L/[L]$,

$$\frac{C_L}{[L]} = \frac{0.0010}{0.1\% \text{ of } 0.0010} = 1.0 \times 10^3$$
$$= 1.4 \times 10^5 \times (C_M - 0.001)$$
$$C_M = 8.14 \times 10^{-3} \text{ M}$$

C. Try the simplified equation (Equation 7.19). From this,

$$[Ag^+] = [L] = (5 \times 10^{-2}/1.4 \times 10^5)^{1/2} = 6.0 \times 10^{-4} \text{ M}.$$

This is about 1% of the total concentration, so the simplification is justified to this degree of accuracy.

9. By Beer's law,

$$A = \varepsilon bC$$

The equilibrium concentration of the complex is calculated at each analytical concentration using Equation 7.18 or the quadratic formula:

$$K'_{ML}[M]^2 + [M] - C_{ML} = 0$$

At $C_{ML} = 1.0 \times 10^{-3}$ M, $[M] = 8.4 \times 10^{-4}$ M, so $[ML] = 1.0 \times 10^{-3}$ M $- 8.4 \times 10^{-4}$ M $= 2 \times 10^{-4}$ M. Absorbance is therefore

$$A = \left(\frac{60.72 \text{L}}{\text{mol} \cdot \text{cm}}\right)(1.0 \text{ cm})(1.6 \times 10^{-4}) = 9.7 \times 10^{-3}$$

C_{MX}	[M]	[MX]	A_{MX}
1.0×10^{-3}	8.4×10^{-4}	1.6×10^{-4}	9.7×10^{-3}
3.5×10^{-3}	2.3×10^{-3}	1.2×10^{-3}	7.3×10^{-2}
5.0×10^{-3}	3.0×10^{-3}	2.0×10^{-3}	1.2×10^{-1}
8.0×10^{-3}	4.1×10^{-3}	3.9×10^{-3}	2.4×10^{-1}
1.0×10^{-2}	4.8×10^{-3}	5.2×10^{-3}	3.2×10^{-1}

The data table and the plot made from it show equilibrium concentrations and absorbances for different values of C_{ML}.

The graph of absorbance versus concentration is not linear, particularly at low concentrations. This is be-

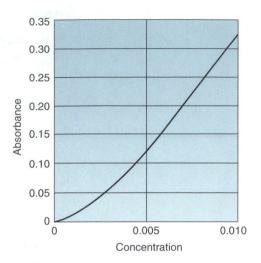

cause the relative concentration of ML increases with increasing analytical concentration in this concentration range. This is an example of a chemical deviation from Beer's law.

10. In many complex systems, more than one ligand can bond to a single coordination center. In most of these systems, the formation constants for the addition of each ligand are less than that for the previous one. The equilibrium constant for the addition of the first ligand is K'_{f1}, for the second, K'_{f2}, and so on. In such stepwise formation systems, the average number of ligands per coordination center increases with increasing ligand concentration.

11.

$$\alpha_M = \frac{1}{1 + K'_1[L] + K'_1K'_2[L]^2}$$

$$\alpha_{ML} = \frac{K'_1[L]}{1 + K'_1[L] + K'_1K'_2[L]^2}$$

$$\alpha_{ML_2} = \frac{K'_1K'_2[L]^2}{1 + K'_1[L] + K'_1K'_2[L]^2}$$

12. Following the line for ML, it is seen to be a maximum at a value of p[L] of 7.6. The concentration of L is thus 3×10^{-8} M.

C. The Effect of [H₃O⁺] on Complex Equilibria

All the development so far has assumed that the complexation reactions were the only reactions that involved the ligand or the coordination center. This is often not the case. A very frequent complication comes from the coordination center complexing OH⁻ ions and/or from the ligand acting as a weak base.

Many coordination centers can form complexes with OH⁻, and many ligands can complex H⁺. These reactions compete with the reaction between the coordination center and other ligands.

HYDROLYSIS OF THE COORDINATION CENTER

If the coordination center forms a complex with OH^- ions, the extent of this reaction will be dependent on the $p[H_3O^+]$ of the solution. In this case, there is a strong similarity between complexation and acid–base reactions, as this reaction shows:

$$M^{n+} + 2H_2O \rightleftharpoons MOH^{(n-1)+} + H_3O^+ \qquad\qquad 7.22$$

This kind of reaction is called a **hydrolysis reaction**, as in a reaction to incorporate part of a water molecule.

Many transition metal ions will form a series of stepwise complexes with OH^-. When their salts are dissolved in water, the solution becomes acidic because of the uptake of hydroxide ions via this reaction. Another effect is that the coordination center is no longer entirely in the form of the species M. This has the effect of reducing the availability of the coordination center for other reactions. An example of this is the cadmium ion (Cd^{2+}). It forms mono- and dihydroxy complexes with log formation constants of 3.9 and 3.8. To consider the reaction of Cd^{2+} with other species, it is helpful to determine what fraction of the cadmium species in solution is in the form of Cd^{2+}. To do this we simply use the equation for $\alpha_{Cd^{2+}}$ for its complexation with OH^-. From the general expressions (Equations 7.21), we obtain a specific equation for the case of the hydrolysis of cadmium:

$$\alpha_{Cd^{2+}} = \frac{1}{1 + 10^{3.9}[OH^-] + 10^{3.9}\,10^{3.8}[OH^-]^2}$$

From the dependence of $\alpha_{Cd^{2+}}$ on the OH^- concentration, it is clear that the availability of Cd^{2+} for other reactions will depend on the $p[H_3O^+]$ of the solution. When $\log \alpha_{Cd^{2+}}$ is plotted as a function of $p[H_3O^+]$ (as shown in Figure 7.5), the extent of this effect can be appreciated. Above $p[H_3O^+] = 9$, the fraction of Cd^{2+} in the unhydrolyzed form drops off precipitously.

This effect of the hydrolysis reaction on the availability of the coordination center for other reactions can be taken into account in either of two ways. The first is to use the calculated concentration of the coordination center in the equilibrium expression for the complexation reaction. For example, if the coordinating center is

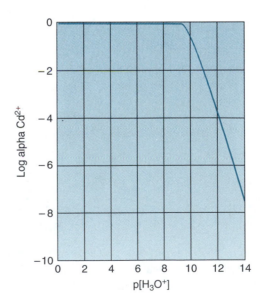

Figure 7.5. Above $p[H_3O^+] = 11$, less than 1% of the dissolved cadmium remains in the unhydrolyzed (Cd^{2+}) form.

to complex with ligand L, we can substitute the concentration of M available from the hydrolysis reaction in the equilibrium constant expression for the reaction of M with L.

$$M + L \rightleftharpoons ML \qquad K'_{ML} = \frac{[ML]}{[M][L]} \qquad\qquad 7.23$$

To calculate the concentration of M in its uncomplexed form, we can start with the equation for α°_M, the fraction of the uncomplexed M that is also unhydrolyzed.

$$\alpha^\circ_M = \frac{[M]}{[M] + [M(OH)] + [M(OH)_2] + \cdots} = \frac{[M]}{[M] + [M]_{hyd}} \qquad\qquad 7.24$$

$$[M] = \alpha^\circ_M ([M] + [M]_{hyd})$$

The term $[M]_{hyd}$ in Equation 7.24 is used to indicate the sum of all the hydrolyzed forms of M. We cannot use the term C_M for the sum of $[M]$ and $[M]_{hyd}$ since M can now form the complex ML. Thus the complete mass balance equation for M is $C_M = [M] + [M]_{hyd} + [ML]$. Equation 7.24 can now be used in the equilibrium expression for the formation of ML.

$$K'_{ML} = \frac{[ML]}{[M][L]} = \frac{[ML]}{\alpha^\circ_M([M] + [M]_{hyd})[L]}$$

Then

$$K'_{MLeff} = \frac{[ML]}{([M] + [M]_{hyd})[L]} = K'_{ML}\alpha^\circ_M \qquad\qquad 7.25$$

An effective complex formation constant, K'_{MLeff}, can be obtained from the product of K'_{ML} and the fraction of the coordination center concentration that is not hydrolyzed (α°_M). This fraction, and thus the effective formation constant, vary with the p[H$_3$O$^+$] of the solution.

In this development, a conditional or **effective equilibrium constant** has been defined in terms of all the forms of M that are not complexed with L. This effective equilibrium constant is dependent on the p[H$_3$O$^+$], but when a particular p[H$_3$O$^+$] is frequently used for the reaction, the use of the effective equilibrium constant for that p[H$_3$O$^+$] simplifies many calculations.

HYDROLYSIS OF THE LIGAND

Complexing agents are very frequently weak bases, that is, they will react with protons as well as with another coordination center. This reaction with water has the form

$$L + H_2O \rightleftharpoons LH + OH^- \qquad\qquad 7.26$$

This reaction can continue stepwise until the maximum number of protons has been added to the ligand. For example, the ligand molecule ethylenediamine can add two protons to the neutral molecule. The protonated forms have a charge equal to the number of protons added. As seen, the ligand is acting as a base in its reaction with water. The relationship between the equilibrium constants for such "base" reactions and the acidic form in which the proton is donated was covered in Chapter 4. In this treatment as well, the K'_a values for the ligand will be used rather than the equilibrium constant for the ligand hydrolysis reaction. This is the most useful approach because not all ligands are unprotonated in their neutral form. The very popular ligand ethylenediaminetetraacetic acid (EDTA) has four replaceable protons in its neutral form. The form most used in the laboratory is the disodium salt of EDTA. When it is dissolved, it may either hydrolyze or lose protons.

The reaction of a ligand with H$^+$ reduces its availability for reaction with the coordination center.

The portion of the ligand concentration that is in the protonated form is not available to the complexation reaction. Again, the alpha relationships can be used. Consider the case of EDTA. The pK'_a values for the reaction of neutral EDTA (H$_4$Y) with water

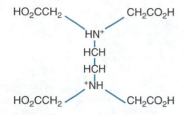

The structure of H_6Y^{2+}. In all but the most acidic solutions, two of the carboxlylic acid protons are absent, giving the neutral H_4Y.

are 1.99, 2.67, 6.16, and 10.26. The fraction of the EDTA that is in the unprotonated form (abbreviated Y^{4-}) is

$$\alpha_{Y^{4-}} = \frac{K'_{a1}K'_{a2}K'_{a3}K'_{a4}}{([H_3O^+]^4 + K'_{a1}[H_3O^+]^3 + K'_{a1}K'_{a2}[H_3O^+]^2 + K'_{a1}K'_{a2}K'_{a3}[H_3O^+] + K'_{a1}K'_{a2}K'_{a3}K'_{a4})} \tag{7.27}$$

$$= \frac{10^{-21.08}}{[H_3O^+]^4 + 10^{-1.99}[H_3O^+]^3 + 10^{-4.66}[H_3O^+]^2 + 10^{-10.82}[H_3O^+] + 10^{-21.08}}$$

A plot of $\log \alpha_{Y_{4-}}$ as a function of $p[H_3O^+]$ is shown in Figure 7.6. From this plot we can see that below $p[H_3O^+] = 12$, the fraction of the EDTA that is in the unprotonated form decreases rapidly.

To form a complex between a ligand L and a coordinating center M, we must consider the available concentration of unprotonated ligand. We will start with the mass balance equation for the ligand, taking into account the protonated and complexed forms (ionic charges have been omitted for simplicity):

$$C_L = [L] + [HL] + [H_2L] + [H_3L] + [H_4L] + [ML]$$

$$\alpha_L^\circ = \frac{[L]}{[L] + [HL] + [H_2L] + [H_3L] + [H_4L]} = \frac{[L]}{[L] + [L]_{prot}} \tag{7.28}$$

$$[L] = \alpha_L^\circ([L] + [L]_{prot})$$

where $[L]_{prot}$ is the sum of all the forms of L that are protonated. The term C_L cannot be used for $[L] + [L]_{prot}$ because, as the mass balance equation shows, C_L now includes the concentration of ML.

As in the case of hydrolysis of the coordination center, we can use the equation for $[L]$ from Equation 7.28 in the equilibrium expression for the formation ML:

$$K'_{ML} = \frac{[ML]}{[M][L]} = \frac{[ML]}{[M]\alpha_L^\circ([L] + [L]_{prot})}$$

$$K'_{MLeff} = \frac{[ML]}{[M]([L] + [L]_{prot})} = K'_{ML}\alpha_L^\circ \tag{7.29}$$

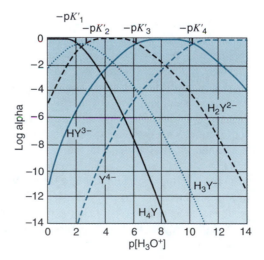

⊙ **Figure 7.6.** The chelating ligand EDTA acts as a base, adding many protons as the pH decreases. Below pH 8, less than 1% of the total EDTA concentration is available for reaction with a coordination center. The effective formation constant for MY is the product of K'_{MY} and α_Y^{4-}.

The effective equilibrium constant here is a function of $p[H_3O^+]$, but it is very convenient as the reaction is normally performed in a buffered solution at a particular $p[H_3O^+]$.

HYDROLYSIS OF LIGAND AND COORDINATION CENTER

The most common situation for the complexation of metal ion coordination centers is that both the coordination center and the ligand are hydrolyzed in solution. In this situation, there are three simultaneous reactions occurring, the two hydrolysis reactions plus the complexation reaction. Seven or more equilibrium constants can be required to solve for the concentrations in such solutions. In this situation, the concept of the effective equilibrium constant is especially helpful. We can begin with the effective equilibrium constant that considers the protonation of the ligand (Equation 7.29) and substitute for [M] from Equation 7.24:

$$K'_{ML} = \frac{[ML]}{[M][L]} = \frac{[ML]}{\alpha^\circ_M([M] + [M]_{hyd})\alpha^\circ_L([L] + [L]_{prot})} \qquad 7.30$$

from which we get

$$K'_{MLeff} = \frac{[ML]}{[M]'[L]'} \qquad 7.31$$

where

$$K'_{MLeff} = K'_{ML}\alpha^\circ_L\alpha^\circ_M \qquad 7.32$$

and $[M]'$ represents the sum of the molar concentrations of all the forms of M except ML, and $[L]'$ is the same for all the forms of L except ML.

The effect of $p[H_3O^+]$ on the values of α°_M and α°_L are opposite. At higher $p[H_3O^+]$ values, α°_L approaches 1 and α°_M is decreasing. These effects are shown in the plots of the effective equilibrium constants for the formation of the Zn^{2+}–EDTA complex in Figure 7.7. The value of $\alpha^\circ_{Zn^{2+}}$ falls off steeply at high $p[H_3O^+]$, while that of α°_Y decreases at lower values of $p[H_3O^+]$. When the two effects are combined, as in the effective formation constant of Equation 7.32, a maximum is observed. This maximum occurs at a $p[H_3O^+]$ of about 9, and even at this $p[H_3O^+]$, the effective formation constant is less than that of K'_{MY} for the Zn^{2+}–EDTA complex ($10^{16.5}$).

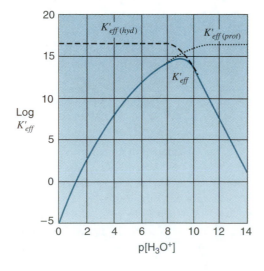

Figure 7.7. In the formation of ZnY^{2-}, Zn^{2+} is hydrolyzed above pH 9 and the EDTA is protonated below pH 10. The overall effective formation constant K'_{eff} has a maximum that is only 1/100 that of the tabular value for K'_{ZnY}.

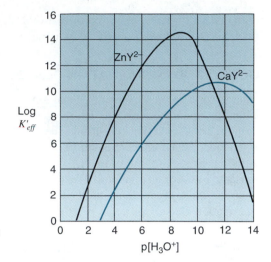

Figure 7.8. The optimum pH for the reaction of various metal ions with EDTA occurs at different $p[H_3O^+]$ values. Controlling the pH can thus be used to selectively enhance or suppress the complexation of specific metals.

The $p[H_3O^+]$ at the maximum for EDTA complexes is a function of the species involved, both from the value of K'_{MY} and from the metal hydrolysis equilibria. The curve of K'_{eff} for the Ca^{2+}–EDTA complex is shown in Figure 7.8 for comparison. These curves illustrate a differentiating characteristic that can be used to advantage in the analysis of either Ca^{2+} or Zn^{2+} when they are present in the same solution. At very high values of $p[H_3O^+]$, the calcium complex will form much more readily than the zinc one. Similarly, at a $p[H_3O^+]$ value of 3, the effective formation constant for the zinc complex is a million times greater than that for the calcium complex.

Study Questions, Section C

13. What is the acidic reaction of many of the coordination centers with a water solvent?

14. How does the reaction described in Question 13 interfere with the reaction of the coordination center with an intentional ligand?

15. How can the amount of the coordination center that is available for the intended complexation reaction be determined?

16. What is the meaning of "effective formation constant," and why does it depend on the pH of the solution?

17. Iron(II) forms complexes with up to two chloride ions by the following reactions:

$$Fe^{2+} + Cl^- \rightleftharpoons FeCl^+ \qquad \log K'_1 = 1.48$$
$$FeCl^+ + Cl^- \rightleftharpoons FeCl_2 \qquad \log K'_2 = 0.65$$

It also reacts with OH^- by the reaction:

$$Fe^{2+} + OH^- \rightleftharpoons FeOH^+ \qquad \log K'_1 = 4.6$$

Calculate $K'_{FeCl_2 eff}$ in a $FeCl_2$ solution buffered at a $p[H_3O^+]$ of (A) 10.0 and (B) 3.5.

18. In what way do many of the ligands react with a water solvent, and how does this interfere with the intended complexation reaction?

19. What is the means to determine the extent of ligand hydrolysis?

20. Does the fraction of unprotonated ligand increase or decrease as the pH increases?

21. K'_{MY} for the complexation of Ca^{2+} and EDTA is 5.0×10^{10}. Calculate K'_{MYeff} for CaY^{2-} in a solution buffered at $p[H_3O^+]$ of (A) 6.5 and (B) 10. The acid equilibrium constants for the acidic groups of EDTA are $K'_{a1} = 1.02 \times 10^{-2}$, $K'_{a2} = 2.14 \times 10^{-3}$, $K'_{a3} = 6.92 \times 10^{-7}$, and $K'_{a4} = 5.50 \times 10^{-11}$. Calcium does not react appreciably with OH^- at either of the $p[H_3O^+]$ values in this problem.

22. What is the form of the effective formation constant when the hydrolysis of both the coordination center and the ligand are significant?

23. Calculate the effective formation constant for EDTA and Ni^{2+} at $p[H_3O^+]$ of (A) 4.0, (B) 7.0, and (C) 12.0. Account for hydrolysis of both EDTA and Ni^{2+}. K'_{MY} for NiY^{2-} is 4.2×10^{18}. For the hydrolysis of Ni^{2+}, $\log K'_{f1} = 4.1$, $\log K'_{f2} = 4.9$, and $\log K'_{f3} = 3.0$. The K'_a values for the hydrolysis of EDTA are listed in Question 21.

13. Hydroxyl ion acts as a ligand for many coordination centers. When dissolved in water, the coordination center can react with the water to complex OH^- ion from the water to produce the hydroxyl complex and hydronium ion.

14. The portion of the coordination center concentration that is complexed with the hydroxide ion is not available for the other complexation reaction. The amount of the intended complex formed will be less.

15. The fraction of the coordination center available for the intended complexation reaction is that fraction that is not complexed by the hydroxide ion. This can be calculated from the alpha equation for the coordination center (the first line of Equation 7.21) considering the complexation by OH^-. The formation constant(s) for the reaction of the coordination center with OH^- must be available.

16. At a given pH, the fraction of the coordination center that is available for the intended complexation reaction is a constant. This fraction, multiplied by the formation constant for the intended complexation reaction, is an effective formation constant in which the total concentration of the coordination center could be used. The effective formation constant is pH dependent because the fraction of the coordination center available for the reaction depends on the concentration of OH^- and thus on the pH.

17. A. Since $K'_{MYeff} = K'_{MY}\alpha^\circ_M$, α°_M must be calculated for the iron at this pH. From Equation 7.21,

$$\alpha^\circ_{Fe^{2+}} = \frac{1}{1 + K'_{f1}[OH^-]}$$

$$[OH^-] = \frac{1 \times 10^{-14}}{[H_3O^+]} = \frac{1 \times 10^{-14}}{1 \times 10^{-10}} = 1 \times 10^{-4}\,M$$

$$\alpha^\circ_{Fe^{2+}} = \frac{1}{1 + (10^{4.6})(10^{-4.0})} = 0.20$$

$$K'_{FeCl_2eff} = (0.200)(10^{1.48})(10^{0.65}) = 27$$

(remember that $K'_{fFeCl_2} = K'_{f1}K'_{f2}$).

B. At pH = 3.5,

$$\alpha^\circ_{Fe^{2+}} = \frac{1}{1 + (10^{4.6})(3.16 \times 10^{-11})} = 1.0$$

$$K'_{FeCl_2eff} = (1.0)(10^{1.48})(10^{0.65}) = 1.3 \times 10^2$$

Comparing parts A and B, we see that at lower pH (lower $[OH^-]$) the effective formation constant for the iron(II) chloride complex is the same regardless of whether we take α°_M into account. At high pH, however, α°_M becomes significantly less than 1 and decreases the overall formation constant considerably.

18. Many ligands are weak bases that will react with the water solvent to create more acidic forms of the ligand group. The extent of such reactions decreases the fraction of the total ligand that is available for the intended complexation reaction.

19. One would use an alpha equation (such as Equation 7.27) to calculate the fraction of the ligand in the unprotonated form. This calculation requires a knowledge of the pH and of the K'_a values the weak acid forms of the ligand.

20. The fraction of the unprotonated ligand increases as the pH increases, since it is the most basic form of the ligand.

21. Given $K'_{MYeff} = K'_{MY}\alpha^\circ_Y$, we need

$$\alpha^\circ_Y = \frac{K'_{a1}K'_{a2}K'_{a3}K'_{a4}}{([H_3O^+]^4 + K'_{a1}[H_3O^+]^3 + K'_{a1}K'_{a2}[H_3O^+]^2 + K'_{a1}K'_{a2}K'_{a3}[H_3O^+] + K'_{a1}K'_{a2}K'_{a3}K'_{a4})}$$

Using the given K'_a values, $\alpha^\circ_{Y^{4-}}$ at a given pH is

$$\alpha^\circ_Y = \frac{8.30 \times 10^{-22}}{([H_3O^+]^4 + 1.02 \times 10^{-2}[H_3O^+]^3 + 2.18 \times 10^{-5}[H_3O^+]^2 + 1.51 \times 10^{-11}[H_3O^+] + 8.30 \times 10^{-22})}$$

A. At pH = 6.5, $\alpha^\circ_{Y^{4-}} = 1.2 \times 10^{-4}$, so

$$K'_{MYeff} = (1.2 \times 10^{-4})(5.0 \times 10^{10}) = 6.0 \times 10^6$$

B. At pH = 10.0, $\alpha^\circ_Y = 0.35$, so

$$K'_{MYeff} = (0.35)(5.0 \times 10^{10}) = 1.8 \times 10^{10}$$

From this example, we see that at higher pH (more basic solution) the fraction of EDTA in the Y^{4-} form is higher and thus the effective formation constant for the EDTA complex is higher.

22. The effective formation constant under these conditions is the product of the α°_M of the coordination center and the α°_L of the ligand times the formal complex formation constant.

23. By Equation 7.32, $K'_{MYeff} = K'_{MY}\alpha^\circ_Y\alpha^\circ_M$
   Question 17 shows how α°_M is calculated at a given pH. For Ni^{2+},

$$\alpha^\circ_{Ni^{2+}} = \frac{1}{1 + K'_{f1}[OH^-] + K'_{f1}K'_{f2}[OH^-]^2 + K'_{f1}K'_{f2}K'_{f3}[OH^-]^3}$$

at a given pH,

$$\alpha^\circ_{Ni^{2+}} = \frac{1}{(1 + 1.3 \times 10^4[OH^-] + 1.0 \times 10^9[OH^-]^2 + 1.0 \times 10^{12}[OH^-]^3)}$$

A. At pH = 4.0, $\alpha^\circ_Y = 3.6 \times 10^{-9}$ and $\alpha^\circ_{Ni} = 1.0$, so

$$K'_{MYeff} = (4.2 \times 10^{18})(3.6 \times 10^{-9})(1.0) = 1.5 \times 10^{10}$$

B. At pH 7.0, $\alpha^\circ_Y = 4.8 \times 10^{-4}$ and $\alpha^\circ_{Ni} = 0.999$, so

$$K'_{MYeff} = (4.2 \times 10^{18})(4.8 \times 10^{-4})(0.999) = 2.0 \times 10^{15}$$

C. At pH 12.0, $\alpha^\circ_Y = 0.98$ and $\alpha^\circ_{Ni} = 1.0 \times 10^{-7}$, so

$$K'_{MYeff} = (4.2 \times 10^{18})(0.98)(1.0 \times 10^{-7}) = 4.1 \times 10^{11}$$

From this example, we see that there is a trade-off between having a high α_Y° at high pH and a high α_M° at low pH. An intermediate pH must be chosen for the titration that gives the highest value for K'_{MYeff}. One consideration not accounted for here is the fact that $Ni(OH)_2$ is a relatively insoluble compound. Precipitation of this compound would be undesirable for the titration, so the pH should be kept as low as possible. Precipitation is discussed in detail in the next chapter.

D. Quantitation by Complexation Titration

The classic method of employing a complexation reaction for quantitative analysis is by titration in a manner similar to that already seen for the use of acid–base reactions. A complexing titrant is chosen that is suitable for the species to be titrated. Frequently, the analyte is a coordination center, and the titrant is the ligand or complexing agent. Several methods for determining the equivalence point of the titration are available. An electrode whose potential is dependent on the coordination center (or ligand) concentration can be used. Indicator compounds have been developed for many complexation titrations. Absorbance spectrometry can also be used if either of the reactants or the complex formed has a significant absorbance. Each of these aspects of complexation titration will be explored in more detail in this section.

STANDARD TITRANTS

The complexing agents most suitable for complexation titration are those that are polydentate. This is because they react with the coordination center in a single step. Complexes formed by stepwise addition of ligands generally do not provide well-defined titration curves owing to the small differences between the successive formation constants. Another advantage is that the formation constants for chelates can be very large, a factor that affects the precision and sensitivity of complexometric titrations. Among the polydentate ligands, ethylenediaminetetraacetic acid (EDTA) is by far the most widely used. Others have been developed but have shown no advantages in application over the very versatile EDTA. Table 7.1 indicates the metal ions with which EDTA forms stable complexes, along with their formal formation constants (for an ionic strength of 0.1).

Standard solutions of EDTA can be made from the primary standard acid form (H_4Y), though that compound is not very soluble in water. (Recall that a primary standard is a material of such pure composition that the mass divided by the molecular weight equals the number of moles within at least 1 ppt.) More commonly used is the disodium salt of EDTA $(Na_2C_{10}H_{14}N_2O_8 \cdot H_2O)$, which is available in nearly primary standard quality. The disodium salt can be gently dried at 80 °C for a short time without losing the

Analytical scheme for Complexation Titration

Differentiating characteristic:
Ability to react to form a complex.

Probe:
Add complexing agent to a solution of the analyte.

Response:
The analyte reacts with the complexing agent to form a complex.

Anticipated response:
The concentration of the analyte will decrease. The concentrations of the complex and the complexing agent will increase.

Measurement of the response:
Follow the changes in the concentration of any of the involved species. Determine the exact volume of complexing agent added at the equivalence point.

Interpretation of data:
Use the equivalence point volume of the complexing agent and its concentration to calculate the amount of analyte originally present.

Table 7.1
Formal Formation Constants for EDTA*

Metal Ion	Log K'_{MY}	Metal Ion	Log K'_{MY}	Metal Ion	Log K'_{MY}
Fe^{3+}	25.1	TiO^{2+}	17.3	Mn^{2+}	14.0
Th^{4+}	23.2	Zn^{2+}	16.5	Ca^{2+}	10.7
Hg^{2+}	21.8	Cd^{2+}	16.5	Mg^{2+}	8.7
Tl^{3+}	21.3	Co^{2+}	16.3	Sr^{2+}	8.6
Cu^{2+}	18.8	Al^{3+}	16.1	Ba^{2+}	7.8
Ni^{2+}	18.6	La^{3+}	15.4	Ag^+	7.3
Pb^{2+}	18.0	Fe^{2+}	14.3		

*At 20 °C, ionic strength 0.1.

water of hydration. If a determinate standard is not prepared, an EDTA solution can be standardized by titration of a weighed-out portion of primary standard quality $CaCO_3$.

TITRATION CURVES

It is a characteristic of useful titration reactions of all types that there is a substantial change in the concentration of the analyte or titration species in the region of the equivalence point of the titration. Putting that another way, when there is not such a significant concentration change in the equivalence point region, the reaction (or the conditions) are not suitable for a successful titration. A plot of the concentration of the analyte or titrant as a function of the titrant volume or the fraction titrated is called a **titration curve**. The form of the titration curve can be solved exactly.

For the complexation reaction $M + Y \rightleftharpoons MY$, the effective formation constant (introduced as Equation 7.31 and repeated here) is

$$K'_{MYeff} = \frac{[MY]}{[M]'[Y]'} \qquad\qquad 7.31$$

The derivation of the titration curve is available on the accompanying CD. If the analyte and titrant are assumed to have roughly equal concentrations, the titration curve equation is

$$\theta = \frac{C_i - [M]' + \dfrac{C_i}{K'_{MYeff}[M]'} - \dfrac{1}{K'_{MYeff}}}{C_i + [M]' + \dfrac{1}{K'_{MYeff}}} \qquad\qquad 7.33$$

With Equation 7.33, the fraction titrated can be calculated for various values of $[M]'$ and the results plotted, as shown in Figure 7.9. This plot is for the titration of several different metal ions, each at an initial concentration of 0.1 M and in a solution of pH = 10. The magnitude of the break at the equivalence point increases as the effective equilibrium constant for complex formation increases.

Another plot (Figure 7.10) has been drawn to illustrate the effect of the initial concentration of analyte. Again, as we have come to expect, the magnitude of the break in the equivalence point region decreases with decreasing concentration of analyte.

Titration curves are generally plotted with the species involved in the end point detection method on the Y-axis and the volume of titrant or fraction titrated on the X-axis. Since the course of complexation titrations could be followed by detection of either M or Y, it is interesting to compare titration plots of $p[Y]'$ with those of $p[M]'$. The equation for a plot of $[Y]'$ versus fraction titrated has also been derived, as pre-

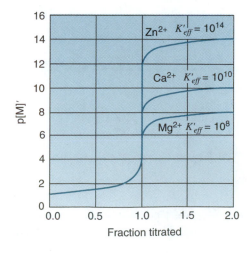

Figure 7.9. The magnitude of the break in $p[M]'$ at the equivalence point volume increases with K'_{eff} for MY.

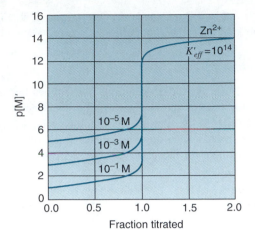

 Figure 7.10. The break in the magnitude of p[M]′ at the equivalence point volume decreases with decreasing initial concentration of M.

sented on the accompanying CD. The result, again assuming that the analyte and titrant concentrations are similar, is

$$\theta = \frac{C_i + [Y]' + \dfrac{1}{K'_{MYeff}}}{C_i + \dfrac{C_i}{K'_{MYeff}[Y]'} - \dfrac{1}{K'_{MYeff}} - [Y]'} \qquad 7.34$$

From Equation 7.34, plots of p[Y]′ for the same conditions as the p[M]′ plots already developed are shown in Figures 7.11 and 7.12. Again, the characteristic break at the equivalence point is seen, with the magnitude of the break increasing with increasing effective equilibrium constant and analyte concentration.

Particular points on the titration curve can be easily determined using equations we have developed earlier. For example, in the titration of M with L, all the points before the equivalence point can be calculated with Equation 7.14, 7.15, or 7.16 since there is an excess of M for these points. At the equivalence point, Equation 7.18 or 7.19 should be used, and after the equivalence point, when L is in excess, Equations 7.11 through 7.13 are most suitable. At the 0% point, $[M]' = C_M$. At the 50% point, $[M]' \approx [ML] \approx \frac{1}{2}C_M$. At the equivalence point, $[M]' \approx (C_M/K'_f)^{1/2}$ (from Equation 7.19). At the 200% point, $[M]' \approx 1/K'_f$ (from Equation 7.13, where $C_L = 2C_M$). The difference in the log of the concentration of M between the 50% point and the 200% point is a good indicator of the feasibility of a titration. A difference of three or more log units will enable good equivalence point detection. (See Example 7.4.)

● ●

Example 7.4

The points on the titration curve of 10^{-5} M Zn^{2+} with EDTA as shown in Figure 7.10 will be obtained mathematically. For ZnY, $K'_{eff} = 10^{14}$.

$$0\%, \quad [M]' = C_M = 1 \times 10^{-5} \text{ M}, \quad p[M]' = 5$$
$$50\%, \quad [M]' = C_M/2 = 5 \times 10^{-6} \text{ M}, \quad p[M]' = 5.3$$
$$100\%, \quad [M]' = (C_M/K'_{eff})^{1/2} = 3.2 \times 10^{-10} \text{ M}, \quad p[M]' = 9.5$$
$$200\%, \quad [M]' = 1/K'_{eff} = 1 \times 10^{-14}, \quad p[M]' = 14$$

● ●

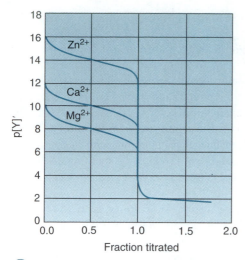

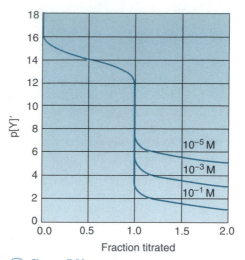

Figure 7.11. The equivalence point volume can be determined by following p[Y]′ as well as p[M]′.

Figure 7.12. The magnitude of the break is the same for a p[Y]′ plot as for a p[M]′ plot and is similarly decreased by decreasing K'_{eff} and C_M.

ACHIEVING SELECTIVITY WITH EDTA TITRATIONS

One of the great advantages of EDTA in complexometric titrations is the great range of metal ions with which it will complex. This certainly makes it a good general-purpose reagent. The other side of this coin is that the ability to react with EDTA is not a highly differentiating characteristic. Many metal ions other than the desired analyte are potential interferents. To minimize this problem, advantage can be taken of the great range of effective formation constants for various metal ions. These are illustrated in Figure 7.13. Recall that the effective formation constant is heavily influ-

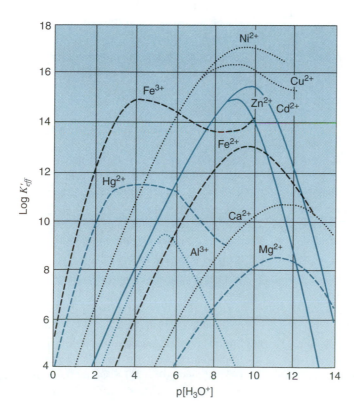

Figure 7.13. The p[H₃O⁺] giving the maximum value of K'_{eff} differs greatly among metal ions (owing to their different degrees of hydrolysis). The use of a buffer to set the pH is important in EDTA titrations.

enced by $p[H_3O^+]$ because of the protonation of the EDTA and the hydrolysis of the metal ions.

Several strategies can be employed. For instance, to titrate Hg^{2+} in the presence of Cd^{2+}, setting the $p[H_3O^+]$ at 2 eliminates the formation of the cadmium complex while keeping the mercuric complex formation constant reasonably high. Cadmium and zinc can be titrated in the presence of calcium and magnesium at pH 7. Where an interfering ion cannot be eliminated by the choice of $p[H_3O^+]$, a $p[H_3O^+]$ can often be found where the effective formation constant is different enough for the titration curve to form separate breaks for each metal. A difference of four or more in $\log K'_{eff}$ is enough for this approach. Successive titration with separately identifiable breaks makes the choice of indicator more critical. In these cases, the use of an electrode to follow the entire titration is better.

Yet another method is to use a **masking agent**. An example is the titration of Cd^{2+} in the presence of Zn^{2+}. If an ammonia and ammonium ion buffer is used, an ammonia–zinc complex is formed, greatly reducing the effective formation constant for the zinc ion. This is a good example of employing a secondary differentiating characteristic to avoid interference. The secondary differentiating characteristic is that the formation constant for the zinc–ammonia complex is much greater that for the cadmium–ammonia complex.

Study Questions, Section D

24. Why are chelates rather than single coordination site ligands chosen for complexometric titration?

25. How is the magnitude of the concentration change in the equivalence point region affected by an increase in the formation constant of the complex and by a decrease in the concentration of analyte?

26. Is there any advantage to monitoring the ligand concentration over the coordination center concentration during a titration?

27. In a titration of M with Y, what are the concentrations of M (relative to the initial concentration, C_i) at the 0%, 50%, 100%, and 200% titration points? Neglect the dilution due to the titrant.

28. What is the characteristic of the metal ion that determines the position of the left side of its $\log K'_{eff}$ curve in Figure 7.13?

29. Estimate the magnitude of the break (50% to 200%) in $p[Fe^{3+}]$ in the titration of 1×10^{-5} M Fe^{3+} with EDTA. The effective formation constant for the Fe^{3+}–EDTA complex is 1×10^{15}.

30. How could one determine the concentration of Fe^{3+} in the presence of Fe^{2+} with an EDTA titration?

Answers to Study Questions, Section D

24. Single coordination site ligands frequently form stepwise complexes, which greatly reduces the sharpness of the concentration changes in the equivalence point region of a titration. The chelates can occupy all the coordination sites of the coordination center with a single ligand, so that a 1:1 complex of generally high formation constant results.

25. An increase in the formation constant of the complex will cause an increase in the magnitude of the concentration changes in the equivalence point region. Conversely, a decrease in the analyte concentration will cause a decrease in the size of the concentration change in this region.

26. The size of the concentration break is unaffected by the choice of species to monitor. The experimenter is free to choose whichever species she can monitor most conveniently.

27. At the 0% point, $[M] = C_i$. At the 50% point, $[M] = C_i/2$. At the 100% point (from Equation 7.19), $[M] = \sqrt{C_i/K'_{MYeff}}$. At the 200% point, $[M] = 1/K'_{MYeff}$.

28. It is the formation constant of the metal ion's reaction with EDTA that determines the position of the left side of the K'_{eff} curve. The lower the formation constant, the more to the right the curve is shifted.

29. At the 50% titration point, $[Fe^{3+}] = C_M/2 = 5 \times 10^{-6}$ M, and $p[Fe^{3+}] = -5.3$. At the 200% point, $[Fe^{3+}] = 1/K'_{eff} = 1 \times 10^{-15}$ M, and $p[Fe^{3+}] = 15.0$. The size of the break, in $p[Fe^{3+}]$ units, is then $15 - 5.3 = 9.7$.

30. From Figure 7.13, we see that at low $p[H_3O^+]$, say 3.0, the formation constant for Fe^{3+} is about 10^{15} while that for Fe^{2+} is about 10^4. EDTA would therefore complex selectively with Fe^{3+} at this $p[H_3O^+]$.

E. Equivalence Point Detection

In this section, equivalence point detection by the now-familiar techniques of indicators, specific electrodes, and optical absorption are explored.

INDICATORS

As in the acid–base titration case, an indicator is a compound that can change form in response to the titrant concentration and for which the two forms of the compound have different colors. In complexometric titrations, the indicator is usually a ligand that has different colors depending on whether it is complexed with a coordination center. An example is the commonly used Eriochrome Black T (EBT). As is often the case, this indicator is a weak acid (see Figure 7.14). The anion of the sodium salt has two protons to donate. Its reaction with water and their equilibrium constants are

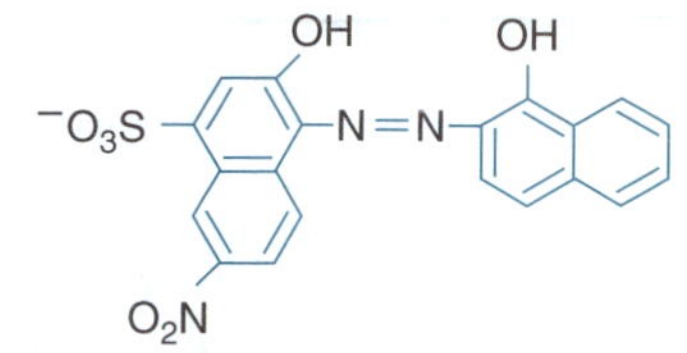

$$H_2In^- + H_2O \rightleftharpoons HIn^{2-} + H_3O^+ \qquad K'_{a2} = 5 \times 10^{-7}$$
$$HIn^{2-} + H_2O \rightleftharpoons In^{3-} + H_3O^+ \qquad K'_{a3} = 2.8 \times 10^{-12}$$

All three of the different conjugate acid forms of the indicator have different colors. The most acidic (H_2In^-) is red, the next (HIn^{2-}) is blue, and the most basic (In^{3-}) is orange. From this we can see that EBT could be used as an indicator in acid–base reactions as well. Concern over its acidity response while acting as a complexometric indicator is nullified by the fact that complexation titrations are almost always carried out in a solution that is buffered to maintain a constant pH. For almost all metals, the color of the complex of EBT is red (the same as its most acidic color). When EBT is used in a solution buffered at pH 10 (as in the case of Ca^{2+} titrations) the uncomplexed color is blue and the complexed color red. Thus, the color of the indicator depends on the metal ion concentration. Assuming the metal is the analyte and the complexing agent the titrant, the indicator is in its metal complex form (red) before the end point (because of the excess of metal ion), and, when an excess of complexing agent has been added, in its uncomplexed form (blue at pH 10). If the pH is below 6.3, the uncomplexed indicator will be predominately in its most acid form, which has the same color as the complex. Thus Eriochrome Black T cannot be used in acidic solutions.

The critical factor in the application of a complexometric indicator is for the color change to occur at that concentration of metal ion that is near its equivalence point value for the titration. The calculation of the metal ion concentration at the point of color change requires knowledge of the protonation and complexation equilibrium constants of the indicator as well as the pH of the solution.

Consider the case of the application of Eriochrome Black T in the titration of calcium with EDTA. The Ca^{2+}–EBT complex formation constant is 2.5×10^5, and the titration is carried out at a pH of 10.00. At this pH, the uncomplexed EBT is almost completely in the blue HIn^{2-} form. The form of the red complex is $CaIn^-$. The reaction with the calcium is thus a combination of the deprotonation reaction of HIn^{2-} and the complexation reaction of In^{3-}:

$$HIn^{2-} + H_2O \rightleftharpoons In^{3-} + H_3O^+$$
$$\underline{Ca^{2+} + In^{3-} \rightleftharpoons CaIn^-}$$
$$Ca^{2+} + HIn^{2-} + H_2O \rightleftharpoons CaIn^- + H_3O^+$$
$$\text{blue} \qquad\qquad\qquad \text{red}$$

for which the formal equilibrium constant is

$$K'_{eq} = \frac{[CaIn^-][H_3O^+]}{[HIn^{2-}][Ca^{2+}]} = K'_{a3}K'_{CaIn} \qquad\qquad 7.35$$

Figure 7.14. The structure of Eriochrome Black T indicator. The protons on the OH groups are lost in the more basic forms. The SO_3^- group is a strong base. For EBT, $pK'_{a2} = 6.3$ and $pK'_{a3} = 11.6$.

Eriochrome Black T indicator is red when complexed with metal ions or H^+. It is blue when uncomplexed in mildly basic solution. Protonation occurs below pH 6.3.

Table 7.2
Selected Metal Complex Indicators

Indicator	Metals
Eriochrome Black T	Ba, Ca, Mg, Zn
Calmagite	Ba, Ca, Mg, Zn
Murexide	Ca, Ni, Cu
Salicylic acid	Fe

To be effective, the indicator must be half complexed by M at the equivalence point value of p[M]. Thus, the optimum K'_{eff} for the indicator will change with different metal ions, concentrations, and reacting ligands.

If the red and blue colors are of equal intensity, the color change occurs when the concentrations of HIn^{2-} and $CaIn^-$ are equal. When this condition is imposed on Equation 7.35,

$$[Ca^{2+}] = \frac{[H_3O^+]}{K'_{a3}K'_{CaIn}}$$

7.36

and with the values for the constants and pH substituted, the $[Ca^{2+}]$ at the end point is $1.0 \times 10^{-10}/(2.8 \times 10^{-12} \times 2.5 \times 10^5) = 1.4 \times 10^{-4}$ M (p[M] = 3.9). From the Ca^{2+} titration curves in Figures 7.9 and 7.11, it can be seen that the end point for this titration at $p[Ca^{2+}] = 3.9$, would be at a much higher $[Ca^{2+}]$ than the equivalence point (more nearly $p[Ca^{2+}] = 5.5$). The full color change would begin 1 p[M] unit before 3.9 and extend to 1 p[M] unit after. This end point is not close enough to the equivalence point to make this an accurate titration.

Fortunately, the end point for the titration of Mg^{2+} is much better. For Mg^{2+}, the EBT complex formation constant is 1.0×10^7. When this value is used in Equation 7.36, the value of $p[Mg^{2+}]$ at the end point is 5.4. From the titration curve for Mg^{2+} shown in Figure 7.9, this is seen to be almost perfect. It is interesting to note that the formation constant of Ca^{2+} with EDTA is greater than that of Mg^{2+}, whereas with EBT, it is the opposite. Thus, when titrating a mixture of Ca^{2+} and Mg^{2+} with EDTA, the EDTA reacts with the Ca^{2+} first; then, when that is almost gone, the EDTA reacts with the Mg^{2+}. Since EBT is a suitable indicator for the Mg^{2+} titration, it can thus be used in the titration of a combination of Ca^{2+} and Mg^{2+} even though it is not suitable for the titration of Ca^{2+} alone.

When a mixture of Ca^{2+} and Mg^{2+} is titrated with EDTA using the Eriochrome Black T indicator, the equivalence point occurs when the amount of EDTA added is equal to the sum of the amounts of Ca^{2+} and Mg^{2+}. Their individual amounts are not determined in this way. The sum of them is useful, however, in the assessment of total water hardness. If the sample contains Ca^{2+}, but very little Mg^{2+}, some Mg^{2+} is added deliberately. To keep from affecting the end point volume, the Mg^{2+} is added in the form of Na_2MgY (already exactly complexed) so that no EDTA titrant is needed to react with it. The red color before the end point will all be due to the magnesium–EDTA complex.

Other indicators have been developed along the lines of Eriochrome Black T. Some of them are listed in Table 7.2. Each has advantages in specific situations, but none enjoys the widespread utility of Eriochrome Black T.

DISPLACEMENT TITRATION

The displacement titration is a way to use the Eriochrome Black T indicator for almost any of the metal ions strongly complexed by EDTA. Already complexed Mg^{2+} is added to the sample solution. The EBT end point occurs when both the metal and the Mg^{2+} are complexed. The amount of EDTA added at the end point is just the amount needed to complex the analyte metal ions.

Since the range of available complexometric indicators is not as great as that for acid–base titrations, it is sometimes difficult to find an appropriate indicator for a particular analyte titration. In this situation, there is a way to use the nearly ideal Mg^{2+}, Eriochrome Black T combination to titrate virtually any metal ion for which the effective formation constant is greater than that for Mg^{2+}. A known amount of MgY^{2-} is added to the analyte solution. The amount of Mg should be greater than the maximum expected amount of the analyte metal ion. The analyte metal ion, having a higher formation constant than the Mg^{2+}, displaces the Mg^{2+} from the EDTA. The equation for this reaction is

$$M^{2+} + MgY^{2-} \rightleftharpoons Mg^{2+} + MY^{2-} \qquad 7.37$$

Assuming a complete reaction, the amount of Mg^{2+} produced is exactly equal to the amount of M^{2+} in the initial solution. The resulting Mg^{2+} is then titrated with EDTA to the Eriochrome Black T end point. This procedure is called a **displacement titration**.

BACK TITRATIONS

In a **back titration**, a known excess of reagent (often the titrant) is added to the analyte solution, and the amount of the excess is determined by titration with a reagent similar to the analyte. For example, an excess of EDTA could be added to a solution of metal ion to form the metal–EDTA complex with some EDTA left over. A standard solution of Mg^{2+} or Zn^{2+} could then be used to determine the amount of excess EDTA present. The amount of the analyte is then equal to the amount of EDTA added less the amount of back titrant.

Back titrations are useful when the analyte may react with the EDTA too slowly to maintain equilibrium concentrations during a normal direct titration. They also offer another way to use an indicator that is best suited to the back titrant. The formation of the metal–EDTA complex in the original solution may also be useful in avoiding the precipitation of the analyte as the oxide or hydroxide at the relatively high pH values sometimes needed for EDTA titration.

SPECIFIC ION ELECTRODES FOR EQUIVALENCE POINT DETECTION

Electrodes for which the voltage output is a function of various metal ions have been developed. These are called specific ion electrodes. Their action (discussed in some detail in Chapter 9) is very similar to that of the pH electrode for H_3O^+. The voltage difference between the specific ion electrode and a reference electrode changes by approximately 0.03 V for each 10-fold change in concentration of a doubly charged ion detected. A normal pH meter is used to convert the electrode output voltage to a number. Most pH meters allow the output to be calibrated in "p" units of the detected ion so that a plot of the meter output versus titrant concentration looks just like the titration curves developed in the previous section. In general, specific ion electrodes are more costly, less specific, and less robust than pH electrodes. However, because they are so potentially useful, much work has been put into their development, and they are steadily improving. One of their great advantages in complexometric titrations is their ability to produce the entire titration curve. This, as in the case of the pH electrode for acid–base titrations, provides a great deal more information about the position of the equivalence point and about the species titrated.

Specific ion electrodes are an excellent way to follow an EDTA complexometric titration. No special indicator conditions need be met, and information regarding the entire course of the titration is available. Electrodes are available for most metal ions. An EDTA electrode that was developed but not commercialized, is described in Chapter 9.

LIGHT ABSORPTION FOR EQUIVALENCE POINT DETECTION

Absorption spectrometers can be used for the detection of the equivalence point in titrimetric procedures if any of the reactants or products has a significant absorptivity. Since many transition metal complexes are colored, this is a very real possibility. A spectrometer with a dip probe is especially convenient for this application. The usual situation is to plot the titration solution absorbance (at a specific wavelength) as a function of the titrant volume. There are several possibilities for the shape of the absorbance titration curve. First, suppose that the complex formed absorbs at the selected wavelength, but the analyte and titrant do not. In this case, the absorbance will be proportional to the concentration of the complex formed at each stage of the titration.

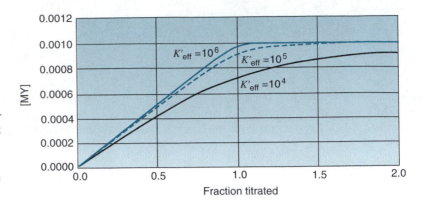

Figure 7.15. Assuming MY is the only species absorbing light of the test wavelength, the plot of [MY] versus fraction titrated will be similar to that of the absorbance versus the fraction titrated. The equivalence point volume is at the intersection of the extensions of the two straight-line portions of the curve. The higher the K'_{eff} for MY, the more precisely this volume can be determined.

A plot of the concentration of MY versus fraction titrated is shown in Figure 7.15. The initial concentration of MY is zero because no complexing agent has yet been added. As the added titrant reacts with the analyte, more complex is formed. At the equivalence point, enough titrant has been added to react with all the analyte, but the completeness of this reaction depends on the initial concentration and the effective complex formation constant. In the case illustrated, the initial concentration is 0.0010 M. For the case of a formation constant of 10^6, the MY concentration increases essentially linearly until all the analyte has reacted. The concentration of MY cannot increase beyond the concentration of the analyte, so it remains constant at higher fractions titrated. In this plot, the dilution effect of the titrant addition has been ignored. As the formation constant decreases, the degree of curvature in the equivalence point region increases because of increasing incompleteness of the reaction at that point. As a rule of thumb, the effective formation constant times the initial concentration should be greater than 100.

The equivalence point volume is determined by plotting the absorbance versus the titrant volume and extrapolating the two straight-line sections toward the equivalence point region. The volume corresponding to the intersection of these lines is taken as the end point volume. As shown in Figure 7.16, the plot can have different shapes depending on which of the species involved in the titration reaction absorbs light at the selected wavelength. For example, if the analyte absorbs, but the titrant and complex do not, the absorbance will decrease steadily to the equivalence point and remain con-

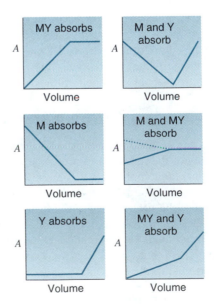

Figure 7.16. Plot of absorbance versus volume of titrant Y will have different shapes depending on the species absorbing and, if more than one absorbs, their relative molar absorptivities. The less obtuse the angle between the straight-line segments, the more precise the end point determination will be.

stant at a low value thereafter. In every case, the titration curve consists of two straight-line sections whose extensions intersect at the end point volume.

Titration with photometric detection offers several advantages over quantitation by the absorbance measurement itself. First, it is a null method, with the amount of analyte given by the amount of titrant at the equivalence point. The titrant volume measurement can be more precise than the absorbance reading. In addition, other absorbing species do not interfere with this determination as long as they do not also react with the titrant. This is because the absolute absorbance is not used in the end point determination, only the change due to the changing concentration of the analyte, titrant, or product. In other words, the characteristics of the potentially interfering species are completely different between direct absorbance determination and a titration with photometric detection.

Study Questions, Section E

31. What are the qualities required of an indicator for a complexation titration?

32. How does the pH of the titration solution affect the choice or requirements of the indicator?

33. In the text, we examined the use of EBT as an indicator in the titration of Mg^{2+} with EDTA. In a similar way, examine the titration of Zn^{2+} using EDTA and EBT.

34. Why is the presence of Mg^{2+} desirable when titrating Ca^{2+} with EDTA using Eriochrome Black T as an indicator?

35. How can Mg^{2+} be added to the solution without affecting the volume of EDTA needed to titrate the Ca^{2+}?

36. Water hardness is defined as the concentration of calcium carbonate equal to the concentration of Ca^{2+}, Mg^{2+}, and heavy metals in the solution. Usually, $[Ca^{2+}]$ and $[Mg^{2+}]$ are much greater than that of other metal cations. If 100.0 mL of a water sample containing Ca^{2+} and Mg^{2+} is titrated with 30.0 mL of standard 0.0180 M EDTA solution containing a small amount of Mg^{2+}, what is the hardness of the water (in ppm $CaCO_3$)? Assume that the Mg^{2+} was added to the EDTA titrant before standardization and that EBT was used as an indicator.

37. Eriochrome Black T as an indicator for the EDTA titration of Mg^{2+} is not only a good example of a complexometric indicator, it is very nearly the only one. Describe the indirect titration techniques that have been devised so that this favorable indicator and metal ion system can be used when the analytes are not Mg^{2+}.

38. A standard solution of MgY^{2-} is added in excess to 50.0 mL of solution containing an unknown concentration of Cu^{2+}. The Mg^{2+} liberated from MgY^{2-} is then titrated with 35.7 mL of standard 0.100 M EDTA. What is the concentration of Cu^{2+} in the unknown solution?

39. In an experiment, 100.0 mL of 0.150 M EDTA is added to 50.0 mL of an unknown solution of Mn^{2+}. The solution is then titrated to the equivalence point with 54.0 mL of standard 0.150 M Mg^{2+} solution using an EBT indicator. What is the concentration of Mn^{2+} in the unknown? Assume that the moles of EDTA added exceed the moles of Mn^{2+} in the unknown solution.

40. Consider the titration curves shown in Figures 7.11 and 7.12. If one had an EDTA electrode, could one develop a method for the titration of a mixture of metal ions? Of those metal ions in Figure 7.12, which ones are the best candidates for individual quantitation by titration of a mixture?

41. Predict the photometric titration curve of absorbance of versus concentration of CuY^{2-} for titration of 100.0 mL of 3.0×10^{-3} M Cu^{2+} with 0.010 M EDTA. Calculate data points after addition of 0 mL, 10 mL, 20 mL, 30 mL, 31 mL, and 35 mL of EDTA. The molar absorptivity of the complex is 60.0 L/mol · cm at 745 nm. Free copper and EDTA do not absorb at this wavelength. Use 1.0 cm as the cell path length.

Answers to Study Questions, Section E

31. An indicator compound for a complexation titration must be a ligand with an effective formation constant for the coordination center smaller that that of the titration ligand (but not too small), and its complexed and uncomplexed forms must be differently colored.

32. The indicator ligand may also hydrolyze, so the protonated forms and the colors of these forms must be considered when choosing an indicator for a specific application.

33. From Figure 7.13, we see that the maximum formation constant for Zn^{2+} with EDTA occurs at a $p[H_3O^+]$ of about 9.0. Below this $p[H_3O^+]$, hydrolysis of Y^{4-} decreases the formation constant. Above it, hydrolysis of the Zn^{2+} ion decreases the effective formation constant. This

p[H_3O^+] is perfectly acceptable for the use of EBT since EBT undergoes a distinct color change above pH 6.3. (Below that p[H_3O^+], it is in its acid form, which is the same color as its complexed form.) The only possible problem with this titration is the insolubility of the hydrolyzed Zn^{2+}. To prevent the precipitation of $Zn(OH)_2$, the titration might be done at a slightly lower p[H_3O^+], say 7.5. The p[H_3O^+] cannot be decreased much more than this without making the color change of the EBT undetectable.

34. The equivalence point for the Mg^{2+} titration is at a much more favorable position for the Eriochrome Black T than it is for the Ca^{2+} titration. When both cations are present, the EBT color changes when both metals are titrated.

35. If the EDTA titrant contains the Mg^{2+} (and was standardized with it present) or if the Mg^{2+} is added as MgY^{2-}, the end point volume is not affected by the Mg^{2+}.

36. At the equivalence point, [Mg^{2+}] + [Ca^{2+}] = [EDTA].

$$\left(\frac{0.0180 \text{ mol EDTA}}{L}\right)(0.0300 \text{ } L) = 5.40 \times 10^{-4} \text{ mol EDTA}$$

$$\left(\frac{0.00540 \text{ mol } Mg^{2+} + Ca^{2+}}{0.100 \text{ L solution}}\right)$$

$$= 5.40 \times 10^{-3} \text{ } M \text{ } Ca^{2+} + Mg^{2+}$$

$$(5.40 \times 10^{-3} \text{ M } Ca^{2+} + Mg^{2+})\left(\frac{100.0 \text{ g } CaCO_3}{\text{mol}}\right)$$

$$\times \left(\frac{1000 \text{ } mg}{g}\right) = \frac{540 \text{ mg}}{L} = 540 \text{ } ppm \text{ } CaCO_3$$

37. Displacement titration starts with a stoichiometric solution of complexed Mg^{2+}. An unknown amount of some other metal ion is added (less than the amount of Mg^{2+}). The metal displaces the Mg^{2+} from the complex. Additional EDTA is added as titrant to regain the equivalence point for the Mg^{2+}.

    In back titration, a known excess of EDTA is added to the unknown metal solution. Mg^{2+} titrant is then added to determine the amount of excess EDTA.

38. The amount of Mg^{2+} is calculated:

$$\text{moles } Mg^{2+} = (0.0357)\left(\frac{0.100 \text{ mol EDTA}}{L}\right)$$

$$\times \left(\frac{1 \text{ mol } Mg^{2+}}{1 \text{ mol EDTA}}\right) = 3.57 \times 10^{-3} \text{ mol } Mg^{2+}$$

Since the amount of Mg^{2+} liberated by the complexation of copper is equal to the amount of copper originally present in the unknown, 3.57×10^{-3} mol Mg^{2+} = 3.57×10^{-3} moles Cu^{2+}, and

$$[Cu^{2+}] = \frac{3.57 \times 10^{-3} \text{ mol}}{0.0500 \text{ L}} = 0.0714 \text{ M } Cu^{2+}$$

39. This is a back titration where the quantity of Mn^{2+} in the solution is equal to the moles EDTA added to the solution minus the amount of uncomplexed EDTA. For moles EDTA,

$$(0.1000 \text{ L})\left(\frac{0.150 \text{ mol}}{L}\right) = 0.0150 \text{ mol EDTA}$$

For the amount of uncomplexed EDTA,

$$(0.0540 \text{ L})\left(\frac{0.150 \text{ mol}}{L}\right)\left(\frac{1 \text{ mol EDTA}}{1 \text{ mol } Mg^{2+}}\right) = 8.10 \times 10^{-3} \text{ mol}$$

The moles Mn^{2+} is $0.0150 - 8.10 \times 10^{-3} = 6.9 \times 10^{-3}$ mol, so the concentration of Mn^{2+} in the unknown is

$$\frac{6.9 \times 10^{-3} \text{ mol}}{0.0500 \text{ L}} = 0.14 \text{ M } Mn^{2+}$$

Note: Back titration was possible in this case because the K_f' of the Mn^{2+}–EDTA complex is higher than that of the Mg^{2+}–EDTA complex.

40. Figure 7.12 shows the break in [Y]′ to be very large at the equivalence point of all the metals titrated. From the plot, it is clear that a mixture of Zn^{2+} with either of the other two metal ions would work well, with the Zn^{2+} titrated first and then with a significant break for either of the other two. Probably, the difference in the formation constants between the Ca^{2+} and Mg^{2+} complexes would not support individual quantitation in a single titration.

41. Since the concentration of the complex is 0 after addition of 0 mL EDTA, the absorbance at this point will be 0 also. At 10 mL,

$$[CuY^{2-}] = [EDTA] = \frac{(0.010 \text{ L})(0.010 \text{ M})}{0.100 \text{ L} + 0.010 \text{ L}}$$

$$= \frac{1.0 \times 10^{-4} \text{ mol}}{0.110 \text{ L}} = 9.1 \times 10^{-4} \text{ M}$$

From Beer's law ($A = \varepsilon bC$)

$$A = \left(\frac{60.0 \text{ L}}{\text{mol} \cdot \text{cm}}\right)(1.0 \text{ cm})\left(9.1 \times 10^{-4} \frac{\text{mol}}{L}\right) = 0.055$$

The table shows the absorbances and concentrations calculated for the rest of the titration. Note: The equivalence

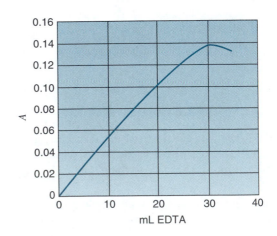

point of the titration occurs where moles EDTA = moles Cu^{2+}, in this case after the addition of 30 mL of 0.010 M EDTA. The concentration of the complex can't increase after this point because there is no more copper to react. The concentration of the complex actually decreases after an excess of EDTA has been added because the EDTA titrant dilutes the solution.

Volume EDTA (mL)	$[CuY^{2-}]$	Absorbance
0	0	0
10	9.1×10^{-4}	0.05
20	1.7×10^{-3}	0.10
30	2.3×10^{-3}	0.14
31	2.3×10^{-3}	0.14
35	2.2×10^{-3}	0.13

F. Log Concentration Plots

The construction of a log concentration plot for complex formation systems can be very instructive in providing quickly attainable approximate solutions of equilibrium concentrations and in easily predicting the critical points in titration curves. As we shall see, log concentration plots are nearly the only easy way to attain exact solutions in systems with stepwise complex formation.

1:1 COMPLEX FORMATION

The log concentration plot for 1:1 complex formation is quite similar to the log alpha plot of Figure 7.2. The principal difference is that the Y-axis is log concentration instead of log α, and the horizontal lines are at the level of log C_M instead of at the top of the plot. The analytical concentration of M (C_M) is the sum of the concentrations of M and ML in the solution. The concentrations of M and ML can be obtained by multiplying their α equations (Equations 7.7 and 7.8) by C_M. The results of this operation are

$$[ML] = \alpha_{ML} C_M = C_M \left(\frac{K_f'[L]}{1 + K_f'[L]} \right)$$

$$\log[ML] = \log C_M - \log\left(1 + \frac{1}{K_f'[L]} \right)$$

7.38

and

$$[M] = \alpha_M C_M = C_M \left(\frac{1}{1 + K_f'[L]} \right)$$

$$\log[M] = \log C_M - \log(1 + K_f'[L])$$

7.39

From these, the log concentration plot is formed, as shown in Figure 7.17. The parameters used for this plot were $-pK_f' = 8$ and $C_M = 1.0 \times 10^{-3}$ M. The construction of this plot is similar to that for the acid–base reactions. A system point $\times$ is placed at the intersection of log C_M and the p[L] value equal to $-pK_f'$. From there, horizontal lines are drawn on the log C_M value, and lines with a slope of 1 and -1 are drawn downward from the system point. The angled and vertical lines are joined by curves that intersect 0.3 log units below the system point. Labels are added knowing that the metal will be in the uncomplexed form (M) at low concentrations (high p values) of L and the ML form will predominate when p[L] is low. Now the line for the ligand concentration is added. This line follows the identity, $-\log[L] = p[L]$.

The log concentration diagram for the complex formation case is simpler than that for the acid–base system since only one equilibrium reaction is plotted. In the acid–base case, the solvent autoprotolysis reaction was also plotted. From the log concentration plot, the concentration of M and ML can be determined for any concentration of L. The plot also provides a graphic illustration of the solution to the

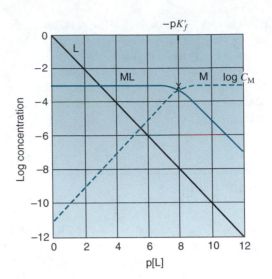

⊚ **Figure 7.17.** The log concentration plot for the reaction
$M + L \rightleftharpoons ML$ reveals all concentrations versus p[L].

question of the concentrations of M and L if 10^{-3} mol of the complex ML were dissolved in a liter of water. To find the solution point, it is necessary to write a **ligand balance equation**. The ligand balance equation is obtained by putting the concentrations of all species with fewer ligands than the starting material on one side of the equation and all with more ligands than the starting material on the other. (The free ligand concentration is on the "more" side since it was zero in the initial solution.)

For the example of dissolving ML in water, the ligand balance equation is

$$[M] = [L]$$

The log concentration plot and a ligand balance equation are used in the graphical solution of complex equilibrium problems.

The ligand balance equation is formed by identifying the starting materials and then equating the concentrations of all species less complexed with those that are more complexed. For each species, the concentration is multiplied by the number of ligands lost or gained relative to the starting material.

This equation could be called the "conservation of ligands" equation; the total concentration of ligands must stay constant. Therefore, for this situation, the concentrations of M and L would be equal. On the plot in Figure 7.17, this is where the lines for these two species intersect. The log of their concentrations would be approximately -5.5; the concentrations, then, are approximately 3×10^{-6} M.

The log concentration plot also facilitates the quick determination of the exact solution. To calculate the equilibrium concentration of M and L in a solution of 1×10^{-3} M ML, one would use one of Equations 7.17 to 7.19. Equation 7.19 is much simpler to implement, but it involves the assumption that [M] is much smaller than C_{ML}. The log concentration plot can reveal at a glance if this assumption is true and therefore whether the simplified or more exact equations must be used. To make this assessment, the values of [M] and C_{ML} are compared at the graphical solution point. They are $[M] = 3 \times 10^{-6}$ M and $C_{ML} = 1 \times 10^{-3}$ M. From this we can conclude that the simplified equation can be used with an error of only 3 ppt. Applying Equation 7.19, $[M] = [L] \approx (1 \times 10^{-3}/1 \times 10^{8})^{\frac{1}{2}} = 3 \times 10^{-6}$ M.

From the log concentration diagram (Figure 7.17), we can also see that the simplified equation used to solve this problem will not always be applicable. If the system point is below (and to the left) of the line for [L], the solution point for the dissolution of ML in water will fall on the part of the [M] line that is horizontal. In other words, [M] will be equal to C_M. This tells us that, in this case, the complex is essentially unformed. This will happen when the complex formation constant is low and/or the total concentration of M in both forms is low. Another possibility is that the system point is near the line for [L]. In this case, the exact solution will require the solution of the quadratic equation, either by the quadratic formula or by successive approximations. Even a crude log plot is a quick way to find out which of the above conditions exists.

STEPWISE COMPLEX FORMATION SYSTEMS

The log concentration plot for stepwise complex formation is very similar to the log alpha plot of Figure 7.4 except for the Y-axis. There will be as many system points as there are stepwise formation constants. These plots are often difficult to construct by hand because the system points are often near each other and because there are many changes in the slopes of the lines. The use of the spreadsheet plotting sheets on the accompanying CD is highly recommended. A log concentration plot for the Tl^{3+} system with Br^- at a concentration of 10^{-2} M, is shown in Figure 7.18. With this plot, the concentrations of all species at any given value of [L] can be easily determined, but that is far from its only virtue. As with the polyprotic acid–base systems, when there are many equilibrium constants acting in a system, exact solutions are complex unless some approximations can be made. The log concentration diagram can provide approximate solutions for many problems, and it can reveal the approximations that can be made in the development of exact solutions. For example, what if you want to know the concentrations of all the species in an 0.01 M solution of $TlBr_2^+$? To find the solution point, it is necessary to write a ligand balance equation. For the example at hand, the ligand balance equation is

$$[ML] + 2[M] = [L] + [ML_3] + 2[ML_4] \qquad 7.40$$

The concentrations of all species with fewer ligands than the starting material (ML_2) are on the left-hand side of the equation, and all those with more ligands than the starting material are on the right. Then each concentration is multiplied by the number of ligands lost or gained relative to the starting material.

To use the ligand balance equation to solve the problem of dissolved ML_2, follow the L line in Figure 7.18 from its highest concentration to its first intersection (ML_4). Since ML_4 is on the same side of the equation as L, switch to the ML_4 line, as it will now have the greater value. The next intersection of the ML_4 line is with ML_3. This too is on the same side of the equation, so follow the ML_3 line to its next intersection (ML_2). This intersection is ignored, as ML_2 is the starting material and therefore not in the ligand balance equation. The ML_3 line next encounters the ML line at p[L] = 5.2. This intersection, then, is the tentative solution point for the problem.

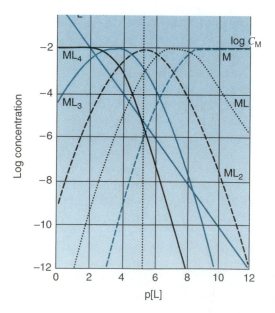

Figure 7.18. The log concentration plot can be used for the graphical solution of complex equilibria. The solution line for the text example is the vertical dotted line. This plot is for the step-wise complexation of Tl^{3+} with Br^-.

Next, look to see whether any of the other species in the proton balance equation have concentrations that are significant relative to $[ML_3]$ and $[ML]$ at the solution point. In this case, we see that $[L]$, $[ML_4]$, and $[M]$ all have values about two orders of magnitude less than $[ML_3]$ and $[ML]$. The values of all the other components at $p[L] = 5.2$ can be read from the plot.

To achieve an exact solution to the problem, we can note from the graphical solution (Figure 7.18) that the only significant species are ML, ML_2, ML_3, and L, and therefore the only relevant equilibrium expressions are those for K'_{f2} and K'_{f3}. We also see that the concentration of ML_2 is not quite 0.01 M. To develop an exact solution to the problem, we can substitute the equilibrium expressions into the abbreviated ligand balance equation:

$$[L] = \frac{[ML_2]}{K'_{f2}[L]} - K'_{f3}[ML_2][L] \qquad\qquad 7.41$$

from which we can obtain

$$[L] = \sqrt{\frac{[ML_2]}{K'_{f2}(1 + K'_{f3}[ML_2])}} \qquad\qquad 7.42$$

Because we know that $[ML_2]$ is somewhat less than 0.01 M, an estimated value of 0.008 M can be used in the above equation to provide $[L] = 5.5 \times 10^{-6}$ M ($p[L] = 5.3$). Because the value of $[ML_2]$ was estimated, the first calculated value of $[L]$ should be used in the equilibrium expressions to obtain a better value for $[ML_2]$. This value can then be used to calculate an improved value for $[L]$, and so on until the iterations converge. In this case, the estimate for $[ML_2]$ gave values for $[ML]$ and $[ML_3]$ of 1.02×10^{-3} M. These confirmed the value of 0.008 M for $[ML_2]$ through the use of the mass balance equation.

PREDICTION OF TITRATION CURVES

Of particular interest to the analyst is the ability to readily assess the practicality of a particular titration. From the titration curves already developed, we can see that with too low an equilibrium constant or too low a concentration, the change in concentration would be too small for an effective determination of the equivalence point. For such an assessment, the logarithmic concentration plot can be very useful. The logarithmic concentration plot for the titration of Ca^{2+} at 10^{-3} M with EDTA is shown next to the titration curve in Figure 7.19. The titration curve with $p[Y]$ as the Y-axis is used because that corresponds to the Y-axis of the rotated log concentration plot. The value for $p[Y]$ is indeterminate at the beginning of the titration because no Y has yet been added to the titration volume. At the half-titrated point ($\theta = 0.5$), the concentrations of M and MY are equal. This occurs at the value of $p[Y]$ corresponding to the system point. The equivalence point is that point where the unreacted M and Y concentrations are equal. Finally, the value of $[Y]$ at $\theta = 2$ is roughly equal to one-third the initial concentration of the analyte. This occurs just 0.5 log units above the intersection of the Y and C_M lines.

From the log plot and the corresponding titration curve shown in Figure 7.19, we can see how the magnitude of the concentration break in the equivalence point region can be estimated. For the current case, the total change in $p[Y]$ from the 50% to 150% titrated points is about 1 log unit less than the difference in $p[Y]$ between the intersections of the M and MY curves and the intersection of the Y and MY curves. On the plot shown, this is 1 less than 10^{-7} or 6 $p[Y]$ units.

The log concentration plot also allows the estimation of the minimum concentration of analyte for which the titration is practical. One simply slides C_m down until the difference between the system point $p[Y]$ and the value of $p[Y]$ at the intersection of Y and MY is too small. Note that this difference decreases only 0.5 log units for every

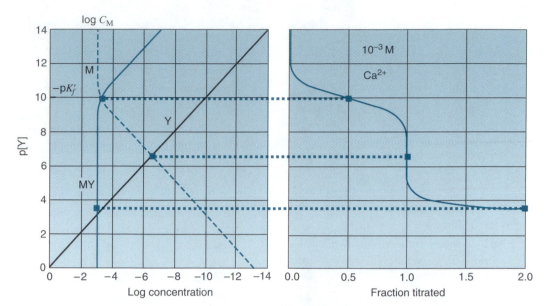

Figure 7.19. Rotating the log concentration diagram 90° counterclockwise reconciles it with the axes of a titration curve. The values of p[L] for the principal points on the titration curve can be located readily on both plots. The 50% point is where the lines for M and ML cross. At the 100% point, the concentrations of L and M are equal. At the 200% point, [Y] approaches C_M.

10-fold decrease in initial concentration, so although the formation constant for the EDTA complex of Ca^{2+} is relatively modest, titrations of micromolar and lower concentrations are possible.

An estimate of the equivalence point break in the analyte can also be obtained from the log plot. This is illustrated in the log plot for the titration of 10^{-6} M Zn^{2+} with EDTA in Figure 7.20. At the beginning (0%), the Zn^{2+} concentration is the initial analyte concentration (10^{-6} M). This concentration is halved at the midpoint when half of the zinc has been complexed. The Y and M concentrations at the equivalence point are seen to be 10^{-10} M. The overall change in the concentration of M from 0% to 200% is seen to be almost eight orders of magnitude—a huge change for such a low concentration of analyte.

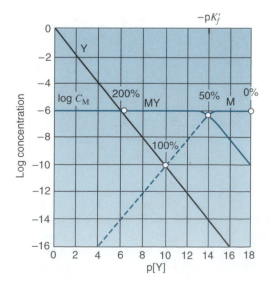

Figure 7.20. On this log concentration plot, the values of p[L] for each of the key points on the titration curve are identified. The effect of C_M and pK_f' on the magnitude of the equivalence point break can be readily seen.

42. Write the ligand balance equation for a solution made by dissolving the complex ML in water.

43. Silver complexes with ethylenediamine in a 1:1 ratio to form $[Ag(en)]^+$ with K_f' of 1.0×10^5. Draw a log concentration diagram for this system. Use 0.010 M as the total concentration of Ag^+ in the complexed and uncomplexed form. You can use the spreadsheet plot for your solution.

44. Using the log concentration diagram from Question 43, estimate the equilibrium concentrations of all species in a 0.010 M solution of $Ag(en)^+$.

45. Write a ligand balance equation for the dissolution of a complex of the form ML_2 in water. The highest order complex formed is ML_4.

46. Estimate the concentrations of all species in a 1.0×10^{-2} M solution of $TlBr^{2+}$ [made by dissolving 0.01 M

Tl$(NO_3)_3$ and 0.01 M KBr in water]. Use the log concentration diagram of Figure 7.18.

47. Aluminum complexes with fluoride by the following stepwise reactions:

$$Al^{3+} + F^- \rightleftharpoons AlF^{2+} \qquad \log K_{f1}' = 7.0$$
$$AlF^{2+} + F^- \rightleftharpoons AlF_2^+ \qquad \log K_{f2}' = 5.6$$
$$AlF_2^+ + F^- \rightleftharpoons AlF_3 \qquad \log K_{f3}' = 4.1$$
$$AlF_3 + F^- \rightleftharpoons AlF_4^- \qquad \log K_{f4}' = 2.4$$

Draw a log concentration diagram for the system and use it to determine the equilibrium concentrations of all the species in a 1.0×10^{-1} M solution of AlF_3.

48. Estimate the equivalence point break for the titration of Co^{2+} with EDTA ($\log K_{MY}' = 16.31$). Also estimate the equivalence point concentrations of Co^{2+} and EDTA. Use 1.0×10^{-6} M as the initial Co^{2+} concentration.

Answers to Study Questions, Section F

42. The list of species present is M, L, and ML. The starting material is ML. The ligand balance equation equates all species less complexed than the starting material with all species more complexed. Thus the ligand balance equation for this case is [M] = [L].

43. The system point is drawn at $p[en] = -pK_f' = 5.0$ and log concentration = log 0.010 = −2. A line representing [en] is drawn based on the identity −log[en] = p[en]. (See the figure.)

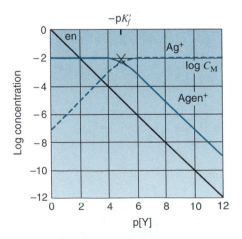

44. We first write a ligand balance equation. All species with fewer ligands than the starting material go on the left side of the equation and all with more go on the right.

$$[Ag^+] = [en]$$

A solution is then found by tracing the Ag^+ line until it crosses that for en. This is the solution point. At this point,

$[en] = [Ag^+] = 10^{-3.5} = 3.2 \times 10^{-4}$ M and $[Ag(en)^+] = 1.0 \times 10^{-2}$ M. Note that since the concentrations of Ag^+ and en are about 3% of 0.010 M, the $[Ag(en)^+]$ could be reestimated as $0.010 - 3.2 \times 10^{-4} = 0.97 \times 10^{-3}$ M.

45. The list of all forms are all the complexes plus L and M. The starting material is ML_2. The ligand balance equation is

$$[ML] + 2[M] = [L] + 2[ML_4] + [ML_3]$$

46. The ligand balance equation for this system is

$$[Tl^{3+}] = [Br^-] + 3[TlBr_4^-] + 2[TlBr_3] + [TlBr_2^+]$$

The line for $[Tl^{3+}]$ is followed until it intersects with one of the other species in the equation. It intersects with $[TlBr_2^+]$ line first, and this is the tentative solution point. All other species (except the starting material $TlBr^{2+}$) have a negligible concentration at this value of $p[Br^-]$ (about 7.1). The concentrations of Tl^{3+} and $TlBr^{2+}$ are about $10^{-3.1}$. At this value of $p[Br^-]$, the concentrations of the other species are $TlBr_3 \approx 10^{-6}$ M and $TlBr_4^- \approx 10^{-10}$ M.

47. The ligand balance equation is written with all species having less ligands than the starting material on one side and all with more on the other.

$$[AlF_2^+] + 2[AlF^{2+}] + 3[Al^{3+}] = [F^-] + [AlF_4^-]$$

The concentrations in the ligand balance equation are multiplied by the number of ligands by which each species differs from the starting material.

Following the line for F^- (L), we first cross the AlF_4^- (ML_4) line. Since this species is on the same side of the equation as F^-, we switch to the AlF_4^- line. The next

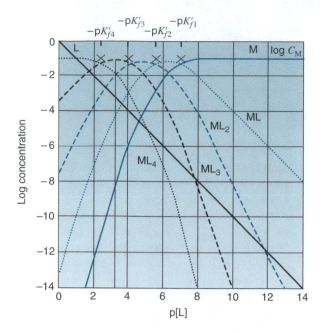

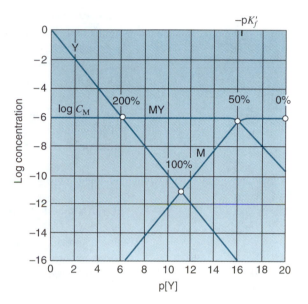

crossing is with the starting material line (not in the equation). Finally, the AlF_4^- line crosses the AlF_2^+ line, and that intersection is the tentative solution point. At the $p[F^-]$ of the solution point (3.3), the concentrations of AlF_2^+ and AlF_4^- are equal at about $10^{-1.8} = 2 \times 10^{-2}$ M. All other concentrations are negligible at this $p[F^-]$ except for $[F^-]$. Its concentration is 5×10^{-4} M or about 3% of the value of $[AlF_2^+]$ and $[AlF_4^-]$. For an approximate solution, $[F^-]$ can be ignored. At the $p[F^-]$ of 3.3, the con-

centrations of the other species are $p[AlF^{2+}] = 4.6$ and $p[Al^{3+}] = 8.5$. The concentration of AlF_3 is $1.0 \times 10^{-1} - 1.6 \times 10^{-2} = 8 \times 10^{-2}$ M.

48. Estimates of the equivalence point break and the concentrations of Co^{2+} and EDTA can be done with a log concentration diagram. From the diagram we see that at the equivalence point (100% titrated) the concentration of EDTA is equal to that of Co^{2+}, about 10^{-11} M. The concentration of Co^{2+} throughout the titration changes from 10^{-6} M to less than 10^{-15} M.

G.   Spot Tests, Test Strips, Flow Injection Analysis, and Immunoassays

Complexation can be a powerful method of identification. For many inorganic complexes the formation and color of the complexes formed can be quite indicative of the reactive species. A significant portion of the classic inorganic qualitative analysis scheme involved complexation reactions. Complexation is often used to form adducts that absorb light where the uncomplexed compound does not. This allows the use of spectrometric measurement for quantitation, detection, and identification. Several of these types of applications are described briefly in this section.

SPOT TESTS

Spot tests are generally procedures for the detection of specific compounds or elements. A plate with a dimple in it to contain the reagents is used. A drop of the test solution is added to a drop of reagent mixture. The color of the resulting solution is observed. Reactions used for spot tests are those that produce a visible color change. Complexation reactions are frequently used for spot tests because a colored complex can often be formed from colorless reagents. The complexing agent 1, 10-phenanthroline is used as a test for ferrous iron (Fe^{2+}). The complex formed has a characteristic orange color. Other reagents for ferrous iron form purple and deep blue purple colored

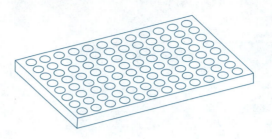

Figure 7.21. A 96-well plate for multiple, simultaneous spot test analysis. Plates with 4 or 8 times as many wells have also been developed. As the number of wells increases, the volume of the reaction mixture gets correspondingly smaller.

complexes. Literally hundreds of spot tests have been devised. Spot tests are a convenient method of screening, they are very handy for field analyses, and they can be applied with a relatively low level of skill.

A modern version of the spot test is staging a literally huge resurgence. Spot plates with 96 small wells, as shown in Figure 7.21, are used. Often the same reagent is placed in all 96 wells, and 96 different samples with different histories are placed in the wells. From the pattern of the reactive response, the effect of the sample history can be observed. Perhaps the samples came from different locations in the field or were synthesized using different parameters. Spectrometers that can quickly obtain the spectrum of all 96 samples have been developed. This method of providing a huge amount of information in a short time is the basis of the new method of discovery called **combinatorial chemistry**.

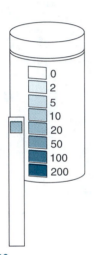

Figure 7.22. A test strip after the test and its container. Comparison of the test square with the color density chart allows a semiquantitative assessment of the analyte concentration.

TEST STRIPS

A variation on the spot test is the test strip. Test strips are like the familiar litmus paper or pH test strips. A narrow band of porous material contains the required test reagents in a dry form, as shown in Figure 7.22. When immersed in the sample, the porous material absorbs a controlled amount of sample, which then reacts with the reagents. Complexation reactions are often used with test strips, where a colored complex indicates the presence of the analyte. In some cases, the density of the color can be used as a rough indication of the amount of analyte present. A key relating color density and concentration is printed on the package containing the test strips. There are test strips for a variety of water quality tests. Complexation reactions are used for most of the tests for metal ions. Complexes with bicinchonic acid and with porphyrin form the basis for tests for copper ions. Test strips for Fe^{2+}, Ni^{2+}, Ag^+, and Co^{2+} are also available.

FLOW INJECTION ANALYSIS

The basic concepts of flow injection analysis were introduced in Section D of Chapter 5. In that discussion, it was assumed that the sample would provide the light absorption on reaching the detector. This method of analysis can be greatly extended by reacting unabsorbing analytes with a reagent that will provide a light absorbing product. Complexation reactions are often used for this purpose. An example is an analysis of Cl^- by FIA.[1] The solvent can be premixed with the required reagents, which are $Hg(SCN)_2$, $Fe(NO_3)_3$, and HNO_3. When the sample in the sample loop is injected into the flow stream, the chloride reacts with the $Hg(SCN)_2$ to displace the SCN^- according to the reaction

$$Hg(SCN)_2 + 2Cl^- \rightleftharpoons HgCl_2 + 2SCN^-$$

The SCN^- in turn, is now free to react with the Fe^{3+} to form the red colored complex $Fe(SCN)_2^+$. This example is interesting from the standpoint of the relative formation

---

[1]G. D. Christian, *Analytical Chemistry*, 5th Ed, p. 760. John Wiley and Sons, New York, 1994.

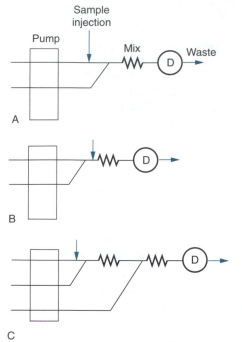

Figure 7.23. Various arrangements of the modules in flow injection systems. In A, a reagent is added after the sample is injected into the carrier solvent. A mixing coil is introduced to allow the sample and reagent to react. The circled D is the detector. In B, a solvent and one or more reagents are mixed prior to injection of the sample. In C, the sample is allowed to mix with the first reagent before introduction of another.

constants of the chloride and thiocyanate complexes of mercuric ion and ferric ion. The formation constant for the ferric thiocyanate complex is much smaller than that for the mercuric thiocyanate complex, which keeps the cyanate complexed with the mercuric ion in the original reagent solution. The addition of chloride ion upsets this equilibrium by the formation of some $HgCl_2$. The thiocyanate released then reacts with the ferric ion, which is present in large excess.

When reactions are used with FIA, it is customary to carry out the reactions in the FIA tubing. This is accomplished by adding more components to the flow system, as shown in Figure 7.23. The symbols used in Figure 7.23 for the FIA components allow the arrangement of each particular system to be described in a readily sketched diagram. For the chloride analysis just described, all the reagents were premixed into the solvent, so the system is that of Figure 7.23A without the lower reagent input. Sometimes, the particular mixture of reagents required for an analysis is not stable enough for long-term storage. When this is the case, the components are mixed in the system. The mixing section is a coil of tubing in which the difference in solution flow rate between the inner coil wall and the outer coil wall provides the mixing. Sometimes the mixing coil is immersed in a temperature-controlled water bath, thus providing control over both the time and temperature of the reaction between mixing and detection.

IMMUNOASSAYS

Immunoassays are an increasingly important part of the analyst's repertoire. The basis of the immunoassay is the formation of a complex between an antibody and an antigen. These complexes are often very highly specific, and immunoassay techniques have been the means to obtain detection and quantitation in systems previously deemed hopelessly complex. The particular chemistry behind and the applications of antibody–antigen interactions are discussed in greater depth in Chapter 12. In that discussion you will find that antibody–antigen reactions can be treated as simple complexation reactions and that all the principles gained in this chapter apply to immunoassay analysis.

49. How do complexation reactions play a role in spot test and test strip analysis?

50. How are the results of the test interpreted in spot tests and with test strips?

51. How would the use of a spectrometer in conjunction with the well plate improve the selectivity and precision of the analysis over visual inspection?

52. Co^{2+} and Ni^{2+} both form complexes with 1-(2-pyridylazo)-2-naphthol (PAN). The cobalt complex has a strong, almost constant absorptivity over the wavelength range from 540 to 640 nm. The nickel complex has a peak absorptivity at 560 mn and no absorptivity above 640 nm. Suggest an FIA method for the analysis of Co^{2+} and Ni^{2+}.

Answers to Study Questions, Section G

49. Complexation reactions are used to form colored complexes when the analyte itself is not colored. The selectivity of the complex forming reaction has an effect on the number of possible interferent species.

50. The results of test strip and spot test analyses are interpreted visually. A change in the color of the test area is an indication of the presence of the sought analyte. The color density can often be used to roughly estimate the analyte concentration. This semiquantitative application requires a table of color density versus analyte concentration to use for comparison.

51. The spectrometer provides more quantitative information. Using the absorbance at specific wavelengths may reduce the number of interfering compounds. The measurement

of light transmittance could provide a more precise measure of color density than the eyeball comparison with the printed chart. The use of the whole spectrum could enable the detection and quantitation of several analytes at once.

52. Use the FIA module arrangement shown in Figure 7.23B. One input to the pump has solvent buffered for the optimum pH for the formation of the complexes and the reduction of interferences. The second input contains a dilute solution of PAN. After mixing, the absorbance of the stream is monitored at both 560 and 640 nm. A working curve for the cobalt can be obtained from the absorbance at 640 nm directly. The absorbance due to the cobalt at 560 nm should be subtracted from the total absorbance to obtain the working curve for nickel.

Practice Questions and Problems

1. Define the following terms:
   A. Coordination center
   B. Ligand
   C. Chelating agent
   D. Hydrolysis
   E. Stepwise complex formation
   F. Formal formation constant
   G. Effective formation constant
   H. Complex

2. A. Explain why stepwise complexation systems are generally unsuitable for analysis by titration.

   B. Sometimes complexation reactions are used to enable an analysis by spectrometry when the complex absorbs light but the uncomplexed species does not. Do you think a stepwise complexation reaction is more suitable for this method of analysis? Why or why not?

3. A. Draw a log concentration plot for the reaction of Ca^{2+} with EDTA complexing agent in a solution buffered at $p[H_3O^+] = 10$. A 1:1 complex is formed. Use 0.001 M

for the total concentration of calcium. The effective formation constant at this $p[H_3O^+]$ is 10^{10}.

   B. Use the log concentration plot constructed in A to determine the equilibrium concentrations that exist in solution when 1×10^{-4} mol of $Ca(EDTA)^{2+}$ is dissolved in 100 mL of water buffered at $p[H_3O^+]$. Be sure to show your ligand balance equation.

   C. Use the log concentration diagram to determine the concentration of EDTA when 0.001 M Ca^{2+} is titrated with an EDTA solution of the same concentration. Find the EDTA concentration at the 50%, 100%, and 200% titration points. Explain your rationale for choosing each point.

4. Ethylenediamine ($H_2NCH_2CH_2NH_2$ is the most basic form) has pK_a' values of 6.8 and 9.9. What is the $p[H_3O^+]$ above which the ethylenediamine is essentially completely unprotonated? (Consider using an acid–base α plot or log plot to get your answer.)

5. A. Why does the indicator Eriochrome Black T not work for EDTA titrations carried out in solutions buffered at $p[H_3O^+]$'s below 6.3?

B. Why do you think so many of the metal ions that can be complexed react with OH^- and so many of the complexing agents react with H^+?

C. Why is the presence of Mg^{2+} important in the titration of Ca^{2+}? If you used an EDTA-sensitive electrode to follow the titration, would the Mg^{2+} be needed?

6. A. Calculate $K'_{f\,eff}$ for the complexation of Co^{2+} with EDTA in water buffered at $p[H_3O^+] = 4.5$. Use 2.0×10^{16} for $K'_{fCoY^{2-}}$. For EDTA, $pK'_{a1} = 1.99$, $pK'_{a2} = 2.67$, $pK'_{a3} = 6.16$, and $pK'_{a4} = 10.26$. Assume that cobalt doesn't react significantly with OH^- at this $p[H_3O^+]$.

B. Would it be possible to perform a titration of Mg^{2+} with EDTA at this $p[H_3O^+]$? ($K'_{MgY^{2-}\,eff} = 4.9 \times 10^8$.)

7. A. Draw a log concentration diagram for 10^{-2} M Co^{2+} and EDTA at $p[H_3O^+] = 4.5$. (Hint: Don't forget to use $K'_{fCoY^{2-}\,eff}$ when finding the system point.)

B. What are the equilibrium concentrations of EDTA (Y^{4-}) and Co^{2+} if 10^{-2} mol of CoY^{2-} is dissolved in one liter of water?

8. From the log concentration diagram of Figure 7.18, what is the $p[Br^-]$ for a solution that is

A. 10^{-2} M in $TlBr_3$?

B. 10^{-2} M in $TlBr^{2+}$?

9. In an analysis, 10 mL of 0.5 M standard EDTA solution is added to 35 mL of a solution containing an unknown concentration of Ba^{2+}. The unreacted EDTA is then titrated with 20 mL of standard 0.1 M Mg^{2+} solution. What is the concentration of Ba^{2+} in the unknown solution?

10. Iron III (ferric ion) forms the complexes $Fe(CN)_6^{3-}$, $Fe(en)_3$ and FeY^- where CN^- is cyanide ion, en is ethylenediamine, and Y^{4-} is EDTA.

A. Explain why the number of ligands is different in each of these complexes.

B. What is the coordination number of Fe^{3+}?

C. Tell which one of these complexes would be better for a complexometric titration and why.

D. Write the formula formation constant expression for the $Fe(CN)_6^{3-}$ complex.

E. Obtain a relationship between the formal and thermodynamic formation constants for the complex $Fe(CN)_6^{3-}$ based on the DHLL. Does the formal formation constant increase or decrease with increasing ionic strength?

11. A. Use the spreadsheet programs to draw the log concentration diagram for a solution that is 0.01 M in Zn^{2+} and is complexed with asparagine. (In the table, the lowest K_1, K_2, and K_3 refer to the Zn complex.)

B. From the plot, give the concentrations of all the species present with an asparagine concentration of 1×10^{-2} M.

C. If the amount of the complex solution is 100 mL, how many moles of asparagines would have to be added to make the solution in part B above?

D. At what concentrations of asparagine are the concentrations of the $Cu(as)$ and $Cu(as)_2$ forms a maximum? (as = asparagine)

12. Fe^{3+} forms a mono complex with asparagine with a formation constant of $1 \times 10^{8.6}$. Will the pH of the solution have an affect on the amount of complex formed? If so, estimate the pH value for which the complex formation will be a maximum.

13. A 0.001 M solution of cupric ion (Cu^{2+}) is titrated with EDTA at a pH for which the effective formation constant is 1×10^{11}. Give the Cu^{2+} and EDTA concentrations for the solutions that have been 0%, 50%, 100%, and 200% titrated. Would eriochrome black T be a good indicator for this titration? Why or why not?

14. Iodide forms a series of complexes with cadmium ion from CdI^+ to CdI_4^{2-}.

A. Use the spreadsheet program to obtain an alpha plot for the cadmium–iodide (Cd–I) system. What is the effect of the closeness of the four complex formation constants?

B. From the log concentration plot for the Cd–I system, determine the concentrations of all the complexes, Cd^{2+}, and I^- in a solution of for which the analytical concentration of CdI_2 is 0.01 M.

D. At what concentration of I^- does the complexation of the Cd^{2+} begin to be significant?

C. Comment on the usefulness of this complexation reaction as a basis for the quantitative analysis of Cd^{2+}.

15. What is the concentration of untitrated equivalence point in an EDTA titration of a mixture of Ca^{2+} and Mg^{2+} in which their initial concentrations were approximately 10^{-3} and 10^{-4} M, respectively? Use 10^{10} and 10^8 for the conditional formation constants of the Ca^{2+} and Mg^{2+} complexes with EDTA and neglect the dilution due to the addition of the titrant.

16. Suppose 20.00 mL of a solution containing both Fe^{2+} and Fe^{3+} ions was titrated with EDTA at pH = 2. The volume of 0.0100 M titrant required to reach the equivalence point was 17.52 mL. When this same titration was performed at pH = 6, the volume of EDTA required to reach the equivalence point was 25.18 mL. What were the concentrations of Fe^{2+} and Fe^{3+} in the original solution?

17. Referring to Figure 7.13, comment on the potential effectiveness of determining aluminum by reaction with excess EDTA and back titration with Mg^{2+} to the eriochrome black T end point.

18. According to Figure 7.20, the magnitude of the break in $p[Y]$ between 50% and 200% titrated is roughly equal to the difference between $-pK'f$ and pCM. For the metal ions

given in Figure 7.13, what is the minimum concentration for an EDTA titration performed at pH 9 for which the break between 50% and 200% will be 3?

19. A. The end point of a complexation titration was followed with light absorption at 478 nm. At that wavelength, the molar absorptivities of the metal, ligand, and complex are 540, 19300, and 83020, respectively. Sketch the shape of the titration curve expected for the titration of the metal with the complexing agent with which it forms a 1:1 complex.

   B. What is the effect of the formation constant of the complex on the shape of the titration curve?

20. A scheme for the flow injection analysis of Cl^- is outlined in the section of FIA at the end of this chapter. In this scheme, a competitive complexation reaction is used to convert the Cl^- into an equivalent amount of thiocyanate ion. The resulting concentration of thiocyanate ion is determined by the absorption of the ferric thiocyanate complex. In such a two step analysis, the differentiating characteristic for each step can be quite different. Analyze this method of Cl^- analysis from the standpoint of the differentiating characteristic(s), probe(s), response(s), measurement(s), and interpretation(s).

Suggested Related Experiments

1. Water hardness determination with EBT visual end point.
2. Water hardness determination with EBT using spectrometer to show the relative concentration of the EBT forms.
3. FIA analysis involving complexation and spectrometric detection.

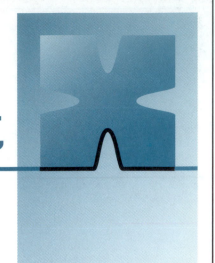

Chapter Eight

ANALYSIS BY PRECIPITATION REACTIVITY

One of the earliest differentiating characteristics to be employed in chemical analysis was chemical reaction to form a precipitate. The consequence of this reaction is an immediately visible product. The formation of a separate phase provides a convenient way to separate the reaction product from the reaction mixture. The ability to form a precipitate with particular reagents may aid in the identification of a species. The nature of the solid material formed (color, crystalline conformation, etc.) may contribute further identification information. Quantitative analysis can be performed with a precipitation reaction by carrying out the reaction as a titration or by adding an excess of precipitating agent (so that the analyte is the limiting reactant), separating the solid product, and determining its quantity by weight. The factors that influence the choice among these options and the procedures that have been developed for their implementation are best understood after a closer look at the precipitation process itself.

A. Precipitation Reactions

A precipitation reaction is one in which a larger amount of product is formed than is soluble in the reaction mixture. This is most readily achieved if the reaction product has a very low solubility in the solvent used. When the solubility limit for a species is exceeded, the material separates from the solution, generally in the solid phase. A specific and familiar example is the reaction between silver ions and chloride ions to form the solid, white precipitate AgCl.

$$Ag^+_{(aq)} + Cl^-_{(aq)} \rightleftharpoons AgCl_{(s)}$$

The general form of a precipitation reaction for ionic reactants is

$$a\text{M}^{m+}_{(aq)} + b\text{X}^{x-}_{(aq)} \rightleftharpoons \text{M}_a\text{X}_{b(s)} \qquad 8.1$$

in which the subscripts indicate the aqueous and solid phases.

Precipitation reactions can be thought of as complexation reactions for which the neutral (uncharged) product has limited solubility. Like complexation products, a precipitate dissociates to form the original reactants.

The precipitate formed must be a neutral species because the solid material formed is essentially neutral. The bonds between the reactants in the product may be covalent or ionic, but the product molecule is always uncharged. For this to be true, $am = bx$.

Many of the species involved in precipitation reactions are familiar from our study of complexation reactions in that most of them have a coordination center and a number of ligands. Another similarity is that the dissociation products are the original reactants rather than some other combination of the atoms. Many precipitates are just the neutral species of limited solubility in a stepwise complex formation. In such cases, the precipitating (also complexing) agent can enable the formation of the next higher complex, which is likely to be ionic and thus dissolved. This is the case of the precipitant reversing its role when too much of it is used.

To carry out a precipitation reaction, one of the reactants is dissolved in a solution, and the other reactant is slowly added (generally with stirring) as the precipitate forms. The amount of precipitate increases with increasing addition of reagent until all of the original reactant is precipitated and an excess of precipitating reagent is present. As we will see, the way in which this reaction is carried out can have a great effect on its usefulness in chemical analysis.

THE EQUILIBRIUM EXPRESSIONS

The precipitation reaction equations given earlier were written with bidirectional arrows to indicate that the reactants and products are in equilibrium. From the reaction, we see that if the solid precipitate is placed in pure solvent, some of it will dissociate to form ions in solution. Furthermore, this process will continue until the ionic concentrations are sufficient to make the rates of precipitation and dissociation equal. For many solids formed from ionic reactants, this is the principal mode of solubility. By convention, the precipitation reaction is usually written in the direction of dissociation rather than precipitation. Thus,

The equilibrium constant for a precipitation reaction is written for the reaction of the precipitate dissolving, not forming.

$$\text{M}_a\text{X}_{b(s)} \rightleftharpoons a\text{M}^{m+}_{(aq)} + b\text{X}^{x-}_{(aq)} \qquad 8.2$$

for which the equilibrium constant equation is

$$K^{\circ}_{sp} = \frac{a^a_{\text{M}^{m+}} a^b_{\text{X}^{x-}}}{a_{\text{M}_a\text{X}_b}} = a^a_{\text{M}} a^b_{\text{X}} \qquad 8.3$$

K°_{sp} is the thermodynamic (activity) equilibrium constant. K'_{sp} is the formal equilibrium constant expressed in terms of concentration. The value of K'_{sp} will change with the solution ionic strength.

In the second expression, the activity of the solid precipitate is taken to be 1 since, if pure, it is in the standard state for a solid substance. The formal equilibrium constant, written in terms of the concentrations, is

$$K'_{sp} = [\text{M}^{m+}]^a [\text{X}^{x-}]^b \qquad 8.4$$

The relationship between K°_{sp} and K'_{sp} is a function of the ionic strength S and can be determined for the range of S over which the Debye–Hückel equation is valid. For the case where the DHLL (Equation 4.23) is adequate,

$$K^{\circ}_{sp} = a^a_{\text{M}} a^b_{\text{X}} = \gamma^a_{\text{M}}[\text{M}^{m+}]^a \, \gamma^b_{\text{X}}[\text{X}^{x-}]^b = \gamma^a_{\text{M}}\gamma^b_{\text{X}}K'_{sp}$$

$$\log K^{\circ}_{sp} = a \log \gamma_{\text{M}} + b \log \gamma_{\text{X}} + \log K'_{sp}$$

$$\log K'_{sp} = \log K^{\circ}_{sp} + aAm^2 \sqrt{S} + bAx^2 \sqrt{S} \qquad 8.5$$

$$\log K'_{sp} = \log K^{\circ}_{sp} + (am^2 + bx^2) A \sqrt{S}$$

For the case of the precipitate AgCl, the thermodynamic K_{sp}° is 1.8×10^{-10}. When only the AgCl and its dissociation products Ag^+ and Cl^- are present, the ions have an equal concentration of 1.3×10^{-5} M. The constants a, m, b, and x are all 1. The value of A in water at 25 °C is 0.5091. The difference between the logs of the formal and thermodynamic solubility products is only 0.0037 log units. However, if there are other salts present such that S is 0.1, K_{sp}' will be larger than K_{sp}° by approximately 0.5 log units.

The subscript sp in the term K_{sp}' is an abbreviation of "solubility product." This constant is called the **solubility product** because if s moles of M_aX_b dissociate in a liter of solution, the concentrations of the ionic products will be as and bs, and

$$K_{sp}' = (as)^a \, (bs)^b \qquad\qquad 8.6$$

From this expression, K_{sp}' is seen to be the "product" of the solubilities.

This limited sense of solubility is not strictly true since insoluble species can dissolve without dissociation. Virtually all substances will dissolve to some extent as the intact molecule. This solubility, which is called the **intrinsic solubility**, is indeed the principal mode of solubility for substances not formed from ionic reactants. The equilibrium constant for the intrinsic solubility of substance M_aX_b is

$$M_aX_{b(s)} \rightleftharpoons M_aX_{b(aq)} \qquad\qquad 8.7$$

for which the formal equilibrium constant is

$$K_{int}' = [\, M_aX_{b(aq)}] \qquad\qquad 8.8$$

This relationship predicts that the equilibrium concentration of the intact molecule is a constant as long as any solid molecules exist in the system. For most precipitates formed of ionic reactants, the intrinsic solubility is less than the ionic solubility. However, there are situations in which the contribution of the intrinsic solubility to the overall solubility can be significant.

THE CRYSTALLIZATION OF PRECIPITATES

Two processes are involved in precipitate formation: **nucleation** and **crystallization**. Because it is easier to describe the process of crystallization, we will begin with it. Crystals of the precipitate A_nX_m have already formed and additional precipitating reagent (X) is being added to the solution. There is still an excess of M in the solution, so the surface of the crystal has more of its "M" sites occupied than its "X" sites. When some of the added X comes to the crystal surface, the lattice for that layer is complete, and more M comes to fill its sites on the next layer. In this way, the added X reacts with the M already at the surface. Notice that both X and M must diffuse to the surface in order to react. This kind of reaction is called a **heterogeneous reaction** (taking place not uniformly throughout the solution, but selectively at the interface between the two phases).

The crystallization process is very dynamic. There is always a tension between the tendency to crystallize and the tendency to dissolve. Therefore, M and X entities will be leaving the surface of the crystal and other M and X entities in solution will continue the formation of the crystal lattice. Two effects can occur during the crystallization process that can affect the composition of the product and its analytical usefulness. They are occlusion and colloid formation.

Occlusion can occur if a second species, N, is present in the solution, where N somewhat resembles M (or X) in size, charge, and conformation. Even if the reaction between N and X would not normally produce an insoluble product under these conditions, it is possible for an N to temporarily take the place of an M on the

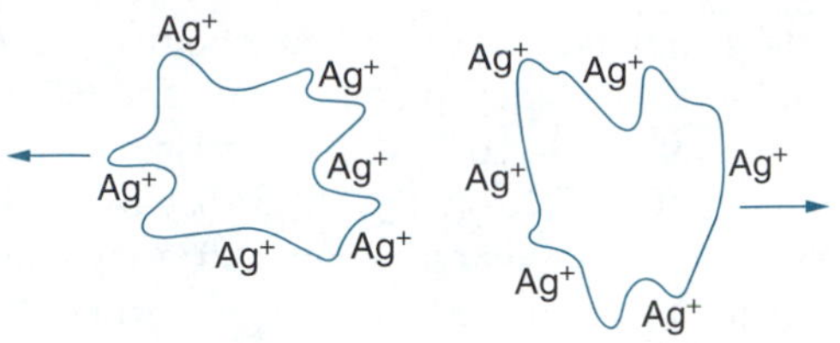

Figure 8.1. The excess positive charge on the particles causes mutual repulsion.

crystal surface and then, under conditions of rapid growth, become trapped there. An example is found in the precipitation of Ba^{2+} with SO_4^{2-}. If there is also some Ca^{2+} in the solution, the resulting precipitate of $BaSO_4$ will have a small $CaSO_4$ impurity. This impurity affects quantitation because the weight of the precipitate formed was not limited by the amount of Ba^{2+} present and because the amount of SO_4^{2-} required to form the precipitate exceeds the stoichiometric amount required by the Ba^{2+}. Occlusion can be reduced by avoiding a too-rapid precipitate growth (by slow addition of dilute reagent and stirring) and by **aging** the precipitate before separation. Aging uses time and the dynamic nature of the crystallization process (often with the help of mild heating) to improve the purity of the precipitate. As the crystal continually dissolves and re-forms, it is increasingly less likely to include disparate species.

Colloid formation comes from the possibility of excess electrical charge on the crystals. If M or X are ionic species, the surface of the crystal, as it is forming, will be charged. For example, in the case of Ag^+ being precipitated by addition of Cl^-, the AgCl crystal will have a positive surface charge because of the excess of Ag^+ taking their place in the lattice before the corresponding Cl^- reagent has arrived. (See Figure 8.1.) This is because, at this point in the precipitate reaction, the concentration of Ag^+ in the solution is higher than that of Cl^-. As more of the Ag^+ in the solution reacts with added Cl^-, the $[Ag^+]$ decreases, and the $[Cl^-]$ increases. This has the effect of also decreasing the amount of excess charge on the crystal surface. When the unreacted Cl^- and Ag^+ concentrations become nearly equal, the surface charge can become zero. If we keep adding Cl^-, this state is short-lived: with an excess of Cl^-, the Cl^- now occupies surface sites and the crystal surface becomes negatively charged. (See Figure 8.2.)

The problem with having charged crystals is that they are mutually repulsive. This interferes with the process of **coagulation**, the joining of small crystals to form larger ones. Instead, the crystals remain very small and can be held in suspension by the charge repulsion effect. This makes it difficult to separate them from the reaction mixture. Repulsion can be reduced by the addition of a salt, not involved in the precipitation reaction, to the reaction mixture. The presence of the salt makes the solution more electrically conductive, which reduces the distance over which the surface charge repulsion effect can occur. (See Figure 8.3.) The added salt does not reduce the *amount* of the surface charge; rather, it just compacts its influence to a degree that allows coagulation to occur.

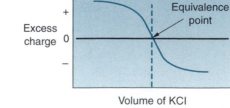

Figure 8.2. At the equivalence point, the particle surface has no excess charge. Beyond the equivalence point, the sign of the excess charge is reversed.

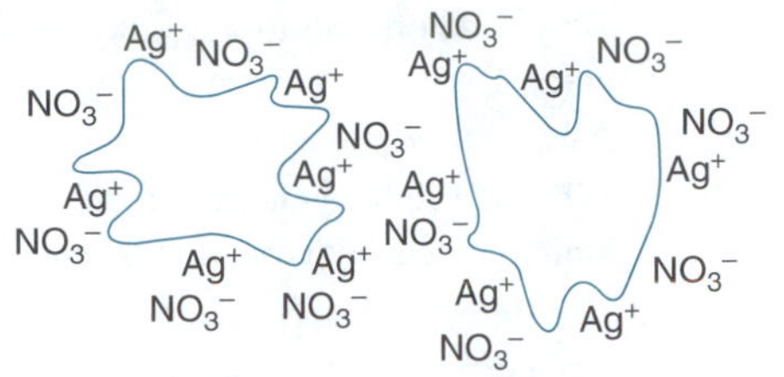

Figure 8.3. The repulsive distance is reduced by the counterions from an added inert salt.

THE NUCLEATION OF PRECIPITATES

The question of how the first crystals of precipitate form (nucleate) is very interesting. If one extrapolates back from the process of crystal growth, one would encounter crystals of very small dimensions and, eventually, very few atoms. Aggregates of relatively small numbers of species are called **clusters**. How does a cluster form, and at what point does a cluster begin to act more like a crystal? One thing we know from experiment is that the solubility of crystals increases significantly as the crystal size decreases below 10^{-3} mm in diameter. Thus, during early growth, a higher concentration of precipitation reactants is required than will eventually be in equilibrium with the larger crystals. The high local concentration of precipitant in the region where the reagent is added to the solution can provide this excess and create the conditions required for nucleation. In some cases, nucleation is difficult to achieve spontaneously in solution, and some "seed" crystals or other catalytic surface are required.

In precipitation separations, it is desirable to reduce the amount of a precipitate that is in the form of very small crystals. Given time, and the avoidance of colloidal suspensions, this will happen naturally. The smaller crystals, having a higher solubility, will dissolve in the solution that is in equilibrium with the larger crystals. New nucleation is unlikely to occur under such conditions, so the material that dissolves from the small crystals will add to the size of the larger ones already formed. When this process is completed, the system will finally have come to equilibrium. In most cases, equilibrium concentration is reached in less than a minute. Full stabilization of the crystal size, however, can take several hours.

Because nucleation occurs in the regions of temporary excess precipitant that results from dropwise addition of the reagent, nucleation can be avoided by increasing the reagent concentration uniformly throughout the solution. This procedure is called **homogeneous precipitation**. An example is the use of sulfamic acid to homogeneously generate SO_4^{2-}. Sulfamic acid is somewhat unstable in water and slowly undergoes the following reaction.

In homogeneous precipitation, the precipitation reaction is still necessarily heterogeneous. The precipitating reagent is formed by a homogeneous reaction to avoid local excess concentrations.

$$H_3NSO_3 + 2H_2O \rightarrow NH_4^+ + H_3O^+ + SO_4^{2-}$$

When the solution is heated to promote the reaction, the sulfate ion is slowly released throughout the solution. This technique is used to precipitate Ca^{2+}, Ba^{2+}, Sr^{2+}, or Pb^{2+}. The advantages of homogeneous precipitation stem from the lack of local excess concentrations, which decreases the generation of small crystals and the slow crystal growth, which in turn reduces occlusion.

Study Questions, Section A

1. What is a precipitation reaction?
2. Write the equilibrium reaction and the formal equilibrium constant expression for the dissolving of the slightly soluble salt $La_2(CO_3)_3$.
3. From Equations 8.5, explain why K'_{sp} is always larger than $K°_{sp}$.
4. Given that $K°_{sp}$ for $CaSO_4 = 2.4 \times 10^{-5}$ at 25 °C, find the value for K'_{sp} of $CaSO_4$ in pure water.

5. What is the intrinsic solubility of a precipitate?

6. Why is a precipitation reaction called a heterogeneous reaction?

7. How does aging relieve some of the impurities due to occlusion?

8. Why do colloids form during precipitation reactions, and how can they be induced to crystallize into larger particles?

9. Why is a concentration of the precipitating reagents in excess of the K'_{sp} product generally necessary to begin the precipitation process?

10. Why is excessive nucleation difficult to avoid when the precipitating reagent is added to the reaction solution dropwise?

11. The term "homogeneous precipitation" sounds like an oxymoron. What is meant by this expression?

Answers to Study Questions, Section A

1. A precipitation reaction is one in which the product has limited solubility.

2.
$$La_2(CO_3)_3 \rightleftharpoons 2La^{3+} + 3CO_3^{2-}$$
$$K'_{sp} = [La^{3+}]^2[CO_3^{2-}]^3$$

3. The solubility product reaction is always written to form ionic products from an ionic reactant. When the DHLL is used to find the relationship between K'_{sp} and $K°_{sp}$, we see that the sign of the difference between them is always positive. Thus, K'_{sp} is always larger than $K°_{sp}$.

4. Before applying Equation 8.5, we must find S, the ionic strength of a solution saturated in $CaSO_4$. By Equation 3.27,

$$S = \frac{1}{2}\sum_i z_i^2 C_i$$

where z is the charge on the ion and C is its concentration. To find the concentrations of the species, we use Equation 8.4:
$$K'_{sp} = [M^{m+}]^a [X^{x-}]^b$$

The reaction for the dissolution of $CaSO_4$ is

$$CaSO_{4(s)} \rightleftharpoons Ca^{2+}_{(aq)} + SO_4^{2-}_{(aq)}$$

So, $[Ca^{2+}] = [SO_4^{2-}]$, and the equilibrium expression is written as

$$K'_{sp} = [Ca^{2+}]^1[SO_4^{2-}]^1 = [Ca^{2+}]^2$$
$$[Ca^{2+}] = \sqrt{K'_{sp}} \approx \sqrt{2.4 \times 10^{-5}} = 4.9 \times 10^{-3}\ M$$

(Note that some error has been introduced here since we used $K°_{sp}$ instead of K'_{sp}.) Now an approximate value for S can be calculated:

$$S = \frac{1}{2}([Ca^{2+}] \cdot 2^2 + [SO_4^{2-}] \cdot 2^2)$$
$$= \frac{1}{2}(4.9 \times 10^{-3} \cdot 4 + 4.9 \times 10^{-3} \cdot 4)$$
$$= 2.0 \times 10^{-2}$$

From Equation 8.5,

$$\log K'_{sp} = \log K°_{sp} + (am^2 + bx^2)A\sqrt{S}$$
$$= \log 2.4 \times 10^{-5} + (1 \cdot 2^2 + 1 \cdot 2^2)$$
$$\times 0.5091\sqrt{2.0 \times 10^{-2}} = -4.1$$
$$K'_{sp} = 8.9 \times 10^{-5}$$

To eliminate the error introduced in the step where we calculated the concentration using $K°_{sp}$ instead of K'_{sp}, we could use this newly calculated value of K'_{sp} instead and redo the calculations. (This is a successive approximations technique for determining the true answer.) The result of this process, after two iterations, is $K'_{sp} = 1.9 \times 10^{-4}$. The reasons for the large difference between $K°_{sp}$ and K'_{sp} are the relatively large solubility of $CaSO_4$ and the double charge on the dissolved ions.

5. Intrinsic solubility is the solubility of the neutral precipitate. This solubility is in addition to the solubility that attends dissociation into ions.

6. A precipitation reaction is heterogeneous because it occurs on the surface of the precipitate. The ionic reactants in the solution have to move to the precipitate surface to precipitate.

7. In the process of aging, the precipitate is constantly dissolving and re-forming owing to the dynamic equilibrium of the precipitation reaction. This is a slow process. Therefore, occluded ions that would not normally be precipitated are released, and they are much less likely to get trapped in crystals as they grow.

8. Precipitates may become colloids from the charge carried by the surface ions that are in excess. Colloids may be helped to coagulate by raising the ionic strength of the solution (i.e., by adding an inert salt).

9. A local excess of reactant concentration is required to initiate nucleation and form a small crystal. This is because the solubility of precipitates below 10^{-3} mm in diameter is greater than that of larger particles. Thus, to form the initial small particles, the final equilibrium concentrations must be exceeded.

10. The solution immediately around the site where the drop entered has a very large excess of precipitating agent. Many precipitate nuclei can form in this region before the stirring dissipates the local excess concentration.

11. In homogeneous precipitation, the precipitating reagent is generated by a homogeneous reaction throughout the solution. This means of introducing the precipitating reagent avoids the local excesses of reagent that result in excessive nucleation and occlusion. The precipitation reaction itself is still a heterogeneous reaction.

B. Equilibrium Concentrations

The bases for the determination of equilibrium concentrations in precipitation systems are the solubility product and intrinsic solubility equilibrium constants introduced in the previous section. Consider again the general reaction

$$M_aX_{b(s)} \rightleftharpoons aM^{m+}_{(aq)} + bX^{x-}_{(aq)} \qquad \text{8.2(repeated)}$$

for which the equilibrium expressions are

$$K'_{sp} = [M^{m+}]^a[X^{x-}]^b \qquad \text{8.4(repeated)}$$

and

$$K'_{int} = [M_aX_{b(aq)}] \qquad \text{8.8(repeated)}$$

These equilibrium expressions assume that there is pure solid precipitate in the system.

SIMPLE SOLUBILITY CALCULATIONS

If the solid M_aX_b is placed in pure water, it will dissociate to some extent according to Equation 8.2. The concentrations of the resulting dissociated ions in solution are related to the ionic solubility s, in moles per liter, of the solid. If there is no other source of M^{m+} or X^{x-}, their concentrations in solution must be $[M^{m+}] = as$ and $[X^{x-}] = bs$. When these values are substituted in Equation 8.4, we get

$$K'_{sp} = (as)^a(bs)^b = a^ab^bs^{(a+b)} \qquad \text{8.9}$$

from which

$$s = \left(\frac{K'_{sp}}{a^ab^b}\right)^{1/(a+b)} \qquad \text{8.10}$$

The total solubility is the sum of the ionic and intrinsic solubilities. For most ionic precipitates, the intrinsic solubility is negligible (see Examples 8.1 and 8.2).

If there is another source for either of the ionic reactants that form the precipitate, Equation 8.10 is not valid. As we will see, it is also not valid if the ionic reactants undergo any other reactions in the system.

The solubility of a precipitate is normally suppressed by the presence of an excess of either reactant. This is called the **common ion effect**. If C_M or C_X is the concentration of the excess reactant, then, in general,

$$K'_{sp} = (C_M + as)^a (bs)^b \qquad \text{or} \qquad K'_{sp} = (as)^a (C_X + bs)^b \qquad \text{8.11}$$

To simplify the solution to this equation, first assume that as or bs is negligible with respect to C_M or C_X. If this proves not to be true, then use successive approximations or the exact formula to calculate the ionic solubility (see Example 8.3).

EFFECT OF HYDROLYSIS ON SOLUBILITY

If the ions formed by the dissociation of a precipitate can undergo a hydrolysis reaction with the solvent, some of them will become protonated or complexed with OH^-. These species do not appear in the solubility product equilibrium expression, so the precipitate will continue to dissolve until the concentrations of M^{m+} and X^{x-} satisfy Equation 8.4. This can greatly increase the effective solubility. The resulting solubility can be calculated by determining the fraction of the M^{m+} or X^{x-} that remains unhydrolyzed.

Example 8.1

The ionic solubility of AgBr, for which $K'_{sp} = 5.2 \times 10^{-13}$, is $s = (5.2 \times 10^{-13})^{1/2} = 7.2 \times 10^{-7}$ M. For nickel arsenate, $Ni_3(AsO_4)_2$, $K'_{sp} = 3.1 \times 10^{-26}$,

$$s = \left(\frac{3.1 \times 10^{-26}}{3^3 2^2}\right)^{1/(3+2)}$$

$$s = \left(\frac{3.1 \times 10^{-26}}{108}\right)^{1/5}$$

$$s = 3.1 \times 10^{-6} \text{ M}$$

Example 8.2

To calculate the solubility of $La(IO_3)_3$ in water, use Equation 8.10. The K'_{sp} is 1×10^{-11}.

$$s = \left(\frac{1 \times 10^{-11}}{1^1 \times 3^3}\right)^{1/4} = (1.1 \times 10^{-12})^{1/4}$$

$$s = 1 \times 10^{-3} \text{ M}$$

Thus, $[La^{3+}] = 1 \times 10^{-3}$ M

$$[IO_3^-] = 3 \times 10^{-3} \text{ M}.$$

Example 8.3

To calculate the solubility of AgBr in a solution to 0.00100 M NaBr (completely dissociated),

$$K'_{sp} = s(s + 0.00100) = 5 \times 10^{-13}$$

First, assume that $s \ll 0.001$ M.
Then

$$s = 5 \times 10^{-10} \text{ M}$$

and the assumption is confirmed.

This approach is similar to that used when hydrolysis reactions were considered in the context of complexation reactions. In fact, the same alpha equations are used. For the protonation of X^{x-}, we use a variation of Equation 4.45, where we have adjusted the number of terms to include the number of equilibria involved.

$$\alpha_{X^{x-}} = \frac{K'_{a1}K'_{a2}K'_{a3}}{[H_3O^+]^3 + K'_{a1}[H_3O^+]^2 + K'_{a1}K'_{a2}[H_3O^+] + K'_{a1}K'_{a2}K'_{a3}} \qquad 8.12$$

and for the complexation of M^{m+} with OH^-, we apply Equation 7.21.

$$\alpha_{M^{m+}} = \frac{1}{1 + K'_{f1}[OH^-] + K'_{f1}K'_{f2}[OH^-]^2 + K'_{f1}K'_{f2}K'_{f3}[OH^-]^3} \qquad 8.13$$

These alpha equations tell us what fraction of the total M^{m+} or X^{x-} concentrations are in the uncomplexed and unprotonated form that make up the precipitate. Thus,

$$[M^{m+}] = \alpha_{M^{m+}}C_{M^{m+}} \qquad \text{and} \qquad [X^{x-}] = \alpha_{X^{x-}}C_{X^{x-}} \qquad 8.14$$

If there are no other sources of M^{m+} or X^{x-} than from the dissociation of the precipitate,

$$C_{M^{m+}} = as \qquad \text{and} \qquad C_{X^{x-}} = bs$$

then,

$$[M^{m+}] = \alpha_{M^{m+}}as \qquad \text{and} \qquad [X^{x-}] = \alpha_{X^{x-}}bs$$

When these quantities are used in Equation 8.4,

$$K'_{sp} = [M^{m+}]^a [X^{x-}]^b = (\alpha_{M^{m+}}as)^a (\alpha_{X^{x-}}bs)^b = \alpha_{M^{m+}}^a \alpha_{X^{x-}}^b a^a b^b s^{(a+b)}$$

from which an equation for s can be obtained.

$$s = \left(\frac{K'_{sp}}{\alpha_{M^{m+}}^a \alpha_{X^{x-}}^b a^a b^b}\right)^{1/(a+b)} \qquad 8.15$$

Equation 8.15 can be compared with Equation 8.10 to see the effect of the ability of M^{m+} or X^{x-} to undergo hydrolysis reactions. Since the alpha values are always less than 1, the effect of the hydrolysis reactions is to increase the quantity in the parentheses and thus the solubility (see Example 8.4).

Example 8.4

Find the solubility of Ag_2CO_3 in a solution at $p[H_3O^+] = 7.0$. Use equilibrium constants from the Appendix tables.

$$\alpha_{Ag^+} = \frac{1}{1 + 10^{2.0} \times 10^{-7} + 10^{2.0} \times 10^{-14}} = 1$$

$$\alpha_{CO_3^{2-}} = \frac{10^{-6.35} \times 10^{-10.33}}{10^{-14} + 10^{-7} \times 10^{-6.35} + 10^{-6.35} \times 10^{-10.33}}$$

$$\alpha_{CO_3^{2-}} = 3.8 \times 10^{-4}$$

$$s = \left[\frac{8.1 \times 10^{-12}}{1^2 \times (3.8 \times 10^{-4}) \times 2^2 \times 1}\right]^{1/(2+1)}$$

$$s = 1.7 \times 10^{-3} \text{ M}$$

EFFECT OF COMPLEXATION ON SOLUBILITY

Hydrolysis is not the only type of reaction in which the ionic products of precipitate dissolution can be involved. A common example is the complexation of the M^{m+} ion as it acts as a coordination center. The complexation can be from the X^{x-} ion or from another ligand species. For these situations, the approach is the same. An expression or value is needed for $\alpha_{M^{m+}}$, and this value or expression is used in Equation 8.15.

Study Questions, Section B

12. Why is the expression in Equation 8.10 called the "ionic solubility" instead of just the "solubility"?

13. Calculate the ionic solubility of $Ni(OH)_2$. Use 6×10^{-16} for K'_{sp}.

14. Why is the solubility of a precipitate decreased when a soluble salt containing one of the precipitate's ions is added to the solution?

15. Calculate the ionic solubility of lanthanum iodate $[La(IO_3)_3]$ in a solution which is 0.010 M in IO_3^-. Use 1.0×10^{-11} for K'_{sp}.

16. Why is the effect of competing reactions (reactions involving the ions formed by dissolution of the precipitate) always to increase the solubility of the precipitate?

17. The K'_{sp} for $FeCO_3$ is 2.1×10^{-11}. Fe^{2+} also complexes with OH^- with log $K'_{f1} = 4.6$. Finally, CO_3^{2-} can be protonated. The K'_{a1} and K'_{a2} values for H_2CO_3 are 4.45×10^{-7} and 4.69×10^{-11}. Calculate the ionic solubility of $FeCO_3$ in aqueous solution buffered at $p[H_3O^+] = 6.0$.

18. Calculate the ionic solubility of AgI in a 1.00×10^{-2} M solution of $K_2S_2O_3$. The formation constants for the complexation of Ag^+ with $S_2O_3^{2-}$ are log $K'_{f1} = 8.82$, log $K'_{f2} = 4.7$, and log $K'_{f3} = 0.7$. The K'_{sp} for AgI is 8.3×10^{-17}. I^- does not hydrolyze because HI is a strong acid in water. Ignore the hydrolysis of Ag^+.

Answers to Study Questions, Section B

12. Equation 8.10 was obtained from the equilibrium expression for the dissolution of the precipitate into its ionic constituents. It is not the total solubility because the intrinsic solubility is not taken into account in the equation.

13. Nickel hydroxide reacts in water as follows:

$$Ni(OH)_2 \rightleftharpoons Ni^{2+} + 2OH^-$$

Using Equation 8.10,

$$s = \left(\frac{K'_{sp}}{a^a b^b}\right)^{1/(a+b)} = \left(\frac{6 \times 10^{-16}}{1^1 \cdot 2^2}\right)^{1/(1+2)}$$
$$= 5 \times 10^{-6} \text{ M}$$

Note that a and b are the stoichiometric coefficients for nickel and hydroxide in the solid nickel hydroxide.

14. From the solubility product expression, one can see that as the concentration of one of the precipitating ions is increased, the concentration of the other one must decrease. Since the only source of the other ion is the dissolution of the precipitate, the solubility must be decreased.

15. The solubility reaction of $La(IO_3)_3$ in water is:

$$La(IO_3)_{3(s)} \rightleftharpoons La^{3+}_{(aq)} + 3IO_3^-_{(aq)}$$

From Equation 8.11, in the case of an excess concentration of X (excess C_X),

$$K'_{sp} = (as)^a (C_X + bs)^b$$

In this case a and b are the stoichiometric coefficients for lanthanum and iodate in the solid. Since K'_{sp} is small, we can assume that $C_X \gg bs$, so Equation 8.11 becomes

$$K'_{sp} = (as)^a (C_X)^b$$

Solving for s gives

$$s = \frac{1}{a}\left[\frac{K'_{sp}}{(C_X)^b}\right]^{1/a} = \frac{1}{1}\left(\frac{1.0 \times 10^{-11}}{0.010^3}\right)^{1/1}$$
$$= 1.0 \times 10^{-5} \text{ M}$$

Note that $0.01 \gg 3.0 \times 10^{-5}$, so the assumption made earlier holds.

16. The competing reactions convert the ions formed by dissolution to another form. Their removal from the solution causes the precipitate to provide more of them so that the concentrations in the K'_{sp} expression can be maintained.

17. Use Equation 8.15. For $FeCO_3$, $a = 1$ and $b = 1$ (see Equation 8.2). From Equation 8.12,

$$\alpha_{CO_3^{2-}} = \frac{(4.45 \times 10^{-7})(4.69 \times 10^{-11})}{(10^{-6.0})^2 + (4.45 \times 10^{-7})(10^{-6.0})}$$
$$+ (4.45 \times 10^{-7})(4.69 \times 10^{-11})$$
$$= 1.4 \times 10^{-5}$$

and from Equation 8.13,

$$\alpha_{M^{m+}} = \alpha_{Fe^{2+}} = \frac{1}{1 + (10^{4.6})(10^{-8.0})} = 1.0$$

Therefore,

$$s = \left[\frac{2.1 \times 10^{-11}}{(1.00)(1.4 \times 10^{-5})}\right]^{1/(1+1)} = 1.2 \times 10^{-3} \text{ M}$$

18. Use Equation 8.15. For AgI, $a = 1$ and $b = 1$. Since I^- does not hydrolyze, $\alpha_{I^-} = 1.0$.
    From Equation 7.21,

$$\alpha_{Ag^+} = \frac{1}{1 + K_{f1}[S_2O_3^{2-}] + K_{f1}K_{f2}[S_2O_3^{2-}]^2 + K_{f1}K_{f2}K_{f3}[S_2O_3^{2-}]^3}$$

Substituting the values given, $\alpha_{Ag^+} = 2.9 \times 10^{-10}$.
So,

$$s = \left[\frac{8.3 \times 10^{-17}}{(2.9 \times 10^{-10})^1(1^1 \cdot 1^1)}\right]^{1/(1+1)}$$
$$= 5.4 \times 10^{-4} \text{ M}$$

Note that without the $S_2O_3^{2-}$ present, the solubility of AgI would be $(8.3 \times 10^{-17})^{1/2} = 9.1 \times 10^{-9}$ M. The greatly increased solubility is due to the reaction of the dissolved Ag^+ to reduce its concentration and shift the dissolution reaction to the right.

C. Quantitation by Precipitation Titration

The process of titration as a means of obtaining quantitative information about a reactant is now becoming familiar. For the precipitation reaction to be used in this way, we need to choose a precipitating reagent for the analyte, make a standard solution of that reagent, and devise a means for determining the equivalence point of the precipitation reaction. The analyte can be either the metallic or nonmetallic participant in the precipitation reaction.

FREQUENTLY USED REACTIONS

The reactions favored for precipitation titrations are those that are fairly specific for the desired analyte and that also produce an exactly stoichiometric precipitate. The precipitate AgCl is a good example meeting these criteria. Relatively few metal chlorides are insoluble (with Hg_2^{2+} and Tl^+ the notable exceptions), which makes chloride a good titrant for the determination of Ag^+. However, although often used for halide ion determinations, Ag^+ is not very specific for halides. Other anions that form insoluble silver salts are arsenate, carbonate, chromate, cyanide, iodate, oxalate, sulfide, and thiocyanate. The problem of having a large number of potential interferents is fairly common among the inorganic metal salts. A few organic precipitating reagents have been developed for precipitation reactions, and some of those are quite specific. As examples, dimethylglyoxime is very specific for Ni^{2+} in weakly alkaline solutions, and 1-nitroso-2-naphthol is selective for Co^{2+}.

TITRATION CURVES

For the sake of equivalence point detection, it is important that the concentration of at least one of the reactants undergo a rapid change in the region of the equivalence point. To determine the shape and extent of this change (and to learn the values of the concentrations at the equivalence point), titration curves are plotted. The Y-axis of the titration curve should be the log of the reactant concentration that is being followed by the equivalence point detection method. The derivations of the titration curves can be found on the accompanying CD. Several families of titration curves obtained from the derived equations are shown here to illustrate the effects of various experimental parameters on their shapes. The first of these is the titration curves for the silver halides (Figure 8.4). For these curves, the analyte and titrant concentrations were 0.10 M, and $[Ag^+]$ is monitored to determine the equivalence point. As we have come to expect, the degree of change in the analyte concentration in the region of the equivalence point is a function of the equilibrium constant for the reaction involved in the titration. In this case, the less soluble the precipitate, the greater the change.

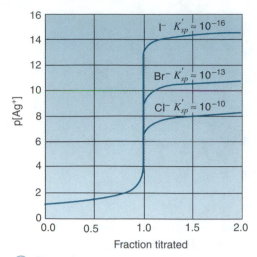

Figure 8.4. The titration curves for 0.10 M Ag^+ titrated with 0.10 M Cl^-, Br^-, and I^-. The magnitude of the break at the equivalence point increases with a decrease in the K'_{sp} of the precipitate formed.

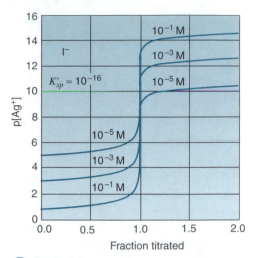

Figure 8.5. The magnitude of the break at the equivalence point also decreases with decreasing initial concentration of analyte. These curves are for three initial concentrations of Ag^+ in a titration with I^- at equal concentration.

The effect of analyte concentration is shown in Figure 8.5. The titrant in this case was I^-, and the concentrations were 1×10^{-1}, 1×10^{-3}, and 1×10^{-5} M. At the lowest concentration, the degree of change is at the marginal edge of usefulness. Note that this is for the most favorable (least soluble) of the halide titrants. If Cl^- were the titrant, only the higher concentrations would yield satisfactory results.

In some cases, we will want to monitor the titrant species to determine the equivalence point. The plots of such titrations are shown in Figures 8.6 and 8.7, both for the effect of K'_{sp} and for the effect of the initial concentration of analyte. Because of the 1:1 stoichiometry of M and X in the precipitate, the plots are the mirror images of the titration curves when [M] is monitored (Figures 8.4 and 8.5).

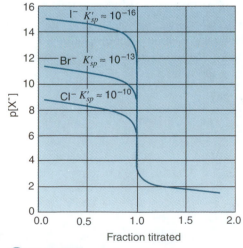

Figure 8.6. Titration curves for halide titrations in which the change in $p[X^-]$ is followed. These curves are complementary to those in Figure 8.4.

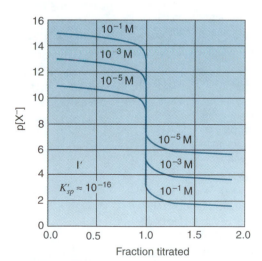

Figure 8.7. Titration curves for three concentrations of I^- when titrated with Ag^+. These curves, for which $p[X^-]$ is followed, complement those in Figure 8.5.

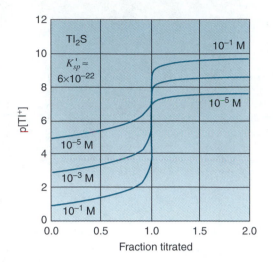

⊙ **Figure 8.8.** The titration curves for unsymmetrical precipitates are also unsymmetrical. In this example for the titration of Tl^+ with S^{2-}, the curvature and concentration dependence are different before and after the equivalence point.

The shape of the titration curves for unsymmetrical reactions (not $1:1$ or $2:2$, etc.) can also be determined. The titration curve for Tl_2S has been derived where S^{2-} is the titrant and Tl^+ is monitored. The resulting plot is shown in Figure 8.8. An obvious observation is that the curve is not symmetrical. Another is that the break at the equivalence point is not as large for the lowest concentration as it was for AgI although the overall K'_{sp} is much smaller. This reminds us that the K'_{sp} is only related to solubility within a given precipitate formula class. The larger the number of entities in the formula, the smaller must be the K'_{sp} for a given solubility.

Titration curves can be estimated by the calculation of the appropriate p[M] or p[X] at key points in the titration curve. We will follow several examples from the titration curves plotted in Figures 8.4 to 8.8. When titrating M^{m+} ions with X^{x-} to form the precipitate M_aX_b, the initial concentrations (0% titrated) are $[M] = C_M$ and $[X] = 0$. At 50% titrated, from Equation 8.4,

$$[M] = C_M/2 \qquad \text{and} \qquad [X] = (K'_{sp}/[M]^a)^{1/b} \qquad 8.16$$

In the case of the titration of 0.10 M Ag^+ with I^-, as in Figure 8.6, $[I^-] = (1 \times 10^{-16}/0.050) = 2 \times 10^{-15}$ M and $p[I^-] = 14.7$. At 100% titrated, the solution is as though the pure precipitate had been dissolved in water. The concentrations of M and X are calculated by first calculating the solubility from Equation 8.10 and then multiplying that by a (for M) or b (for X). The equivalence point concentrations for the titration of 1×10^{-3} M Tl^+ with S^{2-} as in Figure 8.8, are

$$[Tl^+] = 2s = 2\left(\frac{K'_{sp}}{2^2 \times 1^1}\right)^{1/3} = 2(5.3 \times 10^{-8}) = 1.1 \times 10^{-7} \text{ M}$$
$$[S^{2-}] = s = 5.3 \times 10^{-8} \text{ M}$$

from which $p[Tl^+] = 7.0$ and $p[S^{2-}] = 7.3$. At 200% titrated (ignoring dilution),

$$[X] = C_M(b/a) \qquad \text{and} \qquad [M] = (K'_{sp}/[X]^b)^{1/a} \qquad 8.17$$

This relationship is obtained from the stoichiometry and from the fact that this amount of X would exactly titrate the original M once again, thus 200% titrated. The equation for M concentration was obtained from Equation 8.4. In the Tl_2S example for which the equivalence point concentrations were calculated, $[S^{2-}]$ at 200% $= C_M/2 = 5 \times 10^{-4}$ M. The Tl^+ concentration is $(6 \times 10^{-22}/5 \times 10^{-4})^{1/2} = 1 \times 10^{-9}$ M, and $p[Tl^+] = 9.0$.

EQUIVALENCE POINT DETECTION

The determination of the equivalence point volume in a precipitation titration is most generally accomplished by the use of an electrode that responds specifically to one of the reagents in the titration. Such electrodes (discussed in detail in the next chapter) generate an electrical voltage that changes linearly with a change in the logarithm of the concentration. The glass pH electrode itself can be used when one of the reactants is OH^-. (There are a number of insoluble hydroxides and oxides that are formed by reaction with OH^-, but no precipitates that are formed by proton addition.) Specific electrodes are available for many metal ions and for many anions as well. For example, one can use an electrode specific to either Ag^+ or Cl^- in the titration reaction to form AgCl. Specific ion electrodes are particularly advantageous for end point detection in precipitation titrations. They respond to the change in reactant concentrations over the entire titration curve and thus provide more data concerning the equivalence point volume and the species titrated. In the titration of mixtures (such as the mixed halide example) they are essential because the titration curve must be plotted to determine the volume delivered at the point where the next component just begins to react.

Nevertheless, several chemical indicators have found extensive use in precipitation titrations. Chemical indicators are of two varieties. One type undergoes a color changing reaction with excess precipitant. The other is based on the change in the sign of the charge that a colloidal precipitate undergoes at the equivalence point. Three commonly used examples are described below.

In the **Mohr method** for the titration of Cl^-, chromate ion is added as an indicator. The reaction of the silver ion with chromate to form the red precipitate (Ag_2CrO_4) occurs after the equivalence point for the titration of Cl^-. As with all chemical indicators, one must make sure the concentration of titrant at which the color change takes place is at or near the equivalence point concentration. The calculation in Example 8.5 shows that for this to be true, the chromate concentration would have to be 6.7×10^{-3} M. Unfortunately, at this concentration, the intense yellow color of the chromate ion prevents observation of the red Ag_2CrO_4 when it forms. At the low concentrations required to make this indicator practical, the end point volume exceeds the equivalence point volume by a significant amount. Although the error is small at an initial Cl^- concentration of 0.1 M, it becomes much greater at lower concentrations. Methods have been developed to compensate for this error, but in the light of modern alternatives, there seems little point in struggling to make this method more accurate.

The **Volhard method** for the determination of Ag^+ involves titration with thiocyanate ion. The AgSCN precipitate has a K'_{sp} of 1.1×10^{-12}. The first excess of SCN^- reacts with the Fe^{3+} indicator to produce the red complex $FeSCN^{2+}$. Again, the equilibrium constants and concentrations must be just right for the end point and equivalence point volumes to be equal. In this case, it works quite well as far as the titration of Ag^+ is concerned. However, the greatest application of the Volhard method is in the back-titration of excess Ag^+ that has been added to a solution of halide analyte. Here the problem is that the solubility of AgSCN is less than that of AgCl: as the equivalence point is approached, the SCN^- begins to displace Cl^- from the AgCl, causing the end point to fade and the end point volume to exceed the equivalence point for the back-titration. Experienced analysts using this method develop a quickness of technique that minimizes this error. Back-titrations of Br^- and I^- do not share this difficulty because of their lower solubility.

Fluorescein (Figure 8.9) is an example of an **adsorption indicator** in which the adsorption is a function of the sign of the charge on a colloidal precipitate. Fluresceinate anions (formed by reaction of the fluorescein with water) appear yellow-green in solution. They also react with Ag^+ to form an intensely red salt. The formation of the red salt is not used as an indicator, however, because the concentration of Ag^+ required to form it is too high. If a colloid of silver halide has an excess positive charge (due to the Ag^+ on the surface), it will attract anions to its surface, including fluoresceinate anions if they are present. There, the interaction between the Ag^+ and the fluoresceinate

---

■■■ •••••••••••••••••••
■■ **Example 8.5**

Calculation of chromate concentration at the Mohr end point
The K'_{sp} for Ag_2CrO_4 is 1.2×10^{-12}. The $[Ag^+]$ at the Cl^- equivalence point in a Cl^- titration is

$$\sqrt{K'_{sp,AgCl}} = \sqrt{1.8 \times 10^{-10}} = 1.3 \times 10^{-5} \text{ M}$$

The chromate concentration in equilibrium with solid Ag_2CrO_4 at this $[Ag^+]$ is

$$[CrO_4^{2-}] = \frac{K'_{sp,Ag_2CrO_4}}{[Ag^+]^2} = \frac{1.2 \times 10^{-12}}{(1.3 \times 10^{-5})^2}$$
$$= 6.7 \times 10^{-3} \text{ M}$$

•••••••••••••••••••••••••

Figure 8.9. Structure of the fluorescein molecule. The blue H is the acidic proton that reacts with water to form the fluoresceinate anion, the base form of fluorescein.

anions produces the red color. When Cl^- is titrated with Ag^+, the first excess of Ag^+ changes the sign of the colloid charge, and the colloid takes on a red hue. This method is called the **Fajans method** after its originator. The technique is very accurate and reliable. Unfortunately, it is not generally applicable because it requires the formation of a colloidal precipitate and requires an adsorption indicator that changes color when acting as the countercharge of the colloid; only a few of these have been developed.

Photon absorption is not useful for equivalence point detection with precipitation reactions because the precipitate makes the solution cloudy. Reliable absorption measurements cannot be made in the presence of suspended solid particles.

Study Questions, Section C

19. In the titration of M with X, does the concentration of M increase or decrease as the titration proceeds?

20. Comparing Figures 8.5 and 8.7, does it make any difference in the size of the concentration break near the equivalence point whether M is titrated with X or the other way about?

21. For the titration of 1.0×10^{-1} M Hg_2^{2+} with Cl^-, estimate the equivalence point break ($\Delta p[Hg_2^{2+}]$ from 50% to 200% titrated) and the equivalence point concentration of Hg_2^{2+}. Log K'_{sp} for $Hg_2Cl_2 = -16.0$.

22. The Mohr, Volhard, and Fajans methods are based on the use of an indicator that changes color at the first excess of the titrant. What are the reactions involved in the use of these indicators?

23. The minimum detectable $[Fe(SCN)^{2+}]$ in a Volhard titration is about 6.5×10^{-6} M. Calculate the error in $[Ag^+]$ for the titration of 35.00 mL of 0.050 M Ag^+ solution containing Fe^{3+} with 0.150 M KSCN. The K'_f for $FeSCN^{2+}$ is 1.05×10^3.

24. Excess $AgNO_3$ (20.00 mL of 0.0500 M) is added to 25.00 mL of a solution containing Br^-. $Fe(NO_3)_3$ is also added to the solution so that free Fe^{3+} is present. The Ag^+ that does not react with the Br^- is back-titrated with a 0.0100 M solution of KSCN. The color change is observed after addition of 34.03 mL of the KSCN titrant. What is the concentration of Br^- in the unknown solution?

Answers to Study Questions, Section C

19. The concentration of M will decrease as more and more of it reacts with the titrant.

20. For a symmetrical precipitate, the size of the break is independent of which species is the analyte and which is the titrant. It is also the same whichever species is monitored (Y-axis). For an unsymmetrical precipitate (not 1:1 or 2:2), the magnitude of the break does depend on which species is monitored.

21. From Equation 8.16, at the 50% point $[Hg_2^{2+}] = 0.10/2 = 0.050$ M, and $p[Hg_2^{2+}] = 1.30$. From Equation 8.17, at the 200% point $[Cl^-] = 0.10 \times 2 = 0.20$ M. Thus, $[Hg_2^{2+}] = (10^{-16.0}/(0.20)^2) = 2.5 \times 10^{-15}$ M, and $p[Hg_2^{2+}] = 14.6$. Thus, the break is $14.6 - 1.30 = 13.3$ log units.

The concentration of Hg_2^{2+} at the equivalence point can be obtained from Equation 8.10.

$$[Hg_2^{2+}] = s = [10^{-16.0}/(1^1 \times 2^2)]^{1/3} = 3.0 \times 10^{-6} \text{ M}$$
$$p[Hg_2^{2+}] = 5.5$$

22. In the Mohr method, the chromate ion indicator forms a red precipitate Ag_2CrO_4 with excess Ag^+ in halide titrations. With the Volhard method, the ferric ion indicator forms a red complex $FeSCN^{2+}$ with excess thiocyanate in the titration of Ag^+. In the Fajans method, an adsorption indicator forms a colored complex with the ions responsi-

ble for the excess charge on colloidal precipitates. The reactant forming the complex is used as the titrant, so the color formation occurs at the first excess of titrant.

23. To determine the error we must find the $[Ag^+]$ that would be determined by the titration and compare it to the actual concentration of Ag^+ in solution. The $[Ag^+]$ determined by the titration is calculated from the amount of SCN^- required for the titration. This is equal to the amount of SCN^- required to precipitate all of the Ag^+ plus the amount required to form 6.5×10^{-6} M $Fe(SCN)^{2+}$. The amount of SCN^- required to precipitate the Ag^+ is

$$(0.050 \text{ M } Ag^+)(35.00 \times 10^{-3} \text{ L})\left(\frac{1 \text{ mol } SCN^-}{1 \text{ mol } Ag^+}\right)$$
$$= 1.8 \times 10^{-3} \text{ mol } SCN^-$$

The volume of titrant required to get this many moles is

$$(1.8 \times 10^{-3} \text{ mol } SCN^-)\left(\frac{1 \text{ L}}{0.150 \text{ mol } SCN^-}\right)$$
$$= 1.2 \times 10^{-2} L = 12 \text{ mL}$$

To determine the $[SCN^-]$ necessary to give 6.5×10^{-6} M $Fe(SCN)^{2+}$, the complexation equilibrium equation is used.

$$Fe^{3+} + SCN^- \rightleftharpoons Fe(SCN)^{2+} \qquad K'_f = \frac{[Fe(SCN)^{2+}]}{[Fe^{3+}][SCN^-]}$$

By the complexation reaction, $[Fe^{3+}] = [SCN^-]$ so,

$$[SCN^-] = \sqrt{\frac{[Fe(SCN)^{2+}]}{K_f'}}$$

$$= \sqrt{\frac{6.5 \times 10^{-6}}{1.05 \times 10^3}} = 7.9 \times 10^{-5} \text{ M SCN}^-$$

Assuming that the volume of the solution is about the same at the equivalence point and when the actual color change occurs, the moles of SCN^- necessary to bring about the color change is

$$(7.9 \times 10^{-5} \text{ M SCN}^-)(0.012 \text{ L} + 0.03500 \text{ L})$$
$$= 3.7 \times 10^{-6} \text{ mol SCN}^-$$

Comparing the number of moles of SCN^- necessary to complex the Fe^{3+} (3.7×10^{-6} mol) with the number of moles necessary to precipitate the Ag^+ (1.75×10^{-3} mol), the ratio is 2.1×10^{-3}. This means that the error introduced is approximately 2 ppt.

24. The Ag^+ reacts with SCN^- as follows:

$$Ag^+ + SCN^- \rightleftharpoons AgSCN_{(aq)}$$

It also reacts with Br^- to form solid $AgBr$. We assume that this reaction goes to completion since the K_{sp}' for $AgBr$ is so small. Therefore, if we determine the amount of Ag^+ remaining after the reaction with the Br^-, we can subtract this amount from the total amount of Ag^+ added and thus determine $[Br^-]$. The amount of Ag^+ remaining after the reaction with Br^- is

$$(0.03403 \text{ L SCN}^-)(0.0100 \text{ M}) \left(\frac{1 \text{ mol Ag}^+}{1 \text{ mol SCN}^-}\right)$$
$$= 3.40 \times 10^{-4} \text{ mol Ag}^+$$

The number of moles of Ag^+ added is

$$(0.02000 \text{ L})(0.0500 \text{ M}) = 1.00 \times 10^{-3} \text{ mol Ag}^+$$

The amount of Ag^+ that reacted with the Br^- is $1.00 \times 10^{-3} - 3.40 \times 10^{-4} = 6.6 \times 10^{-4}$ mol Ag^+. Thus,

$$[Br^-] = (6.6 \times 10^{-4} \text{ mol Ag}^+) \times$$
$$\left(\frac{1 \text{ mol Br}^-}{1 \text{ mol Ag}^+}\right) \times \left(\frac{1}{0.02500 \text{ L unk. sol'n.}}\right) = \frac{0.026 \text{ mol}}{L}$$

D. Logarithmic Concentration Plots

As we have seen in the acid–base and complexation systems, logarithmic concentration plots can be very useful in obtaining estimates of equilibrium concentrations of species involved in the reactions. This is also true for precipitation reactions. To illustrate the process of constructing a plot for precipitation, we will use the familiar example of $AgCl$. In this diagram (Figure 8.10), $p[Cl^-]$ is plotted against the logarithms of all the concentrations. The line for Cl^- is obtained from the identity $\log [Cl^-] = -p[Cl^-]$. The equation for the Ag^+ line is derived as follows:

$$K_{sp}' = [Ag^+][Cl^-]$$
$$\log K_{sp}' = \log [Ag^+] + \log [Cl^-]$$
$$\log [Ag^+] = \log K_{sp}' - \log [Cl^-]$$

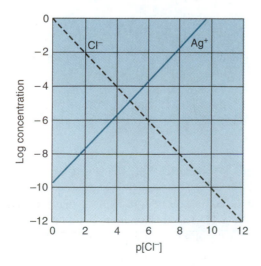

Figure 8.10. The log concentration plot for a precipitation reaction reveals the concentrations of both reactants as a function of the concentration of the precipitating reagent.

where $K'_{sp} = 1.8 \times 10^{-10}$. In practice, the location of the line for Ag^+ is obtained by drawing a line of unit slope from the point where $p[Cl^-] = -\log K'_{sp}$ and log concentration $= 0$.

This plot can be used to solve a variety of problems involving the AgCl dissolution equilibrium. To determine the solubility of AgCl in pure water, find the intersection of the Ag^+ and Cl^- lines. The value of either the Ag^+ or Cl^- concentrations is equal to the solubility. The solubility of AgCl for other values of $[Cl^-]$ can also be found. For example, the solubility of AgCl when $[Cl^-]$ is 1×10^{-2} M is found as the value of $[Ag^+]$ when this is true. (Follow the vertical line at $p[Cl^-] = 2$ upward to get $[Ag^+] \approx 2 \times 10^{-8}$ M.) The solubility of AgCl for a given $[Ag^+]$ can be found the same way, but in this case, the $[Cl^-]$ concentration is equal to the solubility.

This method of determining solubility ignores the solubility that can come from complex formation with an excess of X where X is a ligand for M. For many inorganic precipitates, including AgCl, complex formation can have a significant effect on the solubility. The extent of this effect can be determined graphically by adding the related complexation products to the same log concentration plot.

The following Cl^- complexes are formed with Ag^+.

$$
\begin{aligned}
Ag^+ + Cl^- &\rightleftharpoons AgCl_{(aq)} & K'_{f1} &= 10^{3.0} \\
AgCl_{(aq)} + Cl^- &\rightleftharpoons AgCl_2{}^- & K'_{f2} &= 10^{2.3} \\
AgCl_2{}^- + Cl^- &\rightleftharpoons AgCl_3{}^{2-} & K'_{f3} &= 10^{0.85} \\
AgCl_3{}^{2-} + Cl^- &\rightleftharpoons AgCl_4{}^{3-} & K'_{f4} &= 10^{-0.65}
\end{aligned}
$$

The first product is the dissolved, neutral AgCl. In this case, this reaction is the source of the intrinsic solubility of the precipitate, $[AgCl_{(aq)}]$. In fact, the intrinsic solubility value can be obtained from the value of K'_{f1} and K'_{sp} as follows. The equilibrium constant for the formation of $AgCl_{(aq)}$ is written, and K'_{sp} is substituted for $[Ag^+][Cl^-]$.

$$
K'_{f1} = \frac{[AgCl_{(aq)}]}{[Ag^+][Cl^-]} = \frac{[AgCl_{(aq)}]}{K'_{sp}}
$$

When this expression is rearranged,

$$
[AgCl_{(aq)}] = K'_{f1}K'_{sp} = K'_{int}
$$

As expected from Equation 8.8, the value for the intrinsic solubility, $[AgCl_{(aq)}]$, is a constant. For AgCl, it is $1 \times 10^3 \times 1.8 \times 10^{-10} = 2 \times 10^{-7}$ M.

A line for $AgCl_2^-$ can be constructed from the relationship

$$
K'_{f2} = \frac{[AgCl_2^-]}{[Cl^-][AgCl_{(aq)}]}, \qquad [AgCl_2^-] = K'_{f2}[Cl^-]K'_{f1}K'_{sp}
$$

$$
\log[AgCl_2^-] = \log K'_{f2} + \log[Cl^-] + \log K'_{f1}K'_{sp}
$$

Similarly,

$$
[AgCl_3{}^{2-}] = K'_{f3}[Cl^-][AgCl_2^-] = K'_{f2}K'_{f3}[Cl^-]^2 K'_{f1}K'_{sp}
$$

$$
\log[AgCl_3{}^{2-}] = \log K'_{f2} + \log K'_{f3} + 2\log[Cl^-] + \log K'_{f1}K'_{sp}
$$

and

$$
\log[AgCl_4{}^{3-}] = \log K'_{f2}K'_{f3}K'_{f4} + 3\log[Cl^-] + \log K'_{f1}K'_{sp}
$$

The lines for all the complexes of Ag^+ with Cl^- have been added to the log concentration plot of Figure 8.11. The construction of the lines is much simpler than the algebra used to get their equations would suggest. The line for $AgCl_2^-$ crosses the

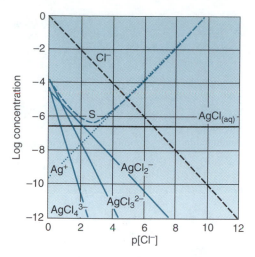

Figure 8.11. When the precipitate is the neutral product in a stepwise complexation series, the precipitation and complexation reactions can share the same log concentration plot. From this plot for the silver–chloride system, the concentration of all species as a function of $p[Cl^-]$ can be seen. The formation of the complex species actually increases the precipitate solubility at values of $[Cl^-]$ greater than 10^{-3} M.

$AgCl_{(aq)}$ line where $\log K'_{f2} = p[Cl^-] = 2.3$. It has a unit slope because the exponent of $[Cl^-]$ in its equation is 1. The line for $AgCl_3^{2-}$ crosses the $AgCl_{(aq)}$ line where $p[Cl^-]$ is equal to half the sum of $\log K'_{f2}$ and $\log K'_{f3}$ ($= 1.6$). It has a slope of $1/2$ because $[Cl^-]$ is squared in its expression. Finally, the line for $AgCl_4^{3-}$ crosses the $AgCl_{(aq)}$ line where $p[Cl^-]$ is equal to one-third $\log K'_{f2} + \log K'_{f3} + \log K'_{f4}$ ($= 0.83$). It has a slope of $1/3$ because $[Cl^-]$ is cubed in its expression.

From the plot including the complexation species (Figure 8.11), it is clear that the concentrations of these species are not negligible at the higher Cl^- concentrations. Our earlier estimate of the solubility of AgCl when $[Cl^-] = 0.01$ M was quite wrong. Although the $[Ag^+]$ at this Cl^- concentration is indeed 2×10^{-8} M, the total solubility is much greater owing to the dissolved silver chloride complexes. The dashed line represents the sum of all the dissolved species. From this plot, one can also obtain the $p[Cl^-]$ value when the solubility is a minimum (about 2.5) and the value of that solubility (a little less than 10^{-6} M). See the spreadsheets on the accompanying CD for automatic plotting of the solubility curves for precipitation reactions complicated by complex formation.

UNSYMMETRICAL LOG PLOTS

The log concentration plots for AgCl are relatively straightforward because of the simple 1:1 stoichiometry between the two ions forming the precipitate. For other ratios, however, only the slopes of some of the lines are changed. An example is the plot for the Ni^{2+}, OH^- system in Figure 8.12. The Ni^{2+} ion forms a series of three complexes with OH^- of which the neutral complex, $Ni(OH)_2$, has limited solubility. The reactions and their K' values are

$$Ni^{2+} + OH^- \rightleftharpoons NiOH^+ \qquad K'_{f1} = 10^{4.1}$$
$$NiOH^+ + OH^- \rightleftharpoons Ni(OH)_{2(aq)} \qquad K'_{f2} = 10^{4.9}$$
$$Ni(OH)_{2(aq)} + OH^- \rightleftharpoons Ni(OH)_3^- \qquad K'_{f3} = 10^{3.0}$$
$$Ni^{2+} + 2OH^- \rightleftharpoons Ni(OH)_{2(s)} \qquad K'_{sp} = 10^{-15}$$

The equations for their lines are

$$\log[Ni^{2+}] = \log K'_{sp} - 2\log[OH^-]$$
$$\log[Ni(OH)^+] = \log K'_{f1}K'_{sp} - \log[OH^-]$$
$$\log[Ni(OH)_{2(aq)}] = \log K'_{f2}K'_{f1}K'_{sp}$$
$$\log[Ni(OH)_3^-] = \log K'_{f3}K'_{f2}K'_{f1}K'_{sp} + \log[OH^-]$$

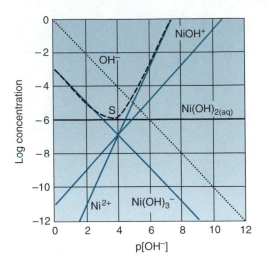

⊙ **Figure 8.12.** The slope of the lines in the combined precipitation– complexation log concentration plot change depending on whether the neutral species is MX, MX_2, or MX_3.
Compare this plot for the nickel–hydroxide system with that in
Figure 8.11 for the silver–chloride system.

The resulting plot is shown in Figure 8.12. The slopes ($\Delta x/\Delta y$) of the Ni-contain-
ing species depend on the coefficient of the $\log[OH^-]$ term in their equation. Since the
plot is of $p[OH^-]$, the signs are changed. The coefficients (and $\Delta x/\Delta y$ slopes) are
then $+2$, $+1$, 0, and -1. The dashed total solubility line has been added. The line
for Ni^{2+} can be readily constructed from the fact that it begins at the point where
$\log[Ni^{2+}] = 0$ and $p[OH^-] = -\log K'_{sp}/2$. At this point, the $p[OH^-]$ increases at twice
the rate of the decrease in $\log[Ni^{2+}]$. From Figure 8.12, we can see that the principal
form of the dissolved precipitate is Ni^{2+} at high values of $p[OH^-]$ and $Ni(OH)_3^-$ at low
$p[OH^-]$ values. The minimum solubility is a bit higher than 1×10^{-6} M and occurs a
$p[OH^-]$ of about 3.8.

All precipitation reaction plots can be constructed similarly. For $Pb_3(PO_4)_2$, the
log concentration is plotted against $p[PO_4^{3-}]$ in Figure 8.13. The $\log[Pb^{2+}]$ concentration intercept (at the left) is where $p[PO_4^{3-}] = \log K'_{sp}/3$ (from the coefficient of
Pb^{2+} in the precipitate). From there it rises to the right with an increase in $\log[Pb^{2+}]$
that is 2/3 times the increase in $p[PO_4^{3-}]$ (from the ratio of the coefficients of Pb^{2+}
and PO_4^{3-}). When important, the related complexation products can be plotted as
well.

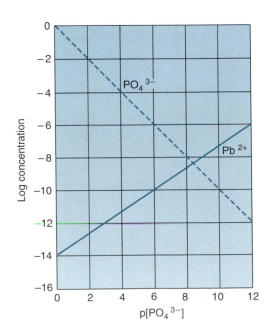

⊙ **Figure 8.13.** Log concentration plot for a precipitate with
the form M_3X_2. The slopes are obtained from the relationship
$\log K'_{sp} = 3 \log[M] + 2 \log [X]$.

ESTIMATION OF TITRATION CURVES

Titration curves can be estimated from the log concentration plots. An example is the titration of mixed halides with Ag^+ when the Ag^+ concentration is monitored. The log concentration plot for this system is shown in Figure 8.14 along with the titration curve for each of the halide ions. The titration curves are for concentrations of 0.100 M for each of the halide ions. A vertical line has been drawn on the log concentration plot at this initial concentration. The $p[Ag^+]$ will change from the value at the intersection of the halide line and the log initial concentration line almost to the intersection of the log initial concentration line and the Ag^+ line. The value of $p[Ag^+]$ at the equivalence point for each analyte is that at which the Ag^+ and halide lines intersect.

Several conclusions can be drawn from a study of the relationship between the log concentration plot and the titration curve. First, we can see that as the concentration of analyte decreases, the change in $p[Ag^+]$ over the titration decreases. (We already knew this, but now we can see how small it gets without having to solve the titration curves.) The Cl^- titration causes a change in $p[Ag^+]$ from roughly 9 to 1.5. If the concentration line is moved to the right, for lower concentrations, the magnitude of this change decreases. If we consider that a change of 3.5 is a minimum for a practical titration, then the minimum concentration of Cl^- for which a silver titration would work is 1×10^{-3} M. For Br^-, this can be extended to 5×10^{-5} M, and for I^-, 1×10^{-6} M. These minimum concentrations for a change of 3.5 $p[Ag^+]$ units are shown in Figure 8.15.

A popular example in analytical texts is the titration of a mixture of halides with silver. The shape of this titration curve can also be visualized with the aid of the log concentration plot (see Figure 8.14). If all three analytes had a concentration of roughly 0.100 M, the Ag^+ titrant would react first with the I^- because it is the least soluble (also demonstrated by the highest value of $p[Ag^+]$ when θ is 0). The titration curve will follow the curve for the titration of I^- until the $p[Ag^+]$ reaches the value at which the Br^- begins to react (the second horizontal line from the top in Figure 8.14). We can see how complete the titration of I^- is by estimating the I^- concentration at this point. We see that it is less than 10^{-4} M, so the fraction untitrated is less than 1 ppt. Now the titration curve will follow the Br^- curve until the $p[Ag^+]$ reaches the value at which the Cl^- begins to react (the third horizontal line from the top). At this point, the Br^- concentration has been reduced to 10^{-4} M, with also just 1 ppt remaining untitrated. The remainder of the titration curve follows that for the Cl^- analyte. Three end points can be obtained for this titration. The first at the point just before the Br^- begins to react, the second at the point just before the Cl^- begins to react and the third at the equivalence point for the Cl^- titration. The volume for the first end point is

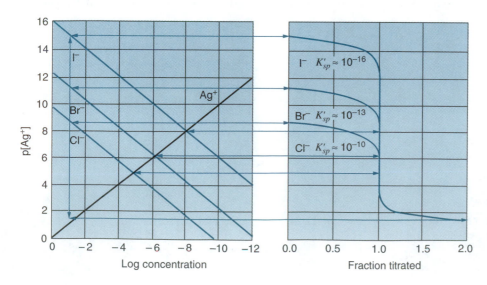

Figure 8.14. The rotated log concentration plot can be used to predict the values of p[M] and p[X] at several key points in the titration curve. Here the magnitude of the break and the value of $p[Ag^+]$ at the equivalence point are compared for the titration of each of the halide ions with Ag^+.

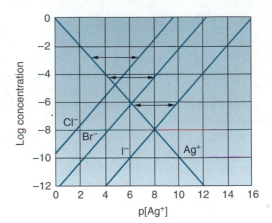

Figure 8.15. These curves illustrate the minimum concentration of halide that can be individually titrated with Ag^+ given the desire for an end point break of 3.5 p units.

related to the amount of I^- present, the volume between the first and second end points to the amount of Br^- present, and the volume between the second and third end points to the amount of Cl^- present.

For this titration to work as favorably as the above example suggests, several conditions need to be met. These can also be obtained from a study of the log concentration plot for the system. First, the fraction of the I^- untitrated when the reaction with the Br^- begins depends on the relative concentrations of these analytes. In Figure 8.16, a dotted square of dimension 3 log concentration units has been drawn from a point where the I^- concentration is 0.10 M. The left side of this box is at the $p[Ag^+]$ for which the I^- is 1 ppt of its original concentration. The $p[Ag^+]$ is also the maximum concentration the Br^- can have if it is not to react until the I^- is titrated to this extent. This Br^- concentration is 0.6 M. In other words, for a successful titration of I^-, $[Br^-] \leq 6 \times [I^-]$. If the initial Br^- concentration exceeds this value, the I^- titration will not be complete at the first end point, and the amount of I^- left untitrated will be added to the volume of titrant ascribed to the Br^- ion. A similar box has been placed at the maximum practical concentration for the Br^- ion. The left line of this box is at the $p[Ag^+]$ for which the Br^- is 1 ppt of its original concentration. The Cl^- concentration at this $p[Ag^+]$ is 0.2 M. Therefore, for an accurate titration of Br^- with Cl^- present, $[Cl^-] \leq 0.33 \times [Br^-]$. This inspection of the plot has demonstrated that mixed halides can be accurately titrated with Ag^+ if they are of nearly equal concentration or if $[Cl^-] \leq 0.33 \times [Br^-] \leq 6 \times [I^-]$. Clearly, this technique cannot be used to determine I^- in seawater, for example. Log concentration diagrams are easy to construct, and they can provide a quick assessment of the practicality of many precipitation titration procedures.

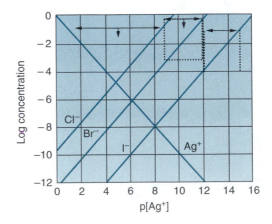

Figure 8.16. These curves illustrate the interactions in the concentrations of mixed halides that can be titrated with Ag^+ in combination. Only a few, rather unlikely combinations of concentrations will work very well.

25. For the log concentration plots, why is it advantageous to plot the logs of the concentrations against p[X] as opposed to p[M]?

26. From Figure 8.11, estimate the concentration of all species when p[Cl⁻] = 3.0. Also estimate the solubility of AgCl in pure water and the concentrations of all other species at the solubility point.

27. Show the steps in deriving the 4 equations given for the lines of the nickel-containing species in Figure 8.12.

28. Cadmium ion reacts with OH^- in the following ways:

$$Cd^{2+} + OH^- \rightleftharpoons CdOH^+ \qquad \log K_1' = 3.9$$

$$CdOH^+ + OH^- \rightleftharpoons Cd(OH)_{2(aq)} \qquad \log K_{f2}' = 3.8$$

$$Cd(OH)_{2(aq)} + OH^- \rightleftharpoons Cd(OH)_3^- \qquad \log K_{f3}' = 0.7$$

The ionic solubility reaction for cadmium hydroxide is

$$Cd(OH)_2 \rightleftharpoons Cd^{2+} + 2OH^- \qquad K_{sp}' = 4.5 \times 10^{-15}$$

Draw a log concentration diagram for $Cd(OH)_2$ and use it to determine the p[OH⁻] at the minimum solubility and the solubility of $Cd(OH)_2$ at that value.

29. For the titration of 1.0×10^{-1} M Hg_2^{2+} with Cl⁻, estimate the equivalence point break and the equivalence point concentration of Hg_2^{2+}. For Hg_2Cl_2, $\log K_{sp}' = -16.0$. Use the spreadsheet plotting program to obtain the plot.

30. Determine the minimum concentration for which the titration in Question 29 is feasible. Consider the titration feasible if there is a break of 3 log units between 50% and 200% titrated.

31. Consider the feasibility of an iodide titration for a mixture of 1.0×10^{-1} M each Pb^{2+}, Hg_2^{2+}, and Ag^+ by drawing the appropriate log concentration diagram. The K_{sp}' values for the cations are 7.9×10^{-9}, 4.7×10^{-29}, and 8.3×10^{-17}, respectively.

25. The real value of the log concentration plots for precipitates comes from the ability to include competing reactions on the same plot. Since the competing reaction is so often the complexation of M by X, having the X-axis in terms of [X] is most convenient.

26. The concentrations can be estimated from Figure 8.11. When p[Cl⁻] = 3.0, [Cl⁻] = 1 × 10⁻³ M; [Ag⁺] ≈ 10⁻⁶·⁴ M = 4.0 × 10⁻⁷ M; [AgCl₂⁻] ≈ 10⁻⁷·⁵ M = 3 × 10⁻⁸ M; [AgCl₃²⁻] ≈ 10⁻⁹·⁵ M = 3 × 10⁻¹⁰ M; and [AgCl₄³⁻] ≈ 10⁻¹³ M. The solubility in pure water is where the Ag⁺ line crosses the Cl⁻ line. At this point, [Ag⁺] = [Cl⁻] ≈ 10⁻⁴·⁸ M = 1.6 × 10⁻⁵ M; [AgCl₂⁻] ≈ 10⁻⁹ M; [AgCl₃²⁻] ≈ 10⁻¹³ M = 1 × 10⁻¹³ M; and [AgCl₄³⁻] ≈ 10⁻²¹ M (found by extending the [AgCl₄³⁻] line). Note that the concentrations of all species besides Ag⁺ and Cl⁻ are very low at the solubility point. This means that complexation of Ag⁺ with multiple Cl⁻ ions does not significantly affect the value of the solubility in pure water.

27. The key is to get equations for all Ni^{2+} species in terms of only constants and [OH⁻]. These equations can then be graphed on the log concentration diagram. When deriving the equations, it is helpful to remember the following properties of logarithms: $\log(ab) = \log a + \log b$, $\log(a/b) = \log a - \log b$, $\log a^b = b \log a$.

    Starting with the solubility reaction,

$$Ni(OH)_{2(s)} \rightleftharpoons Ni^{2+}_{(aq)} + 2OH^-_{(aq)}$$

we write the equilibrium expression as

$$K_{sp}' = [Ni^{2+}][OH^-]^2$$

$$[Ni^{2+}] = K_{sp}'/[OH^-]^2$$

$$\log[Ni^{2+}] = \log(K_{sp}'/[OH^-]^2) = \log K_{sp}' - 2\log[OH^-]$$

The line for $NiOH^+$ is obtained from the expression for K_{f1}'.

$$Ni^{2+} + OH^- \rightleftharpoons NiOH^+$$

$$K_{f1}' = \frac{[NiOH^+]}{[Ni^{2+}][OH^-]}, \qquad [NiOH^+] = K_{f1}'[Ni^{2+}][OH^-]$$

Substituting for $[Ni^{2+}]$ from the K_{sp}' equation ($[Ni^{2+}] = K_{sp}'/[OH^-]^2$), we can get the above equation in terms of [OH⁻].

$$[NiOH^+] = \frac{K_{f1}'K_{sp}'[OH^-]}{[OH^-]^2}$$

$$\log[NiOH^+] = \log K_{f1}' + \log K_{sp}' - \log[OH^-]$$

From the K_{f2}' expression,

$$[Ni(OH)_{2(aq)}] = K_{f2}'[OH^-][NiOH^+].$$

Substituting $K_{f1}'K_{sp}'/[OH^-]$ for $[NiOH^+]$ gives

$$[Ni(OH)_{2(aq)}] = \frac{K_{f1}'K_{sp}'K_{f2}'[OH^-]}{[OH^-]}$$

$$\log[Ni(OH)_{2(aq)}] = \log K_{f1}' + \log K_{sp}' + \log K_{f2}'$$

Finally, the expression for $\log[\mathrm{Ni(OH)_3}^-]$ is determined from the K'_{f3} expression.

$$K'_{f3} = \frac{[\mathrm{Ni(OH)_3}^-]}{[\mathrm{Ni(OH)_{2(aq)}}][\mathrm{OH}^-]}$$

$$[\mathrm{Ni(OH)_3}^-] = K'_{f3}[\mathrm{Ni(OH)_{2(aq)}}][\mathrm{OH}^-]$$

Substituting $K'_{f1}K'_{f2}K'_{sp}$ for $[\mathrm{Ni(OH)_{2(aq)}}]$,

$$[\mathrm{Ni(OH)_3}^-] = K'_{f3}K'_{f1}K'_{f2}K'_{sp}[\mathrm{OH}^-]$$

$$\log[\mathrm{Ni(OH)_3}^-] = \log K'_{f3} + \log K'_{f1} + \log K'_{f2} + \log K'_{sp} + \log[\mathrm{OH}^-]$$

28. The simplest way to make this plot is to use the spreadsheet program on the accompanying CD. However, for the purpose of illustration, the method to obtain the position of each of the lines is given below.

Equations for the Cd species are as follows (derived as in Question 27).

$$\log[\mathrm{Cd}^{2+}] = \log K'_{sp} - 2\log[\mathrm{OH}^-]$$

$$\log[\mathrm{Cd(OH)}^+] = \log K'_{f1} + \log K'_{sp} - \log[\mathrm{OH}^-]$$

$$\log[\mathrm{Cd(OH)_{2(aq)}}] = \log K'_{f1} + \log K'_{f2} + \log K'_{sp}$$

$$\log[\mathrm{Cd(OH)_3}^-] = \log K'_{f1} + \log K'_{f2} + \log K'_{f3} + \log K'_{sp} + \log[\mathrm{OH}^-]$$

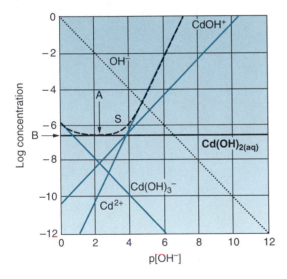

For the Cd²⁺ line: When $\log[\mathrm{Cd}^{2+}] = 0$,

$$\log[\mathrm{OH}^-] = -\log K'_{sp}/2 = 7.17$$

Thus the Cd²⁺ line starts at $\log[\mathrm{Cd}^{2+}] = 0$, $\mathrm{p[OH^-]} = 7.17$ and has a slope of 1/2 (minus 1 over the [OH⁻] coefficient in the $\log[\mathrm{Cd}^{2+}]$ equation).
For the Cd(OH)⁺ line: When $\log[\mathrm{Cd(OH)}^+] = 0$,

$$\log[\mathrm{OH}^-] = -\log K'_{f1} - \log K'_{sp} = -3.9 + 14.34 = 10.4.$$

Thus the Cd(OH)⁺ line starts at $\log[\mathrm{Cd(OH)}^+] = 0$, $\mathrm{p[OH^-]} = 10.4$ and has a slope of 1 (minus 1 over the [OH⁻] coefficient in the $\log[\mathrm{Cd(OH)}^+]$ equation).

For the Cd(OH)₂(aq) line: Since [OH⁻] does not appear in the $\log[\mathrm{Cd(OH)_{2(aq)}}]$ equation, the line has a slope of 0. The value for $\log[\mathrm{Cd(OH)_{2(aq)}}]$ is calculated from the values for the equilibrium constants in the equation. Thus,

$$\log[\mathrm{Cd(OH)_{2(aq)}}] = 3.9 + 3.8 - 14.34 = -6.6$$

and the horizontal line for $[\mathrm{Cd(OH)_{2(aq)}}]$ starts at -6.6 on the log concentration axis (Y-axis).
For the Cd(OH)₃⁻ line: We set $\log[\mathrm{OH}^-] = 0$ to obtain the intercept on the Y-axis. So, $\log[\mathrm{Cd(OH)_3}^-] = \log K'_{f1} + \log K'_{f2} + \log K'_{f3} + \log K'_{sp}$ when $\log[\mathrm{OH}^-]$ is 0. Thus,

$$\log[\mathrm{Cd(OH)_3}^-] = 3.9 + 3.8 + 0.7 - 14.34 = -5.9.$$

This, then, is the value for $\log[\mathrm{Cd(OH)_3}^-]$ when $\log[\mathrm{OH}^-] = 0$. The line has a slope of -1.
Other features: A dashed line is drawn on the graph at the values of the sum of the concentrations of all species at each value of p[OH⁻]. It shows a relatively large region over which the solubility is largely determined by $[\mathrm{Cd(OH)_{2(aq)}}]$. Therefore, the solubility is $10^{-6.6} = 2 \times 10^{-7}$ M (point B). The absolute minimum solubility occurs where $[\mathrm{Cd(OH)}^+] = [\mathrm{Cd(OH)_3}^-]$ (point A). Here p[OH⁻] $= 2.2$. The concentrations of other species at that p[OH⁻] are $\log[\mathrm{Cd}^{2+}] = -10$, $\log[\mathrm{Cd(OH)}^+] = \log[\mathrm{Cd(OH)_3}^-] = -8.2$, $\log[\mathrm{Cd(OH)_{2(aq)}}] = -6.6$, and p[OH⁻] = 2.2.

29. The log concentration diagram was constructed with $\mathrm{p[Hg_2^{2+}]}$ on the Y-axis. The [Cl⁻] line was constructed from a consideration of the solubility equilibrium constant ($= [\mathrm{Hg_2^{2+}}][\mathrm{Cl}^-]^2$), from which $\log[\mathrm{Cl}^-] = \frac{1}{2}\log K'_{sp} - \frac{1}{2}\log[\mathrm{Hg_2^{2+}}]$. When $\log[\mathrm{Hg_2^{2+}}] = 0$, $\log[\mathrm{Cl}^-] = \frac{1}{2}\log K'_{sp}$. When $\log[\mathrm{Cl}^-]$ is 0, $\log[\mathrm{Hg_2^{2+}}] = \log K'_{sp}$. These two points establish the [Cl⁻] line.

Before the titration begins, $\log[\mathrm{Hg_2^{2+}}] = -1$ (0% point). After 50% titration, $[\mathrm{Hg_2^{2+}}] = 0.050$ M and $\log[\mathrm{Hg_2^{2+}}] = -1.3$ (50% point). At the equivalence point, $[\mathrm{Hg_2^{2+}}] = 0.50[\mathrm{Cl}^-]$ (100% point), and when double the required amount of Cl⁻ has been added, $[\mathrm{Cl}^-] = C_{\mathrm{init}}/2 = 0.050$ M (200% point). The break (from 50% to 200%) is estimated at 12 p[Hg₂²⁺] units. The concentration of Hg₂²⁺ at the equivalence point is $10^{-5.3}$ M $= 5.0 \times 10^{-6}$ M.

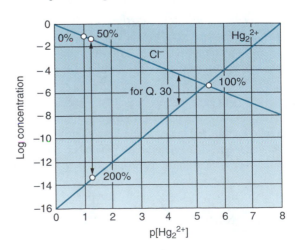

30. Since the titration is only feasible if the change in p[analyte] is greater than about 3, we choose the value for log concentration (on the Y-axis) that gives a difference between the Hg_2^{2+} and Cl^- lines which is no less than this value. From the log–log plot for Answer 29, you see that this occurs at a concentration of about 10^{-4} M.

31. Taking the log of the K'_{sp} expressions yields the following relationships: $\log[Pb^{2+}] + 2\log[I^-] = -8.1$; $\log[Hg_2^{2+}] + 2\log[I^-] = -28.3$; and $\log[Ag^+] + \log[I^-] = -16.1$. The titration is then analyzed as follows: The AgI will precipitate first, the Ag^+ line being the first encountered following the dashed -1 log concentration line from right to left. This will occur at a p[I$^-$] of 15 (point A). The Hg_2^{2+} will precipitate next at a p[I$^-$] of 13.7 (point B). At this point the Ag^+ concentration is $10^{-2.3}$, or 5×10^{-3} M. Thus, at this point 5% of the Ag^+ is still untitrated. That may be an unacceptable error. In addition, the magnitude of the break in p[I$^-$] between the onset of the Ag^+ and Hg_2^{2+} precipitations is only 1.3 log units. This is a marginally useful break even with the most careful data collection and plotting. Thus the titration is only crudely useful in distinguishing between Ag^+ and Hg_2^{2+}. The Pb^{2+} is the last to be precipitated, at a p[I$^-$] of 3.5 (point C). The break in p[I$^-$] between the Hg_2^{2+} and Pb^{2+} precipitations is huge. The equivalence point in the Pb^{2+} titration occurs where $[Pb^{2+}] = 2[I^-]$, or at p[I$^-$] = 2.8 (point D). The

100% excess point in the titration occurs at point E. The break between points C and E is usable if the titration is followed by an electrode, but it is not usable with an indicator. The problem with the Pb^{2+} titration is the relatively high solubility of the PbI_2. In summary, the proposed titration would be useful for the determination of even trace amounts of Ag^+ or Hg_2^{2+} with Pb^{2+}, but it will not accurately distinguish between Ag^+ and Hg_2^{2+}. The Pb^{2+} will only be quantitated at high concentrations.

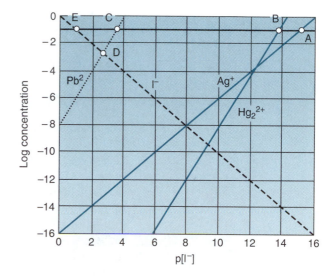

E. Separation by Precipitation

Precipitation can be used to separate an analyte from a mixture to facilitate analysis, or it can be used to remove an interferent from a mixture before analysis. Many analytes or interferents can be separated from each other in this way. Table 8.1 gives a few of the precipitating agents and the ions they precipitate.[1] If separation is the goal, it is important for the separation to be **quantitative**, that is, to be complete to

Table 8.1
Useful Precipitation Reagents

Reagent	Ion Precipitated
$NH_{3(aq)}$	Be^{2+}, Al^{3+}, Sc^{3+}, Cr^{3+}, Fe^{3+}, In^{3+}, Sn^{4+}
H_2S	Cu^{2+}, Zn^{2+}, Ge^{4+}, As^{3+}, Bi^{3+}, Sb^{3+}
H_2SO_4	Sr^{2+}, Cd^{2+}, Pb^{2+}, Ba^{2+}
H_2PtCl_6	K^+, Rb^+, Cs^+
$BaCl_2$	SO_4^{2-}
$MgCl_2$	PO_4^{3-}
8-Hydroxyquinoline	Al^{3+}, Fe^{2+}, Mg^{2+}, many more
Dimethylglyoxime	Only Ni^{2+}, with controlled p[H_3O^+]

[1] W. F. Hillebrand, G. E. F. Lundell, H. A. Bright, and J. I. Hoffman, *Applied Inorganic Analysis*. John Wiley & Sons, New York, 1953.

within 1 ppt. The process of separation by precipitation involves adding an excess of the precipitating agent to the solution (under conditions that discourage colloid formation and encourage crystallization without occlusion), filtering the precipitate from the reaction mixture, and washing the precipitate free of the precipitating solution. If the analyte remains in the solution, all wash solution must be returned to the analyte solution. If the analyte is now the precipitate, it must be completely retained in the filtering and washing steps. Each of these steps will be considered in further detail.

The earlier discussion of the nucleation and crystallization process suggests the procedures desirable for quantitative separation. The precipitating reagent should be added to the reaction mixture slowly in a relatively dilute solution, or homogeneous precipitation can be used. The amount of precipitant added must be carefully considered. As we have seen, there is a concentration of precipitant at which the solubility of the precipitate is a minimum. Too large an excess of precipitant can cause resolution through complexation; without some excess, however, a fraction of the material will not be precipitated.

Often a salt that does not react in the precipitating system is added to aid in the coagulation of any colloidal particles formed. It is common for the solution to be warmed to facilitate the rapid exchange of surface ions and to reduce occlusion. The solution may be kept warm after precipitation to continue the purification of the precipitate. For most precipitates, solubility increases with increased solution temperatures, so the solution will be cooled before filtering. The cooling is often done slowly so that the final growth of the precipitate crystals is gradual.

Filtration is the means by which the precipitate is separated from the reaction mixture. The filtering material can be filter paper held in a filter funnel, or it can be a glass or ceramic container with a fritted (somewhat porous) bottom. This container is fitted over a catch flask. In some cases, a partial vacuum is applied to the catch flask to aid in drawing the reaction fluid through the filter material. Transfer of all the precipitate from the reaction vessel to the filtration system is essential. This is accomplished by rinsing the reaction vessel with clean solution from a squirt bottle. Sometimes particles of precipitate clinging to the wall of the reaction vessel must be dislodged with a tool called a **rubber policeman**. Then the flask and the policeman must be rinsed again, with the rinse solution going into the filter container. The use of too much rinse solution can cause an error from the solubility of the precipitate in the rinse solution. The rinse solution sometimes contains a salt or a wetting agent to prevent loss of coagulation or creeping of the precipitate up the sides of the titration container. Quantitative transfer of a precipitate is an acquired skill that requires care and practice to perfect.

The degree of completeness of a separation by precipitation depends on the relationship between the initial concentration of the analyte and its solubility in the reaction mixture. If the solubility is $\sim 1 \times 10^{-6}$ M as in the case of AgCl described above, a 1×10^{-4} M solution of Ag^+ or Cl^- would only be 99% separated by this process. However, all analyte concentrations above $\sim 1 \times 10^{-3}$ M would be quantitatively separated.

Selectivity of the precipitation reaction can often be enhanced by the control of the $p[H_3O^+]$ of the solution. This is due to the hydrolysis of the cation or the anion, as in the case of complexation reactions. Another possibility is the addition of a complexation agent that is selective for an interferent. This can reduce the alpha value of the interferent below the solubility point and eliminate coprecipitation.

Once separated from its interferents, the analyte in the reaction liquid or in the precipitate can be analyzed in whatever way desired. One way, if the analyte is in the precipitate and the precipitate is pure, is simply to weigh the amount of precipitate produced. The factors that make this a useful method are discussed in the next section.

32. What is meant by quantitative separation, and what steps are required to achieve quantitative separation by precipitation?

33. In an experiment, 3.60×10^{-4} mol of $Pb(NO_3)_2$ is dissolved in 30 mL of a solution buffered at $p[H_3O^+] = 4.0$. Then 20 mL of 0.50 M KI is added to this solution to precipitate the Pb^{2+} as PbI_2. Calculate the amount of lead precipitated and the amount that remains in solution. The K'_{sp} for PbI_2 is 8.7×10^{-9}. The log K'_f for the monohydroxide complex of Pb^{2+} is 7.51. The acid HI is a strong acid in water.

Answers to Study Questions, Section E

32. Quantitative separation is a separation process that is complete to within 1 ppt. The steps for quantitative separation by precipitation include addition of an excess of precipitating reagent, quantitative transfer of *all* the reaction mixture to the filtration apparatus, and washing the precipitate to purity.

33. We must first check to see if either of the species forming the precipitate is significantly hydrolyzed at $p[H_3O^+] = 4.0$. The I^- is not because it is the conjugate base of a strong acid. The fraction of Pb^{2+} hydrolyzed is obtained from Equation 8.13:

$$\alpha_{Pb^{2+}} = \frac{1}{1 + K'_f[OH^-]} = \frac{1}{1 + 10^{7.51} \cdot 10^{-10}} = 0.997$$

which is close enough to 1 to be ignored. The solubility of PbI_2 can be calculated from Equation 8.10, substituting in the appropriate stoichiometric coefficients. Had hydrolysis been involved, Equation 8.15 would be used instead. Thus,

$$s = \left(\frac{K'_{sp}}{1 \cdot 2^2}\right)^{1/(1+2)} = \left(\frac{8.7 \times 10^{-9}}{4}\right)^{1/3} = 1.3 \times 10^{-3} \text{ M}$$

The amount of lead dissolved in the solution is s times the solution volume. The amount of lead not precipitated is 1.3×10^{-3} mol L^{-1} $\times$ 0.050 L $= 6.5 \times 10^{-5}$ mol. Therefore, the amount precipitated is 3.6×10^{-4} mol -6.5×10^{-5} mol $= 2.9 \times 10^{-4}$ mol. Of the total lead present, only $(2.9 \times 10^{-4}$ mol$/3.6 \times 10^{-4}$ mol$) \times 100 = 80\%$ was precipitated. This fraction would increase if the initial concentration of lead were higher. The lower the concentration of analyte in the precipitating solution, the greater must be the K'_{sp} for the separation to be quantitative.

F. Quantitation by Weighing the Precipitate

When the analyte has been separated from the sample mixture by precipitation, in principle, the amount of it can be determined by weighing. This method is one of a class of techniques called **gravimetric analysis**. In most cases, however, we will not be weighing the analyte as it was in the sample. It has now reacted with the precipitating reagent to create a new species. The number of molecules of the new species will be exactly related to the number of atoms of the analyte if the formation of the precipitate is exactly stoichiometric. For example, if Ba^{2+} is separated from a mixture by precipitation with SO_4^{2-}, the precipitate is $BaSO_4$. The number of moles of $BaSO_4$ is equal to the number of moles of Ba in the original sample if the ratio of Ba^{2+} to SO_4^{2-} in the $BaSO_4$ is exactly 1:1. Although we write virtually all chemical reactions with the stoichiometric factors in integer numbers, a surprising number of compounds are not exactly stoichiometric. The next consideration is that we must then be able to determine the number of moles of precipitate by weighing it. This process requires complete removal of the rinse solvent and the achievement of a stable form of the precipitate.

Quantitative transfers of the precipitated material to a weighing vessel are not attempted. Small precipitate particles remain in the pores of the filter, and these create a large error. Instead, the precipitate is filtered in such a way that it can be dried and weighed in the filter medium. This is readily done if a glass or ceramic filter container is used. Filter containers in which the precipitate will be dried are called **crucibles**. The filter container is dried and weighed before the filtration. Once weighed, tongs or

Scheme for gravimetric analysis

Differentiating characteristic:
Ability to react to form a precipitate.

Probe:
Add an excess of precipitating reagent to a solution of the analyte.

Response:
The analyte reacts with the precipitating agent, consuming the analyte and stoichiometrically forming a solid.

Measurement of the response:
Separate the precipitate from the solution, dry, bring to stable and known composition, and weigh.

Interpretation of data:
Use the weight of the precipitate and the stoichiometry of its formation to calculate the amount of analyte in the original solution.

gloves must be used to handle the crucible to avoid adding the weight of fingerprints. Paper filters cannot be returned to their original weight, so they must be burned. This is done by careful transfer of the filter paper, with the precipitate intact, to a crucible in which it will be ashed by placing under a heat lamp. The crucible in which the ashing takes place has previously been weighed. After the entire filter paper has been converted to water vapor and CO_2, only the precipitate remains in the crucible. Paper filters are not often used in industry because of the extra time and inconvenience involved in their use.

A major concern in the determination of the precipitate amount by weighing is the achievement of a stable composition of the precipitate. Many materials cannot be brought to a stable and reproducible composition. As the temperature of the crucible is raised (by placement in a drying oven) to dry the precipitate, the precipitate may begin to lose some waters of hydration. Before the last of these is gone, the material may undergo further decomposition. For example, carbonates lose CO_2 to form oxides, and hydroxides lose H_2O to form oxides. For some materials there is a range of temperatures over which the composition is stable, and they are suitable products for gravimetric analysis.

To calculate the amount of analyte from the weight of the material in the crucible, we must know the ratio of the weights of the substance sought and the substance weighed. For example, if Fe^{3+} ion is precipitated with OH^- ion, the precipitate formed is $Fe(OH)_3$. When it is weighed, it is in the form of Fe_2O_3. The mass of the iron in x g of Fe_2O_3 is

$$\text{grams of Fe} = x \frac{MW_{Fe}}{\frac{1}{2} MW_{Fe_2O_3}} = x \frac{55.847}{79.845} = 0.69943x$$

Only half the molecular weight of Fe_2O_3 is used in the denominator because each mole of Fe_2O_3 contains two moles of Fe. The factor that relates the weight of the Fe_2O_3 to the weight of the iron (0.69943) is called the **gravimetric factor**. It is a handy figure to have for a procedure that is done repetitively. Factors have been calculated for many common gravimetric procedures.

Because of its tedious nature, the gravimetric technique has limited application in modern industry: Moreover, the goals of much industrial analysis have changed over the years. For a long time, purity of a product was determined by finding the fraction of the product weight that could be accounted for by the substances that were supposed to be present. Thus Ivory soap was determined to be 99.44% pure.[2] Now, we are more likely to determine purity by the amount of specifically undesirable material that is present. Thus our attention has turned to trace analysis, where even parts per million of a particular irritant or toxin may be too much. Gravimetric methods have little application in trace analysis. This is because of the relatively large amount of material that is required to obtain an accurately weighable amount. Consider that the analytical balance can weigh to the nearest 0.1 mg. This means that the weighed form of the precipitate must weigh at least 100 mg to be determined quantitatively. If this were Fe_2O_3, the number of moles of Fe would be 0.100 g Fe_2O_3/79.845 g Fe_2O_3 per mole of Fe = 1.25 $\times$ 10^{-3} mol of Fe. If this amount of Fe^{3+} had been dissolved in 100 mL of solution, the concentration would have been 0.0125 M. Refinements of technique could extend the usefulness of gravimetric precipitation analysis by perhaps another two orders of magnitude, but it still could not find significant applicability in the determination of minor components.

---

[2]This famous analysis was done by Sadtler Research Laboratories in Philadelphia. When the standard deviations of the analyses were considered and the result rounded to appropriate significant figures, the result was actually 100% ± 1.2%.

34. For gravimetric analysis, a quantitative separation is required. What additional requirement is there regarding the material that is to be weighed?

35. What must be done to the precipitate to rid it of the last traces of solvent?

36. Precipitates that are hydroxides and carbonates are known to not withstand drying without some decomposition. What is the mechanism for using such precipitates in gravimetric analysis?

37. To determine the amount of Cl^- in 30.00 mL of an unknown solution, an excess of $AgNO_3$ is added. The AgCl precipitate was found to weigh 0.3912 g. What is the concentration of Cl^- (ppt) in the original solution? Assume a density of 1.00 g/mL for the original solution.

38. Compare the minimum concentrations that can be quantitatively analyzed by titration and gravimetry for 100 mL solutions of AgCl and AgI.

34. The material to be weighed must be pure and in a stable, weighable form with an exact and known stoichiometric composition.

35. The precipitate must be thoroughly dried, usually by heating.

36. Hydroxide and carbonate precipitates are heated to carry the decomposition process to a form that is stable and pure. The result is often an oxide. Temperatures much higher than those used for drying are generally required.

37. The reaction of silver and chloride ions is

$$Ag^+_{(aq)} + Cl^-_{(aq)} \rightleftharpoons AgCl_{(s)}$$

The amount of Cl^- in the precipitate is

$$(0.3912 \text{ g AgCl})\left(\frac{1 \text{ mol AgCl}}{143.321 \text{ g}}\right)\left(\frac{1 \text{ mol Cl}^-}{1 \text{ mol AgCl}}\right)$$

$$\times \left(\frac{35.4527 \text{ g Cl}^-}{\text{mol Cl}^-}\right) = 0.09677 \text{ g Cl}^-$$

Since the density of the solution is about 1 g/mL, the amount in parts per thousand is calculated from the milligrams solute per liter of solution.

$$\frac{96.77 \text{ mg Cl}^-}{30.00 \text{ mL}} = 3.226 \text{ ppt Cl}^-$$

38. The requirement for an accurate gravimetric determination is that the precipitate weigh at least 0.100 g (see page 278). The quantity of Cl^- necessary in solution to give 0.100 g AgCl is

$$0.100 \text{ g AgCl}\left(\frac{1 \text{ mol AgCl}}{143.3 \text{ g/AgCl}}\right)\left(\frac{1 \text{ mol Cl}^-}{1 \text{ mol AgCl}}\right)$$

$$= 6.98 \times 10^{-4} \text{ mol Cl}^-$$

In a solution volume of 100 mL, the concentration of Cl^- that can be analyzed gravimetrically is 7×10^{-4} mol/0.100 L = 7×10^{-3} M.

The same calculation for I^- gives

$$0.100 \text{ g AgI}\left(\frac{1 \text{ mol AgI}}{234.8 \text{ g AgI}}\right)\left(\frac{1 \text{ mol I}^-}{1 \text{ mol AgI}}\right)\left(\frac{1}{0.100 \text{ L}}\right)$$

$$= 4.26 \times 10^{-4} \text{ mol I}^-$$

The requirement for a detectable end point in a titration is a change in p[X] of at least 3 p units between 0% and 200% titrated. From Figure 8.15, we see that for the titration of Cl^- with Ag^+, $[Cl^-] \approx 1.0 \times 10^{-3}$ M gives an equivalence point break of 3. For the titration of I^- with Ag^+, the minimum concentration is about 3×10^{-6} M.

The improvement in detection limit of the iodide over the chloride is due to the higher molecular weight of AgI for the gravimetric method and the lower K'_{sp} for the titrimetric method.

G. Spot Tests and Test Strips

Precipitation reactions can be used for quick assessment of the presence of a particular material. For example, if a white precipitate is formed when Ag^+ is added to a solution, it is quite likely that there are halide ions present. Even more specifically, if the addition of dimethylglyoxime produces a bright red precipitate, Ni^{2+} is almost certainly present. The formation of precipitates of various colors, at particular pH values, and the solubility of those precipitates in the presence of specific complexing agents formed the backbone of the inorganic qualitative analysis scheme. Today, they are widely used in the convenient spot tests and test strips introduced in Chapter 7. Some examples of spot and test strip applications will serve to illustrate this point.

Test paper impregnated with a Ag_2CrO_4 solution and dried is used for detection and semiquantitative analysis of halide ions. A drop of the test solution is placed on the test paper. Any halide in the sample will take some of the silver ions from the orange Ag_2CrO_4 and form white silver halide. With some implementations, the size of the white spot produced when a single drop of sample is applied can be correlated (with a chart) to the concentration of halide in the sample. Nickel forms a deep red precipitate with diacetyldioxime. Paper containing some diacetyldioxime makes an effective test strip for Ni^{2+}.

Practice Questions and Problems

1. A. Write the thermodynamic equilibrium constant expression for the solubility of silver carbonate (Ag_2CO_3).

   B. Write the formal equilibrium constant expression for the solubility of silver carbonate.

   C. Write the equation that relates the log of the formal constant to the log of the thermodynamic constant and the ionic strength for the specific case of the K'_{sp} for silver carbonate.

2. A. Explain why occlusion is to be avoided in a precipitate. How can it be avoided? What will be the effect of occlusion in quantitation by titration and in quantitation by weighing the precipitate?

   B. Explain why colloid formation should be avoided and how it can be minimized. Does it affect quantitation by weighing or by titration more?

3. Ferrous sulfide (FeS) has a K'_{sp} of 8×10^{-19}. Ferrous ion forms a monohydroxide complex with a formation constant of 104.6. The K'_a values for H_2S are 9.6×10^{-8} and 1.3×10^{-14}.

   A. What is the solubility of FeS if the hydrolysis reactions are ignored?

   B. Taking the hydrolysis reactions into consideration, what is the solubility of FeS in a solution at $p[H_3O^+] = 3.0$?

   C. What is the solubility of FeS at $p[H_3O^+] = 9.0$?

4. A. Draw a log concentration diagram for the solubility of $BaSO_4$. Use $p[Ba^{2+}]$ for the X-axis. The K'_{sp} for $BaSO_4$ is 1×10^{-10}.

   B. Estimate the values of $p[Ba^{2+}]$ for the titration of 0.01 M Ba^{2+} with SO_4^{2-} titrant at the 0%, 50%, 100%, and 200% points.

   C. What is the magnitude of the titration break (50% to 200%) in $p[Ba^{2+}]$ units?

   D. What is the minimum concentration of Ba^{2+} that could be titrated with a titration break of 3 $p[Ba^{2+}]$ units?

   E. Can you think of any way to determine the equivalence point of this titration?

5. To find the Ce^{4+} content of a solid, 4.37 g was dissolved and treated with excess iodate to precipitate $Ce(IO_3)_4$. The precipitate was collected, washed well, dried, and ignited to produce 0.104 g of CeO_2 (MW 172.114). What was the weight percent of Ce in the original solid?

6. Consider either the gravimetric determination of Cl^- or the titrimetric determination of Ca^{2+}.

   A. What is the discriminating characteristic used in the analysis?

   B. What is the probe for that characteristic?

   C. What is the response to the probe?

   D. How do we measure the response to the probe?

   E. How do we interpret the measurement data to obtain the desired information?

7. Explain why the effect of hydrolysis reactions with the precipitating reagents is always to increase the solubility of the precipitate.

8. Calculate the solubility of AgCl in water with no excess of either Ag^+ or Cl^-. Using Figure 8.11, determine if it is okay in your calculation to ignore the intrinsic solubility of AgCl and the higher complex formation with Cl^-.

9. A. Why is it possible to titrate a lower concentration of I^- with Ag^+ than one can when titrating Cl^- with Ag^+?

   B. Discuss the practicality of determining Tl^+ ion by titrimetric precipitation with Cl^-.

   C. What precipitating anion would work better for Tl^+ than Cl^-?

10. Explain the Fajans end-point determination method for silver halide precipitation in terms of the charge on the precipitate particles.

11. A. In separation by precipitation, why is an excess of the precipitating reagent required?

    B. From Figure 8.12, what concentration of OH^- will result in the minimum solubility of $Ni(OH)_2$?

C. Since a deviation on either side of the value obtained in part B results in a considerable increase in solubility, how could you achieve this optimum $[OH^-]$ in practice?

D. What is the minimum concentration of Ni^{2+} that can be determined to 0.1% by weighing a precipitate of $Ni(OH)_2$? Why?

E. For the precipitation of AgCl and quantitative determination of the amount of Ag^+ in the original sample by weight of the AgCl, the precipitation must be complete, the separation must be complete, all the precipitate must be weighed, and only AgCl must be in the weighed material. Describe the steps that are taken to ensure that all these conditions are met.

12. Limestone contains MgO. In the gravimetric determination of the MgO on limestone, the Mg is precipitated as $MgNH_4PO_4$. This salt cannot be dried without decomposition so the precipitate is ignited (heated to a very high temperature) where its composition stabilizes as $Mg_2P_2O_7$. What is the percent of MgO (by weight) in a limestone sample if 0.9817 g of limestone yielded 0.1238 g of $Mg_2P_2O_7$?

13. A saturated solution of Ag_2SO_4 in otherwise pure water has a silver ion concentration of 3.11×10^{-2} M.

A. What is the solubility product constant for Ag_2SO_4?

B. Is the constant calculated in part A a formal or thermodynamic solubility product?

C. Would the addition of Na_2SO_4 to this solution increase or decrease the solubility?

D. Would the addition of $NaNO_3$ to this solution increase or decrease the solubility?

14. A. What is the solubility of AgCl in a solution that is 1.0 M in NH_3? Remember that Ag^+ forms a complex with ammonia.

B. How many milligrams of AgCl would dissolve in 20 mL of the 1.0 M ammonia solution?

C. What would happen if the amount of AgCl put in the 20 mL of the 1.0 M ammonia solution were 50 mg?

15. In the development of photographic film, a solution of $Na_2S_2O_3$ is used to dissolve all the AgBr remaining on the film. Why should such an insoluble precipitate as AgBr dissolve in this solution? (Hint: See Study Question 18.)

16. A. Calculate the ionic solubility of $Mg(OH)_2$ in pure water. Use 7.1×10^{-12} for K'_{sp}.

B. Determine the equilibrium concentrations of Mg^{2+} and OH^- in a saturated solution $Mg(OH)_2$.

C. What is the $p[H_3O^+]$ of the resulting solution?

D. What is the ionic solubility of $Mg(OH)_2$ in water with a $p[H_3O^+]$ of 12?

17. Calculate the intrinsic solubility of AgI.

18. A. Calculate the ionic solubility of $NiCO_3$ in water buffered at $p[H_3O^+] = 5.0$. Use 1.3×10^{-7} as the K'_{sp} for $NiCO_3$. Ni^{2+} forms complexes with OH^- and CO_3^{2-} is a weak base. $\log K'_{f1} = 4.1$, $\log K'_{f2} = 4.9$, $\log K'_{f3} = 3$, $pK'_{a1} = 6.4$, $pK'_{a2} = 10.3$.

B. Repeat the solubility calculation for water buffered at $p[H_3O^+] = 9.0$.

19. Aluminum can be gravimetrically determined by precipitation as the hydroxyquinolate, $Al(C_9H_6ON)_3$. Calculate the percentage of aluminum in a sample that weighed 0.4827 g and yielded a precipitate weight of 0.3649 g.

20. A sample of a hydrate of copper sulfate ($CuSO_4 \cdot xH_2O$) that weighs 0.2837 g weighs 0.1636 g after being heated to a constant weight. Calculate the value of x assuming that all the waters of hydration were lost by heating.

Suggested Related Experiments

1. Gravimetric and volumetric chloride by precipitation with silver ion.

2. Experiments with test strips and comparison with quantitative analysis.

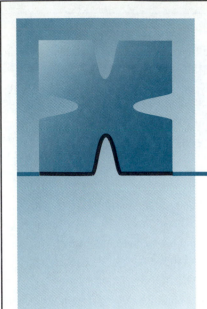

Chapter Nine

ANALYSIS BY ELECTRODE POTENTIAL

The immense success and usefulness of the pH electrode for the measurement of H_3O^+ ion activity has motivated the search for probes that could convert the chemical activity or concentration of other species to a related electrical potential. Coversion of chemical information to an electrode potential is, in many ways, an ideal concept in chemical measurement. Simply insert a probe into the sample (stream, pool, artery, sewage pipe, single cell, etc.) and read the concentration directly from a battery-operated, handheld meter. In this chapter, we will explore the development of such probes and discuss the technique of their application in chemical measurement. Although none of the probes developed has matched the pH probe in universality and ideality of application, probes have been developed that are of great value in many applications. The development of electrode sensors remains an active research area as the search for increased selectivity and decreased detection limits continues.

The discussion below is organized according to the three mechanisms by which the electrode potentials are generated namely, electron exchange at metals, ion reactions on membrane surfaces, and ion diffusion through porous membranes.

A. Electron Exchange at Metals

If a piece of copper metal is immersed in a solution containing cupric ions (from the dissolution of $CuSO_4$, for instance) some of the cupric ions will adopt a position on the copper crystal surface as shown in Figure 9.1. Each cupric ion takes two electrons from the metal to become a part of the solid. To the extent this occurs, the piece of copper will develop a positive electrical charge with respect to the solution. This separation of the positively charged cupric ions from their counterions (SO_4^{2-} in this exam-

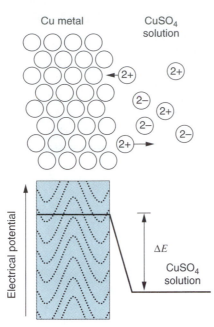

Cu metal

CuSO₄ solution

Electrical potential

ΔE

CuSO₄ solution

Figure 9.1. Copper metal, immersed in a solution with Cu^{2+}, will develop an equilibrium electrical potential difference between the metal and the solution. This potential will be positive if the metal ions have a tendency to add to the metal lattice and negative if the metal has a tendency to dissolve into the solution as ions.

ple) results in the development of an electrical potential between the copper and the solution (see Background Material C) in which the copper is positive with respect to the solution. As more cupric ions, deposit on the copper crystal, the potential difference increases. However, the positively charged cupric ions now have to overcome the electrostatic repulsion of the positively charged metal to take their place on the surface. Surface copper atoms, on the other hand, are increasingly attracted to dissolve into the solution as cupric ions, as this relieves the metal of some of its excess charge. A dynamic exchange is occurring as shown in the following reaction.

$$Cu^{2+}_{(aq)} + 2e^-_{(Cu)} \rightleftharpoons Cu_{(s)} \qquad 9.1$$

This dynamic, bidirectional reaction quickly comes to equilibrium when the reaction proceeds at the same rate in both directions. The electrical potential difference between the metal and solution is also at an equilibrium value.

The electrical potential is constant within the metal and within the solution. The two phases are conductors with no electrical current passing through them, so they must be at the same potential throughout. The entire potential difference between the metal and solution occurs in the immediate vicinity of the interface.

THE EQUILIBRIUM POTENTIAL DIFFERENCE

Recall from Chapter 3 (Equation 3.33) that for a reaction at equilibrium, the sum of the chemical potentials of the reactants is equal to the sum of the chemical potentials of the products. Therefore, for the reaction of Equation 9.1,

$$\overline{\mu}_{Cu^{2+}_{(aq)}} + 2\overline{\mu}_{e^-_{(Cu)}} = \mu_{Cu(s)} \qquad 9.2$$

Notice, however, that the term $\overline{\mu}$ has been used for the cupric ion and the electrons. This quantity, called the **electrochemical potential,** takes into account the electrical energy a charged particle has when it is in a conductor at potential E_{cond}.

The electrochemical potential is related to the chemical potential by a simple relationship.

$$\overline{\mu}_X = \mu_X + zFE_{cond} \qquad 9.3$$

When a reaction involves the movement of a charged particle between regions of different electrical potential, the electrochemical potentials of the reactants and products are equal at equilibrium. The equilibrium condition, then, includes the electrical potential difference between the two phases.

In this equation, z is the number and sign of the unit charges on the species, F is Faraday's constant (96,485 coulombs per mole of electrons), and the subscript cond stands for the conducting medium containing X.

From Equation 3.22, $\mu_X = \mu_X^\circ + RT \ln a_X$, so for the generic ion X in solution,

$$\bar{\mu}_X = \mu_X^\circ + RT \ln a_X + zFE_{\text{sol'n}} \qquad 9.4$$

where μ_X° is the chemical potential of the standard state of X. When these relationships are substituted in Equation 9.2 representing the equilibrium state,

$$\mu_{\text{Cu}^{2+}}^\circ + RT \ln a_{\text{Cu}^{2+}} + 2FE_{\text{sol'n}} + 2\mu_{\text{e}^-\,(\text{Cu})}^\circ - 2FE_{\text{Cu}} = \mu_{\text{Cu}(s)}^\circ \qquad 9.5$$

The term μ° is used for the solid copper because it is in its standard state.

Equation 9.5 can then be solved for the nominal difference in potential between the copper metal and the solution.

$$2F(E_{\text{Cu}} - E_{\text{sol'n}}) = \mu_{\text{Cu}^{2+}}^\circ - \mu_{\text{Cu}(s)}^\circ + 2\mu_{\text{e}^-\,(\text{Cu})}^\circ + RT \ln a_{\text{Cu}^{2+}}$$

$$E_{\text{Cu}} - E_{\text{sol'n}} = \frac{\mu_{\text{Cu}^{2+}}^\circ - \mu_{\text{Cu}(s)}^\circ + 2\mu_{\text{e}^-\,(\text{Cu})}^\circ}{2F} + \frac{RT}{2F} \ln a_{\text{Cu}^{2+}}$$

A term in this equation is composed entirely of constant chemical potentials and Faraday's constant. This collection of constants is given the symbol E°. In addition, since we have invoked the equilibrium condition in writing Equation 9.5, we will call this nominal potential difference E_{eq}.

$$E_{eq} = (E_{\text{metal}} - E_{\text{sol'n}}) \qquad 9.6$$

Thus,

$$E_{eq} = E_{\text{Cu}^{2+},\text{Cu}}^\circ + \frac{RT}{2F} \ln a_{\text{Cu}^{2+}} \qquad 9.7$$

The term E° is called the **standard potential** because it is the value the potential would have if all the reactants were in their standard states. Since the standard potential is different for each metal and ion combination, subscripts are used to identify the particular combination the E° value is for. In this case, it is the combination Cu^{2+},Cu.

Equation 9.7 reveals that the potential difference between the copper metal and the solution should increase with increasing activity of the cupric ion in the solution, as shown in Figure 9.2. This makes sense because the increased chemical activity of the cupric ions makes it possible for them to overcome a higher positive potential barrier.

Equation 9.7 can be generalized for any metal ion in solution in equilibrium with its solid metal:

$$\text{M}^{n+} + ne^- \rightleftharpoons \text{M}_{(s)} \qquad 9.8$$

for which

$$E_{eq} = E_{\text{M}^{n+},\text{M}}^\circ + \frac{RT}{nF} \ln a_{\text{M}^{n+}} \qquad 9.9$$

Every metal, when immersed in a solution containing ions of that metal, will establish an equilibrium potential with respect to the solution. That potential will be different for each metal because of the different values of $E_{\text{M}^{n+},\text{M}}^\circ$, but in every case, the potential difference will increase with increasing activity of the metal ion. Note that this potential difference exists just from the contact between the metal and the solution. It requires some metal ions in the solution but does not need any connection to the

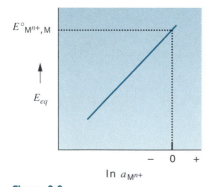

Figure 9.2. The potential difference between a piece of metal and a solution containing ions of the metal (an oxidized form of the metal) depends linearly on the log of the chemical activity of the metal ions. The higher the concentration, the more positive is the potential.

metal. If a wire were attached to the metal, however, we would call it an **electrode,** and the electrode's potential with respect to the solution would respond to the activity of M^{n+} according to Equation 9.9.

METAL–SOLUTION POTENTIALS FROM TWO RELATED IONIC SPECIES

If a solution contains ions that are related to each other by a difference in their oxidation state, a metal immersed in this solution can attain an equilibrium potential with respect to the solution. An example of this is a solution of Fe^{2+} (ferrous ion), Fe^{3+} (ferric ion), and a corresponding number of anions. A piece of metal contacting this solution can serve as a reservoir of electrons for the reaction

$$Fe^{3+} + e^- \rightleftharpoons Fe^{2+} \qquad\qquad 9.10$$

Ferric ions can approach the metal surface, take an electron, and return to the solution as ferrous ions (shown in Figure 9.3). Similarly, ferrous ions can leave an electron with the metal to become ferric ions. This dynamic exchange will come to an equilibrium state. Since the electrons are one of the reactants, their availability (as established by the electrical potential of the metal) is part of the equilibrium.

A relationship for the nominal equilibrium potential can be obtained in a way exactly parallel to that for the case of the copper metal in a solution of cupric ion. We begin by equating the electrochemical potentials of the reactants and products.

$$\overline{\mu}_{Fe^{3+}} + \overline{\mu}_{e^-} = \overline{\mu}_{Fe^{2+}} \qquad\qquad 9.11$$

Then, using Equation 9.5, but for the ferrous and ferric ions,

$$\mu^{\circ}_{Fe^{3+}} + RT \ln a_{Fe^{3+}} + 3FE_{sol'n} + \mu^{\circ}_{e^-(metal)} - FE_{metal} = \mu^{\circ}_{Fe^{2+}} + RT \ln a_{Fe^{2+}} + 2FE_{sol'n}$$

When this equation is solved for E_{eq},

$$E_{eq} = \frac{\mu^{\circ}_{Fe^{3+}} - \mu^{\circ}_{Fe^{2+}} + \mu^{\circ}_{e^-(metal)}}{F} - \frac{RT}{F} \ln \frac{a_{Fe^{2+}}}{a_{Fe^{3+}}} \qquad\qquad 9.12$$

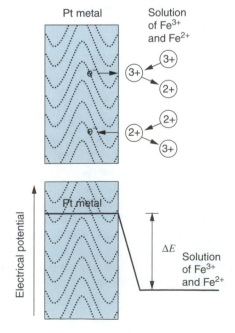

Pt metal

Solution of Fe^{3+} and Fe^{2+}

Electrical potential

Pt metal

ΔE

Solution of Fe^{3+} and Fe^{2+}

Figure 9.3. If a solution contains oxidized and reduced forms of a species, a piece of metal, not related to the oxidized and reduced forms, will develop an electrical potential with respect to the solution. The metal is serving as a reservoir of electrons to be exchanged between the oxidized and reduced forms.

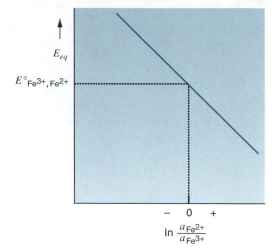

Figure 9.4. In the case of the metal immersed in a solution containing a redox couple, the potential developed between the metal and the solution depends linearly on the log of the ratio of the activities of the oxidized and reduced forms. The greater the relative activity of the more oxidized form, the more positive is the potential difference.

Again, the terms in the left part of the equation are all constants, so we will use the standard potential symbol $E°$ for them.

$$E_{eq} = E°_{Fe^{3+},Fe^{2+}} - \frac{RT}{F} \ln \frac{a_{Fe^{2+}}}{a_{Fe^{3+}}}$$
                                                                                        9.13

From Equation 9.13, we see that the equilibrium potential depends on the ratio of activities of the ferric and ferrous ions. The potential will decrease as the ratio of ferrous ion to ferric ion activity increases. (See Figure 9.4.)

Equation 9.13 can be generalized for the reaction of any pair of species in solution that are related oxidized and reduced forms (here symbolized as Ox and Red).

$$Ox + ne^- \rightleftharpoons Red$$
                                                                                        9.14

The species Ox and Red are called a **conjugate oxidation–reduction pair** or sometimes a **redox couple.** They are related by the gain or loss of n electrons just as a conjugate acid–base pair is related by the gain or loss of a proton. We will refer to reaction 9.14 as an **electron gain reaction** or as a **reduction reaction.**

The equilibrium potential established by the redox couple Ox,Red is

$$E_{eq} = E°_{Ox,Red} - \frac{RT}{nF} \ln \frac{a_{Red}}{a_{Ox}}$$
                                                                                        9.15

Equation 9.15 is called the **Nernst equation.** A consequence of this equation is the general rule that *the more oxidizing the couple (more positive $E°$ and/or greater ratio of a_{Ox} to a_{Red}), the more positive is the equilibrium electrode potential, E_{eq}.*

The metal on which the potential of Equation 9.15 is established acts simply as a reservoir for the electrons exchanged between the oxidized and reduced forms of the couple. Each different redox couple has a different standard potential value. It is important that the metal not be able to produce oxidized forms of itself in solution. The involvement of the metal atoms in the charge exchange would present a competing potential-establishing mechanism. Therefore, we should not use a piece of iron for the electron reservoir when there are Fe^{3+} and Fe^{2+} ions in the solution. It is better to use a nonreactive or **inert electrode** such as a piece of platinum or gold. It also follows that only one redox couple should exist in the solution. Again, note that the potential difference between the metal and the solution exists from the moment the metal contacts the solution. No other reaction or connection to the metal is involved in the establishment of this potential.

The Nernst equation (Equation 9.15) is also the general equation for all cases. We can see that it includes the case of the copper metal in cupric solution where Cu^{2+} is Ox and $Cu_{(s)}$ is Red. The redox couple is then Cu^{2+},Cu. When these species are substituted in the Nernst equation, the exact expression of Equation 9.7 is obtained.

$$E_{eq} = E^{\circ}_{Cu^{2+},Cu} - \frac{RT}{2F} \ln \frac{a_{Cu_{(s)}}}{a_{Cu^{2+}}}$$

$$E_{eq} = E^{\circ}_{Cu^{2+},Cu} - \frac{RT}{2F} \ln \frac{1}{a_{Cu^{2+}}} = E^{\circ}_{Cu^{2+},Cu} + \frac{RT}{2F} \ln a_{Cu^{2+}}$$

The Nernst equation is thus used for all instances of potential establishment on a metal electrode by electron exchange between the oxidized and reduced forms of a redox couple. It is written in the form of a_{Red} over a_{Ox} because that form maintains the convention of $a_{Product}$ over $a_{Reactant}$ that has been accepted for writing equilibrium constants.

Study Questions, Section A

1. Why is an electrical potential developed between a metal and a solution of its ions?

2. Can the electrical potential developed between a metal and a solution of its ions have either sign?

3. In what way does the electrochemical potential differ from the chemical potential?

4. Does the potential of the metal with respect to the solution become more positive or more negative as the concentration of the metal ion in the solution is increased?

5. By what mechanism is the potential difference between an inert metal and an adjoining solution affected by the presence of a redox pair in the solution?

6. If the concentration of the reduced form of the redox couple is increased, does the potential of the inert metal in the solution increase or decrease?

7. How are the forms of a redox couple related to each other, and how is this analogous to an acid–base couple?

Answers to Study Questions, Section A

1. The ions in the solution and the atoms on the metal surface can exchange positions. The species with the higher chemical potential will tend to take the other form. The resulting transfer of charge between the solution and the metal will produce an electrical potential difference between them. The potential difference will come to an equilibrium value when it exactly counteracts the initial chemical potential difference.

2. Yes, E_{eq} may be positive or negative. If the chemical potential of the metal atoms is greater than that of the metal ions in the solution, the resulting potential of the metal will be negative with respect to the solution.

3. If a species is charged, the potential of the medium it is in affects its reactivity with species in a medium of different potential. The electrochemical potential is the chemical potential with this factor included. For neutral species, the chemical and electrochemical potentials are the same.

4. The potential of the metal with respect to the solution increases with increasing metal ion concentration because the

equilibrium is shifted in favor of more metal ions taking a place in the metal lattice. This will make the metal potential more positive. See also Equation 9.9.

5. The oxidized and reduced forms of the redox pair can exchange electrons at the metal surface. This brings the potential of the metal to the oxidizing potential of the redox couple in the solution.

6. The increase in the reduced form concentration makes the solution more reducing (or less oxidizing). Since the more positive potentials are associated with greater oxidizing power, the metal potential would become more negative. See also Equation 9.15.

7. The forms of a redox couple are different from each other by n electrons. If the Ox form gains n electrons, it becomes the reduced form, and vice versa. In the same way, an acid–base couple differs by a proton, with the acid form becoming the base form by losing a proton.

B.   Calculating Redox Equilibrium Electrode Potentials

To calculate the nominal potential between a piece of metal and the solution with which it is in contact, one must first identify the redox couple that is exchanging electrons at the metal surface. Then one must find or calculate the value of the standard potential $E^{\circ}_{\text{Ox,Red}}$ for that couple and use the Nernst equation (Equation 9.15) to calculate the potential, E_{eq}. Tables of standard potential values for redox couples can be found in many sources. Table 9.1 is an abbreviated table of reduction potentials; a more extended table is found in Appendix E. Notice that for some of the entries in the table, other species are involved in the change in oxidation state, especially H_3O^+ and H_2O. (In many tables, H_3O^+ is abbreviated to H^+.) When implementing the Nernst equation for the calculation of E_{eq}, all the species given in the Ox and Red portions of the table must be used, to their appropriate coefficients. As examples,

$$E_{eq} = 0.222 - \frac{RT}{F} \ln \frac{a_{Ag_{(s)}} a_{Cl^-}}{a_{AgCl_{(s)}}} = 0.222 - \frac{RT}{F} \ln a_{Cl^-}$$

$$E_{eq} = 0.334 - \frac{RT}{2F} \ln \frac{a_{U^{4+}} a^6_{H_2O}}{a_{UO_2^{2+}} a^4_{H_3O^+}} = 0.334 - \frac{RT}{2F} \ln \frac{a_{U^{4+}}}{a_{UO_2^{2+}} a^4_{H_3O^+}}$$

There are two difficulties in evaluating such formulas. One is the natural log term, and the other is the use of activities instead of concentrations. The first problem is avoided by the use of the conversion factor between the more familiar base 10 logs

Table 9.1
Standard Reduction Potentials

Ox	n	Red	E°, V
$Cl_{2(g)}$	$+2e^-$	$2Cl^-$	$+1.359$
$O_{2(g)} + 4H_3O^+$	$+4e^-$	$6H_2O$	$+1.229$
$Br_{2(aq)}$	$+2e^-$	$2Br^-$	$+1.087$
Ag^+	$+e^-$	$Ag_{(s)}$	$+0.799$
Fe^{3+}	$+e^-$	Fe^{2+}	$+0.771$
I_3^-	$+2e^-$	$3I^-$	$+0.536$
Cu^{2+}	$+2e^-$	$Cu_{(s)}$	$+0.337$
$VO^{2+} + 2H_3O^+$	$+e^-$	$V^{3+} + 3H_2O$	$+0.337$
$UO_2^{2+} + 4H_3O^+$	$+2e^-$	$U^{4+} + 6H_2O$	$+0.334$
$Hg_2Cl_{2(s)}$	$+2e^-$	$2Hg_{(l)} + 2Cl^-$	$+0.268$
$AgCl_{(s)}$	$+e^-$	$Ag_{(s)} + Cl^-$	$+0.222$
$SO_4^{2-} + 2H_3O^+$	$+2e^-$	$SO_3^{2-} + 3H_2O$	$+0.172$
Sn^{4+}	$+2e^-$	Sn^{2+}	$+0.139$
$TiO^{2+} + 2H_3O^+$	$+e^-$	$Ti^{3+} + 3H_2O$	$+0.099$
$S_4O_6^{2-}$	$+2e^-$	$2S_2O_3^{2-}$	$+0.09$
$2H_3O^+$	$+2e^-$	$H_{2(g)} + 2H_2O$	0.000
Sn^{2+}	$+2e^-$	$Sn_{(s)}$	-0.140
$AgI_{(s)}$	$+e^-$	$Ag_{(s)} + I^-$	-0.151
V^{3+}	$+e^-$	V^{2+}	-0.255
Cr^{3+}	$+e^-$	Cr^{2+}	-0.38
Cd^{2+}	$+2e^-$	$Cd_{(s)}$	-0.403
Fe^{2+}	$+2e^-$	$Fe_{(s)}$	-0.440
U^{4+}	$+e^-$	U^{3+}	-0.63
Zn^{2+}	$+2e^-$	$Zn_{(s)}$	-0.763

and natural logs, $\ln x = 2.303 \log_{10} x$. (See Background Material B.) This allows a modification of the Nernst equation to

$$E_{eq} = E^\circ_{\text{Ox,Red}} - \frac{RT}{nF} 2.303 \log_{10} \frac{a_{\text{Red}}}{a_{\text{Ox}}} \qquad 9.16$$

The term $2.303RT/F$ appears frequently, so we will give it the symbol V_N (V because it has the units of volts and N for Nernst). Therefore, Equation 9.16 can be further simplified as

$$E_{eq} = E^\circ_{\text{Ox,Red}} - \frac{V_N}{n} \log_{10} \frac{a_{\text{Red}}}{a_{\text{Ox}}}$$

$$E_{eq} = E^\circ_{\text{Ox,Red}} = \frac{0.0592}{n} \left(\frac{T}{298}\right) \log_{10} \frac{a_{\text{Red}}}{a_{\text{Ox}}} \qquad 9.17$$

where the parenthetical temperature ratio can be omitted for systems near 25 °C. At 25 °C, $V_N = 0.0592$ V. This is a very useful number to remember.

FORMAL AND THERMODYNAMIC STANDARD POTENTIALS

The difficulty with the use of activities in the calculation of metal–solution potentials can be overcome through the calculation of the activity coefficients involved with the Debye–Hückel equations. An easier alternative, when available, is the use of a concept exactly analogous to the formal equilibrium constants developed in previous chapters. In this case, we use **formal potentials** $(E'_{\text{Ox,Red}})$[1] instead of standard potentials $(E^\circ_{\text{Ox,Red}})$. With formal standard potentials instead of thermodynamic standard potentials, the Nernst equations are written with molar concentrations of dissolved solutes instead of activities. For the examples developed earlier, the Nernst equations using formal potentials are

$$E_{eq} = E'_{\text{AgCl,Ag}} - V_N \log [\text{Cl}^-]$$

$$E_{eq} = E'_{\text{UO}_2^{2+},\text{U}^{4+}} - \frac{V_N}{2} \log \frac{[\text{U}^{4+}]}{[\text{UO}_2^{2+}][\text{H}_3\text{O}^+]^4}$$

There is, of course, a relationship between the formal and standard reduction potentials when the DHLL can be used. For the general reaction

$$a\text{A}^{z_A} + b\text{B}^{z_B} + \cdots + n\text{e}^- \rightleftharpoons p\text{P}^{z_P} + q\text{Q}^{z_Q} + \cdots$$

it is

$$E' = E^\circ - \frac{V_N}{n} \log \frac{\gamma_P^p \cdot \gamma_Q^q \cdots}{\gamma_A^a \cdot \gamma_B^b \cdots}$$

$$E' = E^\circ + \frac{V_N A \sqrt{S}}{n} (pz_P^2 + qz_Q^2 + \cdots - az_A^2 - bz_B^2 - \cdots) \qquad 9.18$$

where, again, the activity coefficients of all the species given in the Ox and Red portions of the table of reduction potentials must be used, to their appropriate coefficients.

Concentrations can be used in the Nernst equation instead of activities if the formal standard potential is used instead of the thermodynamic standard potential. This is analogous to using formal equilibrium constants rather than thermodynamic equilibrium constants for other equilibria. The formal standard potential for a redox couple depends on solution conditions, particularly ionic strength.

[1]Some texts use $E^{\circ\prime}_{\text{Ox,Red}}$ rather than $E'_{\text{Ox,Red}}$ as the symbol for the formal potential. In this text, we are using the super o to indicate thermodynamic values and the prime to indicate formal values, for both equilibrium constants and standard potentials.

For the AgCl,Ag couple,

$$AgCl_{(s)} + e^- \rightleftharpoons Ag_{(s)} + Cl^-$$

$$E'_{AgCl,Ag} = E^\circ_{AgCl,Ag} + \frac{V_N A\sqrt{S}}{1}(1 \cdot 1^2)$$

$$E'_{AgCl,Ag} = E^\circ_{AgCl,Ag} + V_N A\sqrt{S}$$

and for the uranate,uranic ion couple,

$$UO_2^{2+} + 4H_3O^+ + 2e^- \rightleftharpoons U^{4+} + 6H_2O$$

$$E'_{UO_2^{2+},U^{4+}} = E^\circ_{UO_2^{2+},U^{4+}} + \frac{V_N A\sqrt{S}}{2}(1 \cdot 4^2 - 1 \cdot 2^2 - 4 \cdot 1^2)$$

$$E'_{UO_2^{2+},U^{4+}} = E^\circ_{UO_2^{2+},U^{4+}} + V_N A\sqrt{S}\ (4)$$

Formal potentials are available for many couples in the medium in which they are commonly used, but if the formal potential is not available, it can be estimated by the approach illustrated above. The Nernst equation, in terms of the formal potential and concentrations, is

$$E_{eq} = E'_{Ox,Red} - \frac{V_N}{n}\log_{10}\frac{[Red]}{[Ox]} \qquad 9.19$$

where $V_N = 0.0592T/298$

Equation 9.19 is a very useful form of the Nernst equation. However, it is only valid for changes in the ratio of [Red] to [Ox] for which the activity coefficients remain constant. This is because the activity coefficients are now incorporated into the value for E'.

RELATING REDOX CONCENTRATIONS TO EQUILIBRIUM POTENTIALS

Some major points regarding electrode potentials

1. The principal factor determining the potential of an electrode is the E' of the couple acting at the electrode.

2. The more positive the electrode potential, the more oxidizing is the solution it is in.

3. Differences in the ratio of [Ox] to [Red] will cause the electrode potential to differ from the value of E'. The amount of the change is equal to about 60 mV/n for each 10-fold change from a ratio of 1.

Equation 9.19 is the means by which the concentrations of the redox species are related to the equilibrium potential they create in a metal electrode. For example, if a solution at 25 °C contains Fe^{3+} and Fe^{2+} in concentrations of 0.010 M and 0.0010 M, these values along with n and the formal reduction potential are used in Equation 9.19 to calculate the equilibrium potential as follows.

$$E_{eq} = 0.73 - \frac{0.0592}{1}\log\frac{0.001}{0.01} = 0.73 + 0.059 = 0.79\ V$$

This calculation illustrates several important points. One is that the ratio of [Red] to [Ox] changes the potential around the value of the formal reduction potential. The amount of the change from the formal reduction potential is relatively small except in the case of huge or tiny ratios of [Red] to [Ox]. The equilibrium potential is more negative than the formal reduction potential if the reduced form predominates, and it is more positive if the oxidized form predominates.

THE pH DEPENDENCE OF EQUILIBRIUM POTENTIALS

Several of the redox couples in a table of reduction potentials such as Table 9.1, differ in the number of oxygen atoms as well as the number of electrons. In aqueous solutions, the source or acceptor of oxygen atoms is water. Such reactions also involve H_3O^+ or OH^- ions. Because all reactants and products are used in the Nernst equation, the concentrations of H_3O^+ or OH^- will appear in the numerator or denominator of the Nernst equation when it is evaluated. An example is that of the uranate/uranic ion couple developed earlier in this section and repeated here.

$$E_{eq} = E'_{UO_2^{2+},U^{4+}} - \frac{V_N}{2} \log \frac{[U^{4+}]}{[UO_2^{2+}][H_3O^+]^4}$$

Now, with this couple, the H_3O^+ appears on the Ox side of the balanced reduction reaction, so it is in the denominator of the Nernst equation. The equilibrium potential of the couple will be equal to its formal potential if the concentrations of all the species are 1 M. This includes the concentration of H_3O^+. If $[H_3O^+]$ is less than 1 M, the equilibrium potential will be reduced. In other words, the couple becomes less oxidizing. As it turns out, for all reduction reactions, the H_3O^+ is on the oxidized side of the equation, or the OH^- is on the reduced side. The consequence of this is that all couples for which water is involved as a source or acceptor of oxygen are affected by $[H_3O^+]$ the same way. This enables us to make a general rule: *For couples in which H_3O^+ or OH^- are involved in the reduction reaction, as the pH of the solution increases, the equilibrium potential decreases.*

THE HYDROGEN ELECTRODE

The hydrogen electrode exhibits the potential of the couple for which Ox is H_3O^+ and Red is $H_{2(g)}$.

$$2H_3O^+ + 2e^- \rightleftharpoons H_{2(g)} + 2H_2O \qquad 9.20$$

Physically, the reaction is carried out on a platinum electrode in an acidic solution saturated in H_2 gas (formed by bubbling H_2 through the solution), as shown in Figure 9.5. The thermodynamic potential of this couple is defined as 0.000 V to be consistent with the thermodynamic values for the free energies of other reactions. This value can be seen in Table 9.1. Since the reduced form of the couple is a gas, the Nernst equation is written with the H_2 term as a partial pressure. Thus,

$$E_{eq} = 0.000 - \frac{V_N}{2} \log \frac{p_{H_2}}{(a_{H_3O^+})^2} \qquad 9.21$$

Because of the assignment of 0.000 V for the standard reduction potential of the hydrogen electrode, the equilibrium potential of any electrode, as calculated from the Nernst equation, is not the absolute potential of the electrode with respect to the solution; it is the potential of the electrode with respect to the thermodynamically ideal hydrogen electrode (the **standard hydrogen electrode** or SHE). Ideal behavior for the hydrogen electrode is not achieved at a molar concentration for H_3O^+ because the activity coefficient for H_3O^+ at that high concentration is much smaller than 1. At much lower concentrations of H_3O^+, where the activity coefficient can be calculated or is unity, the hydrogen electrode potential (relative to the hypothetical SHE) can be accurately predicted from the Nernst equation.

It is not important whether the potential of the SHE with respect to the solution is really 0 V or not because all potentials are relative. *Electrical potential is a potential difference,* just as altitude is the difference in vertical distance between the location and an arbitrarily assigned "sea level." We will see in the next section that one can only measure the difference in potential between two electrodes, and this difference must correspond to the actual differences in the electrode equilibrium potentials regardless of how the zero level for the table of reduction potentials was chosen. See Figure 9.6 for an illustration of this concept.

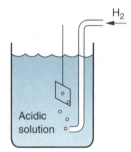

Figure 9.5. A hydrogen electrode. Bubbling H_2 gas through the solution under atmospheric pressure keeps the solution at a concentration of dissolved H_2 that is in equilibrium with an H_2 partial pressure of 1.0 atm.

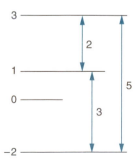

Figure 9.6. A diagram to illustrate how differences between two points on a scale are independent of the position of zero on the scale.

The SHE has been used for the determination of pH since its equilibrium potential is a function of the activity of H_3O^+. As in the case of all the other couples that involve H_3O^+ or OH^-, its potential decreases V_N volts for every unit of pH above pH = 0. The SHE was used extensively in the determination of the pH of the buffer solutions that now constitute the standard definition of pH.

Study Questions, Section B

8. What is the value of V_N, and what is its significance with regard to the effect of varying concentrations of Ox and Red on the equilibrium potential?

9. If the species on the oxidized side of the balanced redox reduction reaction have higher numbers of charges than the species on the reduced side of the equation, will the equilibrium potential increase or decrease with increasing ionic strength?

10. Calculate E_{eq} for a piece of iron immersed in a 0.010 M solution of $FeCl_2$ at 25 °C. Use Table 9.1, assuming the $E°$ value given is E'.

11. Calculate E_{eq} for a piece of Pt immersed in a solution of 0.0970 M $I_{2(l)}$ and 0.240 M NaI at 85 °C. When $I_{2(l)}$ is dissolved in a solution containing I^-, it forms the I_3^- ion according to the reaction $I_{2(l)} + I^- \rightleftharpoons I_3^-$. Use Table 9.1, assuming the $E°$ value given is E'.

12. Calculate E_{eq} for a piece of Pt immersed in a solution buffered at $p[H_3O^+]$ = 4.0 with 0.159 M UO_2^{2+} and 0.086 M U^{4+} at 25 °C. Use $E°$ for E'.

13. Calculate the values of E' for

A. Question 10

B. Question 11

14. In the case of the couple TiO^{2+}, Ti^{3+}, how much is the equilibrium potential changed by operating in a solution of $[H_3O^+] = 1 \times 10^{-3}$ M instead of 1 M?

15. Calculate E_{eq} for the hydrogen electrode in a solution that has a pH of 4.00.

Answers to Study Questions, Section B

8. The value of V_N is 0.0592 V at 25 °C. When divided by n, it is the amount by which the equilibrium potential will change for every 10-fold change in the ratio of [Ox] to [Red].

9. An increase in the ionic strength will cause a greater decrease in activity of the oxidized side species than those on the reduced side. This will make the solution less oxidizing and therefore make its potential less positive. This could also be shown with Equation 9.18, where the negative terms in the parentheses would predominate.

10. We will substitute appropriate values into Equation 9.19 to solve this problem. The reduction reaction is $Fe^{2+} + 2e^- \rightleftharpoons Fe_{(s)}$. In terms of Equation 9.14, Fe^{2+} is Ox and $Fe_{(s)}$ is Red. The value for E' is taken from Table 9.1.

$$E_{eq} = E'_{Fe^{2+},Fe} - \frac{0.0592}{2} \log \frac{1}{[Fe^{2+}]}$$

$$= -0.440 - \frac{0.0592}{2} \log \frac{1}{0.010} = -0.499 \text{ V}$$

Note that the $Fe_{(s)}$ concentration is taken as 1 because it is a pure solid. Also, because T = 25 °C (298 K) the temperature term in Equation 9.17 for V_N cancels out.

11. The concentrations of the species that exist after the reaction between I_2 and I^- are $[I_3^-]$ = 0.0970 M (the original concentration of I_2, assuming the reaction is complete) and $[I^-]$ = 0.240 − 0.0970 = 0.143 M. The reduction reaction to use from Table 9.1 is the one for which I_3^- is Ox and I^-

is Red. The $E°$ value given is +0.536 V. Applying Equation 9.19,

$$E_{eq} = E'_{I_3^-,I^-} - \frac{0.0592}{2} \left(\frac{T}{298}\right) \log \frac{[I^-]^3}{[I_3^-]}$$

$$E_{eq} = +0.536 - \frac{0.0592}{2} \left(\frac{85 + 273}{298}\right) \log \frac{(0.143)^3}{0.097}$$

$$= 0.590 \text{ V}$$

12. From Table 9.1, UO_2^{2+} is Ox and U^{4+} is Red, but other species are involved and must be included in the Nernst equation. The value for $[H_2O]^6$ is unity because it is the solvent, and, from $p[H_3O^+]$, $[H_3O^+] = 1 \times 10^{-4}$ M. Thus,

$$E_{eq} = E'_{UO_2^{2+},U^{4+}} - \frac{0.0592}{2} \log \frac{[U^{4+}]}{[UO_2^{2+}][H_3O^+]^4}$$

$$= +0.334 - \frac{0.0592}{2} \log \frac{(0.086)}{(0.159)(1 \times 10^{-4})^4}$$

$$= -0.13 \text{ V}$$

13. A. First calculate S for 0.010 M $FeCl_2$.

$$S = 0.5([Fe^{2+}] \times 2^2 + [Cl^-] \times 1^2)$$
$$= 0.5(0.010 \times 2^2 + 0.020 \times 1^2) = 0.030$$

Then calculate E' using Equation 9.18.

$$E' = -0.440 + \frac{0.0592 \times 0.5091 \times 0.030}{2}(-1 \cdot 2^2)$$

$$= -0.442 \text{ V}$$

Thus, the error in using -0.440 V as E' was 0.002 V.

B. $S = 0.5([Na^+] \times 1^2 + [I_3^-] \times 1^2 + [I^-] \times 1^2)$
    $= 0.5(0.240 \times 1^2 + 0.097 \times 1^2 + 0.143 \times 1^2) = 0.240$

$E' = 0.536 + \dfrac{0.0592 \times 0.5091 \times 0.240}{2} (3 \cdot 1^2 - 1 \cdot 1^2)$

$= 0.536 + 0.00723 = 0.543$ V

In this case the error caused by ignoring the difference between E' and $E°$ is 7 mV.

14. Since the $[H_3O^+]$ is squared in the denominator of the Nernst equation, its effect on the Red/Ox ratio in the Nernst equation is $(1 \times 10^{-3})^2$ or six powers of ten. This will cause the equilibrium potential to be $6V_N/n = 0.355$ V less positive than it would have been in a solution of 1 M H_3O^+.

15. At pH = 4.00, the activity of H_3O^+ is 1.0×10^{-4}. From Equation 9.21,

$$E_{eq} = 0.000 - \frac{0.0592}{2} \log \frac{1}{(1.0 \times 10^{-4})^2}$$

$$= -0.24 \text{ V}$$

C. Measurement of Electrode Potentials

In the previous section, we have explored one of the mechanisms by which a solid material, a metallic conductor, can develop a potential with respect to the solution in which it is immersed. We have also seen that this potential is a function of the composition of the solution, which makes this phenomenon a potential basis for chemical analysis. For such analysis, the differentiating characteristic of the analyte is that it can affect the electrical potential of a piece of metal with respect to the solution. One would then probe for this characteristic by immersing a piece of that metal in the solution. A typical form for such a probe is shown in Figure 9.7. The metal penetrates a glass container through a watertight seal, and electrical connection to the metal is made inside this container. The complete assembly is called the **test electrode.** Now, to measure the response to the probe, we must be able to measure the potential difference between the test electrode and the solution.

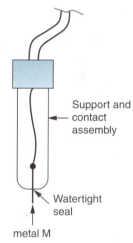

Figure 9.7. A practical metal test electrode can be constructed as shown. The assembly can be easily dipped into the test solution and the wire connected to one terminal of the voltmeter.

THE NECESSITY OF A REFERENCE ELECTRODE

The quantity to be measured is voltage, that is, electrical potential. To make this measurement, a voltage-to-number conversion system is needed. Two such systems were described in Chapter 3 for the measurement of the voltage produced by the pH electrode. (See Figures 3.11 and 3.12.) Since voltage is a relative quantity, we always measure the voltage of one part of a system *with respect to* another. Thus one lead from the voltmeter will be connected to the test electrode. With the other lead, we will have to make contact with the test solution (Figure 9.8). This cannot be done directly because the immersion of the other voltmeter wire into the solution will result in some new potential between this piece of metal and the solution. We will not be able to reliably measure the potential between the electrode and the solution if the other potential developed is not known and constant. For this reason, scientists have developed a type of electrode whose potential with respect to the solution is stable and reproducible. Such an electrode is called a **reference electrode.**

The most widely used reference electrode is based on the $AgCl,Ag_{(s)}$ couple. The equilibrium potential for this couple can be obtained from the Nernst equation of the Ag^+,Ag couple and the K'_{sp} for AgCl as follows. The Nernst equation for the Ag^+,Ag couple is

$$E_{eq} = E'_{Ag^+,Ag_{(s)}} - \frac{V_N}{1} \log \frac{1}{[Ag^+]} \qquad 9.22$$

When Cl^- ion is present, $[Ag^+]$ depends on the chloride ion concentration through the expression for K'_{sp}.

$$[Ag^+] = K'_{sp(AgCl)}/[Cl^-] \qquad 9.23$$

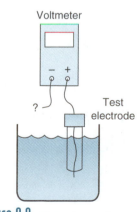

Figure 9.8. The voltmeter has two leads for measuring the potential difference between two points. The positive lead is connected to the test electrode. Some method for connection of the negative lead to the solution must be provided.

When Equation 9.23 is substituted in Equation 9.22, a new relationship involving [Cl⁻] appears.

$$E_{eq} = E'_{Ag^+,Ag_{(s)}} - V_N \log \frac{[Cl^-]}{K'_{sp(AgCl)}}$$

9.24

Splitting the log term in two parts to separate the constant gives

$$E_{eq} = E'_{Ag^+,Ag_{(s)}} + V_N \log K'_{sp(AgCl)} - V_N \log [Cl^-]$$

9.25

The first two terms on the right-hand side of Equation 9.25 are constants. They can be combined into a new constant, which is the formal reduction potential for the reaction $AgCl_{(s)} + e^- \rightleftharpoons Ag_{(s)} + Cl^-$.

$$E_{eq} = E'_{AgCl,Ag_{(s)}} - V_N \log [Cl^-]$$

9.26

where $E'_{AgCl,Ag_{(s)}} = E'_{Ag^+,Ag_{(s)}} + V_N \log K'_{sp(AgCl)}$.

When the values of 0.799 V and 1.82×10^{-10} are substituted for $E'_{Ag^+,Ag_{(s)}}$ and $K'_{sp(AgCl)}$, the calculated value for $E'_{AgCl,Ag_{(s)}}$ is 0.222 V. This is the same value found in Table 9.1 for this reaction. The method used above to combine a related equilibrium constant with the Nernst equation to produce a new couple with its associated $E°$ or E' is a general one.

The physical construction of the AgCl/Ag reference electrode is shown in Figure 9.9. A piece of silver metal is coated with solid AgCl and immersed in a glass tube containing a concentrated solution of KCl (typically 1 M). The nominal equilibrium potential of the silver with respect to the solution, as just calculated and as tabulated in the previous section, is +0.222 V. Contact is made between this reference electrode and the test solution by means of a channel that allows passage of dissolved ions but restricts mixing of the solutions on either side of the reference electrode container. The ionic connection between the solution in the reference electrode and the test solution completes the connection between the other voltmeter test lead and the solution but not without a small complication. Inequalities in the mobilities of the cations and anions through the channel create a small potential difference at this contact. This potential difference is called the **junction potential,** E_j. The junction potential is minimized by the used of concentrated KCl in the reference electrode solution, but its magnitude is difficult to predict and thus results in a measurement uncertainty of several millivolts. For convenience, the test and reference functions can be integrated in a combination electrode, as shown in Figure 9.10.

Since a second electrode is always required in order to measure the test electrode potential with respect to the solution, how can we know what the actual test electrode–solution potential is? The fact is that we can neither calculate nor measure what that actual potential is. That is why we have called it the *nominal* potential thus far. However, it has been confirmed, from theory and experiment, that the potential *difference* between two electrodes can be accurately predicted by the Nernst equations for the couples at each electrode. Thus, the voltage scale is entirely self-consistent, but the potential of the test electrode is far more likely to be measured relative to an AgCl,Ag reference electrode than to the hydrogen electrode. Thus, the practical "zero" on the voltage scale is different from the thermodynamic zero.

CALCULATING THE EXPECTED MEASUREMENT VOLTAGE

The complete potential measurement apparatus is shown in Figure 9.11. The test and reference electrodes are dipped into the test solution, and the potential difference between them is amplified and converted to a number for digital display. When a pH meter is used for this measurement, a calculator contained in the unit will provide a

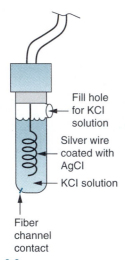

Figure 9.9. A reference electrode is used to probe the solution potential for a measurement of electrode potential. An AgCl,Ag electrode is often used for its convenience and stable potential. The construction of a typical reference electrode is shown.

Fill hole for KCl solution

Silver wire coated with AgCl

KCl solution

Fiber channel contact

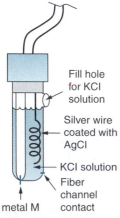

Fill hole for KCl solution

Silver wire coated with AgCl

KCl solution

Fiber channel contact

metal M

Figure 9.10. The reference and test electrodes are sometimes combined in a single unit called a combination electrode.

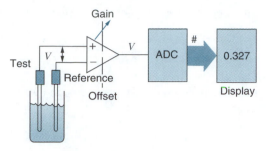

Figure 9.11. The difference in potential between the reference and test electrodes is adjusted for gain and offset by the amplifier before being applied to the analog-to-digital converter. The resulting number is displayed. The transfer functions of the conversion devices are usually adjusted so that the numerical output is in millivolts.

readout in millivolts or in log concentration units for the analyte. The magnitude of the potential difference is the difference in the potentials developed between each of the electrodes and the test solution, as illustrated in the schematic in Figure 9.12. The expected measurement voltage, V, can be calculated by combining the Nernst equations for the test and reference electrodes.

According to the schematic drawn for this measurement, the value of the measured voltage will be

$$V = E_{eq(test)} - (E_{eq(ref)} + E_j) \qquad 9.27$$

Consider the case of a copper test electrode in a solution of cupric ions. An AgCl,Ag reference electrode is used. Now

$$V = E'_{Cu^{2+},Cu_{(s)}} - \frac{V_N}{2} \log \frac{1}{[Cu^{2+}]} - E_{eq(ref)} - E_j$$

$$V = 0.337 - \frac{V_N}{2} \log \frac{1}{[Cu^{2+}]} - 0.222 - E_j \qquad 9.28$$

$$V = 0.115 + \frac{V_N}{2} \log [Cu^{2+}] - E_j$$

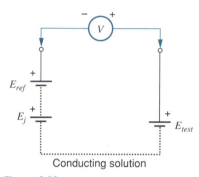

Figure 9.12. The equivalent circuit for the potential measurement system is shown. Summing the voltages clockwise around the loop leads to Equation 9.27.

If E_j is neglected, then the anticipated V can be calculated. For example, if $[Cu^{2+}]$ were 1×10^{-6} M, $V = 0.115 - 0.178 = -0.063$ V.

RELATING ELECTRODE VOLTAGE TO ACTIVITY AND CONCENTRATION

The technique by which concentration or activity information is obtained from the equilibrium potential established at an electrode is called **potentiometry**. In principle, one could measure the equilibrium potential between the test and reference electrodes, calculate the test electrode potential relative to the SHE, and use the Nernst equation to calculate the concentration or activity of Ox or the Ox/Red concentration or activity ratios of species in the test solution. In practice, concentration and activity determinations are rarely done this way. The reason is that an error of only a few millivolts in the measurement can lead to a large error in the calculation of the concentration when a 10-fold change in concentration causes only a 60 mV or 30 mV change in potential. The major sources of error in the calculation come from the uncertainty in E_j, the uncertainty in the absolute accuracy of the reference electrode potential, and the uncertainty that the formal potential for the test electrode couple is accurate. Since all these uncertainties can be eliminated by a simple calibration, that is what is done. A voltage V_{cal} is measured for a test solution of known activity, $a(Cu^{2+}_{std})$. Then, using the same test and reference electrodes, the voltage V of an unknown solution is measured. The following two equations result.

$$V = E°_{Cu^{2+},Cu(s)} - \frac{V_N}{2} \log \frac{1}{a_{Cu^{2+},unk}} - E_{eq(ref)} - E_j \qquad 9.29$$

$$V_{cal} = E°_{Cu^{2+},Cu(s)} - \frac{V_N}{2} \log \frac{1}{a_{Cu^{2+},std}} - E_{eq(ref)} - E_j \qquad 9.30$$

When Equation 9.30 is subtracted from Equation 9.29 (with some rearranging)

$$\log a_{Cu^{2+},unk} = \log a_{Cu^{2+},std} + \frac{V - V_{cal}}{V_N/2} \qquad 9.31$$

This same approach is applicable to all test electrodes and all test electrode couples. The general equations are

$$\log a_{Ox,unk} = \log a_{Ox,std} + \frac{(V - V_{cal})n}{V_N \times o}$$

$$\log a_{Red,unk} = \log a_{Red,std} - \frac{(V - V_{cal})n}{V_N \times r} \qquad 9.32$$

where o and r are the stoichiometric coefficients for the species Ox or Red in the electron transfer reaction $oOx + ne^- \rightleftharpoons rRed$.

The determination of concentration from electrode potentials is not quite so straightforward. One could follow the same treatment of the equilibrium potential equation for formal potentials and concentrations as was done in Equations 9.29 through 9.32. The result would be an equation analogous to Equation 9.32 but expressed in terms of concentrations rather than activities. In making this derivation, when Equation 9.30 is subtracted from 9.29, we would be assuming that the value of E' is the same for both the standard and sample solutions. When the standard solution is an off-the-shelf standard, this is very unlikely to be true. Therefore, *the quantity determined by the difference in the electrode voltage between standard and unknown solutions is activity, not concentration.* When the activity of the analyte has been determined, the analyte concentration can be calculated if the activity coefficient is known.

Another application of electrode potentials is the situation where the electrodes remain in the test solution to follow changes in activity or concentration over time. The change in potential will always be related to the relative change in analyte activity. In cases where the ionic strength does not also change with time, the potential change can also be related to relative change in concentration.

ELECTRODES OF THE SECOND KIND

In the examples of species that affect the potential of a test electrode, we have concentrated on the species that are actually accepting or donating the electrons, such as Cu^{2+} with a copper electrode or H_3O^+, H_2 with a platinum electrode. These are called **electrodes of the first kind.** However, we have also seen that the equilibrium potential of an electrode is affected by all the species that occur in the half reaction and consequently in the Nernst equation. Thus, as shown in Equation 9.26, the potential of an AgCl,Ag electrode is affected by the concentration of Cl^- even though Cl^- is not involved in the electron exchange that establishes the potential. In general, any species that reacts with Ox or Red in the basic couple will affect its activity in the solution and thus the equilibrium potential it establishes. Indirectly, then, the electrode can be used as a probe for the concentration of the secondary reactant species. When probing the concentration of a secondary reactant, the electrode is said to be an **electrode of the second kind.**

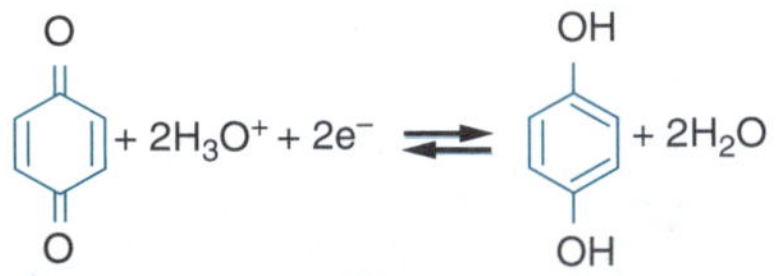

Several ingenious electrodes of the second kind have been developed. An early example makes use of the quinone/hydroquinone couple shown in Figure 9.13. Abbreviating the couple as Q,QH_2, the Nernst equation is

$$E_{eq} = E^\circ_{Q,QH_2} - \frac{V_N}{2} \log \frac{a_{QH_2}}{a_Q a^2_{H_3O^+}} \qquad 9.33$$

Figure 9.13. The quinone, hydroquinone couple involves hydronium ions and water. As such, its equilibrium potential depends on the pH of the solution. A pH sensing electrode can be developed using this couple.

If known and roughly equal amounts of quinone and hydroquinone are added to a solution, the Nernst equation can be rearranged to show the dependence on solution pH.

$$E_{eq} = E^{\circ}_{Q,QH_2} - \frac{V_N}{2} \log \frac{a_{QH_2}}{a_Q} + V_N \log a_{H_3O^+} \qquad 9.34$$

$$E_{eq} = \text{Constant} - V_N \times \text{pH}$$

Since the E° for the Q,QH_2 couple is about 0.70 V and since the activities of the Q and QH_2 in the solution are nearly equal, the value of the constant is about 0.70 V.

A mercury electrode for EDTA analysis was developed[2] based on the complexation of mercuric ion, Hg^{2+}. The Nernst equation for the Hg^{2+},Hg couple is combined with the mercuric ion–EDTA complex formation constant for HgY^{2-} to give a Nernst equation for this electrode.

$$E_{eq} = E'_{Hg^{2+},Hg} - \frac{V_N}{2} \log \frac{1}{[Hg^{2+}]} \qquad 9.35$$

$$[Hg^{2+}] = \frac{[HgY^{2-}]}{K'_{HgY^{2-}eff}[Y^{4-}]} \qquad 9.36$$

$$E_{eq} = E'_{Hg^{2+},Hg} - \frac{V_N}{2} \log K'_{HgY^{2-}eff} - \frac{V_N}{2} \log \frac{[Y^{4-}]}{[HgY^{2-}]} \qquad 9.37$$

$$E_{eq} = E'_{HgY^{2-},Hg} - \frac{V_N}{2} \log \frac{[Y^{4-}]}{[HgY^{2-}]} \qquad 9.38$$

The mercury electrode for EDTA is an example of an electrode of the second kind based on a complexation reaction. Electrodes for many species can be developed on this same principle.

This electrode can be used to follow the titration of metal ions with EDTA. A small amount of Na_2HgY is added to the titration solution. This provides the HgY^{2-} concentration required by Equation 9.38 in a form that does not interfere with the stoichiometric relationship between the EDTA and the analyte metal ion(s). As the titration proceeds, the test electrode potential decreases about 30 mV per unit increase in $\log[Y^{4-}]$. Despite its potential usefulness, this electrode has never been commercially developed.

Study Questions, Section C

16. Why is a reference electrode required to measure the potential developed between a piece of metal and the solution it is in?

17. Why do we not know the actual potential of the solution, and why is this not a problem?

18. What is the junction potential?

19. Calculate the measured potential when a Pt test electrode and an Ag,AgCl reference electrode are dipped into a solution of 1.3×10^{-4} M V^{3+} and 4.2×10^{-2} M V^{2+}.

20. In using electrode potentials to determine the activity of a species, what is the advantage of using a comparison measurement with a known standard instead of a direct calculation from the Nernst equation?

21. A silver test electrode and an Ag,AgCl reference electrode are dipped into a solution that is known to have a pa_{Cl^-} of 4.00. The potential measured is -0.114 V. The same electrodes are then dipped into a sample, and the potential measured is -0.172 V. What is the activity of Cl^- in the sample?

Answers to Study Questions, Section C

16. Electrical potential is always the difference in potential between two points. The voltmeter has two leads (red and black) to connect to these two points. A voltmeter lead dipped into the solution would develop its own potential at the metal–solution interface just as the test electrode does. The potential at this metal–solution interface would not be known or readily predictable, but the voltage developed there would be included in the measured voltage. To avoid this, an electrode with a known and reproducible voltage with respect to the solution is used. The voltmeter can then be connected between the two metallic contacts of the test and reference electrodes.

[2]C. N. Reilley and R. W. Schmid, *Anal. Chem.* **1958**, 30, 947.

17. Since we can only measure the potential difference between two electrodes immersed in a solution, the potential of the solution is not known for sure. The measurement potentials calculated by the Nernst equation and the values of $E°$ or E' are nonetheless accurate because they are based on a consistent scale for which the $E°$ of the $H_{2(g)}$,H_3O^+ couple is assigned a value of 0.000 V.

18. Associated with the reference electrode is a small junction potential E_j, between its inner solution and the test solution.

19. The expected measurement potential can be calculated using Equation 9.27 (neglecting E_j), but first $E_{eq(test)}$ must be calculated. For this, Equation 9.19 can be used. When the temperature is not given, assume it is 298 K. Thus,

$$E_{eq(test)} = E'_{V^{3+},V^{2+}} - \frac{0.0592}{1} \log_{10} \frac{[V^{2+}]}{[V^{3+}]}$$

$$E_{eq(test)} = -0.255 - 0.0592 \log_{10} \frac{4.2 \times 10^{-2}}{1.3 \times 10^{-4}}$$

$$= -0.404 \text{ V}$$

Then $V = -0.404 - 0.222 = -0.626$ V.

20. The measurement of the potential difference between the electrodes dipped in a standard solution and one dipped in the unknown solution cancels out the junction potential and any uncertainty in the reference electrode potential. The voltage difference can then be ascribed to the ratio of the activities of the analyte in the unknown and standard solutions.

21. The reaction that is establishing the potential at the test electrode is

$$AgCl_{(s)} + 1e^- \rightleftharpoons Ag_{(s)} + Cl^-$$

The Nernst equation is

$$E_{eq} = 0.222 - 0.0592 \log a_{Cl^-}$$

Notice that the Cl^- (the varying quantity) is on the reduced side of the equation. A solution can then be obtained using the second of Equations 9.32.

$$\log a_{Cl^-,unk} = -pa_{Cl^-,unk}$$

$$= -4.00 - \frac{[-0.172 - (-0.114)] \cdot 1}{0.0592 \cdot 1} = -3.02$$

$$a_{Cl^-,unk} = 9.6 \times 10^{-4} \text{ M}$$

D. Log Concentration Plots for Redox Couples

In the log concentration plots for previous systems (acid–base, complexation, and precipitation), the concentrations of the various species were related through the equilibrium constant expression. In the case of the redox couples, the relative concentrations are related through the Nernst equation. In view of the reaction relating the species in the redox couple, $Ox + ne^- \rightleftharpoons Red$, we cannot plot the concentration of the electrons, but we can plot a related quantity, the potential of the metal. Thus, the log concentration plot for a redox couple is a plot of E_{eq} versus the logs of the concentrations of Ox and Red. Such a plot for the Fe^{3+},Fe^{2+} couple with a total concentration of 0.001 M is shown in Figure 9.14. Note that the more oxidized form of the couple predominates in the region of more positive potentials and that the more reduced form of the couple predominates at more negative potentials. The potential at which they are equal is where $E_{eq} = E'$.

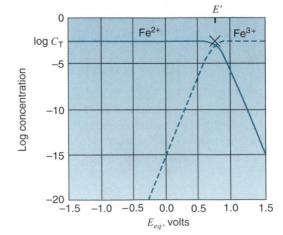

Figure 9.14. The log concentration plot for a redox couple plots the log of the concentrations against electrical potential. We see that at potentials more positive than E' for the couple, the more oxidized form of the couple predominates, and vice versa.

LINES FOR THE CASE WHERE OX AND RED ARE DISSOLVED

The equations for the lines for the log concentrations are obtained as follows. The Nernst equation for the Ox,Red couple (Equation 9.19) is rearranged, and then the mass balance equation $C_T = [Ox] + [Red]$ is invoked.

$$[Red] = [Ox] \times 10^{n(E'-E_{eq})/V_N}, \qquad [Ox] = [Red] \times 10^{-n(E'-E_{eq})/V_N},$$

$$C_T = [Ox] + [Red]$$

$$[Red] = \frac{C_T}{1 + 10^{n(E'-E_{eq})/V_N}}, \qquad [Ox] = \frac{C_T}{1 + 10^{-n(E'-E_{eq})/V_N}} \qquad 9.39$$

The equations for the Ox and Red lines are obtained by taking the logs of Equations 9.39.

$$\log [Red] = \log C_T - \log(1 + 10^{n(E'-E_{eq})/V_N})$$

$$\log [Ox] = \log C_T - \log(1 + 10^{-n(E'-E_{eq})/V_N}) \qquad 9.40$$

In practice, the lines are drawn on a blank plot by the now familiar practice. This is illustrated in Figure 9.14. First, an $\times$ is placed at the intersection of $\log C_T$ and $E_{eq} = E'$. Next horizontal lines are drawn to the right and left of the system point for the concentrations of the dominant species. Then lines are drawn down from the system point with slopes of $\pm V_N/n$ volts per log C. From the log concentration plot, the equilibrium potential can be determined for every [Red]/[Ox] ratio or the concentrations of both species from any equilibrium potential. As this chapter develops, we will see further ways these plots can be used.

LINES FOR THE CASE WHERE ONLY OX OR RED IS DISSOLVED

When one of the species Ox or Red is a solid, as in the case of a metal with its ions, the plot is rather different, as shown in Figure 9.15. In this case, the activity of Red remains constant at $\log a_{Red} = 0$. The equation for the line for [Ox] is obtained from a simple rearrangement of the Nernst equation.

$$E_{eq(M^{n+},M_{(s)})} = E'_{M^{n+},M_{(s)}} - \frac{V_N}{n} \log \frac{1}{[M^{n+}]}$$

$$\log [M^{n+}] = -\frac{n(E' - E_{eq(M^{n+},M_{(s)})})}{V_N} \qquad 9.41$$

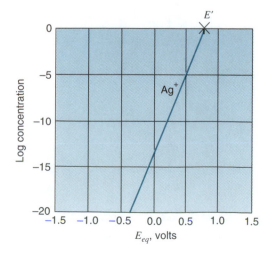

Figure 9.15. The log concentration plot for a redox couple for which the reduced form is a pure metal shows only the oxidized species. The activity of the metal is constant at a value of 1. Note that the potential between the metal and the solution becomes more negative as the concentration of the oxidized form decreases.

The plot in Figure 9.15 was constructed for the system $Ag^+ + e^- \rightleftharpoons Ag_{(s)}$. The system point is placed at the intersection of log $C = 0$ and $E_{eq} = E'$. From the system point, the line for $[Ag^+]$ is drawn to the left and down with a slope of $0.0592/n$ volts per log C. From this plot, the potential for any concentration of metal ion can be estimated, as can the concentration of metal ion in equilibrium with any potential. As expected, the potential increases as the metal ion concentration increases.

LINES FOR SECONDARY REACTION SPECIES

If the metal ion reacts with other species in solution, the concentration lines for these reactants can be added to this log concentration plot. An example is the reaction of Ag^+ with Cl^- to form the solid AgCl. From the solubility product equation for AgCl,

$$\log [Cl^-] = \log K'_{sp} - \log [Ag^+] \qquad 9.42$$

The addition of the Cl^- line to the log concentration plot is shown in Figure 9.16. The construction of this line is similar to that for the plots for precipitation reactions in Chapter 8. When $[Ag^+]$ is equal to $[Cl^-]$, their concentrations are both equal to $\sqrt{K'_{sp}}$. For AgCl, this is $\sqrt{1.82 \times 10^{-10}} = 1.35 \times 10^{-5}$ M. Another point on the Cl^- line is when $\log[Ag^+] = 0$, $\log[Cl^-] = \log K'_{sp}$. The Cl^- line is drawn through these two points.

The Nernst equation for the reaction $AgCl + e^- \rightleftharpoons Ag_{(s)} + Cl^-$ is given in Equation 9.26. When this electrode is in its standard state, the activity of Cl^- is 1. If formal potentials are used, this is $[Cl^-] = 1$ and $\log[Cl^-] = 0$. The potential at which the Cl^- line crosses the log concentration $= 0$ line is thus the formal potential for the AgCl,Ag couple. From Figure 9.16, this is seen to be $+0.22$ V, as predicted with Equation 9.26. Thus, the addition of the chloride line to the log concentration plot of the Ag^+,$Ag_{(s)}$ couple provides the potential relationship for the AgCl,Ag couple.

The Ox and Red forms of the couple may undergo hydrolysis, complexation, and precipitation reactions in the solution. The approach illustrated here allows these reactions to be taken into account and permits the effects of the reactants on the resulting equilibrium potential to be evaluated.

USING LOG PLOTS FOR CELL EQUILIBRIUM POTENTIALS

The equilibrium potential difference between two electrodes can be graphically determined by placing the log plots for each electrode on the same graph. This is illustrated for the case of the measurement of the potential of a Cu^{2+},Cu electrode in 1×10^{-6} M

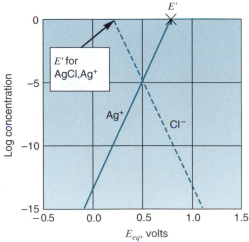

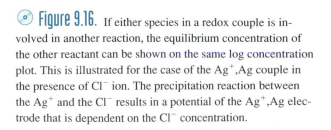

Figure 9.16. If either species in a redox couple is involved in another reaction, the equilibrium concentration of the other reactant can be shown on the same log concentration plot. This is illustrated for the case of the Ag^+,Ag couple in the presence of Cl^- ion. The precipitation reaction between the Ag^+ and the Cl^- results in a potential of the Ag^+,Ag electrode that is dependent on the Cl^- concentration.

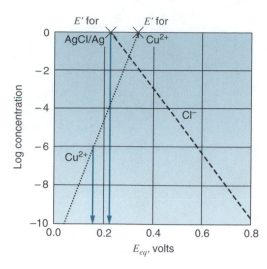

Figure 9.17. Combining the log concentration plots for both electrodes in a potentiometric measurement allows each of their potentials to be determined (arrows). From these, the difference in potential between them can be easily seen.

Cu^{2+} using an AgCl,Ag reference electrode. Both the Cu^{2+},Cu lines and the AgCl,Ag lines have been plotted on the graph of Figure 9.17. The potentials for each couple are determined separately, and the difference between them is calculated. For the reference electrode, the Cl^- line shows how the potential of that electrode would increase with decreasing $[Cl^-]$. However, because the Cl^- concentration in the electrode is constant at 1.0 M, this electrode is assumed to remain constant at the standard value of 0.222 V. The potential of the copper test electrode decreases from its standard potential at a slope of about 0.03 V per log unit. The cupric test electrode, at a $[Cu^{2+}]$ of 1×10^{-6} M, has a potential of about +0.16 V. Thus, the cupric electrode is negative with respect to the reference electrode by about 0.06 V ($E_{test} - E_{ref} = -0.06$ V). The expected measurement voltages for other values of $[Cu^{2+}]$ can be determined in the same way. Note from the plot that as the $[Cu^{2+}]$ decreases from 1 M, the measured voltage goes from +0.115 V through zero (at $[Cu^{2+}] \approx 10^{-4}$ M), to increasingly negative values.

LOG PLOTS OF UNSYMMETRICAL COUPLES

For a few redox couples, the stoichiometric coefficients of Ox and Red in the half reaction are not equal. We will call these **unsymmetrical couples**. The inequality of the stoichiometric coefficients makes these couples more difficult to plot, but the development of the plots for them reveals an interesting and little-mentioned characteristic of these systems. Among the unsymmetrical systems, two are very important in redox chemistry. One is the tetrathionate, thiosulfate couple, and the other is the iodine, iodide couple.

The log plot for the tetrathionate, thiosulfate couple is shown in Figure 9.18. Three aspects of this plot stand out immediately. One is that the maximum concentration lines are not at the same value. This is due to the asymmetry in the stoichiometry in the half reaction.

$$S_4O_6^{2-} + 2e^- \rightleftharpoons 2S_2O_3^{2-} \qquad\qquad 9.43$$

If a given concentration of $S_4O_6^{2-}$ were completely reduced to $S_2O_3^{2-}$, the concentration of $S_2O_3^{2-}$ would be twice the original concentration of $S_4O_6^{2-}$. The plot of Figure 9.18 shows the difference in the concentrations when essentially all the couple is in one form or the other. Another difference from previous plots is that the magnitudes of the slopes of the lines going away from the intersection of the two species are not the same. This is because the ratios of the number of electrons per mole of Ox and Red are different. There are 2 moles of electrons in the half reaction per mole of $S_4O_6^{2-}$, giving rise to a slope of $V_N/2$, whereas the ratio of electrons to $S_2O_3^{2-}$ is 1:1, resulting in a line with a slope of V_N. The third discrepancy is that the potential at which the concentration lines intersect is

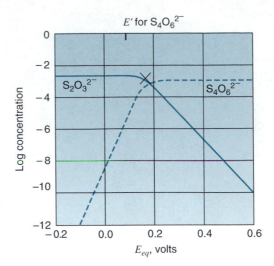

Figure 9.18. Unsymmetrical couples have unequal concentration maxima, unequal slopes in the concentration lines, and a potential at the intersection of Ox and Red that depends on the analytical concentration of the couple.

not the same as the E' for the couple. In some ways, this is the most significant difference because, as we shall see, the oxidizing strength of an unsymmetrical couple is a function of the absolute concentrations, not just the concentration ratios of Ox and Red.

The system point for an unsymmetrical reaction will be at the intersection of C_x and E'_x, where these quantities are functions of C_{Ox} and E' for the couple. These functional relationships are derived in the derivations section of the accompanying CD. The results are given here so that their implications can be appreciated.

For the general reduction half-reaction

$$o\text{Ox} + ne^- \rightleftharpoons r\text{Red} \qquad\qquad 9.44$$

the relationship between C_x and C_{Ox} is

$$\log C_x = \log C_{Ox} + \frac{r}{r+o} \log \frac{r}{o} \qquad\qquad 9.45$$

For the specific example of the couple in Figure 9.18,

$$\log C_x = -3.0 + (2/3)\log 2 = -2.80$$

The relationship for E_x, the potential for the system point, is

$$E_x = E' + \frac{V_N}{n}(o-r)\log C_{Ox} + \frac{V_N}{n}\left(\frac{or}{r+o} - r\right)\log \frac{r}{o} \qquad\qquad 9.46$$

For the specific example of the couple in Figure 9.18,

For unsymmetrical couples where $o > r$, the couple becomes increasingly reducing as the concentration decreases. Conversely, if $r > o$, the potential becomes more oxidizing as the couple is more dilute.

$$E_x = 0.09 + \frac{0.0592}{1}(1-2)\log(1\times 10^{-3}) + 0.0592\left(\frac{1\times 2}{2+1} - 2\right)\log \frac{2}{1}$$

$$E_x = 0.09 + 0.0592 \times 3 - 0.0592 \times 0.40 = 0.17\ \text{V}$$

From this example, we see that a $S_4O_6^{2-},S_2O_3^{2-}$ solution becomes an increasingly strong oxidant as it becomes more dilute. If the stoichiometric coefficient o is greater than r, the effect is reversed. This effect occurs only with unsymmetrical couples. The spreadsheet provided on the accompanying CD will make plots for the couples for which $r = 2$, $o = 1$ and the reverse. Using them enables one to create the log plot without tedious calculation and to easily see the effect of the concentration on the oxidizing strength of the couple.

Study Questions, Section D

22. Why should the X-axis of the logarithmic concentration plot for a redox couple be the equilibrium electrical potential?

23. In the log plot of Figure 9.14, why is the Fe^{2+} form dominant at the less positive potentials? Why are the concentrations of the two forms equal when the equilibrium potential is equal to the formal standard reduction potential?

24. What factor determines the slope of the line as it goes down from the system point?

25. Why does the log plot for a metal ion,metal couple have just one line, and why does it go through the formal reduction potential at a concentration of 1 M?

26. How can the line for Cl^- be added to the plot for the Ag^+,Ag couple as in Figure 9.16 when the Cl^- is not involved in the reduction half reaction?

27. Draw the log concentration plot for the system described in Question 10. Use the value for E' calculated in Question 13. Estimate the value of E_{eq} for a 0.01 M solution of $FeCl_2$.

28. Draw the log concentration plot for the couple Sn^{4+},Sn^{2+} with a total concentration of 0.01 M. Estimate the concentrations of each species if $E_{eq} = -0.10$ V.

29. The solubility product for $CuCrO_4$ (cupric chromate) is 3.6×10^{-6}. Construct a log concentration diagram for the Cu^{2+},$Cu_{(s)}$ system that includes a line for the CrO_4^{2-} ion. Write the electron exchange reaction for the reduction of $CuCrO_4$ to Cu and estimate the E' for this reaction.

30. Use the spreadsheet plotting program to create the log concentration plot for the Hg^{2+},Hg_2^{2+} couple where $[Hg^{2+}] = 1 \times 10^{-4}$ M.

Answers to Study Questions, Section D

22. From the Nernst equation it is the equilibrium potential that is directly related to the ratio of the Red and Ox concentrations. In the acid–base log plots, it was the pH that determined the ratio of acid to base forms.

23. The Fe^{2+} form is the more reduced form of the redox couple. It will be dominant at potentials less positive than the formal standard reduction potential. The concentrations of Ox and Red are equal when the equilibrium potential is equal to the formal reduction potential because, at equal concentrations, the couple is neither more oxidizing nor more reducing than the standard state. These questions can also be answered mathematically using the Nernst equation (Equation 9.19).

24. The slope is V_N/n volts per 10-fold change in the redox concentration ratio. Therefore, the slope is determined somewhat by the temperature, as this has an effect on V_N, but mostly by n.

25. The log plot for a metal ion,metal couple has just one line because there is only one reactant concentration (the metal ion) that can vary. The line for the metal ion goes through the formal reduction potential at a concentration of 1 M because at that concentration, both reactants (metal ion and metal) are in their standard state.

26. The chloride ion reacts with the silver ion through the AgCl precipitation reaction. Through the K'_{sp} expression for AgCl, the Cl^- concentration is related to the Ag^+ concentration. For any given value of $[Ag^+]$ on the plot, $[Cl^-]$ can be determined.

27. Place the system point at the intersection of E' (-0.442 V) and log concentration $= 0$ (point A). The line for the $Fe_{(s)}$ will be a horizontal line at unit activity (log $C = 0$). This

line need not be put on the plot. The line for Fe^{2+} can be obtained from the Nernst equation.

$$E_{eq} = E'_{Fe^{2+},Fe} - \frac{0.0592}{2} \log \frac{1}{[Fe^{2+}]}$$
$$= -0.442 + 0.0296 \log [Fe^{2+}]$$

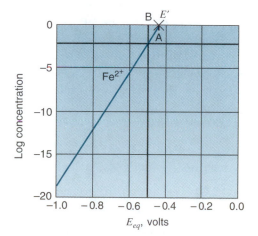

From it we can see that as $[Fe^{2+}]$ decreases from 1.0 M, the potential becomes more negative. For each log concentration decrease in concentration, the potential changes $0.0592/2 = 0.0296$ V. From the log plot, we see that at $\log[Fe^{2+}] = -2$, $E_{eq} \approx -0.50$ (point B). This is the same value of E' calculated in Question 10.

28. The plot is constructed by placing the system point at the intersection of log concentration (-2) and E' (0.139 V). Horizontal lines are drawn for the Sn^{2+} and Sn^{4+}

species, the form representing Ox in the redox couple having the dominant concentration at the higher potentials. The slopes for the Sn^{2+} and Sn^{4+} lines at the lower concentrations are obtained from the Nernst equation.

$$E_{eq} = -0.14 - \frac{0.0592}{2} \log \frac{[Sn^{2+}]}{[Sn^{4+}]}$$

$$= -0.14 - \frac{0.0592}{2} \log [Sn^{2+}] + \frac{0.0592}{2} \log [Sn^{4+}]$$

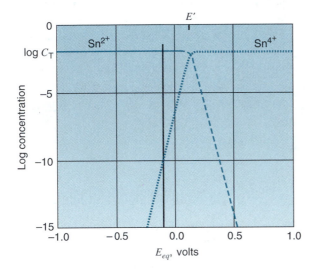

From this we see that the Sn^{2+} line leaves the system point with slope of $0.0592/2$ V per log concentration unit and that the Sn^{4+} has a slope of the same value but in the opposite direction. If the equilibrium potential is -0.10 V, the concentrations of all the species in the solution are obtained along the vertical line for $E = -0.10$ V. Thus $[Sn^{2+}] = 0.01$ M and $[Sn^{4+}] = 10^{-10}$ M.

29. From Equation 9.21, at $\log[Cu^{2+}] = 0$, $E_{eq} = E' \approx E° = 0.337$ (point A). The line for Cu^{2+} is drawn from this point with a slope of $0.0592/2 = 0.0296$. The CrO_4^{2-} line is added using the solubility equation for $CuCrO_4$.

$$K'_{sp} = [CrO_4^{2-}][Cu^{2+}]$$

$$\log [CrO_4^{2-}] = \log K'_{sp} - \log [Cu^{2+}]$$

When $[CrO_4^{2-}] = [Cu^{2+}]$, the concentrations of both species are equal to

$$\sqrt{K'_{sp}} \approx \sqrt{K°_{sp}} = \sqrt{3.6 \times 10^{-6}} = 1.90 \times 10^{-3}$$

This is the point on the graph where the Cu^{2+} line crosses the CrO_4^{2-} line (point B). Another point on the CrO_4^{2-}

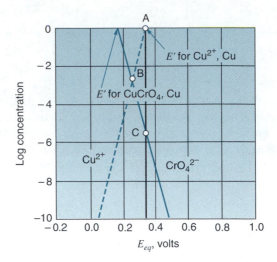

line is where $[Cu^{2+}] = 0$ and $\log[CrO_4^{2-}] = \log K'_{sp} \approx \log K°_{sp} = -5.47$ (point C). The line through points B and C is the CrO_4^{2-} line. The reduction half reaction for the reduction of $CuCrO_4$ to Cu is

$$CuCrO_4 + 2e^- \rightleftharpoons Cu_{(s)} + CrO_4^{2-}$$

E' for this reaction is the value of E_{eq} where $[CrO_4^{2-}] = 0$, about $+0.17$ V. This is considerably different from the value of $E°$ for reduction of Cu^{2+} to $Cu_{(s)}$ of 0.337 V. The addition of chromate ion to the solution has the effect of lowering the standard reduction potential because of its reaction with the cupric ion (the more oxidized form in the redox couple).

30. The resulting graph is shown. Note that the couple is unsymmetrical.

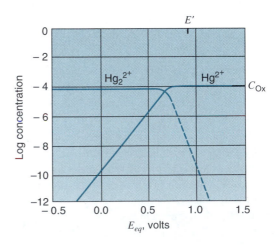

E. Ion Reactions on Membrane Surfaces

In Chapter 8, on precipitation reactivity, we saw how the selective attraction of ions to a surface could produce an electrical potential between a solid surface and the solution containing those ions. This phenomenon is responsible for the generation of colloids. Now we will consider how this potential is related to the concentration of the reacting species in solution and how the interfacial potential developed can be measured. We will use, as our principal example, the selective reaction of protons with a glass surface, the basis of the pH electrode.

H⁺ REACTION WITH A GLASS SURFACE

Silicate glass is a three-dimensional network of silicon atoms, each bonded to five oxygen atoms, two of which are also bonded to other silicon atoms, as shown in Figure 9.19. The stoichiometry of this structure is four oxygen atoms per silicon atom. Given a +4 oxidation state for the silicon and a −2 state for the oxygen, there is a net −4 charge for each silicon atom. To retain electrical neutrality, the spaces in the silicate network contain enough cations to counterbalance this charge. The particular cations contained in the glass depend on the composition of the material used in its fabrication. Monovalent cations within this structure are held only by coulombic forces and are thus mobile enough to impart a modest electrical conductivity to the glass.

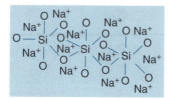

Figure 9.19. The silicate structure provides many sites for the cations that are required to neutralize the negative silicate charge.

At the surface of the glass, monovalent cations such as Na^+ and Li^+ may exchange with ions in the solution. (See Figure 9.20.) For the types of glass used in the pH electrode, the surface absorbs a significant amount of water, and the surface alkali cations have been largely replaced by H^+. The surface, then, has become a kind of solid acid, capable of donating protons to the solution. If some of the glass protons are transferred to the solvent, the negatively charged conjugate base remains part of the glass, and so the potential of the glass will become negative with respect to the solution. Similarly, if H_3O^+ in the solution transfers a proton to the glass surface, the glass will gain a positive potential with respect to the solution. The tendency for protons to be donated or accepted by the glass depends on the activity of hydronium ions in solution according to the reaction

$$GlassH + H_2O \rightleftharpoons H_3O^+ + Glass^- \qquad 9.47$$

CALCULATION OF THE INTERFACIAL POTENTIAL

The equilibrium condition for the interfacial reaction (Equation 9.47) is expressed by equating the electrochemical potentials of the reactants with those of the products.

$$\mu_{GlassH} + \mu_{H_2O} = \overline{\mu}_{H_3O^+} + \overline{\mu}_{Glass^-}$$
$$\mu^{\circ}_{GlassH} + \mu^{\circ}_{H_2O} = \mu^{\circ}_{H_3O^+} + RT \ln a_{H_3O^+} + FE_{sol'n} + \mu^{\circ}_{Glass^-} - FE_{Glass} \qquad 9.48$$

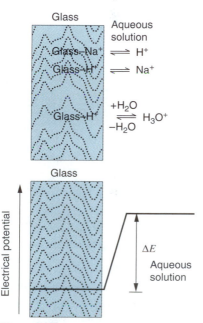

Figure 9.20. An electrical potential will be developed between the glass surface and an aqueous solution from the exchange of hydrogen ions and sodium ions between the glass and the solution.

In this equation, the chemical activities of the GlassH and the H_2O are taken as unity for pure materials. That of $Glass^-$ is also, except that the energy from its charge must be included.

From Equation 9.48, the equation for the potential between the glass and the solution can be obtained.

$$F(E_{Glass} - E_{sol'n}) = \mu^{\circ}_{H_3O^+} - \mu^{\circ}_{H_2O} + \mu^{\circ}_{Glass^-} - \mu^{\circ}_{GlassH} + RT \ln a_{H_3O^+}$$
$$E_{Glass} - E_{sol'n} = \frac{\mu^{\circ}_{H_3O^+} - \mu^{\circ}_{H_2O} + \mu^{\circ}_{Glass^-} - \mu^{\circ}_{GlassH}}{F} + \frac{RT}{F} \ln a_{H_3O^+}$$

Since the entire first term on the right-hand side is a constant, we will call it E°_{Glass}.

$$E_{Glass} - E_{sol'n} = \Delta E_{G-s} = E^{\circ}_{Glass} + \frac{RT}{F} \ln a_{H_3O^+} \qquad 9.49$$

From this equation, we can see that if the activity of H_3O^+ in solution increases, the electrical potential difference between the glass and the solution becomes more positive. This discussion describes the nature of the phenomenon. The challenge for the analytical chemist is how to use this phenomenon as a probe for H_3O^+ activity, how to measure the response to the probe, and how to interpret the response.

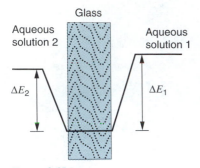

Figure 9.21. The electrical potential between the glass and solution is measured by placing a solution on both sides of a glass membrane. The difference in potential between the two solutions is then measured using two reference electrodes.

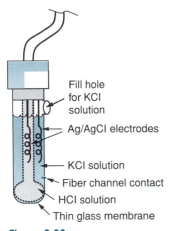

Figure 9.22. The glass electrode part of this assembly is the central section and the exposed glass membrane. A reference electrode has been built around it for the convenience of having a single unit.

MEASUREMENT OF INTERFACIAL POTENTIAL

We have already seen that to measure a potential difference, one must be able to make electrical contact with the media between which the potential is developed. In this case, we must make electrical contact between the glass and the test solution. The problem of contact with the test solution was solved in the case of the metal electrodes by the use of a reference electrode. We will do that again here. The problem is with the glass. The conduction mechanism in the glass is a very modest ion mobility, and this does not allow the creation of a conducting junction with an electronic conductor (a metal wire). A very clever ploy was used. The solution to the problem was to use a thin sheet of glass as the electrode and to contact the other side of the sheet with a second solution as shown in Figure 9.21. Now a potential will be developed on both sides of the glass sheet, each one a function of the H_3O^+ activity in the respective solutions. A second reference electrode can be used to contact the second solution. If the solution on one side has a constant composition, its potential will be constant and can be taken into account.

The resulting probe is shown in Figure 9.22. The thin glass sheet is in the form of a bulb at the end of a tube. The glass is very thin and often fragile. It is made this way to reduce the resistance of the ionic conduction between the two sides of the glass. Even so, glass electrodes have a resistance of the order of $10^8 \ \Omega$, which requires a special high-resistance input amplifier for the voltage measurement. The glass bulb encloses the second solution, which also conveniently contains the second reference electrode. When this electrode is dipped into a test solution that contains the test solution reference electrode, the voltage V between the two reference electrodes can be measured.

The various factors now involved in the measurement are shown in the schematic of Figure 9.23. The measured voltage, V, is obtained by summing (with the appropriate sign) all the voltages in the series starting from the $+$ lead of the voltmeter.

$$V = E_{ref2} - \Delta E_{G-s2} + \Delta E_{G-s1} - E_j - E_{ref1}$$

Now the glass–solution potentials are substituted in from Equation 9.49 and the terms rearranged for convenience.

$$V = E_{ref2} - E_{ref1} - E^{\circ}_{Glass} - \frac{RT}{F} \ln a_{H_3O^+(inner)} + E^{\circ}_{Glass} + \frac{RT}{F} \ln a_{H_3O^+(test)} - E_j$$

In this equation, the E°_{Glass} terms cancel, and all but the last two terms are constants. In practice, the two sides of the glass do not act identically, so there is a finite difference in the two E°_{Glass} values that is called the **asymmetry potential**. Nevertheless, the difference is still a constant. Using the term E_{const} for the combination of constants,

$$V = E_{const} + \frac{RT}{F} \ln a_{H_3O^+(test)} - E_j \qquad 9.50$$

$$V = E_{const} + V_N \log a_{H_3O^+(test)} - E_j$$

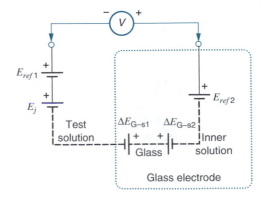

Figure 9.23. The equivalent circuit for the potential measurement with a glass electrode includes all the interfacial potentials in the glass and reference electrodes. Summing the potentials around the loop leads to the equation in the text.

Equation 9.50 is the basic equation for the response of the glass pH electrode when used with a reference electrode for the measurement of pH. The junction potential is generally assumed to be zero or constant. Note that the electrode responds to the activity of the hydronium ions in solution, not the concentration.

USE OF THE MEASURED VOLTAGE TO CALCULATE pH

The previous discussion has demonstrated that the glass electrode potential responds to changes in the activity of the H_3O^+ ion. Since the "p" function is so often used for this quantity, it is useful to rewrite Equation 9.50 in terms of $pa_{H_3O^+}$ rather than $\log a_{H_3O^+}$.

$$V = E_{const} - V_N \, pa_{H_3O^+(test)} - E_j \qquad 9.51$$

We do not know the value of E_{const}, so it is necessary to perform a calibration experiment just as in the case of the electron transfer electrode response. The voltage is first measured when the electrode is dipped into a solution of known pH. (See Chapter 3 for a discussion of pH standards.) Then it can be measured for solutions of unknown pH. The two voltages measured are related by the following two equations.

$$V_{cal} = E_{const} - V_N pH_{(std)} - E_j$$
$$V = E_{const} - V_N pH_{(unk)} - E_j$$

When the difference is taken between these two equations and the terms rearranged to solve for $pH_{(unk)}$,

$$pH_{(unk)} = pH_{(std)} + \frac{V_{cal} - V}{V_N} \qquad 9.52$$

In measurements of electrode potential, the precision is generally much better than the accuracy. For this reason, the difference in the potentials measured in a standard solution and in an unknown solution are used, rather than an absolute potential, when determining activity.

It is important to note that in the combination of the two equations to obtain Equation 9.52, it was assumed that the temperatures and junction potentials were the same for both the calibration and unknown measurements. This must be the case for the calibration process to be accurate.

Though Equation 9.52 and the two measurement results can be used to calculate the pH, this is rarely done when modern pH meters are used. In practice, the pH electrode is dipped into the standard solution, and the pH calibration knob on the pH meter is adjusted to read the pH of the standard. The standard solution should be at the same temperature as that expected for the unknown(s). The temperature of both solutions should be set with the meter temperature adjustment, as it directly affects the slope of the calibration curve. Then the electrode is rinsed and placed in the unknown solution, and the pH is read directly from the display. The operations expressed in Equation 9.52 are carried out within the meter circuitry. In analog meters, the operations are achieved by adjustments to the offset (the calibration) and the slope (the temperature) of the amplifier that interfaces the electrodes to the meter readout. In the case of a pH meter with digital display, the operations of Equation 9.52 can be carried out by digital calculation on the measurement voltages after their conversion to numbers.

TRANSISTOR CHEMICAL SENSORS

Solid-state electronics has provided another way to sense the potential difference between a membrane surface and the solution into which it immersed. These devices are based on the **field-effect transistor.** The heart of a field-effect transistor is a conducting channel between two contacts on a silicon semiconductor base. The conducting channel is a layer of n-type semiconductor between the two contacts, as shown in Figure 9.24. It is called an **n-type semiconductor** because the majority of mobile charge carriers in it are electrons (which have a negative charge). The remainder of the device (the base) is a **p-type semiconductor,** for which the principal conducting species is a positive "hole" or absence of a binding electron.

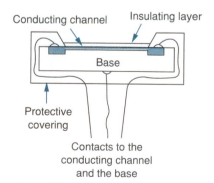

Figure 9.24. A transistor sensor based on the field-effect transistor. The potential between the solution contacting the insulating layer and the base affects the conductivity of the channel. This conductivity can be measured and related to the potential difference.

The surface of the n-type channel is covered with a thin layer of insulating material such as glass, which in turn is brought into contact with a conductor such as an ionic solution. The potential difference between the solution and the base affects the conductivity of the n-type channel. A negative potential between the solution and the base would repel electrons from the conducting channel into the base, decreasing its conductivity. Conversely, a positive potential difference between the solution and the base would attract electrons from the base into the channel and increase its conductivity. The field-effect transistor is thus a potential difference to conductivity converter. To convert the conductivity into an electrical signal, a small voltage is applied between the contacts at the ends of the channel, and the magnitude of the resulting current is detected.

To make this device into a chemical sensor, the insulating layer needs to be, or needs to be covered with, a material with which the solution will establish a characteristic potential. For example, if the insulating layer is glass with which the solution establishes a pH-dependent potential, the conductivity through the channel will be affected by the pH. pH electrodes based on this principle are commercially available and are useful where miniaturization and ruggedness are important. They cannot be used interchangeably with the more traditional glass electrode because the glass electrode converts the pH to a related voltage that is measured by the traditional pH meter. The measuring system for the transistor sensor needs to provide the circuit for converting conductivity to a related current and then convert the current to the output number.

Any mechanism by which the potential of the covering layer can be affected by the composition of the solution is usable with the transistor sensor. In later sections of the book, we will encounter more examples. A general term used for the chemical sensor based on the field-effect transistor is the **CEMFET.**

INTERFERING ELECTRODE REACTIONS

The glass surface is highly specific to reaction with protons. However, the $Glass^-$ sites do have some affinity for alkali metal ions as well. To the extent the alkali metal ions affect the response of the glass electrode, an error in the interpreted pH will result. A modified response equation can be used to quantify this effect. It is based on the concept that the response of the interfering ions is a constant fraction of the response to the ion of choice. This fraction is called the **selectivity coefficient.** The modified response function (starting with Equation 9.50) is

$$V = E_{const} + V_N \log\left(a_{H_3O^+} + k_{H^+,I^+} a_{I^+}\right) - E_j \qquad 9.53$$

The selectivity coefficient k_{H^+,I^+} is the ratio of the response to the interfering ion I^+ to that of the primary ion, H^+. The smaller the value of k_{H^+,I^+}, the less I^+ interferes with the determination by H^+.

For the glass electrode, the selectivity coefficient is extremely small for Na^+, K^+, and Li^+, of the order of 10^{-13} for ordinary electrodes. This would seem to be insignificant, but in a solution of pH = 13.0, the OH^- concentration is 0.1 M. For this to be true, there must also be 0.1 M of some cation, likely Na^+ or K^+. At this point, the alkali metal concentration is 10^{12} times the H_3O^+ concentration and, according to Equation 9.53, some interference will begin to appear. At still higher pH's, the response of the electrode will be primarily determined by the alkali metal ion concentration. To measure pH at high pH values, special electrodes designed for this region should be used. Their selectivity coefficients for the alkali ions can be as low as 10^{-15}.

GAS-SENSING ELECTRODES

A variety of gas-sensing electrodes have been based on the pH electrode (either glass or transistor). In these devices, a thin layer of solution is held next to the pH sensing surface by a membrane. The solution is one with which the sensed gas establishes a pH-dependent equilibrium. The membrane is porous to gaseous components, but it contains the liquid. An example is an ammonia sensor. The solution is NH_4Cl. The pH of the solution is related to the dissolved ammonia from the gas being tested. From the expression for the K'_a of ammonium,

$$[H_3O^+] = \frac{K'_a[NH_4^+]}{[NH_3]}$$

Changing to negative log units since the electrode responds to pH,

$$-\log[H_3O^+] = -\log K'_a - \log[NH_4^+] + \log[NH_3]$$

Since the ammonium concentration is constant, and the $[NH_3]$ is proportional to the partial pressure of NH_3 in the gas,

$$pH \approx p[H_3O^+] = K + \log p_{NH_3} \qquad 9.54$$

A sketch of an ammonia gas electrode based on the pH electrode is shown in Figure 9.25.

   The ammonia electrode described is just one of the electrodes of this type possible. Gas-sensing electrodes can be miniaturized around the transistor pH sensor, where the small solution volume involved can greatly improve the response time. Other gases that can be sensed in this same way are SO_2, NO_2, and H_2S.

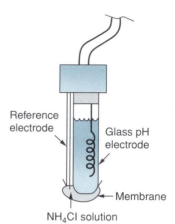

Figure 9.25. An ammonia gas electrode based on a glass pH electrode. This electrode works in the open air while the pH reading can be interpreted in terms of the partial pressure of NH_3 gas in the air.

THE USE OF ION–SURFACE INTERACTIONS FOR THE DETERMINATION OF IONS OTHER THAN H^+

The example developed thus far in this section has been the glass electrode for H_3O^+. Other electrodes have been developed that work on the principle of ion interaction with the solid surface. Glass electrodes have been designed that are less responsive to H_3O^+ and more responsive to various alkali cations. Glass electrodes selective for Na^+, Li^+, and NH_4^+ can be purchased from standard sources. One of the most effective examples is a Na^+ electrode that can be used in solutions with a pH as low as 3. This means that the value of k_{Na^+,H^+} is probably about 1. When the H_3O^+ concentration is below 10^{-3} M, concentrated solutions of Na^+ can be determined. At higher pH's, lower concentrations of Na^+ will work. Besides H_3O^+, the electrode is interfered with by Ag^+, K^+, and NH_4^+. The **selectivity coefficient** relative to K^+ may be as low as 10^{-5}. From this example we can see that although such electrodes can be useful, they do not enjoy the relative selectivity of the pH electrodes, and thus, in their application, all the listed interferents must be carefully considered. The corresponding response equation for electrodes of this type is illustrated for the Na^+ electrode discussed.

$$V = E_{const} + V_N \log\left(\begin{array}{l} a_{Na^+} + k_{Na^+,H_3O^+}a_{H_3O^+} \\ + k_{Na^+,Ag^+}a_{Ag^+} + k_{Na^+,K^+}a_{K^+} \\ + k_{Na^+,NH_4^+}a_{NH_4^+} \end{array}\right) - E_j \qquad 9.55$$

   Reactive electrode surfaces have also been developed for a variety of other ions, principally anions. These electrodes are formed from a pellet of a solid, relatively insoluble crystal of the sensed anion and a metal. For example, a crystal of AgCl is used in an electrode for the sensing of Cl^- ion. The pellet is held at the end of a tube that contains a reference solution of the sensed anion and a reference electrode. The electrode is dipped in the test solution, along with a reference electrode, to make the measurement.

The selectivity of ion selective electrodes

None of the ion selective electrodes has the specificity of the glass electrode for H_3O^+. It is very important to consider possible interferents before applying a specific electrode. Some of the interferents could cause substantial errors in the result. The application of a correction term where the concentration of the interferent is known can help some, but the selectivity coefficients are not completely independent of the composition of the solution.

Silver halide electrodes are available for the selective determination of Cl^-, Br^-, and I^-. As you would suspect, they cannot be responsive to the halide ion at concentrations less than the solubility of the silver halide. Thus, the Cl^- electrode is limited to solutions of 5×10^{-5} M and greater. The Br^- and I^- electrodes work at decreased concentrations since their silver compounds are less soluble. In general, anion electrodes are susceptible to interferences, the worst being other anions that form insoluble compounds with the metal in the pellet crystal. Thus, all the other halides can interfere with the response of any of the halide electrodes. To quantify these effects, selectivity coefficients, as in Equation 9.55, are used.

An electrode made with an Ag_2S pellet is responsive to either Ag^+ or S^{2-}. The Ag_2S has also been mixed with PbS, CdS, and CuS to make electrodes that are selective for Pb^{2+}, Cd^{2+}, and Cu^{2+}. The Ag_2S has been included to take advantage of the mobility of the Ag^+ through the crystal, thus providing the necessary conductivity.

A particularly effective pellet electrode has been made of a crystal of TlF with an impurity of EuF_2 to improve conductivity. The response of this electrode to F^- follows the theoretical model to the micromolar level.

The potential between the pellet material and the solution is developed by the selective attraction of anions to their places in the crystal. The potential difference created by anion adsorption is opposite in sign to that created by cation adsorption. Thus, the sign of the response to these electrodes is negative. The general response function for all ion selective electrodes based on surface reactivity is

$$V = E_{const} + \frac{V_N}{z} \log (a_{A^z} + k_{A^z,X} a_X) - E_j \qquad 9.56$$

When this equation is used, the quantity z should be used with its sign and magnitude. Thus, for a Cl^- electrode, $z = -1$, for a Cd^{2+} electrode, $z = +2$, etc. In this way, z affects both the sign and the slope of the response.

When a calibration measurement is made and the calibration and sample versions of Equation 9.56 are subtracted and rearranged (ignoring the interference term),

$$\log a_{A^z(test)} = \log a_{A^z(std)} + \frac{z(V - V_{cal})}{V_N} \qquad 9.57$$

Equation 9.57 is a general equation that is useful for all reactive-surface specific ion electrodes.

Study Questions, Section E

31. Compare the reaction of the glass electrode with water to that of a weak acid and water.

32. Why is the potential of the glass generally negative with respect to the solution touching it?

33. Why is a standard solution placed on the other side of the glass, and why are two reference electrodes required to measure the glass electrode potential?

34. What is the asymmetry potential?

35. Why is pH measured as the difference in potential between potential measurements in standard and unknown solutions rather than in terms of the absolute glass electrode potential?

36. A glass pH test electrode and a reference electrode are dipped into a solution at 25 °C that is known to have a pH of 6.865. The potential measured is 0.136 V. These same electrodes are then dipped into a sample at 25 °C and the potential measured is 0.245 V.

   A. What is the pH of the sample?

   B. The same standard solution has a pH of 6.838 at 40 °C. If the previous measurements were made at 40 °C, what is the pH of the unknown solution?

37. What kinds of interference can occur with surface interaction-based electrodes?

38. Calculate the error in the pH determination of a solution of 0.01 M NaOH. The value of k_{H^+,I^+} for H^+, Na^+ is 1×10^{-13}.

39. A variety of specific ion electrodes based on an ion–surface interaction have been developed both for cationic and anionic species. Is there any difference in the measured potential differences that result from cationic and anionic analytes?

40. An Ag_2S electrode (with a reference electrode) is dipped into an unknown solution and a potential of 0.442 V measured. This same pair of electrodes is dipped into a solution with a known sulfide ion activity of 4.7×10^{-5} M. In this solution, the measured potential is 0.623 V. What is the activity of sulfide ion in the unknown solution?

Answers to Study Questions, Section E

31. The reaction of Equation 9.47 is identical to that of Equation 4.1 for a weak acid with water. The main difference between these two reactions is that the $Glass^-$ is not soluble in the water and the conjugate base of the weak acid is.

32. The glass, having donated some protons to the solution, has a shortage of positive charge. The solution, on the other hand, has gained positive charge. Thus, the glass is negative with respect to the solution.

33. The glass is not an electronic conductor, so one cannot make an effective connection between the glass and the voltmeter wire. A second solution is put on the other side of the glass to make contact with the glass. To measure the potential difference of the two solutions, a reference electrode must be used in both of them.

34. Asymmetry potential is the difference in the potential between the solutions on either side of the glass electrode when the solution composition on both sides is the same.

35. The measured glass electrode potential includes the asymmetry potential and the reference electrode junction potential. These are not predictable. Under the assumption that they remain constant while the standard and unknown solutions are measured, they cancel out when the difference in the measured potentials is used.

36. A. Using Equation 9.52,

$$pH_{unk} = 6.865 + \frac{0.136 - 0.245}{0.0592} = 5.02$$

B. Again using Equation 9.52,

$$pH_{unk} = 6.838 + \frac{0.136 - 0.245}{0.0592(313/298)} = 5.09$$

37. Any ions that can mimic the reaction of the intended analyte ions with the surface are potential interferents.

38. If the NaOH concentration is 0.01 M, the concentration of the H_3O^+ ion is about $10^{-14}/10^{-2} = 10^{-12}$ M. Thus, the term in the parentheses of Equation 9.53 is approximately $10^{-12} + 10^{-13} \cdot 10^{-2} = 10^{-12} + 10^{-15}$ (when concentrations are substituted for activities). From this we can see that the correction term is one-thousandth the H_3O^+ ion activity and is therefore negligible. However, we can see that it would become significant for NaOH concentrations any greater than 0.01 M.

39. Because of the difference in the sign of the charge, the direction of the potential change as the concentration is varied is opposite for cationic and anionic analytes. This is seen in Equation 9.57, where z represents the magnitude and the sign of the charge on the analyte ion.

40. We will use Equation 9.57.

$$\log a_{S^{2-}(unk)} = -4.328 + \frac{-2(0.623 - 0.442)}{0.0592} = -10.4$$

$$a_{S^{2-}(unk)} = 10^{-10.4} = 4.0 \times 10^{-11} \text{ M}$$

F. Ion Diffusion Through Porous Membranes

Another process by which an electrical potential can be developed across a membrane separating two solutions is by the selective diffusion of ions through the membrane. Consider the case depicted in Figure 9.26. Two solutions of HCl of different concentrations are separated by a membrane through which only H_3O^+ or H^+ ions can pass. This is a possible situation based on size discrimination (because the protons are so small) or based on charge (if the transport mechanism only functions for positive ions).

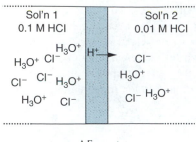

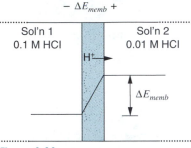

Figure 9.26. The selective diffusion of cations or anions through a porous membrane will develop a potential difference between the solutions on either side of the membrane. The magnitude of the potential difference created depends on the activity difference of the diffusing species between the two solutions and on the selectivity of the diffusion process.

DEVELOPMENT OF ELECTRICAL POTENTIAL DIFFERENCE

Because of the higher concentration of HCl on the left in Figure 9.26, there will be a tendency for the H^+ species to move through the membrane to the right-hand solution. As it does so, it will create a charge imbalance between the two solutions, with the right-hand solution becoming positive with respect to the left-hand solution. The developing positive potential on the right creates a force counter to the concentration difference, and the system will soon come to an equilibrium state. At equilibrium, the chemical potentials of the H_3O^+ ions in both solutions will be equal.

$$\overline{\mu}_{H_3O^+,sol'n2} = \overline{\mu}_{H_3O^+,sol'n1} \qquad 9.58$$

Substitution into this equation from Equation 9.4 gives

$$\mu^\circ_{H_3O^+} + RT \ln a_{H_3O^+,sol'n2} + FE_{sol'n2} = \mu^\circ_{H_3O^+} + RT \ln a_{H_3O^+,sol'n1} + FE_{sol'n1}$$
$$F(E_{sol'n2} - E_{sol'n1}) = RT \ln a_{H_3O^+,sol'n1} - RT \ln a_{H_3O^+,sol'n2}$$

When this equation is solved for the potential difference,

$$E_{sol'n2} - E_{sol'n1} = \Delta E_{memb} = \frac{RT}{F} \ln \frac{a_{H_3O^+,sol'n1}}{a_{H_3O^+,sol'n2}} \qquad 9.59$$

From Equation 9.59, we see that the potential developed across the membrane is affected by the activity ratio of the ions that can move through the membrane. If one of these solutions is of constant concentration, the potential will be a function of the activity of the other.

Membranes have been developed that have selective transport properties for many cations and anions of single and multiple charge. If the term z is included in the above derivation of the membrane potential developed, the resulting membrane potential equation is

$$\Delta E_{memb} = \frac{RT}{zF} \ln \frac{a_{A^z,sol'n1}}{a_{A^z,sol'n2}} \qquad 9.60$$

where, again, the value of z must include its sign.

MEASUREMENT OF MEMBRANE POTENTIAL

The method of developing the phenomenon of selective ionic transport into a probe for concentration follows the approach already seen for the reactive membrane electrodes. Two reference electrodes are used, one in the solution contained in the electrode assembly and the other dipped into the test solution. The schematic of the resulting measurement is shown in Figure 9.27. Adding the potentials around the circuit from the + lead of the voltmeter and substituting for ΔE_{memb} from Equation 9.60,

$$V = E_{ref2} + \Delta E_{memb} - E_j - E_{ref1}$$
$$V = E_{ref2} - E_{ref1} - \frac{RT}{zF} \ln a_{A^z,sol'n2} + \frac{RT}{zF} \ln a_{A^z,sol'n1} - E_j$$

In this equation, the first left three terms on the right-hand side are constants that can be grouped into a single constant, E_{const}.

$$V = E_{const} + \frac{RT}{zF} \ln a_{A^z,sol'n1} - E_j \qquad 9.61$$

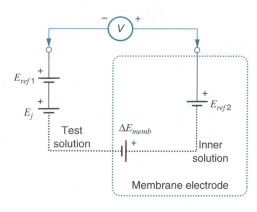

Figure 9.27. An equivalent circuit that includes all the interfacial potentials in the measurement of membrane potentials can be used to develop the equation relating the measured voltage and the membrane potential.

In application, it is necessary to perform a calibration step in order to eliminate the value for E_{const}, which cannot be predicted with sufficient accuracy.

$$V = E_{const} + \frac{V_N}{z} \log a_{unk} - E_j$$

$$V_{cal} = E_{const} + \frac{V_N}{z} \log a_{std} - E_j$$

The difference between these two equations can be taken to obtain an equation into which the measurement data can be inserted.

$$\log a_{unk} = \log a_{std} + \frac{z(V - V_{cal})}{V_N} \qquad 9.62$$

The same precautions apply to the application of selective transport electrodes as to the other types. The calibration must be done at the same temperature as the measurement, and the temperature control of the meter must be set.

ELECTRODE TYPES AND INTERFERENTS

Selective transport electrodes are generally of two types. In one type, the membrane material is the transport medium, and the selected ions move from one site to another through the medium. Selective transport can be achieved through the use of a sheet of cation or anion exchange material. A cation exchange material has immobile anionic sites within it. Only cations will be attracted to these sites and be able to move from one site to the next. To be completely effective, this mechanism of ionic movement must be the only transport mechanism allowed. In other words, there should be no pores for the diffusion of solution from one side to the other. To the extent that the ionic sites can be made to favor particular ions, the ion exchange membranes can be somewhat specific among ions of the same charge type.

The other type of selected transport electrode is made of two thin plastic sheets that are permeable to the ions of interest. Between these two sheets is a solution that contains bulky counterions that can move within the solution but cannot pass through the membrane. The bulky counterions act as transporting agents for the selected ions through the inner solution. Ions not attached to the counterions do not pass through the inner solution because it is composed of a low dielectric medium in which salts, in general, have a very low solubility. If the bulky, transporter ions can be made to have specific affinity to particular ions, the electrode selectivity can be enhanced.

A widely used example of an ion transporter membrane electrode is the Ca^{2+} specific electrode. The transporter anion is dialkyl phosphate. Each of the anions has a single minus charge, so that two anions and the Ca^{2+} are required to provide the neutral species

Though very useful in many specific instances, ion transport specific ion electrodes require more care in storage and more maintenance than electrodes with solid active surfaces. They have a limited shelf life and application lifetime, but they can be readily renewed by exchanging the spent membrane assembly with a new one.

that can permeate the organic liquid layer. The anions can acquire the Ca^{2+} at the test so-lution side of the membrane and release it at the inner solution side. A useful way to think of this process is that the Ca^{2+} is complexed by the dialkyl phosphate while in the organic medium. The higher the formation constant for the complex, the lower is the concentration of Ca^{2+} to which it will be able to respond. The commercial Ca^{2+} elec-trodes are useful to concentrations as low as 5×10^{-7} M. From this example, we can see that the selectivity depends on the differences in the formation constants of the complex-ing agent with the primary ion and potentially interfering ions. In the case of the Ca^{2+} electrode, the selectivity coefficient is 0.02 for Mg^{2+} and 0.001 for Na^+ and K^+. Other interfering ions for the calcium electrode include Pb^{2+}, Hg^{2+}, Sr^{2+}, Fe^{2+}, and Cu^{2+}.

The ion carrier need not be an ion if the charged complex is sufficiently soluble in the organic transport medium. An example of this type of electrode is the K^+ elec-trode in which the antibiotic valinomycin is the carrier. The high specificity of the vali-nomycin for complexing with K^+ makes this a very useful electrode in solutions where sodium ions are also present. The selectivity coefficient for Na^+ relative to the K^+ pri-mary ion is 10^{-4}. Such electrodes are of great value in physiological studies.

LIQUID JUNCTION POTENTIALS

The previous discussion regarding selective transport membranes assumed that the membrane is totally selective for one or the other of the charge states and the move-ment of the analyte ion. If this is not the case (i.e., if both cations and anions can move through the membrane or if significant transport is provided by interfering ions), the situation is very different. Consider the case of a barrier that is designed to prevent large-scale mixing but through which ions of all types can move freely in either direc-tion. Such a barrier is shown in Figure 9.28 for the situation where there is an HCl so-lution on one side and a KCl solution on the other. Initially, the H_3O^+ move through the barrier to the right, the K^+ move to the left, and the Cl^- are free to go either way. However, the mobility of the H_3O^+ is very much greater than that of the K^+, which will cause a positive charge to build up in the right-hand solution. The resulting poten-tial difference between the two solutions will accelerate the rate of K^+ movement to the left and Cl^- movement to the right. It will also slow the movement of H_3O^+ to the right. Soon the system will come to a steady state rate of ion movement in which there is no net movement of charge from one solution to another. The potential will then be constant, and there will be a net flow of HCl to the right and KCl to the left. It is this continued process that makes this a **steady-state process** and not one at equilibrium.

Such a boundary is called a **liquid junction** because it is the electrical connec-tion between two ionic conducting solutions. The potential developed is called the **junction potential.** The example used is exactly the situation that exists in a pH mea-surement between the reference electrode and the test solution. The magnitude of the junction potential is dependent on the difference in the mobilities of the ions carrying the predominance of charge through the barrier. Although the sign of the junction po-tential can often be predicted, the value can be calculated only in the simplest case of the same salt on each side of the barrier.

Liquid junction potentials have been measured for typical situations in which one of the ions is common to both solutions, such as the Cl^- in the example of Figure 9.28. For this case (0.1 M HCl‖0.1 M KCl), the potential is 27 mV. Neglecting this potential would cause an error of nearly half a pH unit in a measurement. In practice, an effort is made to reduce the junction potential to as low a value as possible. This is done by arranging for the predominance of the cations and anions moving through the barrier to have nearly equal mobilities. Fortunately, this can be achieved quite closely by having those ions be K^+ and Cl^-. The mobilities of these ions are equal to within 4%. (The Cl^- mobility is slightly higher.) In addition, the concentration of the KCl should be as large as possible so that it provides the predominance of the ions in the junction. In the case of the junction with 0.1 M HCl, the junction potential is reduced to 3.1 mV when 3.5 M KCl is used.

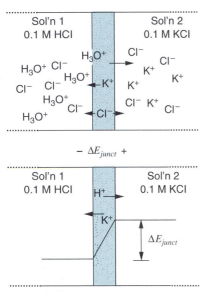

Figure 9.28. The diffusion rates of dif-ferent ionic species can be very differ-ent. Because of this, a potential similar to the porous membrane potential will be developed at the junction of solu-tions that differ in their composition. This liquid junction potential is part of virtually all potentiometric measure-ments.

THE EQUIVALENCE OF RESPONSE FUNCTION FOR ALL ELECTRODES

You may have noticed the remarkable similarity of the response functions of the several electrode types discussed in this chapter. If not, take a moment to compare Equation 9.32 for the $M^{n+}/M_{(s)}$ electrode, Equation 9.57 for the selective surface reaction membrane electrode, and Equation 9.62 for the selective diffusion electrode. Every one of them has the function

$$\log a_{unk} = \log a_{std} + \frac{z(V - V_{cal})}{V_N} \qquad \text{or} \qquad pa_{unk} = pa_{std} + \frac{z(V_{cal} - V)}{V_N} \qquad 9.63$$

On the one hand this seems remarkable, but on the other it is not surprising for two reasons. One was pointed out by Poincaré, a nineteenth century French mathematician, "If a phenomenon admits of a complete mechanical explanation it will admit of an infinity of others which will account equally well for all the peculiarities disclosed by experiment." In other words, the ability of a theory or model to explain the available data does not prove that the model is correct, but it does prove that the observations are consistent and that the model is useful for making predictions. The second reason for the similarity in response functions is that the three mechanisms have something fundamental in common: they are all mechanisms for converting a difference in chemical potential to a difference in electrical potential. From this perspective, it would be surprising if they did not result in the same potential. Failure to recognize that alternate mechanisms may explain the same result perfectly well, and may actually be more correct, can produce a temporary halt in the progress of science. For many years, it was believed that the response of the glass electrode was due to the selective diffusion of protons through the glass. This view prevented the development of a scientific basis for the improvement in glass formulations. It required careful and convincing isotopic studies by one determined doubter to move the scientific community to a new way of viewing the operation of the glass electrode. From this knowledge came better pH electrodes and many other electrode types.

Study Questions, Section F

41. If a phosphate selective electrode were developed using the selective transport of phosphate ion, would the more positive potential be on the side with the higher phosphate concentration or the lower?

42. How is the method of measurement of the membrane potential similar for the ion–surface interaction electrodes and the selective transport electrodes?

43. A calcium combination electrode is dipped into an unknown solution, and a potential of -0.158 V is measured.

This same electrode is dipped into a solution with a known calcium ion activity of 2.3×10^{-4} M. In this solution, the measured potential is -0.094 V. What is the activity of calcium ion in the unknown solution?

44. Why are high concentrations of KCl used in reference electrodes?

45. Why are the equations for the three types of electrodes (redox, ion–surface interaction, and selective transport) all the same for any given ion?

Answers to Study Questions, Section F

41. Some of the phosphate ions would move from the more concentrated side to the less concentrated side. This would produce an excess of negative charge on the lower concentration side and a loss of negative charge on the more concentrated side. Thus, the more concentrated solution would be positive with respect to the less concentrated solution.

42. The methods of measuring potentials using ion–surface interaction electrodes and selective transport electrodes are identical. Two reference electrodes are required to contact the two solutions. The concentrations are determined by measuring the potential in the standard and in the unknown solutions and taking the difference.

43. From Equation 9.62,

$$\log a_{Ca^{2+}(unk)} = -3.64 + \frac{+2[-0.158 - (-0.094)]}{0.0592}$$

$$= -5.800$$

$$a_{Ca^{2+}(unk)} = 10^{-5.800} = 1.6 \times 10^{-6}\ M$$

44. The reference electrode contacts the solution in which it is immersed through a liquid junction. Differences in the mobilities of the ions on either side of this junction can create a potential difference across the junction. The mobilities of K^+ and Cl^- are very nearly equal. When the concentration of K^+ and Cl^- is large, diffusion of the ions from inside the reference electrode to the sample solution predominates, and the junction potential is minimized.

45. The three different types of electrodes are different mechanisms for converting the difference in chemical potential of the analyte ion to an electrical potential. The relationship between the chemical potential and electrical potential is fundamental, so the result must be the same.

G. Selectivity and Detection with Electrodes

The area of potentiometric detection is one of very active research by many groups. The goals of this research are improved sensitivity and selectivity as well as simplicity and miniaturization. We would like to have simple and direct sensors for toxins and pollutants of all types as well as for molecules whose presence is indicative of disease or exposure. Some responsive elements have been combined with the electronic amplification elements in microcircuits to achieve both miniaturization and economy of large-scale manufacture.

As we have seen in this chapter, the main difficulty has been the achievement of sufficient selectivity. In this area, researchers are exploring the implementation of biological reactants, which often exhibit remarkable selectivity in biological functions. Another approach is to acknowledge that a perfectly selective detector is unlikely to be developed and instead use an array of differently selective detectors. If an array of detectors, each with somewhat different selectivity coefficients, was exposed to the same sample, the response of each would be different, depending on the particular mix of analytes to which the detectors are sensitive. The result would be an accumulation of overlapping data from which the individual components could be resolved. This is yet another example of using more data to resolve interfering components.

Practice Questions and Problems

1. A. What is a redox couple? In what way do the forms of the couple differ?

   B. What effect does the presence of a redox couple have on a piece of metal in contact with the solution?

   C. Write the equation that relates the potential of a metal dipped into a solution that contains V^{3+} and V^{2+}.

   D. Calculate the potential of the metal in part C if the concentrations of V^{3+} and V^{2+} are 10^{-5} and 10^{-3} M, respectively.

2. A. Draw a log concentration plot for the electrode where Ox and Red are H^+ and $H_{2(g)}$. Assume that the pressure of H_2 is 1 atm.

   B. The concentration of what species is plotted along the X-axis of this plot?

   C. What is the potential of this electrode at a pH of 7?

   D. Would this be a useful electrode for the measurement of pH?

3. A. Why is a reference electrode required for the measurement of the test electrode potential?

   B. What are the essential qualities of a reference electrode?

   C. Is the potential of the AgCl,Ag reference electrode a function of the Cl^- concentration inside it? Why?

   D. Why is a high concentration of KCl used as the fill solution? (In other words, why is it high, and why is it KCl?)

4. Calculate the potential measured when a AgCl,Ag reference electrode is used with the following test electrodes and solutions.

   A. A silver electrode in a solution of 0.02 M KCl that has been titrated to the equivalence point with 0.01 M $AgNO_3$.

   B. A copper electrode in a solution of 3.6×10^{-3} M Cu^{2+}.

   C. A platinum electrode in a solution that is 3.5×10^{-4} M in Sn^{2+} and 5.8×10^{-3} M in Sn^{4+}.

D. An iron electrode in a solution buffered at pH 9 that is saturated in FeS.

5. The voltage measured with a combination pH electrode in a standard solution of pH = 4.008 is 0.416 V. When the electrode is dipped into the solution of unknown pH, the potential is 0.409 V.

   A. What is the pH of the unknown solution? Assume that both measurements were made at 25 °C.

   B. What kinds of errors would be incurred if the standardization were made at a 30 °C and the measurement of the unknown solution at 45 °C? Assume the temperature adjustment dial on the pH meter remained at 25 °C for both measurements. Refer to Table 3.8. Express the answer in words, not a calculation.

6. A. Briefly describe the mechanisms of the selective adsorption and selective diffusion types of specific ion electrodes.

   B. Why does a Ca^{2+} specific electrode have a different voltage change per $pa_{Ca^{2+}}$ than the pH electrode?

7. A. Why are calibration solutions always used with the determination of solution composition with specific ion electrodes?

   B. What assumption is made when a specific ion electrode measurement is presumed to be measurement of p[ion]?

8. A. On the basis of Table 9.1, what is the sign of the potential difference you would expect if you put an iron nail in a solution containing ferrous ions (Fe^{2+})? Specifically, indicate whether the iron will be positive or negative with respect to the solution and why you think so.

   B. In all the cases of a potential difference between a metal and its dissolved ions, with the formal standard reduction potential be more positive or more negative than the thermodynamic standard reduction potential? Explain your reasoning.

9. A. Write a balanced reduction reaction for the reduction of H_2O to H_2 in basic solution. (Use OH^- in the reaction instead of H_3O^+.).

   B. Derive a value for the standard thermodynamic reduction potential for the reduction reaction written in part A. *Hint:* Start with the reduction of H_3O^+ to H_2 and consider that it is occurring in solution with unit activity of OH^-.

   C. Complete the following sentence: For every ten-fold increase in the ratio of [Ox] to [Red], the equilibrium potential (increases, decreases) by $V_N = 0.0591$ V for a couple with $n = 1$ at 20°C.

10. Based on the thermodynamic standard reduction potentials for $Ag^+ + e^- \rightleftharpoons Ag(s)$ and $AgCl + e^- \rightleftharpoons Ag(s) + Cl^-$, calculate the thermodynamic solubility product for AgCl.

11. Define the terms and briefly describe the use of a:

    A. reference electrode

    B. test electrode

    C. inert electrode

    D. combination electrode

    E. voltmeter

    F. liquid junction

    G. selectivity coefficient.

12. A. In the case of the glass electrode for measurement of pH, how is the electrical potential established between the glass and the solution, and why are two reference electrodes required for its measurement?

    B. When dipped into a solution of unknown pH, the potential of a glass/combination electrode decreases 0.0934 V from that measured with a standard pH = 7 buffer at 20°C. (See Table 3.8). What is the pH of the solution?

13. Write a version of Equation 9.57 for an electrode that responds specifically to SO_4^{2-} ions.

14. A. Why is a liquid junction potential not an equilibrium potential?

    B. Why is the potential across a membrane with selective transport an equilibrium potential?

15. A silver electrode is used in conjunction with a AgCl,Ag reference electrode to follow the titration of 0.001 M Cl^- ion with 0.001 M $AgNO_3$. The voltmeter is connected between the two electrodes with the red lead connected to the silver electrode. This means that if the silver electrode is more positive than the AgCl,Ag electrode, a positive voltage will be indicated. Calculate the voltmeter reading at

    A. 0% titrated

    B. 50% titrated

    C. 100% titrated

    D. 200% titrated.

16. You need to choose between the quantitation of an analyte by electrode potential and titration. Explain which you would choose if you needed to know

    A. the absolute amount of the analyte to a few parts per thousand.

    B. the activity of the analyte.

17. For a calcium-selective electrode, what voltage change would correspond to a change in Ca^{2+} concentration of 1 ppt?

18. All redox couples that involve $[H_3O^+]$ in their half reaction can be written with H_2O on the Ox side of the reaction and the OH^- on the reduced side instead. When the reaction is written in this way, the value of $E°$ changes significantly.

    A. Write the half reaction for the reduction of SO_4^{2-} to SO_3^{2-} using H_2O and OH^- instead of H_3O^+ and H_2O.

B. Calculate the value of $E°$ for the reduction of sulfate to sulfite in basic solution.

C. Would you say that the amount that the $E°$ changes for the sulfate reduction reaction between an acidic solution and a basic solution is the same for all reactions involving H_3O^+ and H_2O or not? Justify your answer.

19. A. Why are all the values of E_{eq} calculated using the Nernst equation (Equation 9.15) values of the equilibrium electrode potential with respect to the standard hydrogen electrode?

B. How does the expected measurement voltage differ from the value calculated from Equation 9.15 when a AgCl,Ag reference electrode is used for the measurement?

20. Suggest a way to use a metallic test electrode to create an electrode of the second kind that is responsive to the activity of:

A. CrO_4^{2-}

B. CN^-

C. Tryptophan

21. A. Write the Nernst equation for the couple $oOx + ne^- \rightleftharpoons rRed$.

B. Use the Nernst equation developed in part A to help explain why the equilibrium potential of equimolar solutions of a symmetrical redox couple is independent of the concentration of the species in the couple.

C. Show why the equilibrium potential of unsymmetrical couples changes with concentration even if the concentration ratio of Ox and Red stay the same.

22. Based on the discussion of the glass electrode and the transistor sensor, determine whether resistance of the conducting channel in a field-effect transistor used as a pH sensor will increase or decrease as the pH of the solution increases.

Suggested Related Experiments

1. Numerous experiments with pH and specific ion electrodes.
2. Analysis of F^- in drinking water.

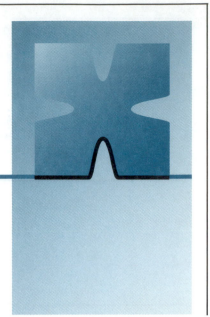

Chapter Ten

ANALYSIS BY OXIDATION–REDUCTION REACTIVITY

Most chemical species can exist in a variety of oxidation states. In the elemental form, the **oxidation state** is defined as zero. When iron rusts, it is being **oxidized** to a higher oxidation state. The oxygen that reacts with the elemental iron to form the oxide is itself **reduced** to a lower oxidation state. Thus, an oxidation reaction always occurs in concert with a reduction reaction. Such reactions are very common in nature. They are central to the conversion of sunlight and food into the energy that sustains plants and animals. They are involved in corrosion and the formation of the ozone hole. It should be no surprise, by now, that they also are used as a differentiating characteristic for chemical analysis. In this chapter, we will explore several ways to probe the quality of oxidizability or reducibility, and we will see how the desired analytical information can be obtained from the response to the probes. We will also see how electrochemical oxidation and reduction can be used as a method of quantitation and separation. The chapter begins with a discussion of the basic qualities of oxidation–reduction reactions.

A. Oxidation–Reduction Reactions

Oxidation–reduction reactions are reactions in which electrons are transferred between the reactants. The reactant that loses electrons to the other is **oxidized,** and the reactant that accepts electrons is **reduced.** Oxidation–reduction reactions are combinations of two electron transfer reactions, one of which is occurring in reverse.

COMBINING ELECTRON HALF REACTIONS

A table of electron transfer reactions was introduced in Chapter 9 and is reprinted here as Table 10.1. Each electron transfer reaction has the form

$$o\text{Ox} + ne^- \rightleftharpoons r\text{Red} \qquad\qquad 10.1$$

Electron transfer reactions are sometimes called **electron half reactions** because two of them are required to make a complete reaction. The Ox form of one redox pair reacts with the Red form of another to make a complete reaction. The overall reaction is the sum of the two electron half reactions, taken with coefficients that make the electrons cancel out. For example, if Fe^{3+} reacts with Sn^{2+}, the overall reaction is

$$2Fe^{3+} + Sn^{2+} \rightleftharpoons 2Fe^{2+} + Sn^{4+} \qquad\qquad 10.2$$

The ferric,ferrous half reaction is multiplied by 2 before combining because the ferric, ferrous couple exchanges only one electron while the stannic,stannous couple exchanges two.

Table 10.1
Standard Reduction Potentials

Ox	n	Red	$E°$, V
Ce^{4+}	$+e^-$	Ce^{3+}	$+1.6$
$Cl_{2(g)}$	$+2e^-$	$2Cl^-$	$+1.359$
$O_{2(g)} + 4H_3O^+$	$+4e^-$	$6H_2O$	$+1.229$
$Br_{2(g)}$	$+2e^-$	$2Br^-$	$+1.087$
$VO_2^+ + 2H_3O^+$	$+e^-$	$VO^{2+} + 2H_2O$	$+1.001$
Ag^+	$+e^-$	$Ag_{(s)}$	$+0.799$
Fe^{3+}	$+e^-$	Fe^{2+}	$+0.771$
I_3^-	$+2e^-$	$3I^-$	$+0.536$
Cu^{2+}	$+2e^-$	$Cu_{(s)}$	$+0.337$
$VO^{2+} + 2H_3O^+$	$+e^-$	$V^{3+} + 3H_2O$	$+0.337$
$UO_2^{2+} + 4H_3O^4$	$+2e^-$	$U^{4+} + 6H_2O$	$+0.334$
$Hg_2Cl_{2(s)}$	$+2e^-$	$2Hg_{(l)} + 2Cl^-$	$+0.268$
$AgCl_{(s)}$	$+e^-$	$Ag_{(s)} + Cl^-$	$+0.222$
$SO_4^{2-} + 2H_3O^+$	$+2e^-$	$SO_3^{2-} + 3H_2O$	$+0.172$
Sn^{4+}	$+2e^-$	Sn^{2+}	$+0.139$
$TiO^{2+} + 2H_3O^+$	$+e^-$	$Ti^{3+} + 3H_2O$	$+0.099$
$S_4O_6^{2-}$	$+2e^-$	$2S_2O_3^{2-}$	$+0.09$
$2H_3O^+$	$+2e^-$	$H_{2(g)} + 2H_2O$	0.000
Sn^{2+}	$+2e^-$	$Sn_{(s)}$	-0.140
$AgI_{(s)}$	$+e^-$	$Ag_{(s)} + I^-$	-0.151
V^{3+}	$+e^-$	V^{2+}	-0.255
Cr^{3+}	$+e^-$	Cr^{2+}	-0.38
Cd^{2+}	$+2e^-$	$Cd_{(s)}$	-0.403
Fe^{2+}	$+2e^-$	$Fe_{(s)}$	-0.440
U^{4+}	$+e^-$	U^{3+}	-0.63
Zn^{2+}	$+2e^-$	$Zn_{(s)}$	-0.763

In some cases, the half reactions to be combined might not be already balanced. Balanced half reactions are required before combining them in the overall reaction. In the case of a simple change in the oxidation state of the only element involved, such as the reduction of Sn^{4+} to Sn^{2+}, one only needs to add the number of electrons to balance the charge. In this case,

$$Sn^{4+} + 2e^- \rightleftharpoons Sn^{2+} \qquad\qquad 10.3$$

If other elements are involved, all elements and the charge need to be balanced. Take the case of the reduction of tantalum (V) oxide to tantalum metal in acid solution. The form the oxygen will take in the product is water. This requires H^+ (or H_3O^+) as a reactant. Balance the elements first, and then use electrons to balance the charge. The result is

$$Ta_2O_{5(s)} + 10H_3O^+ + 10e^- \rightleftharpoons 2Ta_{(s)} + 15H_2O \qquad\qquad 10.4$$

If the solution is basic, one needs to write the reaction with OH^- ions rather than H_3O^+. For example, in the reduction of SO_3^{2-} to $S_2O_4^{2-}$, the half reaction is

$$2SO_3^{2-} + 2H_2O + 2e^- \rightleftharpoons S_2O_4^{2-} + 4OH^- \qquad\qquad 10.5$$

In basic solutions, oxygen atoms on the reactant side are converted to OH^- ions on the product side. The hydrogen required by the OH^- product is supplied by H_2O. Then the charge is balanced by the addition of electrons.

Now we can combine the balanced half reactions into a complete redox reaction. The general form for the reaction that combines any two electron half reactions is shown in Equation 10.6. First, the two reactions are written with an indication of the direction they will take in the combined reaction. One must be forward and the other reverse. Second, each reaction is multiplied by the number of electrons involved in the other reaction. This automatically adjusts the number of electrons in each half reaction so that they will cancel out when the two reactions are summed. (See Example 10.1.) Third, the stoichiometric coefficients (now products of n's with o or r) are reduced to the simplest whole numbers. The first two steps are generalized as follows:

$$n_2(o_1Ox_1 + n_1e^- \rightleftharpoons r_1Red_1) \quad\rightarrow$$
$$n_1(o_2Ox_2 + n_2e^- \rightleftharpoons r_2Red_2) \quad\leftarrow \qquad\qquad 10.6$$
$$n_2o_1Ox_1 + \cancel{n_2n_1e^-} + n_1r_2Red_2 \rightleftharpoons n_2r_1Red_1 + \cancel{n_1n_2e^-} + n_1o_2Ox_2$$
$$n_2o_1Ox_1 + n_1r_2Red_2 \rightleftharpoons n_2r_1Red_1 + n_1o_2Ox_2$$

Redox reactions generally occur homogeneously in solution. There is no piece of metal to serve as the reservoir of electrons for either of the reactions. The electrons must be transferred directly from Red_2 to Ox_1 in order to form the products. Exceptions are redox reactions that are catalyzed at the surface of a solid. For example, the reaction between O_2 as an **oxidant** with H_2 as a **reductant** to form water as the product does not occur in the gaseous phase at room temperature without some spark to initiate the reaction. However, O_2 and H_2 dissolved in water react very vigorously at a platinum surface. The platinum, in this case, is acting as a **catalyst** to facilitate the reaction.

BALANCING COMPLETE REDOX REACTIONS

The result of the addition of two electron half reactions as outlined in the steps of Equation 10.6 is a complete, balanced redox reaction. In some situations, we are not given the balanced half reactions or even the products. In this case, we must first identify the products and then achieve a balanced overall reaction. To identify the products, we identify the possible products for each reactant and then see which of them are en-

Example 10.1

Summing two half reactions
Ferric ion reacts with SO_3^{2-} in acid solution to give Fe^{2+} and SO_4^{2-}. The half reactions are

$$Fe^{3+} + e^- \rightleftharpoons Fe^{2+}$$
$$SO_4^{2-} + 2H_3O^+ + 2e^-$$
$$\rightleftharpoons SO_3^{2-} + 3H_2O$$

The ferric half reaction is multiplied by 2 to give equal numbers of electrons in the two half reactions. The sulfate half reaction is reversed before summing. The result is

$$2Fe^{3+} + SO_3^{2-} + 3H_2O \rightleftharpoons$$
$$2Fe^{2+} + SO_4^{2-} + 2H_3O^+$$

ergetically favored. For example, if the reactants are Sn^{2+} and VO^{2+} in acid solution, there are several possibilities for the products. These can be seen in Table 10.1. Stannous ion can be oxidized to Sn^{4+} or reduced to tin metal. The VO^{2+} can be oxidized to VO_2^+ or reduced to V^{3+}. In the reaction, one of the reactants will be oxidized and the other reduced. If the Sn^{2+} is reduced to Sn, the VO^{2+} is oxidized to VO_2^+. The products are seen to be an oxidant with an $E°$ that is more positive than Sn^{2+} in the table and therefore a stronger oxidant than the oxidizing reactant. Similarly, the Sn metal produced is in a couple with a more negative potential than the reducing reactant, VO^{2+}. This reaction (to produce products that are stronger oxidants and reductants than the reactants) is not energetically favored. Therefore, the reaction that will occur is the oxidation of Sn^{2+} to Sn^{4+} by the VO^{2+}, which is reduced to V^{3+}. A favored reaction involves an oxidant (species on the left side of Table 10.1) with a reductant (species on the right side of Table 10.1) and which is in a couple below that of the oxidant.

Choose reaction products so that Ox_1 reacts with a Red_2 that is below and to the right of it in Table 10.1.

Now that we have the products, we can proceed to balance the reaction. The unbalanced reaction is $Sn^{2+} + VO^{2+} \rightleftharpoons Sn^{4+} + V^{3+}$. To balance this reaction, we first need to balance the changes in oxidation state. For the tin, it is clear that the oxidation state changes from $+2$ to $+4$, a change of $+2$ overall. For the vanadium, it is less clear. The oxidation state of V in the VO^{2+} is obtained by considering the O to be -2. If the overall charge on the ion is $+2$ and the O is -2, the oxidation state of V must be $2 - (-2) = +4$. Therefore, the V is reduced from the $+4$ oxidation state to $+3$ for a change of -1. To balance the oxidation state changes, two VO^{2+} will have to be reduced for every Sn^{2+} that is oxidized. Now the equation for the reaction is $Sn^{2+} + 2VO^{2+} \rightleftharpoons Sn^{4+} + 2V^{3+}$. It remains to balance the O by the use of H_3O^+ and H_2O. To convert one O to H_2O, two H_3O^+ are needed. Therefore, the final, balanced reaction is

The oxidation state of an element is obtained from its net charge if the other atoms with it are assigned:

$$O = -2$$
$$H = +1$$
$$halide = -1$$

$$Sn^{2+} + 2VO^{2+} + 4H_3O^+ \rightleftharpoons Sn^{4+} + 2V^{3+} + 6H_2O \qquad 10.7$$

Confirmation of the equation balance is made by calculating the charge balance ($+10$ on both sides).

The challenge in this method of balancing redox reactions is determining the correct species being oxidized or reduced and then the correct change in its oxidation state (see Example 10.2). Uncertainties in this approach can be avoided by writing and balancing the two half reactions and then summing them as in Equation 10.6.

Example 10.2

Finding oxidation states
The oxidation state is the charge on the ion or molecule less the oxidation states of all the other elements. For As in H_3AsO_3,

$$0 - 3(-2) - 3(+1) = +3$$

For Br in BrO_3^-

$$-1 - 3(-2) = +5$$

For Al in $AlCl^{2+}$,

$$2 - (-1) = +3$$

EQUILIBRIUM CONSTANTS FOR REDOX REACTIONS

Redox reactions will proceed to an equilibrium state at which the rates of the forward and reverse reactions are equal. The equilibrium constant for redox reactions is formed in the usual way, taking the product of the activities of the products over the product of the activities of the reactants raised to the powers of their stoichiometric coefficients. For example, for the reaction between triiodide and uranate ions, the reaction and equilibrium expressions (in both thermodynamic and formal forms) are

$$I_3^- + U^{4+} + 6H_2O \rightleftharpoons 3I^- + UO_2^{2+} + 4H_3O^+ \qquad 10.8$$

$$K_{eq}^° = \frac{a_I^3 \, a_{UO_2^{2+}} a_{H_3O^+}^4}{a_{I_3^-} a_{U^{4+}}} \qquad 10.9$$

$$K_{eq}' = \frac{[I^-]^3 [UO_2^{2+}][H_3O^+]^4}{[I_3^-][U^{4+}]} \qquad 10.10$$

One does not look up an equilibrium constant for a redox reaction in the same way as for other reactions; one must calculate it from the standard reduction potentials of the two couples involved. The equilibrium constant is related to the difference between the standard reduction potentials for the two electron half reactions by Equation 10.11.

$$\ln\ K^{\circ}_{eq} = \frac{nF}{RT}(E^{\circ}_f - E^{\circ}_r)$$

$$\log\ K^{\circ}_{eq} = \frac{n}{V_N}(E^{\circ}_f - E^{\circ}_r) \qquad\qquad 10.11$$

E°_f and E°_r are the reduction potentials for the half reactions that were written forward and reverse to obtain the complete reaction, and n is the number of electrons that were canceled in creating the complete reaction ($n_1 n_2$ in the general example of Equation 10.6 and 2 in the case of the reaction in Equation 10.7). The Nernst factor, V_N, was defined after Equation 9.16. It is equal to $0.0592 \times (T/298)$ V. (See Example 10.3.)

If the formal potentials are used, the formal equilibrium constant is calculated. Thus,

$$\log\ K'_{eq} = \frac{n}{V_N}(E'_f - E'_r) \qquad\qquad 10.12$$

If the reaction is written with the stronger oxidant on the left, the reaction will proceed predominantly to the right and the equilibrium constant will be greater than 1. This is because the reaction is written so that the stronger oxidant reacts with the stronger reductant to produce the relatively weaker products.

FORMAL AND THERMODYNAMIC EQUILIBRIUM CONSTANTS

The relationship between the formal and thermodynamic equilibrium constants has the now familiar form (Equation 3.49, accurate when the DHLL is valid),

$$\log\ K'_{eq} = \log\ K^{\circ}_{eq} + A\sqrt{S}\,(pz^2_P + qz^2_Q + \cdots - az^2_A - bz^2_B - \cdots) \qquad 10.13$$

For the reaction of triiodide with uranate ions (Equation 10.8),

$$\log\ K'_{eq} = \log\ K^{\circ}_{eq} + A\sqrt{S}\,(3\cdot 1^2 + 1\cdot 2^2 + 4\cdot 1^2 - 1\cdot 1^2 - 1\cdot 4^2) \qquad 10.14$$

$$\log\ K'_{eq} = \log\ K^{\circ}_{eq} - 6A\sqrt{S}$$

Another option for this calculation is to calculate the values of the separate formal reduction potentials using Equation 9.18 and calculate the formal equilibrium constant with Equation 10.12.

▪▪▪ Example 10.3

Calculation of K°_{eq} from standard potentials

Calculate the K°_{eq} for the reaction

$$2Fe^{3+} + Sn^{2+} \rightleftharpoons 2Fe^{2+} + Sn^{4+}$$

at 25 °C.

From Table 10.1, the E° values for the forward and reverse reactions are $+0.771$ and $+0.139$ V. Using Equation 10.11.

$$\log K^{\circ}_{eq} = \frac{2}{0.0592}\left(\frac{298}{298}\right)(0.771 - 0.139)$$

$$= 21.35$$

$$K^{\circ}_{eq} = 2 \times 10^{21}$$

Study Questions, Section A

1. How are electron exchange half reactions combined to make a complete redox reaction?

2. Write a balanced reaction for the oxidation of Fe^{2+} by $O_{2(g)}$. From the E° values, predict the direction of the spontaneous reaction.

3. A. Write the thermodynamic and formal equilibrium expressions for the reaction obtained in Question 2.

   B. Find the relationship between the thermodynamic and formal equilibrium expressions.

   C. Calculate the thermodynamic equilibrium constant and estimate the formal equilibrium constant for a solution with an ionic strength of 0.10.

4. If a reaction is considered complete when the equilibrium constant is 100 or more, what should the difference in the standard reduction potentials of the half reactions used to make the complete reaction be? Assume that $n = 1$ for both half reactions.

5. Write a complete, balanced reaction for the reaction that occurs when a solution containing Ag^+ is mixed with a solution containing V^{2+}.

1. The two electron exchange half reactions that include the redox reaction reactants and products are chosen. The one for which the Ox form is the product of the overall reaction is reversed. A coefficient is applied to one or both reactions to make the number of electrons exchanged in each half reaction equal. The reactions are summed, canceling the electrons and any other species appearing on both sides of the final reaction.

2. From the list of reduction potentials (Table 10.1), the two half reactions involved are

$$O_2 + 4H_3O^+ + 4e^- \rightleftharpoons 6H_2O$$
$$Fe^{3+} + e^- \rightleftharpoons Fe^{2+}$$

The Fe^{3+}, Fe^{2+} reaction is reversed to put both reactants on the lefthand side of the resulting redox equation. All the stoichiometric coefficients in the Fe^{3+}, Fe^{2+} reaction are then multiplied by 4 so that the number of electrons involved in both equations are equal. Then the two reactions are summed.

$$O_2 + 4H_3O^+ + 4e^- \rightleftharpoons 6H_2O$$
$$\underline{4Fe^{2+} \rightleftharpoons 4Fe^{3+} + 4e^-}$$
$$O_2 + 4H_3O^+ + 4Fe^{2+} \rightleftharpoons 6H_2O + 4Fe^{3+}$$

Since O_2 is a stronger oxidant than Fe^{3+} (i.e., it has a higher value of $E°$), the reaction will proceed spontaneously to the right.

3. A. For the reaction

$$O_2 + 4H_3O^+ + 4Fe^{2+} \rightleftharpoons 6H_2O + 4Fe^{3+}$$

the thermodynamic equilibrium expression is

$$K°_{eq} = \frac{a_{H_2O}^6 a_{Fe^{3+}}^4}{p_{O_2} a_{H_3O^+}^4 a_{Fe^{2+}}^4} = \frac{a_{Fe^{3+}}^4}{p_{O_2} a_{H_3O^+}^4 a_{Fe^{2+}}^4}$$

in water solvent. The $O_{2(g)}$ is included as its partial pressure, and the water activity is unity in dilute aqueous solutions. The formal equilibrium expression is

$$K'_{eq} = \frac{[H_2O]^6 [Fe^{3+}]^4}{p_{O_2}[H_3O^+]^4 [Fe^{2+}]^4} = \frac{[Fe^{3+}]^4}{p_{O_2}[H_3O^+]^4 [Fe^{2+}]^4}$$

in water solvent.

B. Using Equation 10.13,

$$\log K'_{eq} = \log K°_{eq} + A\sqrt{S}\,(4 \cdot 3^2 - 4 \cdot 1^2 - 4 \cdot 2^2)$$
$$\log K'_{eq} = \log K°_{eq} + 16A\sqrt{S}$$

C. From Equation 10.11,

$$\log K°_{eq} = \frac{4}{0.0592}(1.229 - 0.771) = 30.9$$
$$K°_{eq} = 10^{30.94} = 9 \times 10^{30}$$

From the equation derived in part B,

$$\log K'_{eq} = 30.94 + 16 \times 0.509\sqrt{0.10} = 33.5$$
$$K'_{eq} = 3.2 \times 10^{33}$$

4. From Equation 10.11,

$$K°_{eq} > 100 \qquad \text{if} \qquad \frac{n(E°_f - E°_r)}{V_N} > 2$$
$$(E°_f - E°_r) > 2V_N/n$$
$$(E°_f - E°_r) > 0.118 \text{ V}$$

Thus, for a reaction involving the exchange of one electron to be complete, the difference in standard reduction potentials should be 0.118 V or greater.

5. From the location of these reactants in Table 10.1, it is clear that Ag^+ is the oxidant and V^{2+} is the reductant (because the V^{2+} is below and to the right of the Ag^+). The Ag^+ will be reduced to Ag metal, and the V^{2+} will be oxidized to V^{3+}. However, we see that V^{3+} is also a reductant that is below and to the right of Ag^+, so the V^{3+} formed will be oxidized to VO^{2+}. (Note that the VO^{2+} will not be oxidized to VO_2^+ because that product is above Ag^+ in Table 10.1.) Thus the overall reaction (unbalanced) is

$$Ag^+ + V^{2+} \rightleftharpoons Ag_{(s)} + VO^{2+}$$

The vanadium oxidation state is increased by 2 (from +2 to +4), as can be determined by adding the electrons required to oxidize V^{2+} to VO^{2+} or from calculating the oxidation state of V in V^{2+} and VO^{2+}. The oxidation state of Ag is reduced only by 1. Thus, two Ag^+ reduction reactions are required for each V^{2+} oxidized to VO^{2+}. Now the reaction is

$$2Ag^+ + V^{2+} \rightleftharpoons 2Ag_{(s)} + VO^{2+}$$

After balancing the O, we get

$$2Ag^+ + V^{2+} + 3H_2O \rightleftharpoons 2Ag_{(s)} + VO^{2+} + 2H_3O^+$$

This reaction is confirmed by checking the charge balance (+4 on each side).

B. Equilibrium Concentrations

As in all previous reaction types, the equilibrium concentrations of the redox reactants and products are related through the formal equilibrium constant expressions. To obtain the equilibrium constant, we must first identify the two oxidation–reduction pairs that are involved in the reaction. Then we combine them into a complete reaction as in

Equation 10.6 and calculate the formal equilibrium constant as in Equation 10.12 or 10.13. As an example, consider the case where 50 mL of 0.20 M Cr^{3+} (as $CrCl_3$) is mixed with 50 mL of 0.02 M U^{3+} [as $U(NO_3)_3$]. These two reactants are located on Table 10.1. One must be on the left and the other on the right for a complete redox reaction. The only entries that meet the criteria are the Cr^{3+}/Cr^{2+} couple at -0.38 V and the U^{4+}/U^{3+} couple at -0.63 V. We can now write the complete balanced reaction

$$Cr^{3+} + U^{3+} \rightleftharpoons Cr^{2+} + U^{4+} \qquad 10.15$$

and the equilibrium constant expression.

$$K'_{Cr^{3+},U^{4+}} = \frac{[Cr^{2+}][U^{4+}]}{[Cr^{3+}][U^{3+}]} \qquad 10.16$$

The value for K' can be calculated from the formal potentials (if they are available) or can be estimated from the thermodynamic potentials. Here, we have used the thermodynamic potentials from Table 10.1, so the approximately equals sign is used in the calculation.

A rule of thumb is that log K'_{eq} increases by 1 for every V_N/n difference in the E' values of the couples involved.

$$\log K'_{Cr^{3+},U^{4+}} \approx \frac{1}{0.0592}[-0.38 - (-0.63)] = 4.22 \qquad 10.17$$

$$K'_{Cr^{3+},U^{4+}} \approx 10^{4.22} = 1.7 \times 10^4$$

The equilibrium constant expression tells us that the concentration of each species is dependent on the concentration of the three other species involved. The rather high value for $K'_{Cr^{3+},U^{4+}}$ indicates that the reaction would proceed substantially to form the products. There are two approaches to the numerical solution of the equilibrium concentrations: the approximation method and the rigorous algebraic method.

APPROXIMATION METHOD FOR EQUILIBRIUM CONCENTRATIONS

After obtaining the balanced chemical reaction and the formal equilibrium constant, check to see that the equilibrium constant is much greater than 1. Then determine the analytical concentrations of the reactants. In the example of 50 mL of 0.20 M Cr^{3+} with 50 mL of 0.02 M U^{3+}, the analytical concentrations are $C_U = 0.01$ M and $C_{Cr} = 0.1$ M. Normally, one reactant will be present in excess (more is present than is required to react with the other reactant completely), and the other reactant is limiting (it imposes an upper limit on the amount of product that can be formed). Identify which reactant is in excess and which is limiting. Consider the stoichiometry of the reaction in deciding which reactant is in excess. In the example, the Cr^{3+} is in excess and the U^{3+} is the limiting reactant.

Assume that the reaction goes to completion, that is, that virtually all the limiting reagent is converted to product. From the stoichiometry of the reaction, this will make $[U^{4+}] = 0.01$ M and $[Cr^{2+}] = 0.01$ M as well. Now calculate the concentration of the excess reagent remaining unreacted. $C_{Cr} = 0.1$ M $= [Cr^{3+}] + [Cr^{2+}]$ so $[Cr^{3+}] = 0.10 - 0.01 = 0.09$ M. Now we have concentrations for all the species except for the very small amount of the limiting reagent that is left unreacted. This is obtained from the equilibrium constant expression, Equation 10.16.

$$[U^{3+}] = \frac{[Cr^{2+}][U^{4+}]}{K'_{Cr^{3+},U^{4+}}[Cr^{3+}]} \qquad 10.18$$

$$[U^{3+}] = \frac{(0.01)(0.01)}{1.7 \times 10^4(0.09)} = 7 \times 10^{-8} \text{ M}$$

By comparing the initial and final concentrations of U^{3+} (0.01 M and 7×10^{-8} M), we can determine if the assumption about the reaction being complete was valid. (It was,

Steps in the Approximation Method for Equilibrium Concentration Calculations

1. Write a balanced equation for the reaction.

2. Calculate the equilibrium constant and make sure that it is much greater than 1.

3. Identify the excess and limiting reactants.

4. Assume that the reaction is complete (i.e., the limiting reactant is essentially consumed).

5. Calculate the concentrations of the products for the stoichiometry and the concentration of limiting reactant.

6. Calculate the concentration of the excess reactant remaining.

7. Calculate the concentration of the unreacted limiting reactant using the formal equilibrium constant expression.

8. Use the answer to step 7 to validate the assumption concerning the completeness of the reaction.

9. Iterate steps 5–8 if necessary.

to 7 parts per million.) If the assumption was not valid to the level required, the first calculated value can be used to correct the concentrations in the equilibrium expression to obtain a closer value, and so on, until the calculation converges.

ALGEBRAIC METHOD FOR EQUILIBRIUM CALCULATIONS

Equations for the concentrations of all species in the mixture can be calculated in the following way. Mass balance equations are formed for each of the reactants. Again using the example of the reaction between Cr^{3+} and U^{3+} in Equation 10.15, we have

$$C_{Cr} = [Cr^{3+}] + [Cr^{2+}] \quad\text{and}\quad C_U = [U^{3+}] + [U^{4+}] \qquad 10.19$$

from which we can get

$$[Cr^{3+}] = C_{Cr} - [Cr^{2+}] \quad\text{and}\quad [U^{3+}] = C_U - [U^{4+}] \qquad 10.20$$

From the initial conditions, C_U and C_{Cr} are 0.01 and 0.1 M, respectively.

From the stoichiometry of the reaction, we can see that for every Cr^{2+} formed, there will also be a U^{4+}. Thus $[Cr^{2+}] = [U^{4+}]$. We now have four equations with which to solve the four unknown concentrations. Substitution of the above expressions into the equilibrium constant expression enables the derivation of an equilibrium expression with only $[Cr^{2+}]$ and constants.

$$K'_{Cr^{3+},U^{4+}} = \frac{[Cr^{2+}]^2}{(C_{CR} - [Cr^{2+}])(C_U - [Cr^{2+}])} \qquad 10.21$$

When Equation 10.21 is rearranged, the result is a equation that is quadratic in $[Cr^{2+}]$.

$$(K'_{Cr^{3+},U^{4+}} - 1)[Cr^{2+}]^2 - K'_{Cr^{3+},U^{4+}}(C_U + C_{Cr})[Cr^{2+}] + K'_{Cr^{3+},U^{4+}}C_U C_{Cr} = 0 \quad 10.22$$

This equation can be solved with the quadratic formula introduced with Equation 4.12. You might also observe that if $K'_{Cr^{3+},U^{4+}}$ is much greater than 1, the K'''s cancel out so that $[Cr^{2+}]^2 - (C_U + C_{Cr})[Cr^{2+}] + C_U C_{Cr} = 0$. When the values for the example are substituted in Equation 10.22, the solution is $[Cr^{2+}] = 0.01$ M. Thus, we have confirmed, algebraically, that the reaction is complete. From this, $[U^{4+}] = 0.01$ M and $[Cr^{3+}] = 0.10 - 0.01 = 0.09$ M. Again, use the formal equilibrium expression to calculate the value of $[U^{3+}]$.

The algebraic approach has its greatest value for situations where the reaction is not complete. In that case, one would get values for both products in one step. Then the mass balance and/or formal equilibrium constant expressions can be used to get the reactant concentrations.

Study Questions, Section B

6. Consider a mixture of 20 mL of 0.03 M Fe^{3+} solution with 20 mL of 0.02 M SO_3^{2-} solution.

   A. What would the products of the resulting redox reaction be?

   B. Which reactant is limiting, and which is in excess?

   C. Write the formal equilibrium constant for the reaction.

   D. Estimate the value of the formal equilibrium constant.

   E. Based on the estimated K'_{eq}, can we assume that the reaction is essentially complete?

   F. Calculate the equilibrium concentrations of all species but the limiting reactant.

   G. Calculate the equilibrium concentration of the limiting reactant. Use the estimated K'_{eq} derived in part D. Assume the reaction occurs in 1 M acid.

7. Air is bubbled through a solution that was originally 0.004 M $FeCl_2$. The solution is buffered at pH = 2.0 with a buffer that has an ionic strength of 0.1. What are the concentra- tions of all the species when the system has come to equi- librium? Assume that the air is 20% O_2 and use the value of K'_{eq} calculated in Question 3.

Answers to Study Questions, Section B

6. A. From Table 10.1, the Fe^{3+} is the oxidant and the SO_3^{2-} is the reductant. The products of the reaction are Fe^{2+} and SO_4^{2-}.

B. Because two electrons are required to oxidize SO_3^{2-} to SO_4^{2-}, two Fe^{3+} are required for each SO_3^{2-} oxidized. However, there is only 1.5 times as much Fe^{3+} as SO_3^{2-}. Therefore, the Fe^{3+} is limiting and the SO_3^{2-} is in excess.

C. From the balanced reaction

$$2Fe^{3+} + SO_3^{2-} + 3H_2O \rightleftharpoons 2Fe^{2+} + SO_4^{2-} + 2H_3O^+$$

the equilibrium constant expression is

$$K'_{eq} = \frac{[Fe^{2+}]^2[SO_4^{2-}][H_3O^+]^2}{[Fe^{3+}]^2[SO_3^{2-}]}$$

D. The difference in $E°$'s is $0.771 - 0.172 = 0.599$ V. This voltage divided by V_N/n (~0.03 V) is ~20. There- fore, K'_{eq} is approximately 10^{20}.

E. Yes, the reaction is complete because 10^{20} is very much larger than 1.

F. The initial concentrations of Fe^{3+} and SO_3^{2-} are 0.015 M and 0.01 M. Since Fe^{3+} is the limiting reactant, the concentration of the SO_4^{2-} produced will be half that of the original Fe^{3+}, so $[SO_4^{2-}] = 0.0075$ M. The Fe^{3+} is almost completely converted to Fe^{2+}, so $[Fe^{2+}] = 0.015$ M. The SO_3^{2-} concentration is equal to the initial con- centration less the concentration of SO_4^{2-} produced, so $[SO_3^{2-}] = 0.01 - 0.0075 = 0.003$ M.

G. From the equilibrium constant expression derived in part C,

$$[Fe^{3+}]^2 = \frac{[Fe^{2+}]^2[SO_4^{2-}][H_3O^+]^2}{K'_{eq}[SO_3^{2-}]}$$

$$[Fe^{3+}] = \left(\frac{(0.015)^2(0.0075)(1)^2}{10^{20}(0.0025)}\right)^{1/2} = 10^{-12} \text{ M}$$

7. The formal equilibrium constant from the answer to Ques- tion 3 is 3×10^{33}. With such a large equilibrium constant favoring the products, we can assume that essentially all of the Fe^{2+} is oxidized to Fe^{3+}. Thus, $[Fe^{3+}] = 0.004$ M. Now we can substitute into the equilibrium constant expression to solve for $[Fe^{2+}]$.

$$3 \times 10^{33} = \frac{[Fe^{3+}]^4}{p_{O_2}[H_3O^+]^4[Fe^{2+}]^4}$$

$$[Fe^{2+}] = \left[\frac{(0.004)^4}{3 \times 10^{33} \times 0.2(1 \times 10^{-2})^4}\right]^{1/4}$$

$$[Fe^{2+}] = 3 \times 10^{-9} \text{ M}$$

From this result, we see that the Fe^{2+} concentration is negli- gible compared to 0.004 M, so our assumption is justified. If it were not, a corrected value for $[Fe^{3+}]$ could be used in a successive approximations approach. Note that the equi- librium concentration of Fe^{2+} increases as the pH of the so- lution increases.

C. Quantitation by Redox Titration

Titration is now a familiar method by which quantitation can be achieved in cases where the differentiating characteristic is some sort of chemical reactivity. Redox reac- tivity was one of the earliest to be used in this way. Two oxidizing reagents were first used as titrants over 150 years ago, namely, the permanganate ion (MnO_4^-) and the dichromate ion ($Cr_2O_7^{2-}$). Redox titrations are still among the most widely applied titrimetric techniques.

For a redox reaction to be a suitable basis for a titration, it must be rapid, stoi- chiometric, and amenable to a suitable equivalence point detection method. In addi- tion, the titrant material should be readily standardized and reasonably stable. Among the reagents that meet these criteria, the oxidants predominate. Therefore, the analytes for redox titration are generally reductants or if not, the analysis is performed by **back titration.** As we shall see, reductors are often used to assure that all the analyte is in the reduced state before its titration.

This section is arranged according to the titrant employed because not all oxi- dants will react quickly and stoichiometrically with all reductants even when the equi-

Back Titration

To analyze an oxidant with a standard titrant, a known excess of standard re- ductant is added to the sample solution. The excess reductant (and possibly the analyte) is then titrated with the stan- dard oxidant.

librium constant is very high. In general, if the number of electrons in the two half re-actions that are added to obtain the titration reaction is not the same, the reaction may be mechanistically complex. Unless it is directly reversible, it is likely to be slow. Ex-perience has shown which titrants are most suitable for each analyte and under what conditions the titration is best carried out.

TITRATION WITH Ce(IV)

The ceric ion (Ce^{4+}) is a very strong oxidizing agent. Its formal reduction potential ranges from $+1.4$ V to $+1.9$ V depending on the acid used in making the titration solu-tion acidic. It has the highest potential in perchloric acid solutions and the lowest in sulfuric acid solutions. As you can see from Table 10.1, this potential is high enough for Ce^{4+} to oxidize water to O_2 gas. The reason it does not is that it normally lacks the mechanism to do so at any significant rate. A determinate preparation of Ce(IV) titrant is possible because of the existence of a very stable compound $(NH_4)_2Ce(NO_3)_6$ called ammonium hexanitratocerate(IV). This complex salt is available in primary standard quality. It can be dried for 1 to 6 hours at 85 °C without decomposition. Weighed out $(NH_4)_2Ce(NO_3)_6$ is dissolved in sulfuric acid solution and diluted to volume in a volu-metric flask. The sulfuric acid is important because the complexation of the Ce(IV) with SO_4^{2-} is sufficient to stabilize the solution. When perchloric or nitric acids are used, the Ce(IV) is reduced at the rate of a few tenths of a part per thousand per day, and this rate is considerably increased if the solution is exposed to light. If standardiza-tion of the Ce(IV) solution is needed, a procedure involving the titration of primary standard sodium oxalate has been developed.

The analytes that have been successfully titrated with standard Ce(IV) titrant (besides oxalate ion) are Fe(II), H_2O_2, As(III), I^-, Sb(III), Mo(V), Pu(III), Sn(II), $S_2O_3^{2-}$, Tl(I), U(IV), and V(IV). Some of the factors involved are illustrated with the determination of Fe(II). The overall titration reaction is

$$Fe^{2+} + Ce^{4+} \rightleftharpoons Fe^{3+} + Ce^{3+} \qquad 10.23$$

ASSURING ANALYTE OXIDATION STATE BEFORE TITRATION

Before the titration is begun, we must make sure all the iron we want to determine is in the ferrous form. A look at Table 10.1 shows that Fe^{2+} can be oxidized by O_2 gas. Dis-solved oxygen will be a part of all aqueous solutions that have been exposed to the at-mosphere. Thus, unless precautions have been taken, some of the ferrous iron will already be in the ferric form. This will have to be reduced back to Fe^{2+} before the titra-tion begins. There are several ways to do this. One is to use a Zn metal **reductor.** The reaction is

$$2Fe^{3+} + Zn_{(s)} \rightleftharpoons 2Fe^{2+} + Zn^{2+} \qquad 10.24$$

The zinc is often in the form of an amalgam with mercury. A wand of the material can be swirled in the solution, or the solution can be passed over the zinc reductor before titration. It is essential that the solution be free of metallic Zn before the titration be-gins. It is also desirable that the solution be freed of O_2 before the reduction and subse-quent titration. This is accomplished by bubbling pure N_2 or H_2 gas through the solution for several minutes.

What fraction of the Fe^{3+} is reduced to Fe^{2+} by the Zn reductor? This is an im-portant consideration if exact quantitation is to be achieved. The equilibrium constant for the reaction of Equation 10.24 is

$$K'_{eq} = \frac{[Fe^{2+}]^2[Zn^{2+}]}{[Fe^{3+}]^2} \qquad 10.25$$

Analytical Scheme for Redox Titration

Differentiating characteristic:
Ability to be oxidized or reduced.

Probe:
Titrate by gradually adding an oxidiz-ing or reducing reagent to a solution of the analyte.

Response to the probe:
The analyte reacts with the oxidizing or reducing reagent. The concentra-tion of the analyte is reduced, the con-centration of the titrant is increased, and the oxidizing potential of the so-lution changes.

Measurement of the response:
Follow the decrease in analyte con-centration, the increase in titrant con-centration, or the change in oxidizing potential of the solution. Determine the exact volume of the titrant added at the equivalence point.

Interpretation of data:
Use the equivalence point volume of titrant and its concentration to calcu-late the amount of analyte originally present.

The value for K'_{eq} can be estimated from the thermodynamic $E°$ values.

$$\log K'_{eq} \approx \frac{2}{0.0592}(0.771 + 0.763) = 51.8 \qquad 10.26$$

Equation 10.25 can be rearranged to give the ratio of $[Fe^{3+}]$ to $[Fe^{2+}]$ in the equilibrium solution.

$$\frac{[Fe^{3+}]}{[Fe^{2+}]} = \left(\frac{[Zn^{2+}]}{K'_{eq}}\right)^{1/2} \qquad 10.27$$

From Equation 10.27, it can be seen that the ratio of $[Fe^{3+}]$ to $[Fe^{2+}]$ in the solution depends on the concentration of Zn^{2+} produced by the Fe^{3+} reduction. Even in the worst case, for example, where the initial solution has a concentration of Fe^{3+} as high as 0.1 M,

$$\frac{[Fe^{3+}]}{[Fe^{2+}]} = \left(\frac{(0.05)}{10^{51.8}}\right)^{1/2} = 3 \times 10^{-27} \qquad 10.28$$

The ratio of $[Fe^{3+}]$ to $[Fe^{2+}]$ is extremely small.

Clearly, this reductor is totally effective for the reduction of all the ferric ion present. In fact, the potential of the Zn^{2+},Zn system is so low that it will reduce many other ions as well. A look at Table 10.1 reveals that all species in the Ox column could be reduced by Zn, the lowest entry. Ag^+, Hg_2^{2+}, and Cu^{2+} (among others) would be reduced to their metallic form. If their reactions were not mechanistically hindered, Fe(II), Co(II), and Ni(II) would be, too (from the $E°$ values in Appendix E). If the Zn could reduce Fe^{2+} to Fe, this would make the Zn reductor useless for the prior reduction of Fe^{3+} to Fe^{2+} because it would reduce Fe^{3+} to metallic iron, instead. Because of its great reducing power, the Zn reductor is very nonselective. If the unknown solution has many species present, several of them may be reduced by the Zn reductor and then be reoxidized during the titration with the Ce(IV) titrant. Since virtually the entire reduction potential table falls between these two reagents, almost any species that can be oxidized or reduced becomes a potential interferent in the analysis.

A reductor that is more selective is the silver reductor. It is silver metal and is always used in the presence of hydrochloric acid. Since there is a large concentration of Cl^-, the potential of the system is essentially that of the AgCl,Ag system ($+0.222$ V). The ferric ion concentration can be seen to be less than 10^{-9} M in the presence of the silver reductor. Thus this reductor, too, is completely effective for the prior reduction of Fe^{3+}. It has the advantage of being much more selective than the Zn reductor. Only species between its potential and that of the Ce(IV) are potential interferents in this determination.

Effectiveness of the Silver Reductor for Fe^{3+}

The reaction of $Ag_{(s)}$ with Fe^{3+} is

$$Ag_{(s)} + Fe^{3+} + Cl^- \rightleftharpoons AgCl_{(s)} + Fe^{2+}$$

for which the equilibrium constant equation and value are

$$K'_{eq} = \frac{[Fe^{2+}]}{[Fe^{3+}][Cl^-]}$$

$$\log K°_{eq} = \frac{1}{0.0592}(0.771 - 0.222)$$

$$= 9.27$$

$$K'_{eq} \approx 10^{9.3}$$

Thus, if the Cl^- concentration is roughly 1 M, the remaining Fe^{3+} in the solution is less than 1 ppb of the Fe^{2+} concentration.

TITRATION CURVES

The titration curve is generally plotted as E_{eq} versus the volume of titrant. The curve follows the familiar sigmoidal shape. The initial solution contains only Fe^{2+} and a very small concentration of Fe^{3+}. If the silver reductor were used, the relative Fe^{2+} and Fe^{3+} concentrations would correspond to an E_{eq} of $+0.222$ V. This is a reasonable estimate of the E_{eq} at the start of the titration. An exact determination of this potential is not necessary because with the first drop of Ce^{4+} titrant, the potential will rise rapidly from the sudden increase in the Fe^{3+} concentration. At the 33% titrated point, the Fe^{2+} concentration is twice that of Fe^{3+}, and the potential is just a few millivolts negative from the value of E' for the Fe^{3+},Fe^{2+} couple. At the 50% titrated point, $[Fe^{2+}] = [Fe^{3+}]$, and E_{eq} is equal to E' for the Fe^{3+},Fe^{2+} couple. At the 66% titrated point $[Fe^{2+}] = 0.5[Fe^{3+}]$, just a few millivolts above E'. Thus, all through the middle of the titration, the E_{eq} remains close to the E' for the analyte system. For all points in this region, the equilibrium potential is calculated using the fraction of Fe^{2+} titrated and the Nernst equation for the Fe^{3+},Fe^{2+} couple.

At the equivalence point there is exactly enough titrant present to react with any unreacted analyte. Thus, $[Ce^{4+}] = [Fe^{2+}]$. Also, from the stoichiometry of the titration reaction,

$$Fe^{2+} + Ce^{4+} \rightleftharpoons Fe^{3+} + Ce^{3+} \qquad \qquad 10.29$$

the product concentrations, $[Fe^{3+}]$ and $[Ce^{3+}]$, are also equal. These substitutions can be made in the equilibrium constant expression.

$$K'_{eq} = \frac{[Fe^{3+}][Ce^{3+}]}{[Fe^{2+}][Ce^{4+}]} = \frac{[Fe^{3+}]^2}{[Fe^{2+}]^2} \qquad \qquad 10.30$$

from which

$$\log \frac{[Fe^{2+}]}{[Fe^{3+}]} = -\frac{1}{2} \log K'_{eq} \qquad \qquad 10.31$$

From Equation 10.12,

$$\log K'_{eq} \approx \frac{1}{V_N}(E'_{Ce^{4+},Ce^{3+}} - E'_{Fe^{3+},Fe^{2+}}) \qquad \qquad 10.32$$

Substituting Equations 10.31 and 10.32 into the Nernst equation for the Fe^{3+},Fe^{2+} couple,

$$E_{equiv.pt.} = E'_{Fe^{3+},Fe^{2+}} + \frac{V_N}{2} \frac{1}{V_N} (E'_{Ce^{4+},Ce^{3+}} - E'_{Fe^{3+},Fe^{2+}}) \qquad \qquad 10.33$$

from which

$$E_{equiv.pt.} = \frac{(E'_{Ce^{4+},Ce^{3+}} - E'_{Fe^{3+},Fe^{2+}})}{2} \qquad \qquad 10.34$$

If the formal potentials for the Ce^{4+},Ce^{3+} and Fe^{3+},Fe^{2+} couples are 1.4 and 0.77 V, the equivalence point potential for the titration is 1.1 V.

After the equivalence point, there is an excess of Ce(IV). At the 200% titrated point, this excess is equal to the amount of the original Fe^{2+}. The amount of Ce(III) is also equal to the amount of the original Fe^{2+}. Thus at the 200% point, $[Ce^{4+}] = [Ce^{3+}]$ and $E_{eq} = E'$ for the Ce^{4+},Ce^{3+} system (which we have taken to be 1.4 V for the titration conditions).

The overall titration curve, estimated from the calculated points, is shown in Figure 10.1. The amount of the titration break from 50% to 200% titrated, in volts, is then

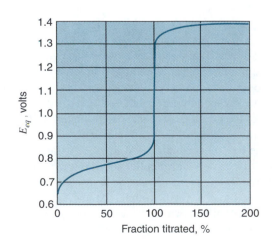

Figure 10.1. Sketch of the response curve for the titration of Fe^{2+} with Ce^{4+}. The values were obtained from the calculations of E_{eq} at 50, 100, and 200% titrated.

equal to the difference in the formal reduction potentials of the analyte and the titrant. In this case, that is about 0.63 V. This is a very large break, so the equivalence point for this titration can be readily determined by an electrode potential or by an indicator as described in a later section. It is very interesting to note that the magnitude of the break is not dependent on the concentration of the original species. This will always be true if neither of the couples is unsymmetrical. The lowest practical analyte concentration for this titrimetric determination is determined by the levels of other reactive components in the solution and by the concentrations required to produce a response in the equivalence point determination technique.

MULTIPLE EQUIVALENCE POINT TITRATIONS

If Ce(IV) were used to titrate Sn^{2+}, the difference in the formal reduction potentials for the two systems would be even larger then it is with Fe^{2+}; $1.4 - 0.14 = 1.3$ V. This break is so great that both Fe^{2+} and Sn^{2+} can be readily differentiated in the titration. The Ce(IV) would react first with the Sn^{2+} because of the greater difference in the formal reduction potentials. The E_{eq} for the 50% titrated point would be 0.14 V. As the titration proceeds, the $[Sn^{4+}]/[Sn^{2+}]$ ratio increases, as does E_{eq}. As E_{eq} increases, so also does the concentration of Fe^{3+}. This occurs, of course, as a result of the oxidation of Fe^{2+} by the Ce(IV).

The equivalence point in the titration of the Sn^{2+} with the Ce(IV) occurs when the concentration of unreacted Sn^{2+} is equal to half the combined concentrations of the Fe^{3+} and the Ce(IV), that is,

$$[Sn^{2+}] = \tfrac{1}{2}([Fe^{3+}] + [Ce^{4+}]) \qquad 10.35$$

Because the Ce^{4+},Ce^{3+} reduction potential is so much more positive than the Fe^{3+},Fe^{2+} reduction potential, the concentration of Ce^{4+} is very much less than that of Fe^{3+}, so to a good approximation, $[Sn^{2+}] = [Fe^{3+}]$. Note that this is the same condition we would have at the equivalence point if the Sn^{2+} were being titrated with Fe^{3+}. This, in effect, is what is happening because the equilibrium potential cannot rise above the potential of the Fe^{3+},Fe^{2+} couple until half the Fe^{2+} has been titrated. The potential at the equivalence point for the Sn^{2+} titration will then be between that of the Sn^{4+},Sn^{2+} couple and the Fe^{3+}, Fe^{2+} couple. It is not the simple average as in the case of Equation 10.34 for the Ce^{4+},Fe^{2+} titration because there is a different number of electrons involved in the half reactions for the Sn^{4+},Sn^{2+} and Fe^{3+},Fe^{2+} couples. It is weighted toward the potential with the higher number of electrons in the reduction half reaction. Specifically,

$$E_{equiv.pt.} = \frac{2E'_{Sn^{4+},Sn^{2+}} + E'_{Fe^{3+},Fe^{2+}}}{3} = \frac{2(0.14) + 0.77}{3} = +0.35\,V \qquad 10.36$$

The general equation for the equilibrium potential at the equivalence point (when both couples are symmetrical) is

$$E_{equiv.pt.} = \frac{n_1 E'_1 + n_2 E'_2}{n_1 + n_2} \qquad 10.37$$

When the value of n for both couples is 1, Equation 10.37 reduces to Equation 10.34. It should be remembered that this often given formula for the equivalence point potential of a redox reaction is only valid when neither of the couples involved is unsymmetrical. For titrations involving unsymmetrical couples, the equivalence point potential is dependent on the species concentrations in a complex way. Such systems are most easily evaluated using log concentration plots, as shown in Section E.

A plot of the titration curve is shown in Figure 10.2. The E_{eq} remains in the region of the E' for Sn^{4+},Sn^{2+} until the equivalence point for the Sn^{2+} titration ap-

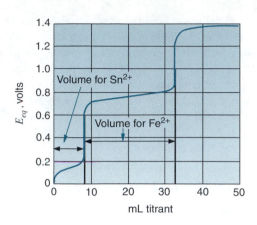

Figure 10.2. This titration curve for the mixed titration of Sn^{2+} and Fe^{2+} with Ce^{4+} illustrates the method for obtaining the portion of the titrant volume that was used by each of the analytes. Where analyte reduction potentials are sufficiently different, mixed titrations are very practical. End point determination by electrode is much easier than finding separate indicators for each end point region.

proaches. Over the greatest part of the Fe^{2+} titration, E_{eq} is around the formal reduction potential for the Fe^{3+},Fe^{2+} system. The equivalence point and 200% point are determined as before. The total titration curve for the two analytes would have two distinct breaks, the first for the volume required for the Sn^{2+} and the second at the volume equivalent to the sum of the Sn^{2+} and Fe^{2+}. To determine the volume required for the Fe^{2+} alone, the difference in the two equivalence point volumes is used.

An interesting application of this mixture titration is the use of the Sn^{2+} as a reductant for any Fe^{3+} in the original sample solution. Then, when the mixture is titrated with the Ce(IV), the excess Sn^{2+} is titrated before the Fe^{2+}. If a known amount of Sn^{2+} reductant was added, the titration reveals how much of it was used to reduce the Fe^{3+} in the original solution. The Fe^{2+} titration then gives total iron. In this way, the fraction of the total iron in the ferrous and ferric forms is determined.

EQUIVALENCE POINT DETECTION

By now, we are familiar with the various techniques for equivalence point detection. They include the measurement of electrode potential, the color change of an indicator, and the use of spectroscopy to follow changes in the solution composition. With redox titrations, the use of an electrode is very straightforward. It is not necessary to find an electrode that has a specific response to one of the reactants or products since the entire oxidizing level of the solution is changing during the titration. We have been characterizing the magnitude of the end point break in terms of the change in the equilibrium potential between the 50% and 200% titration points. The potential of an inert test electrode dipped in the solution would change by this same amount. The test electrode potential is measured with the aid of a reference electrode. Suitable materials for the inert test electrode include platinum, gold, mercury, and carbon. The exact potential between the test and reference electrodes at the equivalence point can be determined theoretically or empirically, and then the titration can be stopped at that potential. Alternatively, the curve of potential versus volume of titrant can be plotted, with equivalence point determination by noting the inflection point or by more sophisticated mathematical analysis.

A number of redox indicators are also available. They are compounds that can be reversibly oxidized or reduced and for which the oxidized and reduced forms are differently colored. When the solution reaches the formal potential for the indicator redox couple, the color will change. A change in the solution equilibrium potential from V_N/n above E' for the indicator to V_N/n below E' will change the Ox/Red concentration ratio from $10:1$ to $1:10$.

An example of a redox indicator is the phenanthroline complex with iron. Phenanthroline forms a complex containing three phenanthrolines with both ferric and ferrous iron.

Redox Equivalence Point Detection with a Potentiometric Electrode

1. One can titrate to a fixed potential that has been predetermined to be the equivalence point potential for the system. This is similar to using an indicator except that the resolution of the potentiometric end point detector is better.

2. One can record the titration curve and analyze the shape of the curve to obtain the equivalence point. This method could include either derivative techniques to find the inflection points of the curve (which are not equal to the equivalence points if n for the titrant and analyte couples is different) or curve fitting to a theoretical curve.

3. Potentiometric equivalence point detection is especially useful for the titration of multiple analytes. The same response system is used at each equivalence point.

All redox indicators must be redox couples for which the two forms (Ox and Red) are differently colored. Since the color transition will occur when the two forms of the couple have roughly equivalent concentrations, the potential for the color change is very near E' for the indicator couple.

The ferrous complex is red, and the ferric, pale blue.

$$Fe(phen)_3^{3+} + e^- \rightleftharpoons Fe(phen)_3^{2+} \qquad E' = +1.06 \text{ V} \qquad 10.38$$

From Figure 10.1, this is clearly an excellent choice for the titration of Fe^{2+} with Ce(IV). On the other hand, you can see from Figure 10.2 that it would be completely unsuitable for the titration of Sn^{2+} with Fe^{3+}. Substitutions on the phenanthroline rings have produced indicator compounds that have somewhat different potentials, so some fine-tuning is possible. With the availability of inexpensive electrodes and meters to follow redox reactions, efforts to develop new redox indicators have waned considerably.

Some redox reactants themselves are differently colored in their oxidized and reduced forms. Iodine is an example. The dissolved I_2 is reddish brown, and the I^- is colorless. For any such situations, spectrometric end point detection will work well. If the color is sufficiently intense, the eye may substitute adequately for the spectrometer. The color of the iodine does not appear to the eye until a significant excess is present, but when starch is added to the solution, an intense blue color appears with the first excess of I_3^-, which is adsorbed on the starch. The starch–iodine system is very convenient when the iodine–iodide system is involved in the titration. It can also be used by adding starch and a trace of KI to any redox solution. Care should be taken, however, since the I_3^-,I^- couple is unsymmetrical so the equivalence potential is a function of the iodine indicator concentration.

Redox reactions can be very sensitive. For all symmetrical systems, the magnitude of the titration break, in volts, does not decrease with decreasing concentration of reactants. Therefore, the response function does not impose a fundamental limit on the lowest concentration determinable. The factors that do determine the least practical concentration are the presence of interfering oxidants and reductants and the response of the equivalence point detection system. In principle, electrodes will respond at any concentration level, but their action depends on the active exchange of electrons at the surface of the electrode. If the concentrations of the titrated species are so low that there is little exchange, the electrode response will be sluggish, or some other minor redox couple may compete for the control of the electrode potential. In practice, titrations to the micromolar level are possible for some well-behaved systems.

Study Questions, Section C

8. Quite remarkably, the magnitude of the break at the equivalence point for redox titrations with symmetrical couples does not depend on the concentration of the analyte. Why is this the case?

9. What weight of $(NH_4)_2Ce(NO_3)_6$ should be used to prepare 1 liter of 0.10 M Ce(IV) titrant?

10. In Figure 10.1 you can see that the equivalence point for the titration of Fe^{2+} with Ce^{4+} is midway between the E' values for the Fe^{3+},Fe^{2+} and Ce^{4+},Ce^{3+} couples. On the other hand, the break between the Fe^{3+},Fe^{2+} and Sn^{4+},Sn^{2+} couples in Figure 10.2 is not midway between their E' values. What is the reason for this difference?

11. Consider the reduction of 50.0 mL of a solution containing both Fe^{3+} and Fe^{2+} by the addition of 10.0 mL of 0.100 M $SnCl_2$. The resulting solution is titrated with 0.100 M Ce(IV). Two end points are observed, one at 7.32 mL and a second at 34.83 mL (cumulative total). Calculate the concentrations of Fe^{3+} and Fe^{2+} in the original solution.

12. What qualities are required of a redox indicator?

13. The starch indicator for titrations involving iodine is very widely used. What is the basis of its operation?

14. What will limit the lower level of concentration for which a redox titration can be used?

Answers to Study Questions, Section C

8. The potentials of the couples involved in the titration are a function of the ratio of their concentrations, not their absolute values. Even though the oxidizing potential of a low concentration couple is undiminished, the amount of reductant it can react with certainly is less.

9. To prepare 1 liter of 0.10 M $(NH_4)_2Ce(NO_3)_6$, use the equation, moles = molarity × volume = 0.10 × 1 = 0.1 mol. The molecular weight is 548.222 g/mol. The required weight is

$$0.10 \text{ mol} \times 548.222 \text{ g/mol} = 54.8 \text{ g}$$

10. The reason for the difference is that in the Ce^{4+} titration of Fe^{2+}, both couples involve the exchange of the same number of electrons. In the case of the Fe^{3+},Fe^{2+} and Sn^{4+},Sn^{2+} systems, the number of electrons is 1 and 2, respectively. This difference in n results in an asymmetry in the titration curve. Equation 10.37 shows the equivalence point potential to be the weighted average of the E' values for the two couples, with the values for n being the weighting factors.

11. The original Fe^{3+} was reduced by the added Sn^{2+} according to the reaction

$$2Fe^{3+} + Sn^{2+} \rightleftharpoons 2Fe^{2+} + Sn^{4+}$$

This left a solution with $[Fe^{2+}]$ equal to the total iron in the sample, an amount of Sn^{4+} produced by the reduction of the Fe^{3+}, and the excess Sn^{2+} reducing reagent. When the solution was titrated, the Sn^{2+} was titrated first, then the Fe^{2+}. The reaction of the Ce(IV) with the Sn^{2+} is

$$2Ce^{4+} + Sn^{2+} \rightleftharpoons 2Ce^{3+} + Sn^{4+}$$

Thus, mmol Sn^{2+} = 0.5 × mmol Ce^{4+} to first end point = 0.5 × 7.32 mL × 0.100 mol/L = 0.366 mmol Sn^{2+}. The original mmol Sn^{2+} = 10.0 mL × 0.100 mol/L = 1.00 mmol Sn^{2+}. Therefore, mmol Sn^{2+} that reacted with the Fe^{3+} = 1.00.− 0.366 = 0.634 mmol Sn^{2+}. The mmol Fe^{3+} in original solution = 2 × mmol reacted Sn^{2+} = 2 ×

0.634 = 1.268 mmol Fe^{3+}. The mmol Ce^{4+} that reacted with the total iron as Fe^{2+} = (34.83 − 7.32) mL × 0.100 mol/L = 2.751 mmol Ce^{4+}. The total iron is mmol Fe^{3+} + mmol Fe^{2+} in sample = 2.751 mmol. Thus, mmol Fe^{2+} in sample = mmol total iron − mmol Fe^{3+} = 2.751 − 1.268 = 1.483 mmol Fe^{2+}. In the original solution, $[Fe^{3+}]$ = 1.268 mmol/50.0 mL = 0.0254 M, and $[Fe^{2+}]$ = 1.483 mmol/50.0 mL = 0.0297 M.

12. A redox indicator must be a redox couple that is differently colored in its oxidized and reduced forms. For an indicator to be useful in a redox titration, the E' for the indicator couple must be near the equivalence point potential for the titration.

13. The starch forms an intensely blue complex with I_3^-. Therefore, in a titration with I_2, the first excess of titrant appears as the blue starch–I_3^- complex.

14. Since the theoretical magnitude of the break in the oxidation potential at the equivalence point is not a function of concentration for redox reactions (except for those that have different stoichiometric coefficients for Ox and Red), the limitation does not come from the decrease in the size of the break. Possible limiting factors are the ability to add sufficiently small amounts of titrant, the weakening of the response of potentiometric electrodes at low concentration, or the presence of redox-active interferents in the sample.

D. Iodine and Thiosulfate

The iodine reduction half reaction we considered earlier may be found in tables in three forms.

$$I_{2(s)} + 2e^- \rightleftharpoons 2I^- \qquad E° = +0.535 \text{ V} \qquad 10.39$$

$$I_{2(aq)} + 2e^- \rightleftharpoons 2I^- \qquad E° = +0.615 \text{ V} \qquad 10.40$$

$$I_3^- + 2e^- \rightleftharpoons 3I^- \qquad E° = +0.536 \text{ V} \qquad 10.41$$

These three half reactions correspond to three different forms of the iodine. In Equation 10.39, the solution is saturated with iodine, as evidenced by the existence of solid iodine. At room temperature, the solubility of iodine in water is 0.002 M. Equation 10.40 is for solutions in which the concentration of iodine is below the saturation level. Equation 10.41 is for solutions in which there is an excess of iodide ion. In such solutions, the iodide ion (I^-) and iodine (I_2) react to form a complex, the triiodide ion (I_3^-). The complexation of iodine with iodide greatly increases its solubility, but it reduces its oxidation strength as seen by the lower reduction potential for Equation 10.41 relative to that for Equation 10.40.

The Nernst equations for the three half reactions reveal the differences in the way in which we describe the iodine activity in the denominator of the log term.

$$E_{eq} = E'_{I_{2(s)},I^-} - \frac{V_N}{2} \log \frac{[I^-]^2}{1} \qquad 10.42$$

$$E_{eq} = E'_{I_{2(aq)},I^-} - \frac{V_N}{2} \log \frac{[I^-]^2}{[I_2]} \qquad 10.43$$

$$E_{eq} = E'_{I_3^-,I^-} - \frac{V_N}{2} \log \frac{[I^-]^3}{[I_3^-]} \qquad 10.44$$

Comparing the standard potentials for expressions 10.42 and 10.43, we can get an estimate of the concentration of a saturated solution of I_2.

$$0.535 = 0.615 - \frac{0.0592}{2} \log \frac{1}{[I_2]_{sat'd}} \qquad 10.45$$

$$\log [I_2]_{sat'd} = \frac{(0.535 - 0.615)2}{0.0592} = -2.7 \qquad 10.46$$

$$[I_2]_{sat'd} = 2 \times 10^{-3} \text{ M}$$

Comparing the standard potentials for expressions 10.43 and 10.44, we can get an estimate of the formation constant of the triiodide complex.

$$K_f = \frac{[I_3^-]}{[I_2][I^-]} \qquad \text{so} \qquad K_f[I_2] = \frac{[I_3^-]}{[I^-]} \qquad 10.47$$

Substituting Equation 10.47 into Equation 10.44,

$$E_{eq} = E'_{I_3, I^-} + \frac{V_N}{2} \log K'_f - \frac{V_N}{2} \log \frac{[I^-]^2}{[I_2]} \qquad 10.48$$

$$0.615 = 0.536 + \frac{0.0592}{2} \log K'_f \qquad 10.49$$

$$\log K'_f = 2.7$$

The iodine,iodide half reaction, because of its intermediate potential, is useful as either an oxidant or a reductant. We will illustrate its use here as an oxidant for the thiosulfate ion ($S_2O_3^{2-}$). The redox log plot for this system has been shown previously in Figure 9.18. The thiosulfate half reaction is

$$S_4O_6^{2-} + 2e^- \rightleftharpoons 2S_2O_3^{2-} \qquad E' = +0.09 \text{ V} \qquad 10.50$$

When thiosulfate is used to titrate an iodine-containing solution, the overall reaction is

$$2S_2O_3^{2-} + I_3^- \rightleftharpoons S_4O_6^{2-} + 3I^- \qquad 10.51$$

When iodine is the analyte, there is usually an excess of I^- to form the complex. If the iodine analyte is at roughly 0.01 M concentration and there is enough excess I^- so that its concentration is 0.1 M, the equilibrium potential at 0% titrated would be

$$E_{eq} = 0.536 - \frac{0.0592}{2} \log \frac{(0.1)^3}{(0.01)} = 0.57 \text{ V} \qquad 10.52$$

At 50% titrated, the I_3^- concentration has dropped to 0.005 M, so

$$E_{eq} = 0.536 - \frac{0.0592}{2} \log \frac{(0.1)^3}{(0.005)} = 0.56 \text{ V} \qquad 10.53$$

At the equivalence point, we need to use the equilibrium constant expression and value

$$\log K'_{eq} \approx \frac{(0.536 - 0.09)2}{0.0592} = 15 \qquad 10.54$$

$$K'_{eq} = \frac{[S_4O_6^{2-}][I^-]^3}{[S_2O_3^{2-}]^2[I_3^-]} = 10^{15} \qquad 10.55$$

Choosing the Appropriate Half Reaction for Iodine Reduction

From these interrelationships, it is seen that the three formalisms for the I_2 reduction to I^- are all equivalent when the solubility of I_2 and the complexation are considered. In application, it is important to use the expression and E' that matches the experimental conditions. When excess I^- is present, use the I_3^- form of the iodine. When solid I_2 is present, use that form. The dissolved I_2 form is most often used for the oxidation of I^- to I_2.

Assuming the reaction is complete, $[S_4O_6^{2-}] = 0.01$ M. From the condition of the equivalence point, $[S_2O_3^{2-}] = 2[I_3^-]$. The $[I^-]$ remains constant at 0.1 M. Now we can calculate $[I_3^-]$ and with that, we can calculate the equilibrium potential.

$$K'_{eq} = \frac{(0.01)(0.1)^3}{(2[I_3^-])^2[I_3^-]} = \frac{10^{-5}}{4[I_3^-]^3} = 10^{15} \qquad 10.56$$

$$[I_3^-] = 10^{-7} \text{ M}$$

$$E_{eq} = 0.536 - \frac{0.0592}{2} \log \frac{(0.1)^3}{10^{-7}} = 0.42 \text{ V} \qquad 10.57$$

If Equation 10.37 were used to calculate the equivalence point potential, the result would be 0.313 V, which is very different from the result of 0.42 V in Equation 10.57. This is because of the asymmetry of these reactions.

At the 200% point, there is an excess of unreacted $S_2O_3^{2-}$. Its concentration is 0.02 M while that of the $S_4O_6^{2-}$ remains at 0.01 M. The equilibrium potential can be obtained from the Nernst equation for the $S_4O_6^{2-}, S_2O_3^{2-}$ couple.

$$E_{eq} = 0.09 - \frac{0.0592}{2} \log \frac{(0.02)^2}{(0.01)} = 0.13 \text{ V} \qquad 10.58$$

The titration curve is shown in Figure 10.3.

The basic titration described above has been adapted to a variety of situations. For example, the thiosulfate titrant may be standardized against an exactly weighed amount of potassium iodate or pure metallic copper. For the first case, the potassium iodate is dissolved and reacted with an excess of potassium iodide. The reaction

$$IO_3^- + 8I^- + 6H_3O^+ \rightleftharpoons 3I_3^- + 9H_2O \qquad 10.59$$

is quick and stoichiometric. The resulting I_3^- is then titrated with the thiosulfate solution.

For standardization against copper, a weighed piece of copper is oxidized to cupric ion (Cu^{2+}) by dropping it in a small amount of nitric acid. The excess nitric acid is replaced with a mixture of sulfuric and phosphoric acids by first neutralizing the nitric acid with ammonia and then adding the other acids. The resulting cupric ion is then reacted with an excess of KI. The reaction is

$$2Cu^{2+} + 5I^- \rightleftharpoons 2CuI_{(s)} + I_3^- \qquad 10.60$$

Now the I_3^- can be titrated with the thiosulfate solution and the concentration of the thiosulfate solution calculated. This same procedure can be used to determine the amount of copper in an ore or alloy.

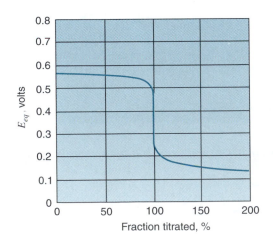

Figure 10.3. The resulting curve for the iodine–thiosulfacte titration has a break of about 0.4 V.

15. Why is the iodine redox couple I_3^- and I^- when there is an excess of I^- in the solution?

16. What are the reactants and products if the iodine couple is to be used as a reductant, and which species in Table 10.1 could be reduced by it?

17. In an experiment, 1.652 g of a copper mineral was dissolved in 100 mL of acid. The copper was brought quantitatively to the Cu(II) oxidation state. Excess KI was then added to form $CuI_{(s)}$ and I_3^-. The latter was then titrated with standard thiosulfate, and 28.74 mL of 0.1000 M titrant was required to reach the end point. What is the percentage of the mineral that is copper (by weight)?

15. The I^- reacts with I_2 to form I_3^-.

16. I^- would be the reactant and I_2 the product. There cannot be any significant I_3^- product because there would not be an excess of I^- at the same time there was a significant amount of I_2. All species above the iodine, iodide couple in Table 10.1, in the left-hand column, could be reduced by I^-.

17. First, mmol thiosulfate = 28.74 mL $\times$ 0.1000 mmol/mL = 2.874 mmol $S_2O_3^{2-}$. From Reaction 10.51, mmol I_3^- = 0.5 $\times$ mmol $S_2O_3^{2-}$ = 0.5 $\times$ 2.874 = 1.437 mmol I_3^-. From Reaction 10.60, mmol Cu^{2+} = 2 $\times$ mmol I_3^- = 2.874 mmol Cu^{2+}. The mass of copper is 2.874 $\times$ 10^{-3} mol Cu $\times$ 63.546 g/mol Cu = 0.1826 g Cu. Thus,

$$\%Cu = (0.1826/1.652) \times 100 = 11.05\%$$

E. Log Concentration Plots

The electrode potential log concentration plots developed in Section C of Chapter 9 can be used to solve for equilibrium concentrations in redox reactions. To do so, the log plots of both couples are placed on the same graph. This is illustrated for the oxidation of U^{3+} with Cr^{3+} as analyzed in Section B (Equations 10.15 through 10.22). In the mixture, the original concentration of U^{3+} is 0.01 M and Cr^{3+} is 0.10 M, so C_U and C_{Cr} are 0.01 M and 0.10 M. System points for each of the redox couples are placed on the plot as shown in Figure 10.4. Then lines are drawn from the system points at slopes of V_N/n (0.0592/n at 25 °C) volts per log concentration unit. Labels are then added. The line with the greater concentration at the more positive potential is always the more oxidized form of the redox couple.

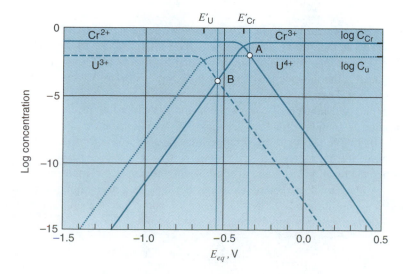

Figure 10.4. The log concentration diagram for a redox reaction is simply the combination of the log concentration plots for each couple, as developed in Chapter 9. With a stoichiometric equation obtained from the starting conditions, the solution point is readily found.

The solution point for the mixture of 0.01 M U^{3+} and 0.10 M Cr^{3+} is found by generating another equation that expresses the concentration relationships of the resulting solution. This could be an **electron balance equation** in which the concentrations of the species with more electrons than the starting material is equated with the concentrations of the species with fewer electrons than the starting material. For this example, the electron balance equation is

$$[U^{4+}] = [Cr^{2+}]$$ 10.61

This equation is the same as that obtained previously from the reaction stoichiometry.

To find the solution point, follow the U^{4+} line until it meets the Cr^{2+} line. The intersection is the tentative solution point (marked A on the plot). A vertical line going through this point would intersect all the other concentration lines at their log concentrations. Thus, $\log[Cr^{3+}]$ is slightly less than -1, $\log[U^{4+}] = \log[Cr^{2+}] = -2$, and $\log[U^{3+}] = -7$. The mathematical solution obtained in Equation 10.18 was $\log[U^{3+}] = -7.17$. Further, the potential of an inert metal electrode immersed in this solution would be about -0.33 V (with respect to 0.00 V on the $E°$ scale). Since the log plots can be constructed automatically using the spreadsheet templates on the accompanying CD, they offer a quick means to approximate solutions of redox equilibria, and they provide the means to determine which assumptions are appropriate for more accurate algebraic solutions.

The plot of Figure 10.4 could also be used to obtain the equilibrium concentrations reached in a mixture of 0.10 M Cr^{2+} with 0.01 M U^{4+}. For these starting materials, the electron balance equation is $[U^{3+}] = [Cr^{3+}]$ which occurs at solution point B. For this mixture, the starting materials maintain nearly their starting concentrations, and the product concentrations have values of $10^{-3.7}$ M. An electrode immersed in this solution would have a potential of -0.53 V. The equilibrium constant for this reaction is clearly less than 1, as would also be predicted from the relationship of the redox couples in Table 10.1 and from a calculation using Equation 10.11.

REDOX LOG PLOTS WITH A CONSTANT ACTIVITY REACTANT

For many redox couples, the activity of one of the reactants is constant. Examples are the reduction of metal ions to the metal, the reduction of $AgCl_{(s)}$ to $Ag_{(s)}$, and the reduction of continuously supplied O_2 to H_2O. For these systems, too, the log plots for the reactions are just the combination of the log plots for both couples involved. To illustrate this point, we will consider a mixture of equal volumes of 0.1 M Sn^{2+} and 0.002 M Cu^{2+}. A look at Table 10.1 reveals that the products will be copper metal and Sn^{4+}.

$$Sn^{2+} + Cu^{2+} \rightleftharpoons Cu_{(s)} + Sn^{4+}$$ 10.62

From the relative potentials of the two couples, we see that the equilibrium constant is greater than 1 and therefore the reaction will favor the formation of the products.

The log concentration plot is constructed as shown in Figure 10.5. A system point is placed at the intersection of E' for Sn^{4+},Sn^{2+} ($+0.139$ V) and its total concentration (0.05 M). Then lines are drawn from it with a slope of 0.0592/2 V per log concentration unit and labeled. Then a system point is placed at the intersection of log concentration = 0 and E' for Cu^{2+},Cu ($+0.337$ V). The line for the Cu^{2+} also has a slope of 0.0592/2 V per log concentration unit. Finally, a line for the log of the initial concentration of Cu^{2+} is put in (at a constant -3.0).

When the reaction has come to equilibrium, the number of moles of copper metal and Sn^{4+} produced will be equal. This does not help us get an equation relating concentrations at equilibrium because the copper has come out as a solid. However, from the mass balance equation and the equality given above,

$$[Sn^{4+}] = C_{\text{init,Cu}^{2+}} - [Cu^{2+}]$$ 10.63

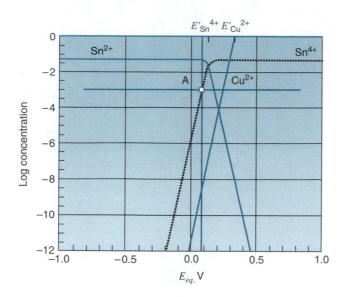

⊚ **Figure 10.5.** In this redox log concentration diagram, one of the redox couples is the Cu^{2+},Cu couple for which the log concentration plot is one line.

To apply Equation 10.63, we first note that the reaction is expected to proceed to the right and that the Cu^{2+} is the limiting reactant. Therefore, we could expect that only a negligible amount of the initial Cu^{2+} will remain unreacted. Thus we can put a tentative solution point at the intersection of $[Sn^{4+}] = C_{init,Cu^{2+}}$. This is marked as point A in Figure 10.5. A vertical line through this point indicates a $\log[Cu^{2+}]$ of -8.5, so our assumption is valid. Along this line, we also see that $[Sn^{2+}]$ is only slightly less than its original value and that a Cu electrode dipped in this solution would have a potential of $+0.08$ V.

LOG PLOTS WITH ASYMMETRY AND pH COMPLICATIONS

The reaction of $S_4O_6^{2-}$ (tetrathionate ion) with Ti^{3+} (titanic ion) illustrates two important aspects of redox log plots. One is the asymmetry of the $S_4O_6^{2-},S_2O_3^{2-}$ couple, and the other is the pH dependence of the TiO^{2+},Ti^{3+} couple. Assume the initial concentrations of the $S_4O_6^{2-}$ and Ti^{3+} reactants are 1.0×10^{-5} M and 1.0×10^{-3} M, respectively, in a solution buffered at pH $= 2$. A look at the reduction potentials in Table 10.1 shows that the reaction will be

$$S_4O_6^{2-} + 2Ti^{3+} + 4H_2O \rightleftharpoons 2S_2O_3^{2-} + 2TiO^{2+} + 4H_3O^+ \qquad 10.64$$

and that the equilibrium will somewhat favor the reactants. This last conclusion turns out to be wrong, as we have ignored some important factors. First, the TiO^{2+},Ti^{3+} couple involves H_3O^+, and the concentration of H_3O^+ is not 1 M. We must first find the **equivalent formal potential** for this half reaction at $p[H_3O^+] = 2$. This is done by writing the Nernst expression for the half reaction, separating the $[H_3O^+]$ term, and evaluating the formal potential at the given $p[H_3O^+]$.

$$E_{eq} = E'_{TiO^{2+},Ti^{3+}} - \frac{V_N}{1} \log \frac{[Ti^{3+}]}{[TiO^{2+}][H_3O^+]^2} \qquad 10.65$$

$$E_{eq} = E'_{TiO^{2+},Ti^{3+}} + 2 \times 0.0592 \ \log [H_3O^+] - 0.0592 \ \log \frac{[Ti^{3+}]}{[TiO^{2+}]}$$

$$E_{eq} = E'_{eff\,TiO^{2+},Ti^{3+}} - 0.0592 \ \log \frac{[Ti^{3+}]}{[TiO^{2+}]}$$

$$E'_{eff\,TiO^{2+},Ti^{3+}} = E'_{TiO^{2+},Ti^{3+}} + 2 \times 0.0592 \ \log [H_3O^+] \qquad 10.66$$

$$E'_{eff\,TiO^{2+},Ti^{3+}} = 0.099 - 4 \times 0.0592 = -0.138 \ V$$

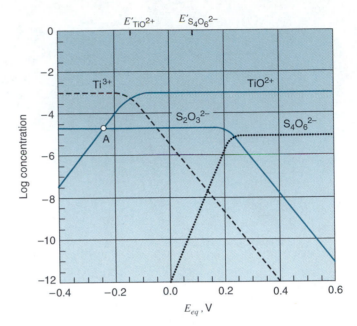

⊙ **Figure 10.6.** When the stoichiometric coefficients for the Ox and Red species are not the same, there is no single value for C_T. The concentrations at which the Ox and Red species predominate are different. Further, the system potential depends on the absolute concentration of the couple. This is shown here for the $S_4O_6^{2-}$, $S_2O_3^{2-}$ couple in contrast with the TiO^{2+}, Ti^{3+} couple.

Note that the potential of this couple is shifted considerably to more negative values when it is carried out at $[H_3O^+]$ less than 1 M.

Now the TiO^{2+}, Ti^{3+} system can be put on the log plot (which is part of Figure 10.6). Place a system point at the intersection of log $C = -3$ and $E'_{eff} = -0.138$ V. From this system point, draw lines at slopes of 0.0592 V per log concentration unit away from the system points.

The method of obtaining the log plot for the $S_4O_6^{2-}$, $S_2O_3^{2-}$ system was presented in Chapter 9. (See Figure 9.18.) The plot is placed on the combined plot so that the concentration of $S_4O_6^{2-}$ (if it were all in that form) is the initial concentration of 10^{-5} M.

The solution point is found by finding an equation relating component concentrations. This can be obtained from an electron balance equation or from the reaction stoichiometry. The reaction occurring is given in Equation 10.64, from which we see that two $S_2O_3^{2-}$ are produced for every two TiO^{2+} produced. The solution point (marked A on Figure 10.6) is where $[S_2O_3^{2-}] = [TiO^{2+}]$. An inert electrode dipped in this solution would have a potential of -0.24 V (with respect to 0.00 V on the table of reduction potentials).

THE IODINE, IODIDE SYSTEM

The iodine reduction reaction has several forms, as given in Equations 10.39 to 10.41. The consequence of this is that each of the forms of the iodine reduction half reaction will produce different log concentration plots. All three of these have been plotted in the graph of Figure 10.7. The curves were derived from Equations 10.42 to 10.44. The system point for the case of $I_{2(s)}$ was placed at the intersection of log $C = 0$ and $+0.535$ V. The concentrations of $I_{2(aq)}$ and I_3^- were taken as 10^{-4} M and 0.1 M, respectively.

Quite a bit can be learned from a careful look at this combined log plot. The system point for the $I_{2(aq)}$, I^- system falls on the I^- line from the $I_{2(s)}$ system point. This is expected, since the activity of iodine has changed from 1 to the value of its concentration. The E' for the $I_{2(aq)}$, I^- system can be obtained from the intersection of the I^- line and the saturation concentration for I_2 (0.002 M). Also, the I^- line appears to follow a different course from the I_3^-, I^- system point than from the $I_{2(aq)}$, I^- system point. Then we realize that the right-hand portion of the I_3^-, I^- system plot cannot be valid because in this region the I^- concentration is much lower than the I_3^- concentration. This condition will preclude the formation of the complex. For this reason, those lines were drawn with dashes. If I^- were being oxidized to I_2, the product would be in the form of

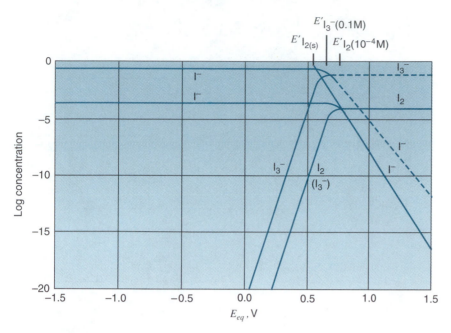

Figure 10.7. Each of the three methods of expressing the iodine,iodide couple results in a different log concentration plot. By combining them, we can better understand their differences. In most analytical applications, the I_3^-,I^- couple is the most appropriate.

dissolved or solid I_2, not I_3^-. If the system concentration is greater than the I_2 solubility (0.002 M), I_2 will precipitate out. In either case, the I^- concentration will follow the line that goes through the $I_{2(aq)}$,I^- and $I_{2(s)}$,I^- system points.

In the case of I_2 reduction to I^-, the usual situation would be to have excess I^-. The I^- would then be at a constant high concentration. To the left of the system points, where the I^- concentration exceeds that of the oxidized form, the I_2 will be complexed by the I^- and will thus be in the form of I_3^-, as indicated in Figure 10.7.

POINTS ON TITRATION CURVES FROM LOG CONCENTRATION PLOTS

Equilibrium concentrations at various points on redox titration curves can be predicted from redox log concentration plots. They are particularly useful for evaluating end point detection means because they provide the related redox electrode potentials for those points at a glance. This process will be illustrated for several titrations.

The first example will be that of the titration of Fe^{2+} with Ce^{4+} introduced with Equation 10.29 and Figure 10.1. For titration redox plots, we will assume that the analytical concentration of the titrant is the same as that of the analyte. The redox plots for 0.1 M Ce^{4+} and Fe^{3+} are shown in Figure 10.8.

As discussed before, the potential at 0% titrated is placed at the potential of the silver reductor (+0.22 V). The log plot is useful in giving the concentration of unreduced Fe^{3+} at that potential (about 10^{-28} M). At the 50% titrated point, E_{eq} is at the system point where $[Fe^{2+}] = [Fe^{3+}]$.

At the equivalence point there is exactly enough titrant present to react with any unreacted analyte. Thus, $[Ce^{4+}] = [Fe^{2+}]$. The intersection of these two lines on the log plot in Figure 10.8 represents the equivalence point condition. The E_{eq} at the equivalence point is about +1.09 V. Because the slopes of the lines for the Fe^{2+} and the Ce^{4+} are equal, E_{eq} is equal to the average of the two system point potentials, as given in Equation 10.34.

After the equivalence point, there is an excess of Ce(IV). At the 200% titrated point, this excess is equal to the amount of the original Fe^{2+}. The amount of Ce(III) is also equal to the amount of the original Fe^{2+}. Thus at the 200% point, $[Ce^{4+}] = [Ce^{3+}]$ and $E_{eq} = E'$ for the Ce^{4+},Ce^{3+} system. The titration break from 50 to 200% titrated, in volts, is the difference in the formal reduction potentials of the analyte and the titrant. The log plot makes it very clear that the magnitude of the break is not dependent on the concentration of the analyte. The whole curve just shifts down as the concentration decreases.

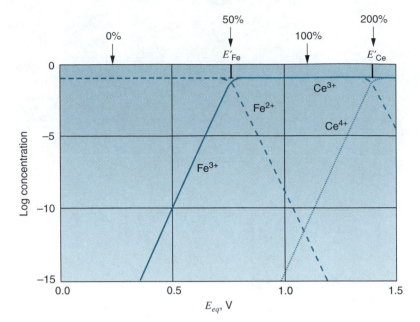

Figure 10.8. The log concentration plots for a redox reaction carried out as a titration enable the prediction of the equilibrium potentials at several key points in the titration curve. This is illustrated here for the titration of Fe^{2+} with Ce^{4+}.

If Ce(IV) were used to titrate a mixture of Sn^{2+} and Fe^{2+}, all three couples can be placed on the plot, as in Figure 10.9. The E_{eq} for the 50% titrated point for Sn^{2+} would be 0.14 V. The 100% point occurs where Equation 10.35, $[Sn^{2+}] = 1/2([Fe^{3+}] + [Ce^{4+}])$, is satisfied, very near the intersection of the Sn^{2+} and Fe^{3+} lines. The location of this point on the plot is where $[Sn^{2+}] = 1/2[Fe^{3+}]$ because, at that point, the Ce(IV) concentration is negligible. The plot also makes it clear that the equivalence point potential is not the simple average of the E' values of the couples. Also, the concentration of the Fe^{3+} at the Sn^{2+} equivalence point can be seen to be about 10^{-8} M. Thus the amount of Fe^{2+} titrated at this point is negligible. The second part of this mixed titration (titration of the Fe^{2+}) is the same as that in Figure 10.8.

The final titration for which the log plots will be developed here is the oxidation of thiosulfate with iodine. This is a combination of two unsymmetrical half reactions and as such is complicated to solve algebraically; we will see, however, that the graphical solution is simple. The combined plot of the I_3^-,I^- and $S_4O_6^{2-}$,$S_2O_3^{2-}$ systems is shown in

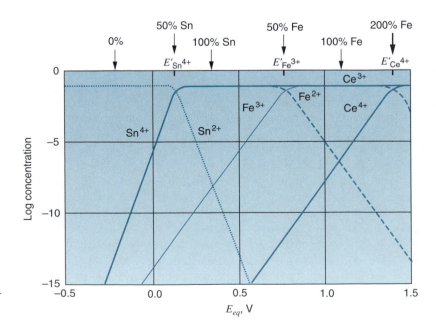

Figure 10.9. This log concentration plot contains three couples a the same value of C_T. From it, one can obtain the points in the titration of a mixture of Sn^{2+} and Fe^{2+} with Ce^{4+}. The plot is very useful for predicting the equilibrium potentials at the 100% points.

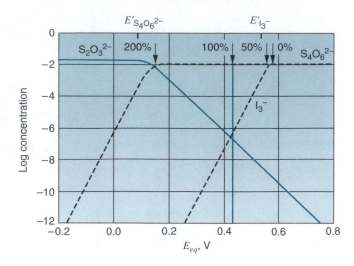

Figure 10.10. This log concentration plot illustrates the system in which iodine is being titrated by thiosulfate. The log plot enables the prediction of the equilibrium potentials at key points in the titration curve.

Figure 10.10. Initial concentrations of $[I_3^-] = 0.010$ M and $[I^-] = 0.10$ M were chosen to match the conditions studied algebraically in Equations 10.52 through 10.58.

When thiosulfate is used to titrate an iodine-containing solution, the overall reaction is

$$2S_2O_3^{2-} + I_3^- \rightleftharpoons S_4O_6^{2-} + 3I^- \qquad\qquad 10.67$$

The triodide, iodide couple is established by having a relatively high concentration of I^- in the initial solution. At the beginning of the titration, the equilibrium potential would be determined by the ratio of iodide and triiodide concentrations. Assuming they are roughly equal, this would be near the system point potential (calculated in Equation 10.52 to be 0.57 V). When the I_3^- is half titrated, its concentration is one-half the initial concentration while the I^- concentration has increased. This will cause the equilibrium potential to decrease somewhat (generally less than V_N). (This point was calculated in Equation 10.53 to be 0.56 V.) The condition for equivalence (from the overall reaction) is $2[S_2O_3^{2-}] = [I_3^-]$. This is marked on the plot as 100%. (The calculated value for this point from Equation 10.57 is 0.421 V.) When a 100% excess of titrant has been added, the concentration of thiosulfate equals half the initial analyte concentration. (A value of 0.13 V was obtained from Equation 10.58 for this point.)

These cases illustrate the value of log concentration plots for the determination of critical points in titration curves, in assessing the feasibility of titrations, and in informing the choice of end point detection methods and devices. A series of algebraic calculations is replaced by the inspection of a pair of plots that have been generated automatically from the concentration and E' data for the couples involved.

Study Questions, Section E

18. Use a log concentration plot to find the concentrations of all species when a 2×10^{-3} M solution of $FeCl_3$ is mixed with an equal volume of a 6×10^{-2} M solution of $SnCl_2$. Use the potentials in Table 10.1 as though they were formal potentials.

19. Calculate the equilibrium concentrations for a mixture of 1×10^{-2} M $Hg(NO_3)_2$ and 3×10^{-3} M $Tl(NO_3)_3$. The reduction half reactions are

$$2Hg^{2+} + 2e^- \rightleftharpoons Hg_2^{2+} \qquad E' = +0.920 \text{ V}$$
$$Tl^{3+} + 2e^- \rightleftharpoons Tl^+ \qquad E' = +1.25 \text{ V}$$

Use a spreadsheet log plot for your solution.

20. What will happen to the oxidizing power of a solution that has equal concentrations of V^{3+} and V^{2+} when a complexing agent is added that reacts preferentially with the V^{3+}?

21. Cyanide ion (CN^-) complexes both Fe^{2+} and Fe^{3+}, forming the hexacyano complex for each. The formation constants are 10^{24} and 10^{31}, respectively.

    A. For equal concentrations of ferric and ferrous iron in a solution of excess CN^-, will the equilibrium potential be greater or less than 0.771 V, and why?

    B. Calculate the effective potential of the ferric, ferrous couple in a solution of 1 M CN^- and make a log concentration plot for this couple at $C_T = 10^{-3}$ M.

22. Use algebra, reasoning, or the spreadsheet program to determine whether the potential at the equivalence point for the titration of I_3^- with $S_4O_6^{2-}$ (in excess I^-) will increase, decrease, or stay the same as the concentration of the reactants is decreased.

Answers to Study Questions, Section E

18. From Table 10.1, the two half reactions involved are

$$Fe^{3+} + e^- \rightleftharpoons Fe^{2+} \quad \text{and} \quad Sn^{4+} + 2e^- \rightleftharpoons Sn^{2+}.$$

The overall reaction is

$$2Fe^{3+} + Sn^{2+} \rightleftharpoons 2Fe^{2+} + Sn^{4+}$$

The initial concentrations of the Fe^{3+} and Sn^{2+} are 1×10^{-3} M and 3×10^{-2} M, respectively. The system points are placed on the plot as shown. From the system points, the lines are drawn at a slope of V_N/n per log concentration unit. Thus, on the bare plot, the lines for the ferric,ferrous system have a slope of 1, and the lines for the stannic,stannous system have a slope of $\frac{1}{2}$. According to the reaction, the concentration of Fe^{2+} produced will be twice that of the Sn^{4+} ion that is, $[Fe^{2+}] = 2[Sn^{4+}]$. The point that satisfies this equation is marked with a circle on the plot, and a line is drawn at the E_{eq} corresponding to the solution point. From this line, we can see that the ferrous concentration is 10^{-3} M, the Sn^{2+} concentration has been reduced a little (by 0.5×10^{-3} M) from its original 3×10^{-2} M, and the ferric ion concentration is about 10^{-14} M.

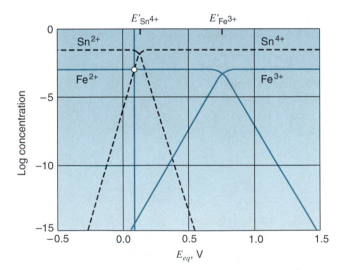

19. We see that the reduction potential for the Tl^{3+},Tl^+ couple is more positive than that for the Hg^{2+},Hg_2^{2+} couple. Therefore, the equilibrium constant should favor the formation of the products Hg^{2+} and Tl^+. In other words, the thallic ion will oxidize the mercurous ion to mercuric ion and form thallous ion. In the log plot containing the two systems, the dashed lines are for the Hg^{2+},Hg_2^{2+} system, and the solid lines are for the Tl^{3+},Tl^+ system. The reaction is

$$Tl^{3+} + Hg_2^{2+} \rightleftharpoons Tl^+ + 2Hg^{2+}$$

Thus, the concentration of Hg^{2+} produced will be twice the amount of Tl^+ produced. The solution point should be where $[Hg^{2+}] = 2[Tl^+]$. This point is marked with a vertical line through it. The Tl^{3+} is the limiting reagent, and it is essentially used up in the oxidation of the Hg_2^{2+}. If all the Tl^{3+} is reduced, $[Tl^+]$ must be 3×10^{-3} M. The $[Hg^{2+}]$ is then twice that, namely, 6×10^{-3} M, and the $[Hg_2^{2+}]$ is the original concentration less 3×10^{-3} M, or 7×10^{-3} M. From the plot, the Tl^{3+} concentration is seen to be far less than 10^{-12} M, thus justifying the assumption that it had been essentially all consumed.

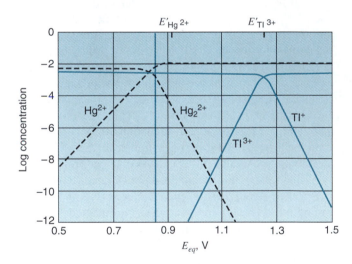

20. The oxidizing power of the solution would begin at a potential near the E' for the V^{3+},V^{2+} couple (about -0.25 V). The addition of a complexing agent that would react more with the V^{3+} than with the V^{2+} would reduce the effective activity of the V^{3+} and thus make the solution less oxidizing.

21. A. First write the formation constant expressions for the complexes and arrange them to solve for the uncomplexed iron ions.

$$[Fe^{3+}] = \frac{[Fe(CN)_6^{3-}]}{K'_{Fe(CN)_6^{3-}}[CN^-]^6}$$

$$[Fe^{2+}] = \frac{[Fe(CN)_6^{4-}]}{K'_{Fe(CN)_6^{4-}}[CN^-]^6}$$

Next, substitute these expressions into the Nernst equation for the Fe^{3+}, Fe^{2+} couple.

$$E_{eq} = E' - \frac{V_N}{1} \log \frac{[Fe^{2+}]}{[Fe^{3+}]}$$

$$= E' - \frac{V_N}{1} \log \frac{[Fe(CN)_6^{4-}]}{K'_{Fe(CN)_6^{4-}}[CN^-]^6} \frac{K'_{Fe(CN)_6^{3-}}[CN^-]^6}{[Fe(CN)_6^{3-}]}$$

$$E_{eq} = E' + V_N \log \frac{K'_{Fe(CN)_6^{4-}}}{K'_{Fe(CN)_6^{3-}}} - V_N \log \frac{[Fe(CN)_6^{4-}]}{[Fe(CN)_6^{3-}]}$$

Now evaluate the second term and combine it with E' to obtain the effective E' for Fe^{3+}, Fe^{2+} in excess CN^- which is calculated to be 0.357 V. Thus,

$$E_{eq} = 0.357 - 0.0592 \log \frac{[Fe(CN)_6^{4-}]}{[Fe(CN)_6^{3-}]}$$

From this result, it can be seen that the standard potential for the Fe^{3+}, Fe^{2+} couple has moved from 0.771 to 0.357 V. Thus, the equilibrium potential with equal concentrations of ferric and ferrous will be less. This is because the formation constant for the CN^- complex is greater for Fe^{3+} than for Fe^{2+} thus reducing the activity of the oxidized form more than that of the reduced form.

B. The resulting plot is made by placing the system point at the effective E' and the total concentration. The plot is shown with a plot for the normal Fe^{3+}, Fe^{2+} couple for comparison.

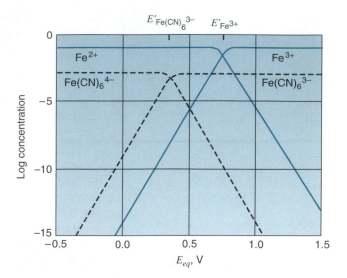

22. From Figure 10.10 or equation 9.46, it can be seen that the oxidizing power of the $S_4O_6^{2-}, S_2O_3^{2-}$ couple increases with decreasing concentration. Thus, the thiosulfate couple will shift to more positive potentials (closer to the I_3^-, I^- couple) as the concentration decreases. As the concentration of I_3^- decreases, the initial potential of the solution to be titrated decreases (moves toward the thiosulfate potential). Thus, as the concentration decreases, these couples move toward each other, which will decrease the size of the break at the equivalence point.

F. Karl Fischer Titration for Water

A very interesting variation of the redox titration is one that uses a species involved in a redox reaction as the limiting reagent. For example, you will note that several of the redox couples listed in Table 10.1 involve water without the water being oxidized or reduced. Even though the water is not being oxidized or reduced in these reactions, it is unavoidably involved. The **Karl Fischer titration** for water takes advantage of this fact. The reaction chosen is the reduction of sulfate ion. The half reaction for this couple is written in Table 10.1 as

$$SO_4^{2-} + 2H_3O^+ + 2e^- \rightleftharpoons SO_3^{2-} + 3H_2O \qquad 10.68$$

When there is a very limited amount of water, the reaction can be written

$$SO_3 + 2H^+ + 2e^- \rightleftharpoons SO_2 + H_2O \qquad 10.69$$

If water is to be the limiting reactant, this reaction should go to the left so SO_2 will be oxidized. The oxidizing agent chosen for this reaction is I_2. The overall reaction (combining the I_2 reduction with the SO_2 oxidation) is

$$SO_2 + I_2 + H_2O \rightleftharpoons SO_3 + 2H^+ + 2I^- \qquad 10.70$$

Since the goal is the determination of water, the solvent used and all the reagents must be water-free. An alcohol is used as the principal solvent. A base is added to enhance the solubility of the acidic reactants and products. The overall reactions (using B to represent the base and ROH as the alcohol) are

$$BSO_2 + BI_2 + B + H_2O \rightleftharpoons B^+SO_3^- + 2BH^+I^- \qquad \text{10.71}$$
$$B^+SO_3^- + ROH \rightleftharpoons BH^+ROSO_3^-$$

The application of this reaction for the titration of water involves filling the buret with the Karl Fischer reagent (the I_2, SO_2, and base dissolved in the alcohol in the mole ratio of $1:3:10$). The alcohol is often methanol. The base in the original Karl Fischer formula was pyridine. More recently, this has been replaced by imidazole or dimethanolamine. This is added to the sample, which has been dissolved in a nonaqueous solvent. The products of the reaction are slightly yellow, but the complex formed between the base and I_2 is deep brown. Therefore, the first excess of the titrant begins to color the flask solution brown. With some practice, this visual end point can be quite effective. The Karl Fischer reagent is not very stable, so it must be restandardized frequently. A common standard is a methanol solution with a known concentration of water (see Example 10.4).

The Karl Fischer titration remains an important analytical technique, extensively used in industry for a variety of applications. The water content in various foods or other products is often determined in this way. In addition, other materials that react to produce or consume water can be determined. For example, the esterification of alcohols and the neutralization of carboxylic acids both produce water. Other reactions have been developed for the production or consumption of water from aldehydes, ketones, primary amines, nitriles, acid anhydrides, and peroxides. However, just as with other redox reactions, it is necessary to avoid sample constituents that may react with any of the reagents. Carbonyls react with the alcohol to yield water. Aldehydes and ketones, when used as part of the solvent, may bind the SO_2.

Trace determinations of water are made more difficult by the need to keep the reagents and sample free of atmospheric water. The performance of buret titrations in a closed system requires special equipment. For this reason, many Karl Fischer titrations are performed with electrochemical generation of the titrant. This procedure is described in the next section.

Example 10.4

Karl Fischer titration

In a titration 14.81 mL of Karl Fischer reagent was required to titrate 10.00 mL of a 0.100 M solution of water in methanol. Titration of 1.00 g of a methanol–water mixture required 38.23 mL of this same reagent. What is the mole % of water in the mixture?

$$\text{mol } H_2O \text{ in mixture} = \frac{38.23 \text{ mL}}{14.81 \text{ mL}} \left(\frac{10.00 \text{ mL} \times 0.100 \text{ mol L}^{-1}}{1000 \text{ mL L}^{-1}} \right) = 2.58 \times 10^{-3} \text{ mol}$$

$$\text{g } H_2O \text{ in mixture} = 2.58 \times 10^{-3} \text{ mol} \times 18.00 \text{ g mol}^{-1} = 0.0465 \text{ g}$$

$$\text{g methanol in mixture} = 1.00 - 0.0465 = 0.954 \text{ g}$$

$$\text{moles methanol in mixture} = \frac{0.954 \text{ g}}{32.00 \text{ g mol}^{-1}} = 29.8 \times 10^{-3} \text{ mol}$$

$$\text{mole \% water in mixture} = \frac{2.58}{29.8 + 2.58} \times 100 = 7.97\%$$

23. What is the differentiating characteristic used in the Karl Fischer determination of water?

24. What is the specific reaction that is at the heart of the Karl Fischer determination of water?

25. Why must all water, including atmospheric humidity, be excluded in the Karl Fischer procedure?

26. What oxidant is used for the Karl Fischer reaction?

27. A 10.00 g sample of an alcohol suspected of having a water impurity was titrated with the Karl Fischer reagent, and 3.27 mL of the reagent was required. When 5.00 mL of a standard methanol/water mixture (0.100 M water in MeOH) was titrated with the same reagent, 7.67 mL were required. Calculate the parts per thousand of water (by weight) in the alcohol sample.

23. The differentiating characteristic needs to be some quality of water because that is the analyte. In this case the quality is the requirement of some redox reactions for water as a reactant.

24. The central reaction is the oxidation of SO_2 to sulfur trioxide. The oxygen required for this reaction is supplied by water, with the release of two protons as products.

25. Any water entering the reaction vessel other than from the sample will interfere with the determination and cause a positive error.

26. The oxidant chosen is I_2.

27.
$$\text{mol } H_2O = 3.27 \text{ mL} \times \frac{5.00 \times 10^{-3} \text{ L} \times 0.100 \text{ mol/L}}{7.67 \text{ mL}}$$

$$= 2.13 \times 10^{-4} \text{ mol } H_2O$$

$$\text{ppt } H_2O = \frac{2.13 \times 10^{-4} \text{ mol} \times 18.00 \text{ g/mol}}{10.00 \text{ g}} \times 1000$$

$$= 0.384 \text{ ppt } H_2O$$

G. Quantitation by Electrolytic Redox Reaction

In the previous chapter, it was shown how the presence of a redox couple could establish a related equilibrium potential between a solution containing that couple and a piece of metal immersed in it. In the measurement of the electrode potential, we were careful to avoid the passage of any electrical current through the voltmeter that connects the test and reference electrodes. A net flow of current to or from an electrode must accompany a net oxidation or reduction reaction at that electrode. Such a process would very likely affect the potential being measured because this potential is no longer the equilibrium potential. However, just as electrode current is deleterious to the technique of potentiometry, it is desirable whenever we want to carry out an oxidation or reduction reaction at an electrode surface. Such electrode oxidation and reduction reactions have many applications.

ELECTROLYTIC REACTIONS

An apparatus that is designed to carry out a redox reaction is shown in Figure 10.11. The two compartments are separated by a porous divider that allows ionic conductivity, but not intermixing between the compartments. The solutions in each compartment contain a redox couple and an electrode suitable for that couple. A system specifically designed for electrolytic reactions is often called a **cell.** If the potential between the two electrodes is measured, the result will be the voltage E_{eq}. The calculation of the voltage expected follows Equation 9.27; in other words, it is the difference in the equilibrium potentials of the two redox half reactions (neglecting the junction potential).

$$E_{eq,cell} = E_{eq,Ox,Red2} - E_{eq,Ox,Red1} \qquad\qquad 10.72$$

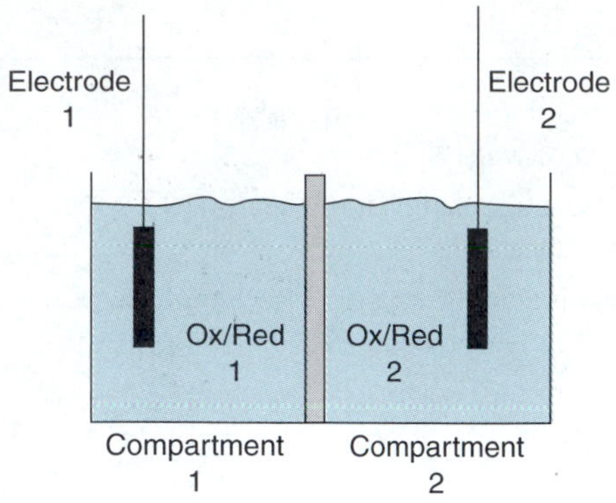

Figure 10.11. An electrochemical cell is composed of two electrodes, often in compartments separated by a porous membrane that prevents mixing. Each electrode will have an equilibrium potential relative to the solution and relative to each other. The potential difference between the two electrodes is $E_{eq,cell}$.

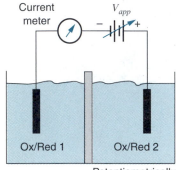

$$V_{app} < E_{eq}, i$$
$$V_{app} = E_{eq}, i = 0$$
$$V_{app} > E_{eq}, i$$

Figure 10.12. If an external voltage source is applied to the cell, a complete circuit exists between the cell electrodes, and a current will pass through the circuit. The direction of the current depends on whether the applied voltage is greater or less than $E_{eq,cell}$.

The terms anode and cathode have specific meanings in electrochemistry that may differ from their use in other contexts. In electrochemistry, the cathode is always the electrode at which reduction is taking place. Its potential with respect to the anode may be positive or negative.

Now an adjustable voltage source and a current meter are connected between the electrodes, as shown in Figure 10.12. For this illustration, it is assumed that the right-hand electrode has the more positive potential at equilibrium. The voltage V_{app} is shown opposing the equilibrium potential. The direction of the current in the circuit, and therefore through the electrodes and solutions, depends on the relative values of V_{app} and E_{eq}. If V_{app} is greater than E_{eq}, the net voltage is positive and the direction of the current is clockwise around the circuit in Figure 10.12. The converse is true if V_{app} is less than E_{eq}. Adjusting V_{app} to the exact value of E_{eq} will result in a zero net potential around the circuit and thus a zero current through it.

Take first the case where V_{app} is greater than E_{eq}. The current is flowing clockwise through the right-hand electrode, through the interface between the electrode and the solution, through the solution, through the porous membrane, through the other solution, through the interface to the left-hand electrode, through the current meter, and back to the external voltage source. According to one of Kirchoff's laws, the current is the same at every point in the circuit. In each conductor, the current can be the result of the movement of positive charge in the clockwise direction, negative charge in the counterclockwise direction, or both. The same is true for each contact between conductors. Thus, at the interface of the right-hand electrode and the solution, there must be a motion of positive ions from the metal into the solution or negative electrons from the solution into the metal. An example of the first case could be copper atoms in a copper electrode becoming copper ions in solution. The second case could be ferric ions at the electrode giving up an electron to become ferrous ions. In either case, the net reaction occurring at the electrode is the conversion of the reduced form of the couple to the oxidized form. In other words, at this electrode, an **electrolytic oxidation** is occurring.

At the left-hand electrode, the direction of the current requires the movement of positive ions from the solution to the metal or the movement of electrons from the metal to the solution. Examples of this are the deposition of silver ions on a silver electrode or the transfer of two electrons from the metal to a dissolved iodine molecule (I_2) to form $2I^-$. Both of these are examples of reduction reactions. Therefore, at this electrode, an **electrolytic reduction** is occurring. It will always be the case that when an oxidation is occurring at one electrode a reduction is occurring at the other. The electrode at which the oxidation is occurring is called the **anode.** The electrode where the reduction takes place is called the **cathode.** An arrangement of two electrodes separated by an ionic conductor that is designed for an electrolytic reaction is called an **electrolytic cell** or just **cell.**

Now consider the case where V_{app} is less than E_{eq}. In this case, the current will flow in a counterclockwise direction through the circuit shown in Figure 10.12. Positive ions must flow from the left electrode to the left compartment, or electrons must flow from the solution into the left electrode. In either case, the electrode reaction will be an oxidation. Similarly, a reduction process will be occurring at the right-hand electrode. Note that the compartments in which the oxidation and reduction are occurring

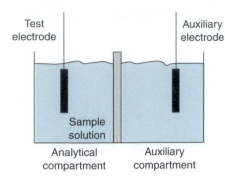

Test electrode

Auxiliary electrode

Sample solution

Analytical compartment

Auxiliary compartment

Figure 10.13. A cell intended for analytical application has a test electrode in the compartment with the sample solution and an auxiliary electrode that is used to supply the desired electrolysis current.

are reversed from the first case when V_{app} was greater than E_{eq}. This is the result of the difference in the direction of the current. When the current is reversed, the electrode formerly called the cathode is now the anode, and vice versa. When the current is zero, as in potentiometry, no reaction occurs at either electrode, and neither is properly called the anode or cathode.

In analytical applications of oxidation or reduction at an electrode, one of the electrodes will be in the compartment with the sample solution as shown in Figure 10.13. It is at this electrode that the analysis reaction will take place. This electrode is called the **test electrode** or **working electrode.** So that an oxidation or reduction can take place at the test electrode, there must be another electrode. This electrode, called the **auxiliary electrode** or **counter electrode,** is in another compartment separated from the analytical compartment by a porous membrane.

COULOMBS AND MOLES

When two electrodes are immersed in solutions that have an ionic connection and there is also a metallic connection between the electrodes, a complete circuit exists. Current can flow in this circuit according to the sign of the difference between the equilibrium and applied voltages. When a net current does flow in either direction, there will be an oxidation at one electrode and a reduction at the other. The rate of these reactions depends on the amplitude of the current. The amplitude of the current in amperes is the rate of charge flow in coulombs per second. (See Background Materials C for a review of electrical quantities.) For each coulomb of charge that crosses the boundary between an electrode and its solution, a proportionate number of moles of material will be oxidized or reduced. From the oxidation and reduction half reactions given in Table 10.1, the stoichiometric relationship between the number of electrons supplied or consumed by the electrode and the number of moles of Ox converted to Red (or vice versa) is clear. For example, if the electrode reaction is the reduction of I_3^- as shown in the margin, two electrons are required to convert one I_3^- to three I^-. If exactly 1 mol of electrons passed from the electrode to the solution containing the I_3^-, exactly 0.5 mol of I_3^- would be reduced and exactly 1.5 mol of I^- would be produced.

$$I_3^- + 2e^- \rightleftharpoons 3I^-$$

The unit most commonly used for the quantity of electrons is not moles; it is coulombs. A **coulomb** (C) is the amount of charge that is passed in 1 second when the electrical current is 1 ampere. Fortunately, Michael Faraday determined the number of coulombs in 1 mol of electrons. It is 96,485 C/mol. This number, symbolized F, is called **Faraday's constant.** Faraday obtained his constant by measuring the mass of silver metal reduced from a solution containing Ag^+ when a measured number of coulombs of charge had been used. From the atomic weight of silver and the stoichiometry for the reduction of Ag^+ to $Ag_{(s)}$ the number of coulombs of electrons required to reduce a mole of silver was determined. Please note that since the current must always be the same at both electrodes in the circuit, the amount of reduction oc-

Faraday's constant, Avogadro's number, and the charge on an electron are all interrelated.

$$e\,\frac{\text{coulombs}}{\text{electron}} = F\,\frac{\text{coulombs}}{\text{mole}} \times \frac{1}{N_A}\,\frac{\text{mole}}{\text{electrons}}$$

$$F = 96{,}485\,\frac{\text{coulombs}}{\text{mole of electrons}}$$

Example 10.5

Moles of reactant consumed

A cell is arranged so that the reaction at one electrode is the reduction of I_3^- according to the reaction

$$I_3^- + 2e^- \rightleftharpoons 3I^-$$

How much I_3^- was consumed if a current of 20.0 mA passed for 50.0 s?

$$\text{mol } I_3^- \text{ consumed} = \frac{(0.0200 \times 50.0) \text{ C}}{\dfrac{96{,}485 \text{ C}}{\text{mol } e^-} \times \dfrac{2 \text{ mol } e^-}{\text{mol } I_3^-}} = 5.18 \times 10^{-6} \text{ mol } I_3^-$$

curring at one electrode is exactly equal to the amount of oxidation occurring at the other. The amount of reactant consumed or product formed at each electrode can be determined from the redox stoichiometry and the relationship

$$\text{moles of } e^- = \frac{Q \text{ coulombs}}{F \text{ coulombs/mol } e^-} = \frac{i \text{ amperes} \times t \text{ seconds}}{96{,}485} \qquad 10.73$$

where Q is the charge passed in coulombs (ampere-seconds). (See Example 10.5.) The only caveat in the application of the relationship between charge and reaction quantity is that the reaction in the calculation must be the only one occurring at the test electrode.

COULOMETRY

The ability of a species to be electrolytically oxidized or reduced can be used for quantitative analysis in either of two ways. One is to transform all the analyte from its oxidized form to its reduced form or vice versa and then *measure the amount of product* thus produced. This is called **electrogravimetry.** An example of this is determining the amount of cupric ions in a solution by the complete electrolytic reduction of them to copper metal and then weighing the copper metal produced. This method requires that the product of the electrolytic reaction be readily separated from the other species in the solution. Electrogravimetry is discussed further in Section H. The other method is to *measure the number of coulombs* required to completely oxidize or reduce the analyte present. This use of the relationship between the total amount of charge passed and the amount of oxidation or reduction that has occurred is called **coulometry.** In direct coulometry, the analyte itself is oxidized or reduced at the electrode. An indirect method, in which the analyte reacts with an electrolytically generated redox reactant, is called **coulometric titration** and is discussed in Section I.

To accomplish direct coulometry, it is necessary to impose a potential at the test electrode that is sufficient to reduce the analyte concentration to the required value. For example, in the reduction of Ag^+ to silver metal, the silver electrode potential must be negative relative to the equilibrium potential for the Ag^+,Ag couple. As seen in Figure 9.15, this potential depends on the concentration of Ag^+ in the solution. If the initial concentration of Ag^+ were 1×10^{-3} M, the final concentration would have to be 1×10^{-6} M to have reduced 99.9% of the initial Ag^+ present. At a Ag^+ concentration of 1×10^{-6} M, the equilibrium potential is 0.44 V. This value can be obtained by inspection of Figure 9.15 or from a substitution in the Nernst equation for the half reaction, as shown in Example 10.6. As long as the potential of the silver electrode is maintained at a potential that is 0.44 V negative (relative to the SHE), the silver ion will be reduced, and the final concentration will be no greater than 1×10^{-6} M.

Example 10.6

Electrode potential required to reduce $[Ag^+]$ to 10^{-6} M

From Equation 9.19,

$$E_{eq} = 0.799 - 0.0592 \log(1/[Ag^+])$$

where $[Ag^+] = 1 \times 10^{-6}$ M,

$$E_{eq} = 0.799 - 0.36 = 0.44 \text{ V}$$

This E_{eq} is relative to the SHE. If an AgCl,Ag reference electrode is used ($E_{eq} = 0.22$ V with respect to the SHE), the desired potential of the test electrode relative to the reference electrode would be $0.44 - 0.22 = +0.22$ V.

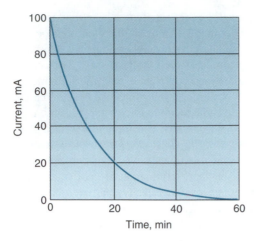

Figure 10.14. The current in a cell is proportional to the rate of the electrolytic reactions going on at the electrodes. As the reactant is depleted, the rate decreases so that total consumption is approached asymptotically.

For a coulometric determination of the initial amount of Ag^+ in the solution, there must be a way to measure the number of coulombs used in this process. For a constant potential electrolysis, the rate of the reaction decreases as the reactant concentration decreases. This is because the rate of diffusion of the reactant to the test electrode surface decreases. The current in the electrolysis circuit (a direct measure of the rate of the reaction) then decreases exponentially with time, as shown in Figure 10.14. Two difficulties are seen from this plot. One is that the reaction rate approaches zero slowly. Thus, in the plot shown, even after 60 minutes, roughly 1% of the reactant remains in the solution. For the rate shown, 100 minutes would be required to reduce the reactant concentration to 0.1% of its initial value. The second difficulty is that the number of coulombs used in the process is the integral of the current taken over the time of the electrolysis (the area under the curve). Integration can be accomplished by recording the current–time curve (as shown in Figure 10.14) and finding the integral by computation or by replacing the current meter in the circuit with an integrating device (sometimes called a **coulometer**).

To control the voltage at the test electrode, it is necessary to measure it with respect to an electrode of known potential. In potentiometry, we used a reference electrode for this purpose. Reference electrodes, however, cannot be used as the auxiliary electrode because they do not maintain their potential if a significant current is passing through them. Thus, two other electrodes must be used: a reference electrode to measure the test electrode potential and an auxiliary electrode to supply the electrolysis current to the solution. The resulting system is shown in Figure 10.15. A circuit called a **potentiostat** (literally, potential controller) keeps the test electrode at the desired value.

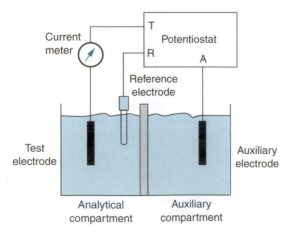

Figure 10.15. To ensure a constant potential between the test electrode and the solution during electrolysis a potentiostat is used. This device compares the measured test electrode potential (relative to the reference electrode) and applies whatever current is required (through the auxiliary electrode) to keep the test electrode at the desired voltage.

The potentiostat includes the power supply that is applied between the test electrode and auxiliary electrode. It also includes a circuit that measures the potential between the test and reference electrodes. The power supply voltage is automatically and continuously adjusted so that the potential between the test and reference electrodes remains at the desired value. For coulometry, a coulometer can be used in place of the current meter.

Direct coulometry has limited value in the analysis of larger sample volumes. This is because of the amount of time required for complete reaction, the difficulty of reducing the remaining analyte to trace levels, and the general lack of specificity among other species that can also react at the electrode. **Microcoulometry,** however, in which the sample volumes are in the microliter range, can achieve complete electrolysis very quickly, and it is useful for the determination of very small amounts of analyte. Fortunately, coulometric methods are very effective with very small amounts of analyte. Remember that a coulomb of charge is one ampere-second. One can readily measure currents at the microampere level or less, and one can certainly measure electrolysis times in the millisecond range. Thus, one could imagine an accurate determination of 10^{-9} coulomb. Using Faraday's constant, 10^{-9} coulomb would electrolyze roughly 10^{-14} mol of analyte. Great advantage is taken of this in modern electroanalytical techniques.

Study Questions, Section G

28. What is an electrolytic reaction?

29. Why does the presence of one electrolytic reaction require a second electrolytic reaction to occur simultaneously?

30. In an electrolytic cell, what determines the direction of the current in the circuit and thus the electrode at which the reduction reaction will occur?

31. A cell is made up of one compartment with a platinum electrode immersed in a solution of 0.10 M KI in which 0.01 M I_2 has been added (resulting in $[I_3^-] = 0.01$ M and $[I^-] = 0.09$ M). The other compartment contains a Ag electrode in 0.1 M KI. An external power supply of 0.8 V is connected to the cell (+ side connected to the Pt electrode).

    A. Determine the direction of the flow of positive current.

    B. Identify the cathode and anode and the reactions occurring at them.

    C. Over what range of power supply voltages would the direction of the current be reversed? In this case, give the reactions occurring at the cathode and anode.

32. What is the relationship between the amount of charge passed through a cell and the number of moles of electrolytic products formed?

33. Calculate the moles of $S_4O_6^{2-}$ reduced and moles of $S_2O_3^{2-}$ produced when a current of 20 mA is passed through a Pt cathode at which this reaction is occurring for a period of 15 minutes.

34. Calculate the charge on one electron (in coulombs) from Faraday's constant and Avogadro's number.

35. What is coulometry?

36. It is desired to analyze Sn^{4+} by reducing it quantitatively to Sn^{2+} at a platinum electrode. If the original solution is roughly 0.01 M in Sn^{4+}, what potential (relative to the SHE) would the Pt electrode have for the final concentration of Sn^{4+} to be 1ppt of the original value? Will H_2 gas be evolved at this potential? Will Sn^{2+} be reduced to Sn at this potential?

37. What is a potentiostat, and what is a coulometer?

Answers to Study Questions, Section G

28. An electrolytic reaction is an oxidation or reduction reaction that takes place at an electrode via a transfer of electrons to or from the electrode material.

29. To pass charge from the electrode to the solution, one must have a complete circuit. The second electrode used to complete this circuit must pass the same charge. If an electrolytic oxidation is occurring at one electrode, an electrolytic reduction is occurring at the other.

30. The direction of the current in an electrolytic cell is determined by the sign of the difference between the applied voltage and the equilibrium voltage of the cell. If the ap-

plied voltage is the larger (and of the same sign), oxidation will occur at the electrode to which the positive side of the applied voltage is connected.

31. A. The equilibrium potential for the cell is obtained from the difference in the equilibrium potentials for the two half reactions involved,

$$E_{eq(Pt-Ag)} = 0.536 - \frac{0.059}{2} \log \frac{(0.09)^3}{0.01}$$

$$- \left(-0.151 - \frac{0.059}{1} \log 0.10\right) = 0.662 \text{ V}$$

with the Pt electrode being the more positive. The external supply is connected to oppose the equilibrium potential and is greater than it. Therefore, the sign of its potential determines the direction of the current through the cell. With the positive lead connected to the Pt electrode, the positive current will flow to that electrode.

B. The Pt electrode receives the positive current from the power supply. Its reaction must be oxidation ($3I^- \rightleftharpoons I_3^- + 2e^-$). Therefore, it is the anode. The Ag electrode is the cathode, and a reduction reaction occurs there ($AgI + e^- \rightleftharpoons Ag + I^-$).

C. The direction of the current will be reversed if the applied voltage is less than the equilibrium cell potential rather than greater. This would be any voltage less than 0.662 V when the + side is connected to the Pt electrode. If the direction of the current is reversed, the directions of the reactions are reversed, as are the designations of cathode and anode. Specifically, at the Ag electrode the reaction is $Ag + I^- \rightleftharpoons AgI + e^-$, and it is the anode. At the Pt electrode, the reaction is $I_3^- + 2e^- \rightleftharpoons 3I^-$, and it is the cathode.

32. The moles of reactant consumed or product formed in an electrolytic reaction can be calculated from the number of coulombs passed through the cell. The relationship is given in Equation 10.73. The ratio of moles of reactant or product to the moles of electrons is obtained from the reduction half reaction.

33. First, calculate the number of coulombs passed.

$$Q = i(\text{amperes}) \times t(\text{seconds})$$
$$Q = 0.020 \text{ A} \times 15 \text{ min} \times 60 \text{ sec/min} = 18 \text{ C}$$

According to the reduction half reaction (see Table 10.1), one-half mole of $S_4O_6^{2-}$ is consumed for every mole of electrons passed. Therefore,

$$\text{mol } S_4O_6^{2-} \text{ consumed} = (18 \text{ C})(0.5 \text{ mol/mol e}^-)/$$
$$(96,485 \text{ C/mol e}^-) = 9.3 \times 10^{-5} \text{ mol } S_4O_6^{2-}$$

Since twice as many moles of $S_2O_3^{2-}$ are produced for each mole of $S_4O_6^{2-}$ consumed,

$$\text{mol } S_2O_3^{2-} \text{ produced} = 2 \times 9.3 \times 10^{-5}$$
$$= 1.9 \times 10^{-4} \text{ mol } S_2O_3^{2-}$$

34. The charge on one electron is

$$\frac{\text{coulombs}}{\text{e}^-} = F \frac{\text{coulombs}}{\text{mol e}^-} \div N_A \frac{\text{e}^-}{\text{mol e}^-}$$
$$= \frac{96,485}{6.022 \times 10^{23}} = 1.602 \times 10^{-19} \text{ C}$$

35. Coulometry is the determination of chemical quantity by measurement of the coulombs of charge necessary to completely oxidize or reduce the analyte.

36. The final concentration of Sn^{4+} is 1/1000 the original, or 1×10^{-5} M. The concentration of Sn^{2+} will be 1×10^{-2} M if all the Sn^{4+} is reduced. The equilibrium potential of the Sn^{4+},Sn^{2+} couple for these concentrations is

$$E_{eq(Sn^{4+},Sn^{2+})} = 0.139 - \frac{0.059}{2} \log \frac{[Sn^{2+}]}{[Sn^{4+}]}$$
$$= 0.139 - \frac{0.059}{2} \log \frac{1 \times 10^{-2}}{1 \times 10^{-5}}$$
$$E_{eq(Sn^{4+},Sn^{2+})} = 0.05 \text{ V}$$

This potential is still positive with respect to the H_3O^+ reduction potential, though it is wise to keep the pH from getting too close to 0. No H_2 gas will be evolved at this potential. The equilibrium potential for the Sn^{2+},Sn couple at 0.01 M Sn^{2+} is -0.199 V, so the reduction of stannous ion to tin metal cannot happen.

37. A potentiostat is a device that applies the electrical current to a cell for electrolysis while keeping the electrode potential equal to a set value. A coulometer is a device for measuring the total charge that has passed in a circuit over the measurement time. In effect, it integrates the current over the measurement time.

H. Electrogravimetry

In the electrogravimetric technique, the amount of analyte that has undergone electrolytic conversion is determined by separating it from the sample compartment and weighing the amount of product produced. This approach eliminates the need to determine the coulombs used in the electrolysis. This is particularly desirable when the analytical reaction is not the only one occurring at the test electrode. A standard example of electrogravimetry is the determination of copper in a copper solution. The test electrode is a relatively large piece of supported platinum mesh that is carefully weighed before the electrolysis begins. The electrolysis apparatus is shown in Figure 10.16. In this case, there is no need to separate the sample and auxiliary compartments because the product of the auxiliary electrode reaction (oxygen) does not interfere with the deposition of the copper on the test electrode. The test electrode and auxiliary electrodes

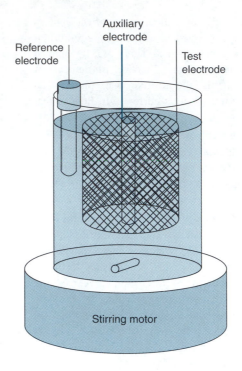

Figure 10.16. The apparatus for electrogravimetry consists of an inert test electrode with a large area along with the auxiliary and reference electrodes. The sample solution is placed in the compartments, and the metal ions are then reduced to the metallic state on the test electrode. From the weight gain of the test electrode, the amount of the metal ions in the sample can be determined.

are immersed in the solution containing Cu^{2+}, and the electrolysis is begun. Stirring is used to help bring the Cu^{2+} nearer to the test electrode. A potentiostat is used to control the test electrode potential.

     Completion of the electrolysis is indicated by the decrease in the electrolysis current to a small fraction of its original value. The setting of the test electrode potential is important in achieving a complete electrolysis and avoiding interferences. Consider the determination of silver in copper ore. The analyte solution will contain both Ag^+ and Cu^{2+}. We would like to deposit the silver ions without depositing any of the copper. This is possible, to a point, because of the much more positive potential of the Ag^+,Ag couple. Assume that the Cu^{2+} concentration is 0.01 M. The potential at which the copper will begin to deposit is then

$$E_{dep} = E'_{Cu^{2+},Cu} - \frac{V_N}{n} \log \frac{1}{[Cu^{2+}]}$$ 10.74

$$E_{dep} = 0.337 - \frac{0.0592}{2} \log \frac{1}{0.01} = 0.28 \text{ V}$$

This is the potential with respect to the SHE. If an AgCl,Ag reference electrode were used ($E_{eq} = 0.222$ V), the potentiostat voltage setting at which copper would first start to deposit would be $0.28 - 0.222 = +0.06$ V with respect to the reference electrode. To be on the safe side, one would probably use a slightly more positive setting, such as $+0.100$ V. This would keep the test electrode at a potential of $+0.100 + 0.222 = +0.322$ V with respect to the SHE. One can now calculate the concentration of Ag^+ that would be in equilibrium with a silver electrode at this potential using the Nernst equation for the Ag^+,Ag couple.

$$E_{dep} = E'_{Ag^+,Ag} - \frac{V_N}{n} \log \frac{1}{[Ag^+]}$$

$$0.322 = 0.799 - \frac{0.0592}{1} \log \frac{1}{[Ag^+]}$$ 10.75

$$[Ag^+] = 8.8 \times 10^{-9} \text{ M}$$

This sounds like a very low concentration until one considers that at the end of the electrolysis the remaining silver concentration is just under one-millionth of the copper ion concentration. Thus the silver content in the copper ore would have to be considerably higher than one-millionth the copper content for this technique to work. In other circumstances (with more nearly equal concentrations of analyte), selective electrolysis has been used effectively. The species with the most positive reduction potential will deposit on the test electrode first. The potential should be set at a voltage just positive of the reduction potential of the next most positive analyte. The closer the reduction potentials of the two analytes, the less complete the deposition can be without interference.

To determine the amount of the analyte deposited, the electrode is removed after each electrolysis, dried, and weighed. The amount of analyte is then determined from the difference in the electrode weight before and after the electrolysis. (See Example 10.7.) The detection limits for gravimetric techniques were discussed earlier in Section F of Chapter 8. The amount of material required for accurate weighing precludes using the gravimetric technique for trace analysis.

Example 10.7

Electrogravimetric determination of silver

The electrode weight before and after the deposition of the Ag was 35.7262 and 36.7294 g. The weight of the silver deposited was the difference (1.0032 g). The moles of silver deposited is then 1.0032 g/107.9 g/mol $= 9.298 \times 10^{-3}$ mol Ag.

Study Questions, Section H

38. In electrogravimetry, what quantity is measured, and how is it related to the amount of analyte in the sample?

39. Some electrolytic reactions may interfere with the quantitation in electrogravimetry, and some may not. Give an example of each type.

40. A 10.0 g sample of copper ore is dissolved, and all the copper contained in the ore is converted to Cu^{2+}. The resulting solution was diluted to 200 mL and subjected to controlled potential electrolysis with a preweighed Pt test electrode. After the electrolysis current dropped to less than 0.1% of its original value, the electrolysis was stopped, and the test electrode was dried and weighed. The

weight gained by the test electrode in electrolysis was 0.148 g.

A. What is the apparent percentage of copper, by weight, in the original sample?

B. If the potential of the potentiostat was set to keep the potential of the test electrode 0.00 V with respect to the AgCl,Ag reference electrode, what is the Cu^{2+} concentration in the final solution? What percentage of the copper was left in the solution?

C. Among the other possible metallic constituents in the sample are Ag, Mn, and Zn. Which of these might interfere with the described determination of copper?

Answers to Study Questions, Section H

38. The quantity measured in electrogravimetry is mass. The difference in the mass of an electrode before and after exhaustive electrodeposition of the unknown species is related to the amount of the analyte in the original sample.

39. An interfering electrolytic reaction would be one that results in adding to the deposit on the electrode. An example would be the deposition of Ag in the determination of Cu. An electrolytic reaction that would not interfere is one that does not affect the amount of deposit. An example is the reduction of water to form hydrogen gas.

40. A. A simple calculation gives the percentage Cu in the ore:

$$\frac{0.148 \text{ g Cu}}{10.00 \text{ g sample}} \times 100 = 1.48\% \text{ Cu}$$

B. The potential of the AgCl,Ag reference electrode is 0.222 V with respect to the SHE. Thus the final equilibrium potential of the test electrode is 0.222 V. At this potential the copper concentration is obtained

from the Nernst equation for the Cu^{2+},Cu half reaction.

$$0.222 = 0.337 - \frac{0.0592}{2} \log \frac{1}{[Cu^{2+}]}$$

$$\log[Cu^{2+}] = -3.89, \qquad [Cu^{2+}] = 1.3 \times 10^{-4} \text{ M}$$

The concentration of weighed copper in the original solution was

$$\frac{0.148 \text{ g}}{63.5 \text{ g/mol}} \times \frac{1.00 \text{ L}}{0.200 \text{ L}} = 1.17 \times 10^{-2} \text{ M}$$

Thus, the amount of copper left in solution, relative to the weighed copper, is

$$\frac{1.3 \times 10^{-4}}{1.17 \times 10^{-2}} \times 100 = 1.1\%$$

C. The reduction potentials of Ag^+, Mn^{2+}, and Zn^{2+} (from Appendix E) are $+0.799$ V, -1.18 V, and -0.763 V. Of these, only the silver potential is more positive than the test electrode potential, and Ag is the only interferent.

I. Coulometric Titration

Coulometric titration is a very elegant method by which a redox reaction can be electrolytically generated. Instead of requiring the analyte to react at the test electrode, a titrant species is generated electrolytically. The titrant then reacts with the analyte. Many useful titrant species can be generated effectively by electrolytic oxidation or reduction reactions. The principal limitation is that all the current through the test electrode must go to generating the desired titrant. Avoiding unwanted reactions involves using electrode reactions that occur readily and then making sure the supply of reactant to the electrode is sufficient to keep up with the electrode current. Table 10.2 lists some of the titrant species that can be electrolytically generated and the test electrode reaction involved in their generation. In this list, you will recognize reagents that are useful in acid–base, precipitation, complexation, and redox titrations. Virtually all the chemical reaction distinguishing characteristics studied thus far can be probed with electrolytically generated titrant.

Electrolytic generation of the reactant species takes the place of the standardized reagent and buret in the traditional titration. This offers several great advantages. Titrant standardization is not required since the amount of titrant delivered to the reaction is calculated from the number of coulombs used to generate it. Very small amounts of titrant can be accurately generated since, as discussed earlier, measurement of the charge equivalent to 10^{-14} mol electrons is quite practical. There is no dilution of the reaction volume owing to the addition of electrolytically generated titrant. Finally, it is possible to use titrants that are not stable for very long since they react so quickly after their generation. Several examples of this are given in Table 10.2. The titrants Ag^{2+} and Fe^{2+} react slowly with water or air and thus are not practical as titrants for buret delivery.

A typical coulometric titration apparatus is shown in Figure 10.17. The generator electrode (often a piece of platinum) is placed in the analytical solution. The auxiliary electrode and its solution are separated from the analytical solution by a glass frit at the bottom of the auxiliary electrode compartment. A stirrer is used to keep the solution

Table 10.2

Coulometrically Generated Titrants*

Titrant	Test Electrode Reaction
OH^-	$2H_2O + 2e^- \rightleftharpoons 2OH^- + H_2$
H^+	$3H_2O \rightleftharpoons 2H_3O^+ + \frac{1}{2}O_2 + 2e^-$
Ag^+	$Ag \rightleftharpoons Ag^+ + e^-$
Hg_2^{2+}	$2Hg \rightleftharpoons Hg_2^{2+} + 2e^-$
HY^{3-}	$HgNH_3Y^{2-} + NH_4^+ + 2e^- \rightleftharpoons Hg_{(l)} + 2NH_3 + HY^{3-}$
Br_2	$2Br^- \rightleftharpoons Br_2 + 2e^-$
Cl_2	$2Cl^- \rightleftharpoons Cl_2 + 2e^-$
I_3^-	$3I^- \rightleftharpoons I_2 + I^- + 2e^-$
Ce^{4+}	$Ce^{3+} \rightleftharpoons Ce^{4+} + e^-$
Mn^{3+}	$Mn^{2+} \rightleftharpoons Mn^{3+} + e^-$
Ag^{2+}	$Ag^+ \rightleftharpoons Ag^{2+} + e^-$
Fe^{2+}	$Fe^{3+} + e^- \rightleftharpoons Fe^{2+}$
Ti^{3+}	$TiO^{2+} + 2H_3O^+ + e^- \rightleftharpoons Ti^{3+} + 3H_2O$
U^{4+}	$UO_2^{2+} + 4H_3O^+ + 2e^- \rightleftharpoons U^{4+} + 6H_2O$

*The reagent from which the titrant is generated is in blue. In the case of a metallic reagent such as Ag or Hg, the generator electrode is made of this metal. Y = EDTA.

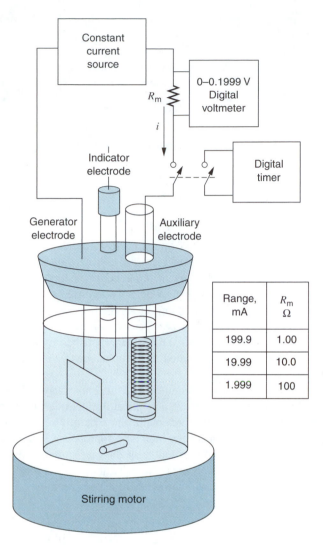

Range, mA	R_m Ω
199.9	1.00
19.99	10.0
1.999	100

Figure 10.17. Titration with electrons is performed by this apparatus. A constant electrolysis current is supplied to the compartment with the sample and the coulometric reagent. The total electrolysis time at the end point is related to the amount of analyte present.

mixed. The titrant is generated by passing a current through the generator and auxiliary electrodes. In this case, a constant current source is used. The use of a constant current supply greatly simplifies the measurement of the coulombs delivered in the titrant generation process. At constant current, the coulombs passed through the electrode is simply the product of the current and the time.

$$Q \text{ (coulombs)} = i \left(\frac{\text{coulombs}}{\text{second}} \right) \times t \text{ (seconds)} \qquad 10.76$$

Constant current electrolysis could not work for direct electrolysis of the analyte because eventually the analyte concentration would be too low to sustain the applied current and some other process would occur at the electrode. The complete absence of processes other than the one desired is essential for the validity of the relationship between coulombs delivered and amount of analyte consumed. In the case of coulometric titration, the reactant in the electrolysis reaction is the reagent from which the titrant is generated. The concentration of this reagent is made large enough so that its diffusion to the electrode can keep up with the current applied. The constant current generator circuit (sometimes called an **amperostat** or **galvanostat**) is adjustable over a range from a few microamperes to a few tenths of an ampere. Commercial units include an indication of the current being delivered. Otherwise, a digital voltmeter can be used for the current measurement, as shown in the diagram of Figure 10.17. The value of the current-measuring resistor R_m depends on the current range desired, according to the table provided in Figure 10.17.

Setting the current from the constant current source sets the rate at which titrant will be generated in the analytical compartment. This rate is obtained from Faraday's constant according to the equation

$$\frac{\text{moles titrant}}{\text{sec}} = \frac{i\left(\dfrac{\text{moles titrant}}{\text{mole of e}^-}\right)}{F} \qquad 10.77$$

For example, if the current were 100 mA, the titrant generation rate (for one molecule of titrant per electron) would be 1.04×10^{-6} mol/sec. In approximately 5 seconds, one would generate the amount of titrant contained in 1 drop (0.05 mL) of 0.1 M titrant. For titrant delivery at a slower rate, one would choose a smaller current.

The constant current source is connected to the system through a switch so that the current can be turned on and off. This switch acts as the stopcock on the buret in a traditional titration. A push-button type of switch is often used so that brief periods of current can readily be applied (corresponding to adding very small amounts of titrant). To know the total coulombs applied, it is necessary to know the total time that the switch in the current circuit has been on. This is accomplished by using a dual switch so that a timer and the current are switched on simultaneously. The timer should read to the closest tenth of a second.

Equivalence point detection is accomplished the same way in coulometric and traditional titrations. That is, one can use an indicator electrode (as shown in Figure 10.17), a spectroscopic probe, or an indicator. In the case of an indicator, the current switch is closed until the end point for the indicator approaches. Thereafter, there is a pause to allow complete mixing and reaction between brief applications of the current. When the indicator has changed color, the titration is stopped, and the cumulative time is read. The amount of analyte is then obtained from the equation

$$\text{moles analyte} = \frac{\text{moles titrant}}{\text{sec}}\, t\,(\text{sec})\, \frac{\text{moles analyte}}{\text{moles titrant}} \qquad 10.78$$

$$\text{moles analyte} = \frac{it\left(\dfrac{\text{moles titrant}}{\text{mole of e}^-}\right)}{F}\, \frac{\text{moles analyte}}{\text{moles titrant}}$$

In the case of a potentiometric or spectroscopic indicator, the response of the electrode or spectrometer can be recorded as a function of the time during which the current has been applied so that a complete titration curve can be produced. From it, the time of equivalence can be obtained and the moles of analyte calculated.

Because coulometric titrant generation affords excellent control over very small amounts of titrant, it is an excellent way to perform analysis on very small amounts of analyte or on very dilute solutions. For this purpose, it is important to consider the practical concentration limits for the titration you would like to pursue. For many of the chemical differentiating characteristics, the change in the titrant or analyte concentration in the equivalence point region decreases dramatically as the concentration of analyte decreases. Only for redox reactions was this not the case. Therefore, microanalysis using redox reactivity as the differentiating characteristic can be accomplished by coulometric titration.

An excellent example of a frequently used coulometric titration is the Karl Fischer determination of water described in Section F. (See Example 10.8.) All the reagents required except the I_2 are placed in the analyte compartment. KI is also added to supply the reagent (I^-) for the electrolytic generation of the I_2 titrant which appears

Example 10.8

Coulometric titration

A coulometric Karl Fischer apparatus is set up and the reagents installed in the titration compartment. The coulometric generation current of 10.00 mA is applied until the end point indication is reached. This ensures that there is no initial water, and it creates the amount of excess I_3^- required for end point indication.

Now a sample is added to the titration volume and the current applied again. The end point is reached in 289.4 s. How many moles of water were in the sample?

The stoichiometry is that 2 mol of electrons are required to generate 1 mol of I_2, and 1 mol of I_2 reacts with 1 mol of water. From Equations 10.77 and 10.78,

$$\frac{\text{moles titrant}}{\text{sec}} = \frac{0.0100 \times \frac{1}{2}}{96,485}$$

$$= 5.18 \times 10^{-8}\ \text{mol s}^{-1}$$

moles $H_2O = 5.18 \times 10^{-8}$ mol s^{-1} × 289.4 s $= 1.50 \times 10^{-5}$ mol H_2O

in Table 10.2 as I_3^-. The equivalence point detection can be either by the brown color of the first excess of I_2 or by a Pt potentiometric electrode. The residual water in the reagents and in the analytical compartment air is first titrated by adding current until the equivalence point is reached. It is important not to overshoot this end point because the purpose of this titration is to bring the solution just to the end point composition. Then the timer is reset, and the sample is added to the analytical compartment (usually by a syringe through a septum in the compartment cover). Current is added until the end point is reached and the time recorded. A closed system is very desirable in this case because it avoids the interference of atmospheric water. The coulometric measurement of titrant addition avoids the frequent standardization required of solution-based titrants. Many successive samples can be run in the same solution until the reagents are exhausted. With this technique, there is a tendency to overestimate the water content by about 50 μg. Precision for samples with 10 mg of water is of the order of 1% but improves with larger amounts.

Study Questions, Section I

41. What differentiating characteristics can be used with a coulometric titration?

42. Why is it important to have a large concentration of coulometric reagent?

43. How is the amount of coulometric titrant generated determined?

44. How is the equivalence point determined in a coulometric titration?

45. In a Karl Fischer titration, a 20.0 mA current was applied to the reagent until the first appearance of the brown I_2–base complex. The time required was 23.67 s. Then the timer was reset, the sample was introduced, and the current was again applied to the first appearance of the color. The time required was 369.32 s. Calculate the moles of water in the sample.

Answers to Study Questions, Section I

41. Coulometry is used to generate a titrant. The titrant can be an acid, a complexing agent, a precipitant, or a redox reagent. Therefore, the differentiating characteristics that are amenable to coulometric titration are virtually any kind of chemical reactivity.

42. It is essential that all the current applied to the cell be used to generate the coulometric titrant. A large concentration of reagent is desired to avoid exceeding the rate with which it can get to the electrode to react.

43. The product of the constant current and the generation time is equal to the number of coulombs passed. This, in turn, is related to the amount of titrant generated through Faraday's constant. The value of the current is sometimes

determined by the titration of a known amount of sample.

44. Just as with other titrations, there are many ways to determine the equivalence point. Indicators, specific electrodes, and photometry are a few of the possibilities.

45. The moles of electrons used to generate the I_2 titrant was 369.32 s $\times$ 0.0200 A/(F C mol^{-1}) = 7.66 $\times$ 10^{-5} mol. Two moles of electrons are required to generate one mole of I_2 (which is given as I_3^- in the half reaction in Table 10.2). Thus the moles of I_2 generated are 7.66 $\times$ 10^{-5}/2 = 3.83 $\times$ 10^{-5} mol. From the titration reaction stoichiometry given in Equation 10.70, one mole of water reacts with one mole of I_2. Thus, at the equivalence point, moles water = moles I_2 = 3.83 $\times$ 10^{-5} mol.

J. Diffusion-Limited Electrodes

Many analytical techniques involving redox reactions at an electrode occur under conditions where the rate of the electrode reaction is limited by the rate at which the analyte can diffuse to the electrode surface. Such techniques are at the heart of the field of **electroanalytical chemistry**.

When an electrode reaction is carried out under conditions of low reactant concentration, the process of bringing the reactant to the electrode surface can limit the

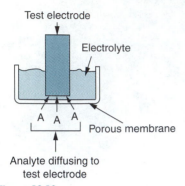

Test electrode

Electrolyte

A A A

Porous membrane

Analyte diffusing to
test electrode

Figure 10.18. A porous membrane can serve to limit and control the rate of diffusion of an analyte to the test electrode surface.

rate of the reaction. When the region between the source of the reactant and the electrode surface is carefully controlled, the rate of the reaction (thus, the electrode current) can be proportional to the concentration of the reactant. Consider a test electrode separated from a thin, porous membrane by a very thin region of solution, as shown in Figure 10.18. This assembly is then dipped into the fluid containing the analyte. To react at the electrode, the analyte must diffuse through the porous membrane and through the thin layer of solution.

The potential of the electrode will be controlled to have a potential for which the concentration of analyte at the electrode surface is negligible compared to its concentration in the sampled fluid. In this case, the rate of diffusion of the analyte (A) to the electrode will be proportional to the concentration of analyte in the sample:

$$\frac{\text{moles A}}{\text{sec}} = KC_A \qquad 10.79$$

where K is the analyte diffusion constant in L s^{-1}. Since the electrode potential is such that the concentration of A at the electrode surface remains very small, all the analyte reaching the electrode is being reduced (or oxidized.). This gives rise to an electrode current that can be calculated from a variation of Equation 10.77 where, instead of the rate of titrant generated with a given current, we are calculating the current resulting from a given rate of diffusion.

$$i = \frac{\text{moles A}}{\text{sec}} \times \frac{F}{\left(\dfrac{\text{moles A}}{\text{mole of e}^-}\right)} \qquad 10.80$$

Now, combining Equations 10.79 and 10.80,

$$i = KC_A \times \frac{F}{\left(\dfrac{\text{moles A}}{\text{mole of e}^-}\right)} \qquad 10.81$$

from which we see that the electrode current will be directly proportional to the concentration of A in the sample.

To keep the test electrode at the proper potential and to allow the measurement of the test electrode current, another electrode must be added to the internal solution. This is often the familiar AgCl,Ag electrode. An implementation of such a diffusion-limited electrode is the Clark electrode[1] shown in Figure 10.19. The reaction at the test electrode is the reduction of O_2 to H_2O. The membrane material must be permeable to O_2 but impermeable to the electrolyte solution that makes up the interior electrolytic cell. An external voltage is applied between the Pt and Ag electrodes such that the Pt electrode is -0.6 V relative to the Ag electrode. This voltage is sufficient to keep the O_2 concentration in the inner electrolyte nearly zero and to reduce the O_2 diffusing through the membrane as soon as it comes in contact with the test electrode. The resulting current is related to the O_2 concentration through Equation 10.81. This electrode can be miniaturized to measure blood oxygen in arteries, and it can also be used to measure atmospheric oxygen. (See Example 10.9.) It is also important in this study because it is used in many sensors based on biochemical reactivity, as we will see in Chapter 12.

The principle behind the Clark electrode can be used for many other analytes and situations. The test electrode material and the membrane must be suitable for the reaction and transport of the analyte, and the membrane must effectively contain the inner electrolyte.

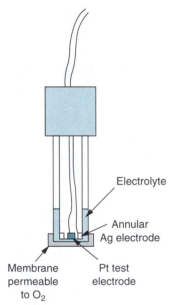

Electrolyte

Annular
Ag electrode

Membrane
permeable
to O_2

Pt test
electrode

Figure 10.19. A Clark oxygen sensor electrode. A voltage applied between the Ag and Pt electrodes keeps the concentration of O_2 in the inner electrolyte near zero. The current at the test electrode is proportional to the rate of diffusion of O_2 through the membrane, which in turn is proportional to the concentration of O_2 in the medium contacting the porous membrane.

[1]M. L. Hitchman, *Measurement of Dissolved Oxygen.* John Wiley & Sons, New York, 1978.

Example 10.9

The Clark electrode

The current measured with a microsized Clark electrode is 14.3 μA when exposed to normal air (20% O_2). When the electrode is placed in the artery of a patient with lung problems, the current is 3.7 μA. If the blood O_2 concentration were in equilibrium with the air, the current would be the same as it was in the air. What fraction of the optimum concentration is the blood O_2 of this patient?

$$3.7/14.3 = 0.26$$

Thus, the patient's lungs are working at roughly one-quarter capacity.

Study Questions, Section J

46. What two conditions are required for the oxidation or reduction current at an electrode to be proportional to the concentration of the reactive species?

47. For the Clark electrode, what is the equivalent partial pressure of O_2 at the test electrode surface if the electrolyte is buffered at $p[H_3O^+] = 7$ and the concentration of Cl^- (for the AgCl,Ag reference electrode) is 1? The electrode potential is -0.60 V relative to the SHE. Use $+1.2$ V as the approximate value for the formal reduction potential for O_2 being reduced to H_2O in acid solution.

48. For a current of 10 μA with a Clark electrode, what is the rate of O_2 diffusion to the electrode surface?

Answers to Study Questions, Section J

46. The electrode potential must be such that the surface concentration of the reactant is negligible compared to that in the sample medium, and the diffusion of the analyte to the electrode must be through a controlled thin layer.

47. From the Nernst equation (Equation 9.19) for the O_2 reduction reaction,

$$E_{eq} = 1.2 - \frac{0.059}{4} \log \frac{1}{p_{O_2}[H_3O^+]^4}$$

$$-0.60 + 0.22 = 1.2 - 0.015 \log \frac{1}{p_{O_2}[10^{-7}]^4}$$

$$105 = \log \frac{1}{p_{O_2}} + \log \frac{1}{[10^{-7}]^4}$$

$$p_{O_2} = 10^{-77} \text{ atm}$$

48. From Equation 10.80,

$$\frac{\text{moles A}}{\text{sec}} = 10 \times 10^{-6} \times \frac{\text{moles A/mole of } e^-}{F}$$

$$= 1.0 \times 10^{-5} \times \frac{1/4}{96,485} = 2.6 \times 10^{-11} \frac{\text{moles A}}{\text{sec}}$$

Thus, the rate of diffusion, of O_2 is 2.6×10^{-11} mol s^{-1}.

K. Test Strips, Spot Tests, and FIA

Many of the reactions that form the basis of spot test and test strip applications are redox reactions. A good example is the test for free chlorine in water. When chlorine is dissolved in water, it reacts with the water to form hydrochloric and hypochlorous acids according to

$$Cl_2 + 2H_2O \rightarrow H_3O^+ + Cl^- + HOCl$$

Thus the test for "chlorine" is actually a test for the HOCl present in the solution. Both Cl_2 and hypochlorous acid will oxidize the molecule N,N-diethyl-p-phenylenediamine

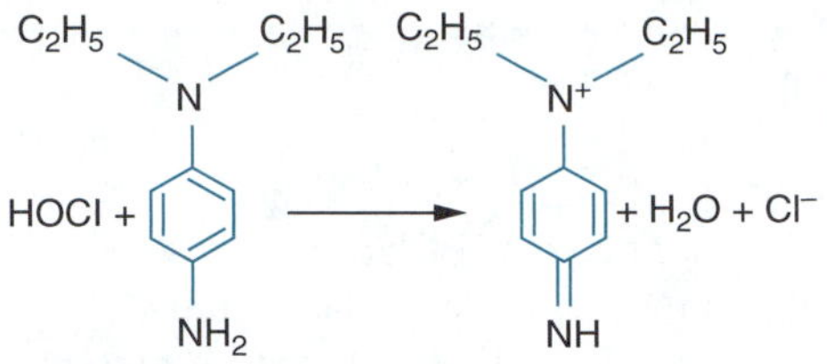

Figure 10.20. Redox reaction of DPD with HOCl to form the magenta amine ion. The intensity of the color is related to the amount of HOCl introduced into the test solution.

(DPD) to form a magenta-colored ion. The reaction is shown in Figure 10.20. The reaction is pH-dependent, so it is carried out in a buffered solution. This reaction can be used in spot tests with spectrometric determination of the intensity of the red color, in FIA systems as part of a water analysis system, or in test strips. In the case of the test strip, the absorbing material on the support must contain the buffer and the DPD. Chlorine test strips are popular devices for pool and spa operators. The procedure involves dipping the strip into the test liquid for just one second in order to control the amount of the test solution that is sampled.

As you might expect, selectivity can be a problem with redox-based test strips. All oxidants with a more positive reduction potential than that of the reductant will oxidize the reductant. Thus, the test strip for Cl_2 is also sensitive to Br_2, permanganate, and ozone. Biochemically catalyzed redox reactions are often much more specific since they add a mechanistic selectivity to that of the sign of the difference in oxidizing strength. Examples of such systems are described in Chapter 12.

Practice Questions and Problems

1. A. $Sb_2O_{5(s)}$ is reduced to SbO^+. Write a balanced half reaction for this reduction.

   B. The $E°$ for the reaction in part A is $+0.581$ V. What will happen if some VCl_3 is dissolved in a solution in which some $Sb_2O_{5(s)}$ is present? Write a balanced reaction.

   C. Write the equilibrium constant expression for the reaction in part B.

   D. Calculate the thermodynamic equilibrium constant for the reaction in part B.

   E. Calculate the formal equilibrium constant for the reaction in part B. Assume a temperature of 25 °C and an ionic strength of 0.01.

2. A. Draw a log concentration plot for the system in Question 1. Take the initial concentration of V^{3+} to be 1.0 $\times 10^{-3}$ M. Also, assume that the $[H_3O^+]$ is 1.0 M.

   B. From the log concentration plot, estimate the concentrations of the reactants and products in the equilibrium solution.

   C. Using the equilibrium constant expression, calculate the concentrations of the reactants and products in the equilibrium solution.

3. A. Draw a log concentration plot for the system in Question 1 assuming the $p[H_3O^+]$ is 4.0.

   B. Using results from part A and from Question 2, describe the difference in the reaction carried out at the two values of $[H_3O^+]$.

4. Consider a titration of V^{2+} with Ce(IV).

   A. What reactions will be involved? Write balance equations for each.

   B. Sketch the titration curve you expect, taking the titration to the point where the Ce(IV) concentration is approximately equal to that of the original V^{2+}. Estimate the values of E_{eq} for each midpoint and equivalence point and for the final solution. Assume that the $[H_3O^+]$ is 1.0 M. You may use a log plot to help with this answer or not, as you choose.

   C. Why do we not need to know the concentrations of the titrant and reactant (assuming they are approximately equal) to answer part B?

   D. If the amount of titrant required to reach the second equivalence point is greater than twice that required to reach the first, what does that indicate about the composition of the original solution?

   E. Calculate the original solution concentrations (in terms of vanadium species) if 18.42 mL of 0.0100 M titrant is required to reach the first equivalence point and 41.74 mL is required to reach the second.

F. What would happen, in the titration of part E, if some Sn^{2+} was also present in the original solution?

5. A. Comment on the relative advantages of using indicators versus the Pt test electrode for the titration of Question 4. What qualities would the indicators have to have?

B. What effect would changing the pH have on the titration in Question 4?

C. Do you think you could determine the amount of Fe^{2+} in the same solution as the V^{2+} by the titration of Question 4? Explain why or why not.

6. What is the equilibrium potential of a copper electrode in a solution that is 1×10^{-3} M in copper(II) and which is also buffered at a pH of 10? The formation constant of the $Cu(OH)^+$ complex is $10^{6.5}$.

7. A. Write a balanced reaction for the reaction of UO_2^{2+} with $S_2O_3^{2-}$.

B. Calculate the equilibrium constant for this reaction.

C. If 50 mL of a solution that is 0.02 M in UO_2^{2+} is mixed with 50 mL of a solution that is 0.002 M in $S_2O_3^{2-}$, what will the equilibrium concentrations of the reactants and products be? Assume that the reaction is being carried out in 1 M strong acid. Also assume that the $E°$ is approximately equal to E'.

D. What are the reactant and product concentrations if this reaction is carried at pH 4 instead?

8. A solution that is 0.001 M in both Fe^{3+} and Fe^{2+} also has 0.01 M EDTA at pH = 6.

A. What is the equilibrium potential of a Pt electrode in this solution? *Hint:* Use the effective formation constants from Figure 7.13.

B. If the pH is increased to 8, will the potential increase or decrease? Explain.

9. A. Describe how you would make a titration out of the reaction studied in Question 7 above. Assume that the UO_2^{2+} is the analyte and $S_2O_3^{2-}$ is the titrant. How would you standardize the titrant? What method of end point detection would you use?

B. If the equivalence point volume for the titrant is 34.87 mL and the titrant is 0.0103 M, what was the weight of uranium in the sample?

10. An electrolytic cell is made up with a copper electrode in a solution of 0.01 M $CuSO_4$ and a Zn electrode in a solution of 0.1 M $ZnSO_4$.

A. Calculate the potential difference between these two electrodes at equilibrium.

B. What voltage would you apply to this cell to oxidize Zn from the electrode into the solution? (Be explicit about which metal electrode is connected to the positive pole of your voltage source.)

C. In part B, when the Zn is being oxidized, what is happening at the copper electrode?

D. In parts B and C, which electrode is the cathode and which the anode?

E. How long will it take to oxidize 7.2 mg of Zn at a current of 43.3 mA?

11. A. In a coulometric titration, why must the auxiliary electrode be separated from the main titration volume?

B. Name three advantages in coulometric titration over addition of titrant from a buret.

12. Sn^{4+} is mixed with U^{3+} in a solution.

A. Write a balanced equation for the overall reaction.

B. Calculate K_{eq} for the reaction.

C. Will the reaction proceed as written?

D. If the initial amounts of Sn^{4+} and U^{3+} were 0.010 mol and 0.006 mol and the final solution volume were 100 mL, what are the approximate concentrations of Sn^{4+}, U^{3+}, Sn^{2+}, and U^{4+} in the final solution (assuming the formal equilibrium constant is roughly equal to the thermodynamic constant)?

13. Ferric ion is used to titrate V^{2+}.

A. What are the products of the titration reaction?

B. What is the equilibrium potential (relative to SHE) at the 50%, 100% and 200% titration points?

C. What is the concentration of V^{2+} at the equivalence point assuming an initial concentration of 0.010 M? (Neglect dilution.)

14. A solution contains Cr^{2+} and V^{2+} in roughly comparable concentrations. Discuss the feasibility of a redox titration procedure by which these two species could be independently quantitated.

15. A cell is made up with a Pt electrode in one compartment immersed in a solution of 0.09 M I^- and 0.01 M I_3^-. The other compartment has a Cd electrode in a solution of 0.030 M Cd^{2+}.

A. Calculate E_{eq} for the cell, clearly indicating which of the two electrodes is the more positive.

B. If the two leads of the cell are connected directly, in which direction is the current through the connecting wires?

C. A voltage source is connected between the two electrodes so that its positive pole is connected to the more positive electrode. What must its value be in order to reverse the direction of the current from that in part B?

D. In the case where the positive lead of the voltage source is connected to the more positive electrode, and $V_{app} > E_{eq}$, which electrode is the anode and which is the cathode?

16. A. Write a balanced equation for the reaction that occurs when UO_2^{2+} is mixed with Ti^{3+}.

 B. Consider how the equilibrium constant for this reaction might change with the pH of the solution. Consider, also, how the potential of an inert electrode following this reaction might change with the pH of the solution.

17. What would be the effect of using hydroiodic acid instead of hydrochloric acid with the silver reductor? Are there species that would be reduced when hydroiodic acid is used that wouldn't be when hydrochloric acid is used or vise versa?

18. Balance the following reactions and calculate their equilibrium constants.

 A. $MnO_4^- + Fe^{2+} \rightleftharpoons Mn^{2+} + Fe^{3+}$

 B. $Fe^{2+} \rightleftharpoons Fe^{3+} + Fe(s)$.

19. The half reaction and standard reduction potential for arsenate ion are

$$AsO_4^{3-} + 2H_3O^+ + 2e^- \rightleftharpoons AsO_3^{3-} + 3H_2O, E° = 0.575 \text{ V}.$$

 An excess of I^- is added to a solution derived from the dissolution of 1.348g of arsenate pesticide in acid. The I_3^- formed required 26.87 mL of 0.100 M $S_2O_3^{2-}$. What is the percentage, by weight, of arsenate in the original sample?

20. As shown in Table 10.2, the reagent TiO^{2+} can be used for the coulometric generation of Ti^{3+}. What are some of the species that might be usefully titrated with Ti^{3+}?

21. The times of a coulometric Karl Fischer titration system are read after the blank and each successive replicate sample without resetting to zero. The current was set to 10.00 mA. The times were 2.72, 38.25, 73.79, 109.27, 144.80, and 180.36 s. Calculate the average mg of water in the samples and the confidence intervals at the 95% confidence level.

Suggested Related Experiments

1. Determination of copper with iodine.
2. Iron titration with cerium.

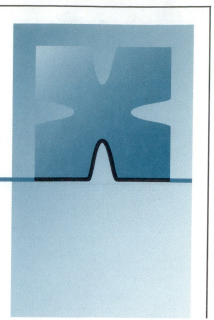

Chapter Eleven

ANALYSIS BY INTERPHASE PARTITION

In a system with two phases, mobile species in either phase may choose to become part of the other phase. Molecules in a liquid can choose to become part of the gaseous phase above the liquid. Molecules in an aqueous phase may choose to dissolve in an adjacent phase of immiscible organic liquid. The capability of a molecular species to distribute or **partition** between two phases is used as a differentiating characteristic in chemical analysis. The analytical goal achieved by interphase partition is separation. A compound that partitions between the two phases is separated from those that have a lower tendency to do so. An analyte that can evaporate from a liquid mixture can be separated from those with a low vapor pressure. Such separation procedures are often used before the application of detection or quantitation techniques in order to remove potential interferents from the sample.

   Interphase partitioning is the phenomenon in which molecules may exist in either of two phases that are in contact with each other. It is assumed that there is no chemical or electrochemical reaction required for the molecules to move between the phases (except for the change in solvation). In other words, the molecule has essentially the same chemical composition in both phases. No electrical effects are involved since the molecules that move between the phases must be electrically neutral. This is true even if the species is significantly dissociated into ions in one of the phases.

   In general, if a species can exist in either of the phases, and if a method of transport between the phases exists, there will be a dynamic exchange of this species at the interface between the two phases. This process is sketched in Figure 11.1. The distribution of the species will come to equilibrium when the chemical potential of the

If a species is free to cross the boundary between two phases, some of that species' molecules will move across the boundary in an attempt to establish an equal chemical potential in both phases. For two different species, the equal chemical potential may be achieved at very different relative concentrations. This fact is the basis of the partition separation techniques such as distillation, extraction, and adsorption. It is also the basis of most modern chromatographic techniques.

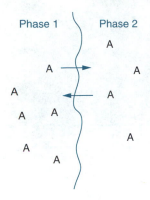

The reaction

$$A_{(Phase\ 1)} \rightleftharpoons A_{(Phase\ 2)}$$

is in equilibrium when

$$\mu_{A(Phase\ 1)} = \mu_{A(Phase\ 2)}$$

Figure 11.1. The species A, free to cross the phase boundary, will cross predominately into the phase in which it has the lower chemical potential. When the concentration in that phase reaches the equilibrium value (the point at which the chemical potentials are equal), A will continue to cross the phase boundary, but at an equal rate in both directions.

species is the same in both phases. Following the approach used in Section D of Chapter 3,

$$\mu^{\circ}_{A(Phase\ 1)} + RT \ln a_{A(Phase\ 1)} = \mu^{\circ}_{A(Phase\ 2)} + RT \ln a_{A(Phase\ 2)} \qquad 11.1$$

$$\frac{\mu^{\circ}_{A(Phase\ 1)} - \mu^{\circ}_{A(Phase\ 2)}}{RT} = \frac{-\Delta G^{\circ}}{RT} = \ln \frac{a_{A(Phase\ 2)}}{a_{A(Phase\ 1)}} = \ln K^{\circ}_{p}$$

The thermodynamic equilibrium constant K°_{p} for this process (called the **partition constant** or **distribution coefficient**) is thus the ratio of the molecular activities in each phase (product phase over reactant phase).

$$K^{\circ}_{p} = \frac{a_{A(Phase\ 2)}}{a_{A(Phase\ 1)}} \qquad 11.2$$

Depending on the specific phases involved, the activity of the species may be related to the concentration, the partial pressure, or the mole fraction. The three phases involved in partitioning are solid surfaces, liquids, and gases. We will take these up two at a time, discussing the partitioning phenomenon that can occur at the interface between them.

When the process of interphase partition is carried out in discrete steps such as extraction or distillation, it is assumed that the materials to be separated differ greatly in their partition characteristics, for example, solubility in the organic phase or volatility. A major advance in chemical analysis was the development of continuous methods of interphase partition in which species differing only by degree in the partitioning tendencies could be separated from each other. Continuous partition methods are referred to by the name **chromatography.** The combination of chromatographic separation techniques with methods of chemical detection and/or identification has become a cornerstone of modern analytical practice.

A. The Liquid–Liquid Interface: Extraction

When two immiscible liquids are in contact with each other, solutes in each can partition between them. The same qualities that make the solvents immiscible create very different environments for the solutes. The distinction is generally between polar solvents and nonpolar solvents. Water is a polar solvent. Each water molecule is a small electrical dipole, that is, its electrical charge is not distributed evenly throughout the molecule. The more positive end of one molecule is attracted to the more negative end of another, and so on. Similarly, the dipolar environment provides an excellent environment for the solubilization of other polar molecules and ionic species. Ions are stabilized by attracting the oppositely charged ends of the solvent molecules. Polar solvents can be recognized by their relatively high dielectric constant.

In nonpolar solvents, the electrical charge is distributed more evenly through the molecule. An example is benzene. Nonpolar molecules have a much higher solubility in nonpolar solvents than in those with a polar nature. Salts have virtually no solubility in nonpolar solvents.

Solvent molecules may be more or less polar; in other words, they differ in the degree of unevenness of charge distribution within the molecule. The dielectric constants for a variety of solvents are given in Table 11.1, which lists the solvents in order of increasing polarity.

Just as nonpolar molecules have low solubility in polar solvents, polar and nonpolar solvents are not miscible. The greater the difference in the dielectric constant, the

Table 11.1
Common Solvents

Solvent	Dielectric Constant, D
Nonpolar	
n-Hexane	1.89
n-Decane	1.99
Cyclohexane	2.02
Carbon tetrachloride	2.23
Benzene	2.27
Toluene	2.31
Diethyl ether	4.30
Polar	
Ethanol	24.3
Methanol	32.63
Formic acid	58
Water	78.54
Formamide	109

smaller is the mole fraction each solvent will have in the other when they are brought into contact. Thus, cyclohexane and water, stirred in the same flask, form two distinct phases, with the less dense solvent (the cyclohexane) above the water.

Solute molecules in either solvent can cross the phase boundary to become solutes in the other. They will preferentially move to the solvent in which they have the lower chemical potential until they come to equilibrium concentrations in each. For dissolved solutes, the concentration is related to the activity by the relationship

$$a_A = \gamma_A [A] \qquad 11.3$$

Therefore

$$K_p^\circ = \frac{\gamma_{A(Phase\ 2)}[A]_{Phase\ 2}}{\gamma_{A(Phase\ 1)}[A]_{Phase\ 1}} = \frac{\gamma_{A(Phase\ 2)}}{\gamma_{A(Phase\ 1)}} K_p' \qquad 11.4$$

where $K_p' = [A]_{Phase\ 2}/[A]_{Phase\ 1}$. Since the solutes that can partition are all uncharged, their activity coefficients at relatively low concentrations are near 1. For this reason, thermodynamic and formal partition constants for many solutes are nearly equal. By convention, the partition equilibrium constant is generally given such that Phase 2 is the nonpolar phase and Phase 1 is the polar phase. On the basis of their usual composition, these are also referred to as the organic and aqueous phases. To continue the convention used throughout this book, we will use the formal constant symbol when we use concentrations in the equilibrium expression. Thus,

K_p° is the thermodynamic partition constant expressed in terms of chemical activity.

K_p' is the formal partition constant expressed in terms of concentration.

$$K_p' = \frac{[A]_{nonpolar}}{[A]_{polar}} = \frac{[A]_{org}}{[A]_{aq}} \qquad 11.5$$

Consider the extraction of species A from a water solution with an organic solvent. This is accomplished by placing both the aqueous and organic solutions in a separatory funnel and shaking vigorously to create a lot of interfacial area. This facilitates the prompt achievement of equilibrium. If the volumes of the aqueous and organic phases are V_{aq} and V_{org}, the fraction of A that remains in the aqueous phase is

In liquid–liquid extraction, the fraction of the analyte extracted depends on the partition constant and on the volume ratio of the two solvents. Two extractions with half the volume of extractant is seen to be more efficient than a single extraction.

$$\alpha_{A,aq} = \frac{[A]_{aq}V_{aq}}{[A]_{aq}V_{aq} + [A]_{org}V_{org}} = \frac{1}{1 + K_p'(V_{org}/V_{aq})} \qquad 11.6$$

From this we see that the larger the value of K_p' and the greater volume of organic solvent used, the larger is the fraction of A that gets extracted. It is not good to use too large a volume of extracting solvent, however, because the species extracted can become too dilute.

To make extraction with a particular volume of extractant more efficient, it is sometimes divided up into several portions of equal volume, with each portion used sequentially. If the total desired volume of extractant V_{org} is divided into n portions, the fraction remaining after all n extractions is

$$\alpha_{A,aq,n} = \left[\frac{1}{1 + K_p'(V_{org}/nV_{aq})} \right]^n \qquad 11.7$$

With this equation, one can readily demonstrate how much more effective 10 extractions of 20 mL each is than one extraction of 200 mL. (See Example 11.1.) One can also predict the volume of extractant needed to achieve a desired level of separation.

Batch extraction is an effective way of removing organic solutes from an aqueous solution. It is a frequently used technique in preparation for the analysis of the organic constituents in biological or environmental fluids. It can be very selective for

Example 11.1

Multiple extractions

In an extraction, 100 mL of organic solvent was used to extract an analyte with a K_p' of 10 from 50 mL of sample solution. Five equal extraction volumes were used. The fraction of the analyte remaining is

$$\alpha_{aq} = \left[\frac{1}{1 + 10(100/250)} \right]^5$$
$$= 3.2 \times 10^{-4}$$

Thus the extraction is complete to within 320 ppm.

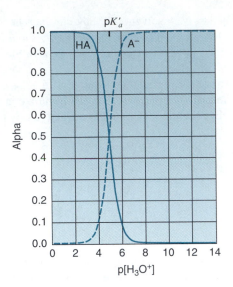

Figure 11.2. The pH has a great effect on which form of a weak acid will predominate in solution. In the case illustrated, the $p[H_3O^+]$ must be below 5 for the majority of the species to be in the neutral form.

nonpolar compounds, in general, but not very selective among the various nonpolar constituents that may be present.

An interesting aspect of extraction concerns species in the aqueous phase that may or may not be in the ionic form. If the conditions in the aqueous phase favor the ionic form, the species will not be extracted because only neutral species can cross the interface between the solutions. On the other hand, extraction may occur when the aqueous conditions favor the neutral form. For example, if the solute is a weak acid HA in the aqueous phase, its extraction to the organic phase will not be favored if the $p[H_3O^+]$ in the aqueous phase is greater than pK_a' for the acid. The reason for this can be seen in the alpha plot of Figure 11.2, repeated from Figure 4.1. When the $p[H_3O^+]$ in the aqueous phase is greater than pK_a' for the acid, the anionic basic form of the species predominates. At $p[H_3O^+]$ values below the pK_a' for the acid, the neutral form predominates, and extraction can occur. Another example of this occurs with complexes in the aqueous phase. The alpha plot in Figure 11.3 is for the complexation of Tl^{3+} with Br^- (from Figure 7.3). For this system, the ML_3 form $(TlBr_3)$ is

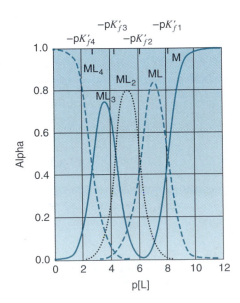

Figure 11.3. Only electrically neutral species can cross the phase boundary in significant quantity. In the case of the thallium and bromide complexes illustrated here, where the neutral form is ML_3, the optimum ligand concentration for extraction is $10^{-3.5}$ M.

the neutral species. The extraction of thallium into the organic phase will be maximized at a Br^- concentration of $10^{-3.5}$ M, where the fraction of Tl in the $TlBr_3$ form is the greatest.

For systems in which the species to be extracted may not be entirely in the neutral, extracting form, a **practical distribution ratio** has been defined. It has the symbol D_C and is the ratio of the concentrations of the solute in all its forms. That is,

$$D_C = \frac{[\text{all forms of A}]_{org}}{[\text{all forms of A}]_{aq}}$$ 11.8

Consider the case where the solute to be extracted is the weak acid HA with a total initial concentration in the aqueous phase of C_T. Assume that in the organic phase, all the solute is in the neutral form HA. D_C is equal to $[HA]_{org}/C_T$. In Equation 4.6, a relationship was given for the fraction of the total A that is in the HA form. It is

$$\alpha_{HA} = \frac{1}{1 + K_a'/[H_3O^+]}$$ 11.9

We can use the concept of alpha with the tabulated value of K_p' to obtain the more practical D_C.

$$K_p' = \frac{[HA]_{org}}{[HA]_{aq}} = \frac{[HA]_{org}}{\alpha_{HA}C_T} = \frac{D_C}{\alpha_{HA}}$$

$$D_C = \alpha_{HA} K_p'$$ 11.10

If Equation 11.9 is combined with Equation 11.10,

$$D_C = \frac{K_p'}{1 + K_a'/[H_3O^+]}$$ 11.11

This relationship will work well for the extraction of a monoprotic acid or base. (See Example 11.2.) Relationships for other acid and complex systems can be readily obtained. Equation 11.10 is a general form, independent of the type of equation used to obtain the value for alpha.

The practical distribution ratio is used in place of K_p' in Equations 11.6 and 11.7 when only a fraction of the species to be extracted is in the neutral, extractable form.

Example 11.2

Extraction of a weak acid

If the partition coefficient for the weak acid HA is 7.4 and the K_a' for the acid in water is 1×10^{-6}, what $p[H_3O^+]$ range is required for 99% of the acid to be in the neutral form, and what fraction of the acid will be unextracted at $p[H_3O^+] = 3$ and at $p[H_3O^+] = 8$? Assume a single extraction with equal volumes of organic and aqueous phases.

From Figure 4.1 and related equations, the $p[H_3O^+]$ at which 99% of the acid is in the neutral (acid) form is $2p[H_3O^+]$ units less than the value of pK_a'. In this case, that is any $p[H_3O^+]$ less than $6 - 2 = 4$. From Equation 11.11, at $p[H_3O^+] = 3$, $D_C = 7.4/(1 + 0.001) = 7.4$. At $p[H_3O^+] = 8$, $D_C = 7.4/(1 + 100) = 0.074$. From Equation 11.6, and using D_C for K_p', the fraction unextracted at $p[H_3O^+] = 3$ is $\alpha_{A,aq} = 1/(1 + 7.4) = 0.12$. At $p[H_3O^+] = 8$, the fraction unextracted is $\alpha_{A,aq} = 1/(1 + 0.074) = 0.93$.

Study Questions, Section A

1. Why must a species be electrically neutral to take part in an interphase partition?

2. What is the condition of equilibrium for a species involved in interphase partition?

3. Why would the species involved in interphase partition have different concentrations in the two phases if the chemical potentials are equal?

4. What is the partition constant (or distribution coefficient)?

5. What are the qualities of the two liquid phases used in liquid–liquid extraction, and why must this be so?

6. What is the form of the formal distribution coefficient for liquid–liquid extraction?

7. Why is an extraction more efficient if the extractant liquid is used in several fractions instead of all at once?

8. You wish to extract I_2 from 100 mL of aqueous phase solution. The K_p' of I_2 for extraction with CCl_4 is 85.

   A. What fraction of I_2 remains in the aqueous phase after a single extraction with 100 mL of CCl_4?

   B. What fraction of I_2 remains in the aqueous phase after 10 extractions with 10 mL of CCl_4 (total CCl_4 volume of 100 mL)?

   C. Using five extractions, what is the minimum volume of CCl_4 required to achieve a 99.9% extraction of the I_2?

9. What is a "practical distribution ratio"?

10. A weak acid HA is to be extracted into an organic solvent, and $K_p' = 3$ for the system. The pK_a' of the acid in the aqueous phase is 1×10^{-5}. Calculate D_C at $p[H_3O^+]$ values of 4 and 9.

11. You are going to attempt to extract an aluminum(III) fluoride complex into diethyl ether. The log of the formation constants for the aluminum complexes with F^-, beginning with $\log K_{f1}'$, are 7.0, 5.6, 4.1, and 2.4.

    A. What is the optimum concentration of F^- for the aqueous phase?

    B. What fraction of the Al is in a neutral form at the optimum $[F^-]$ and also at $[F^-] = 0.1$ M?

Answers to Study Questions, Section A

1. A charged species crossing the phase boundary will create a potential difference between the phases that will impede more ions of that charge crossing the boundary. The amount of charge that is necessary to achieve the impeding voltage is extremely small.

2. The equilibrium condition is that the chemical potential of the species is the same in both phases.

3. Because the species is in a different environment in each phase, the amount of that species required for a given chemical potential will be different. If there is a large difference in the degree of interaction between the species and each environment, the resulting concentration difference can be large.

4. The partition constant is the ratio of the activities of a species in the two phases it partitions between.

5. The qualities of the two liquid phases are that one is relatively polar and the other is relatively nonpolar. This must be the case because this is the basis of the nonmiscibility of two liquids. If both were polar or both were nonpolar, they would have substantial solubilities in each other, and the result would be a solvent mixture with no phase boundary.

6. The formal distribution coefficient for extraction has the form $K_p' = [A]_{org}/[A]_{aq}$, where the concentration in the less polar phase is always the numerator.

7. Each fresh batch of extractant begins with zero concentration of the extracted species. The fraction of each extraction is taken to the power of the number of fractions used. The fraction extracted per volume of extraction liquid is higher with the multiple extractions than with a single extraction of the same total volume.

8. A. Using Equation 11.6,

$$\alpha_{I_2,aq} = \frac{1}{1 + 85\left(\dfrac{100}{100}\right)} = \frac{1}{86} = 0.012$$

   B. Using Equation 11.7,

$$\alpha_{I_2,aq,10} = \left(\frac{1}{1 + 85(100/1000)}\right)^{10}$$

$$= \left(\frac{1}{9.5}\right)^{10} = 1.7 \times 10^{-10}$$

   C. Again, from Equation 11.7,

$$10^{-3} = \left(\frac{1}{1 + 85V_{org}/(5 \cdot 100)}\right)^5 = \left(\frac{1}{1 + 0.17V_{org}}\right)^5$$

$$10^{-3/5} = \frac{1}{1 + 0.17V_{org}} = 0.25$$

$$1 = 0.25(1 + 0.17V_{org})$$

$$V_{org} = \frac{1 - 0.25}{0.043} = 17 \text{ mL}$$

9. A practical distribution ratio takes into account the possibility that not all of the species to be extracted is in the

neutral form in the polar phase. It is simply the formal distribution coefficient multiplied by the fraction of the species that is in the neutral form.

10. Calculate α_{HA} for each $p[H_3O^+]$ value using Equation 11.9. For $p[H_3O^+] = 4$, $\alpha_{HA} = 1/(1 + 10^{-5}/10^{-4}) = 0.91$. For $p[H_3O^+] = 9$, $\alpha_{HA} = 1/(1 + 10^{-5}/10^{-9}) = 10^{-4}$. For $p[H_3O^+] = 4$, $D_C = 0.91 \times 3 = 2.7$. For $p[H_3O^+] = 9$, $D_C = 3 \times 10^{-4}$.

11. A. The neutral complex of Al^{3+} with F^- is AlF_3. The easiest way to find the fraction in this form is to construct a log concentration plot as in Figure 7.4. Also, the maximum $[AlF_3]$ will be at a $p[F^-]$ that is the average of log K'_{f3} and log K'_{f4}. Thus $[AlF_3]$ is maximized at $p[F^-] = \dfrac{4.1 + 2.4}{2} = 3.3$. Therefore the optimum $[F^-]$ is 5×10^{-4} M.

B. To find the fraction of total Al that is in the form AlF_3, use Equation 7.21.

$$\alpha_{AlF_3} = \frac{K'_{f1}K'_{f2}K'_{f3}[F^-]^3}{1 + K'_{f1}[F^-] + K'_{f1}K'_{f2}[F^-]^2 + K'_{f1}K'_{f2}K'_{f3}[F^-]^3 + K'_{f1}K'_{f2}K'_{f3}K'_{f4}[F^-]^4}$$

At $p[F^-] = 3.3$,

$$\alpha_{AlF_3} = \frac{10^{6.8}}{1 + 10^{3.7} + 10^6 + 10^{6.8} + 10^{5.9}} = 0.78$$

At $p[F^-] = 1$,

$$\alpha_{AlF_3} = \frac{10^{13.7}}{1 + 10^6 + 10^{10.6} + 10^{13.7} + 10^{15.1}} = 0.038$$

B. The Gas—Liquid Interface: Distillation

The composition of the vapor that is in equilibrium with a liquid mixture is richer than the solution in the more volatile components of the mixture. If this vapor is then condensed, the process is called **distillation.** Distillation is a method of separating solution components according to their relative volatility. Solutions undergoing distillation are those of liquids that are miscible over the concentration range employed.

The partition that occurs in distillation is between molecules in the liquid and gaseous phases. For these two phases, the activity is expressed as the vapor pressure in the gaseous phase and as the mole fraction in the liquid phase. These choices correspond to the choice of pure vapor at 1 atm pressure for the gaseous phase and pure liquid for the liquid phase. Thus for the vaporization step, the equilibrium equations (Equations 11.2 and 11.4) are written as

$$K_v^\circ = \frac{a_{A(g)}}{a_{A(l)}} = \frac{\gamma_{A(g)}P_A}{\gamma_{A(l)}X_A}, \qquad K_v' = \frac{P_A}{X_A} \qquad 11.12$$

In distillation situations, the concentrations are often quite high from the standpoint of molarity. This is one reason they are often expressed as the mole fraction. A consequence of the high concentrations, however, is that the activity coefficients are not likely to be unity as they are for dilute solutions of neutral species. In the vapor phase, the activity coefficient will be unity if the ideal gas law is obeyed. This will be true if there are no chemical interactions among the gas phase components. In the liquid phase, the activity coefficient will be unity if the solutions are ideal. An ideal solution obeys Raoult's law, which says that the vapor pressure of each component is directly proportional to its mole fraction in the solution. That is, $P_A = P_A^\circ X_A$, where P_A° is the vapor pressure of pure liquid A. Then, for an ideal solution, $K_v^\circ = K_v' = P_A^\circ$.

Solutions are ideal if the intermolecular interactions among all species are essentially identical. This condition is rarely met, so most solutions exhibit deviations from Raoult's law. Ideal and nonideal behaviors for a binary (two-component) solution are shown in Figure 11.4. As shown, the deviations may be either positive or negative depending on whether the two components are more or less attractive to each other than they are to themselves.

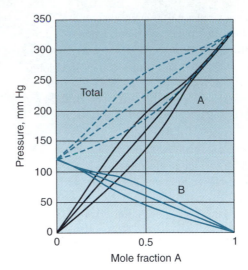

Figure 11.4. Ideal (straight line) and nonideal relationships between mole fraction and vapor pressure. The positive deviations from ideal indicate that the molecules of the two solvents are less attracted to each other than they are to themselves.

The vapor pressure of a species is a strong function of temperature, as shown in Figure 11.5. In this plot, we see that the vapor pressure rises very steeply as the temperature approaches the boiling point. The boiling point, of course, is the temperature at which the vapor pressure equals that of the atmosphere. Species that are more volatile have higher vapor pressures at any temperature, and they have lower boiling points. At any given temperature, the **relative volatility** RV of two species A and B is the ratio of the vapor pressures of the pure substances:

$$RV = P^\circ_A / P^\circ_B \qquad\qquad 11.13$$

where A is always chosen to be the more volatile component so that $RV \geq 1$.

The ratio of the mole fractions of A and B in the vapor phase is equal to the ratio of the parts of the pressure due to each component (called its **partial pressure**). So

$$\frac{P_A}{P_B} = \frac{Y_A}{Y_B} = \frac{P^\circ_A X_A}{P^\circ_B X_B} = RV \frac{X_A}{X_B} \qquad\qquad 11.14$$

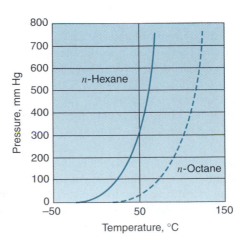

Figure 11.5. As this plot indicates, the volatility of compounds increases steeply with increasing temperature, but at different temperatures for different compounds.

where Y_A and Y_B are the mole fractions, and P_A and P_B are the partial pressures of A and B in the vapor phase. The substitution of $P_A^\circ X_A$ for P_A assumes that the solutions are ideal. Since RV is greater than 1, the mole fraction of A in the gas phase is greater than it is in the liquid phase. The relative volatility is somewhat a function of the temperature, but Equation 11.14 is quite useful for providing approximate solutions. The value of RV can be estimated by an equation derived by Rose.[1] It is

$$\log RV = 8.9 \frac{T_{boiling,B} - T_{boiling,A}}{T_{boiling,B} + T_{boiling,A}} \qquad 11.15$$

This equation, too, is based on a number of assumptions, including the assumption that the components are chemically similar.

In performing a distillation, the solution is boiled to enhance the production of the vapor. Boiling occurs when the vapor pressure of the total solution is equal to the atmospheric pressure.

$$\sum_{i=1}^{n} P_i = 1 \text{ atm} \qquad 11.16$$

The boiling point of the solution depends on its composition, as shown in the diagram in Figure 11.6 for a binary mixture of hexane and octane. At $X_A = 1$, the boiling point is that for pure hexane; at $X_A = 0$, it is that for pure octane. The curve is bowed to lower temperatures in the middle because of the higher vapor pressure of the hexane. If one considers a solution with $X_A = 0.1$, the boiling point will be 112 °C. Because of the higher volatility of hexane, the mole fraction of hexane in the vapor phase will be larger than in the liquid phase. The vapor phase composition at each temperature is plotted as the dotted line. Thus a boiling solution with $X_A = 0.1$ will have a vapor phase mole fraction $Y_A = 0.47$. This is determined by drawing a horizontal line at the boiling point temperature to the vapor phase composition line, then dropping a vertical to determine the mole fraction.

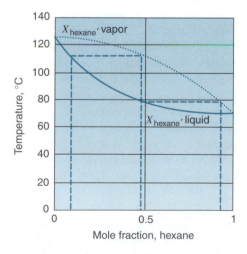

Figure 11.6. At any boiling point temperature, the vapor is richer in the more volatile component than is the liquid. The vapor will therefore condense at a lower temperature. The vapor above this condensate will be still richer in the more volatile component.

[1]A. Rose, *Ind. Eng. Chem.* **1941,** 33, 594.

Example 11.3

Enrichment achieved in multiple-plate distillation

The relative volatility of two organic liquids is 1.3. Starting with an equal mole fraction solution of each, what will be the mole fraction and enrichment of the more volatile compound in the first distillate from a 5-plate distillation column?

From Equation 11.18,

$$\frac{Y_A}{1 - Y_A} = (1.3)^5 \frac{0.5}{1 - 0.5} = 3.7$$

$$Y_A = (1 - Y_A)3.7$$

$$Y_A = 3.7/4.7 = 0.79$$

The degree of enrichment is

$$Y_{A(final)}/X_{A(original)} = 0.79/0.5 = 1.6$$

From this example, we can see that enrichment in the more volatile component has been achieved, but not a separation. Complete separation by a single distillation step would require a much greater difference in the boiling points of the components. Further enrichment could be achieved by condensing the vapor from the first distillation and then distilling it. The result of the second distillation is shown by extending the boiling point of the first distillate to the vapor composition line and dropping a vertical from it (see Figure 11.6). The second distillation has improved the mole fraction of hexane from 0.47 to 0.9. Third, fourth, fifth, and so on distillations could be performed to continue to increase the mole fraction of the more volatile component in the final distillate.

The degree of enrichment can be estimated for a binary mixture by substituting $1 - Y_A$ for Y_B and $1 - X_A$ for X_B in Equation 11.14.

$$\frac{Y_A}{Y_B} = RV \frac{X_A}{X_B} = \frac{Y_A}{1 - Y_A} = RV \frac{X_A}{1 - X_A} \qquad 11.17$$

The term on the right is the starting composition, and the term in Y_A is the composition of the distillate. If the first distillate is redistilled, and so on for n times, the enrichment is then

$$\frac{Y_A}{1 - Y_A} = RV^n \frac{X_A}{1 - X_A} \qquad 11.18$$

(See Example 11.3.)

Multiple distillations would be very cumbersome if it were not for the development of distillation columns that can perform the sequential stages in a single apparatus. A device for this purpose is the bubble-cap column shown in Figure 11.7. In it, the first distillate is condensed to a liquid at its boiling point T_1. The vapor from this liquid is condensed in the second cap, and so on for as many stages as are built into the column. In this apparatus, each stage is called a **plate.** The temperature of the liquid at each higher plate decreases as the liquid becomes increasingly enriched in the more volatile component.

Columns with less explicit stages have also been developed. In these, the column is loosely packed with glass beads, short glass tubes, etc., that can serve as condensation surfaces. The effectiveness of these columns is determined by the use of Equation 11.18. The composition of the emerging distillate is compared with that of the initial solution, and the exponent for RV is determined. This value, n, is called the number of **theoretical plates.** Even with the explicit plates, the calculated value for n is often smaller than the number of explicit plates, indicating that each plate is not completely effective. This can be due to the temperature at each plate not being at exactly the boiling point for the solution it contains. It is also important to note that the equations developed above here only apply to the very first distillate produced. As the distillation progresses, the liquid in the stillpot and subsequent stages increases in the mole fraction of the less volatile components. The boiling point temperature at each higher stage rises, and the mole fraction of the less volatile components in the distillate increases.

In analytical work, distillation is used for separation rather than purification. In purification, significant quantities of the desired component are left impure in order to obtain a sufficiently pure product. In analysis, we would greatly prefer a quantitative separation. This is achieved by distillation only for components that differ greatly in their volatility. Nevertheless, distillation is widely used to eliminate substances that might interfere with the subsequent analysis. It is also important for the conceptual framework it provides for the understanding of gas–liquid chromatography.

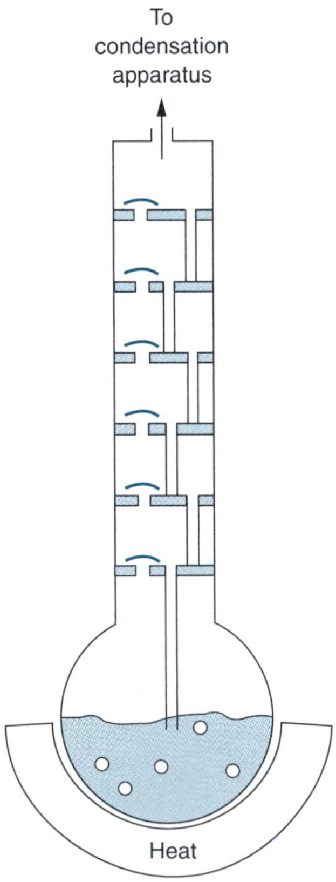

Figure 11.7. An apparatus for carrying out a fractional distillation has a number of explicit or effective "plates" within which the vapor from the lower plate can condense and revolatilize onto the plate above.

12. What is the differentiating characteristic used in a distillation separation?

13. What is the form of the formal partition coefficient for distillation?

14. For the distillation of a binary mixture, how are the degree of enrichment and the relative volatility related?

15. How does multistage distillation improve the degree of enrichment?

16. What is a "plate," and what is a "theoretical plate"?

17. Consider a binary solution of 40% A and 60% B, where the relative volatility of A over B is 4.2.

   A. Calculate the partial pressure of A after a single stage of distillation.

   B. Calculate the number of theoretical plates in a distillation column for which the mole fraction of A in the distillate is 0.99.

12. The differentiating characteristic used in distillation separation is degree of volatility. This is the principal quality that determines the partition coefficient between the liquid and the gas phase.

13. The formal equilibrium constant for a solute in equilibrium with the vapor from the solution is

$$K_v' = \frac{P_A}{X_A}$$

   where P_A is the pressure of A in the vapor and X_A is the mole fraction of A in the solution.

14. The relative volatility, RV, of two components in a binary liquid mixture is defined as the ratio of their equilibrium vapor pressures. The degree of enrichment of the more volatile component in the vapor when a binary mixture is boiled is

$$\frac{Y_A}{Y_B} = RV \frac{X_A}{1 - X_A}$$

   where the Y terms are the mole fractions in the vapor phase.

15. The second stage of distillation condenses the enriched vapor from the first and then develops the vapor from this enriched liquid. The vapor is further enriched in the more volatile component. Each succeeding stage carries this process further toward increasing purity of the more volatile component.

16. A plate in distillation is a name for the site at which a stage of distillation can occur. If the site is not physically recognizable, but rather a section of the distillation column that acts like a stage, it is called a theoretical plate.

17. A. We will use Equation 11.17.

$$\frac{Y_A}{1 - Y_A} = RV \frac{X_A}{1 - X_A} = 4.2 \frac{0.4}{0.6} = 2.8$$

$$Y_A = 2.8(1 - Y_A), \qquad Y_A = \frac{2.8}{3.8} = 0.74$$

   B. We will use Equation 11.18.

$$\frac{0.99}{0.01} = 4.2^n \frac{0.4}{0.6}, \qquad \frac{0.99}{0.01} \times \frac{0.6}{0.4} = 4.2^n = 148.5$$

$$n \log 4.2 = \log 148.5, \qquad n = \frac{\log 148.5}{\log 4.2} = 3.5$$

C. The Gas—Solid Interface: Adsorption

There is a natural tendency for gases to adsorb on solid surfaces. The surface molecules or atoms of a solid are not surrounded by other molecules as they are in the interior of the solid. Consequently, they have a bonding capability that is not being used. This tendency toward bond formation is at least partially satisfied by the adsorption of a gas molecule. The solid on which the adsorption takes place is the **adsorbant,** and the molecule being adsorbed is the **adsorbate.** The reaction that occurs is

$$A_{(g)} \rightleftharpoons A_{(ads)} \qquad\qquad 11.19$$

for which the equilibrium constant is

$$K_{ads}^\circ = \frac{a_{A_{(ads)}}}{a_{A_{(g)}}} = \frac{a_{A_{(ads)}}}{P_{A_{(g)}}} \qquad\qquad 11.20$$

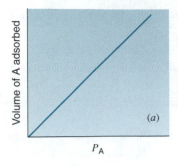

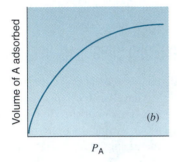

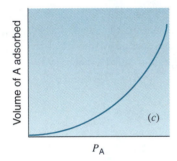

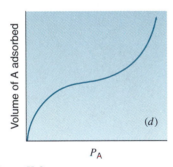

Figure 11.8. The relationship between the amount of vapor adsorbed and the vapor pressure can take many forms depending on the interaction of the absorbate with the surface and with already adsorbed molecules. See text for details.

The magnitude of K_{ads}° depends on many factors. One is the type of bonding involved in the adsorption. All gases and surfaces exhibit van der Waals bonding. Some specific combinations of gases and surfaces may form covalent bonds as well. In the case of covalent bond formation, the adsorption is much stronger. However, the adsorption reaction may be slower and may not be reversible. In the absence of covalent bonding, the order of K_{ads}° for a series of gases is independent of the adsorbant. Another factor is the temperature. For all gases, K_{ads}° increases with decreasing temperature. This corresponds to the increasing tendency of the gas to condense. The increase in K_{ads}° can be quite dramatic as the temperature decreases toward the boiling point of the gaseous material.

ADSORPTION ISOTHERMS

The activity of the gaseous material is equal to its partial pressure, as shown in Equation 11.20. For the adsorbed material, the activity will generally increase with the amount of gas adsorbed, but the relationship between activity and surface concentration varies with the particular combination of adsorbants and adsorbates. The relationship is generally shown as an experimental **isotherm.** The isotherms for only a few gas–surface combinations have been fundamentally rationalized. Several typical adsorption isotherms are shown in Figure 11.8. The isotherm in Figure 11.8a is called the ideal isotherm because the adsorbate activity is proportional to the amount adsorbed. Figure 11.8b represents an isotherm in which saturation occurs. Saturation might be expected since there is only so much room on the surface for the adsorbate. After the surface is filled, the adsorbant surface is no longer available. Successive adsorption must take place on the already adsorbed material. This is certain to have different characteristics.

It is interesting to calculate how much adsorbate can be adsorbed before monolayer coverage is achieved. A number of assumptions are required, so the result is a ballpark estimate. First, assume a spherical molecule. Its volume will be

$$\frac{V\,(\text{cm}^3)}{\text{molecule}} = \frac{MW\left(\dfrac{g}{mol}\right)}{\rho\left(\dfrac{g}{cm^3}\right)N_A\left(\dfrac{\text{molecules}}{mol}\right)} \qquad 11.21$$

where ρ is the density of the solid or liquid adsorbate and N_A is Avogadro's number. The radius of a sphere with this volume is $r = (3V/4\pi)^{1/3}$. The closest arrangement for spheres on a surface is hexagonal packing exemplified by billiard balls in the rack. The area occupied by each spherical molecule is then a regular hexagon with each side equal to r. The area of such a hexagon is $3r^2$. Combining these equations yields the area each molecule occupies on the surface.

$$\frac{\text{area}\,(\text{cm}^2)}{\text{molecule}} = 3\left(\frac{3}{4\pi}\frac{MW\left(\dfrac{g}{mol}\right)}{\rho\left(\dfrac{g}{cm^3}\right)\cdot N_A\left(\dfrac{\text{molecules}}{mol}\right)}\right)^{2/3} \qquad 11.22$$

For a molecule with a MW of 50 and a density of 1, the area occupied is 2.2×10^{-15} cm^2. There are then 4.6×10^{14} molecules/cm^2. This sounds like a lot, but when divided by Avogadro's number it is only 7.6×10^{-10} mol/cm^2. This amount of ideal gas at STP would occupy a volume of 1.7×10^{-5} mL, just over a hundredth of a microliter. These figures are rough, but they serve to convey the point that the amount of material adsorbed per square centimeter of surface is extremely small.

The material in the isotherm shown in Figure 11.8c has approximately the same tendency to adsorb on itself as on the surface. Thus, the adsorption continues with increased gas pressure. The very steep rise indicates that the pressure of condensation is being approached. The isotherm in Figure 11.8d approaches a nearly level region after the achievement of a monolayer of coverage. At still higher pressures, deposition of multiple layers and finally condensation can occur.

Adsorption has a variety of applications, particularly in gas analysis. Air to be analyzed for trace organic molecules can be pumped through a tube filled with adsorbant to concentrate the species that will adsorb in the tube. The adsorbant might be cooled to enhance the effect, especially if low boiling analytes are sought. Then the tube with adsorbates is taken to the lab and heated to release the adsorbates into the analysis apparatus. A part of the analysis apparatus is likely to be a gas chromatograph.

SOLID PHASE EXTRACTION

A method of extraction in which the extractant is the surface of solid particles has become a very powerful method of analyte separation. The way it works is that the surface of the solid particles is coated with a very thin layer of material to which the analyte(s) will have an affinity. The solid particles are then mixed with the sample fluid, allowing the analyte(s) to partition to the solid surface (step 1 in Figure 11.9). If the analyte partition coefficient is high enough, the fraction of the total analyte on the solid can be very high. The solid is then rinsed with pure solvent to complete the separation of the analyte(s) from the other sample components (step 2). The solid is then rinsed with a different solvent for which the partition coefficient for the analytes greatly favors the solvent. In this step, called **elution**, the analytes desorb from the solid surface (step 3). The solvent, now containing the purified analyte(s), is washed from the region containing the solid particles.

This separation is seen to be a combination of extractions, first from the sample solution to the solid surface and second from the solid surface to the dissolving solution. In a typical application, the sample solution will be aqueous, the solid surface will be coated with a nonpolar hydrocarbon that will attract all the nonpolar components of the solution, and the dissolving solution will be a strong organic solvent such as methanol. Since the volume of the dissolving solution can be much less than the sample volume from which the analytes were extracted, the analytes can actually be more concentrated in the extract than they were in the original sample. This **preconcentration** step makes the subsequent quantitation or detection of the analytes that much easier.

To achieve the separation of a useful amount of analyte, the total surface area of all the solid particles used must be very large. To accomplish this with a reasonable volume of solid particles, the particles are very small, typically spheres roughly 40 μm in diameter. One cubic centimeter of such particles would have an area of roughly 300 cm^2. From our previous calculation of molecules per area, this amount of surface could contain up to 0.25 μmol of analyte.

Several techniques have been developed for carrying out solid phase extraction. The general method is to arrange for solutions to flow through the region containing the solid phase extractant. Thus, the solid phase extractant can be in the form of a filter disk, adsorbant particles placed in a filter funnel, or particles in an enclosed cartridge through which the solutions are forced to flow. One of the more common implementations is one in which the cartridge is designed for a syringe to be used to introduce the sample, elution solvent, and wash fluids as well as to provide the pressure to force the liquids through the volume containing the solid phase extractant. Figure 11.10 is a sketch of such a cartridge.

Solid phase extractants are made with a variety of surface molecules attached to them in order to provide selective adsorption of analytes with different differentiating characteristics. A surface with C_{18} aliphatic hydrocarbons selectively adsorbs nonpolar analytes from polar solutions, whereas particles with an alumina (AlO or AlO$^-$) surface selectively adsorb polar molecules from nonpolar solutions. Other surfaces de-

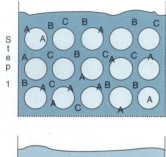
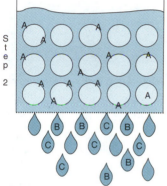
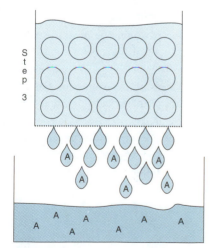

Figure 11.9. The process of solid phase extraction. In step 1, the analyte A is adsorbed on the particle surface. The remainder of the sample is washed off in step 2. The analyte is then eluted with a strong solvent for A in step 3.

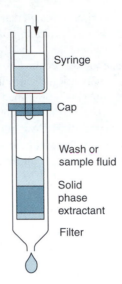

Figure 11.10. Solid phase extraction cartridge designed to work with a syringe for fluid introduction and pressure.

signed for specific ionic or complexation-based interactions with the analyte have also been developed. A similar technique based on biochemical interactions, called **affinity chromatography,** is introduced in Chapter 12.

Study Questions, Section C

18. What is the form of the partition coefficient for the partition between a gas and a solid surface?

19. What is an adsorption isotherm?

20. Which way does an isotherm curve as the partial pressure of the adsorbate approaches the condensation pressure?

21. Calculate the surface area required to contain 10^{-5} mol of adsorbed dichloroethane ($C_2H_4Cl_2$). Assume the adsorption is only one layer thick. Assume a density of 0.8 g/cm^3.

Answers to Study Questions, Section C

18. The partition coefficient in adsorption is the ratio of the activity of the adsorbed gas to the partial pressure of the adsorbate.

19. An adsorption isotherm is the relationship between the volume of the gas adsorbed and the partial pressure of the gas.

20. The isotherm curves upward as the adsorbate's partial pressure approaches the condensation pressure, indicating that the volume adsorbed can increase greatly without very much further increase in the pressure.

21. $$\frac{\text{area}}{\text{mole}} = N_A \frac{\text{area}}{\text{molecule}}$$

$$= 6 \times 10^{23} \times 3\left(\frac{3 \times 99}{4\pi \times 0.8 \times 6 \times 10^{23}}\right)^{2/3}$$

$$= 2.4 \times 10^9 \frac{\text{cm}^2}{\text{mol}}$$

For 10^{-5} mol, the area equals 2.4×10^4 cm^2.

D.   Continuous Partition: Chromatography

In the previous section, interphase partition was shown to be a method by which species differing greatly in their partition constants could be separated. It is one of the most impressive accomplishments of modern chemical analysis that this relatively crude phenomenon could be transformed into a technique by which species differing only slightly in their partitioning characteristics could be quantitatively separated and independently characterized. The basis for this transformation is the creative application of another dimension to the partitioning, one that distributes the partitioning components over space and time. This section is an introduction to the principles of how this is accomplished.

THE FRACTION OF TIME SPENT IN EACH PHASE

A partitioning species is divided between the two phases according to the equilibrium constant expression. To remain completely general, these will be called Phases 1 and 2. The partitioning reaction is then

$$A_{(Phase\ 1)} \rightleftharpoons A_{(Phase\ 2)} \qquad 11.23$$

The fraction of the total amount of material in Phase 1 (expressing the amounts as molarities and volumes) is

$$\frac{A\ in\ Phase\ 1}{total\ A} = \frac{[A]_1 V_1}{[A]_1 V_1 + [A]_2 V_2} = \frac{1}{1 + \dfrac{[A]_2}{[A]_1}\dfrac{V_2}{V_1}} = \frac{1}{1 + K'_p \dfrac{V_2}{V_1}} \qquad 11.24$$

This relationship reveals that the fraction in Phase 1 depends on the equilibrium constant and the relative volumes. As K'_p increases, the fraction of the molecules of A in Phase 1 decreases.

Sometimes we forget that reactions at equilibrium are dynamic, that is, that the reaction is proceeding at equal rates in both directions. Any given molecule of A will spend part of its time in Phase 1 and the rest of its time in Phase 2. The fraction of the time it spends in Phase 1 is the same as the fraction of A molecules that are in Phase 1. For example, if there are twice as many molecules of A in Phase 2 as in Phase 1 (fraction in Phase 1 = 1/3), any given molecule of A is twice as likely to be found in Phase 2 as in Phase 1. Therefore, it spends two-thirds of its time in Phase 2 and one-third of its time in Phase 1. Repeating Equation 11.24 in terms of the fractional time in Phase 1,

$$\frac{t_{A,Phase\ 1}}{t_{total}} = \frac{1}{1 + K'_p \dfrac{V_2}{V_1}} \qquad 11.25$$

MOVING ONE OF THE PHASES

From the relationship in Equation 11.25, we can see that the fraction of the time component A spends in Phase 1 depends on K'_p. If there were a way to separate components according to this time fraction, species with minor differences in K'_p might be separable. This is accomplished elegantly by making Phase 1 move with respect to Phase 2. In this way, the distance a species moves in a given time will depend on the fraction of its time it spends in Phase 1. The phase that is moving is called the **mobile phase** or the **eluant**; the other is called the **stationary phase**. A simple model for such a system is that of a tube lined with the stationary phase, as shown in Figure 11.11. The mobile phase then flows through the tube. If the velocity of the mobile phase is u, the average velocity $\bar{v}_A$ of the A molecules will be u times the fraction of the time each A molecule is in the mobile phase.

$$\bar{v}_A = u \times \frac{t_{A,phase\ 1}}{t_{total}} = u\frac{1}{1 + K'_p \dfrac{V_2}{V_1}} = u\frac{1}{1 + k'} \qquad 11.26$$

where k' is called the **capacity factor.** As seen, $k' = K'_p(V_2/V_1)$. If all the components were placed in the mobile phase at the same time, they would separate along the tube

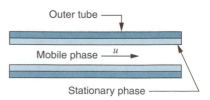

Figure 11.11. A common arrangement for the movement of one phase with respect to the other is to flow the mobile phase through a tube coated with the sationary phase. Here we see a cutaway of such a tube.

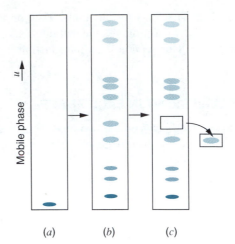

Figure 11.12. A thin strip of porous material can be used for the stationary phase. In such a case, the sequence is as follows: (*a*) sample is placed at bottom and flow of mobile phase is started; (*b*) mobile phase is stopped, and components are located; (*c*) components of interest are removed.

according to their respective velocities like runners in a race. This is the basic process employed in chromatographic separation.

The differences in component velocities can be used for separation in two basic ways. In one, the movement of the mobile phase is stopped before the fastest component has passed beyond the last of the stationary phase. In this form, the components are then arrayed in sections along the path of mobile phase motion, as shown in Figure 11.12. If the components have a color, colored bands will be seen at the component locations along the stationary phase. This effect gives the technique its name of "chromatography." If the components are not colored, the mobile phase may be sprayed with a material that reacts with the components to form a colored product. Sections of the stationary phase containing the components of interest can then be removed for further analysis or reaction. This is the method used with paper and thin-layer chromatography.

The other method of performing chromatographic separation is to place a component detector at the exit of the stationary phase and to keep the mobile phase moving until the slowest component has gone through the detector. This process, shown in Figure 11.13, is called **elution.** The record of detector response as a function of time is an indication of the components present and their relative velocities in the mobile phase.

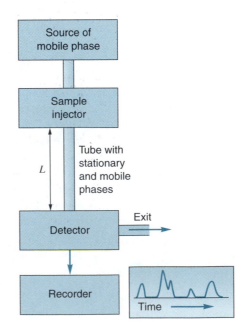

Figure 11.13. When a tube is used to contain the stationary and mobile phases, the distance each component must move to be detected is the length of the tube. The time required to do that will depend on the velocity of each component.

The detector signal may also be used to initiate collection of the exiting mobile phase at the time components of interest appear.

What is measured, in the case of full elution of all components, is the time between sample injection and component appearance at the detector. This time is called the **retention time,** t_R. As expected, the retention time is related to the partition equilibrium constant K_p' and the capacity factor k'. The velocity of component A in the mobile phase has been derived in Equation 11.26. It is also equal to $L/t_{R,A}$ where L is the column length as in Figure 11.13. Likewise, the mobile phase velocity u can be measured. Often the sample injected contains some components such as solvent or nitrogen gas for which k' is essentially zero; that is, they are not retained at all, and their velocity is that of the mobile phase. The time of their appearance at the detector has the symbol t_M and the velocity of the mobile phase is $u = L/t_M$. When both these values for the component and mobile phase velocities are substituted into Equation 11.26, the result is

$$\bar{v}_A = u \frac{1}{1 + k_A'}, \qquad \frac{L}{t_{R,A}} = \frac{L}{t_M} \frac{1}{1 + k_A'} \qquad 11.27$$

Equation 11.27 can be rearranged to

$$k_A' = \frac{t_{R,A} - t_M}{t_M} = \frac{t_{R,A}'}{t_M} \qquad 11.28$$

where $t_{R,A}'$ is the retention time after the unretained components have eluted. Equation 11.28 provides an easy way to determine k' experimentally.

The value of k' has some practical boundaries. If it is less than 1, the component will appear at a time when the unretained component(s) are still present. If the value of k' is too large, the retention time for the component may be impractically long. Values of k' between 1 and 5 are considered ideal.

MEASURES OF EFFECTIVENESS

The equations developed here can provide information on the difference in retention time that can be expected for a given difference in k'. This, by itself, does not allow an evaluation of the separation of the two components because not all the molecules of a given component arrive at the detector at the same instant. One of the factors causing the spread in the arrival times among identical analyte molecules is the equilibrium distribution of analyte between the two phases. One can think of the stationary phase as divided into a number of sections, each of which comes to equilibrium with the analyte in the mobile phase at that time. The section at the front edge of the analyte peak is just retaining the first part of the analyte, and the section at the rear edge of the analyte peak is just releasing the last of the analyte into the mobile phase that now has very little analyte in it. As this process continues along the stationary phase, the analyte peak naturally spreads out. The shape of the peak is Gaussian, so the peak width is expressed in units of the standard deviation σ, as shown in Figure 11.14. The width of the peak may be expressed in length units while the component is still on the column or in time units as the component exits the column and enters the detector.

The width of the peak has been modeled mathematically in terms of the equivalent number N of equilibration sections over which the analyte has passed:

$$N = \frac{t_R^2}{\sigma_t^2} = \frac{l_R^2}{\sigma_l^2} \qquad 11.29$$

where σ_t is the standard deviation of peak width in time. If the peak width is measured in length, l is the distance the peak has traveled, and σ_l is the standard deviation of peak width in length. The nomenclature of multistage distillation was borrowed, so N

Analytical Scheme for Chromatography

Distinguishing characteristic:
Ability to partition between the stationary and mobile phases.

Probe:
Introduce analytes into the mobile phase that then flows past a length of stationary phase or at one spot on the stationary phase prior to the application of the mobile phase.

Response to the probe:
The velocity of the peak concentration of analyte along the chromatographic path is related to its partition constant between the stationary and mobile phases.

Measurement of response:
Stop the elution after a fixed time and locate the analyte peak concentrations on the stationary phase or continue the elution measuring the analyte concentration at the end of the stationary phase as a function of elution time. These measurements require a detector that is responsive to some differentiating characteristic of the analyte.

Interpretation of data:
Relate the area under the peak in a plot of the detector response versus distance or time and relate that to the amount of analyte using an internal standard or a working curve.

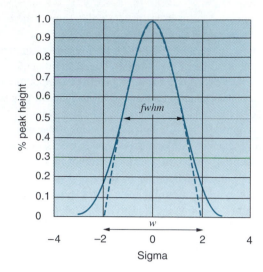

Figure 11.14. The concentration of the analyte as a function of distance or time roughly follows the Gaussian error curve.

is called the number of **theoretical plates.** A measure of the efficiency of a chromatographic column is the number of theoretical plates it has per length of column. This value is called the **height-equivalent theoretical plate,** HETP, or just H. It is found by dividing the column length L by N.

$$H = \frac{L}{N} = \frac{\sigma_t^2 L}{t_R^2} = \frac{\sigma_l^2 L}{l_R^2} \qquad 11.30$$

With Equations 11.29 and 11.30, it is possible to determine H or N by measuring the width and retention time or length of any peak. To avoid having to determine the value of σ, relationships between σ and the width of the peak at its base and at its half-height have been developed. They are

$$\sigma = \frac{w}{4} = \frac{fwhm}{2.355} \qquad 11.31$$

Example 11.4

Calculation of *H* and *N* from peak width

A chromatogram has a peak with a retention time of 5.0 min. The time of the unretained peak is 0.7 min. Its peak width (*fwhm*) as it elutes is 10 s. The column length is 30 m. What are the values of H and N for this column under these conditions? What is the value of k' for the compound?

From Equation 11.32,

$$N = \frac{5.54(5.0 \times 60)^2}{10^2} = 5.0 \times 10^3 \text{ plates}$$

$$H = \frac{L}{N} = \frac{30 \times 100}{5.0 \times 10^3} = 0.60 \text{ cm}$$

From Equation 11.28,

$$k_A' = \frac{5.0 - 0.7}{0.7} - 1 = 6.1$$

where w is the width of the peak at its base and *fwhm* is the full width at half-maximum. If the peak width is measured in time units, the standard deviation calculated is σ_t. The width is determined from the peak shape as shown in Figure 11.14.

For time-based chromatograms, expressions for N and H in terms of the retention time and the peak width can be derived by combining Equation 11.31 with Equations 11.29 and 11.30,

$$N = \frac{16t_R^2}{w_t^2} = \frac{5.546t_R^2}{(fwhm)_t^2} \qquad 11.32$$

$$H = \frac{w_t^2 L}{16t_R^2} = \frac{(fwhm)_t^2 L}{5.546t_R^2} \qquad 11.33$$

The values for w_t and $(fwhm)_t$ are measured in the same time units used to measure t_R.

The peak width in length and time are related through the velocity of the component.

$$w_{t,A} = w_{l,A}/\bar{v}_A \qquad 11.34$$

When Equation 11.26 is used for the velocity of the component in Equation 11.34, we see that the relationship between the peak width in time and the peak width in length depends on the capacity factor of the component.

$$w_{t,A} = \frac{w_{l,A}(1 + k_A')}{u} \qquad 11.35$$

It is desirable for the peak width to be as small as possible because a narrow peak width means that the peaks of species differing only slightly in k' will be less likely to overlap. A measure of the ability to resolve two components A and B is the resolution, R_s.

$$R_s = \frac{t_{R,A} - t_{R,B}}{(w_{t,A} + w_{t,B})/2} = \frac{\Delta t}{w_t} \qquad 11.36$$

With a resolution of 1, the adjacent peaks would overlap at their 2σ points, and the separation would not be complete. As we can see from the plot of the Gaussian distribution in Figure 11.14, it would be desirable for Δt to be at least 6σ. Since w is 4σ, R_s should be at least 1.5 for good separation.

The values for H and N obtained from these calculations are approximate because the model used to obtain the equations has some assumptions that are not entirely valid in the experimental situation. For instance, the system is dynamic so that true equilibrium is rarely reached under normal operating conditions. Furthermore, different peaks in the same chromatogram are likely to give somewhat different values for H and N. Nevertheless, these are useful concepts and measurements when developing new columns and when finding the optimum conditions for a chromatographic experiment.

EFFECT OF MOBILE PHASE FLOW RATE

The choice of flow rate for a chromatographic experiment is quite important. If the flow rate is too slow, there is a tendency for the peak to spread out owing to diffusion of the component from the higher concentrations at the center of its location to the lower concentrations on either side. If the flow rate is too fast, it is difficult to achieve equilibrium partition between the two phases. There must be time for the molecules in the mobile phase to get to the surface where the interphase transition can occur before the portion of the mobile phase has moved on. The effect of flow rate on separation effectiveness is determined by measuring the peak width and retention time, calculating H, and plotting H versus flow rate.

A typical plot of H versus flow rate is shown in Figure 11.15. As expected, there is a flow rate for which H is a minimum since there are factors that cause it to increase at flow rates that are too high or too low. A great deal of effort has gone into studies to understand each of the factors that lead to an increase in H (also sometimes called **band broadening**). The principal factors are given in an equation known as the **van Deemter equation**.

$$H = A + \frac{B}{u} + Cu \qquad 11.36$$

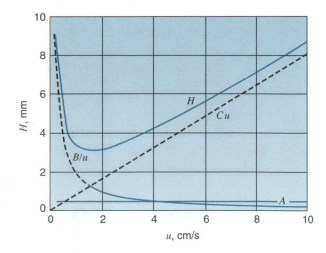

Figure 11.15. The effective value for the "height" of a theoretical plate is a function of the flow rate of the mobile phase. At lower flow rates, diffusion along the column causes H to increase, while at higher flow rates, the increase is caused by a decreasing ability to achieve equilibrium at each plate.

Each of the terms in this equation corresponds to a different mechanism of band broadening. The A term is related to **eddy diffusion.** This is the extent of turbulence in the flow that exists in the mobile phase. It is approximately independent of u. The B/u term is called the **longitudinal diffusion** term. This term describes the effect of the diffusion of analyte in the mobile phase due to the concentration gradients on either side of the peak concentration. At the lowest flow rates, the analyte motion due to diffusion can become significant with respect to the mobile phase velocity. The Cu term derives from the kinetics of the transfer of the analyte to and across the interface between the mobile and stationary phases. It is called the **kinetic transfer** term. Some time is required for the concentrations in each section of the solution to approach their equilibrium values. As the flow rate increases, there is less chance for that to occur, and the theoretical plate height increases.

The steady improvement in the performance of chromatographic separation is due to continued efforts to understand the factors that contribute to band broadening. For example, the quantity C can be reduced by a reduction in the cross-sectional dimension of the mobile phase so that the analyte does not have so far to diffuse. This reduces the increase in H that occurs at higher flow rates and allows faster separations without serious loss of resolution. This understanding was followed by the development of columns with smaller diameters. Making a plot of H versus the flow rate for a given experimental system allows the operator to choose the operating conditions that will provide the maximum resolution.

The discussion in this section applies generally to all types of chromatography. Current chromatographic science has many widely differing modes of implementation, and new ones are being developed regularly. The common element is the partition of solutes between a mobile phase and a stationary phase. The mobile phase is necessarily fluid, so it must be a gas, a liquid, or a supercritical fluid (one which is above its critical temperature and pressure). The stationary phase may be a nonvolatile or immiscible liquid or an adsorbant. The support for the stationary phase may be a metal or glass tube, a piece of paper, or a thin layer of porous solid. The force that causes the mobile phase to move can be a pump, gas pressure, gravity, capillary action, or electrophoretic impulse. The two implementations of column chromatography that are considered in this chapter are gas chromatography (in which the stationary phase may be a nonvolatile liquid or adsorptive surface) and liquid chromatography (in which two immiscible liquids make up the stationary and mobile phases).

Study Questions, Section D

22. How is it that moving one of the phases in the partition converts a difference in partition coefficient to a difference in velocity?

23. What is the capacity factor, and how is it related to the partition coefficient?

24. What is the relationship between the capacity factor and the fraction of the mobile phase velocity exhibited by the portioning component?

25. How is the retention time of a component related to the capacity factor?

26. In a chromatographic separation, the mobile phase velocity is 4.0 cm/s. For a length of stationary phase of 25 m, calculate t_M and the retention time of components with k' values of 1.3 and 9.0.

27. A. If a component in Question 26 has a retention time of 3,782 s, what is its k' value?

B. If the mobile phase volume is 5.3 times that of the stationary phase, what is the value of K'_p for the component?

28. What is a height-equivalent theoretical plate, and how is it related to band broadening?

29. Calculate the values of N and H for a separation in which a peak has a retention time of 923 and a *fwhm* of 3.5 s. The length of the stationary phase is 10 m.

30. Calculate the number of theoretical plates a separation system would need to resolve ($R_s = 1.5$) two peaks whose retention times were 824 s and 829 s.

31. Describe the three principal factors in the van Deemter equation for band broadening.

32. Which of the band broadening effects is worst at low flow rates, and which is worst at higher flow rates?

22. Molecules of the species with the higher partition coefficient will spend a larger fraction of their time in the stationary phase and consequently spend less time being moved along with the mobile phase. Thus, the fraction of the time spent in the mobile phase affects the average velocity of the compound relative to the stationary phase.

23. The capacity factor k' is the partition coefficient multiplied by the ratio of the volumes of the stationary and mobile phases. It is also the ratio of the elution time after that of the unretained components to the unretained elution time.

24. The fraction of the mobile phase velocity achieved by the component is $1/(1 + k')$, where k' is the capacity factor. The higher the value of k', the smaller is the fraction.

25. The retention time of a component is equal to the capacity factor multiplied by the time of the unretained peak.

26. The mobility of unretained components can be calculated from $u = L/t_M$:

$$t_M = \frac{25 \text{ m}}{4.0 \text{ cm s}^{-1}} \frac{100 \text{ cm}}{\text{m}} = 625 \text{ s}$$

The velocities of the components can be calculated from Equation 11.27.

$$\bar{v}_A = u \frac{1}{1 + k'} = 4.0 \frac{1}{1 + 1.3} = 1.8 \text{ cm s}^{-1}$$

$$\bar{v}_B = 4.0 \frac{1}{1 + 9} = 0.40 \text{ cm s}^{-1}$$

Their retention times are then

$$t_{R,A} = \frac{2500 \text{ cm}}{1.8 \text{ cm s}^{-1}} = 1.4 \times 10^3 \text{ s}$$

$$t_{R,B} = \frac{2500 \text{ cm}}{0.40 \text{ cm s}^{-1}} = 6.3 \times 10^3 \text{ s}$$

27. A. From Equation 11.28,

$$k'_A = \frac{t_{R,A} - t_M}{t_M} = \frac{3,782 - 625}{625} = 5.05$$

B. From Equation 11.26,

$$u \frac{1}{1 + K'_p \dfrac{V_2}{V_1}} = u \frac{1}{1 + k'}$$

$$k' = K'_p \frac{V_2}{V_1}$$

$$K'_p = k' \frac{V_1}{V_2} = 5.05 \frac{5.3}{1} = 27$$

28. The height-equivalent theoretical plate is related to the theoretical plate in distillation. It is the length in the direction of mobile phase flow that represents an equilibration of the analyte between the two phases. The more theoretical plates in the column, the narrower will be the peaks relative to the total retention time.

29. From Equations 11.32 and 11.30,

$$N = \frac{5.54 t_R^2}{(fwhm)_t^2} = \frac{5.54(923)^2}{(3.5)^2} = 3.9 \times 10^5$$

$$H = \frac{L}{N} = \frac{1,000 \text{ cm}}{3.9 \times 10^5} = 2.6 \times 10^{-3} \text{ cm}$$

30. From Equation 11.36, w would have to be $\Delta t/R_s = 5/1.5 = 3.3$ s. From Equation 11.32, and using the average of the retention times,

$$N = \frac{16 t_R^2}{w_t^2} = \frac{16(826)^2}{(3.33)^2} = 984,000$$

31. Longitudinal diffusion (B/u) is caused by the concentration gradient within the peak. It worsens the longer the component is in the system and therefore is decreased with increasing flow rate.

   Eddy diffusion (A) is due to the possibility of multiple paths for the mobile phase along the separation axis. Flow rate has little effect on this term.

   Kinetic transfer (Cu) relates to the time required for each plate to reach the equilibrium concentrations in the stationary and mobile phases. As the flow rate increases, less time is available for this process, and the separation efficiency decreases.

32. As seen in Figure 11.15, at low flow rates, the longitudinal diffusion term dominates. At higher flow rates, it is the kinetic transfer term.

E. Gas Chromatography

The term **gas chromatography** is applied to all chromatographs in which the mobile phase is a gas. The partitioning occurs between the gas and either a solid surface or a thin layer of liquid coated on a solid surface. In the latter case, the technique is sometimes called **gas–liquid chromatography.** The components of a gas chromatograph are the source of the mobile phase (called the **carrier gas**), a means of measuring the

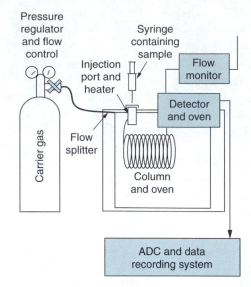

Figure 11.16. Elements of a chromatographic instrument. The column is contained in an oven for accurate temperature control. Methods of sample injection and component detection are included.

gas flow rate, a method for injecting the sample into the flow stream, the column through which the analytes move according to their values of k', a detector to indicate the amount of analyte in the carrier gas exiting the column, and a recorder from which information regarding retention time and analyte quantity can be obtained. A block diagram of the complete apparatus is shown in Figure 11.16.

THE INSTRUMENT

In modern gas chromatographs, an **open tubular capillary column,** generally made of fused silica (glass), is used. The inner diameter (i.d.) may be between 0.1 and 0.5 mm. The inner surface of the glass is lined with the stationary phase or with a deactivator. The earliest columns were made of stainless steel and had much larger inner diameters, sometimes 0.25 in. These tubes were filled with solid particles covered with a thin layer of the stationary phase. These columns were called **packed columns.** The flow volumes were necessarily much greater than with the capillary columns. Also, owing to the variety of paths available for the carrier gas through the particles, the A term in the van Deemter equation was large and the resolution correspondingly poor.

The chromatograms in Figure 11.17 illustrate the improved resolution obtained with modern capillary columns. The sample was peppermint oil in each case. The chromatogram in Figure 11.17a was with a 6-foot, 0.25-in. packed column. A 0.03-in.-i.d. stainless steel capillary, 500 feet long, was used for the chromatogram in Figure 11.17b. The column used for the chromatogram in Figure 11.7c was a 50-m, 0.25-mm-i.d. glass capillary column. In each case, the stationary phase was the same. The improvement in resolution is evident with each decrease in capillary diameter and increase in capillary length.

The carrier gas used is generally H_2 or He. After pressure regulation and flow control, the carrier gas is split into two paths. One path goes through the column; the other bypasses the sample injection port and column but does go through the oven. Some detectors operate on the difference in response of the carrier gas with and without analyte present. The gas in the column path first goes through the injection port volume. Here it will pick up vaporized sample that has been injected into this volume. The injection volume is heated so that sample in a liquid solvent will be completely volatilized for inclusion in the carrier gas. A major limitation of gas chromatography is that the analyte molecules must be able to be vaporized. If, on heating, they decompose rather than vaporize, they cannot be analyzed or separated directly by gas chromatography.

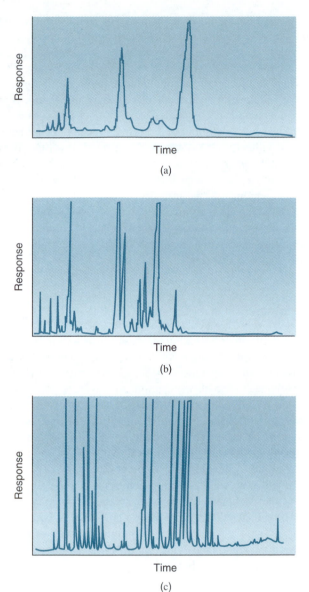

Figure 11.17. Representative chromatograms of the same sample taken with columns that have different resolutions. Many more compounds are seen as the resolution increases from (*a*) to (*c*). (From W. Jennings, *J. Chromatogr. Sci.* **1979,** *17,* 637, with permission.)

The temperature of the column is carefully controlled by the column oven. This is because the value of k' for all the components is greatly affected by temperature. Controlling the temperature allows k' to be adjusted to a value favorable for chromatographic separation.

SAMPLE INJECTION

Samples are generally injected into a gas chromatograph by insertion of a syringe containing the sample through a rubber disc called a **septum** and then quickly emptying the syringe into the injection volume. The sample in the syringe can be either gas or liquid. If it is a liquid, the solvent containing the analyte should be volatile and the injection port temperature should be well above the boiling point of the solvent. The solvent is thus converted to gas and enters the column along with the analyte. It is desirable that the solvent gas molecules have a k' on the column that is much less than 1, so they will come out first and not interfere with the other, much less abundant components of the sample.

The time required for the injected sample to be swept from the injection volume onto the column can be a limiting factor on peak width. Traps based on specific adsorption or freezing have been used to concentrate samples prior to chromatographic analysis. In these, large volumes of sample may be passed through the trap, which is then suddenly heated to release the accumulated analyte into the gas stream and onto the column. This sampling approach is called the **purge and trap** method.

A solid phase extractor is sometimes used in a syringe container. With this approach, a stiff fiber is coated with the solid phase extractant. This fiber can be exposed at the syringe tip or withdrawn into the syringe barrel. The syringe is inserted into the sample container and the fiber exposed to the sample. Analyte is adsorbed on the solid phase extractant, which is then withdrawn into the syringe barrel. Next the syringe is inserted through the septum into the injection volume of the gas chromatograph and the fiber exposed to the injection volume. The hot gas within the injection volume vaporizes the sample components and carries them onto the column. This is called **solid phase microextraction.** The processes of extraction, concentration, and injection are all completed in a single, simple step.

THE COLUMN

Capillary columns come with a variety of stationary phases. They can provide a surface for adsorption partition or a thin liquid phase for gas–liquid partition. The former is called a **porous layer open tubular** column or PLOT column. These have particular application for highly volatile species such as the components of air and very light organics such as methane. Columns with liquid coatings are called **wall-coated open tubular** columns or WCOT columns. The coatings used vary in the highest temperature at which they are stable and in their chemical characteristics. Regardless of the coating used, the principal factor governing the k' value of any analyte is the temperature of the column. This is because the principal factor governing the gas–liquid partition is the relationship of the temperature to the boiling point of the analyte. However, when compounds with very similar boiling points need to be separated, a secondary distinguishing characteristic may be used to advantage. For example, a somewhat polar liquid phase may be selected. This will increase the k' for polar analytes with respect to nonpolar analytes with the same boiling point. A good example is the use of β-cyclodextrin, a chiral compound with complexing capabilities, on the stationary phase. With such a column, optically active isomers or enantiomers can be separated. This capability has been of great value in the study of biological chemistry and in the preparation of optically active drugs.

Bare capillary columns are unsuitable for gas chromatography. The relatively high adsorption energies involved result in k' values that are impracticably high, and the slowness of the analyte to release from the surface results in chromatographic peaks with serious tailing (a stretching of the peak in the direction of increasing time). All bare surfaces in the chromatograph must be deactivated before use. A variety of compounds and techniques are used for this purpose, all of which have resulted in surfaces that are greatly reduced in adsorption activity.

THE DETECTOR

The output of the column goes directly into the detector assembly. Several types of detectors for gas chromatography have been developed. Following the theme of this book, they are based on some characteristic of the analyte molecules that is not shared by the carrier gas. The three types of characteristics that have been used extensively in detectors are thermal conductivity, ionizability, and electron capture. Quite likely, other possibilities await development. The **thermal conductivity detector** shown in Figure 11.18 is based on the extraordinarily high thermal conductivity of He gas. Thermal conductivity of a gas is measured by the use of a thermistor.

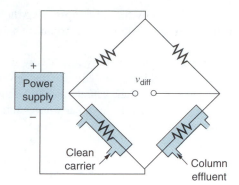

Figure 11.18. Arranging two detectors in a Wheatstone bridge allows the output of the detector in the column stream to be compared with detector that is exposed only to the pure mobile phase. The difference between them is then due principally to the effect of the analyte on the detector.

The electrical resistance of a thermistor is strongly dependent on temperature. The circuit used to measure its resistance causes some current through the thermistor that, in turn, results in the dissipation of some power in the thermistor. The temperature of the thermistor will then be slightly greater than that of the carrier gas around it. How much greater depends on the thermal conductivity of the gas. Helium is used as the carrier gas so that, when an analyte is present, the thermal conductivity of the carrier is reduced and the temperature of the thermistor goes up. A measurement of the difference in the resistance between a thermistor in the pure helium flow stream and that in the column effluent provides a response to the presence of analyte. The circuit that enables this difference measurement is the venerable Wheatstone bridge. Its output voltage v_{diff} is related to the difference in the resistances in the two flow paths.

The thermal conductivity detector is based on a characteristic of the carrier gas that is diminished by the presence of the analyte. For maximum sensitivity, it is better to test for an analyte characteristic that is not shared by the carrier. One such characteristic is ionizability. When substantial energy is absorbed by most organic molecules, they will fragment or react to produce positive ions. Several energy deposition devices have been used, including impact by energetic electrons and ultraviolet photons. A very simple and highly effective method is the combustion of the molecules in a small, controlled hydrogen/air flame. The carbon in the organic molecules reacts to form CHO^+ ions plus an equal number of electrons. This process gives the detector its name, **flame ionization detector** or FID. As shown in Figure 11.19, the positive ions are collected by the negative charge on the collector elec-

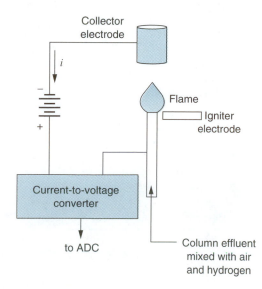

Figure 11.19. The flame ionization detector, shown here, is a very popular gas chromatography detector. Its sensitivity (response per molecule of analyte) is proportional to the number of carbon atoms in the analyte molecule.

trode, and the electrons are collected by the tube supporting the flame. The resulting current is proportional to the concentration of the organic analyte molecules in the effluent at that moment.

The FID is marked by very high sensitivity and a very wide dynamic range. In the chromatograms shown in Figure 11.17, you will see that several of the peaks have exceeded the vertical scale in the graph; their actual amplitudes are off scale. If the detector has sufficient dynamic range, this scale expansion enables the observation of very minor components. The dynamic range of the FID has been reported to be as large as 10^6.

The electron-capture detector uses a natural emitter of beta rays (high-energy electrons) that ionize the carrier gas and produce a small, steady current between two polarized electrodes (similar to those in the FID). When analyte molecules that have a high electron affinity are present, this process is diminished and the current drops. The electron-capture detector is among the most sensitive chromatographic detectors available, but it is only responsive to halogenated analyte molecules.

THE GENERAL ELUTION PROBLEM

The **general elution problem** in chromatography is that it is often impossible to find a single set of conditions for which the values of k' for all the analytes fall within the most desired range of 1–5. This problem is illustrated with the chromatograms in Figure 11.20. The sample is gasoline, a complex mixture of hydrocarbons. When the column temperature is set at 100 °C as in Figure 11.20a, the most volatile components all come off very early and are too crowded together to be distinguished. Setting the temperature lower helps this problem (see Figure 11.20c), but after an hour and a half, the least volatile compounds still have not eluted. You will also notice that the late eluting peaks are very broad. This is because their velocity along the column is so slow that it takes them a long time to exit the column.

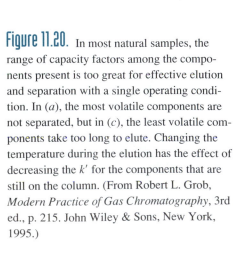

Figure 11.20. In most natural samples, the range of capacity factors among the components present is too great for effective elution and separation with a single operating condition. In (a), the most volatile components are not separated, but in (c), the least volatile components take too long to elute. Changing the temperature during the elution has the effect of decreasing the k' for the components that are still on the column. (From Robert L. Grob, *Modern Practice of Gas Chromatography*, 3rd ed., p. 215. John Wiley & Sons, New York, 1995.)

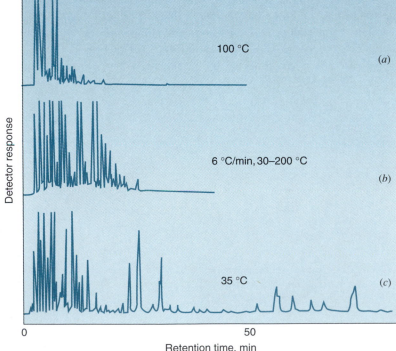

The answer to the general elution problem is to change the conditions during the elution so that the k' value for compounds that have not yet eluted is reduced. We can reduce k' in gas chromatography by raising the temperature. To accomplish this, the column oven temperature can be programmed to follow a specified temperature profile over the course of the experiment. This is called **programmed temperature elution.** Separation at a fixed temperature is called **isocratic elution.** Some chromatographs are capable of following quite complex temperature profiles, and the development of the optimum profile is a skill acquired by experienced chromatographers.

The chromatogram in Figure 11.20b was performed with a simple temperature profile, namely, a linear ramp with a constant 6-degree per minute change beginning at 30 °C and ending at 200 °C. An advantage of starting the column at a relatively cool temperature is that all the sample components flushed out of the inlet port by the carrier gas are stopped at the first bit of the column because the velocity of all of them is very low at that temperature. This has the effect of focusing all the sample in one place at the beginning of the elution. Another advantage of ramping the temperature is that the later eluting components have relatively low values of k' at the time they are coming off the column. This keeps the peak widths narrow for all components. A disadvantage is that it is necessary to allow time for the oven to cool to the starting temperature before the next chromatogram can be run.

PEAK OVERLAP

An interesting point can be gleaned from the example chromatograms shown in Figure 11.17. As the resolution of the chromatography improves, more components are seen. Clearly, in the more poorly resolved chromatograms, each apparent peak contains many compounds. In such a complex sample, how do we know how much resolution is required to resolve all the components? In a landmark paper, Cal Giddings and Joe Davis[2] applied Poisson statistics to the appearance of peaks in a chromatographic experiment. They reasoned that the retention time of a given peak was a random occurrence in that in an uncontrived sample, the k' values of each component are independent of the others. The answer is that to be 90% sure that a given peak represents a single analyte, *the chromatogram would have to be 95% baseline;* peaks would be present only 5% of the elution time. For the chromatograms illustrated in this section, the probability is very high that almost all of the peaks contain two, three, or more components. This startling but irrefutable finding has significant implications in the interpretation of the chromatograms produced.

It has been proved that virtually all natural samples have many more components than can fit in one chromatogram. Even when a peak appears to result from a single component, coelutants that have a much lower abundance are highly likely. Peak overlap can be reduced by more selective detection and by better chromatographic resolution.

INFORMATION OBTAINED FROM GAS CHROMATOGRAMS

The data produced from a gas chromatograph is a plot or record of the detector intensity versus time. Information is contained in the area and arrival time of the peaks. The peak area is related to the amount of the eluting compound that was injected, and the retention time is related to the k' of the compound, the column type, the flow rate, and the temperature. To obtain quantitative information from the peak area, one must determine the sensitivity factor for the detector. A working curve can be constructed by injecting increasing amounts of analyte and plotting the peak area as a function of analyte quantity.

The sensitivity (the slope of the working curve) is a function of the specific compound for all detector types. Therefore, the sensitivity determined for one compound is not usable for another. In addition, the sensitivity may or may not be dependent on the

Chromatographic detectors are generally either concentration or mass flow sensitive. For the mass flow sensitive detector, the area under the peak is proportional to the amount of analyte and is independent of flow rate. For the concentration sensitive detector, the peak amplitude is independent of flow rate, but the peak area increases with decreasing flow rate.

---

[2]J. Davis and C. Giddings, *Anal. Chem.* **1983,** 55, 418–424.

flow rate. If the detector is sensitive to the concentration of the analyte in the detector volume (as in the case of the thermal conductivity detector), the area will increase as the flow rate is decreased because the peak will get broader but remain the same amplitude. If the detector is destructive (as in the case of the FID), a decrease in flow rate reduces the rate of delivery of the analyte to the detector and thus lowers the peak height. The peak also broadens, so the area remains constant. Such a detector is said to be **mass flow sensitive.**

Quantitation in chromatography is most often accomplished with an internal standard against which the analyte response is compared. Peak area is the most reliable form of response for quantitation, but determination of the areas of peaks that are partially overlapped can lead to significant error.

The reproducibility of duplicate runs done sequentially can be within a few percent. This limits the precision of quantitation using a working curve. Better quantitation can be obtained if an **internal standard** of a very similar compound is added to the original sample and carried through all the preparative steps. The peak areas of the analyte and internal standard are then compared, the sensitivity ratio between them having been previously determined.

Identification in chromatography is frequently done by comparing the retention time of the analyte with that for known and expected compounds. Since the number of compounds that could have that same retention time is huge, this is not a positive identification in most cases involving natural samples. For confirmation, spectrometry of the separated components is highly desirable.

It is very tempting to use the retention time as a means of identifying analytes. When a particular compound has been observed to appear at a particular retention time repeatedly, it is natural to assume that a peak at that position is due to the known compound. This assumption may be justified when the number of sample components is limited and when the full range of possible sample compositions is well characterized. An example of this might be a manufacturing process where the expected components are known. There is a great danger in compound identification by retention time alone for natural samples that could contain any of thousands of components. In that case, all we learn from the detector is that the concentration of something at that retention time was significant. Many attempts have been made to use relative retention time as a fundamental quality, but they are incompatible with the concurrent use of temperature programming and so are little used. The solution to the problem of identification is to choose a detector that has greater characterizing power, such as ultraviolet spectrometry or mass spectrometry.

Study Questions, Section E

33. Comment on the usefulness of a table of boiling points in predicting the elution order of compounds separated by gas chromatography.

34. In chromatography, why is the differentiating characteristic used for the separation different from that used for detection?

35. What is the differentiating characteristic used in the flame ionization detector?

36. What is the general elution problem, and what is the most desirable range for k'?

37. What is the solution to the general elution problem for gas chromatography?

38. Explain the appearance of new peaks in the chromatograms of Figure 11.17 as the resolution increases, using Giddings' and Davis' theory of peak overlap.

39. Why is peak area better related to analyte amount than peak height for the flame ionization detector?

40. Why is an internal standard so often used with gas chromatography?

Answers to Study Questions, Section E

33. Boiling point temperature is related to volatility, and volatility is related to the partition of the analytes between the mobile phase and the stationary phase. The higher the volatility (the lower the boiling point), the greater is the fraction of time in the mobile phase and the shorter is the retention time. Therefore, the elution order should be in approximate order of the boiling points, with the lowest boiling compounds eluting first.

34. The differentiating characteristic used for separation is the degree of portioning between the stationary and mobile

phases. This has not yet provided a good characteristic for detection. For detection, some physical or chemical property of the analytes must be exploited.

35. The differentiating characteristic used in the flame ionization detector is the ability to form an ion when exposed to a flame. This is a chemical reactivity type of differentiating characteristic, which will not be shared by all possible analytes.

36. The general elution problem is that compounds with a value of k' over 5 have a very slow velocity. They take a

long time to elute and have broad peaks when they do. Compounds with a k' of less than 1 have a velocity only a little slower than the mobile phase. Little differentiation is achieved among compounds with these k' values. Most samples have compounds that, for a given column temperature, fall outside the ideal range for k', which is between 1 and 5.

37. Since the value of k' for a compound decreases with increasing temperature, the solution to the problem of a wide range of k' values is to change the column temperature as the chromatographic separation proceeds.

38. The chromatogram of Figure 11.17a has very little baseline. Therefore, we could assume that there is considerable overlap of coeluting compounds throughout the chromatogram. This is confirmed by the chromatogram of Figure 11.17b, taken at higher chromatographic resolution. It, too, has little baseline so more peaks should be revealed with still higher resolution (as seen in Figure 11.17c). Since the last chromatogram is still not mostly baseline, we can assume that this chromatogram is not completely resolved, either.

39. The total number of ions produced in the detector by an analyte is the best measure of its amount. The measurement most closely related to the total ions is the area under the current–time peak.

40. The use of an internal standard cancels out many variables in the detector response.

F. High-Performance Liquid Chromatography

In liquid chromatography, the mobile phase is a liquid. The stationary phase may be either an immiscible liquid or a solid. In liquid–liquid chromatography, partitioning occurs between the two liquids. Generally, one of the liquids is organic (nonpolar) and the other is aqueous (polar). In liquid–solid chromatography, the partitioning process is adsorption. In the first implementations of liquid chromatography (LC), the stationary phase consisted of solid particles, which could be coated with a layer of immiscible liquid if liquid–liquid partition was desired. The particles were packed into a tube, which had a stopcock at the bottom and was open at the top. (See Figure 11.21.) The mobile phase liquid was dripped into the open top of the column and collected under the stopcock at the bottom. The number of theoretical plates that could be achieved by this system was modest, but it offered a great improvement over the batch extraction techniques.

It was soon noticed that the HETP decreased with the size of the solid particles due to a reduction in the eddy diffusion and kinetic transfer terms. Unfortunately, the rate of flow of the mobile phase, impelled only by gravity, decreased as well. The solution to this problem was to force the mobile phase through the column with a pump. This began a cyclic development process in which pumps capable of higher pressures encouraged the use of smaller stationary phase particles and vice versa. This cycle is still ongoing. For this reason there is some confusion over whether the HP in the acronym HPLC refers to "high performance" or "high pressure."

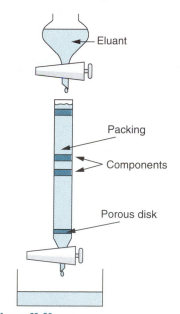

Figure 11.21. Sketch of the original liquid chromatographic apparatus. The stationary phase was adsorbed on the inert packing. Colored components could be seen, hence the name "chromatography."

THE HPLC SYSTEM

In the present development of HPLC, the stationary phase is tiny silica beads 3 to 10 μm in diameter. A short column (5–30 cm) is packed with such beads, whose surface may be coated or treated according to the type of separation mechanism desired. As shown in Figure 11.22, the liquid phase solvents are drawn by the pump from one or more reservoirs. Between the pump and the column are the sample injection mechanism and a precolumn filter (sometimes called a **guard column**). The detector and data recorder follow the column. The pump is a special device, as pressures as high as 15 MPa are often needed.

The point at which the sample must be introduced is then also at this very high pressure, so the simple syringe injection used in gas chromatography is thus not practi-

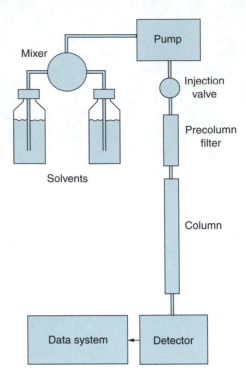

Figure 11.22. Diagram of the HPLC apparatus. A high pressure pump is required to force the mobile phase through the very fine particles in the column.

cal. Instead, a loop of tubing must first be filled with the sample and this loop then valved into the solvent stream. The valve that accomplishes this task is shown in cross section in Figure 11.23. In the injection position, the sample is introduced into the sample loop through the central opening (port 1). Sample flows through the loop and again through the valve (in port 5 and out port 4) and then to waste until the sample loop is filled. When the injector valve is switched to the injection/run position, the solvent from the pump goes through the sample loop, back through the valve (in port 2 and out port 7), and on to the column and detector. The valve remains in this state throughout the run.

The detectors used for liquid chromatography are based on a differentiating characteristic shared by the components of interest. They include light absorbance, fluorescence (see Chapter 6), refractive index, coulometry and other electrochemical techniques, and mass spectrometry. Important characteristics of the detector are a volume no larger than the mobile phase volume in a single theoretical plate (to avoid loss of resolution) and sufficient sensitivity to respond to the very small amounts of analyte eluting from the column.

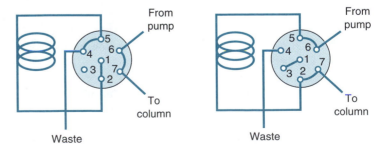

Figure 11.23. Detail of the injection valve. A loop of tubing is first filled with sample and then switched into the flow of mobile phase from the pump.

With such a system, a wide variety of separations are possible. The various techniques are categorized by the type of reaction the analytes undergo between the stationary and mobile phases as discussed in the sections below.

PARTITION CHROMATOGRAPHY

The term **partition chromatography** is generally used for chromatography based on the partition of the analytes between two liquid phases. In this context, partition has a meaning much narrower than we have been using until now. To minimize kinetic transfer broadening, the thickness of the liquid coating on the stationary phase beads and the space between the beads must be as small as possible. Early systems involved the adsorption of the immobile liquid phase on the silica bead surface. This resulted in nonuniform coating thickness and poor adhesion of the coating to the beads. Column deterioration with time was a major concern. Chromatographers have since learned to chemically bond the desired molecular environment to the surface atoms of the silica beads. The reaction that provides this custom surface is shown in the margin. The composition of R determines the solubilizing nature of the stationary phase, now called a **bonded phase.** Because of their greater stability, bonded phases are used almost exclusively.

The R group in a bonded phase may be either relatively nonpolar (saturated or aromatic hydrocarbons) or relatively polar (amino, cyano, or alcohol termini). Nonpolar stationary phases are used with more polar mobile phases and vice versa. In the first implementation of liquid–liquid chromatography, the more polar phase, was the stationary phase, and the relatively nonpolar phase was the mobile phase. This was considered "normal." Thus, when a column packing was devised from a low-polarity phase and the mobile phase was the more polar, the new type of chromatography was called **reverse phase chromatography.** These completely nondescriptive names (normal phase and reverse phase) of the two liquid modes continue to be used.

When **normal phase chromatography** is used, the more polar solutes have the greatest affinity for the polar stationary phase. Since these components spend the least time in the mobile phase, they elute last. From this, we can see that the differentiating characteristic that is probed in partition chromatography is the polarity of the analyte. Compounds will come off the column in order of increasing polarity with normal phase operation. Table 11.2 lists rough orders of polarity for various organic functional groups.

The general elution problem exists for liquid chromatography, too. Analytes in the same mixture can have a wide range of k' values for any given stationary phase and solvent combination. The solution to the general elution problem with liquid chromatography follows the same idea as for gas chromatography: decrease the values of k' for the analytes while the elution is occurring. In normal phase liquid chromatography, the affinity for the mobile phase is increased by making the mobile phase more polar. Thus, as the elution continues, an increasing amount of a more polar solvent is mixed into the mobile phase. This is called **gradient elution.**

The majority of liquid chromatography performed today is with reverse phase bonded phase columns. With these, the least polar compounds are the more highly retained and come off last. Gradient elution begins with the most polar solvent composition and proceeds with the addition of greater amounts of nonpolar solvent. Common solvents are water (perhaps with a buffer), methanol, acetonitrile, and tetrahydrofuran. Solvents can be listed in order of their ability to dissolve polar compounds. Such a list is shown in Table 11.3 including the **polarity index** P assigned to the solvent. The higher the polarity index of a solvent is, the greater is the solubility of polar compounds in it. To a first approximation, solvent mixtures are assumed to have a polarity index that is the sum of their component polarity indices times their mole fraction.

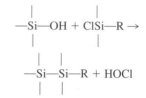

Reaction to form bonded phase

Table 11.2
Functional Groups, in Order of Increasing Polarity

Aliphatic hydrocarbons
Olefins
Aromatic hydrocarbons
Halides
Sulfides
Ethers
Nitro compounds
Esters, aldehydes, ketones
Alcohols, amines
Sulfones
Sulfoxides
Amides
Carboxylic acids
Water

Table 11.3
Solvent Characteristics

Solvent	Polarity Index, P	Elution Strength, ϵ°
Fluoroalkane	<-2	-0.2
Cyclohexane	0.04	0.03
n-Hexane	0.1	0.01
Carbon tetrachloride	1.6	0.11
Diisopropyl ether	2.4	0.22
Toluene	2.4	0.22
Diethyl ether	2.8	0.38
Dichloroethylene	3.1	0.34
Tetrahydrofuran	4.0	0.35
Chloroform	4.1	0.26
Ethanol	4.3	0.68
Dioxane	4.8	0.49
Methanol	5.1	0.73
Acetonitrile	5.8	0.50
Nitromethane	6.0	0.49
Water	10.2	

Study Questions, Section F

41. In reverse phase liquid chromatography, is the polar solvent or the nonpolar solvent the mobile phase?

42. In reverse phase liquid chromatography, will the more polar analytes or the less polar analytes elute first?

43. How is the general elution problem solved in liquid chromatography?

44. What are some of the distinguishing characteristics that have been used for detection in HPLC?

Answers to Study Questions, Section F

41. In reverse phase liquid chromatography, the more polar solvent is the mobile phase.

42. In reverse phase liquid chromatography, the stationary phase is the less polar. Thus, the less polar compounds will be more retained, and the more polar compounds will elute first.

43. The general elution problem is solved by gradually increasing the quality of the solvent that will decrease the k' value of the components still on the column. In the case of reverse phase liquid chromatography, this means gradually decreasing the polarity of the solvent.

44. HPLC detectors have been based on UV absorption, fluorescence, refractive index, and mass spectrometry.

G. Variations on the Chromatographic Theme

The great success of gas and liquid chromatography has given rise to many variations of the basic theme. Some variations are based on the differentiating characteristic used by the reaction of the analyte between the stationary and mobile phases. Other variations are based on the mobilities of the analyte in the mobile phase or on the method of pumping the mobile phase. A few of these variations are briefly described in the paragraphs below.

In **adsorption chromatography,** the stationary phase is a solid surface on which adsorption can take place. Silica gel and alumina were the first to be employed. They both have quite highly polar surfaces, so they serve as the stationary phase for normal phase chromatography. Because of the nonpolar nature of the mobile phase, adsorption chromatography has its greatest application in the separation of species that would have very little solubility in an aqueous phase. Mixtures of nonpolar analytes can be separated quite readily with this approach. The method of adjusting the k' values of the analytes is by adjusting the **eluent strength** $\epsilon°$ of the mobile phase. The values of $\epsilon°$ for various solvents are also given in Table 11.3. A comparison of the values for elution strength and polarity index in Table 11.3 shows that they are closely-related properties.

Ion exchange chromatography, sometimes just called **ion chromatography,** uses an ion exchange resin as the stationary phase. Mixtures of cations can be separated with a cation exchange resin and mixtures of anions with an anion exchange resin. The active surface of the cation exchange resin is a sulfonic acid group $—SO_3^-$ or carboxylic acid group $—COO^-$. These groups are normally protonated, but a cation can displace the proton and thus become part of the stationary phase (temporarily). Analytes elute in the order of their affinity for the resin. The values of k' can be adjusted by changing the pH of the aqueous solvent used.

Capillary electrophoresis is the term used to describe the separation of charged species in a column through which solute is moving because of electroosmotic flow. The mechanism of electroosmotic flow is beyond the scope of this text, but its characteristic is that the flow velocity is absolutely uniform from one edge of the tube to the other. This is in contrast to normal tubular flow, which is greatest at the center and nearly zero at the tube wall. Electroosmotic flow is the result of a combination of specific ion attraction to the tubular surface and a strong electric field along the direction of flow. The field strength is several hundred volts per centimeter. This same field imparts a force on the ions in the solution, with the force on cations being in the direction of the solvent flow and that on the anions being in the upstream direction. The actual velocity of the ions (relative to the mobile phase velocity) will be proportional to their charge over their mobility. The absence of a stationary phase with which to partition and the advantage of an open tubular column in eliminating eddy diffusion reduce the van Deemter equation to just the longitudinal diffusion term. Separations with a measured value of up to 500,000 theoretical plates have been achieved with this technique.

Supercritical fluid chromatography uses a mobile phase material that is above its **critical temperature.** This is the temperature above which the liquid state cannot exist. For liquids at normal temperatures, this temperature is very high, and the pressures at these temperatures are also high. For normally gaseous materials, however, the critical temperature need not be experimentally excessive, and the corresponding pressures can be reasonable. A common example is carbon dioxide. The solubilizing properties of supercritical fluids increase rapidly with increasing pressure. Packed or open tubular columns are used. An advantage of supercritical fluid chromatography over gas chromatography, which it most resembles, is that high temperatures are not required in order to get analytes into the mobile phase. The system must have the ability to automatically regulate the pressure of the supercritical fluid because gradient elution is accomplished by increasing the supercritical fluid pressure.

PLANAR CHROMATOGRAPHY

Thin-layer chromatography is a very common example of planar chromatography. It is called planar chromatography because it takes place on an almost two-dimensional surface. Thin-layer chromatography is most similar to liquid chromatography and enjoys widespread use in biomedical research. The stationary phase is a thin layer of fine particles of the types used in HPLC columns spread uniformly on a glass plate. A drop of sample material is placed near one edge of the plate, or several drops of samples and standards can be placed along the same edge of a wide plate. The solvent from the sample application is dried. The edge of the plate with the sample spots is then immersed in a shallow pool of the mobile phase. The top edge of the pool must be between the bottom edge of the plate and the spots of sample. The mobile phase then moves up the plate by capillary action. This process is illustrated in Figure 11.12. The pool and plate(s) are in an enclosed container so that the mobile phase liquid can saturate the atmosphere with its vapor. This high mobile phase vapor pressure prevents evaporation of the mobile phase as it is moving up the plate. The least retained components of the sample move upward most rapidly. This process, called **plate development,** is stopped before the components of interest have reached the top of the plate.

The location of the sample components on the plate after development is done by some form of indicator or spectroscopy. The indicator solution that reacts with the analyte to form a colored compound may be sprayed on after development. The components may be located by optical spectroscopy scanners or by fluorescence. An advantage of thin-layer chromatography is that the separated components can be removed from the plate by scraping that section into a beaker and dissolving the sample to be used for further experiments. Another advantage is its great simplicity. Disadvantages are that the number of theoretical plates is low, so resolution is poor. Identification is generally done by the application of a standard solution on a parallel track. Spots that appear at the same elution distance are assumed to be the same compound.

H. Mass Spectrometric Detection

The mass spectrometer has become one of the most powerful detectors for chromatography. The mass information it provides is a fundamental property of the compound and is often sufficient for positive identification. In addition, the mass spectrometer can be operated in a mode that is selective for only certain analyte mass values. The capability of variable selectivity enables the operator to make a trade-off between general detection and targeted compound analysis at very low detection levels.

The major components of a mass spectrometer are shown in Figure 11.24. The heart of the mass spectrometer is the mass analyzer. To sort out the sample components by mass, the molecules of the sample must first be converted to ions. This is because all mass analyzer devices operate on the differences in the path that charged particles of different mass will take through electric or magnetic fields. These differences in trajectory allow the mass analyzers to disperse ions according to their mass-to-charge ratio. Thus, strictly speaking, the differentiating characteristic being used in a mass spectrometer is m/z, where m is the exact mass of the ion in atomic mass units (amu) and z is the electronic charge on the ion.

The sample is introduced onto the mass spectrometer inlet, and the molecules of the sample are ionized in the ion source. Using an electric field, the ions are transferred to the mass analyzer, where they are given different paths according to the ion m/z value. The ion detector detects the ions that have traversed a particular path and presents the ion intensity data for that m/z value to the data system. To obtain a complete mass spectrum (ion intensity over a range of m/z values), the ion detector sequentially detects the ions that have traversed a series of trajectories. The data are typically plotted in histogram form as the intensity at each m/z value. Such a spectrum is shown in Figure 11.25.

There are many different types of mass spectrometers owing to the several different types of mass analyzers and ionization mechanisms that have been developed. All types of mass analyzers have been adapted for chromatographic detection, but they differ greatly in the mass range, mass resolution, spectral generation rate, and ion detection efficiency available. Ionization methods can be categorized as either **fragmenting ionization** or **soft ionization.** Electron ionization (EI) bombards the vaporized sample molecules with energetic electrons. An example of an EI mass spectrum is seen

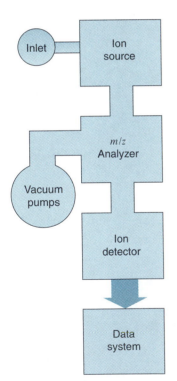

Figure 11.24. Elements of a mass spectrometer. Analyte must first be converted to ions for the mass analyzer to disperse them according to their mass-to-charge ratio. Each m/z value of ions is detected separately and the mass spectrum constructed in the data system.

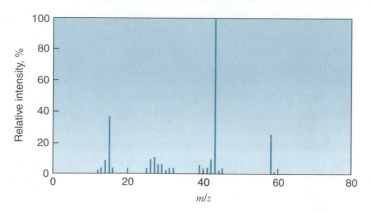

Figure 11.25. Mass spectrum of acetone, CH_3COCH_3. The molecular mass of acetone is 58 amu. The other m/z values in the spectrum result from ion fragments produced in the ionization process.

in Figure 11.25. All the ions in this spectrum have a single charge. The molecular ion is seen at $m/z = 58$. The small peak at $m/z = 59$ is due to the small fraction of acetone ions that contain a ^{13}C atom. The remaining peaks in the spectrum are due to ion fragments produced by the energetic ionization process. The pattern of these fragment masses and intensities is extremely useful for the positive identification of the analyte molecules.

Soft ionization techniques have a much lower production of fragment ions. They frequently involve proton or electron transfer to the analyte molecule. Soft ionization often includes a gain or loss of a H atom, in which case the resulting ion is 1 mass unit different from the analyte molecule. The mass spectra produced by soft ionization are much less complex. They are preferable when the sample contains a mixture of analytes, but they provide little information beyond the molecular mass.

For gas chromatography detection, the principal ionization methods employed are electron ionization (fragmenting) and chemical ionization (soft). The production of vapor phase ions from a liquid matrix has proved a difficult task. Among the most successful of the methods developed is electrospray ionization (ESI), a soft technique. This recent development has sparked a revolution in liquid chromatography–mass spectrometry because of its ability to detect peptides and other large biopolymers as they elute from liquid chromatography and electrophoresis columns.

The combination of mass spectrometry (MS) and chromatography has resulted in some of the most powerful analytical tools ever developed. The discriminating power of the chromatography–MS techniques is a combination of the separating power of the chromatography and the characterizing power of the mass spectrometer. Fortunately, MS is also a very sensitive technique. Detection limits into the femtomole (10^{-15} mol) region are now being achieved routinely for many applications.

I. Multichannel Chromatographic Detection

One of the major advances of the last few decades has been the emergence of multichannel detectors for chromatography. In chromatographic applications, there is an inherent conflict between a desire to see a peak for every compound in the sample and the problem of peak overlap that exists when the sample has more than a few dozen components. If the detector is sensitive to most of the components in the sample (a general detector), the resulting chromatogram will be complex and will almost certainly contain many multicomponent peaks. One solution is to use a more selective detector. An example is the electron-capture detector, which is very selective for halogenated compounds. This detector simply ignores the nonhalogenated components of the sample and thus produces a much simpler chromatogram. The selectivity of this detector (the reduction in the number of interfering compounds) is one of the reasons for its very low detection limit. To take this direction to its extreme, imagine a detector that is specific for just one compound of interest. That compound would be perfectly detected and quantitated even if in the midst of a very complex matrix.

From that example, we see that selectivity, chromatographic simplicity, and reliable quantitation are increased at the expense of decreasing information about the sample composition. The solution to this trade-off is to use a detector that is separately sensitive to more than one quality of the eluting analytes. Each quality to which it is sensitive can be considered a separate channel of the detector. Such a detector would have multiple, simultaneous outputs. In practice, this is achieved by using a spectrometer of some sort as the detector. If it is an optical spectrometer, it could take the entire absorption spectrum of the effluent at frequent intervals along the chromatographic time axis. This could provide data for positive identification, indication of multicomponent peaks, and separate quantitation for components that absorb at wavelengths not shared by coeluting components. To date, optical spectrometers and mass spectrometers have been used for chromatographic detection.

The acquisition of full spectra at frequent intervals presents two challenges. One is that the spectrometer must be capable of producing spectra at the desired rate (10 to 50 spectra per *fwhm*), and the data collection system must be capable of sorting, storing, and presenting all the data thus acquired. With most modern chromatographic methods, the amount of sample is measured in picomoles and less. Scanning spectrometers cannot achieve such a low detection limit at the desired spectral generation rate. Thus, diode array detector spectrometers are used exclusively for this application. Mass spectroscopists are currently working on new generations of instruments that will have the needed capability.

A great advantage of multichannel detection for chromatography is that the response of each detector element or channel can be plotted as a function of time. In effect, this gives as many simultaneous chromatograms as there are channels in the detection system. The division of the sample component's responses among the many channels decreases the density of peaks in all the channels, thus greatly decreasing the problem of peak overlap and increasing the effective dynamic range. The pattern of responses to an analyte among the many channels can also provide data for positive identification.

Practice Questions and Problems

1. Name the three classes of phases that exist in matter. For each combination of two phases (there are six combinations altogether), briefly describe its potential usefulness for separation by interphase partition.

2. What is the minimum partition constant necessary if four extractions of 25 mL each are to extract 99% of the solute from 100 mL of aqueous solution?

3. It is desired to extract Cd^{2+} into an organic solvent as the $Cd(CN)_2$ complex. The logs of the formation constants for Cd^{2+} with CN^- are (in order from 1 to 4) 5.48, 5.14, 4.56, and 3.58. Assume that the extraction coefficient (with tributyl phosphate in benzene) is 4.5.

   A. What is the $[CN^-]$ for which the practical distribution ratio is highest? (It is suggested that you estimate this value from a log alpha plot.)

   B. What is the value of the maximum practical distribution ratio? (Again, an estimation from the log plot will be fine.)

   C. Is the pH of the aqueous solution a concern, and if so, in what range should it be and why?

4. Many amino acids have a pK'_{a1} between 2 and 3 and a pK'_{a2} between 9 and 10 (e.g., glutamine). Over what pH range will quantitative (at least 99%) extraction be possible?

5. At 1 atm pressure, a mixture of ethanol and water (90% mole ratio water) boils at 91.5 °C. The composition of the vapor is 49% ethanol. The boiling points of water and ethanol are 100 and 79 °C, respectively. Calculate and compare the values of RV obtained from Equations 11.14 and 11.15.

6. A distillation column is set up and tested with the water–ethanol mixture of Problem 5. The first distillate has a composition of 82% ethanol. How many theoretical plates has the column? Use the value of RV obtained from Equation 11.14.

7. Some crystal sensors respond to the mass of an adsorbate by the change in the natural frequency of vibration caused by the weight of the adsorbed material. In terms of conversion devices, they are mass-to-frequency converters. Calculate the mass of an adsorbate of density 0.9 g/cm^3 and molecular weight 253 if the surface of the sensor is 1.0 cm^2 and it is half covered with a single layer of adsorbate.

8. In a chromatographic separation, two components appear at retention times of 136 and 138 s later than an unretained compound. The unretained compound appeared 24 s after the sample injection. What is the ratio of their partition constants?

9. A compound is known to have a partition coefficient between the mobile and stationary phases of 4.7. When the compound was injected into a chromatographic column, the unretained peak and the compound peak retention times were 14 and 159 s, respectively. What is the ratio of the volumes of the stationary and mobile phases?

10. A specific chromatographic medium is characterized as achieving up to 400 theoretical plates. When operated under these optimum conditions, what will be the peak width, (*fwhm* in seconds), of a component eluting at a retention time of 729 s?

11. In gas chromatography, how will the capacity factor change as the temperature is increased? What is the general elution problem, and how does programmed temperature gas chromatography help to solve it?

12. In the van Deemter equation, it is the *Cu* term that often contributes the most to the band broadening at higher flow rates. This term has been reduced by column designs that lessen the maximum distance analytes in the mobile phase need to diffuse to get to the stationary phase (e.g., narrower bores, finer packing). Discuss the advantages of such developments and the desirability of working at higher flow rates.

13. In gas chromatography with a flame ionization detector, identify the conversion devices used to obtain a number related to amount of analyte in the detector at each sampling time. A block diagram is recommended. Be sure to identify the forms of the data that are the input and output of each device.

14. Is the flame ionization detector a mass detector or a concentration detector? Explain your answer.

15. The longitudinal diffusion coefficient is generally greater in the mobile phase than in the stationary phase, and it is much greater in gas chromatography than in liquid chromatography. Provide an explanation for this observation.

16. In liquid chromatography with an UV absorbance detector, describe the analytical scheme by which the analyte concentration is determined. By scheme, we mean the differentiating characteristic employed, the nature of the probe, the response to the probe, the measurement of the response, and the interpretation of the data.

17. What is the solution to the general elution problem in liquid chromatography? For your explanation, use the example of a reverse phase column.

18. A. What are the common names given to the partitions that occur at a liquid–gas interface, a liquid–liquid interface, and a gas– or liquid–solid interface?

    B. What is the required charcteristic of two liquids involved in a liquid–liquid partition? How is this characteristic usually achieved? (That is, what types of liquids are used for each phase?)

    C. Which approach provides greater discrimination among analytes with similar partition coefficients, batch or continuous partition? Why?

    D. How is the capacity factor related to the partition coefficient?

    E. How is the velocity of an analyte in chromatography related to its capacity factor?

    F. Which of the four goals of chemical analysis is best met by partition techniques?

19. Why don't all the analyte molecules of a given species emerge from a chromatographic column at the same time? Name and describe three phenomena that contribute to the distribution of retention times.

20. A gas chromatogram is given in the figure below. The carrier is helium at 40 cm/sec. The oven is constant at 90°C. The components of the sample are numbered. The offscale peak early in the chromatogram is from unretained solvent introduced with the sample.

    A. The column is 30 m in length. Estimate HETP and the number of theoretical plates for this column under the operating conditions used.

    B. Explain why the peaks get broader with increasing retention time.

    C. Calculate the capacity factors for peaks 1 and 4.

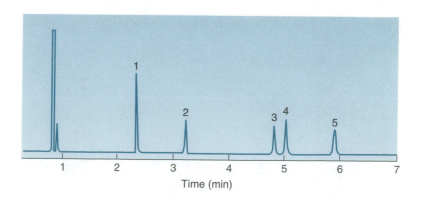

21. Another gas chromatogram is given in the figure below.

    A. In this case, why do the peak widths not increase with increasing retention time?

    B. Compare the HETP values for peak 2 and peak 17.

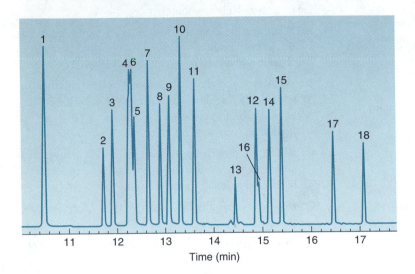

Time (min)

22. For gradient elution, reverse-phase HPLC:

    A. In what way does the solvent change as the chromatographic time increases?

    B. Will the more polar or the less polar analytes emerge from the column first?

    C. Comment on the types of analytes what would be observed when a UV absorption detector is used with HPLC. How is this different from the case where fluorescence detection is used?

Suggested Related Experiments

1. Measure the fraction of a weak acid extracted versus the pH of the solution.

2. Perform a solid phase extraction and test the effectiveness.

3. Use gas chromatography to analyze breath or another interesting, complex example.

4. Analyze a mixture of amino acids or small peptides with liquid chromatography.

5. Perform trace analysis with gas chromatography–mass spectrometry.

Chapter Twelve

ANALYSIS BY BIOCHEMICAL REACTIVITY

An important quality of a differentiating characteristic is the degree to which it can differentiate the intended analyte(s) from the rest of the material in the sample. All of the differentiating characteristics studied thus far are shared by many potential components in the sample. These components can then interfere with the chemical analysis. In identification, they may give a false positive; in quantitation, they can produce an inaccurately high result; in every situation, they raise the level of the detection limit. Therefore, the more discriminating the differentiating characteristic is, the fewer are the possible interferences and the lower the detection limit. In fact, it is usually true that the detection limit in a chemical analysis is set by the level of the interferences present. In other words, the detection limit is more often a function of the selectivity of the differentiating characteristic than of the sensitivity of the measurement system used.

The chemical reactions we have studied thus far can very effectively differentiate one class of compounds from another, for example, those that can react with a base from those that do not. Full spectroscopic analysis and chromatographic methods provide a greater degree of discriminating power in many situations, but the lower the concentration of the analyte we wish to characterize, the greater the degree of selectivity must be. One of the many remarkable aspects of the chemical reactions that occur in biological systems is their incredible degree of selectivity, that is, the extremely limited number of compounds with which they will carry out a given reaction. This is accomplished by adding molecular shape and electron density distribution to the qualities the reactant must have in order to undergo a reaction.

The two types of biochemical reactions most often used in chemical analysis are enzyme reactions and antigen–antibody reactions. As we shall see, they have been used for quantitation, detection, and separation in a variety of ingenious ways. Altogether, this is one of the most rapidly developing areas of chemical analysis, and it involves virtually every one of the principles studied thus far.

A. Enzyme Reactivity

glucose + O_2 + H_2O $\xrightarrow{E}$ gluconic acid + H_2O_2

(E = glucose oxidase)

$(NH_2)_2CO$ + H_2O $\xrightarrow{E}$ $2NH_3$ + CO_2

(E = urease)

CH_3CH_2OH + NAD^+ $\xrightarrow{E}$ CH_3COH + NADH + H^+

(E = alcohol dehydrogenase)

(NAD^+ is a coenzyme)

L-histidine $\xrightarrow{E}$ histamine + CO_2

(E = histidine decarboxylase)

Enzymes are proteins that serve as catalysts for reactions that occur in biological systems. Each one is designed to facilitate a very specific type of reaction with a very small range of reactant molecules. Examples of some reactions facilitated by enzymes are shown in the margin. For example, the enzyme glucose oxidase facilitates the oxidation of glucose, and the enzyme histidine decarboxylase removes the carboxylic acid function (COO^-) from histidine (see the structure in Appendix B) to form histamine. From these examples, we can see that the suffix "ase" is used for enzyme names, often attached to the chemical function they perform, and that the name frequently includes the molecule for which they facilitate the reaction. The molecule that is being acted on is called the **substrate.** The dehydrogenase enzyme requires another reactant (NAD^+) to aid in the conversion of the alcohol to an aldehyde. The ion NAD^+ (nicotinamide adenine dinucleotide) is reduced, by the addition of an H atom, to NADH. In this situation, NAD^+ is called a **coenzyme** (or sometimes a **cofactor**). The NADH formed is often reoxidized to NAD^+ by a separate enzymatic reaction to complete the cycle and reconstitute the dehydrogenation reaction catalysts.

If one wished to oxidize glucose ($C_6H_{12}O_6$) to gluconic acid ($C_6H_{12}O_7$) with a general oxidizing agent, one would simply add an oxidizing agent that had an E' greater than that for the glucose/gluconic acid couple. When this is done, every oxidizable species in the solution with an E' less than that of the oxidizing agent can also be oxidized. On the other hand, if the oxidizing agent were O_2, there may be no reasonable mechanism by which the oxidation reaction could occur under normal conditions. In that case, the reaction would not occur even if it were energetically favorable. Glucose oxidase provides a practical mechanism for the reaction of O_2 with the glucose. Its function, in this reaction, is to act as a **catalyst.**

A catalyst cannot enable a reaction that is not energetically favorable, so all the relationships we have studied regarding relative acidities, oxidation potentials, and other reaction equilibria apply to enzyme-catalyzed reactions. What the catalyst does is provide a lower energy path for the reaction to occur so that its rate is increased. To appreciate the degree of rate enhancement achieved by enzymes, consider the reaction of fumarate (the basic form of fumaric acid) with water shown below. The water adds across the double bond to form malate (the basic form of malic acid). The rate for this reaction in neutral solution with the enzyme catalyst fumarase is 3.5 × 10^{15} times faster than when the catalyst is not present. The amount of product formed in one minute with the fumarase present would take 6.6 billion years to form without it.[1]

Specificity in enzyme reactions is achieved by complexation, which is strongly affected by conformation. For an oxidase to catalyze the oxidation of a substrate molecule, the substrate must complex with the oxidase at a particular location on the enzyme called the **active site.** The active site is three-dimensional, occurring in clefts, crevices, or holes in the enzyme protein. For complexation, the specific atoms involved in the reaction must be in the correct orientation to corresponding atoms in the

---

[1]A. Radzicka and R. Wolfenden, *Science* **1995**, 267, 90.

active site of the oxidase. Thus, glucose oxidase provides a highly selective mechanism for the oxidation of glucose. Other oxidizable species in the solution, lacking the enzymatic oxidation mechanism, will not be oxidized.

VARIETIES OF ENZYMES AND REACTIVITY

Enzymes are proteins, sometimes with the addition of particular functional groups. The folding of the amino acid chain produces surface conformations that are attractive to specific organic molecules through a combination of coulombic attraction, hydrogen bonding, van der Waals forces, and hydrophobic environment. The variety of enzymes that are found in nature is astounding, and the range of chemical reactions they can catalyze includes virtually all the reactions of biological significance. Hundreds of naturally occurring enzymatic reactions have been discovered. Enzymes are often categorized according to the type of reaction they catalyze. The six principal reaction types as defined by the Enzyme Commission of the International Union of Biochemistry are shown in Table 12.1.

An enzyme-catalyzed reaction begins with the binding of the reactants to specific sites on the enzyme. One of these reactants is the substrate. The other (if required) is O_2, water, NAD^+, or others. The bound enzyme–substrate is a complex and follows the equilibria developed in Chapter 7. The reaction then occurs between the substrate and the other reactant while bound to the enzyme. The final step is the release of the products to the surrounding solution. Some enzyme reactions require the presence of a cofactor such as NAD^+. When necessary, the cofactor must be bound to the enzyme for the reaction to occur. This bond may be covalent or noncovalent. The cofactor is also sometimes called the coenzyme or **prosthetic group.** Specificity for the cofactor can be very high as in the case of dehydrogenases, or the cofactor may be any of a variety of compounds as in the case of glucose oxidase. The complete enzyme–coenzyme complex is called a **holoenzyme.** The enzyme without the coenzyme is called an **apoenzyme** and is generally completely inactive. Many enzymes require two substrates and may produce more than one product. The second substrate is sometimes called a **cosubstrate.** Glucose oxidase, for example, can employ a variety of cosubstrates other than glucose.

Many enzymes are formed of more than one polypeptide chain, with subunits bound together through hydrogen bonding, coulombic forces, hydrophobic interactions, and/or disulfide linkages. If the subunits are identical peptide chains, the enzyme is a **homomultimer;** if not, it is a **heteromultimer.** The physical arrangement of the subunits determines the **quaternary structure** of the complete enzyme protein. (The **primary, secondary** and **tertiary structures** are the sequence of amino acids in the peptide chain, the interchain bonding structure,[2] and the folding, respec-

A Quick Summary of Enzyme Reactivity

Enzymes are large proteins, made up of several subunits, that can serve as catalysts for highly specific reactions.

At least one of the reactants, called the substrate, forms a complex with the enzyme at an active site in the enzymes.

A different enzyme is required for each reaction type, and enzymes are often highly specific with regard to the reactant molecules for which they are effective catalysts.

Enzymes may have to be activated through the binding of cofactors, coenzymes, effectors, or cosubstrates. When not activated, the enzyme has no catalytic activity. The nature of the activators can be highly specific or more general.

Enzyme activity can be inhibited by the complexation of inhibitor molecules. The inhibitor can interfere with the binding of the reactant substrate or with the mechanism of the reaction. The inhibition effect can be induced by only specific molecules or by a wider variety of inhibitors.

All specific reactions, whether the substrate reaction, the activation complexation, or the inhibition complexation, can be used as the basis for chemical analysis.

Table 12.1
Main Heading for Enzyme Classification

Enzyme Class	Function
1. Oxidoreductases	Transfer of hydrogen, oxygen, or electrons
2. Transferases	Transfer of specific groupings
3. Hydrolases	Hydrolytic reactions
4. Lyases	Bond cleaving by nonhydrolytic reactions
5. Isomerases	Intramolecular rearrangements
6. Ligases	Bond formation requiring ATP

---

[2]Interchain bonding structures include α-helix, β-pleated sheet, etc.

tively.) The quaternary structure is often essential to the enzyme activity and is involved in the mechanism of the enzyme reaction. For example, the enzyme lactate dehydrogenase is made up of four subunits, which can be either of two types, A or B. The subunits are enzymatically inactive, but a tetramer composed of the subunits A or B will be active. Each of the five possibilities (A_4, A_3B, A_2B_2, AB_3, and B_4), which are called **iso-enzymes**, differs somewhat from the others in its activity and substrate specificity. At the most complex end of the scale, human cytochrome oxidase has 13 subunits.

ENZYME ACTIVATION AND INHIBITION

Every enzyme has what is called an **active site**. This is the place on the enzyme surface that binds with the substrate molecule. We have seen that enzyme activity depends on composition and structure, and in some cases it depends on the presence of a coenzyme. In addition, some enzymes require an additional substance called an **activator** to function effectively. In some cases, the necessary activator is an inorganic ion. The enzyme amylase is activated by the presence of chloride, but other anions can serve to activate it as well. Some enzymes require specific cations such as Mg^{2+} or Cu^{2+} to become active. Another kind of activating molecule can be one that stabilizes the enzyme structure in the conformation that optimizes the effectiveness of the active site. Enzymes such as glycogen phosphorylase that require structural stabilization are called **allosteric** enzymes. Glycogen phosphorylase is critically involved in the synthesis of glycogen. In its normal state it is inactive, but it can be activated by the allosteric **effectors** AMP and IMP (adenosine monophosphate and inosine monophosphate). These effectors are not involved in the catalytic reaction, but their binding to the enzyme is thought to affect the nature of the active site.

As you might expect, there are also many ways by which an enzyme can be made less effective. These processes come under the general heading of **inhibition.** One of the most direct forms of inhibition is **competitive inhibition.** In this form, a molecule somewhat like the substrate molecule, but nonreactive, binds in the active site and prevents the enzyme from binding the substrate. One of the possible competitive inhibitors might be the product of the enzymatic reaction. In this case, the rate of the reaction would be slowed as the concentration of product increases. Other inhibitors operate at sites distinct from the active site. A **noncompetitive inhibitor** binds to the enzyme in such a way as to distort its active site or prevent activation. Either of these effects will cause a decrease in the activity of the enzyme. In another mechanism, called **uncompetitive inhibition,** the inhibitor molecule binds to the enzyme–substrate complex in such a way as to inhibit the completion of the catalytic reaction or the release of products.

The activation and inhibition of enzyme activity are important functions in nature, where a means of regulating the chemical activities of each enzyme is essential to the maintenance of chemical balance for optimum biological functioning. In the analytical application of enzymatic reactivity, the presence of necessary activators must be provided for, and the loss of function due to the presence of inhibitors may result in errors in quantitation and detection.

THE EFFECT OF pH

Enzymes are sensitive to the pH of their environment. This is no surprise since the amino acids that the enzyme proteins are made of are weak acids and thus have different forms at high and low values of pH. The most severe form of this pH dependence is the denaturation of the protein that occurs in very low pH and very high pH solutions. **Denaturation** is the loss of tertiary and quaternary structure and possibly an alteration of secondary structure as well. Clearly, this will be accompanied by a complete loss of enzyme activity. Within the pH range of overall enzyme structural integrity, there can

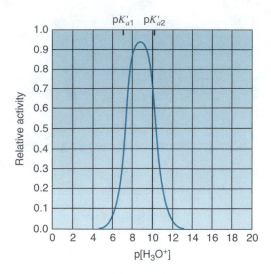

Figure 12.1. Relative activity of a typical enzyme versus pH. If the limiting pK_a' values are closer together, the useful range will be narrower. Correspondingly, a greater difference in the limiting pK_a' values results in a broader useful pH range.

be other pH effects. pH can affect the charge state of acidic species in the active site region. If the substrate is a weak acid, its form will be dependent on the pH of the solution. The specificity of the enzymatic reaction is such that it will only react with a single acidic form of the substrate. The pH may have an effect on the conjugate form of amino acids that can affect the tertiary structure of the peptide and thus the conformation of the active site. Finally, the pH may affect the extent of activation or inhibition reactions.

Because of all these factors through which the enzyme activity depends on the pH, it is common to experimentally determine the degree of enzyme activity as a function of the pH of the solution. The general form of this function is shown in Figure 12.1. In many cases, this curve can be fit to the alpha curve for a particular acid form in a polyprotic acid system. This suggests that two acidic moieties with different pK_a' values are critical to the enzyme activity.

Since many enzymes function at physiological pH ($\sim$7.4), one would expect them to have significant activity at that pH. This pH, however, is not the optimal pH for many enzymatic reactions. When using enzymatic reactions in the laboratory, the activity of the enzyme is maintained by putting it in a buffer solution in the optimal pH range. Good buffer strength is often required because many hydrolytic enzyme reactions involve protons as reactants or products. For the greatest buffer strength with the least concentration of buffer, one wants to choose a conjugate acid–base pair with a pK_a' close to the desired solution pH. Biochemists have developed a set of "favorite" buffer systems that span the useful range for optimizing enzyme reactions. Several widely used buffer systems are shown in Table 12.2. Since the buffering species may

Table 12.2
Some Buffers Used with Biochemical Reactions

Abbreviation	Acid	pK_a'
MES	2-(*N*-Morpholino)ethanesulfonic acid	6.1
ADA	*N*-(2-Acetamido)iminodiacetic acid	6.6
PIPES	Piperazine-*N*,*N'*-bis(2-ethanesulfonic acid)	6.8
ACES	*N*-(2-Acetamido)-2-aminoethanesulfonic acid	6.8
MOPS	(3-*N*-Morpholino)-propanesulfonic acid	7.2
HEPES	*N*-2-Hydroxyethylpiperazine-*N'*-2-ethanesulfonic acid	7.5
HEPPSO	*N*-2-Hydroxyethylpiperazine-*N'*-3-propanesulfonic acid	8.0
TRIS	Tris(hydroxymethyl)aminomethane	8.1
Bicine	*N*,*N*-Bis(2-hydroxyethyl)glycine	8.3

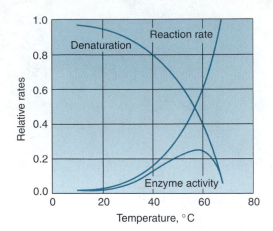

Figure 12.2. Enzyme activity as a function of temperature. Increasing the temperature increases the reaction rate but also reduces activity by denaturation.

interact with the biochemically active compounds, results may vary considerably between systems with different buffers, even when the buffered pH is the same.

ENZYME STABILITY

As we have seen so far, maintaining optimum enzyme activity depends on a number of independent factors. The storage of enzyme reactants requires careful investigation into the solution conditions that will minimize degradation of the enzyme structure or activity. Long-term storage is frequently accomplished by addition of 50% glycerol to the enzyme solution and storing at $-70\,°C$. Shorter term storage is best in a solution of the optimum pH, ionic strength, and anionic/cationic composition. Cooling, short of freezing, generally enhances enzyme life, but repeated freeze–thaw cycles cause denaturation. The conditions for optimum storage are not necessarily those for optimum enzyme activity, and this is especially true of temperature. Enzyme activity will increase exponentially with increasing temperature (roughly doubling for each 10 °C increase in temperature, as with virtually all chemical reactions). However, too high a temperature induces denaturation and total, sometimes irreversible, loss of activity. These opposing trends are illustrated in Figure 12.2. Most laboratory enzymatic reactions are run at a controlled temperature that enables a compromise between enzyme activity and enzyme lifetime. Another reason for carefully controlling the temperature of enzyme reactions used analytically is that the property of the reaction measured is the rate, which is highly dependent on the temperature.

Study Questions, Section A

1. What is an enzyme, and what is a substrate?

2. Is the enzyme consumed in carrying out the enzyme-catalyzed reaction? Is the cofactor altered by the catalyzed reaction?

3. Reactions may not occur for either of two reasons: The products are in a higher energy state than the reactants, or there is no suitable mechanism by which the reaction can take place. In which of these situations will a catalyst enable a reaction, and why or why not?

4. What is the active site of an enzyme?

5. List all the species that might have to be bound to an enzyme for the catalytic reaction to occur.

6. The pH of a solution can affect the activity of enzymes in it in two ways. One is generally reversible; the other may not be. What are the two actions of pH on enzyme activity?

7. Which of the buffer systems in Table 12.2 would be suitable for maintaining a pH of 7.5? How would you choose among them?

8. Is there any disadvantage in working at the temperature of maximum activity as shown in the curve of Figure 12.2?

1. An enzyme is a protein that catalyzes a chemical reaction. One of the reactants is the substrate, an organic molecule that is being chemically altered by the enzymatic reaction.

2. The enzyme is a true catalyst in that it is not changed or consumed by the reaction. When it has catalyzed the conversion of one substrate molecule, it is free to facilitate the reaction of another. The cofactor or coenzyme is not necessarily a true catalyst and may be changed by the reaction. In this case, a second reaction (often enzyme-catalyzed) is needed to return the coenzyme to its original form so that it can participate in further reactions.

3. Since the catalyst is unchanged by the reaction, it cannot compensate for an unfavorable energy imbalance between the reactants and products. However, the catalyst can provide a reaction pathway or mechanism that is practical under the reaction conditions and thus enable a mechanistically hindered reaction.

4. The active site on an enzyme is where the substrate is bound in the enzyme–substrate complex.

5. In addition to binding the substrate, the enzyme may need a cofactor to which it must be bound. If another reactant such as O_2 is required, it must also be bound to the enzyme. In some cases, the formation or conformation of the active site requires the formation of a multimer of peptide chains. In addition, some enzymes require an activator ion or effector molecule.

6. At very high and very low pH's, enzymes will lose their tertiary and quaternary structure. The enzyme may not recover its active structure after a more reasonable pH is established. Some of the amino acids in the peptide structure can change their acid form as a function of pH. This can cause a change in the shape and binding properties of the active site. This kind of deactivation is generally reversible.

7. Any buffer system with a pK_a' value within 0.5 p units of the desired pH can be used for effective pH control. From Table 12.2, possible buffers would include MOPS, HEPES, and HEPPS. The ones selected will have to be tested with the biochemical reaction to determine whether they inhibit the desired function. This is because the weak acid anion in the buffer may complex metal ions that are part of the enzyme and critical to its activity.

8. Yes, there is a disadvantage in working at the temperature of maximum enzyme activity. The denaturation process continually decreases the concentration of active enzyme. The longer one works at the elevated temperature, the less active the system will become. If the loss of activity through denaturation is reversible, one might cycle through warm and cool cycles to regenerate the activity between experiments.

B. Kinetics of Enzyme Reactions

To apply the differentiating characteristic of enzymatic reactivity in chemical analysis, it is necessary to carry out the enzymatic reaction with the analyte as the limiting reagent in the reaction. The classic method of using chemical reactivities for analysis is by titration. Recall, however, that for a reaction to be suitable for titration, it must be perfectly stoichiometric and attain equilibrium rapidly relative to the addition of titrant. It is frequently the case that enzymatic reactions do not achieve equilibrium quickly, if at all. In addition, a significant part of the selectivity of enzymatic reactions has to do with the relative *rates* of the reaction of the enzyme with all possible substrates. Waiting for equilibrium would increase the interference from other reactants, but using the rate of the reaction as the differentiating characteristic maximizes the selectivity of the analysis. Therefore, the application of enzymatic reactions in chemical analysis is usually accomplished by techniques that are called **kinetic methods of analysis**. To understand these methods better, we will first look at some of the principles of the kinetics of enzyme reactions.

The process of an enzymatic reaction (as proposed by Victor Henri and later by Leonor Michaelis and Maude Menten) takes place in two steps. The first is the reversible formation of the enzyme–substrate complex. The second step is the catalytic conversion of the complex to reaction products, and it is generally not reversible. The overall reaction scheme is thus represented as

$$E + S \underset{k_{-1}}{\overset{k_1}{\rightleftharpoons}} ES \overset{k_2}{\rightarrow} E + P \qquad 12.1$$

Kinetic Methods of Analysis

Kinetic methods of analysis are based on the fact that for most reactions, the rate of the reaction increases with increasing concentration of reactants.

Developing an analytical procedure around this quality of the reaction provides an alternative to the equilibrium concentration method of analysis (titration) when chemical reactivity is the differentiating characteristic. Kinetic methods are particularly valuable when the reaction rate is slow (so that equilibrium is not quickly attained) and when even slower interfering reactions make the equilibrium method less discriminating.

Application of the kinetic method requires that the reactants be quickly mixed, that the analyte concentration limits the rate of the reaction, and that there is a method to follow the progress of the reaction with time.

where E is the enzyme, S is the substrate, ES is the complex (sometimes called the **Michaelis complex**), and P is the product. The rate constants are k_1 for the formation of the complex, k_{-1} for the dissociation of the complex, and k_2 for the catalytic conversion of the complex to the catalyst (E) and the product(s) (P). The ratio k_1/k_{-1} is equal to the formal formation constant of the complex, K_f'.

The rate of the catalytic conversion of the complex to catalyst and products will be dependent on the concentration of the complex. It is a simple, first-order reaction, which has the form

$$v = k_2[\text{ES}] \qquad\qquad 12.2$$

where v is the rate of the reaction in moles per liter per second of product formed or reactant consumed. As the reaction proceeds, the substrate is converted to product, and so the concentration of S decreases. This decreases the concentration of ES, which in turn decreases the reaction rate v.

FIRST-ORDER REACTION RATES

If this reaction were purely first order, that is, if there were no way to replace ES as it is converted to product, the concentration of ES and the product as a function of time would follow the relationship

$$[\text{R}] = C_\text{R}e^{-kt} \qquad \text{and} \qquad [\text{P}] = C_\text{R}(1 - e^{-kt}) \qquad 12.3$$

where R is the reactant, P is the product, C_R is the initial concentration of reactant, and k is the rate constant for the first-order reaction R $\rightarrow$ P. A plot of this function is shown in Figure 12.3. The concentrations are continually changing as the reactant concentration approaches zero and the product concentration approaches the initial reactant concentration. When measuring reaction rates, the rate of change of concentration as the reaction is just beginning is often measured (see Example 12.1). This rate, v_0, is

$$v_0 = kC_\text{R} \quad \text{mol/L s} \qquad\qquad 12.4$$

ENZYME-CATALYZED REACTIONS

With the enzyme-catalyzed reaction, the overall reaction is not simply first order because the complex ES can be replaced after it reacts to form product by the reaction of the released enzyme with further substrate. To assess the rate of the reaction through Equation 12.2, we need an equation for [ES]. If we assume that the complex formation reaction is fast compared to the catalytic reaction, the concentration of ES will be in equilibrium with the concentration of E and S in solution, as with other complexation

Example 12.1

Calculation of initial reaction rate and time to completion

Using the data in the reaction rate plot of Figure 12.3 and Equation 12.4, the initial reaction rate is

$$v_0 = 1 \times 10^{-4}\,\text{mol L}^{-1} \times 0.02\,\text{s}^{-1}$$
$$= 2 \times 10^{-6}\,\text{mol/L s}$$

To calculate the time to completion, consider the reaction complete when only 1% of the reactant remains. From Equation 12.3,

$$\ln[\text{R}] = \ln C_\text{R} - kt$$
$$t = \frac{\ln\dfrac{C_\text{R}}{[\text{R}]}}{k} = \frac{\ln 100}{0.02\,\text{s}^{-1}} = 230\,s$$

where ln is the natural logarithm.

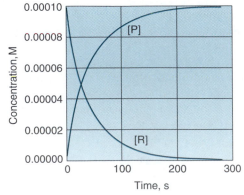

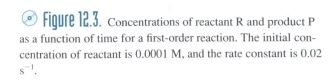
Figure 12.3. Concentrations of reactant R and product P as a function of time for a first-order reaction. The initial concentration of reactant is 0.0001 M, and the rate constant is 0.02 s^{-1}.

reactions. We can therefore write an expression for the fraction of the enzyme that is in the complex form using Equation 7.8

$$\alpha_{ES} = \frac{[ES]}{C_E} = \frac{K_f'[S]}{1 + K_f'[S]} \tag{12.5}$$

where C_E is the analytical concentration of the enzyme. Solving Equation 12.5 for [ES],

$$[ES] = \frac{C_E K_f'[S]}{1 + K_f'[S]} \tag{12.6}$$

gives us an expression for [ES]. Combining this equation with Equation 12.2 will yield an equation for the rate of the reaction.

First, we will modify Equation 12.6 to incorporate the common usage of enzyme chemists who express the complex formation equilibrium as a dissociation, that is, $ES \rightleftharpoons E + S$, for which the equilibrium constant is

$$K_S = \frac{[E][S]}{[ES]} = \frac{1}{K_f'} \tag{12.7}$$

Now, in terms of K_S,

$$[ES] = \frac{C_E K_f'[S]}{1 + K_f'[S]} = \frac{C_E[S]}{1/K_f' + [S]} = \frac{C_E[S]}{K_S + [S]} \tag{12.8}$$

Combining Equations 12.8 and 12.2, we get

$$v = \frac{k_2 C_E[S]}{K_S + [S]} \tag{12.9}$$

from which we can see that the rate of the reaction is proportional to the concentration of the enzyme and depends on the concentration of substrate.

The initial rate of the reaction, v_0, will have the same relationship, where $[S]_0$ is the initial concentration of substrate. If the initial concentration of substrate is larger than C_E, then only a small fraction of it can be complexed, so $[S]_0 \approx C_S$. Now,

$$v_0 = \frac{k_2 C_E C_S}{K_S + C_S} \tag{12.10}$$

The relationship between the initial reaction rate and the analytical substrate concentration is shown in Figure 12.4. Three regions of the curve have been identified.

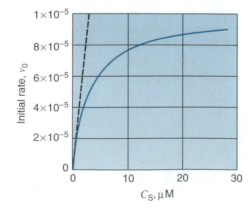

⊙ Figure 12.4. Initial reaction rate versus the molar concentration of substrate for an enzyme reaction following Equation 12.10. The dashed line shows the slope of the rate versus concentration curve at 0 concentration of substrate. For this figure, $k_2 = 1000 \text{ s}^{-1}$, $C_E = 1 \times 10^{-7}$ M, and $K_S = 3 \times 10^{-6}$ M.

In the first region, at low concentration of substrate, the initial rate increases approximately linearly with increasing concentration of substrate. This can be verified from Equation 12.10 for the case where C_S is much less than K_S. In this region, the rate equation becomes

$$v_0 = \frac{k_2 C_E C_S}{K_S} \qquad \text{when } C_S \ll K_S \qquad\qquad 12.11$$

Recall from Equation 12.8 that K_S is the inverse of the formation constant for the complex. Thus, if the formation constant is $3.3 \times 10^5 \text{ M}^{-1}$ (as in the case illustrated in Figure 12.4), $K_S = 1/K_f' = 3 \times 10^{-6}$ M. For all substrate concentrations less than micromolar, the measured initial rate will be proportional to the substrate concentration. A dashed line has been included in Figure 12.4 showing the slope of this linear relationship and the concentration region over which linearity is achieved.

Figure 12.4 also shows that the initial reaction rate approaches a limit as the substrate concentration increases. This occurs when the reaction becomes limited by the rate at which the enzyme can turn the substrate into product. It will occur when virtually all the enzyme is in the complexed form. From our study of complexes in Chapter 7 (and from Equation 12.5), this condition will occur when the product $K_f'[S]$ is greater than 1. In terms of K_S and Equation 12.10, the initial rate will reach a limiting value when C_S is greater than K_S. This value, called v_{max}, will be

$$v_{max} = k_2 C_E \qquad\qquad 12.12$$

THE MICHAELIS–MENTEN EQUATION

In the development of Equation 12.10, it was assumed that the complex formation reaction is at equilibrium. For this to be true, the rate constants k_1 and k_{-1} in the mechanism of Equation 12.1 must be greater than k_2. When this is not the case, the time required to re-form the complex from the enzyme newly released from the products will slow down the reaction. A kinetic derivation of the reaction mechanism in Equation 12.1 yields an equation for the initial rate in which the K_S term in the denominator of Equation 12.10 is replaced by $(k_2 + k_{-1})/k_1$.[3] This term is called K_M, the **Michaelis constant.** Notice that K_M has the units of concentration. In terms of K_S and the rate constants,

$$K_M = K_S + \frac{k_2}{k_1} \qquad\qquad 12.13$$

If the catalytic step is much slower than the complex formation, k_2 will be very much smaller than k_1, and K_M will equal K_S. If the rates of the catalytic and complex formation steps are comparable, K_M will be larger than K_S, and the reaction rate will be decreased.

The reaction mechanism of Equation 12.1 is the simplest possible form of the enzyme reaction. In some cases, a coenzyme has been oxidized or reduced, and a second enzymatic reaction needs to occur to regenerate the original, active enzyme–coenzyme combination. Such complications affect the overall rate of the catalytic step. In the general case, where the catalytic step may be more complex, a generic rate constant k_{cat} is used instead of the more specific k_2. In no case can the value of k_{cat} be greater than the value of k_2. When these changes are made, the resulting equation for the initial rate is

$$v_0 = \frac{k_{cat} C_E C_S}{K_M + C_S} \qquad\qquad 12.14$$

---

[3] A. Fersht, *Structure and Mechanism in Protein Science*, pp. 106–107. Freeman, New York, 1999.

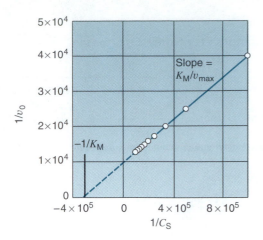

⊚ **Figure 12.5.** Lineweaver–Burk plot to determine values of K_M and v_{max} from a limited data set. The data points are linearly extrapolated to the X-axis, where the intercept is $-1/K_M$, and the slope is K_M/v_{max}.

Equation 12.10 can now be modified to incorporate v_{max} from Equation 12.12, where v_{max} is now equal to $k_{cat}C_E$.

$$v_0 = \frac{v_{max}C_S}{K_M + C_S} \qquad\qquad 12.15$$

This equation is known as the **Michaelis–Menten equation.** It is the fundamental equation of enzyme kinetics and is virtually universally used in characterizing enzyme reactions. The two constants, K_M and k_{cat}, are the factors most frequently used to compare the effectiveness of enzyme reactions. They are determined empirically by measuring the initial reaction rate for different concentrations of substrate with a known, constant concentration of enzyme. The resulting plot is similar to the one shown in Figure 12.4. The value of K_M is equal to the concentration of substrate for which the initial rate is half of v_{max}. (Make $K_M = C_S$ in Equation 12.15, and $v_0 = 1/2\ v_{max}$.) The value of k_{cat} is obtained by dividing v_{max} by C_E.

Sometimes it is impractical to determine the initial rate over a wide enough range of substrate concentration to obtain a good value for v_{max}. When dealing with a limited data set, workers use a curve fitting program where K_M and k_{cat} are the adjustable parameters, or they use a graphical technique in which $1/v_0$ is plotted against $1/C_S$. Such a plot for the data shown in Figure 12.4 is shown in Figure 12.5. This method of determining K_M and v_{max} is called the Lineweaver–Burk plot.

ANALYTICAL SIGNIFICANCE OF K_M and k_{cat}

As we saw in the previous subsection, the value of K_M depends on the formation constant for the ES complex and on the rate of its formation relative to the catalytic conversion of the ES complex to products. The lower the formation constant of the complex and the slower its formation, the larger is the value of K_M and the slower the overall reaction. Similarly, the lower the rate constant for the catalytic conversion, the slower is the overall reaction. Consider the case where C_S is small compared to K_M. When this is true, Equation 12.14 reduces to

$$v_0 = \frac{k_{cat}}{K_M}C_E C_S \qquad\qquad 12.16$$

Thus, it is the ratio of k_{cat} to K_M that determines the enzyme-catalyzed reaction rate. Recall that the differentiating characteristic for the analytical application of enzyme

Table 12.3

Characteristic Constants for Some Enzymes

System	K_M, M	k_{cat}, s^{-1}	k_{cat}/K_M, M^{-1} s^{-1}
Acetylcholinesterase with acetylcholine substrate	9.5×10^{-5}	1.4×10^4	1.5×10^8
Chymotrypsin with N-acetylglycine ethyl ester substrate	4.4×10^{-1}	5.1×10^{-2}	1.2×10^{-1}
Chymotrypsin with N-acetyltyrosine ethyl ester substrate	6.6×10^{-4}	1.9×10^2	2.9×10^5
Fumarase with fumarate substrate	5.0×10^{-6}	8.0×10^2	1.6×10^8
Fumarase with malate substrate	2.5×10^{-5}	9.0×10^2	3.6×10^7
Urease with urea	2.5×10^{-2}	1.0×10^4	4.0×10^5

reactions is the rate of the enzyme reaction relative to the rates of the reaction with possible interferents. Equation 12.16 provides us with a factor with which to evaluate the selectivity of an analysis based on enzyme reaction rate. If there are two substrates that can potentially react similarly through the enzyme pathway, the relative initial reaction rates for these species will be

$$\frac{v_{0,A}}{v_{0,B}} = \frac{(k_{cat}/K_M)_A C_A}{(k_{cat}/K_M)_B C_B} \qquad 12.17$$

For this reason, the term k_{cat}/K_M is sometimes called the **specificity constant** for the catalytic reaction. The catalytic rate constant k_{cat} has the units reciprocal seconds (s^{-1}). It is equal to the maximum number of times per second that each active site can convert a substrate molecule to products. This factor is sometimes called the **turnover number.** Values of k_{cat}, K_M, and k_{cat}/K_M can vary greatly. Values of these constants for a few enzymes are given in Table 12.3.[4] From Table 12.3, we can see that malate would be an interferent in the analysis of fumarate using fumarase but that chymotrypsin could be used to determine tyrosine without interference from glycine.

In this development, we have assumed that the value of C_E, the analytical concentration of active enzyme, is known. In many cases, this is not practical. The enzyme might not be separately weighable, it might not all be active, and there might be factors in the implementation of the assay that inhibit the enzyme activity. Because of this, it is easier to use the actual enzyme activity as a measure of the amount of effective enzyme present. The SI unit of activity is called the **katal.** One katal of enzyme will convert 1 mol of substrate to product in 1 s. A katal of enzyme is a huge amount, so units of nanokatals are used.

A similar but less frequently used measure is the **international unit** (IU) of activity. One international unit of enzyme will convert 1 μmol of substrate to product in 1 minute. The international unit of activity is the activity defined under specific conditions of pH, ionic strength, presence of certain ions, etc. Other conditions, of course, will yield different rates. One nanokatal is equal to 0.06 IU. Related to the IU measure of activity is the **specific activity,** which is the number of international units of activity per gram of the enzyme formulation. In many cases, the specific activity is the most practical measure of activity. Applying these units of activity to Equation 12.16, one can see that the initial reaction rate will be proportional to the number of activity units of enzyme present.

---

[4]D. Voet and J. G. Voet, *Biochemistry*, 2nd ed., p. 353. John Wiley & Sons, New York, 1995.

9. For enzyme reactivity, why is the rate of the reaction rather than the extent of the reaction used as the differentiating characteristic?

10. Prove that a plot of ln[R] versus time will be a straight line for the first-order kinetics described in Equation 12.3.

11. What is the principal assumption involved in the derivation of Equation 12.9?

12. How is the constant K_S related to the complex formation constant K_f'?

13. What justification is there for using the analytical concentration of S for the actual concentration of S in Equation 12.10?

14. From the data given in the caption of Figure 12.4, calculate v_{max} for the reaction plotted.

15. In what way is the Michaelis constant K_M different from K_S?

16. What are the parameters in the Michaelis–Menten equation that are used to characterize an enzyme?

17. Why are the units of K_S mol L^{-1}?

18. From Equation 12.13, deduce the units for the rate constant k_1.

19. Why is the ratio of k_{cat} to K_M of particular importance in the analytical context?

20. If an enzyme preparation has an activity of 30 nanokatal mg^{-1}, how much product will be formed in 2.0 min using 10 mg of the enzyme preparation?

9. Enzyme reactions often do not proceed to complete equilibrium. Even when they can approach equilibrium closely, the time required is often many minutes. This rate of reaction precludes practical titration. At equilibrium, all interfering substrates would also have reacted, thus increasing the degree of their interference.

10. Take the natural logarithm of both sides of the [R] equation in Equation 12.3.

$$\ln[R] = \ln C_R + \ln e^{-kt} = \ln C_R - kt$$

This equation has the form of a straight line ($y = mx + b$) in which the intercept on the ln C_R axis at $t = 0$ is ln C_R and the slope of the line is $-kt$.

11. In the derivation of Equation 12.9, the assumption is that [ES] can be solved from the equilibrium constant and the concentrations of E and S. This assumes that the complex formation reaction is at equilibrium. For this to be true, the rate of the catalytic reaction of ES to products must be very slow compared to the rate of complex formation. In other words, $k_2 \ll k_1$ in the reaction mechanism of Equation 12.1.

12. The constant K_S is the equilibrium constant for the dissociation of the complex. As such it is the reciprocal of the complex formation constant K_f.

13. In Equation 12.10 it is assumed that the initial concentration of substrate is large compared to the concentration of enzyme. In this case, the fraction of substrate that is complexed at the beginning of the reaction will be small. Therefore, the initial concentration of uncomplexed substrate will be very nearly equal to the analytical concentration of substrate. This will not remain true as the reaction proceeds, but Equation 12.10 applies only for the rate at the beginning of the reaction.

14. Using Equation 12.12,

$$v_{max} = k_2 C_E$$
$$= (1 \times 10^3 \text{ s}^{-1})(1 \times 10^{-7} \text{ M}) = 1 \times 10^{-4} \text{ M s}^{-1}$$

15. The Michaelis constant takes into account the possibility that the concentration of ES is not in equilibrium with the concentrations of E and S. The difference between these two constants is given in Equation 12.13. The Michaelis constant is always equal to or larger than K_S.

16. The enzyme characterizing parameters in the Michaelis–Menten equation are v_{max} (or k_{cat}) and K_M.

17. K_S is the equilibrium dissociation constant for the Michaelis complex, ES. Substituting units in Equation 12.7 for this dissociation constant gives

$$K_S = \frac{[E]\text{mol } L^{-1}[S]\text{mol } L^{-1}}{[ES]\text{mol } L^{-1}}$$

Thus, the units of K_S are mol L^{-1}.

18. Because only terms with like units can be added, the units of k_2/k_1 in Equation 12.13 must be mol L^{-1}. The units of k_2, the first-order rate constant, are s^{-1}. Therefore, from Equation 12.13,

$$\text{mol } L^{-1} = \frac{s^{-1}}{x}, \qquad x = \frac{L}{\text{mol s}}$$

and the units for k_1 are L mol^{-1} s^{-1}.

19. The ratio of k_{cat} to K_M is called the specificity constant. It is proportional to the initial reaction rate for the enzyme-catalyzed reaction. The ratio of specificity constants for two substances indicates the degree to which they will interfere in an enzymatic analysis.

20. One milligram of the preparation will produce 30 nmol product per second. Therefore, the total moles produced is equal to

$$30 \times 10^{-9} \text{ mol s}^{-1} \text{ mg}^{-1} \times 120 \text{ s} \times 10 \text{ mg} = 3.6 \times 10^{-5} \text{ mol}$$

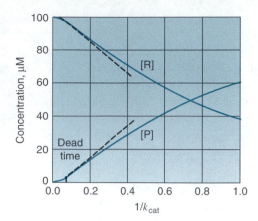

Figure 12.6. Time course of reactants and products for a first-order reaction. The dead time is the time for the reaction to reach its predicted rate. The dotted lines have the slopes of the theoretical initial reaction rate.

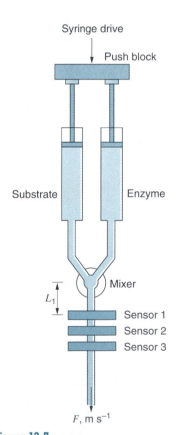

Figure 12.7. An apparatus for continuous flow kinetic measurement. The push block is moved downward at a steady rate, producing the mixture flow rate, F. Each sensor position samples the reaction mixture composition at a different time after mixing.

C. Kinetic Methods of Analysis

To obtain the best selectivity (differentiation) and the most accurate quantitation with enzyme reactivity, it is desirable to measure the initial rate of the enzyme reaction. To do so, we must mix the enzyme and substrate and then determine the rate of disappearance of the substrate or the rate of appearance of the product. The time course for the reactant and product concentrations for a typical, first-order kinetic determination is shown in Figure 12.6. Time zero on this plot is the instant the reactants are brought in contact with each other. In any practical system, some time is required for the mixing of the reagents. During this time, the reaction is proceeding at less than its theoretical rate. This time is called the **dead time,** the time between mixing and the earliest time a valid concentration measurement can be made. Dotted lines for the limiting initial rate of this reaction are shown in Figure 12.6 superimposed on the concentration profiles. We can see that the concentration profile follows the initial rate for only a short period after the dead time.

The time scale for this plot is in units of $1/k_{cat}$. On this scale, the dead time is 0.07, and the concentrations deviate from the initial rate values after 0.2. Values for k_{cat} fall in the range of 1 to 10^7 s^{-1}, so the plot time scale is between 10^{-7} s (100 ns) and 1 s. To obtain the initial rate, the dead time must be less than $0.1/k_{cat}$ s. Therefore, one of the goals of kinetic methods developers has been the invention of techniques and devices for quickly and thoroughly mixing the reactants. The other development of kinetic analysis is the technique used to determine the change in concentration of the substrate or product with time. For this, one must choose a differentiating characteristic for one of the product or reactant species. The most commonly used ones are photon absorption, photon emission, and electrode potential.

Quantitative analysis is achieved by calculating the initial reaction rate from the change in the reactant or product concentration with time for the very early part of the reaction. From Equation 12.16, we know that this initial rate is proportional to the substrate concentration. The usual method is to develop a working curve by measuring the initial rate for several known concentrations of substrate.

CONTINUOUS FLOW METHODS

In **continuous flow** methods, the enzyme and substrate solutions are continuously pumped into the mixing chamber. The effluent from the mixing chamber then flows through a tube while the reaction proceeds. As the distance along the tube increases, the concentration of product will increase and the concentration of reactant will decrease. If the pumping rate for the solutions into the mixing chamber is constant, the concentrations of the reactants and products will achieve a steady-state value at each point along the reaction tube. A sketch of this apparatus is shown in Figure 12.7. Turbulent mixing is achieved when a sufficient flow rate into and out of the mixer is maintained.

For a flow rate of F m s^{-1}, the solution has been mixed for L_1/F seconds when it passes sensor 1. If L_1 is 1 cm and the flow rate is 20 m s^{-1}, the reaction time at sensor 1 is 500 μs. If the sensors are spaced every centimeter down the reaction tube, the composition will be sampled at 500 μs intervals along the reaction time coordinate. This would be perfect for a reaction for which k_{cat} is about 100 s^{-1}. For reactions with a higher value of k_{cat}, higher flow rates and shorter measurement distances are used. Even though this technique was introduced over 75 years ago (by Hartridge and Roughton[5]), it is still being used and improved. In one recent apparatus using photon absorption, a dead time of about 10 μs was achieved.[6] This raises the accessible value for k_{cat} to 10,000 s^{-1}. Individual elements in an array detector provided the sensors for an equivalent of over 1000 sensor positions along the reaction tube.

For lower reaction rates, the automated flow injection techniques introduced in earlier chapters can be used for kinetic analysis. The approach is the same as that shown in Figure 12.7, except that standard flow method tubing, pumps, and reagent mixing methods are used. A sensor is placed downstream from the mixer, and the decrease in reactant or increase in product is determined. As long as the reaction conditions remain constant and first-order reaction conditions prevail, the amount of the reaction at a given position (time) will be proportional to the concentration of the substrate. A frequent application of a variation on this technique is the determination of any substrate that reacts with one of the "oxidase" enzymes. A product of an oxidase reaction is hydrogen peroxide. As we saw in Section C of Chapter 6, peroxide reacts with luminol to produce chemiluminescence. This reaction is made to go to completion by having a long reaction tube. Then the luminol is added in a second mixer. The luminescence intensity at a given point after mixing is related to the amount of hydroxide present, which is related to the amount of glucose originally reacted. In this way, glucose and many other compounds can be routinely analyzed with simple, automated apparatuses. An enzyme reactor in which the enzymes are immobilized is often used for routine analysis.

STOPPED FLOW METHODS

The **stopped flow** technique is a variant of the continuous flow technique described above. It, too, was introduced by Roughton[7] and then improved by Chance.[8] In it, a steady state is established and then the flow is suddenly stopped. The solution in the reaction tube then continues to react, but it is no longer moving. This allows a single detector to follow the remaining time course of the reaction. The apparatus is very similar to that of the continuous flow system of Figure 12.7, except that the reaction mixture goes on to fill a third syringe that starts out nearly empty. (See Figure 12.8.) The steady-state condition is established while the collection syringe is filling. The plunger of this syringe, however, will hit a stop block before it is full. When this happens, further flow is blocked. The collection syringe is called the **stop syringe**. For the example developed in the previous sub-section, in the steady-state condition, the reaction time sampled by the first detector is 500 μs. When the flow stops, the solution at this detec-

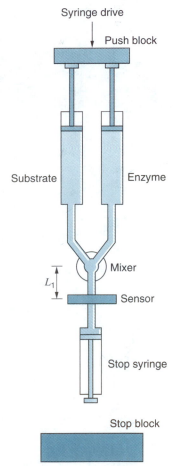

Figure 12.8. An apparatus for stopped flow kinetic measurement. The reaction solution fills the stop syringe, but the flow suddenly stops when the plunger of the stop syringe hits the stop block. The sensor follows the continuation of the reaction.

[5]H. Hartridge and F. J. W. Roughton, *Proc. R. Soc.* **1923**, *A104*, 376.

[6]M. C. Ramachandra Shastry and H. Roder, *Nature Struct. Biol.* **1998**, *5*, 385.

[7]F. J. W. Roughton, *Proc. R. Soc.* **1934**, *B115*, 475.

[8]B. Chance, *J. Franklin Inst.* **1940**, *229*, 445, 613, 637.

tor stage continues to react, and the detector output can reflect the composition changes going on during this time. Therefore, a single detector can sample the entire reaction time course from the steady-state time onward. A trigger attached to the stop block initiates the data collection from the sensor.

For the stopped flow method, the steady-state reaction time at the detector sets an upper limit on the rate of reaction that can be successfully followed. Commercial stopped flow analyzers routinely achieve 1 ms for this parameter. For the case where the initial reaction rate must be determined, this limits the value of k_{cat} to less than 100 s^{-1}. However, if the reaction kinetics are well established, the entire measurable time course of the reaction can be used to calculate a value for the substrate concentration. This has been a very active area of research for many years. With numerical methods, the effective k_{cat} range of reactions amenable to stopped flow analysis is 0.01 to 1000 s^{-1}.

The versatility of the stopped flow technique has led to its commercial availability and continued development. Modern systems are simple to operate, rugged in application, and require a reactant volume of only a few hundred microliters per analysis.

QUENCHING METHODS

For the rapid mixing methods of continuous flow and stopped flow, the system that monitors the changes in solution composition must also be rapid. The time required for each composition determination must be less than the dead time of the mixing system. For most systems, the concentration determination must be completed in 1 ms or less. This poses little difficulty for the technique of photon absorption. For photon emission methods, the emitted photon flux must be high enough to provide the needed measurement precision in this time frame. The photon flux available from chemiluminescence is so low that flux measurements of 1 s or more are required. In this case, the composition monitoring system becomes the limiting factor in the reaction rates that can be used. Many other powerful analytical tools do not provide their compositional information on the millisecond time scale. Among these are chromatography, nuclear magnetic resonance, and mass spectrometry. If it is desired to use any of these slower, batch methods as monitors for a kinetic analysis, it is necessary to halt the reaction at a specific time along the reaction course. The technique of suddenly stopping a reaction is called **quenching.**

The mechanism most often used to quench an enzymatic reaction is to add a reagent that deactivates the enzyme. The simplest one to implement is a solution of strong acid. A quenching system is shown in Figure 12.9. The reaction mixture flows from the first mixer into the second mixer, where it is quenched. The reaction time is equal to the time required for the reaction solution to flow between the two mixers. In operation, the drive is turned on, and collection is delayed until a steady-state effluent is achieved. This delay is just longer than the flow time of the reaction solution from the first mixer to the collection outlet. The collected solution has the composition of the reaction solution for the given reaction time. Since it is now stable with this composition, any analysis method can be used to determine the extent of the reaction. The reaction time can be varied on subsequent collections by changing either the flow rate or the distance between the two mixers. The collected solutions can then be analyzed at any time.

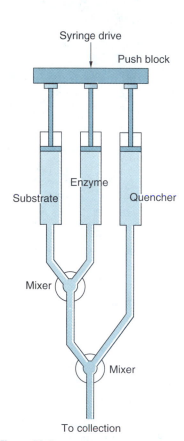

Figure 12.9. An apparatus for quenched kinetic measurement. When a steady state is reached, the reaction mixture is quenched in the second mixer. The quenching time is equal to the distance between the two mixers times the flow rate.

APPLICATIONS OF KINETIC ANALYSIS WITH ENZYMES

A representative example of the use of enzyme kinetic analysis is the determination of blood alcohol. The enzyme used is alcohol dehydrogenase, and the product of the reaction is acetaldehyde. In this reaction, NAD^+ is a cofactor that is reduced to NADH.

The extent of this reaction can easily be followed by UV absorption at 340 nm, where the NADH adsorbs strongly and the NAD^+ has negligible absorption. As the reaction proceeds, the absorbance at 340 nm increases.

The conversion of lactic acid to pyruvic acid by lactic acid dehydrogenase is used in the confirmation of heart attack. Again, NAD^+ is a cofactor, and the reaction can be followed the same way as for the blood alcohol assay. This specific method for the analysis of lactate can also be used as a diagnostic for hypoxia because of the related buildup of lactate in the blood caused by this condition.

Urea and uric acid can be determined with the aid of the enzymes urease and uricase, which facilitate their oxidation by oxygen. In the case of urea, the products are ammonia and CO_2. Uric acid is oxidized to allantoin ($C_4H_6O_3N_4$), CO_2, and hydrogen peroxide. The urea, which is usually in relatively high concentration in urine, is determined by following the ammonia released. The ammonia is used in another reaction that produces a colored product. An example is the reaction of ammonia, hypochlorite, and phenol to produce the blue indophenol. The uric acid has a characteristic absorption in the UV at 292 nm and could be analyzed directly. Here the absorbance change before and after the enzyme reaction is measured because of the interferences and the very low concentrations of uric acid normally encountered. The loss of absorbance is then taken to be due to the destruction of the uric acid by the uricase reaction.

Glucose oxidase catalyzes the reaction of glucose to gluconic acid. Because diabetics need to monitor their blood glucose levels often, an active area of research is the development of more accurate, convenient, and noninvasive ways to perform this test. (See the discussion of approaches to this problem in the side bar.[9])

In addition to the determination of substrates, these reactions are used to detect and quantitate the enzymes themselves. Enzyme levels in the blood can sometimes be used as indicators of various kinds of trauma. A suitable substrate for the enzyme is added to the sample solution, and the rate of product formation is related to the concentration of the enzyme present. An example is creatinine phosphokinase, which removes a phosphate group from the substrate and adds it to the cofactor ADP (adenosine diphosphate) to form ATP (adenosine triphosphate). The ATP production is then followed by a colorant reaction and absorption spectrometry. Elevated levels of creatinine phosphokinase-MB (one of three creatinine phosphokinase isoenzymes) are an indicator of myocardial infarction.

From these examples, it follows that enzyme kinetic reactions can be used in the determination of all the factors that can affect the rate of the enzyme reaction. This includes the cofactors and the inhibitors as well as the enzymes and substrates. As more is learned about the reactions of specific enzymes, more analytical possibilities are opened up.

The Search for the Ideal Glucose Monitor

The oxidation of glucose to gluconic acid via the enzyme glucose oxidase provides a very specific test for glucose. The variety of approaches that can be taken makes it a good example of the way in which analysts can adapt a differentiating characteristic into an effective method of analysis. The overall reaction is

$$\text{glucose} + H_2O + O_2 \xrightarrow{\text{glucose oxidase}} \text{gluconic acid} + H_2O_2$$

1. Choose an equilibrium or a rate method. Equilibrium (titration) requires no temperature control, but the reaction is slow and subject to interferences.

2. If rate is chosen, measure the disappearance of O_2 or the appearance of H_2O_2. A Clark electrode can be used for O_2, but for trace amounts, the change in O_2 could be very small.

3. Appearance of H_2O_2 can be determined by its oxidation of o-toluidine in the presence of horseradish peroxidase. The o-toluidine dye changes color on oxidation.

4. Appearance of H_2O_2 can be determined by electrochemical oxidation and measurement of the resulting current. A miniaturized electrode capable of being implanted in humans for continuous monitoring has been developed. Selectivity for the specific oxidation of H_2O_2 was achieved by surrounding the electrode with a thin layer of polymer that has selective permeability.

5. Other approaches have been developed. Still more are needed to achieve an ideal, noninvasive, portable, and inexpensive method.

Study Questions, Section C

21. In a first-order reaction, is the rate of change of the reactant and product concentrations linear with time?

22. What is the dead time of a reaction?

23. In making a kinetic measurement, why must the dead time be much less than $1/k_{cat}$?

24. For the reaction of β-lactamase with benzylpenicillin, the value of k_{cat} is 2000. For this reaction, what is the time scale for Figure 12.6?

25. A continuous flow system has five sensors spaced at 1 mm intervals along the reaction tube. The distance from the

[9]S-K Jung and G. S. Wilson, *Anal. Chem.* **1996**, *68*, 591.

mixer to the first sensor is 2 mm. For a flow rate of 52 m s^{-1}, what are the reaction times sampled by the sensors?

26. What reaction rate would be ideal for the system described in Question 25?

27. What flow rate should be used with the system described in Question 25 if the reaction is the hydrolysis of fumarate with the catalyst fumarase? Please see Table 12.3.

28. Justify the statement that values of k_{cat} greater than 100 s^{-1} cannot be used with a stopped flow system that has a dead time of 1 ms if the initial rate is to be directly observed.

29. For a first-order reaction to go to completion, a time equal to $3/k_{cat}$ is needed. For a glucose oxidase reaction to be complete, how long a reaction tube is needed if k_{cat} for the glucose oxidase reaction is 100 s^{-1} and the flow rate is 15 m s^{-1}?

30. For the quenching system of reaction rate measurement shown in Figure 12.9, why does the material that appears at the collection tube come to a constant concentration with time? How can you change the reaction time represented by the material collected?

Answers to Study Questions, Section C

21. For the first 10–20% of the reaction, the rate of change of the reactant and product concentrations is approximately constant. The overall time course of the reaction is not linear; it is exponential.

22. The dead time is the shortest time after mixing at which the composition of the solution will be related to the reaction rate. It is due to the time required to thoroughly mix the reactants.

23. If the reactants were mixed instantly, after a time equal to $1/k_{cat}$, the reaction would be 63% complete. To use the first, approximately linear, region of the reaction curve to advantage, the mixing time must be less than 10% of $1/k_{cat}$.

24. The time scale in Figure 12.6 goes from 0 to 1.0 $1/k_{cat}$ s. The reciprocal of the k_{cat} for benzylpenicillin, 2000 s^{-1}, is 500 μs, so the time scale would be 0 to 500 μs.

25. The reaction time at the first sensor is 2×10^{-3} m/52 m s^{-1} = 39 μs. The increase in reaction time to the next sensor is 1×10^{-3} m/52 m s^{-1} = 19 μs. Therefore, the sampling times, in microseconds, are 38, 57, 76, 95, and 114.

26. It would be good to have the last sensor in the region of 15% reacted. From Figure 12.6, this would be at a time equal to $0.2/k_{cat}$ s. Therefore, the ideal value for k_{cat} would be

$$1.14 \times 10^{-4} \text{ s} = \frac{0.2}{k_{cat} \text{ s}^{-1}}$$

$$k_{cat} = \frac{0.2}{1.14 \times 10^{-4}} = 1{,}750 \text{ s}^{-1}$$

27. The value of k_{cat} for the reaction of fumarate with fumarase (from Table 12.3) is 800 s^{-1}. The reciprocal of this value is 1.25 ms. The time at the last sensor should be about 15–20% of $1/k_{cat}$, or 2.5×10^{-4} s. The last sensor is 6 mm from the mixer, so the flow rate should be

$$F = \frac{6 \times 10^{-3} \text{ m}}{2.5 \times 10^{-4} \text{ s}} = 24 \text{ m s}^{-1}$$

28. If the dead time is 1 ms, and this is to be less than 1% of $1/k_{cat}$, then $1/k_{cat}$ must be 10 ms or more. The reciprocal of 0.01 s is 100 s^{-1}, so k_{cat} can be no greater than 100 s^{-1}.

29. The time $3/k_{cat}$ is 3×10^{-2} s. The length of tubing required is then

$$L = 15 \text{ m s}^{-1} \times 3 \times 10^{-2} \text{ s} = 45 \text{ cm}$$

30. After the first substrate and enzyme have reached the second mixer, the material in the tube between the mixers has a composition dependent on the distance from the first mixer, just as in the case of the continuous flow analyzer. Thus, the solution that enters the second mixer has a constant composition. In this mixer, the reaction is stopped, so the collection composition is very nearly equal to that which enters the second mixer. To sample the reaction at different reaction times, one has to vary the time required for the solution to get from the first mixer to the second mixer. This is done by changing the flow rate or the distance between the two mixers.

D. Antigen–Antibody Reactivity

Antibodies are proteins that are produced by an animal to protect it against the activity of foreign substances that have been introduced into its tissues. They act by binding to the large molecules in the foreign substance in order to deactivate, destroy, and/or remove them. Each antibody generated is extremely specific for particular intruder molecules. In fact, in most cases, an antibody has been designed and produced specifically in response to each invader. A molecule that can trigger the production of an antibody is called an **antigen** or an **immunogen.** Usually, the molecule targeted by the antibody produced is the antigen itself. Antigens are always very large molecules such as proteins, polysaccharides, or oligonucleotides. The section of the antigen molecule that triggers the production of the antibody is called the **haptenic determinant.** This determinant usually contains an organic functional group such as dinitrophenyl that is essential to the antibody triggering process. When the organic group exists as a separate organic molecule, it is called a **hapten.** However, the hapten, by itself, is not an effective trigger.

The bonding between the antibody and the antigen is made up of the same forces that we saw between the enzyme and its substrate. In other words, the bond is a combination of hydrogen bonding, van der Waals forces, and coulombic and hydrophobic interactions. Since antibodies and antigens are very large molecules, only a portion of the antibody and antigen are involved in this bonding. These portions are called the **epitope** (for the antigen) and the **paratope** (for the antibody). The bonding to form the antibody–antigen complex is reversible, so the reaction is generally written as

$$Ab + Ag \rightleftharpoons AbAg \qquad\qquad 12.18$$

where Ab and Ag are the antibody and antigen, respectively. The formation constant for this complex is generally quite large (of the order of 10^8 to 10^{10}) and extremely specific. The great specificity comes from the large area of overlap between the epitope and paratope. This means the quality of being able to participate in an antibody–antigen reaction can be a highly differentiating characteristic and therefore of potentially great analytical value. Antigen–antibody interactions have been used for detection and rough quantitation in biological systems for several decades. Their application in the broader context of chemical analysis has been one of the most fertile fields of development in the last decade.

ANTIBODIES AND THEIR GENERATION

The response of an animal to inoculation with an antigen follows a complex series of steps that are beyond the scope of this text. Of interest, however, are the types of antibodies produced and the time scale of their production. One type of antibody, immunoglobulin M, is produced in the first few days, followed by another type, immunoglobulin G, which is produced in larger quantity over the next few weeks. These antibodies are part of a group of proteins called **immunoglobulins.** There are five classes of immunoglobulins with symbols IgA, IgD, IgE, IgG, and IgM. In animal blood serum, the IgG antibodies are present in the largest concentrations.

All the immunoglobulins are composed of similar structural units. The overall structure of the IgG antibody is shown in Figure 12.10. It is composed of three subunits, each having a molecular weight of about 50,000 g mol^{-1}. The two Fab subunits contain the antigen binding sites. The intersection of the three subunits is quite flexible, so the overall structure can readily assume a Y or T shape or any angle in between. In each class of immunoglobulins, the majority of the protein sequence is the same for all the antibodies produced. In the area of the antigen binding site, however, the sequence is variable. It is this part of the molecule that confers its specificity for each particular antigen target. These same substructures are found in different combinations in all the immunoglobulins. The five immunoglobulin classes and some of their characteristics are listed in Table 12.4.

Immunoglobulins are manufactured in cells called **lymphocytes.** If an antigen has more than one haptenic determinant or if the intrusion contains more than one antigen, different lymphocytes may generate antibodies that target the different determinants contained in the triggering intrusion. The serum produced is called **polyclonal** because it contains antigens specific for more than one haptenic determinant. The development of the technique for generating **monoclonal** antibodies was a major break-

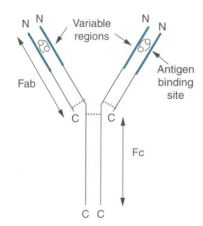

Figure 12.10. Structure of goat immunoglobulin G. The two Fab subunits have the antigen binding sites near their tips. The Fc subunit affects the biological function of the antibody. C and N termini of the proteins are indicated. The dotted lines are disulfide linkages.

Table 12.4
Characteristics of Immunoglobulins

Class	Amount in Blood Serum, %	Approximate MW, g mol^{-1}	Features
IgG	80	1.5×10^5	Crosses cell membranes and diffuses into extravascular spaces
IgA	13	3.6×10^4	Secreted in saliva, tears, and nasal fluids as a dimer with an additional chain
IgM	6	1.0×10^6	Tetramer of units similar to IgG with an additional chain; antigen precipitating agent found principally in the bloodstream
IgD	1	1.6×10^5	Function not yet known
IgE	0.002	1.9×10^5	Parasite protection and allergic reactions

through in biochemistry and paved the way for a substantial improvement in the specificity of antibody reactions available to the analyst. Monoclonal antibodies are all produced by the same gene and therefore have the same amino acid sequence.

As mentioned earlier, small organic molecules do not act as antigens. A molecule must have a molecular weight of at least 4000 g mol^{-1} to trigger the immune response. However, if a small organic molecule is covalently bonded to the surface of a protein such as serum albumin, a haptenic determinant may be produced on the albumin molecule. When the bound albumin is injected into an animal, antibodies against this antigen will be produced. Isolation of the system that produces a particular antibody is the key to producing an antibody solution that is monoclonal.

DETECTING THE ANTIBODY–ANTIGEN REACTION

The complex formed between an enzyme and its substrate is not catalytically active. Therefore, to use this reactivity as a differentiating characteristic, a change induced by the formation of the complex must form the basis of the measured response to the antibody probe. The method most often used to accomplish this is to tag one of the reactants (usually the antigen) with a subgroup that can be detected by radioactive emission or fluorescence. The unreacted antigen is then separated from the antigen–antibody complex, and the amount of the tagged substance remaining is determined. The methods of separation and complex determination are treated in later sections of this chapter.

In addition to forming a complex, some antibody–antigen reactions form a precipitate. This reaction occurs because multiple haptenic determinants often occur on an antigen and because multiple antigen binding sites occur on the antibodies. This enables a precipitation process in which an insoluble matrix of antigens and antibodies is formed through cross-linking, as shown in Figure 12.11. This process is called **immunoprecipitation.** When immunoprecipitation is performed in solution, the antigen is added to an excess of antibody. The precipitate formed creates a turbid solution. The degree of turbidity can be measured by measuring the loss of light transmission through the sample or by determining the fraction of incident light scattered by the precipitate particles. Note that drying and weighing, the usual method of quantitation with precipitation, are not used. This is because drying to a constant weight is not achievable with these precipitates. They decompose before complete dryness. Since high accuracy is not achievable by weighing, the quicker methods involving light absorption or light scattering are used.

The most elegant and widely used methods of determining the antibody–antigen reaction are the various forms of immunoassay. These are described in a later section of this chapter.

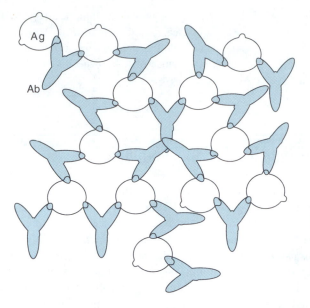

Figure 12.11. Formation of a precipitate through the cross-linking of antigen–antibody complexes. The Y shapes are the antibodies, and the circular shapes are the antigens, with bumps indicating the haptenic determinants.

Study Questions, Section D

31. In an animal under invasion by a foreign protein, which comes first, the antibody or the antigen?

32. What is the difference between a haptenic determinant and a hapten?

33. What is the nature of the bond formed between the epitope and the paratope regions of the antibody and the antigen? Which regions belong to which molecules?

34. Is it true that antigens have only one haptenic determinant per molecule?

35. What is the function of the Fab section of an immunoglobulin?

36. What is the time scale and function of the IgM antibody, and what makes it so large?

37. Is the serum normally produced in response to an intrusion polyclonal or monoclonal? Why?

38. Each paratope region on an antibody can complex only one epitope region of an antigen. The complex remains soluble. How then does a precipitate of the antigen and antibody molecules form?

39. In immunoprecipitation, how is the amount of precipitate formed determined?

Answers to Study Questions, Section D

31. The antigen triggers the formation of the antibody.

32. The hapten is the organic substructure that triggers the production of an antibody when it is covalently bonded to a large molecule. The region of the large molecule containing the hapten group is called the haptenic determinant.

33. An epitope region of an antigen bonds to a paratope region of the antibody designed for it. The bond is a combination of van der Waals forces, hydrogen bonding, coulombic interactions, and hydrophobic interactions.

34. No, it is not true that an antigen molecule has only one haptenic determinant. Antigens usually have several repetitions of each type of haptenic determinant. They may also have several types of haptenic determinants.

35. The Fab sections contain the antigen binding sites. They contain a variable region that has been tailored to the haptenic determinant.

36. The IgM antibody is released into the bloodstream shortly after the introduction of the antigen (within the first few days). According to Table 12.4, its role is to precipitate the antigen out of the bloodstream. It is large because it contains four of the units that make up the more common IgG antibody.

37. The serum normally produced is polyclonal because of the number of different haptenic determinants that have been introduced by one or more antigens.

38. Each antigen and each antibody have several epitope and paratope regions. This allows each one to react with sev-

eral other molecules. In this way, the individual complexes can be cross-linked together into a much larger cohesive mass that is eventually large enough to precipitate.

39. The precipitate produces turbidity in the solution that reduces the transmittance of light through the sample and increases the amount of light scattered by the sample. The more precipitate that is formed, the lower is the transmittance and the greater the scattering. The scattering or transmittance of unknown samples is compared with that of standards.

E. Immobilized Enzymes and Antibodies

As the use of enzymes and antibodies for their specific chemical reactivity increased, the desire to reuse these costly reactants grew. The attachment of these reactants to some form of solid support makes it easier to separate them from the soluble reactants and products. As we shall see, such immobilization also enables other applications in sensing and separation. The two general approaches by which enzymes and antibodies can be immobilized are containment and bonding. In containment, the biochemical reagent is physically separated from the solution phase. In bonding, the reagent is bonded to a solid phase but is immediately accessible to the solution.

BONDING BIOCHEMICAL REAGENTS

Enzymes and antibodies can be bonded to solid supports by either adsorption or covalent bonding. Adsorption is easily achieved, as proteins adsorb readily on a number of solid surfaces such as alumina, charcoal, silica gel, and glass. This method has often been used for developing new methods. Unfortunately, desorption and deactivation occur quite rapidly with this method of attachment, so the useful lifetime of adsorbed reagents is of the order of one day. For this reason, much effort has gone into the development of methods of covalently binding the biochemical reagents to a solid support. For covalent binding, the solid surface is generally glass or an organic polymer. Reactive groups at the solid surface can be used to provide a covalent link between the atoms in the solid and those in the enzyme or antibody protein.

Many chemical schemes for bonding the protein to the solid surface have been developed. Most commonly, the amino groups that form the N terminus of proteins are reacted with an acidic function at the solid surface. Alternately, a basic group on the solid can react with the carboxylic acid group at the C terminus of the protein. Chemistries have been developed for modifying and bonding to surfaces on which the reactivity is provided by hydroxyl, sulfhydryl, and siloxyl groups as well as amines, carboxylic acids, and esters.[10,11] The synthetic steps involved must occur under mild conditions of temperature and pH to avoid disturbing the protein structure and therefore its activity. An example of a reaction of a surface carboxylic acid group and the N terminus of a protein is shown in Figure 12.12. The achievement of bonding without distorting the active sites has been one of the more difficult aspects of this development. In current practice, the active sites are protected by a competitive inhibitor or hapten during the attachment synthesis. The protecting groups are then displaced to reactivate the sites. Covalently bound enzymes and antibodies can maintain their activity for months with careful treatment and storage.

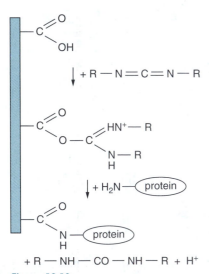

Figure 12.12. One of many synthetic schemes for the covalent bonding of proteins to a solid surface. The amino end (N terminus) or carboxylic end (C terminus) are most often used to form the covalent bond.

[10]R. Kellner, J.-M. Mermet, M. Otto, and H. M. Widmer, Eds., *Analytical Chemistry*, pp. 381–383. Wiley-VCH, Weinheim, 1998.

[11]B. Eggins, *Biosensors: An Introduction*, pp. 35–37. John Wiley & Sons, Chichester, 1996.

CONTAINING BIOCHEMICAL REAGENTS

Containment methods for biochemical reagents take advantage of the great size difference between the biochemical reagent and the reactants and products in the techniques for which they are designed. Two methods predominate, namely, membrane encapsulation and polymer entrapment. In both approaches, small molecules and ions can diffuse through the containment material to gain access to or depart from the biochemical reagent. The pores of the containment material, however, are too small for diffusion of the reactive proteins they contain. Therefore, this method of immobilization is not useful for reactions that involve other large biomolecules.

Membrane materials used for encapsulation include cellulose acetate, polycarbonate, collagen, Teflon, Nafion, and polyurethane. Each of these has some selectivity for the types of molecules that can diffuse through it. For example, Teflon is selectively permeable to gases such as oxygen. This type of immobilization is used when the product of the reaction, such as oxygen or hydrogen peroxide, can be detected as it exits the encapsulated reaction region. Examples of membrane applications are found in section H on sensors in this chapter. The active regions of the biochemical reactants are not well protected in this type of immobilization, and thus they lose their activity over time. Generally, the active material needs to be replaced every week.

Physical retention of the biochemical reactant in a polymer matrix can extend the useful life of the reagent to up to a month. In this approach, the biochemical reagent is mixed with one of the polymer-forming reagents. The other reagent is then added, and the polymer forms around the biochemical reagent, effectively trapping it. Polyacrylamide gel is the most often used, though gelatin, agar, and many other polymers have been used. Many of the biochemically mediated reactions to be employed involve water, so the hydration properties of the polymer can be very important. Recently developed film-forming emulsion polymers allow microencapsulation of the biochemical reagent and optimization of the pH and hydration properties of the encapsulating film for each application.[12]

Immobilized biochemical reagents are rapidly gaining applications in the areas of quantitation, separation, and detection. Many important applications of these materials are found in the next three sections.

Study Questions, Section E

40. Which of the methods of immobilization enables the longest period of activity?

41. Which of the immobilization techniques will allow the reagent to be used for reactions with large biomolecules, and which will not?

42. How are the activity and shape of proteins maintained through the chemical steps involved in immobilization?

Answers to Study Questions, Section E

40. The method of covalent bonding provides activity for up to a year. Next in longevity are the polymer entrapment, with a lifetime of one month, and membrane containment, which lasts one week. Adsorption is the easiest to implement but lasts only about one day.

41. The containment immobilization techniques depend on the small pore size of the containment material to prevent the loss of the bioactive reagent. These same pores will prevent large molecules from reaching or leaving these reagents. The immobilization techniques of adsorption and covalent bonding leave the biochemical reagent free to interact with all species in the adjacent solution.

42. The chemical processes in immobilization of bioactive reagents are carried out under conditions that do not destabilize the protein, and the active sites are protected by prior complexation with a reactant that can later be removed.

---

[12]N. Martens and E. A. H. Hall, *Anal. Chim. Acta* **1994**, *292*, 49–63.

F. Separation with Biochemical Reactions

There are two significant aspects of separation when applied to reactions involving biomolecules. One is the method used to separate the reaction products from the reaction mixtures when that is a necessary part of the determination. The other is the use of the selective bonding of enzymes and antibodies to aid in the separation of various molecules according to the degree of their bonding to the biomolecules. The great size and mass of the proteins involved in biochemical reactivity make them amenable to techniques based on these physical characteristics. All these processes will be introduced in this section.

SEPARATION ON THE BASIS OF SIZE

The simplest separation based on size is filtration using a filter material with a pore size that will differentiate among the materials to be separated. The size ranges of the ions, molecules, and particles of interest in systems of biochemical reactivity are shown in Table 12.5. Cellophane, which is a thin film of cellulose acetate, has pore sizes in the range of 4–8 nm. Macromolecules with a molecular weight of greater than 10,000 (for which the diameter would be roughly 2 nm) have difficulty passing through it. If a cellophane membrane is used to separate two solutions, the solvent and electrolyte ions will pass freely through it, but the proteins will not. When the driving force for the net exchange of ions and solvent through the membrane is the concentration differences between the two solutions, the process is called **dialysis.** For example, a volume of solution containing proteins and an undesirable electrolyte can be placed in a dialysis container, which in turn is placed in a beaker containing a much larger volume of the desired solvent buffer, as shown in Figure 12.13. In a few hours, the buffer and electrolyte concentrations will have equilibrated, removing most of the undesired electrolyte and replacing it with the desired buffer solution. If further improvement in the solvent composition is desired, the process can be repeated with fresh buffer solution.

In filtration, a solution is forced through the filter material. Since only the solvent and particles smaller than the pore size will get through, the larger ions and molecules remain on the inlet side of the filter material. Polymeric materials with a variety of pore sizes have been developed for this purpose. Since the pores are necessarily very small, the force of gravity is insufficient to drive the solution through the membrane at a reasonable rate. When a pump is used to provide the force, the process is called **ultrafiltration.**

You will recognize filtration as a batch process that can only discriminate between ions and molecules that are much larger or much smaller than the pore size. When finer degrees of discrimination are required, a continuous process is needed, as we saw in Chapter 11. Thus, a type of chromatography has been developed for separation based on particle size. It is called **gel permeation chromatography.** In this technique, the chromatographic column is packed with beads of a porous gel. The pore size of the gel material has been carefully controlled in manufacture. The mobile phase is usually an aqueous buffer solution in which the macromolecules are stable.

As the portion of the mobile phase containing the sample encounters the bead packing, the ions and molecules that can diffuse into the pores of the beads will do so. The extent of the diffusion into the pores depends on the size and shape of the particle relative to the pore dimensions. While a particle is in the pore, its velocity through the column is zero; while it is in the mobile phase, it has the mobile phase velocity. Thus, just as in partition chromatography, the relative velocity of the component depends on the fraction of the time it spends in the stationary phase. In this situation, there are two limits. Particles that are too large to enter the pores will remain in the mobile phase, and all will have the mobile phase velocity. The mass value above which this is true is

Table 12.5
Particle Sizes

Particle	Diameter, nm
Small molecules and ions	<1.0
Proteins, viruses	1.0–100
Bacteria	100–1000

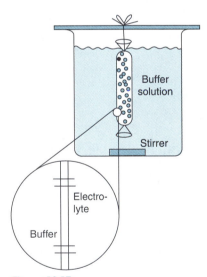

Figure 12.13. In the process of dialysis, the electrolytes, solvents, and buffer components can equilibrate (come to equal concentrations) across the membrane. The proteins and other macromolecules remain inside the dialysis container.

called the **exclusion limit.** At the other extreme, particles that are so small that they can readily diffuse through the entire pore structure of the beads will all have the minimum velocity. They enter new bead material as the portion of them in the mobile phase moves through the column, and they diffuse out of the equilibrated beads as they encounter the fresh mobile phase solution behind them. Between these two extremes, particles elute in order of size, with the largest particles eluting first.

The resulting chromatograms are a series of peaks, just as those in Chapter 11. The time scale of the chromatogram is in hours, with peak widths of the order of 30 min. People working in gel permeation chromatography characterize the elution of the components in terms of the volume units of the mobile phase. The volume of the space between the beads is called the **void volume,** V_0. The total available volume (void volume plus pore volume) is the bed volume, V_t. The void volume is typically one-third of the bed volume. After injection of the sample, the volume of mobile phase at which the molecules above the exclusion limit will appear is the void volume, V_0. The smallest molecules have access to the total bed volume, so the mobile phase volume at which they will appear is the bed volume, V_t. Since the bed volume is about 3.3 times the void volume, the retention volume, V_e, of all molecules will appear between V_0 and $\sim 3.3 V_0$.

A plot of appearance volume versus the log of mass for a few very large molecules is shown in Figure 12.14.[13] The mass units are kilodaltons (kD). The **dalton** (also known as amu, or atomic mass unit) is the unit used to express molecular mass. The dalton is often used by people working with very large molecules. As expected, the curve flattens out at a value of 1.0 for all molecules over the exclusion limit, which in this case is almost 10^6 daltons. Similarly, the curve flattens out for low mass molecules at an elution volume just over 3 times the void volume. The shape of the curve is determined by the distribution of the pore sizes in the gel beads used. A narrower range of pore sizes will make the curve steeper in the useful working range, but the mass range covered by the sloping part of the curve will be smaller. Several manufacturers supply beads in a variety of pore sizes and pore size distributions. For the Sephadex G-200 beads used for the data in Figure 12.14, the mass range over which separation can be expected is given as 5 to 600 kD. Some of the other gel permeation materials available are listed in Table 12.6. A selection among them is made to just barely cover the mass range needed.

Gel permeation chromatography is used not only as a means of separation, but also as a means to estimate the molecular weight of macromolecules. The molecular mass estimation works because most macromolecules fall close to the same curve relat-

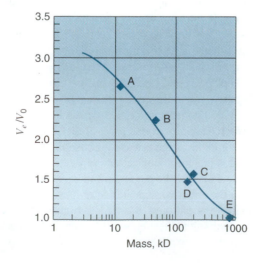

Figure 12.14. Retention volume versus mass for a Sephadex G-200 gel permeation column. Most molecules fall relatively close to the line. The specific examples in the data points are as follows: A, cytochrome c; B, ovalbumin; C, fumarase; D, γ-globulin; E, α-crystallin.

[13]P. Andrews, *Biochem J.* **1965**, *96*, 597.

Table 12.6
Gel Permeation Materials

Name and Supplier	Mass Range, kD
Sephadex by Pharmacia*	
G-10	0.05–0.7
G-25	1–5
G-50	1–30
G-100	4–150
G-200	5–600
Bio-Rad**	
P-2	0.1–1.8
P-6	1–6
P-10	1.5–20
P-30	2.4–40
P-100	5–100
P-300	60–400

*Sephadex is made from dextan by Parmacia Fine Chemicals AB.

**The P series is made from polyacrylamide by Bio-rad Laboratories.

ing mass to relative retention volume. To determine mass, one would elute the unknown species with several species of known mass. The known species would be used to construct the curve, and the unknown mass could then be determined by the intersection of its retention volume with the curve. As one can see from the specific proteins in Figure 12.14, not all proteins fall on the curve. This is due to differences in density, shape, and hydration among proteins of different types. The differentiating characteristic being probed by this technique is the bead volume to which the analyte has access.

SEPARATION ON THE BASIS OF COMPLEXATION REACTIVITY

The specific binding of enzymes, antibodies, and other proteins to specific molecules can be used as a method of separation, as shown in Figure 12.15. In this method, a molecule that binds to the analyte that is to be separated is immobilized on a solid support. This material is then used in a column through which a solution containing the analyte material is passed. The analyte will complex selectively with the immobilized ligand and thus remain attached to the solid support. The rest of the material in the sample is eluted in a single void volume. The mobile phase composition is then changed to a formulation that will cause the analyte to unbind from the immobilized ligand and thus be eluted in a highly purified form. This process is called **affinity chromatography** because the method of separation is based on the affinity of the analyte to the stationary phase.

If there were several components in the sample that had an affinity for the immobilized ligand, they could be released in reverse order of their formation constant by a gradual increase in the concentration of the releasing constituent in the mobile phase. This approach has been exploited in a technique called immobilized metal affinity chromatography or IMAC.[14] For this approach, the binding affinities of the analytes need to be in an intermediate range so that the effective capacity factor is not too large or too small.

In a single-step affinity separation, a high and highly specific binding affinity is desired. The other sample components are simply washed off the column, and then the analyte is released in a single step. In this sense, the technique is a batch rather than

---

[14]J. Porath, J. Carlsson, I. Olsson, and G. Belfrage, *Nature* **1975**, *258*, 598–599.

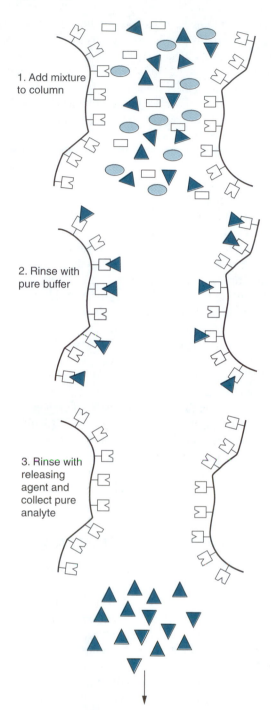

1. Add mixture to column

2. Rinse with pure buffer

3. Rinse with releasing agent and collect pure analyte

Figure 12.15. Steps involved in affinity chromatography. The sample solution is brought into contact with the immobilized ligand. The analyte adheres to the ligand, and the remaining solution components are washed off. Finally, the purified analyte is eluted with a solution that releases the analyte–ligand bond.

continuous process and therefore not strictly chromatography. In the analysis of 2,4-dichlorophenoxyacetic acid (the pesticide 2,4-D) in water, a single-step purification of 30,000 : 1 is achieved by affinity chromatography.[15]

There are two difficulties in implementing affinity chromatography. One is the effective immobilization of the ligand in such a way that it is still free to react with the analyte. The second is the dissociation of the analyte after the rest of the sample has been washed away. The immobilization is accomplished by techniques discussed in section E of this chapter. The unbinding is accomplished by competitive binding or by

---

[15]T. Dombrowski, G. Wilson, and E. Thurman, *Anal. Chem.* **1998**, *70*, 1969–1978.

changing the solution conditions to make the binding less favorable. In the former case, the elution solution contains a molecule that has a higher complex formation constant for the ligand than does the bound analyte. This molecule will displace the analyte from the ligand and allow it to be washed out. The more often used technique is to change the pH or the ionic strength of the solution to distort the active site of the protein and reduce its formation constant. This process must be tested for its reversibility if the original form and activity of the protein need to be regained.

Biochemical species that have been used in affinity chromatography include enzymes, antibodies, lectins, receptor proteins, and nucleic acids. **Lectins** are a class of proteins that have specific binding to certain carbohydrate groups. In the case of enzymes, the substrate, coenzyme, or competitive inhibitor is most often immobilized, and the desired enzyme is selectively removed from the sample solution. With antibodies, either the antigen or the antibody can be immobilized. Immobilized monoclonal antibodies can then be used to select for their antigens or haptens.

Immobilized haptens or antigens can be used for purification of the related antibodies. They are generally the immobilized component used for the separation of carbohydrates or certain carbohydrate-rich proteins. Receptor proteins are involved in the hormone recognition activity of cells. Such receptor proteins can be specifically recovered from solution by immobilizing the compounds they recognize. Nucleic acids will specifically bind complementary strands of DNA. The immobilization of a particular polynucleotide provides a means of separating the sample molecules that contain a complementary sequence.

Affinity chromatography is playing an increasingly important role in chemical analysis. It provides highly selective separation along with a great reduction in the sample volume. As such, its application has vastly improved the detection limits of many instrumental techniques. It can accomplish truly difficult separations from extremely complex mixtures, including the separation of optical isomers. New methods and materials are continually being developed for affinity chromatography. In working with analytes that are found in biological fluids, such specific methods are essential because of the great complexity of the sample. For a review of the applications of affinity methods, see Larive *et al.*[16]

SEPARATION ON THE BASIS OF MASS AND DENSITY

A mainstay of biochemical research is the **ultracentrifuge.** With it, macromolecules are forced to form a sediment roughly in order of their mass, with the most massive molecules at the bottom of the centrifuge tube. This is achieved by placing the sample solution in a special centrifuge tube that is then placed in one of the tube holders of the centrifuge. The centrifuge spins these tubes very rapidly, and the bottom of the tube spins outward from the center of rotation. The centrifugal force on a molecule in the tube is related to its mass and to the rotational velocity of the centrifuge. This force is counteracted in solution by the need for the molecule to displace an equivalent volume of the solution. The net sedimentary force provided by the rotation of the tube is then

$$F_{sed} = (m_m - m_s)\omega^2 r \qquad\qquad 12.19$$

where m_m is the mass of the molecule, m_s is the mass of the equivalent volume of solution, ω is the rotational speed of the centrifuge in radians per second, and r is the distance of the molecule from the center of rotation in centimeters. The mass of an equivalent volume of solution is equal to the volume of the particle, V_m, times the density of the solution ρ_s. In addition, the volume of the molecule is equal to the mass of

[16]C. Larive, S. Lunte, M. Zhong, M. Perkins, G. Wilson, G. Gokulrangan, T. Williams, F. Afroz, C. Schöneich, T. Derrick, C. Middaugh, and S. Bogdanowich-Knipp, *Anal. Chem.* **1999**, *71*, 396R–398R.

the molecule times the **partial specific volume** $\overline{V}_m$ of the molecule in cubic centimeters per gram. These two equations are

$$m_s = V_m \rho_s \qquad\qquad 12.20$$

and

$$V_m = m_m \overline{V}_m \qquad\qquad 12.21$$

When Equations 12.19, 12.20, and 12.21 are combined,

$$F_{sed} = m_m(1 - \rho_s \overline{V}_m)\omega^2 r \qquad\qquad 12.22$$

from which we see that the sedimentation force is related to the molecular mass. The product of the solution density and the partial specific volume of the molecule must be less than 1 for sedimentation to occur. Another way to say this is that the density of the molecule must be greater than that of the solution. This point is even clearer if Equation 12.22 is rearranged in terms of the relative densities of the molecule and the solution:

$$F_{sed} = m_m(\rho_m - \rho_s)\overline{V}_m\omega^2 r \qquad\qquad 12.23$$

where ρ_m is the density of the molecule.

The sedimentation force accelerates the molecule toward the end of the tube. As the molecule gains velocity through the solution, the frictional force fv impeding its motion increases. This force is proportional to the friction coefficient of the molecule f and the velocity v. When the velocity has increased to the point where the frictional force is equal to the sedimentation force, the velocity becomes constant. The equation for the limiting velocity, now called the **sedimentation rate** v_{sed}, is

$$v_{sed} = \frac{m_m(1 - \rho_s \overline{V}_m)\omega^2 r}{f} \qquad\qquad 12.24$$

Recall that m_m is the actual mass of the molecule. To express this in terms of the molecular weight of the molecule, MW/N_A (where N_A is Avogadro's number) must be substituted in Equation 12.24 for m_m.

$$v_{sed} = \frac{MW_m(1 - \rho_s \overline{V}_m)\omega^2 r}{N_A f} \qquad\qquad 12.25$$

From these equations, we can see that the sedimentation rate is proportional to $\omega^2 r$. This rate can be measured by noting the rate at which the position of a species changes with time. This rate is expressed as $v_{sed} = dr/dt$. With this substitution (and an integration),

$$\frac{1}{r}\frac{dr}{dt} = \frac{MW_m(1 - \rho_s \overline{V}_m)\omega^2}{N_A f} = \frac{\Delta \ln r}{\Delta t} \qquad\qquad 12.26$$

where $\Delta \ln r$ is the change in the natural log of the position (in centimeters) over the time Δt.

The sedimentation rate measured is proportional to ω^2. To characterize the sedimentation rate without this experimental variable included, Equation 12.26 is rearranged.

$$\frac{MW_m(1 - \rho_s \overline{V}_m)}{N_A f} = \frac{2.303 \Delta \log_{10} r}{\omega^2 \Delta t} = s \qquad\qquad 12.27$$

where s is called the **sedimentation coefficient** and has the units of seconds. (See Example 12.2.) The value of s is a characteristic of the molecule being centrifuged (if the density of the solution is taken as a constant). Values of s have been measured for many biological molecules in the buffer solutions most commonly used with them. For

Example 12.2

How fast do they spin?
Suppose we wished the sedimentation rate to be such that a protein with an s value of 10×10^{-13} s would move 0.1 cm (from an r of 6.0 to 6.1 cm) in 5 hours. We can calculate what rotational speed would be required from Equation 12.27.

$$\omega^2 = \frac{2.303(\log 6.1 - \log 6.0)}{1.0 \times 10^{-12}\,\text{s} \times 1.8 \times 10^4\,\text{s}}$$
$$= 9.2 \times 10^5\,\text{s}^{-2}$$
$$\omega = 960\,\text{rps}$$

To get the rotational speed in rotations per minute (the usual rpm), multiply the rps by 60 to get 57,500 rpm.

This value is toward the upper end of rotational speeds available in commercial ultracentrifuges. Lower speeds or lower sedimentation coefficients will require more patience.

proteins, they vary from 1×10^{-13} s to 100×10^{-13} s. For convenience, sedimentation coefficients are given in units of 10^{-13} s called Svedbergs (S) after the person who developed ultracentrifugation.

An approximate relationship between the sedimentation coefficient and the molecular weight can be obtained from Equation 12.27. The reason that it is not the simple relationship suggested by this equation is that the frictional coefficient is a function of the molecular radius, which is related to the molecular weight. If one assumes the proteins are spherical and of equal density, the radius of the molecule will be proportional to the cube root of the molecular weight. According to Stokes law, the frictional coefficient is proportional to the radius of the particle. Therefore, the frictional coefficient should be roughly proportional to the one-third power of the molecular weight.

$$f = K(MW)^{1/3} \qquad\qquad 12.28$$

Also, since we have assumed that the density of the proteins is approximately constant, the term $(1 - \rho_s \overline{V}_m)$ will also be a constant. Substituting Equation 12.28 into Equation 12.27 and lumping the constants, we get

$$s = K'(MW_m)^{2/3} \qquad\qquad 12.29$$

From this derivation, we can expect the sedimentation coefficient to increase roughly with the two-thirds power of the molecular weight.

Figure 12.16 is a plot of the sedimentation coefficients in Svedbergs of several proteins as a function of their molecular weight. A log–log plot was used to obtain a line with a slope of 2/3. To understand the deviations of the actual points from the theoretical line, we can look at the assumptions involved in this derivation. One was that the density of all proteins is the same. This assumption is quite good. The partial specific volume only varies from 0.70 to 0.75 among different proteins. A greater relative variation, however, is found in the way the friction coefficient f varies with the mass of the molecule. The actual frictional coefficient f can be compared with the calculated frictional coefficient f_0 assuming a spherical molecule of the protein's molecular weight. The ratio of f to f_0 is called the **frictional ratio.** This ratio varies from just a little over 1 to well over 2 among various proteins. It is this substantial variation that is the basis for most of the deviation from the linear relationship plotted in Figure 12.16. In addition, the fact that the frictional ratio tends to increase slightly with molecular weight causes the slope of the theoretical line to be slightly greater than the general trend of the points.

Structural information can be obtained from the frictional ratio, as the increase in value over unity is generally due to an increase in hydration or to a deviation from

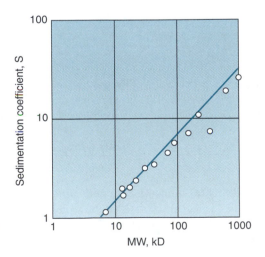

Figure 12.16. Plot of the sedimentation coefficient versus molecular weight for a variety of proteins. The straight line follows the theoretical relationship, $s = K'(MW)^{2/3}$.

spherical shape. Globular proteins have frictional ratios of less than 1.5, but fibrinogen, which has an elongated shape, has a frictional ratio of over 2.3. Denatured proteins, which are no longer folded, have much higher frictional ratios than when their secondary and tertiary structures are intact.

There are many useful analytical applications of the ultracentrifuge. One is the ability to obtain an approximate molecular weight from an observed sedimentation rate. This technique, however, requires that the type of the compound be known and that a curve similar to that of Figure 12.16 is available for that compound type. The average value of the partial specific volume of DNA molecules, for instance, is only about 0.5. Since they are more dense than proteins, DNA molecules would all fall on a line that is well above that for proteins in Figure 12.16.

Several methods of fractionation based on relative sedimentation rate have been developed. In the **moving boundary** method, one begins with a homogeneous mixture. During the centrifugation, the molecules with the greater sedimentation rate clear the upper regions of the solution faster. For each species, there will be a level above which it is absent at any given time. This boundary moves down the tube as the rotation continues. The centrifuge is stopped when the desired component has reached the bottom. The supernatant liquid is then poured off and replaced with fresh solvent, and the tube contents are mixed. The process is then repeated, stopping when the boundary of the next heavier component reaches the bottom. The liquid fraction is now enriched in the desired component. In the **moving zone** method, the species to be separated are added to the top of the tube just before the rotation is started. During rotation, they will move down the tube at different rates. The rotation is stopped before the heaviest desired component reaches the bottom of the tube. The solution is drained from the bottom into fractions, with the heaviest components occurring in the first drawn fractions.

Another technique, called the **density gradient** technique, is based on Equation 12.23. In it, tubes of solvent that have an increasing density from top to bottom are prepared. The sample is then added to the top of the tube just prior to rotation. The molecules will move down the tube until they come to the region where their density is equal to that of the solvent. At this point, the sedimentation force is zero, and no further sedimentation can occur. When all the components have reached the layer corresponding to their density, the process is stopped and the fractions withdrawn.

An analytical ultracentrifuge, as opposed to a preparative ultracentrifuge, will include a method for following the positions of the analytes while the centrifuge is in operation. The methods employed are generally light absorption, light scattering, or light refraction. A blank solution is placed in one of the tube positions for reference. For each rotation, a series of detector responses corresponding to each tube in the rotor will be obtained. These are electronically sorted out to produce a record of the optical response for each tube versus time.

Although ultracentrifugation is an indispensable part of the biochemist's and biologist's set of tools, the long times required and the relatively poor resolution of the separations achieved make the ultracentrifuge a last-resort technique for most analysts.

Study Questions, Section F

43. What are the three differentiating characteristics on which the separations introduced in this section are based?

44. Comment on the desirability of having very uniform pore sizes for the materials used in filtration and in gel permeation chromatography.

45. Why does affinity chromatography require an immobilized ligand?

46. What must be done to release the analyte from the immobilized ligand in affinity chromatography?

47. Are the immobilized ligands in affinity chromatography limited to antigens or antibodies?

48. How is the partial specific volume of a macromolecule related to its density?

49. What factors cause the sedimentation coefficient, s, to deviate from the theoretical two-thirds power dependence on the molecular weight?

50. It is desired to completely separate all the molecules that have a sedimentation coefficient of 30 S from a homoge-

neous solution. The ultracentrifuge tubes are 5 cm in length and are filled to within 1 cm of the top. The bottom of the tube is 11 cm from the center of rotation, and the rotational velocity is 50,000 rpm. How long should you leave the centrifuge spinning to ensure complete settling of the desired molecules?

43. The differentiating characteristics discussed in this section are size, complexation reactivity, and a combination of mass, frictional ratio, and density.

44. A uniform pore size in the membrane material used for filtration will create a sharp cutoff between the sizes of particles that will pass the filter and those that won't. Since this is a batch operation, the high degree of discrimination can be a good thing. For gel permeation chromatography, a uniform pore size will produce a narrow range of sizes between the unretained and total access sizes. This will give better sensitivity for size dispersion in this range, but it provides a smaller range of sizes over which separation occurs.

45. If the ligand were not immobilized, it would not be possible to rinse away the sample components that had not become bound to it without rinsing the complex away as well.

46. The analyte is released by introduction of a component with a still higher affinity to the ligand than the analyte (displacement) or by changing the solvent pH or ionic strength so that the active site of the ligand is distorted and the formation constant for the complex reduced.

47. No, ligands in affinity chromatography are not limited to antigens or antibodies. Any biochemical species with the property of selective binding can suffice. Enzymes and other proteins are often used.

51. Of the three fractionation methods discussed for separation by ultracentrifuge, which one(s) will produce bands of molecules of different s along the length of the centrifuge tube?

Answers to Study Questions, Section F

48. The partial specific volume has the units of $cm^3 \, g^{-1}$. It is the reciprocal of the density, which has the units of $g \, cm^{-3}$.

49. Several factors cause s to deviate from the theoretical two-thirds power dependence on molecular weight. Hydration of the macromolecule increases the mass and size of the particle moving through the solvent. A nonspherical shape will have a different relationship between the frictional coefficient and mass. The partial specific volume is not the same for all molecules of the same type.

50. Find the spin time by rearranging Equation 12.27.

$$\Delta t = \frac{2.3(\log 11 - \log 7)}{30 \times 10^{-13} \, s \left(50,000 \, min^{-1} \times \dfrac{1 \, min}{60 \, s} \right)^2}$$

$$\Delta t = \frac{0.451}{30 \times 10^{-13} \times 6.9 \times 10^5} = 2.2 \times 10^5 \, s$$

This many seconds is two and a half days.

51. The moving zone method will produce bands because all the compounds start together at the top of the tube and move down at a rate that is related to the s values of the molecules. The density gradient method also produces bands because the compounds come to a steady-state position where their density is the same as the density of the solvent at that distance.

G. Immunoassay

An **immunoassay** is a method in which the specific binding of antibodies is used for detection and quantitation. The basic selective reaction of the several forms of immunoassay is that of either a direct binding between the antibody and antigen or the displacement of one antigen by the analyte antigen. Following the selective reaction, immunoassay methods sometimes include ways to separate the reactants from the products, and they always include a method to determine the extent to which the selective reaction has proceeded. We will first discuss the chemistry behind the direct binding approach.

DIRECT BINDING REACTIONS

The direct binding reaction can be written as

$$Ab + Ag \rightleftharpoons AbAg \qquad\qquad 12.30$$

where Ab and Ag are the antibody and antigen and AbAg is the complex formed between them. You will recognize this as a complex formation reaction, as studied in Chapter 7. The extent of this reaction can be taken as the fraction of the original antibody that has been complexed by the analyte antigen. We will use the formation constant expression and mass balance equations to obtain an equation for this fraction. The formation constant expression is

$$K_f' = \frac{[AbAg]}{[Ag][Ab]} \qquad\qquad 12.31$$

and the mass balance equations are

$$C_{Ag} = [Ag] + [AbAg], \qquad [Ag] = C_{Ag} - [AbAg]$$
$$C_{Ab} = [Ab] + [AbAg], \qquad [Ab] = C_{Ab} - [AbAg] \qquad 12.32$$

where C_{Ab} and C_{Ag} are the analytical concentrations of antibody and antigen. If the antibody has more than one active site per molecule, C_{Ab} should be taken as the concentration of active sites. When the mass balance equations of Equation 12.32 are substituted into the formation constant expression of Equation 12.31 and the terms rearranged, the following equation results.

$$[AbAg]^2 - \left(\frac{1}{K_f'} + C_{Ag} + C_{Ab}\right)[AbAg] + C_{Ag}C_{Ab} = 0 \qquad 12.33$$

This quadratic equation forms the basis for the plot shown in Figure 12.17. Here we see the fraction of bound antibody as a function of the relative antigen concentration. When the concentrations are relatively large compared to the reciprocal of the formation constant (for this plot, $K_f' = 10^8$, so $1/K_f' = 10^{-8}$), the fraction of bound antibody increases nearly linearly with the amount of antigen added. Since the maximum fraction of the antibody that can be bound is 1, the curve shows a limit at that value. This limit puts an upper boundary on the amount of antigen that can be quantitated. In general, the amount of antibody used as the probe should always exceed the amount of antigen in the sample. However, some methods used to detect the extent of the reaction work best when a significant fraction of the antibody sites have reacted with the antigen, so in those cases, too large an excess should be avoided. To achieve an optimum antibody concentration, the antiserum containing the antibodies is generally diluted until roughly half the sites have reacted with a known amount of antigen in the expected working range.

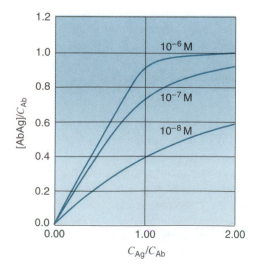

Figure 12.17. Plot of the fraction of antibody bound by the antigen as a function of the relative antigen concentration. For this plot, $K_f' = 10^8$. Values are shown for three different concentrations of antibody.

As the concentration of the antibody decreases toward $1/K_f'$, the extent of the reaction for a given ratio of antigen to antibody decreases. The curvature of the working curve decreases as well. From Figure 12.17, we can see that the minimum concentration of analyte for which this technique will work is approximately equal to $1/K_f'$. Fortunately, the formation constants for many antibody–antigen complexes are very large. Many methods have been developed for determining the extent of the reaction. They will be discussed in a later subsection.

COMPETITIVE BINDING REACTIONS

A particularly elegant method of applying the antibody–antigen complexation reaction for analysis is the competitive binding approach. In it, the reagent antibody is already stoichiometrically bound with an antigen. The analyte is an antigen that can also bind with the antibody. When the sample is added to the system with the antibody–antigen complex, it reacts with the antibody to displace some of the previously bound antigen. The reaction can be written as follows

$$\text{AbAg*} + \text{Ag} \rightleftharpoons \text{AbAg} + \text{Ag*} \qquad\qquad 12.34$$

where Ag* is the original antigen and Ag is the analyte. To measure the extent of this reaction, one can determine the fraction of the antibody that is still bound to the original antigen or the amount of free antigen that is released.

The equilibrium constant for the reaction is

$$K'_{\text{Ag/Ag*}} = \frac{[\text{AbAg}][\text{Ag*}]}{[\text{Ag}][\text{AbAg*}]} = \frac{K'_{f\text{AbAg}}}{K'_{f\text{AbAg*}}} \qquad\qquad 12.35$$

which we can see is the ratio of the formation constants of the analyte and antigen for the antibody.

The relationship between the concentration of the original AbAg* complex and the concentration of analyte added can be derived using appropriate mass balance equations. We will also assume that the test reagent solution is composed of the complex AbAg* without an excess of either Ab or Ag*. The mass balance equations are then

$$C_{\text{Ag}} = [\text{Ag}] + [\text{AbAg}]$$
$$C_{\text{Ab}} = [\text{AbAg*}] + [\text{AbAg}] \qquad\qquad 12.36$$
$$C_{\text{Ag}}* = C_{\text{Ab}} = [\text{AbAg*}] + [\text{Ag*}]$$

where the second mass balance equation assumes that there is negligible free antibody. This will be true as long as the concentration of antibody is substantially greater than $1/K_f'$ for the antibody–antigen complex.

When the mass balance expressions of Equation 12.36 are substituted into Equation 12.35, the following quadratic equation results.

$$(1 - K'_{\text{Ag/Ag*}})[\text{AbAg*}]^2 + \{K'_{\text{Ag/Ag*}}(C_{\text{Ab}} - C_{\text{Ag}}) - 2C_{\text{Ab}}\}[\text{AbAg*}] + C^2_{\text{Ab}} = 0 \quad 12.37$$

Equation 12.37 has been used to obtain the curves in Figure 12.18. In this graph, we can see the effect of the relative formation constant of the displacement antigen (the analyte). If K_f' of the analyte–antibody complex is the same as that of the antigen–antibody complex, the curve marked $K = 1$ is obtained. As the formation constant for the analyte–antibody complex becomes greater than that for the antigen-analyte complex, the linear portion and the dynamic range increase.

Other representations of this curve are also instructive. In Figure 12.19, the fraction of the initial antigen–antibody complex is plotted against the analyte con-

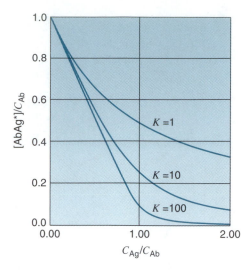

⊙ **Figure 12.18.** Plot of the fraction of antibody bound by the initial antigen (Ag*) as a function of the relative analyte antigen concentration. Values are shown for three different ratios of formation constants as in Equation 12.35.

centration for three different concentrations of the antibody. Surprisingly, the sensitivity (change in output for a given change in input) of the technique increases as the concentration of antibody decreases. Remember, though, that this only works as long as the reciprocal of the formation constant is equal to or larger than the concentration. There is also an advantage in using conditions that, according to our analysis, would seem to be unfavorable. As the formation constant of the analyte–antibody complex decreases relative to that of the original antigen–antibody complex, displacement of the original antigen will require more of the analyte. This approach can be used to extend the working curve over a very wide concentration range. In Figure 12.20, the fraction of the initial antigen bound to the antibody is plotted against the log of the analyte concentration. A log plot is used to encompass the wide range of analyte concentrations over which the response is useful. In this plot, the value of $K'_{Ag/Ag*}$ was only 0.08. This mode of operation is quite common in practical immunoassay systems.

Often, the initial antigen will be labeled for easy quantitation. The antigen that has been displaced from the antibody can be separated from the antibody complexes by washing (if the antibody has been immobilized) or by filtration or ultracentrifugation. The amount of the labeled antigen left with the antibody or the amount of antigen in the wash solution can then be determined.

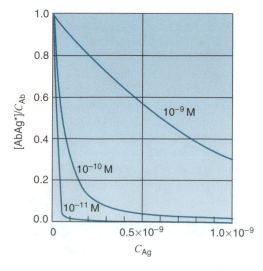

⊙ **Figure 12.19.** Plot of the fraction of antibody bound by the initial antigen as a function of the analyte concentration. Values are shown for three different concentrations of antibody.

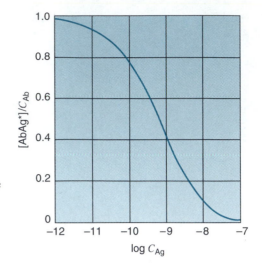

Figure 12.20. Plot of the fraction of antibody bound by the initial antigen as a function of the antigen concentration when the formation constant for the analyte complex is relative low. A wide dymanic range is enabled with a concomitant loss in precision.

ENZYME-LINKED IMMUNOSORBENT ASSAY (ELISA)

Having achieved such exceptional selectivity with antibody–antigen reactivity, the natural desire is to use the very low level of interferences to lower the detection limit. In many cases, the detection limit is now determined by the detection limit for the labeled antigen. Even though the detection limit for radioactive and fluorescence labels can be very low, a method to enhance the sensitivity could lead to better quantitation and shorter measurement times. This has been accomplished by combining the selectivity of antibody–antigen reactions with the ability of an enzyme to facilitate the production of many product molecules.

In one of the implementations of the method of enzyme-linked immunosorbent assay (ELISA), an antibody is immobilized in the reaction vessel (often just a depression or **well** in a plate that can have many such wells). A known amount of an antigen that is labeled with an enzyme and the sample containing the analyte are then added sequentially or simultaneously. The antigen and analyte compete for the antibody sites, reaching the same equilibrium state described earlier for the competitive binding immunoassay. The amount of bound antigen is inversely related to the amount of analyte in the sample. The unbound antigen and analyte are then washed off. Now the enzyme is put to work. The substrate for the enzyme is added to the well. The rate of production of the product of the enzymatic reaction is proportional to the amount of enzyme present. The amount or concentration of product is generally measured after a specific reaction time has elapsed, and a working curve is developed. A typical working curve has the appearance of that in Figure 12.20. The response can be rate of radio emissions or fluorescence intensity. Photon absorption has also been used.

In another variation on the ELISA technique shown in Figure 12.21, the analyte antigen is added to the well with the immobilized antibody. The antigen binds to the antibody (of which there is an excess). Then a secondary antibody that has an affinity for a different epitope of the antigen is added. The secondary antibody now binds to the analyte antigens that are also bound to the immobilized antibody. This creates a kind of sandwich effect, with the analyte antigen between the immobilized and secondary antibodies. The solution in contact with the antibodies now has no analyte. If the analyte–antibody reaction were to come to equilibrium, some of the analyte would dissociate from the antibody and the sandwich would be destroyed. Fortunately, the rate of this dissociation reaction (called the **off-rate**) is very slow, so the structure remains bound and intact for the time of the analysis. To implement the ELISA approach, the secondary antibody has been labeled with an enzyme. After the excess

Antibodies ⬠ with analyte △ added

Secondary antibody △ with enzyme label
added and excess washed off

Substrate S added, and enzyme catalyzes
reaction of substrate S to product P

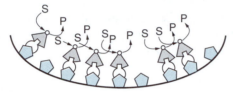

Figure 12.21. Sandwich scheme for antigen analysis using ELISA. The catalytic reaction of the enzyme yields many product molecules per enzyme, thus amplifying the amount of analyte that can be detected.

labeled antibody has been washed off, the substrate for the enzyme is added. In this case, the amount of labeled antibody is equal to the amount of analyte bound to the immobilized enzyme, and the rate of product production is directly related to the amount of analyte in the sample.

A third approach is used when the analyte is an antibody rather than an antigen. In this case, an excess of antigen for the analyte antibody is immobilized in the well. The sample containing the antibody is then added and the excess washed off. Then a secondary antibody that can bind to the analyte antibody is added. The secondary antibody has an enzyme label that can then be used as in the previous examples.

ENZYME MULTIPLIED IMMUNOASSAY TECHNIQUE (EMIT)

A type of enzyme immunoassay for haptens has been developed, called **enzyme multiplied immunoassay technique** (EMIT[17]), that does not require separation of the free and bound enzyme label. The assay depends instead on the effect the binding of the hapten has on the enzyme activity. Generally the hapten binding inhibits the enzyme activity, but the technique can be used for hapten activators as well. The technique is centered on an antibody that has specific binding sites for the hapten. An antigen containing the analyte hapten is bound to the label enzyme at or very near its substrate active site. The binding of the antigen to the antibody covers the enzyme active site or distorts it enough to inhibit enzyme activity. Therefore, the addition of the antibody to a solution containing the antigen will decrease the catalytic activity of the enzyme. This can be seen as a change in the rate of production of the product of the enzyme reaction. Now if the analyte hapten is added to the solution, it will compete with the antigen for the sites on the antibody. The enzyme on the released antigens will now be active, and the enzymatic activity will increase. The greater the concentration of hapten analyte, the greater is the increase in catalytic activity.

---

[17]EMIT is a registered trade name of the Syva Corporation, 3181 Porter Drive, Palo Alto, CA 94304.

APPLICATIONS OF IMMUNOASSAY

Immunoassay has become a mainstay in the analysis of biological molecules. Its great specificity allows the detection and quantitation of species at the very low levels found in natural fluids. Many kits have been developed, based on immunoassay, that can be kept on the shelf until needed and then implemented with a minimum of additional equipment and expertise. One example is the test for the placental hormone chorionic gonadotropin in urine. This hormone is present in the urine of pregnant women and is the basis for the easily implemented pregnancy test. Methods for many other hormones have been implemented. The first of these was an insulin assay, for which the developers were awarded the Nobel Prize.[18]

Most of the attention in immunoassay development in recent years has been with the enzyme-linked techniques because of their great sensitivity and ease of application once the kits have been developed. A 1976 review article[19] lists dozens of hormones, proteins, and viruses for which methods have been developed. In addition, kits based on EMIT have been developed for a variety of small organic molecule haptens including several drugs of abuse, antiepileptic drugs, and toxins. Detection limits for many of these species are generally in the range of micrograms per liter, though much lower levels can be detected in certain instances.

Study Questions, Section G

52. Which of the differentiating characteristics studied earlier in this text forms the basis for the analysis of the immunoassay methods?

53. What determines the lower analyte concentration limit for an immunoassay technique?

54. What is the principal advantage of the competitive binding technique over the direct method?

55. In the competitive binding technique, does the response increase or decrease with increasing analyte concentration if the amount of bound, labeled antigen is determined?

56. Why should the sensitivity of the immunoassay technique increase with decreasing antibody concentration?

57. Why is it necessary to separate the reaction products from the initial complex solution in immunoassay techniques?

58. Why does the technique of enzyme linking so greatly increase the sensitivity of the immunoassay technique?

59. Is it true that only antigens and antibodies can be analyzed by immunoassay?

60. Of the four goals of chemical analysis, which are enabled by immunoassay techniques?

Answers to Study Questions, Section G

52. The immunoassay techniques are based on the formation of complexes between antibodies and antigens. The complex formation reaction studied in Chapter 7 forms the basis for the mathematical analysis of this technique.

53. The lower concentration limit is that for which a significant fraction of the analyte will be complexed with the antibody (or antigen). This fraction decreases as the concentrations of the analyte and antibody decrease. The lower limit is roughly equal to the reciprocal of the formation constant of the complex.

54. In the competitive binding technique, the initial antigen can be prelabeled for easy detection by radioemmision, photon absorption, or fluorescence.

55. The amount of labeled antigen still bound to the antibody decreases with increasing analyte. Therefore, the response decreases with increasing analyte concentration.

56. When the antibody concentration decreases, the amount of analyte required to react with it also decreases. Therefore, the fraction of the label remaining will be smaller when a smaller amount of antibody is used.

57. The labeled material will respond to the probe for the label whether it is bound or not. Therefore, in order to determine the extent of the reaction, it is necessary to separate the bound material from that which is free.

58. When the label is an enzyme, the enzyme can be probed by the addition of its substrate. One enzyme molecule can

---

[18]The developers were Berson and Yalow, who reported their findings in *J. Clin. Invest.* **1960**, *39*, 1157. Berson died before the Nobel Prize was granted, so it went to Yalow alone.

[19]G. B. Wisdom, "Enzyme-immunoassay," *Clin. Chem.* **1976**, *22*, 1243.

catalyze the conversion of many substrate molecules to detectable product molecules. Therefore, there are many more product molecules than enzyme molecules, and the response has been amplified as a result.

59. It is not true that the immunoassay technique is limited to antigen and antibody molecules. Any molecule that can act as a hapten can also be determined. This ability to respond to a variety of small organic molecules is being used as the basis for their detection at very low levels.

60. Immunoassay techniques provide both detection (because of the high specificity of the reactions) and quantitation (through the determination of the extent of the reaction). They are also useful for separation when the complex can be made to precipitate or when it can be separated on the basis of size. The goal of identification is possible, but only over a limited set of possibilities for which the appropriate reagents have been prepared.

H. Biochemically Based Sensors

We have studied a number of chemical sensors in previous chapters. The ideal chemical sensor could be placed in contact with a liquid or gaseous sample. It would then provide a response that is related to the amount or concentration of a single component in the sample. The closest we have come to this ideal in the systems studied thus far is the glass electrode for the determination of solution pH. Its response is very specific for the proton, and it has a huge dynamic range over which it can operate. Other potentiometric sensors are less specific and have a narrower dynamic range. Probes based on optical absorption or fluorescence can be made of light fibers. They, too, are subject to more interference than we would like. These sensors are, however, very convenient to apply, and thus ways to improve their selectivity have been the subject of continuous research.

In 1962, Clark and Lyons[20] devised the first sensor in which the selectivity of a biochemical reaction was combined with the convenience of an ordinary chemical sensor. They used the Clark electrode for oxygen to make a sensor for glucose in which the selectivity was provided by glucose oxidase. Since that time, the field has seen continued active development and remarkable advances. This section is organized by the differentiating characteristic used in the sensor that monitors the results of the biochemical reaction employed for the primary selectivity.

PHOTON EMISSION SENSORS

Photon emission sensors are based on either of two differentiating characteristics. One is fluorescence, and the other is chemiluminescence (or bioluminescence). A photon emission sensor (often called an **optode**) is a fiber optic assembly that carries the emitted light to the spectrometric detector. The end of the fiber is also in contact with the biochemical reactant that provides the specificity.

An example of a chemiluminescent optode is an optical fiber coated with luminol. (See the reaction scheme for luminol in Section D of Chapter 6.) Luminol reacts with H_2O_2 to produce chemiluminescence. This probe can then be used in conjunction with any biochemical reaction that produces peroxide. An obvious case is any of the peroxidase enzymes. They react with the substrate and oxygen to form the oxidized substrate and peroxide. The released peroxide can then react with the luminol to produce the detected chemiluminescence. A sketch of such a sensor is shown in Figure 12.22. In a specific example of such a sensor, glucose oxidase was used to make a sensor specific for glucose. The resulting sensor had a linear range of 0.15 to 1.5×10^{-3} M. In another variation on this theme, the probe can contain a tiny photodiode, light

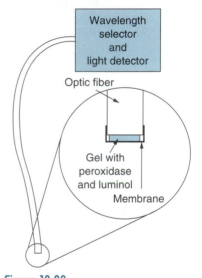

Figure 12.22. Luminescence sensor for the substrate of the peroxidase enzyme. The substrate molecules diffuse through the membrane and are catalytically oxidized, and the peroxide byproduct is detected by the luminescent reaction with luminol.

---

[20]L. C. Clark, Jr., and C. Lyons, "Electrode systems for continuous monitoring in cardiovascular surgery", *Ann. N. Y. Acad. Sci.* **1962**, *102*, 29.

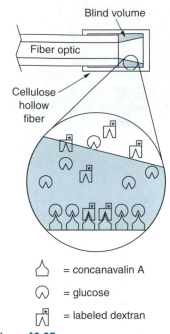

= concanavalin A

= glucose

= labeled dextran

Figure 12.23. Glucose sensor based on fluorescence. When glucose diffuses through the cellulose, it displaces some of the labeled dextran, which is then free to move out of the blind volume and to be excited by the excitation light from the fiber optic. The greater the concentration of glucose, the greater is the intensity of fluorescence radiation detected.

sensor, and filter at the tip. This assembly is then covered with the gel and membrane. The body of the probe is then a pair of wires rather than the optical fiber. Chemiluminescent immunoassay devices have been developed for human IgG, testosterone, thyroxine, biotin, hepatitis B, rabbit IgG, and cortisol.[21]

There are actually three differentiating characteristics employed in the glucose sensor just described. One is the reaction of glucose catalyzed by the glucose oxidase to produce peroxide. The second is the reaction of the peroxide with luminol to give the chemiluminescence. The third is the particular wavelength characteristic of the luminol chemiluminescence. All three of these differentiating characteristics must be met to produce a response at the detector. The resulting selectivity of this sensor is thus greater than the selectivity of any of the three separate processes alone. This multiplication of selectivities is a common feature of biochemically based sensors.

A probe for glucose concentration has been made using a competitive immunoassay and a fluorescent tag.[22] The protein concanavalin A, which can bind to glucose, is attached to the walls of a cellulose hollow fiber. As shown in Figure 12.23, one end of the hollow fiber is closed, and an optical fiber is inserted in the other end. This fiber optic introduces the excitation radiation and returns the collected fluorescence radiation to a detector. A labeled, competitively binding molecule is introduced into the enclosed volume. In this case, the molecule is dextran, and the label is fluorescein isothiocyanate. The glucose molecule is small enough to diffuse readily through the cellulose fiber walls. The labeled dextran is large and thus trapped inside. Note that the fiber optic is arranged so that the volume closest to the walls does not receive the excitation radiation. This part of the volume is called the **blind volume.** The labels on the bound dextran are not excited since they are in the blind volume.

When there is no glucose in the fiber interior, a small fraction of the dextran is unbound and thus in the solution. Some fluorescence radiation is returned to the detector through the optic fiber. When the probe is introduced into a solution containing glucose, the glucose diffuses through the cellulose fiber and displaces some of the dextran from the immobilized concanavalin A. The resulting increase in the concentration of the dextran in the excited volume of the solution produces an increase in the intensity of the fluorescence.

This glucose sensor is an example of a biochemical reaction using a label in which the unbound material did not have to be washed from the reaction mixture. The requirement for complete separation of the bound and unbound labeled molecules is avoided by immobilizing the concanavalin A in the blind volume of the sensor.

PHOTON ABSORPTION SENSORS

A probe for UV–visible photon absorption is similar to that shown in Figure 5.7. Such a probe could potentially be used for any reaction that can cause a change in the concentration of a photon absorbing species. In practice, such probes are difficult to implement with biochemical reagents because of the need to include some of those reagents in the light path and to avoid having them adsorb on the mirror. In one mode, an optical cell is used with the walls coated with an immobilized receptor. In this case, it is necessary for the absorbance spectrum of one of the reactants to change between the bound and unbound states. A much more commonly used approach uses attenuated reflectance spectroscopy as shown in Figure 5.39.

In the normal implementation of attenuated total reflectance (ATR), the bioreceptor is immobilized at the surface of the waveguide or in a gel that is on the wave-

---

[21]E. A. H. Hall, *Biosensors*, Open University Press, Milton Keynes, 1990.

[22]J. S. Shultz, "Design of fibre optic biosensors based on bioreceptors," in A. P. F. Turner, I. Karube, and G. S. Wilson (Eds.), *Biosensors: Fundamentals and Applications*, Chap. 33, pp. 655–679. Oxford University Press, Oxford, 1987.

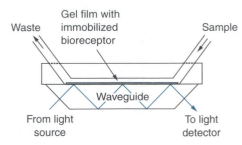

Figure 12.24. Optical waveguide with a flow-through sampling arrangement. The immobilized receptor concentrates the analyte in the region of the evanescent wave.

guide surface, as shown in Figure 12.24. The **bioreceptor** is a biomolecule that binds selectively to the analyte. It can be an antibody, an antigen, an enzyme, or other protein. If the analyte absorbs in the wavelength region of the illumination, the absorption spectrum will be affected by the presence of the bound analyte. The bioreceptor acts as an agent to concentrate the analyte in the region very close to the waveguide surface, within reach of the evanescent wave. Unbound molecules are dispersed throughout the solution and, even if they absorb light, will have little effect on the spectrum. Here, then, is another example of an optical technique in which the required separation is accomplished by the spectroscopic technique selecting between the bound and unbound portions of the system. If the analyte is not absorbing at the test wavelengths, a competitive assay, which uses another binding species that does absorb, can be implemented.

A related technique called **surface plasmon resonance** is also used. This technique is sensitive to the refractive index of the species within the evanescent wave. A change in the refractive index of the layer within the evanescent wave results in a change in the resonance angle of a laser light beam reflected off the top surface of the waveguide. Since refractive index is a function of the composition of the material, the resonance angle is a function of composition. Only the material within the evanescent wave portion of the waveguide surface is sampled.

ELECTRODE POTENTIAL SENSORS

Sensors based on electrode potentials use an electrode to determine the extent of a biochemical reaction. As described in Chapter 10, electrode potentials can be affected by the concentrations of the ions of the electrode metal, by the oxidation level of the solution in which an inert metal electrode is immersed, or by a specific ion electrode based on selective adsorption or diffusion of ions on or through a conducting membrane. The use of biochemical reactions in conjunction with such electrodes greatly increases the range of materials to which they can respond, and it can provide sensors of exquisite selectivity. The electrodes used most successfully for biochemically based sensors are the glass pH electrode and ion-selective electrodes for NH_3 and CO_2.[23] Iodide and sulfide electrodes have also been used, but with less success.

Examples of potentiometric biosensors based on the pH electrode include sensors for penicillin, glucose, and urea. The typical electrode is arranged as shown in Figure 12.25. The biochemical reagent is sandwiched between two membranes through which it cannot diffuse. This assembly is then attached to the contact area of the electrode. A very thin layer of electrolytic gel coats the surface of the electrode to ensure contact. The substrate (analyte) diffuses from the test solution through the outer membrane to undergo the enzyme-catalyzed reaction. The reaction product (in this case solvated protons) can then diffuse through the inner membrane to bring the pH of the contact electrolyte into equilibrium with that of the reaction volume. The pH elec-

---

[23]B. Eggins, *Biosensors: An Introduction*, pp. 68, 167–170. John Wiley & Sons, Chichester, 1996.

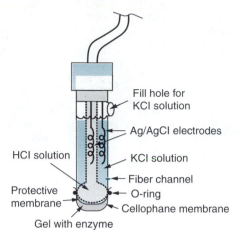

Figure 12.25. Arrangement for the containment of a biochemical reagent at the surface of a potentiometric electrode. This example is for a pH electrode. Other specific ion electrodes can also be used.

trode then responds to the changes in the pH of the gel reaction region caused by the reaction. This same arrangement can be used with a field-effect transistor based electrode.

In one specific application of an enzyme/pH electrode combination, penicillinase catalyzes the conversion of penicillin to penicilloate with a release of H_3O^+. The pH thus decreases with increasing concentration of penicillin in the solution. Similarly, the glucose oxidase catalysis of the oxidation of glucose produces gluconic acid, which lowers the pH.

The ammonia electrode is particularly useful. Deaminase enzymes catalyze the removal of NH_3 from their substrate. With creatinase, ammonia is removed from creatinine. Adenine deaminase produces inosine and ammonia from adenosine. Aspartame is deaminated with L-aspartase. The action of the decarboxylase enzymes to produce CO_2 can similarly be monitored by a CO_2-sensitive electrode. Oxalate is the substrate for both oxalate decarboxylase and oxalate oxidase enzymes, both of which produce CO_2.

ELECTRODE CURRENT SENSORS

Oxygen is a common reactant or product of many enzyme-catalyzed reactions. For this reason, the Clark electrode (see Section J of Chapter 10), that produces an electrical current proportional to the O_2 concentration in solution is a common sensor element. As mentioned in the introduction to this section, the use of the oxygen electrode to follow the reduction in O_2 caused by the catalytic oxidation of glucose was the first example of a biosensor. Another example that has been commercialized is a sensor for blood alcohol. Enzymes in the microbes *Acetobacter xylinum* and *Trichosporon brassicae* catalyze the oxidation of ethanol to acetic acid. Again, the decrease in oxygen caused by this reaction is monitored by a Clark electrode. This last example is interesting because the whole microorganism is used without isolation of the enzyme(s) responsible for the reaction. Such a device is sometimes called a **microbial sensor**.

A significant number of sensors have been developed using plant and animal tissues without prior purification of the enzymes involved in the reactions.[24] Among the first of these **tissue-based sensors** was the **bananatrode**.[25] As the name suggests, an

[24]M. A. Arnold and G. A. Rechnitz, "Biosensors based on plant and animal tissues," in A. P. F. Turner, I. Karube, and G. S. Wilson (Eds), *Biosensors: Fundamentals and Applications*, Chap. 3, pp. 30–59. Oxford University Press, Oxford, 1987.

[25]J. S. Sidwell and G. A. Rechnitz, "Bananatrode—an electrochemical sensor for dopamine," *Biotechnol. Lett.* **1988**, 7, 419.

enzyme contained in banana tissue catalyzes the reaction with the dopamine or catechol substrate sensed. These substrate molecules have similar structures as shown in Figure 12.26. Dopamine is found in the brain and is commonly used as an indicator of brain activity. The enzyme polyphenol oxidase, found in banana tissue, catalyzes the oxidation (with O_2) of several catechol-type molecules to the corresponding quinones. Quinones are very active electrochemically, being readily reduced at an electrode back to the catechol. The electrode is made of carbon paste mixed with banana tissue. A voltage sufficient to reduce the quinone is applied between this electrode and a suitable reference electrode. The resulting current is related to the concentration of the catechol-type compounds in the test medium. Because of the cyclic process possible (the catechol produced by the quinone reduction at the electrode can be reoxidized at the enzyme, and so on), this electrode can measure catechol or dopamine at the 10 μM level.

A variation of the catechol electrode is being developed for the determination of aspirin levels in blood.[26] This is important because of the small difference between the therapeutic and toxic levels for this commonly used drug. The enzyme salicylate hydrolase catalyzes the oxidation of salicylate to catechol.

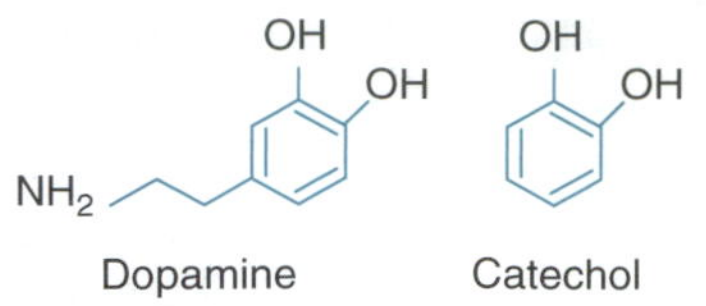

Figure 12.26. Chemical structures of dopamine and catechol. The enzyme polyphenol oxidase catalyzes the oxidation of both these molecules to the quinones (both —OH groups are replaced by ═O).

Study Questions, Section H

61. Justify the statement that biochemically based sensors use at least two differentiating characteristics in their operation.

62. What is the blind volume, and why is it so important in fluorescence biosensors?

63. When a waveguide is used with attenuated total reflectance (ATR), what portion of the sample is probed by the source radiation?

64. How is separation between the bound and unbound complexing agents achieved in the method of optical absorption with ATR? Which species are sampled, the bound or the unbound?

65. Is the potential achieved by the biochemical reagent/pH electrode combination shown in Figure 12.25 an equilibrium potential or a steady-state potential?

66. Many biochemical reactions involve ammonia, so when an ammonia electrode is used in a biochemically based sensor, what is it that provides the selectivity?

67. Is it true that the current measured in a sensor based on electrode oxidation or reduction current is related to the rate of diffusion of the analyte through the outer membrane? Why or why not?

Answers to Study Questions, Section H

61. The first differentiating characteristic used is the reaction of the analyte with (or catalyzed by) the biochemical reagent. The second differentiating characteristic is the one that detects the formation of the reaction product or the depletion of a reactant. The detection method may involve more than one level of selectivity.

62. The blind volume is that part of the reaction volume that is not illuminated by the excitation radiation. It is important because it is necessary to determine only the unbound molecules that have the fluorescent tag. The immobilized reagent is located principally in the blind volume, thus keeping the bound species from fluorescing.

63. The portion of the sample probed by the radiation transmitted through the waveguide is that part closest to the surface of the waveguide.

64. Separation is achieved by immobilization of the biochemical complexing agent in the penetration depth region of the waveguide. Since this is the only region that is optically sampled, the species probed are those bound to the immobilized agent.

65. The potential of the pH electrode is in equilibrium with the pH of the conducting gel with which it is in immediate contact. The pH of the gel, however, depends on the pH of the reaction volume. This pH depends on the rate of diffusion of the analyte though the outer membrane, the rate of the biochemical reaction, and the rate of movement of the protonated solvent back through the outer membrane. When the rate of production of protonated solvent is equal to the rate of diffusion away from the electrode, the pH will come to a steady-state value. Thus, the electrode po-

[26]B. Eggins, *Biosensors, An Introduction*, p. 166. John Wiley & Sons, Chichester, 1996.

tential overall is a steady-state value, not an equilibrium one.

66. While it's true that many biochemical reactions involve ammonia, the reaction that can occur within the biosensor is that for which the enzyme is specific. Therefore, the only species that can react will be those that can act as a substrate for the biochemical reagent(s) included in the sensor reaction volume.

67. The current measured by the Clark electrode is related to the rate at which the O_2 can diffuse to the electrode surface. This O_2 comes through the outer membrane and the reaction volume to reach the electrode surface. Some frac-

tion of it also gets consumed by the catalytic oxidation process when it encounters an enzyme to which a substrate (analyte) molecule is already attached. When the substrate is converted to product, it must be replaced by diffusion of the substrate through the outer membrane. Thus, the decrease in current of the Clark electrode is related to the rate of diffusion of the analyte through the outer membrane and into the reaction volume. The same arguments can be made for all the other current-based sensors because the quantity measured is current (rate of charge flow), and this rate is limited by the rate of supply of fresh reactant.

Practice Questions and Problems

1. Which of the following species are changed to another chemical form (not just complexed) by an enzyme-catalyzed reaction?

   A. Enzyme

   B. Coenzyme (or cofactor)

   C. Substrate

   D. Activator ion

   E. Effector

   F. Inhibitor

   G. Reactant

2. In the instructions for the preparation of biological buffer solutions, one is to calculate the relative concentrations of the acid and base forms desired from an equation like that of Equation 4.3 and a table of pK_a's such as Table 12.2. Then one is to measure the pH of the resulting buffer with a pH meter and adjust the pH by adding more of the base or acid form. If it is desired to have the pH controlled to within a few hundredths of a pH unit, will adjustment normally be required or not? Why do you think so?

3. What is denaturation, why is it a problem with enzyme-catalyzed reactions, and how is it minimized?

4. Why is the initial rate used in the measurement of catalytic activity rather than a later value?

5. In what way is the Michaelis constant for an enzyme related to the formation constant of the ES complex?

6. If 23.6 mg of an enzyme preparation produces 7.2×10^{-4} mol of product in 8.3 min, what is the specific activity of the preparation in nanokatals per milligram?

7. In Table 12.3, the value of k_{cat} for the reaction of fumarase with fumarate is $8.0 \times 10^2\ s^{-1}$. What is the longest dead time for which the initial rate of this reaction can be measured? Is this achievable with the usual commercial apparatus?

8. What is the purpose of the stop block in the stopped flow apparatus for kinetic measurements?

9. A continuous flow system has 20 sensors spaced at distances of 5 to 24 mm from the mixing chamber. It will be used to monitor a reaction for which k_{cat} is 190. It is desired to have the reaction be about 40% complete at the last sensor. What flow velocity should be used for this system? What will be the fraction complete at the first detector? You may use the plot of Figure 12.6 to help obtain the answer.

10. Antibodies and enzymes have very different functions and a quite different kind of chemical reactivity. Briefly, describe each and compare them.

11. What is the difference between monoclonal and polyclonal antibody preparations?

12. In single radial immunodiffusion, does the sensitivity of the method (in terms of the change in radius for a given change in concentration) increase or decrease as the concentration increases? Why?

13. Name the two major categories for the immobilization of enzymes and antibodies. Give at least one difficulty encountered with each.

14. In chemical bonding for immobilization, how are the active sites maintained during the chemical synthesis of the bonding?

15. Why is it easier to separate molecules of nearly equal size with gel permeation chromatography than with ultrafiltration?

16. The technique of chromatography is often assumed to include the separation in time or space of components that differ in the degree to which they exhibit the differentiating characteristic. Explain why the procedure described in Figure 12.15 is not a chromatographic technique in this sense. Could a chromatography be based on the affinity approach?

17. Since the density of all proteins is roughly the same, what phenomenon causes the layering of proteins of decreasing molecular weight in an ultracentrifuge tube?

18. What would the value of the sedimentation coefficient have to be in order to separate an analyte by ultracentrifuging at 60,000 rpm? The time allotted is 64 hours (Friday afternoon to Monday morning), and the tube is 11 cm from the center of rotation at the bottom and 8 cm at the top of the solution.

19. An antibody–antigen complexation is used in a direct binding assay. The formation constant for the complex is 5×10^7 L mol^{-1}. What is the minimum concentration of antigen for which this assay will work effectively? How many moles of antigen does this represent if the reaction volume is 100 μL?

20. In a competitive binding immunoassay, does the formation constant of the displacement antigen need to be larger than that of the antigen it is displacing?

21. In the enzyme-linked immunoassay, how is the enzymatic reaction linked to the amount of analyte, and how is it initiated?

22. Describe the analytical scheme for the determination of glucose based on the fluorescence sensor shown in Figure 12.23. Your answer should include the distinguishing characteristic used, the method of probing the characteristic, the anticipated analyte response, the method of measurement, and the way the measurement data are interpreted to obtain the desired information.

23. Why is it desirable to have the thickness of the gel layer containing the immobilized biocomplexing agent comparable to the penetration depth of the evanescent wave?

24. Horseradish peroxidase catalyzes the cleavage of carbon–fluorine bonds to form F$^-$. Suggest a method for the sensing of fluorocarbon compounds based on this biochemical reaction.

Suggested Related Experiments

1. Stopped flow or continuous flow enzyme kinetics.
2. Gel permeation chromatography.
3. Affinity chromatography.
4. An immunoassay or ELISA with fluorescence detection.
5. Glucose analysis with an ion selective electrode and glucose oxidase.
6. Catechol analysis with a bananatrode.
7. ATR analysis of an analyte complexed by an immobilized bioreceptor.
8. Determination of nitrate in drinking water by the reaction $NADH + NO_3^- \xrightarrow{\text{nitrate reductase}} NO_2^- + NAD^+$.

Background Materials

This material is intended to provide a convenient review of topics that many people using this book have already studied in other contexts. Examples and exercises are provided to aid in this review. Three background topics are covered: A, Stoichiometric Ratios; B, Logarithms and Exponents; and C, Electrical Quantities.

A. Stoichiometric Ratios in Chemical Compounds and Reactions

MOLECULAR FORMULAS

The atomic composition of most chemical compounds can be expressed as the **chemical formula.** An example is the molecule glucose (the sugar found in blood). Its formula is $C_6H_{12}O_6$. Every molecule of glucose has exactly 24 atoms: 6 carbons, 12 hydrogens, and 6 oxygens. In normal molecular matter, these formula numbers are all relatively small integers. Thus, in a sample of pure glucose, the ratio of hydrogen atoms to carbon atoms is exactly 2.000000 (to as many zeros as the purity of the glucose will allow).

REACTION STOICHIOMETRY

Furthermore, if the glucose is burned in oxygen so that the products are CO_2 and H_2O, we can predict exactly how many molecules of CO_2 and H_2O will be produced per molecule of sugar. The first step in this process is to write a **balanced equation** for the reaction.

$$C_6H_{12}O_6 + 6O_2 \rightarrow 6CO_2 + 6H_2O$$

The reaction is said to be **balanced** when the coefficients of the reactants and products are such that there are exactly the same number of atoms of each element on both sides of the equation. This is based on the assumption that this is the only reaction occurring and that every atom consumed as a reactant must appear in one of the products. The relationships of these small whole numbers of reactant and product molecules is called the **stoichiometry** of the reaction. Calculations based on these stoichiometric relationships are frequently used in quantitative analysis.

Here are some more reactions you can use to practice your equation balancing skills. (Answers to all the practice problems are found at the end of this section.)

1. _NaCl + _SO$_2$ + _H$_2$O + _O$_2$ → _Na$_2$SO$_4$ + _HCl

2. _B$_2$O$_3$ + _C → _B$_4$C + _CO$_2$

IONIC REACTIONS

In many of the reactions in this book, the reactants or produces are **ions** in solution. The charge that is part of the formula indicates the ionic state. An example is SO_4^{2-}, the sulfate ion that has a charge of -2. Negatively charged ions are called **anions;** positively charged ions are **cations.** In nature, the sulfate anions are always neutralized by an equal charge of cations, so sulfate anions would have to be introduced into the reaction as H_2SO_4 or Na_2SO_4 or $NaHSO_4$. However, if the species of the neutralizing ion is not important to the reaction involving the sulfate anions, it is often left out of the reaction equation. An example is the reaction

$$2Ag^+ + CrO_4^{2-} \rightarrow Ag_2CrO_4(s)$$

In this case, the silver cation may have been added as $AgNO_3$ (which dissolves into Ag^+ and NO_3^- in water), and the chromate anion may have been added as K_2CrO_4 (which dissolves into $2K^+$ and CrO_4^{2-} in water). Since the nitrate and the potassium ions are not involved in the reaction and since the same reaction would occur if other neutralizing ions were present, their omission simplifies the reaction and focuses attention on the chemically significant species.

When balancing an ionic reaction, the charges on both sides of the equation must be equal as well as the number of atoms of each element. For instance, in the balanced reaction

$$As_2O_3 + 4OH^- \rightarrow 2HAsO_3^{2-} + H_2O$$

the elements balance, and there are 4 negative charges on each side of the reaction $[4\times(-1) = 2\times(-2)]$. There do not have to be the same number of charges of each sign on each side of the equation, but the *net charge* on each side must be equal. An example of a reaction that has charges of both signs is

$$6Cu^{2+} + 15I^- \rightarrow 6CuI + 3I_3^-$$

In this case, there is a net negative charge of -3 on each side of the equation $[6\times(+2) + 15\times(-1) = 3\times(-1)]$.

Here are some practice examples involving ions.

3. $\_MnO_4^- + \_Mn^{2+} + \_H_2O \rightarrow \_MnO_2(s) + \_H^+$

4. $\_S_2O_8^{2-} + \_H_2O \rightarrow \_SO_4^{2-} + \_O_2 + \_H^+$

5. $\_H_4IO_6^- + \_I^- \rightarrow \_IO_3^- + \_I_2 + \_OH^- \_H_2O$

COMBINING QUANTITIES

An advantage to knowing the exact molar ratios of reactant and product species is that knowing the amount of one reactant or product can enable the prediction of the amounts of the other species involved. For example, you may wish to know how much Na_2CO_3 to add to a sample containing up to 50 mg of calcium (as Ca^{2+}) in order to convert all the Ca^{2+} to $CaCO_3$. We first write the reaction to get the combining ratios.

$$Ca^{2+} + CO_3^{2-} \rightarrow CaCO_3(s)$$

The reaction indicates that 1 mole of CO_3^{2-} will be required for every mole of Ca^{2+}. The maximum number of moles of Ca^{2+} is 0.050 g/40.078 g mol^{-1} = 1.3 $\times$ 10^{-3} mol, where 40.078 is the formula weight of Ca. Therefore, it will take at least 1.3 $\times$ 10^{-3} mol of CO_3^{2-} to make sure all the Ca^{2+} is precipitated. If the carbonate is in the form of sodium carbonate, this will be 106 g mol^{-1} × 1.3 × 10^{-3} mol = 0.14 g = 140 mg Na$_2$CO$_3$.

6. If the Na$_2$CO$_3$ were in the form of a 0.10 M solution, how many milliliters would be required?

Now, you may wish to know how much Ca there was in the original sample by weighing the CaCO$_3$. This will not work very well, as the CaCO$_3$ precipitate is hydrated and the temperature required to drive off the last of the water is very close to the decomposition temperature of the CaCO$_3$. What is done is to decompose the CaCO$_3$ until it is all converted to CaO(s) and CO$_2$(g). The purer and more stable CaO is then weighed. If the CaO weighed 47.3 mg, how could you determine the amount of Ca in the original sample? Again, we must write the balanced reaction to get the ratios.

$$CaCO_3 \rightarrow CO_2 + CaO$$

From the reaction, we can see that the number of moles of CaO produced are equal to the number of moles of CaCO$_3$ decomposed. In this case, 0.0473 g/56.1 g mol^{-1} = 8.43 × 10^{-4} mol CaO. The number of moles of Ca is also 8.43 × 10^{-4}. The weight of Ca in the original sample is 8.43 × 10^{-4} mol × 40.78 g/mol = 0.0344 g.

From this example, we can see that the balanced reaction gives us the exact combining ratios in terms of the *moles* of reactants and products and that the formula weight of the materials must be used to calculate the amounts in terms of grams.

In most analytical reaction situations, the reactant that is the analyte reacts with an excess of the other reactants. This is to ensure that the analyte will be entirely transformed by the reaction. The analyte, in this case, is the **limiting reactant.** It is the amount of the limiting reactant that determines the amounts of the other reactants consumed and the amounts of all the products formed.

Here are some additional reactions with which to polish your ability with stoichiometric calculations.

7. Exactly 29.37 mL of 0.02000 M NaOH is required to react with the acetic acid (FW = 60.053) in a 50.00 mL sample of vinegar. Calculate the concentration and the grams of acetic acid in the sample. The reaction is

$$OH^- + HAc \rightleftharpoons H_2O + Ac^-$$

8. Ferric ion (Fe^{3+}) is reduced to ferrous ion (Fe^{2+}) by stannous ion (Sn^{2+}). The reaction (not balanced) is

$$Fe^{3+} + Sn^{2+} \rightleftharpoons Fe^{2+} + Sn^{4+}$$

How many grams of SnCl$_2$ must be added to 100 mL of 0.023 M Fe^{3+} to provide a 50% excess of Sn^{2+} reactant?

9. A solution of Ni^{2+} has been obtained from dissolving 0.147 g of steel. The nickel has been precipitated with dimethylglyoxime (DMG) (FW 116.12) in the ratio of two DMG per Ni^{2+}. The dried precipitate weighs 0.0274 g. What is the weight percent of Ni in the steel?

The answers to all the numbered exercises in this section are given below. If you found these exercises difficult, it is highly recommended that you undertake further review using an introductory chemistry text. The concepts reviewed in this background section are essential for an understanding of the material in this text.

Balanced Reactions

1. $4NaCl + 2SO_2 + 2H_2O + O_2 \rightarrow 2Na_2SO_4 + 4HCl$

2. $2B_2O_3 + 4C \rightarrow B_4C + 3CO_2$

Balanced Ionic Reactions

3. $2MnO_4^- + 3Mn^{2+} + 2H_2O \rightarrow 5MnO_2(s) + 4H^+$

4. $2S_2O_8^{2-} + 2H_2O \rightarrow 4SO_4^{2-} + O_2 + 4H^+$

5. $H_4IO_6^- + 2I^- \rightarrow IO_3^- + I_2 + 2OH^- + H_2O$

Stoichiometric Calculations

6. $\dfrac{1.25 \times 10^{-3} \text{ mol Na}_2\text{CO}_3}{0.10 \text{ mol/L}} = 1.3 \times 10^{-2}\text{ L} = 13\text{ mL}$

7. $(29.37 \times 10^{-3}\text{ L})(0.02000\text{ mol/L}) = \text{moles NaOH} = \text{moles}$
HAc present $= 5.874 \times 10^{-4}\text{ mol}$

$$\frac{5.874 \times 10^{-4}\text{ mol}}{0.05000\text{ L}} = 0.01175\text{ M HAc}$$

$(5.874 \times 10^{-4}\text{ mol})(60.053\text{ g/mol}) = 0.03528\text{ g}$

8. The balanced equation is

$$2Fe^{3+} + Sn^{2+} \rightleftharpoons 2Fe^{2+} + Sn^{4+}$$

$$(0.023\text{ mol/L})(0.10\text{ L}) = 2.3 \times 10^{-3}\text{ mol Fe}^{3+}$$

To equal moles $Sn^{2+} = 1/2$ moles $Fe^{3+} = 1.2 \times 10^{-3}$ moles. For 50% excess, moles $Sn^{2+} = 1.15 \times 10^{-3}$ mol $\times 1.5 = 1.73 \times 10^{-3}$ mol.

$$(1.73 \times 10^{-3}\text{ mol})(189.61\text{ g/mol}) = 0.328\text{ g}$$

9. The formula weight of $Ni(DMG)_2$ is 290.93.

$$\frac{0.0274\text{ g}}{290.93\text{ g/mol}} = 9.41 \times 10^{-5}\text{ mol Ni(DMG)}_2$$

$(9.41 \times 10^{-5}\text{ mol})(58.69\text{ g/mol Ni}) = 5.523 \times 10^{-3}\text{ g Ni}$

$$\frac{5.523 \times 10^{-3}\text{ g Ni}}{0.147\text{ g steel}} \times 100 = 3.76\%\text{ Ni}$$

B. Logarithms and Exponents

For many chemical quantities, the range of useful values is so great that the use of ordinary numbers is awkward. For example, the concentration of H_3O^+ (hydronium ions) in pure water is approximately 0.000 000 1 M, and the number of H_3O^+ in 1 mL of this water is 60,220,000,000,000. Numbers that are so large or so small are unwieldy because of the long strings of zeros involved. Scientists have developed a kind of shorthand called **scientific notation** for using numbers of these magnitudes. In scientific notation, the example numbers are 1×10^{-7} and 6.022×10^{13}. Here the superscript numbers (called **exponents**) reveal the orders of magnitude (factors of 10) larger or smaller than the decimal part of the number the actual number is. So, to get the number 0.000 000 1 you would have to multiply 1 by one-tenth seven times. To obtain 60,220,000,000,000 you would multiply 6.022 by 10, thirteen times. The exponent is referred to as the **power** to which 10 should be taken to obtain the desired value.

Write the following numbers in scientific notation:

1. 0.0000427

2. 89650000

Exercise 2 above is an example of another value of scientific notation. As the number is written, it is not clear if the four zeros at the right of the number are significant figures. They are necessary to give the number 8965 their proper weight. In scientific notation, the zeros that are not significant can be eliminated from the expression.

LOGARITHMS

We have seen that it is convenient to use scientific notation (powers of 10) when expressing large or small numbers. Another way to express 6.022×10^{13} is as the power to which one should take 10 to express the entire value. For this example, the number is larger than 10^{13}, but not as large as 10^{14}, so the power of 10 that will equal 6.022×10^{13} is between 13 and 14. In fact, it is 13.780. Therefore, the numbers $10^{13.780}$ and 6.022×10^{13} are equal. The number 13.780 is called the **logarithm** of 6.022×10^{13}. Some more examples are as follows:

$$\log 4.73 \times 10^{-5} = -4.325$$
$$\log 10.0 = 1.000$$
$$\log 1.000 = 0.0000$$

Practice obtaining the logarithms of these numbers with the "log" command on your calculator.

3. 437

4. 0.0038

Antilogarithms

Many times, when the logarithm of a number is given, it is desired to express the number in scientific notation. This is done by finding the antilogarithm of the log value. The **antilogarithm** (sometimes just called "antilog") of a number x is the number that is equal to 10 to the x power. Some examples of this operation are as follows:

$$\text{antilog } 6.93 = 8.5 \times 10^6$$
$$10^{-2.48} = 3.3 \times 10^{-3}$$
$$\text{antilog } 1.000 = 10.0$$

The calculator command for obtaining the antilog of a number is sometimes labeled "10^x". Practice converting these numbers to scientific notation.

5. $\log x = 8.625$

6. $\log x = -5.1$

"p" Units

As a result of the definition of pH as related to the negative logarithm of H_3O^+ activity in solution, chemists have found the use of negative log values convenient. The "p" function is then defined as $-\log$. Therefore $p(x) = -\log x$. To obtain the "p" value for a number, find the log of that number and change the sign. Some examples of this operation are

$$p(4.72 \times 10^{-7}) = 6.326$$
$$p(9.38 \times 10^4) = -4.972$$
$$p(10) = -1.0$$
$$p(1.00) = 0.000$$

If the "p" value of a number is given and the scientific notation value is desired, change the sign of the "p" value and take the antilog of the result. For instance

$$\text{If } p(x) = 3.85, x = 1.4 \times 10^{-4}$$
$$\text{If } p(x) = -1.74, x = 55$$
$$\text{If } p(x) = 10.00, x = 1.0 \times 10^{-10}$$
$$\text{If } p(x) = 1.000, x = 0.100$$

It is helpful not only to gain a facility with the conversions of numbers to these several forms, but to develop a "feel" for the magnitudes of numbers when they are expressed in their logarithmic or "p" forms. Try these exercises:

7. $p(2.83 \times 10^{-4}) =$

8. $p(7.4136 \times 10^{7}) =$

9. If $p(x) = 23.8527, x =$

10. If $p(x) = -4.6, x =$

Logarithmic Scales

Another consequence of the large range of useful numbers in chemical studies is that these numbers are often plotted as their log values. (This is the same as using log or semilog graph paper.) Interpolation of values from logarithmic plots is not the same as with a linear scale. The safest method, until some experience is gained, is to read the log value and then calculate the number it represents. This is because the relationship between the value and its logarithm is nonlinear, as shown in Figure BB.1. The log of 0.01 is -2.0 and the log of 0.001 is -3.0, but the log of 0.005 is not midway between -2.0 and -3.0, it is -2.3. The value corresponding to $\log x = -2.5$ is actually closer to 0.002. This nonlinear aspect of log plots can be a bit tricky until you get used to them.

NATURAL LOGARITHMS

The logarithmic values we have discussed so far have been the power to which the number 10 should be raised to equal the number. In this case, 10 is called the **base** of the logarithm. In principle, any number can serve as the base for a logarithmic system. In practice, only two bases are in common use: 10 and e. The constant e is an irrational number with a value of 2.71828. . . . It is the base of the **Naperian** or **natural** logarithms. Thus if $a = \ln x$, $e^a = e^{\ln x}$.

To distinguish between the two logarithm bases, we use $\log x$ or $\log_{10} x$ when we mean the base 10 logarithm and $\ln x$ or $\log_e x$ when we mean the natural logarithm. It is sometimes necessary to convert the values of logarithms between the two bases. This is accomplished by the following relationship.

$$\ln x = 2.3026 \log_{10} x$$

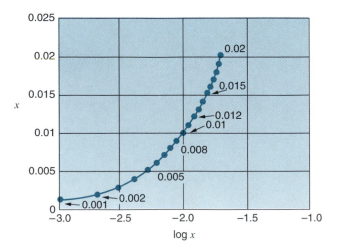

Figure BB.1. A plot of $\log x$ versus x is not linear, which makes interpolation on a log scale difficult.

To find the values of the following numbers (in scientific notation), divide $\ln x$ by 2.303 to obtain $\log_{10} x$. Then take the antilog of $\log_{10} x$. Try these.

11. $\ln x = 8.274$

12. $\ln x = -36.2$

OPERATIONS WITH EXPONENTS

Many of the relationships used in chemistry involve exponents. In deriving new relationships from existing ones and in the evaluation of relationships, we frequently need to perform operations on numbers or symbols for values that are in the exponential form. This section reviews the forms of the results of performing the common mathematical operations on numbers in the exponential format.

Addition and Subtraction

Two values in exponential form can only be added or subtracted when their exponents are the same. Thus

$$a10^x + b10^x = (a + b)10^x$$

To add 3.87×10^3 to 6.29×10^4, it is first necessary to make their exponents identical. Then the numerical parts can be added. $0.387 \times 10^4 + 6.29 \times 10^4 = 6.68 \times 10^4$.

Multiplication and Division

The product of two values in exponential form is obtained by multiplying the preexponential parts and adding the exponents. Thus

$$a10^x \times b10^y = ab10^{(x+y)}$$

The product of 3.87×10^3 and 6.29×10^4 is 24.3×10^7 or 2.43×10^8.

When values in exponential form are divided, the preexponential parts are divided and the exponents are subtracted. Thus

$$\frac{a10^x}{b10^y} = \frac{a}{b}10^{(x-y)}$$

If 3.87×10^3 is divided by 6.29×10^4, the result is 0.615×10^{-1} or 6.15×10^{-2}.

Here are some exercises:

13. $3.1 \times 10^{-4} + 4.653 \times 10^{-3} =$

14. $5.95 \times 10^5 - 4.72 \times 10^3 =$

15. $8.37 \times 10^4 \times 6.24 \times 10^{-3} =$

16. $3.631 \times 10^{-6}/5.942 \times 10^4 =$

OPERATIONS WITH LOGARITHMS

Since the logarithm of a number is the exponent to which 10 must be raised to equal that number, the operations with logarithms are similar to those of exponents. In this case, there is no preexponential part.

$$\log a + \log b = \log(ab)$$

$$\log a - \log b = \log\left(\frac{a}{b}\right)$$

$$\log \frac{1}{b} = -\log b$$

For example

$$x = \log 3.41 + \log 2.83 = \log 9.65 + 0.98$$
$$x = \log 3.41 - \log 2.83 = \log 1.20 = 0.081$$
$$x = \log \frac{1}{2.83} = -\log 2.83 = -0.45$$

Here are some for you to try:

17. $x = \log 34.9 + \log 3.86$

18. $x = \log 1.63 - \log 9.475$

19. $x = \log 1/84.56$

Answers

Scientific Notation
1. 4.27×10^{-5}
2. 8.965×10^7

Logarithms
3. $\log 437 = 2.640$
4. $\log 0.0038 = -2.42$

Antilogarithms
5. $10^{8.625} = 4.22 \times 10^8$
6. $10^{-5.1} = 8 \times 10^{-6}$

"p" Units
7. $p(2.83 \times 10^{-4}) = 3.548$
8. $p(7.4136 \times 10^7) = -7.87003$
9. If $p(x) = 23.8527, x = 1.404 \times 10^{-24}$
10. If $p(x) = -4.6, x = 4 \times 10^4$

Natural Logs
11. If $\ln x = 8.274, x = 3.92 \times 10^3$
12. If $\ln x = -36.2, x = 2 \times 10^{-16}$

Operations with Exponents
13. $3.1 \times 10^{-4} + 4.653 \times 10^{-3} = 0.31 \times 10^{-3} + 4.653 \times 10^{-3} = 4.96 \times 10^{-3}$
14. $5.95 \times 10^5 - 4.72 \times 10^3 = 5.95 \times 10^5 - 0.0472 \times 10^5 = 5.90 \times 10^5$
15. $8.37 \times 10^4 \times 6.24 \times 10^{-3} = 5.22 \times 10^2 = 522$
16. $3.631 \times 10^{-6}/5.942 \times 10^4 = 6.111 \times 10^{-11}$

Operations with Logarithms
17. $x = \log 134.7 = 2.1294$
18. $x = \log 0.172 = -0.764$
19. $x = -\log 84.56 = -1.9272$

C. Electrical Quantities and Their Relationships

All electrical phenomena are the result of the existence of species such as ions and electrons that can carry an electrical charge. In normal matter, the quantity of positively charged species is equal to the quantity of negatively charged species so that the entity is uncharged or **electrically neutral.** However, it is possible to induce the motion of charged species within a medium, and it is possible to transfer charged species from one entity to another. These two simple possibilities enable all the electrical processes that have become such an essential part of our science and, indeed, our life.

ELECTRICAL CONDUCTORS AND INSULATORS

Materials are electrically **conducting** if they have charged species that can move within the material. Such species are called **mobile charge carriers.** Electrical conductivity, then, is the net motion of the mobile charge carriers through the material. The mobile charge carriers are often electrons, as in a metallic conductor. But they can also be positive and negative ions, as in an ionic solution. **Conductivity** in solid materials requires that there be more or fewer bonding electrons than are needed for the interatomic bonding. If there are more, the excess bonding electrons are the mobile charge carriers. If there are fewer, there are some unoccupied bonding sites. Electrons can then skip from one bonding site to an adjacent unoccupied bonding site and thus move the unoccupied site through the material. In this case, the mobile charge carriers are the unoccupied bonding sites, called **holes.**

If all the bonding electrons in the atoms making up the material are used in bonding, there are no holes or free electrons. With no mobile charge carriers, the material does not conduct charge and is called an **insulator.** Insulators are used extensively to avoid electrical conductivity when it is not wanted.

SEPARATION OF CHARGE

Normally, materials in the world are considered uncharged. That is, they contain no excess charge of either sign. Consider two such pieces of material placed near each other, but separated by air. They are depicted as sheets of metal in Figure BC.1. Air is an insulator, so there is no conductivity between them. By some mechanism, an electron from one of the pieces of metal is removed and placed in the other one. It requires work to do this. One piece now has an excess negative charge and the other an excess positive charge. This difference in charge results in an **electrical potential** difference between them. If more electrons are moved from the first piece to the second, this electrical potential difference increases. This potential difference, v, is proportional to the difference in charge and is measured in the units of **volts,** V. Note that potential difference is always measured between two objects: it is the voltage of one *with respect to* the other.

Scientists have developed many devices for the separation of charge and thus the charging of one piece of metal with respect to another. They include electromagnetic generators and electrochemical batteries. The electrochemical battery has a symbol that depicts the charged metallic plates as shown in Figure BC.2. The end with the longer lines is always positive with respect to the other. Batteries and power supplies are rated for the electrical potential difference they can develop between their output contacts. The cylindrical flashlight batteries produce 1.5 V, an automobile battery 12 V, and so forth.

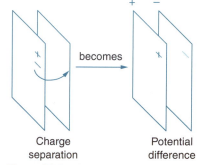

Figure BC.1. This sequence illustrates the work involved in separating electrical charge.

Charge separation — Potential difference

Figure BC.2. The symbol for a battery has a series of alternating long and shorter lines.

ELECTRICAL CURRENT

At the positive contact of a battery, there is a deficiency of electrons compared to the negative contact. If a conductor is connected between the two battery contacts, the positive connector of the battery would attract negative mobile charge carriers in the conductor toward it. If there were positive mobile charge carriers in the conductor (like holes in a semiconductor) they would be attracted toward the negative connector. In this way, the electrical potential difference from the battery causes a net motion of the mobile charge carriers in the conductor connected to it. This motion of mobile charge carriers is called an electrical **current.**

The magnitude of the current, i (the rate of charge flow past a cross section of the conductor), is measured in amperes, A. One **ampere** is the flow of one coulomb of charge per second. A **coulomb** (C) of charge is 6.2383×10^{18} electron charges.

When a conductor is connected between the battery contacts, the current that results in that conductor depends on the density of mobile charge carriers available, the dimensions of the conductor, and the ease with which they can move through the material. The ease of conducting charge is an intrinsic property of a conductor called its **conductance,** the opposite of which is called **resistance.** The less dense the mobile charge carriers in the conductor, the longer or narrower their path, or the more difficult their motion through the conductor, the higher the resistance of the conductor will be. The resistance, R, of a conducting device is measured in **ohms,** Ω. The symbol for a resistive conductor, or **resistor** (an intentionally poor conductor, as opposed to a wire) is shown in Figure BC.3.

Figure BC.3. The symbol for a resistor has zig-zag lines to indicate resistance to the flow of current.

The magnitude of the current, i, through a conductor also depends on the voltage applied across it. The higher the voltage, the greater is the attractive force on the mobile charge carriers in the conductor. The relationship among the voltage, v, the current, i, and the resistance, R, is called **Ohm's law.**

$$i = \frac{v}{R} \qquad \text{or} \qquad v = iR \qquad\qquad \text{BC.1}$$

From Ohm's law we see that a current of 1 A will result from the application of 1 V to a conductor with a resistance of 1 Ω.

The circuit diagram for a resistor connected to a voltage source (often symbolized as a battery) is shown in Figure BC.4. The lines connecting the battery and resistor represent the connecting wires. The arrow represents the direction of the flow of current i (always taken as positive to negative).

Here are some exercises involving Ohm's law:

1. A resistor that has a resistance of 4.5 kΩ is connected between the contacts of a 1.5 V battery. What is the value of the current through the resistor?

2. The current through a 100 Ω resistor is 9.2 μA. What is the potential difference between the contacts of the resistor?

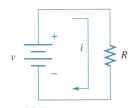

Figure BC.4. The resistor provides a conductive path for current to flow from the positive terminal of the battery to the negative.

ELECTRICAL CAPACITANCE

In the above discussion on the separation of charge, it was said that the greater the charge difference developed between the conducting plates, the greater is the electrical potential difference between them. The form of that relationship is

$$C = \frac{q}{v} \qquad \text{or} \qquad v = \frac{q}{C} \qquad\qquad \text{BC.2}$$

where C is the **capacitance** of the two conductors. In this relationship, q is the magnitude of the charge difference in coulombs, C. The unit for capacitance is the Farad, F. The quality of capacitance is that it requires a certain amount of charge difference to maintain an electrical potential difference between the two conductors. The amount of that charge difference is given by q in Equation BC.2. If it is desired to accumulate charge, a device called a capacitor is used. The symbol for a capacitor is shown in Figure BC.5. In a capacitor, the two conductors are arranged to have some specific value of capacitance.

Try these exercises in the relationship between voltage, capacitance, and charge.

Figure BC.5. The symbol for a capacitor represents the two parallel conductive sheets from which it is formed.

3. How many coulombs of charge are required to change the potential of a 0.010 μF capacitor by 3.6 mV?

4. A capacitor with 4.7 μC of charge has a potential difference between its contacts of 0.094 V. What is the value of the capacitance?

ELECTRICAL SIGNALS

The relevance of the electrical quantities discussed in this section to the study of chemical analysis is that the information we obtain from our experiments is so often encoded in the form of an electrical signal. We use electrodes that produce a voltage (electrical potential) that is related to the composition of the solution into which they are dipped. We use light detectors that produce a current that is related to the intensity of the light falling on them. We cannot correctly decode this information (converting it to a related number) unless we appreciate the fundamental nature of the electrical quantities from which the electrical signal is composed.

Voltage is the electrical potential difference caused by the separation of charge. It is a kind of "electrical pressure" in that it is the motivating force behind the concerted movement of charge carriers. Voltage is always measured as the potential difference between two conductors. The ideal measurement of a signal voltage is one that does not require the voltage source to induce *any* current through the measurement device. In other words, we would like to have a volt-meter with an infinite resistance so that the current through the meter, resulting from the measurement of v, would be zero.

Current is the rate of the flow of charge through a conductor. To have current in a conductor, one must have an electrical potential difference across it. In some chemical and light sensors, the phenomenon sensed produces the mobile charge carriers so that the *current* through the device is related to the process under study, and therefore the *current* must be measured to decode the signal. Current is best measured with a device that will not impede the current at all; in other words, an ideal current meter has zero resistance.

Current and voltage signals are both **analog** signals in that they are continuously variable. However, it is important to remember that they are different quantities and their decoding requires different conversion devices.

Answers

1. The current through the resistor is obtained by the use of Equation BC.1.

    $$i = v/R = 1.5 \text{ V}/4.5 \times 10^3 \, \Omega = 3.3 \times 10^{-4} \text{ A}$$

2. If a current of 9.2 µA passes through a 100 Ω resistor, the potential difference developed across the resistor is $v = iR = 9.2 \times 10^{-6}$ A $\times$ 100 $\Omega = 9.2 \times 10^{-4}$ V, or 920 µV.

3. From Equation BC.2, the charge required to change the potential of a 0.010 µF capacitor by 3.6 mV is $q = C v = 1.0 \times 10^{-8}$ F $\times$ 3.6 $\times 10^{-3}$ V $= 3.6 \times 10^{-11}$ C.

4. Again, Equation BC.2 is used to determine the value of the capacitance from the charge it contains and the voltage across it.

    $$C = 4.7 \times 10^{-6} \text{ C}/0.094 \text{ V} = 5.0 \times 10^{-5} \text{ F} = 50 \text{ µF}$$

Tables of Activity Coefficients and a_x

Table 1 gives values of the activity coefficient, γ_x, for several values of the ionic strength, S, and ionic charge, z_x. The values given were calculated by the use of the Debye–Hückel equation (DHE), Equation 3.28. The calculations are for an aqueous solvent at 25 °C. To use this table, one must know the ion size parameter, a_x. This is given for many ions in Tables 2 and 3.

1. Calculated Values of Activity Coefficient Using the DHE

a_x, nm	$z_x = \pm 1$ and $S =$			$z_x = \pm 2$ and $S =$			$z_x = \pm 3$ and $S =$			$z_x = \pm 4$ and $S =$		
	0.001	0.01	0.1	0.001	0.01	0.1	0.001	0.01	0.1	0.001	0.01	0.1
1.10	0.967	0.918	0.841	0.875	0.709	0.501	0.741	0.461	0.211	0.587	0.252	0.063
1.00	0.967	0.916	0.834	0.874	0.703	0.483	0.739	0.452	0.195	0.584	0.244	0.055
0.90	0.967	0.914	0.826	0.873	0.696	0.465	0.737	0.443	0.179	0.581	0.235	0.047
0.80	0.966	0.911	0.817	0.872	0.690	0.445	0.735	0.434	0.162	0.578	0.227	0.039
0.70	0.966	0.909	0.807	0.871	0.683	0.424	0.733	0.424	0.145	0.575	0.218	0.032
0.60	0.966	0.907	0.796	0.870	0.676	0.401	0.731	0.414	0.128	0.572	0.209	0.026
0.50	0.965	0.904	0.784	0.869	0.669	0.377	0.728	0.404	0.111	0.569	0.200	0.020
0.45	0.965	0.903	0.777	0.868	0.665	0.364	0.727	0.399	0.103	0.567	0.195	0.018
0.40	0.965	0.902	0.770	0.867	0.661	0.351	0.726	0.394	0.095	0.566	0.191	0.015
0.35	0.965	0.900	0.762	0.867	0.657	0.337	0.725	0.388	0.087	0.564	0.186	0.013
0.30	0.965	0.899	0.754	0.866	0.653	0.323	0.724	0.383	0.079	0.563	0.181	0.011
0.25	0.965	0.897	0.745	0.865	0.648	0.308	0.722	0.377	0.071	0.561	0.177	0.009
0.20	0.964	0.896	0.736	0.865	0.644	0.293	0.721	0.372	0.063	0.559	0.172	0.007

2. Table of Ion Size Parameters, a_x, for Inorganic Ions*

Ion	a_x, nm	Ion	a_x, nm	Ion	a_x, nm	Ion	a_x, nm
Ag^+	0.25	HCO_3^-	0.45	K^+	0.3	S^{2-}	0.5
Al^{3+}	0.9	Cr^{3+}	0.9	La^{3+}	0.9	HS^-	0.35
Ba^{2+}	0.5	CrO_4^{2-}	0.4	Li^+	0.6	$S_2O_3^{2-}$	0.4
Be^{2+}	0.8	Cu^{2+}	0.6	Mg^{2+}	0.8	SCN^-	0.35
Br^-	0.3	F^-	0.35	Mn^{2+}	0.6	Sn^{2+}	0.6
BrO_3^-	0.35	$Fe(CN)_6^{3-}$	0.4	MnO_4^-	0.35	Sn^{4+}	1.1
$C_2O_4^{2-}$	0.45	$Fe(CN)_6^{4-}$	0.5	Na^+	0.4	SO_3^{2-}	0.45
Ca^{2+}	0.6	Fe^{2+}	0.6	NH_4^+	0.25	HSO_3^-	0.4
Cd^{2+}	0.5	Fe^{3+}	0.9	Ni^{2+}	0.6	SO_4^{2-}	0.4
Ce^{3+}	0.9	$H_2AsO_4^-$	0.4	NO_2^-	0.3	Sr^{2+}	0.5
Ce^{4+}	1.1	H_3O^+	0.9	NO_3^-	0.3	Th^{4+}	1.1
Cl^-	0.3	$HCOO^-$	0.3	OH^-	0.35	Tl^+	0.25
ClO_3^-	0.35	Hg^{2+}	0.5	Pb^{2+}	0.45	Zn^{2+}	0.6
ClO_4^-	0.35	Hg_2^{2+}	0.4	PO_4^{3-}	0.4	Zr^{4+}	1.1
CN^-	0.3	I^-	0.3	HPO_4^{2-}	0.4		
Co^{2+}	0.6	IO_3^-	0.4	$H_2PO_4^-$	0.4		
CO_3^{2-}	0.45	IO_4^-	0.35	Rb^+	0.25		

*Values found in J. Kielland, *J. Am. Chem. Soc.* **1973**, *59*, 1675.

3. Table of Ion Size Parameters, a_x, for Organic Ions*

Ion Name	Formula	ax, nm
Phthalate	$C_6H_4(COO)_2^{2-}$	0.6
Acetate	CH_3COO^-	0.45
Tetrabutylammonium	$(C_4H_9)_4N^+$	0.8
Benzoate	$C_6H_5COO^-$	0.6
Tetraethylammonium	$(C_2H_5)_4N^+$	0.6
Dichloroacetate	$CHCl_2COO^-$	0.5
Malonate	$H_2C(COO)_2^{2-}$	0.5
Dibenzylacetate	$(C_6H_5)_2CHCOO^-$	0.8
Citrate	$OOC(OH)C(CH_2COO)_2^{3-}$	0.5
Hydrogen citrate	$HOOC(OH)C(CH_2COO)_2^{2-}$	0.45
Dihydrogen citrate	$HOOC(OH)C(CH_2COO)_2H^-$	0.35

*Values found in J. Kielland, *J. Am. Chem. Soc.* **1937**, *59*, 1675.

K_a° Values for Some Weak Acids in Water

Listed are the equilibrium constants for the protonation of water by the acid given. The values given for K_a° are for solutions at 25 °C and are the thermodynamic equilibrium constants. Where the acid can lose more than one proton to the water, the additional equilibrium constants are listed. The value of pK_a' depends on the ionic composition of the solution. From Equation 3.50, pK_a' can be estimated from a knowledge of the ionic strength, S, in situations where the Debye–Hückel limiting law (DHLL) is valid. The number given in the last column is the factor N in the equation $pK_a' = pK_a^\circ - NA\sqrt{S}$. The value of A at 25 °C is 0.5091. The method of calculating the ionic strength is given in Equation 3.27. The K_a° and pK_a° values in the table were taken from A. E. Martell and R. M. Smith, *Critical Stability Constants*, Volumes 1–5. Plenum Press, New York, 1974.

Name	Most acidic formula or structure	K_a°	pK_a°	N
Acetic acid	CH_3CO_2H	1.75×10^{-5}	4.757	2
Alanine	NH_3^+ $\|$ $CHCH_3$ $\|$ CO_2H	4.49×10^{-3} 1.36×10^{-10}	2.348 (CO_2H) 9.867 (NH_3)	0 2
Ammonia	NH_4^+	5.70×10^{-10}	9.244	0
Arginine	NH_3^+ $\|$ $CHCH_2CH_2CH_2NHC$ NH_2^+ $\|$ CO_2H NH_2	1.50×10^{-2} 1.02×10^{-9} 3.3×10^{-13}	1.823 (CO_2H) 8.991 (NH_3) 12.48 (NH_2)	-2 0 2
L-Ascorbic acid	HO OH $HOCH_2CH$ O O $\|$ OH	9.33×10^{-5} 4.57×10^{-12}	4.03 11.34	2 4
Asparagine	NH_3^+ O $\|$ $\|\|$ $CHCH_2CNH_2$ $\|$ CO_2H	7.2×10^{-3} 1.9×10^{-9}	*2.14 (CO_2H) (0.1) 8.72 (NH_3) (0.1)	0 2
Aspartic acid	NH_3^+ $\|$ $CHCH_2CO_2H$ $\|$ CO_2H	1.02×10^{-2} 1.26×10^{-4} 9.95×10^{-11}	1.990 (terminal CO_2H) 3.900 (alkyl CO_2H) 9.90 (NH_3)	0 2 4
Benzoic acid	$\bigcirc$—CO_2H	6.28×10^{-5}	4.202	2

Name	Most acidic formula or structure	K_a°	pK_a°	N
Butanoic acid	$CH_3CH_2CH_2CO_2H$	1.52×10^{-5}	4.819	2
Butylamine	$CH_3CH_2CH_2CH_2NH_3^+$	2.29×10^{-11}	10.640	0
Carbonic acid	$HO-\overset{\displaystyle O}{\overset{\displaystyle \|}{C}}-OH$	4.45×10^{-7} 4.69×10^{-11}	6.352 10.329	2 4
Chloroacetic acid	$ClCH_2CO_2H$	1.36×10^{-3}	2.865	2
Chlorous acid	$HOCl{=}O$	1.12×10^{-2}	1.95	2
Chromic acid	$HO-\overset{\displaystyle O}{\underset{\displaystyle O}{\overset{\displaystyle \|}{\underset{\displaystyle \|}{Cr}}}}-OH$	1.6 3.1×10^{-7}	-0.2 (20 °C, 0) 6.51	2 4
Citric acid	$HO_2CCH_2\underset{\displaystyle OH}{\overset{\displaystyle CO_2H}{CCH_2CO_2H}}$	7.44×10^{-4} 1.73×10^{-5} 4.02×10^{-7}	3.128 4.761 6.396	2 4 6
Cysteine	$\underset{\displaystyle CO_2H}{\overset{\displaystyle NH_3^+}{CHCH_2SH}}$	1.95×10^{-2} 4.4×10^{-9} 1.70×10^{-11}	1.71 (CO_2H) 8.36 (SH) 10.77 (NH_3)	0 2 4
Dichloroacetic acid	Cl_2CHCO_2H	5.0×10^{-2}	1.30	2
Diethylamine	$(CH_3CH_2)_2NH_2^+$	1.17×10^{-11}	10.933	0
Dimethylamine	$(CH_3)_2NH_2^+$	1.68×10^{-11}	10.774	0
EDTA	$\overset{\displaystyle HO_2CCH_2}{\underset{\displaystyle HO_2CCH_2}{}}N CH_2CH_2 N\overset{\displaystyle CH_2CO_2H}{\underset{\displaystyle CH_2CO_2H}{}}$	6.31×10^{-3} 5.01×10^{-3} 4.79×10^{-7} 9.68×10^{-12} 1.0×10^{-2} 2.09×10^{-3} 7.76×10^{-7} 6.76×10^{-11}	2.2 (20 °C, 1.0) 2.3 (20 °C, 1.0) 6.320 (20 °C, 0) 11.014 (20 °C, 0) 2.0 (0.1) 2.68 (0.1) 6.11 (0.1) 10.17 (0.1)	2 4 6 8
Ethylamine	$CH_3CH_2NH_3^+$	2.31×10^{-11}	10.636	0
Formic acid	HCO_2H	1.80×10^{-4}	3.745	2
Glutamic acid	$HO_2CCH_2CH_2\overset{\displaystyle NH_3^+}{CHCO_2H}$	5.89×10^{-3} 3.80×10^{-5} 1.12×10^{-10}	2.23 (CO_2H) 4.42 (CO_2H) 9.95 (NH_3)	0 2 4
Glutamine	$H_2N\overset{\displaystyle O}{\overset{\displaystyle \|}{C}}CH_2CH_2\overset{\displaystyle NH_3^+}{CHCO_2H}$	6.76×10^{-3} 9.77×10^{-10}	2.17 (CO_2H) (0.1) 9.01 (NH_3) (0.1)	0 2
Glycine	$\underset{\displaystyle CO_2H}{\overset{\displaystyle NH_3^+}{\underset{\displaystyle \|}{\overset{\displaystyle \|}{CH_2}}}}$	4.47×10^{-3} 1.67×10^{-10}	2.350 (CO_2H) 9.778 (NH_3)	0 2

Name	Most acidic formula or structure	K_a°	pK_a°	N
Guanidine		2.9×10^{-14}	13.54 (27 °C, 1.0)	0
Histidine		2×10^{-2}	1.7 (CO_2H) (0.1)	-2
		9.5×10^{-7}	6.02 (NH) (0.1)	0
		8.3×10^{-10}	9.08 (NH_3) (0.1)	2
Hydrogen cyanate	$HOC\equiv N$	3.3×10^{-4}	3.48	2
Hydrogen cyanide	$HC\equiv N$	6.2×10^{-10}	9.21	2
Hydrogen fluoride	HF	6.8×10^{-4}	3.17	2
Hydrogen peroxide	HOOH	2.2×10^{-12}	11.65	2
Hydrogen sulfide	H_2S	9.5×10^{-8}	7.02	2
		1.26×10^{-14}	13.9	4
Hydroxybenzene	—OH	1.05×10^{-10}	9.98	2
8-Hydroxyquinoline		1.12×10^{-5}	4.95 (NH)	0
		1.70×10^{-10}	9.77 (OH)	2
Hypochlorous acid	HOCl	3.0×10^{-8}	7.53	2
Iodic acid		0.17	0.77	2
Isoleucine		4.80×10^{-3}	2.319 (CO_2H)	0
		1.76×10^{-10}	9.754 (NH_3)	2
Leucine		4.69×10^{-3}	2.329 (CO_2H)	0
		1.79×10^{-10}	9.747 (NH_3)	2
Lysine		9.1×10^{-3}	2.04 (CO_2H) (0.1)	-2
		8.3×10^{-10}	9.08 (terminal NH_3) (0.1)	0
		2.0×10^{-11}	10.69 (alkyl NH_3) (0.1)	2
Malonic acid	$HO_2CCH_2CO_2H$	1.42×10^{-3}	2.847	2
		2.01×10^{-6}	5.696	4
Methionine		6.3×10^{-3}	2.20 (CO_2H) (0.1)	0
		8.9×10^{-10}	9.05 (NH_3) (0.1)	2

Name	Most acidic formula or structure	K_a°	pK_a°	N
Methylamine	$CH_3\overset{+}{N}H_3$	2.3×10^{-11}	10.64	0
Nicotinic acid		8.91×10^{-3} 1.55×10^{-5}	2.05 (CO_2H) 4.81 (NH)	0 2
Nitrous acid	$HON{\equiv}O$	7.1×10^{-4}	3.15	2
Oxalic acid	HO_2CCO_2H	5.60×10^{-2} 5.42×10^{-5}	1.252 4.266	2 4
1,10-Phenanthroline		1.38×10^{-5}	4.86	0
Phenylalanine		6.3×10^{-3} 4.9×10^{-10}	2.20 (CO_2H) 9.31 (NH_3)	0 2
Phosphoric acid		7.11×10^{-3} 6.32×10^{-8} 4.49×10^{-13}	2.148 7.199 12.35	2 4 6
Phosphorous acid		3×10^{-2} 1.62×10^{-7}	1.5 6.79	2 4
Phthalic acid		1.12×10^{-3} 3.90×10^{-6}	2.950 5.408	2 4
Proline		1.12×10^{-2} 2.29×10^{-11}	1.952 (CO_2H) 10.640 (NH_2)	0 2
Propylamine	$CH_3CH_2CH_2NH_3^+$	2.72×10^{-11}	10.566	0
Pyridine		5.90×10^{-6}	5.229	0
Pyruvic acid		2.82×10^{-3}	2.55	2
Serine		6.50×10^{-3} 6.18×10^{-10}	2.187 (CO_2H) 9.209 (NH_3)	0 2
Succinic acid	$HO_2CCH_2CH_2CO_2H$	6.21×10^{-5} 2.31×10^{-6}	4.207 5.636	2 4

Name	Most acidic formula or structure	K_a°	pK_a°	N
Sulfuric acid	$\begin{array}{c} O \\ \parallel \\ HO-S-OH \\ \mid \\ O \end{array}$	1.02×10^{-2}	1.99 (pK_2°)	4
Threonine	$\begin{array}{c} OH\ NH_3^+ \\ \mid\ \ \mid \\ CH_3CHCHCO_2H \end{array}$	8.17×10^{-3} 7.94×10^{-10}	2.088 (CO_2H) 9.100 (NH_3)	0 2
Trichloroacetic acid	Cl_3CCO_2H	0.22	0.66 (0.1)	2
Triethylamine	$(CH_3CH_2)_3NH^+$	1.93×10^{-11}	10.715	0
Tryptophan	(structure: indole with $\begin{array}{c} NH_3 \\ \mid \\ CH_2CHCO_2H \end{array}$)	4.47×10^{-3} 4.68×10^{-10}	2.35 (CO_2H) (0.1) 9.33 (NH_3) (0.1)	0 2
Tyrosine	$\begin{array}{c} NH_3^+ \\ \mid \\ CHCH_2-\bigcirc-OH \\ \mid \\ CO_2H \end{array}$	6.8×10^{-3} 6.5×10^{-10} 3.4×10^{-11}	2.17 (CO_2H) (0.1) 9.19 (NH_3) 10.47 (OH)	0 2 4
Valine	$\begin{array}{c} NH_3^+ \\ \mid \\ CHCH(CH_3)_2 \\ \mid \\ CO_2H \end{array}$	5.18×10^{-3} 1.91×10^{-10}	2.286 (CO_2H) 9.718 (NH_3)	0 2

*Formal equilibrium constant for (temperature, ionic strength) conditions given. If temperature is not specified, it is 25 °C.

Table of Complex Formation (K_f° and K_f') Values

The log K_f values in this table were taken from A. E. Martell and R. M. Smith, *Critical Stability Constants,* Volumes 1–5. Plenum Press, New York, 1974. Some of the values are for products of specific formation constants. This occurs most often when the stepwise formation constants do not decrease in value with each step. The term N is from Equation 3.50, log $K_f' = $ log $K_f^\circ + NAS^{1/2}$. It is only given for those formation constants determined at 0 ionic strength.

Ligand and Metal	log K_f	Conditions (Temperature, °C, and Ionic Strength)	N
Ammonia (NH_3)			
Ag^+	$K_1 = 3.31$	25, 0	0
	$K_2 = 3.91$	25, 0	0
Cu^{2+}	$K_1 = 4.04$	25, 0	0
	$K_2 = 3.43$	25, 0	0
	$K_3 = 2.80$	25, 0	0
Hg^{2+}	$K_1 = 8.8$	25, 2.0	
	$K_1 \cdot K_2 = 17.4$	23, 2.0	
	$K_3 = 1.0$	25, 2.0	
	$K_4 = 0.7$	25, 2.0	
Mg^{2+}	$K_1 = 0.23$	23, 2.0	
Zn^{2+}	$K_1 = 2.21$	25, 0	0
	$K_2 = 2.29$	25, 0	0
	$K_3 = 2.36$	25, 0	0
	$K_4 = 2.03$	25, 0	0
Asparagine ($C_4H_8O_3N_2$)			
Cu^{2+}	$K_1 = 7.83$	25, 0.1	
	$K_2 = 6.53$	25, 0.1	
Fe^{2+}	$K_1 = 3.40$	20, 1.0	
Fe^{3+}	$K_1 = 8.6$	20, 1.0	
Zn^{2+}	$K_1 = 5.07$	25, 3.0	
	$K_2 = 4.36$	25, 3.0	
	$K_3 = 2.87$	25, 3.0	
Bromide (Br^-)			
Ag^+	$K_1 = 4.30$	25, 0.1	
	$K_2 = 2.34$	25, 0.1	
	$K_3 = 1.46$	25, 0.1	
	$K_4 = 0.80$	25, 0.1	
Hg^{2+}	$K_1 = 9.00$	25, 0.5	
	$K_2 = 8.10$	25, 0.5	
	$K_3 = 2.30$	25, 0.5	
	$K_4 = 1.6$	25, 0.5	

Ligand and Metal	$\log K_f$	Conditions (Temperature, °C, and Ionic Strength)	N
Chloride (Cl⁻)			
Ag^+	$K_1 = 3.70$	25, 5.0	
	$K_2 = 1.92$	25, 5.0	
	$K_3 = 0.78$	25, 5.0	
Ca^{2+}	$K_1 = 0.08$	25, 0.7	
Cu^{2+}	$K_1 = 0.4$	25, 0	-4
Fe^{3+}	$K_1 = 1.48$	25, 0	-6
	$K_2 = 0.65$	25, 0	-4
Hg^{2+}	$K_1 = 6.74$	25, 0.5	
	$K_2 = 6.48$	25, 0.5	
	$K_3 = 0.88$	25, 0.5	
	$K_4 = 1.00$	25, 0.5	
Zn^{2+}	$K_1 = 0.30$	25, 4.0	
	$K_2 = -0.30$	25, 4.0	
	$K_3 = 1.0$	25, 4.0	
Cyanide (CN⁻)			
Ag^+	$K_1 \cdot K_2 = 20.48$	25, 0	-2
	$K_3 = 0.92$	25, 0	2
	$K_4 = 0.5$	30, 1.0	
Fe^{2+}	$K_1 \cdot K_2 \cdot K_3 \cdot K_4 \cdot K_5 \cdot K_6 = 35.4$	25, 0	6
Fe^{3+}	$K_1 \cdot K_2 \cdot K_3 \cdot K_4 \cdot K_5 \cdot K_6 = 43.6$	25, 0	-6
Hg^{2+}	$K_1 = 17.00$	25, 0	-4
	$K_2 = 15.75$	25, 3	
	$K_3 = 3.56$	25, 4	
	$K_4 = 2.66$	25, 4	
Zn^{2+}	$K_1 = 5.3$	25, 3.0	
	$K_1 \cdot K_2 = 11.07$	25, 0	-6
	$K_3 = 4.98$	25, 0	0
	$K_4 = 3.57$	25, 0	2
EDTA ($C_{10}H_{16}O_8N_2$)			
Ag^+	$K_1 = 7.32$	20, 0.1	
Al^{3+}	$K_1 = 12.6$	20, 0	-24
Ca^{2+}	$K_1 = 11.00$	20, 0	-16
Cu^{2+}	$K_1 = 18.70$	25, 0.1	
Fe^{2+}	$K_1 = 14.27$	25, 0.1	
Fe^{3+}	$K_1 = 25.0$	25, 0.1	
Hg^{2+}	$K_1 = 21.5$	25, 0.1	
K^+	$K_1 = 0.8$	25, 0.1	
Mg^{2+}	$K_1 = 9.12$	20, 0	-16
Na^+	$K_1 = 1.64$	25, 0.1	
Zn^{2+}	$K_1 = 16.44$	25, 0.1	
Ethylenediamine ($C_2H_8N_2$)			
Ag^+	$K_1 = 5.06$	25, 1.0	
	$K_2 = 2.64$	25, 1.0	
Cu^{2+}	$K_1 = 10.48$	25, 0	0
	$K_2 = 9.02$	25, 0	0
Fe^{2+}	$K_1 = 4.34$	25, 1.4	
	$K_2 = 3.32$	25, 1.4	
	$K_3 = 2.06$	25, 1.4	
Hg^{2+}	$K_1 = 14.3$	25, 0.1	
	$K_2 = 8.94$	25, 0.1	
Mg^{2+}	$K_1 = 0.37$	30, 1.4	

Ligand and Metal	$\log K_f$	Conditions (Temperature, °C, and Ionic Strength)	N
Ethylenediamine ($C_2H_8N_2$) (*cont.*)			
Zn^{2+}	$K_1 = 5.66$	25, 0.1	
	$K_2 = 4.98$	25, 0.1	
	$K_3 = 3.25$	25, 0.1	
Hydroxide (OH^-)			
Ag^+	$K_1 = 2.0$	25, 0	-2
	$K_2 = 1.99$	25, 0	0
Al^{3+}	$K_1 = 9.01$	25, 0	-6
	$K_2 = 9.69$	25, 0	-4
	$K_3 = 8.30$	25, 0	-2
	$K_4 = 6.00$	25, 0	0
Ca^{2+}	$K_1 = 1.3$	25, 0	-4
Cu^{2+}	$K_1 = 6.3$	25, 0	-4
	$K_1 \cdot K_2 = 12.8$	25, 1.0	
	$K_3 = 1.70$	25, 1.0	
	$K_4 = 1.10$	25, 1.0	
Fe^{2+}	$K_1 = 4.5$	25, 0	-4
	$K_2 = 2.9$	25, 0	-2
	$K_3 = 2.6$	25, 0	0
Fe^{3+}	$K_1 = 11.81$	25, 0	-6
	$K_2 = 10.49$	25, 0	-4
	$K_3 \cdot K_4 = 34.4$	25, 0	-2
Hg^{2+}	$K_1 = 10.6$	25, 0	-4
	$K_2 = 11.2$	25, 0	-2
Mg^{2+}	$K_1 = 2.58$	25, 0	-4
Zn^{2+}	$K_1 = 5.0$	25, 0	-4
	$K_2 = 6.1$	25, 0	-2
	$K_3 = 2.50$	25, 0	0
	$K_4 = 1.20$	25, 0	2
Iodide (I^-)			
Ag^+	$K_1 = 6.58$	25, 0	-2
	$K_2 = 5.12$	25, 0	0
	$K_3 = 1.40$	25, 0	2
Hg^{2+}	$K_1 = 12.87$	25, 0.5	
	$K_2 = 10.95$	25, 0.5	
	$K_3 = 3.78$	25, 0.5	
	$K_4 = 2.20$	25, 0.5	
Thiocyanate (SCN^-)			
Ag^+	$K_1 = 4.8$	25, 0	-2
	$K_2 = 3.43$	25, 0	0
	$K_3 = 1.27$	25, 0	2
	$K_4 = 0.20$	25, 0	4
Al^{3+}	$K_1 = 0.42$	20, 0	-6
Cu^{2+}	$K_1 = 2.33$	25, 0	-4
	$K_2 = 1.32$	25, 0	-2
Fe^{2+}	$K_1 = 1.31$	25, 0	-4
Fe^{3+}	$K_1 = 2.21$	25, 3.0	
	$K_2 = 1.43$	25, 3.0	
	$K_3 = 1.36$	25, 3.0	
	$K_4 = 1.30$	25, 3.0	

Ligand and Metal	log K_f	Conditions (Temperature, °C, and Ionic Strength)	N
Thiocyanate (SCN^-) (*cont.*)			
Hg^{2+}	$K_1 = 9.08$	25, 1.0	
	$K_1 \cdot K_2 = 17.26$	25, 0	-6
	$K_3 = 2.71$	25, 0	0
	$K_4 = 1.83$	25, 0	2
Zn^{2+}	$K_1 = 1.33$	25, 0	-4
	$K_2 = 0.58$	25, 0	-2
	$K_3 = 0.09$	25, 0	0
Tryptophan ($C_{11}H_{12}O_2N_2$)			
Cu^{2+}	$K_1 = 8.71$	25, 3.0	
	$K_2 = 7.95$	25, 3.0	
Fe^{2+}	$K_1 = 3.92$	25, 3.0	
	$K_2 = 3.47$	25, 3.0	
	$K_3 = 2.11$	25, 3.0	
Fe^{3+}	$K_1 = 9.0$	20, 1.0	
Zn^{2+}	$K_1 = 5.01$	25, 3.0	
	$K_2 = 4.17$	25, 3.0	
	$K_3 = 3.72$	25, 3.0	

K_{sp}° and K_{sp}' Values for Some Precipitates

The log K_{sp} values in this table were taken from A. E. Martell and R. M. Smith, *Critical Stability Constants,* Volumes 1–5. Plenum Press, New York, 1974. The term N is from Equation 3.50, $\log K_{sp}' = \log K_{sp}^{\circ} + NAS^{1/2}$.

Formula	$\log K_{sp}$	Conditions (Temperature, °C, Ionic Strength)	N
Bromide			
AgBr	−12.30	25, 0	2
CuBr	−8.3	25, 0	2
$HgBr_2$	−18.9	25, 0.5	6
Hg_2Br_2	−22.25	25, 0	6
$PbBr_2$	−5.68	25, 4.0	6
TlBr	−5.44	25, 0	2
Carbonate			
BaCa	−8.30	25, 0	
MgCa (aragonite)	−8.22	25, 0	
MgCa (calcite)	−8.35	25, 0	
$MgCO_3$	−7.46	25, 0	8
SrCa	−9.03	25, 0	8
Chloride			
AgCl	−9.74	25, 0	2
CuCl	−6.73	25, 0	2
Hg_2Cl_2	−17.91	25, 0	6
$PbCl_2$	−4.78	25, 0	6
TlCl	−3.74	25, 0	2
Chromate			
Ag_2CrO_4	−11.92	25, 0	6
$BaCrO_4$	−9.67	25, 0	8
$CuCrO_4$	−5.44	25, 0	8
Hg_2CrO_4	−8.70	25, 0	6
Tl_2CrO_4	−12.01	25, 0	6
Cyanide			
AgCN	−15.66	25, 0	2
$Hg_2(CN)_2$	−39.3	25, 0	6
$Zn(CN)_2$	−15.5	25, 3.0	
Fluoride			
BaF_2	−5.76	25, 0	6
CaF_2	−10.41	25, 0	6
LaF_3	−18.9	25, 0.1	12
LiF	−2.77	25, 0	2

Formula	log K_{sp}	Conditions (Temperature, °C, Ionic Strength)	N
Fluoride (*cont.*)			
MgF_2	−8.18	25, 0	6
PbF_2	−7.44	25, 0	6
SrF_2	−8.54	25, 0	6
ThF_4	−28.3	25, 3.0	20
Hydroxide			
$Ba(OH)_2 \cdot 8H_2O$	−3.6	25, 0	6
$Ca(OH)_2$	−5.19	25, 0	6
$Co(OH)_2$	−14.9	25, 0	6
$Cu(OH)_2$	−19.32	25, 0	6
$Fe(OH)_2$	−15.1	25, 0	6
$Fe(OH)_3$	−38.8	25, 0	12
$Mg(OH)_2$	−11.15	25, 0	6
$Mn(OH)_2$	−12.8	25, 0	6
$Ni(OH)_2$	−15.2	25, 0	6
Iodide			
AgI	−16.08	25, 0	2
Hg_2I_2	−28.33	25, 0	6
PbI_2	−8.10	25, 0	6
Sulfide			
$BaSO_4$	−9.96	25, 0	8
$CaSO_4$	−4.62	25, 0	8
$SrSO_4$	−6.50	25, 0	8

The $E°$ values in this table were taken from S. G. Bratsch, *Journal of Physical and Chemical Reference Data,* 18, 1. American Chemical Society, Washington, D.C., 1989. The value of N is for Equation 9.18 in which $E' = E° + \dfrac{V_N A \sqrt{S}}{n} N$.

Half Reaction	$E°$, V	N
Bromine		
$Br_{2(l)} + 2e^- \rightleftharpoons 2Br^-$	+1.087	2
Cadmium		
$Cd^{2+} + 2e^- \rightleftharpoons Cd_{(s)}$	−0.403	−4
Cerium		
$Ce^{4+} + e^- \rightleftharpoons Ce^{3+}$	+1.72	−5
Chlorine		
$Cl_{2(g)} + 2e^- \rightleftharpoons 2Cl^-$	+1.359	2
Chromium		
$Cr_2O_7^{2-} + 14H^+ + 6e^- \rightleftharpoons 2Cr^{3+} + 7H_2O$	+1.33	0
$Cr^{3+} + e^- \rightleftharpoons Cr^{2+}$	−0.38	−5
$Cr^{3+} + 3e^- \rightleftharpoons Cr_{(s)}$	−0.74	−9
$Cr^{2+} + 2e^- \rightleftharpoons Cr_{(s)}$	−0.89	−4
Cobalt		
$Co^{3+} + e^- \rightleftharpoons Co^{2+}$	+1.92	−5
$Co^{2+} + 2e^- \rightleftharpoons Co_{(s)}$	−0.282	−4
Copper		
$Cu^{3+} + e^- \rightleftharpoons Cu^{2+}$	(+2.4)	−5
$Cu^+ + e^- \rightleftharpoons Cu_{(s)}$	+0.518	−1
$Cu^{2+} + 2e^- \rightleftharpoons Cu_{(s)}$	+0.337	−4
$Cu^{2+} + e^- \rightleftharpoons Cu^+$	+0.161	−3
Gold		
$Au^+ + e^- \rightleftharpoons Au_{(s)}$	+1.69	−1
Hydrogen		
$2H^+ + 2e^- \rightleftharpoons H_{2(g)}$	0.000	−2
Iodine		
$2I^+ + 2e^- \rightleftharpoons I_{2(s)}$	+1.35	−2
$3I_2 + 2e^- \rightleftharpoons 2I_3^-$	+0.789	2
$I_2 + 2e^- \rightleftharpoons 2I^-$	0.620	2
$I_3^- + 2e^- \rightleftharpoons 3I^-$	0.536	2
Iron		
$Fe^{3+} + e^- \rightleftharpoons Fe^{2+}$	+0.771	−5
$Fe^{2+} + 2e^- \rightleftharpoons Fe_{(s)}$	−0.44	−4
Lead		
$Pb^{2+} + 2e^- \rightleftharpoons Pb_{(s)}$	0.126	−4

Half Reaction	$E°$, V	N
Manganese		
$MnO_4^- + 4H^+ + 3e^- \rightleftharpoons MnO_{2(s)} + 2H_2O$	+1.692	-5
$Mn^{3+} + e^- \rightleftharpoons Mn^{2+}$	+1.56	-5
$MnO_4^- + 8H^+ + 5e^- \rightleftharpoons Mn^{2+} + 4H_2O$	+1.507	-5
$MnO_4^- + e^- \rightleftharpoons MnO_4^{2-}$	+0.56	+3
$Mn^{2+} + 2e^- \rightleftharpoons Mn_{(s)}$	-1.182	-4
Mercury		
$2Hg^{2+} + 2e^- \rightleftharpoons Hg_2^{2+}$	+0.908	-4
$Hg_2^{2+} + 2e^- \rightleftharpoons 2Hg$	+0.789	-4
$Hg_2Cl_{2(s)} + 2e^- \rightleftharpoons 2Hg_{(l)} + 2Cl^-$	+0.268	2
Nickel		
$Ni^{3+} + e^- \rightleftharpoons Ni^{2+}$	(2.3)	-5
$Ni^{2+} + 2e^- \rightleftharpoons Ni_{(s)}$	-0.236	-4
Oxygen		
$H_2O_2 + 2H^+ + 2e^- \rightleftharpoons 2H_2O_{(l)}$	+1.763	-2
$O_{2(g)} + 4H^+ + 4e^- \rightleftharpoons 2H_2O_{(l)}$	+1.229	-4
Potassium		
$K^+ + e^- \rightleftharpoons K_{(s)}$	-2.936	-1
Silver		
$Ag^+ + e^- \rightleftharpoons Ag_{(s)}$	+0.799	-1
$AgCl + e^- \rightleftharpoons Ag_{(s)} + Cl^-$	+0.222	1
$AgI_{(s)} + e^- \rightleftharpoons Ag_{(s)} + I^-$	-0.151	1
Sodium		
$Na^+ + e^- \rightleftharpoons Na_{(s)}$	-2.714	-1
Sulfur		
$S_4O_6^{2-} + 2e^- \rightleftharpoons 2S_2O_3^{2-}$	+0.024	4
Thalium		
$Tl^{3+} + 2e^- \rightleftharpoons Tl^+$	+1.280	-8
$Tl^{3+} + 3e^- \rightleftharpoons Tl_{(s)}$	+0.741	-9
$Tl^+ + e^- \rightleftharpoons Tl_{(s)}$	-0.336	-1
Tin		
$Sn^{2+} + 2e^- \rightleftharpoons Sn_{(s)}$	-0.140	-4
$Sn^{4+} + 2e^- \rightleftharpoons Sn^{2+}$	0.139	-12
Titanium		
$TiO^{2+} + 2H^+ + e^- \rightleftharpoons Ti^{3+} + H_2O$	+0.099	3
Uranium		
$UO_2^+ + 4H^+ + e^- \rightleftharpoons U^{4+} + 2H_2O$	+0.39	11
$UO_2^{2+} + 4H^+ + 2e^- \rightleftharpoons U^{4+} + 2H_2O$	+0.334	8
$UO_2^{2+} + 3H^+ + 2e^- \rightleftharpoons UOH^{3+} + H_2O$	+0.254	2
$UO_2^{2+} + e^- \rightleftharpoons UO_2^+$	+0.16	-3
$UOH^{3+} + H^+ + e^- \rightleftharpoons U^{3+} + H_2O$	-0.539	-1
$U^{4+} + e^- \rightleftharpoons U^{3+}$	-0.63	-7
Vanadium		
$VO_2^+ + 2H^+ + e^- \rightleftharpoons VO^{2+} + H_2O$	+1.001	1
$VO^{2+} + 2H^+ + e^- \rightleftharpoons V^{3+} + H_2O$	+0.337	3
$V^{3+} + e^- \rightleftharpoons V^{2+}$	-0.255	-5
$V^{2+} + 2e^- \rightleftharpoons V_{(s)}$	-1.125	-4
Zinc		
$Zn^{2+} + 2e^- \rightleftharpoons Zn_{(s)}$	-0.763	-4

International Atomic Weights

Atomic masses are 1987 IUPAC values. Numbers in parentheses indicate the mass number of the most stable or best known isotope.

Element	Symbol	Atomic Number	Atomic Weight	Element	Symbol	Atomic Number	Atomic Weight
Actinium	Ac	89	(227)	Curium	Cm	96	(247)
Aluminum	Al	13	26.98154	Dysprosium	Dy	66	162.50
Americium	Am	95	(243)	Einsteinium	Es	99	(252)
Antimony	Sb	51	121.76	Erbium	Er	68	167.26
Argon	Ar	18	39.948	Europium	Eu	63	151.965
Arsenic	As	33	74.92159	Fermium	Fm	100	(257)
Astatine	At	85	(210)	Fluorine	F	9	18.9984
Barium	Ba	56	137.327	Francium	Fr	87	(223)
Berkelium	Bk	97	(247)	Gadolinium	Gd	64	157.25
Beryllium	Be	4	9.0122	Gallium	Ga	31	69.723
Bismuth	Bi	83	208.9804	Germanium	Ge	32	72.61
Boron	B	5	10.811	Gold	Au	79	196.9665
Bromine	Br	35	79.904	Hafnium	Hf	72	178.49
Cadmium	Cd	48	112.411	Helium	He	2	4.0026
Calcium	Ca	20	40.078	Holmium	Ho	67	164.930
Californium	Cf	98	(251)	Hydrogen	H	1	1.00794
Carbon	C	6	12.01115	Indium	In	49	114.82
Cerium	Ce	58	140.115	Iodine	I	53	126.90447
Cesium	Cs	55	132.9054	Iridium	Ir	77	192.22
Chlorine	Cl	17	35.4527	Iron	Fe	26	55.847
Chromium	Cr	24	51.9961	Krypton	Kr	36	83.80
Cobalt	Co	27	58.9332	Lanthanum	La	57	138.9055
Copper	Cu	29	63.546	Lawrencium	Lw	103	(260)

Element	Symbol	Atomic Number	Atomic Weight	Element	Symbol	Atomic Number	Atomic Weight
Lead	Pb	82	207.19	Rhodium	Rh	45	102.9055
Lithium	Li	3	6.941	Rubidium	Rb	37	85.4678
Lutetium	Lu	71	174.967	Ruthenium	Ru	44	101.07
Magnesium	Mg	12	24.3050	Samarium	Sm	62	150.36
Manganese	Mn	25	54.9380	Scandium	Sc	21	44.9559
Mendelevium	Md	101	(258)	Selenium	Se	34	78.96
Mercury	Hg	80	200.59	Silicon	Si	14	28.0855
Molybdenum	Mo	42	95.94	Silver	Ag	47	107.8682
Neodymium	Nd	60	144.24	Sodium	Na	11	22.9898
Neon	Ne	10	20.1797	Strontium	Sr	38	87.62
Neptunium	Np	93	(237)	Sulfur	S	16	32.066
Nickel	Ni	28	58.69	Tantalum	Ta	73	180.9479
Niobium	Nb	41	92.9064	Technetium	Tc	43	(98)
Nitrogen	N	7	14.00674	Tellurium	Te	52	127.60
Nobelium	No	102	(259)	Terbium	Tb	65	158.9253
Osmium	Os	76	190.2	Thallium	Tl	81	204.3833
Oxygen	O	8	15.9994	Thorium	Th	90	232.0381
Palladium	Pd	46	106.42	Thulium	Tm	69	168.9342
Phosphorus	P	15	30.9738	Tin	Sn	50	118.710
Platinum	Pt	78	195.08	Titanium	Ti	22	47.88
Plutonium	Pu	94	(244)	Tungsten	W	74	183.85
Polonium	Po	84	(209)	Uranium	U	92	238.0289
Potassium	K	19	39.0983	Vanadium	V	23	50.9415
Praseodymium	Pr	59	140.9076	Xenon	Xe	54	131.29
Promethium	Pm	61	(145)	Ytterbium	Yb	70	173.04
Protactinium	Pa	91	(231)	Yttrium	Y	39	88.90585
Radium	Ra	88	(226)	Zinc	Zn	30	65.39
Radon	Rn	86	(222)	Zirconium	Zr	40	91.224
Rhenium	Re	75	186.207				

Constants and Prefixes

Useful Constants and Conversions

Term	Symbol	Value
Avogadro's number	N_A	6.022×10^{23} mol^{-1}
Gas constant	R	8.314 J mol^{-1} K^{-1}
		8.206×10^{-2} L atm mol^{-1} K^{-1}
Faraday constant	F	9.6485×10^4 C mol^{-1}
Coulomb	C	A s
Joule	J	10^7 erg
		6.2415×10^{18} eV
		4.184 cal
Speed of light (vacuum)	c	2.998×10^8 m s^{-1}
Planck's constant	h	6.626×10^{-34} J s
Electron charge	e	-1.602×10^{-19} C
Atomic mass unit	u	1.661×10^{-24} g
Frequency	Hz	s^{-1}
Electrical resistance	Ω	V A^{-1}
Electrical potential	V	J C^{-1}
Electrical current	A	C s^{-1}
Acceleration of gravity	g	9.806 m s^{-2}
Boltzmann's constant (R/N_A)	k	1.3807×10^{23} J K^{-1}
Nernst factor	V_N	$0.0592(T/298)$ V

Prefixes for Units and Their Multiplication Factors

Factor	Prefix	Symbol	Factor	Prefix	Symbol
10^{18}	exa	E	10^{-2}	centi	c
10^{15}	peta	P	10^{-3}	milli	m
10^{12}	tera	T	10^{-6}	micro	μ
10^{9}	giga	G	10^{-9}	nano	n
10^{6}	mega	M	10^{-12}	pico	p
10^{3}	kilo	k	10^{-15}	femto	f
10^{2}	hecto	h	10^{-18}	atto	a
10	deka	da	10^{-21}	zepto	z
10^{-1}	deci	d			

Technique	Chapters and Sections
Acid-Base	3, 4
Activity	3C, D
Adsorption	11C
Affinity Chromatography	11C, 12F
Amperometry	10J
Atomic Spectrometry	6F
Attenuated Total Reflectance Spectrometry	5F, 12H
Chemiluminescence	6D
Chromatography	11
Complexation	7
Continuous Flow methods	12C
Coulometric Titration	10I
Coulometry	10G, I
Distillation	11B
Electrochemistry	9, 10G–J
Electrogravimitry	10H
Equilibrium	3D, 4A,B E, 7B, C, F, 8B, D, 9B, D, 10B, E, 11A–C
Extraction	11A
Flow Injection Analysis	5D, 7G, 10K
Fluorescence	6B
Gas Chromatography	11D, E
Gel Permeation Chromatography	12F
Gravimetry	8A, B, E
Immunoassay	12G
IR Spectrometry	5B, D, F, H
Karl Fischer Titration	10F

Technique	Chapters and Sections
Kinetic Methods	12C
Liquid Chromatography	11D, F
Mass Spectrometry	11H
Neutralization Titrations	4C, F
Phosphorescence	6B
Potentiometry	9
Precipitation	8
Raman Spectrometry	6C
Sensors	5F, 9E, F, 10J, 12H
Solid-Phase Extraction	11C, E
Solubility	8
Specific Ion Electrodes	9E, F
Spectrometric Titration	7E
Spectrometry	5, 6
Spot Tests	7G, 8G, 10K
Statistics	2H-J
Stopped Flow Methods	12C
Test Strips	7G, 8G, 10K
Thin-layer chromatography	11G
Titrimetry	4D-F, 7D-F, 8C, D, 10C-F, I
Ultracentrifugation	12F
UV-Visible Spectrometry	5B-I

Page references followed by *t* indicate material in tables.